The Beginning of the Use of Metals and Alloys

The MIT Press
Cambridge, Massachusetts
London, England

The Beginning of the Use of Metals and Alloys

Papers from the
Second International Conference
on the Beginning of the
Use of Metals and Alloys,
Zhengzhou, China,
21–26 October 1986

Edited by
Robert Maddin

This book was set in Palatino by Asco Trade Typesetting Ltd., Hong Kong, and printed and bound by Halliday Lithograph in the United States of America.

Library of Congress Cataloging-in-Publication Data
International Conference on the Beginning of the Use of
Metals and Alloys (2nd: 1986: Cheng-chou shih, China)
The beginning of the use of metals and alloys.

Includes index.
1. Metallurgy—History—Congresses. I. Maddin, Robert. II. Title.
TN16.I58 1986 669'.009 87-29391
ISBN 0-262-13232-X

Contents

Foreword

Cyril Stanley Smith

The contents of this volume are of unusual interest. Previous conferences on metallurgical history have tended to overemphasize the scientific techniques used for studying artifacts without much input from professional archaeologists and anthropologists. In this volume about a third of the papers use scientifically disclosed detail as the basis for a reexamination of some of the broadest aspects of historical change.

The chapters embody a far richer appreciation of the importance of cultural diversity than would have appeared in contributions to the field even as little as a decade ago. Not only have East and West discovered in each other much more than geographic and mercantile opportunity, but a more intimate view of history has become possible because of the ability to infer the history of objects on the basis of their internal structure. This microscopic supplement to the macroscopically visible records on which historians have had to depend in the past makes it increasingly clear that technology has played a central role in permitting and sometimes necessitating changes in human values and social organizations. Technology has enabled an improved physical standard of living for many people by increasing the speed and cheapness of making a diversity of objects, but its greatest effects will probably arise from the enhanced availability of information. The world has both shrunk in size and enormously increased in complexity. This has brought both new problems and new means of understanding and responding to them.

Structural change can be seen to underlie changes in the zeitgeist. The frame within which history can be, indeed must be, studied has vastly enlarged. Both scholars and the general public are becoming more aware of the value of cultural differences. An equally important change has happened in science. Not long ago complex fields such as biology, metallurgy, and natural history were regarded as being in some way less "scientific" than mathematical physics, and it was rarely appreciated that, although mathematical physics could handle with beauty and exactitude those properties that were insensitive to structure, it required that the real nature of matter be ignored.

In the last half century or so the study of those levels of structure that are intermediate between the atom and objects on a human scale has led to exciting advances in explaining biological diversity and in the virtually new field of materials science. Among many other opportunities that have been opened up by materials science is the ability to reconstruct the technical history of artifacts and so to learn the individual experiences of their makers. This richly supplements the verbal and stylistic evidence from which human activities have been reconstructed in the past—and it suggests that some questions regarding environmental influences can never be answered in a rigidly quantitative way. The approach of the humanist and the artist is paramount.

One must constantly remember that "scientific" evidence itself has some uncertainty. This arises both in sampling and analytical errors and in the interpretation of the data, for it is hard to avoid thinking that an ancient artisan "must have" done something or other in the way that seems natural to a modern technologist.

Unlike most professional historians whose standard is the understanding in depth of human behavior at a given place and time, the technical historian is likely to be unrestrained by geographic and temporal limits and to follow a single thread of curiosity into any culture, seeing the richness of history in the diversity of ways in which matter, with its constant properties, has been handled. This narrow-minded view is not without value, though of course it takes meaning only within the larger level of cultural history. Nevertheless one cannot help but notice that the history of human societies is rather similar to the history of the way in which atoms aggregate to form molecules and cells and crystals and landscapes.

Philosophers have long speculated about the relations between microcosm and macrocosm: At last it is beginning to be possible to study the origins of similarity and differences. No one level can be understood without some knowledge of those levels immediately above and below it in scale and complexity; it is the very differences that exist on one scale that give coherence and entity at another, yet there is a repetitive hierarchy and a similarity of pattern of interaction across boundaries regardless of scale; it is the inclusion of structureless space that defines structure.

The nuances of style in the way materials are treated or used can be uncovered by the use of laboratory techniques; these nuances throw much light on the cultural environment within which they originated. Metalsmiths today can have sensory experiences identical with those of craftsmen of earlier times, though they cannot, of course, know the earlier craftsmen's thoughts. The study of structure on the scale of the optical microscope provides a record of those changes that were a direct consequence of human intervention, such as heating, mixing, casting, forging, annealing, or other treatment. Other laboratory techniques based on trace elemental analysis, isotope analysis, and so on do not give such personal information but nevertheless are useful in establishing dates or provenance to aid the studies of cultural interchange.

It is interesting to note the clear preferences that metalworkers in different cultures have for shaping by hammer or by casting or for various methods of treating surfaces by thermal, mechanical, or chemical means and the way in which the technical possibilities interact with the uses of material objects for ceremony, enjoyment, or purposeful activity. On practically every page in this volume there can be found some hint of the way in which the properties of materials have interacted with the properties of a human body and mind. The record preserved in the structure of artifacts makes it evident that the sensually acquired knowledge of the nature of things as felt and as seen by artisans long preceded and supposedly inspired the first recorded attempts of philosophers to provide intellectual explanations. [Some discussion of this is found in my book *A Search for Structure* (MIT Press, 1981) and in the article "Notes on the prehistory of solid state physics," in *Solid State Physics: Past, Present and Predicted*, edited by D. L. Weaire and C. G. Windsor (Bristol: Adam Hilger, 1987, 11–42).]

Workers in materials use the materials' properties to make history, but the experience changes them. Familiarity with changes in structure in materials (for example, those responsible for the reduction of ore to metal, the work-hardening of bronze and the hardening or softening of steel by thermal treatment) inevitably influences scientists' ideas on the nature of change in general: The transition between small and large things and small and large interactions in small and large intervals of time is not dissimilar to the changes in human history. But, and it is a big but, this gives only a general viewpoint, in no way carrying exact understanding of detail; indeed rather the reverse, for a structural change that might be desirable within a small aggregate can be restrained or accelerated by the structure of the interface with the environment. Both are historically generated. The rules are inescapable, but they are like thermodynamic principles that apply to whole assemblies, not to local structures at any one time. Nevertheless the structural metaphor is useful, and technicians can sometimes see important facets of interaction between peoples that better historians might not perceive among the multitude of other factors. Indeed one can see a tendency among scholars in all fields to appreciate the importance of what is called "indeterminancy" in quantum physics and the inevitability of something like uncertainty and art in all biological and human interactions. The only aspects of the universe that we can perceive are those that are reducible to patterns processible by the human senses and brain—and these are dependent on the same inertial hierarchies of electron interaction that give rise to atoms and their assemblies.

I was unable to attend the conference at which these papers were presented, but I was privileged to be at the one in Beijing in 1981 and remember vividly the excitement of learning what the collaboration between Chinese archaeologists and metallurgists had achieved, then little known in the West: the casting techniques used in making ceremonial bronze vessels of exactly known provenance, putting on sounder base much that had previously only been inferred; the construction of ancient furnances and kilns; molds of both clay and metal used for mass production; the unsuspectedly vast scale of cast iron production and use; the copper mines at Tonglüshan (Mount Verdigris); and perhaps most striking, the great diversity of techniques used to generate beautiful detail in metallic and nonmetallic objects recovered from the fifth century B.C. tomb of the Marquis Yi of Zeng in the Hubei Provincial Museum.

The 1986 conference, for the organization of which we owe a great debt to the joint chairmen Ko Tsun and Robert Maddin, went some steps beyond the earlier one. The papers published here take a world view. They contain much new information and important reevaluation of older knowledge of the technical archaeology of materials in Africa, northern and eastern Europe, the Middle East, and Southeast Asia, as well as China and Japan. Focused studies by what has come to be called "archaeometry" are interspersed with interpretative essays dealing with the intangible aspects of cultural change and with others suggestively examining the way in which techniques have served as metaphor to generate both popular myth and philosophical speculation about cosmological principles. This combination of views makes this volume unusually valuable. It is likely to be seminal.

Preface

Robert Maddin

It was during a 1978 visit to the Beijing University for Iron and Steel Technology (BUIST) that I first became aware of the full range of the studies being carried out by the Ancient Metals Group directed by Tsun Ko. I had come across their papers in the journal *Kaogu Xuebao*, which was available outside China (albeit with some delay), under the collective nom de plume Li Chung or under the names of the individual authors. What I found was that many important papers were being published in much harder to find journals of provincial museums and historical societies, and there was a whole range of scientists and scholars whose work was completely new to me. The Ancient Metals Group was perhaps the largest interdisciplinary gathering of scientists, engineers, archaeologists, and historians anywhere, and it was carrying out studies on a variety of archaeological and historical questions that I knew would be of great interest to people pursuing similar research problems in other parts of the world. It was for this reason that I started corresponding with Ko about the possibility of organizing a small symposium that would initiate further contact and communication with the scholarly community outside China.

We came to call that symposium BUMA-I (the First International Conference on the Beginning of the Use of Metals and Alloys). It was held in Beijing in the fall of 1981 and was attended by about eighty Chinese and twelve foreign scholars and scientists. Most of the papers were presented in Chinese with concurrent translation. We intended to publish the proceedings, but obtaining suitable English renderings of the Chinese papers proved more difficult than we had expected, and the idea was eventually dropped.

The success of BUMA-I led to the planning of a follow-up conference that would survey the beginning of the use of metals and alloys in all parts of the world. We hoped that such a conference might reveal common threads to the patterns that were emerging and might also allow a useful interchange of techniques and approaches. We agreed that progress in understanding early metallurgy would come only when scholars and scientists from many disciplines focused their talents on the same problems.

Some sixty invitations were sent out from China to individual scholars, and initially all but two accepted. At the same time, announcements and a call for papers were sent to journals. Economic realities and some political constraints subsequently forced about half of the initial invitees to withdraw, and as a result there were some important unplanned gaps in the coverage of the conference and the present volume (among these I would include many Eastern European and some European countries, the Levant, some parts of Southeast Asia, and some important areas of the Orient). Nevertheless the gathering was to include a remarkable array of respected scholars and scientists with active interests in early metallurgy.

Our Chinese hosts set the symposium for 21–26 October 1986 in Zhengzhou, Henan Prefecture, a region that has been called "the cradle of Chinese civilization." It was here that bronze was developed around 2000 B.C. and that the making of cast iron and steel from pig iron (on a large scale) was initiated between the sixth century B.C. and the first century A.D. Sponsors for the symposium included the Government of Henan Province, the Henan Association of Science and Technology, the Henan Academy of Social Science, the Institute of History of National Science (Academia Sinica), the Chinese Society of the History of Science and Technology, the Henan Institute of Cultural Relics Research, the Department of History of Zhengzhou University, the Chinese Society of Natural Dialectics of Nature, and the Archaeometallurgy Group of the Beijing University of Iron and Steel Technology.

Topics discussed at the conference included recent studies on ore deposits and mining, the beginning of metallurgy, ancient alloys and alloy development, the development of early iron and steel, and archaeometric techniques. Some thirty-six papers were presented. Unfortunately the Chinese papers were not available in time for inclusion in this volume. We have, however, included three papers from BUMA-I: those by Pieter Meyers (whose data were used in the analysis by I. L. Barnes et al. at BUMA-II), William Rostoker, and Michael Notis. I have generally tried to arrange the papers chronologically and geographically.

BUMA-II was felt to be an important meeting by those who attended, and it is hoped that the present volume will convey some of the spirit of that meeting. The encouragement and support of C. C. Lamberg-Karlovsky, Director of the Peabody Museum of Archaeology and Ethnology, Harvard University, and a partial travel grant from the American Council of Learned Societies are gratefully acknowledged.

The Beginning of the Use of Metals and Alloys

1
The Beginnings of Metallurgy in the Old World

James D. Muhly

Dedicated to the memory of Xai Nai (1910–1985), the founder of modern Chinese archaeology.

We are still a long way from being able to document the development of metallurgical technology in the Old World, from the first use of metal (c. 7000 B.C.) to the full development of a copper smelting and alloying technology (c. 2000 B.C.). Developments took place in different ways and at different rates in various parts of the world, but a detailed accounting of these developments is hampered by the random nature of archaeological discovery, by the chronological problems that still make comparative study a hazardous undertaking, and by factors beyond our control, such as the state of preservation of material remains.

The idea of a unified sequence of steps or stages in technological progress—from cold-hammered native copper to the melting and casting of native copper to the smelting of first simple and then complex copper ores to, finally, the alloying of copper, first with arsenic and then with tin—seems to be gone forever. It is doubtful that we will ever return to the concept that, because the physical properties of metals remain constant regardless of the historical or cultural setting in which they were being used, metallurgical discovery must follow an inevitable sequence. For better or for worse humans have always had the ingenuity to impose their will on the physical environment and to develop ways often unique to time and place of working and manipulating that environment.

From this state of affairs modern scholarship has drawn the corollary that metallurgy also developed independently in different parts of the world. The old idea of stimulus diffusion, working in ways that brought technological developments out of the ancient Near East across Anatolia into the Balkans and Central Europe, a sequence of events that came to be summed up in the phrase "Ex oriente lux," has now been replaced by a reconstruction positing six or more independent centers of metallurgical development.

Migration versus Diffusion

Local invention is the hallmark of current archaeological research. The rejection of stimulus diffusion, or of any form of outside influence or inspiration, has been followed by "the retreat from migrationism" (Adams et al. 1978), and retreat it is—full-scale, pell-mell retreat. I would argue that, as usual, opinion has swung from one extreme to the other, and we are now at the far side of the swing of the pendulum. It is hard to believe that the famous book by Sir Grafton Elliot Smith, *The Migration of Early Culture,* in which all of European civilization was derived from that of ancient Egypt, was published as late as 1929. To go from Elliot Smith to Colin Renfrew in little more than the space of one generation is a breathtaking intellectual journey. It is time, therefore, to pause for a breath and to take stock of where we are and where we want to go. The emphasis here is on the origins of metallurgy and on what can now be said with profit regarding the first use of metals and the ways in which metallurgy first began in the Old World, from the British Isles to Japan.

I want to make it clear from the outset that I still believe that all these developments are, in ways however indirect, somehow related. I believe that the discovery whereby a hard, intractable rock is turned into a soft, pliable, and malleable metal, was a unique discovery, not one miraculously repeated in much the same way at different times in different parts of the world. In defending such a thesis I make no attempt here to deal with the development of metallurgy in the New World, especially among the Inca of modern Peru. Such developments constitute a complex set of problems best dealt with in a different context.

Any discussion of origins must first face the fact that such investigations are no longer popular. There is a growing feeling among anthropologists that origins are not important, that there are better things to do than attempting to determine who came first, and that such research is, more often than not, a thinly veiled cover for nationalistic puffery and chauvinistic propaganda. According to Richard Pearson, "tracing diffusion, however gratifying the nationalistic sentiment, is no longer of much interest anthropologically" (Pearson 1976, p. 395).

As Gordon Willey put it over twenty-five years ago: "It may be ... that the search for the very well-springs of origin and cause is meaningless, and that the limits of anthropology are to appraise and understand the continuum of process as it is disclosed to us rather than to fix its ultimate beginnings" (Willey 1962, p. 10).

I would argue, on the other hand, that origin is part and parcel of understanding. To appreciate how things developed, we have to understand how they started. Perhaps what is really at issue here is what Willey calls "the very well-springs of origin." I do not believe that we will ever uncover the first copper artifact fashioned by an ancient artisan or the first copper smelting furnace ever constructed. In that sense I would argue that origins will forever elude us and to search for them is to chase after a will-o'-the-wisp. What is within our grasp is a correct understanding of beginnings insofar as they are preserved in the existing archaeological and historical record. That is what I mean by a study of origins. I would contend that such a study is both necessary and essential for any proper understanding of the development of metallurgical technology.

China: Diffusion or Independent Invention?

We need only look to China to see the truth of this observation. For years the bronze industry of Shang Dynasty China was seen as the classic example of an industry imported wholesale from abroad. There were no formative or incipient stages within China, but such stages certainly did exist within the archaeological record of older, more highly developed cultures in Mesopotamia. For Carl Whiting Bishop, writing in the *Origin of the Far Eastern Civilizations*, the implications of this were obvious:

Thus in Babylonia, writing, wheeled vehicles, the ox-drawn plow, wheat, and all the common domestic animals except the horse, with a complete mastery of bronze working, all existed well before 3000 B.C. *In China, on the other hand, there seems to have been no knowledge of the metals before around 2000* B.C. *Yet only something like 500 years later we find the Yellow River basin occupied by an already well-developed Bronze Age civilization which had most (though not quite all) of the elements known to the Near East a thousand years or more earlier. This civilization must therefore have appeared in northern China during the first half of the second millennium* B.C. *(Bishop 1942, p. 14)*

For similar observations, see also Creel (1935, pp. 43–69).

As far as comparative levels of culture are concerned, Bishop's summary remains unchanged, and David Keightley made essentially the same point in 1977: "The fact remains that the Neolithic Chinese of the third millennium B.C. were still living under conditions that can only be characterized as primitive by comparison with those of contemporary Mesopotamia or Egypt" (Keightley 1977, p. 399). What is at issue here is not the basic set of facts but the interpretation of and conclusions to be drawn from those facts. If we take the three most visible characteristics of the Shang Dynasty in China—writing, wheeled vehicles (chariots), and bronze metallurgy—we find that the first two appear in Mesopotamia (Uruk IVb) at least 2,000 years before they do in China and that bronze metallurgy is about 1,000 years earlier in Mesopotamia (EDIII: Royal Cemetery of Ur) than in China. Furthermore, if we move from bronze to simple copper metallurgy, we find Western Asia (Çayönü) some 5,000 years ahead of China.

But, as I have already indicated, the real problem here concerns the conclusions to be drawn from these comparative observations. No one worries about the origin of these developments in Mesopotamia, and that is because a fairly detailed formative sequence could be reconstructed, from the Hassuna and Halaf cultures down to the culmination of Sumerian civilization as represented in the finds from the Royal Cemetery of Ur. It apparently just never occurred to anybody that a comparable sequence was missing in China because no one had ever bothered to look for it. Neolithic cultures in China were known chiefly from the work of J. G. Andersson, whose extreme positions should have been quickly discredited, and the only Shang Dynasty site really excavated was that at Anyang, the site of the dynasty's capital only during the final phase of its existence. Nothing at Anyang was earlier than c. 1300 B.C. This was not the place to look for the origins of Chinese metallurgy.

The lack of a formative phase for Chinese metallurgy was, in reality, the lack of archaeological evidence for the Early Shang and pre-Shang phases of the Chinese Bronze Age. What was being studied, what constituted the material evidence for the Chinese Bronze Age, were the sophisticated ceremonical and ritual vessels that had been looted from Chinese tombs and sold on the antiquities market to European museums and private collectors. This was hardly the stuff from which to reconstruct the formative stages of the Chinese bronze industry![1]

The decades since 1948 have seen a complete transformation of the archaeological scene in China. With good reason the noted scholar An Zhimin, of the Institute of Archaeology, Chinese Academy of Social Sciences in Beijing, has claimed that "in the past thirty years ... archaeology has exhibited a totally new face; from surveys to excavations to scientific studies, the magnitude of its manpower, the wealth of its discoveries, and the significance of its results, all are rarely equalled in the history of world archaeology" (An Zhimin 1979–1980, p. 35).

One of the major consequences of this information explosion has been a great swing of the pendulum. Whereas as late as 1974 it was still possible for such a noted scholar of the "old school" as Max Loehr to state that "some of the Shang people's most significant cultural possessions were paralleled in the west and ultimately may have come thence" (Loehr 1974, p. 300), today most of the leading scholars of Chinese archaeology and Chinese bronzes, including all scholars in China as well as such figures as Noel Barnard and Kwang-Chih Chang, stoutly support all aspects of the indigenous autonomy of Chinese civilization.

Writing about the catalogue of the great 1980 show of Chinese bronzes in the United States, the art historian Michael Sullivan felt confident that "any remnants of the notion that the Chinese got their bronze technology from western Asia—the denial of which was taken by some Western scholars as merely Chinese chauvinism—are demolished once and for all" (Sullivan 1980, p. 11). In many respects the culmination of this trend was the publication, in 1975, of Ping-ti Ho's monumental work *The Cradle of the East: An Inquiry into the Indigenous Origins of Techniques and Ideas of Neolithic and Early Historic China* (Ping-ti Ho 1975).

Such conclusions, based on simplistic black and white, either/or approaches, quite ignore the complex nature of the problem. Recent archaeological work in south Asia, especially Baluchistan, in central Asia and Southeast Asia, especially in Thailand and Vietnam, is beginning to uncover evidence that will one day enable us to deal in a meaningful way with the many elements that came together to create what we have come to recognize as ancient Chinese bronzes.

The very question of what could properly be designated as "China" and "Chinese" in ancient times is far more complicated than scholars once believed. K. C. Chang once tried to discount the significance of the emerging regional variations in Chinese archaeology, maintaining that "all local cultures in prehistoric China that, in their entirety or in large part, became part of the historical Chinese civilization must be referred to as *Chinese* or *proto-Chinese*" (Chang 1977, p. 640). That is like saying that what is now China has always been Chinese!

Such a position contrasts sharply with that of Donn Bayard who maintains that, before the Han Dynasty, the southern two-thirds of the country was not Chinese in the sense that its inhabitants spoke not Chinese but some form of Austro-Thai (Bayard 1975, pp. 76–77). As we will see later, a proper understanding of developments in south China is the key to any future attempts to relate what happens in China to what is going on in surrounding lands.

Before we can understand the beginnings of metallurgy in China, we must take a look at what is now known about the early stages of copper metallurgy throughout the Old World and what, on the basis of existing evidence, can be said about where it all seems to have begun.

Early Metallurgy in the Ancient Near East

In 1944 the distinguished prehistorian V. Gordon Childe published the influential paper "Archaeological Ages as Technological Stages" (Childe 1944). The main thrust of the paper was in its study of the impact of metal technology on society. In plotting human progress in the use of metals, Childe set up a series of four stages, or "modes," taking care to emphasize that progress through these modes was neither uniform nor consistent. Childe used the term "homotaxis" to designate the chronological variations in such developments around the world, borrowing a term from Thomas Huxley that, fortunately, failed to be picked up by other scholars, thus sparing us at least one terminological monstrosity.

I am concerned here only with the first use of metal, a period sometimes designated as a Copper Age and one that Glyn Daniel at one time proposed to designate as an "eochalcic episode" (another unsuccessful label, if anything, even uglier than "homotaxis"). For Childe this was Mode 0, a name that made clear the insignificance Childe attached to the period. Childe chose to emphasize the "limited significance and doubtful relevance" (Childe 1944, p. 10) of the period because he was more concerned with archaeology than with technology. For Childe the period before Mode 1, his "Early Bronze Age," a time when copper and its alloys were used for weapons and ornaments but when tools and implements were still made of stone, was a period when metal was not prominent in the archaeological record and obviously of little significance in contemporary material culture [for Childe's concept of technology, see Trigger (1986)].

Yet Mode 0 was the very period when all the critical first steps were taken in learning what metal was, how metals behaved, how metal had to be worked (in ways quite different from the familiar techniques used on stone, wood, and bone), and finally in learning all the complex skills connected with the mining and smelting of copper ores, with the casting and hammering of metallic copper, and then the alloying of copper with arsenic and with tin. All this had to be in place in order to make possible Childe's Mode 1. Clearly Childe was not interested in metal production but only in the stages where that production had reached the point of being able to influence the social, economic, and even political organization of society at large. It must be admitted that the limited use of metal, especially copper but also lead and even some gold, during Childe's Mode 0 meant that the social impact of the newly developing technology was quite insignificant. Nevertheless those cautious beginnings were of crucial significance for everything that was to follow, right to the "advanced metals" being created today in the never-ending search for new and better materials (Kear 1986).

The earliest example of the use of copper is often taken to be the small oval-shaped pendant from the Shanidar Cave in Iraqi Kurdistan, dated to the early ninth millennium B.C. (Solecki 1969). It is more likely that this pendant, said to be completely mineralized, was in fact never metallic copper but malachite carved as a semiprecious stone. It is unfortunate that there still is considerable confusion between mineral and metal. As recently as 1984 we find a hematite macehead from Korucutepe, southeastern Anatolia, being presented as evidence for the early use of iron and the beginnings of iron metallurgy (Yakar 1984, pp. 67, 73).

The first uncontested use of metallic copper dates to the late eighth millennium B.C. Joint Turkish-American excavations at the aceramic Neolithic site of Çayönü Tepesi in southeastern Turkey, under the direction of Halet Çambel and Robert Braidwood, have uncovered well over fifty artifacts made of metallic copper and numerous beads carved directly from malachite. Many of the copper and malachite finds come from the so-called Intermediate Layer, between the two building levels known as the Grill Phase and the Cell Phase. A series of sixteen radiocarbon dates place this early phase of occupation at Çayönü within the period 7250–6750 B.C. (Braidwood et al. 1981; Çambel and Braidwood 1983; Schirmer 1983).

Preliminary work on the copper finds from Çayönü indicates that they were all made of native copper, most likely native copper collected at the great Ergani Maden mines located only some 20 km north of the site. Of special interest is the fact that several of the Çayönü artifacts show evidence of recrystallization, indicating that the objects had been annealed during the working or hammering stage of fabrication. Evidence for this annealing can be seen in an "awl" and two "hooks" from Çayönü (ÇT 70.I.5, ÇT 72.I.14, ÇT 84.I.19; figures 1.1 and 1.2).

This application of pyrotechnology to copper artifacts that are, in fact, the earliest such artifacts yet excavated, is quite astounding. I would also argue that the evidence shows that the smiths at Çayönü did not yet understand the effects of annealing and did not really know what they were doing. They reheated the artifact to prevent it from cracking or splitting under continued hammering. Such cracking can be seen in some of the heavily cold-worked artifacts from Çayönü, including one of the awls we have studied (ÇT B1.I.6; figure 1.3). But after annealing the "hooks" and "awls" being made at Çayönü, the smith then left them in the softened state, obviously not fully understanding the effects of recrystallization. This is not surprising as we are here on the very threshold of copper metallurgy.

Another interesting aspect of the copper artifacts

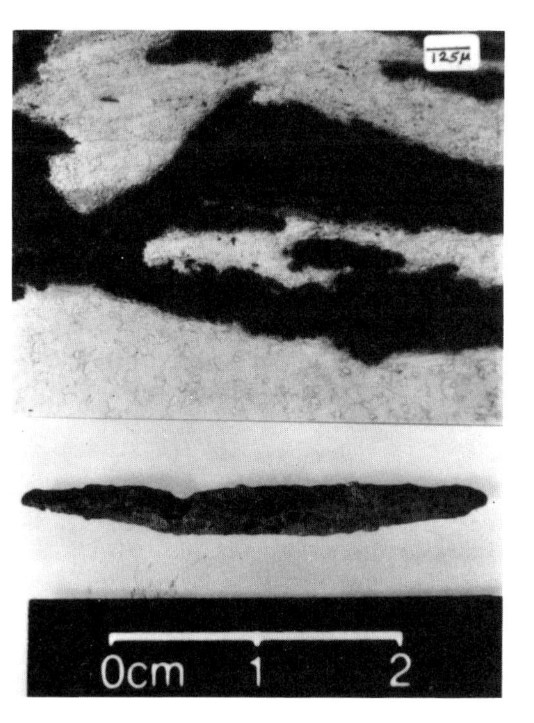

Figure 1.1 Awl from Çayönü Tepesi (ÇT 70.I.5) showing evidence of recrystallization produced by annealing. The presence of nonmetallic inclusions is readily visible. PIXE analysis gave 98.91% Cu, 0.025% Ni, 0.875% As, 0.019% Zn, and 0.014% Fe.

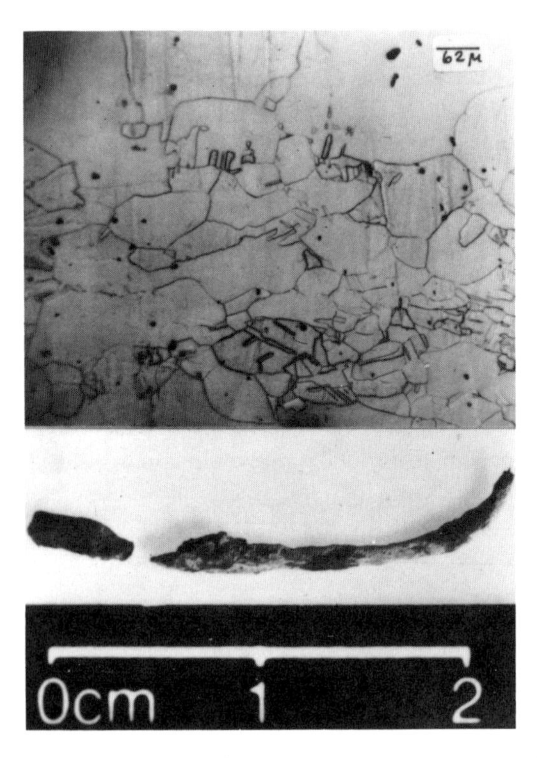

Figure 1.2 Hook from Çayönü Tepesi (ÇT 72.I.14) showing evidence of recrystallization produced by annealing. The presence of nonmetallic inclusions is also readily visible. It is likely that there was some further cold-working after annealing. PIXE analysis gave 99.69% Cu, 0.02% Ni, 0.04% Zn, 0.011% Pb, and 0.017% Fe.

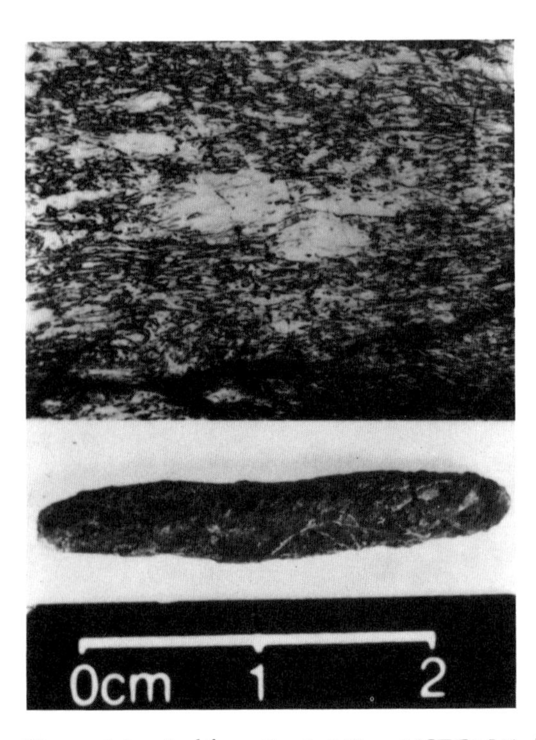

Figure 1.3 Awl from Çayönü Tepesi (ÇT B1.I.6) showing customary cracking resulting from extensive cold-working. PIXE analysis gave 99.6% Cu, 0.03% Ni, 0.028% As, 0.031% Zn, 0.035% Ag, and 0.014% Pb.

from Çayönü is that several of them show evidence of significant amounts of arsenic. The first analysis of one of the awls or reamers from Çayönü, by the German team of Junghans, Sangmeister, and Schröder, reported 0.8% arsenic (Esin 1969, p. 130; JSS 18 431). This was in turn taken as evidence for the smelting of arsenic-bearing copper ores already in this incipient stage of copper metallurgy (Selimchanow 1977, p. 4; Esin 1976, pp. 212ff). We now realize, thanks to the studies of George Rapp, that native coppers from around the world can contain as much as 5.0% arsenic and that a specimen from the Anarak-Talmessi mines in Iran contained 3.0% arsenic (Rapp 1982, p. 35). There is no need to reconstruct a copper smelting technology at seventh millennium B.C. Çayönü.

I have discussed the material from Çayönü at such length not just because I am working on the material myself, together with Robert Maddin and Tamara Stech, but because at Cayönü we are looking at the earliest metal artifacts yet known from any archaeological context. At Çayönü we are the closest we have ever come to the origins of copper metallurgy. The high level of technological achievement in metallurgy is repeated in other aspects of the site's material culture. The rooms in the buildings at Çayönü had true terrazzo floors made of pieces of grayish white and reddish orange limestone bonded in a lime mortar and then carefully polished.

What makes the material from Çayönü all the more surprising is that nearby contemporary sites, showing many similar traits in material culture, have produced no evidence for the use of metal. A French team excavating at Cafer Höyük, located just to the northwest of Çayönü (figure 1.4), has uncovered a site having many close parallels with Çayönü, but in four seasons of excavation they have not yet found a single object of copper (Cauvin 1985). The abundance of such finds at Çayönü remains a great enigma. It is fair to say that the more we learn about Çayönü, the more baffling the site becomes.

The extensive modern copper mines at Ergani Maden have never been investigated for possible traces of ancient mining activity. For the seventh and sixth millennia B.C., when exploitation amounted to no more than collecting lumps of native copper, it is hard to imagine what sort of physical evidence could possibly survive to attest to the utilization of a particular ore deposit. Nevertheless there are geographical factors that support the identification of Ergani Maden as the source of much of the copper found at Neolithic sites in Anatolia and in northern Mesopotamia. The Tigris was navigable, by raft or hide boat, from Çayönü down the length of Mesopotamia. It is entirely possible that the small number of copper artifacts from seventh and sixth millennia B.C. levels at such Mesopotamian sites as Tell Maghzaliyeh, Tell Sotto, Yarim Tepe, Tellul eth-Thalathat, Tell es-Sawwan, and even the

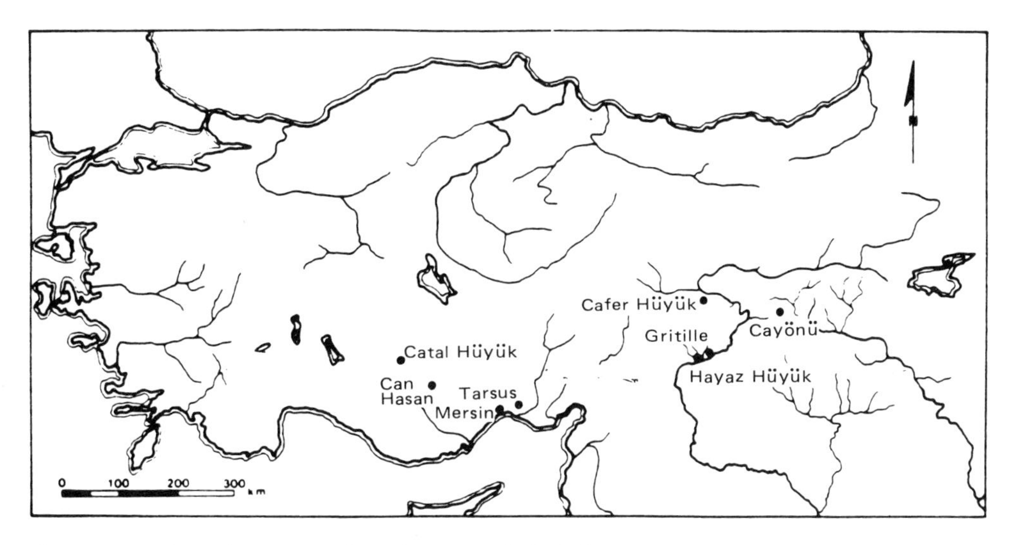

Figure 1.4 Anatolian sites with early copper finds (before 5000 B.C.). After A. M. T. Moore, in *Advances in World Archaeology*, (Academic Press, 1985), vol. 4, 28, fig. 1.16.

nearby Iranian site of Ali Kosh (figure 1.5) were all made of Ergani copper. [For finds, see Moorey (1982, pp. 17–18) and Muhly (1983, pp. 350–351).]

A recent publication of the awl from Tell Maghzaliyeh (figure 1.6), which is dated in the seventh millennium B.C. and is thus the earliest copper artifact from Mesopotamia, tries to show that the awl was made of copper from the Anarak-Talmessi mines in Iran (Ryndina and Yakhontova 1985). The argument is based on evidence from the trace element composition of the copper and, like all such arguments, simply cannot be accepted. Although still not widely appreciated, it has long been recognized that the various processes of smelting, remelting, refining, mixing, and even alloying so altered the chemical composition of every ancient metal artifact that attempts to establish provenance on the basis of chemical composition are doomed to failure (Muhly 1973, pp. 339–342; Gale and Stos-Gale 1982, pp. 11–12).

Early Smelting Technology

All of the copper artifacts discussed up to this point presumably were made of unalloyed native copper; no evidence exists for the smelting of copper from its ores before c. 5000 B.C. The earliest evidence for a smelting technology involves not copper but artifacts of metallic lead. As native lead is exceedingly rare in a geological context, it is reasonable to assume that all existing lead objects came from the smelting of lead ores, notably galena (PbS). Thus the necklace with thirteen lead beads from Level VI A at Çatal Hüyük, in the Konya Plain of south-central Anatolia, and the lead bracelet from Level XII at Yarim Tepe, in the Sinjar Valley of northern Mesopotamia—both finds dating to the early sixth millennium B.C.—must represent the earliest known examples of the use of a metal smelted from its ore [see Muhly (1983, p. 351)].

The beginnings of copper smelting technology are still poorly understood. A piece of slag from Level VI A at Çatal Hüyük, the same level that produced the lead beads, has been identified as a piece of copper smelting slag (Neuninger et al. 1964), but this identification has been contested (Tylecote 1976, pp. 5, 19). It is more likely that what was found at Çatal Hüyük, if it is a slag at all, is a crucible or melting slag, having nothing to do with any smelting operation [for this distinction, see Cooke and Nielsen (1978, p. 182)]. Evidence for copper smelting has often been identified in the remains of copper workshops excavated at Tepe Ghabristan in northwestern Iran, west of Tehran (Majidzadeh 1979), and at Tall-i-Iblis in southern Iran, near Kerman (Caldwell 1968)—both sites dating to the fifth millennium B.C.—but again the evidence is ambiguous and more likely relates only to the melting and casting of native copper.

The macehead from Level 2B at Can Hasan in southern Anatolia, dating to c. 5000 B.C. (French 1962, pp. 33; de Jesus 1980, vol. 2, pl. 22:4), is perhaps the most substantial copper artifact produced before c. 4000 B.C. Spectrographic analysis proved it to be made of a pure copper having only 0.05% silver as a trace element (Esin 1969, p. 130; JSS 17 635). It could well represent the use of native copper. Silver, as a sole trace element in copper, has in fact been said to be one of the identifying characteristics of native copper (Ottaway 1973, pp. 303ff). It is unlikely, however, that the simple presence or absence of trace element silver can identify the use of native copper.

It may turn out that the scale of the metal industry will prove to be one of the best indicators for the use of smelted copper. The argument here rests on

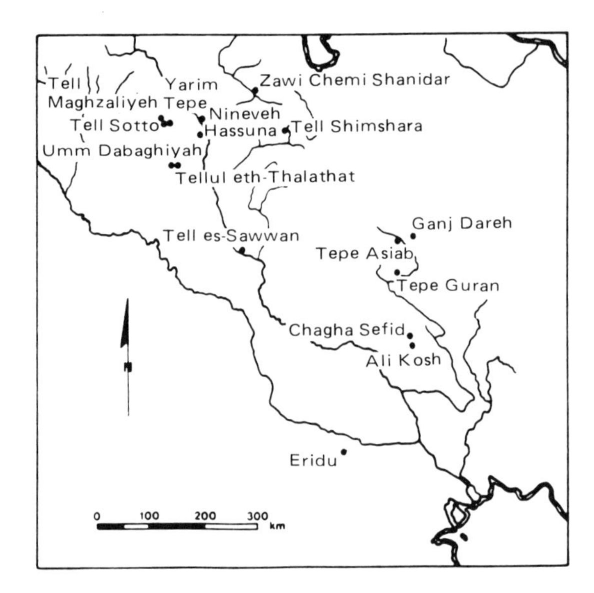

Figure 1.5 Mesopotamian sites with early copper finds (before 5000 B.C.). After A. M. T. Moore, in *Advances in World Archaeology* (Academic Press, 1985), vol. 4, 39, fig. 1.21.

the assumed relative scarcity of native copper in Old World geological contexts. According to the most complete study yet made on the distribution of native metals, it was only in the New World that native copper existed on a scale sufficient to support a metal industry producing something more than the odd trinket (Patterson 1971, pp. 198–199). The surviving fifty-five copper axes from graves near Susa, contemporary with the founding of Susa in the late fifth millennium B.C. (Susa A or I) indicate the existence of a metal industry that must have been based on the use of copper smelted from simple oxide or even from complex sulfide copper ores [for the axes, see Amiet (1979, p. 198)].

Beginnings of Metallurgy in Europe

The pattern of metallurgical development in Europe, although taking place somewhat later in time, followed the western Asiatic pattern. From southeastern Europe the earliest object of metal that has yet come to light is a massive copper awl (14.3 cm in length) from the site of Balomir in Romania dated to the Early Neolithic period. On the basis of calibrated radiocarbon dates the Early Neolithic period covers the centuries c. 5900–5300 B.C. This awl remains an isolated find in an Early Neolithic (Starčevo, Karanovo I–II) context, but a small corpus of copper finds can be assembled from the following Middle Neolithic period (5300–4700 B.C.; Karanovo III–IV). Although no analytical evidence is available, it is reasonable to assume that all these copper finds from Neolithic Europe were made of native copper. No evidence for actual metalworking has come to light: no crucibles,

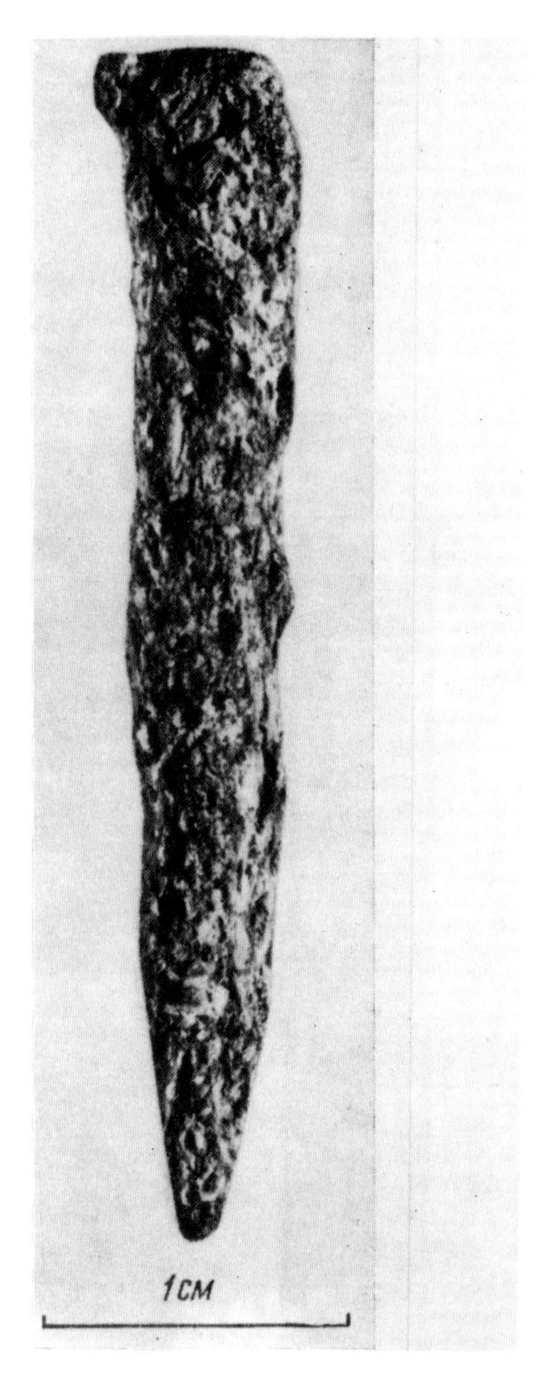

Figure 1.6 Awl from Tell Maghzaliyeh, northern Mesopotamia [from Ryndina and Yakhontova, *Sovetskaya Arkheologiya* (1985), 2; 157, fig. 1].

tuyeres, or molds and no evidence for the mining of copper ores. These are exactly the same circumstances that surrounded the first use of metal in western Asia [for early copper finds in Europe, see Bognár-Kutzián (1976), Chapman and Tylecote (1983), and Muhly (1985, pp. 109–112)].

The Middle Neolithic period also saw the first appearance of copper in the Aegean world. From two sites in northern Greece, Dikili Tash and Sitagroi (also known as Photolivos), come several small artifacts of copper—including beads from Sitagroi and a needle from Dikili Tash—all dating to the late sixth millennium B.C. The evidence from Sitagroi is of special interest because the following period, Sitagroi III or Late Neolithic (4700–4300 B.C.), produced not only awls and pins of copper but also slags and crucibles, the first evidence from Europe attesting to the melting and casting of copper, perhaps even to the smelting of copper ores [for references, see Muhly (1985, pp. 109–112); for Dikili Tash, see also Deshayes (1972)].

By the Late Neolithic period copper metallurgy had spread to central Greece, a directional development that points to southeastern Europe as the most likely place to look for the origins of Aegean metallurgy. From Level II at the Kitsos Cave, in the Laurion district of southern Attica, comes a small copper pin that, on the basis of six calibrated radiocarbon dates, is to be placed in the mid–sixth millennium B.C. Spectrographic analysis of this pin has shown it to contain 0.03% iron, 0.03% zinc, 0.01% gold, and 0.05% silver. According to the Stuttgart group who did the analysis and to Nicole Lambert, the excavator of the Kitsos Cave, this means that the pin was made of copper imported from Yugoslavia (Lambert 1973, pp. 510–511). It is entirely possible that this pin was made of imported copper, but there is nothing in the spectrographic analysis that would either prove or disprove this attribution.

What we do have from Yugoslavia, as well as from Bulgaria, is the first solid evidence for the mining of ores of copper in a Late Neolithic context. Excavations at Rudna Glava, East Serbia, Yugoslavia, and at Ai Bunar in southern Bulgaria have uncovered evidence for shaft mining (Rudna Glava) and for a combination of open cast and shaft mining (Ai Bunar) that can be dated, on the basis of pottery found in the mine shafts as well as the associated artifacts, to the Late Neolithic period (4700–4300 B.C.; Karanovo V–VI, Gumelnitsa).

Archaeologists working in the Balkans designate the Late Neolithic period the Eneolithic (or Aeneolithic), further subdivided into early, middle, and late phases. They also tend to use a somewhat lower chronology than the one advocated here, putting the entire Eneolithic at c. 4700–3900 B.C. What followed was a transitional period or *Übergangsperiode* of c. 600 years, thought by some scholars to be the result of a nomadic incursion, or *Steppeninvasion*. On the basis of this scheme the beginning of the Early Bronze Age in the Balkans would be put at c. 3300 B.C. (Quitta 1986), contemporary with the date now assigned for this on Crete (Warren 1980; figure 1.7).

Copper Ores and Arsenical Copper

These chronological distinctions are essential to any understanding of this crucial period in the prehistory of Old World metallurgy. The mining operations at Rudna Glava are placed early in this sequence; Borislav Jovanović, the excavator of Rudna Glava, now regards all early mining activity there as belonging to the Early Eneolithic period (Jovanović 1985). This is said to explain the general lack of associated finds in the excavation. Ai Bunar, being somewhat later, proved to be a much richer site with a nearby settlement (and cemetery) for the miners, something missing at Rudna Glava. E. N. Chernykh, the excavator of Ai Bunar, estimates that some 20,000 to 30,000 tons of material were dug out of the ground during Eneolithic mining operations at the site, including some 2,000 to 3,000 tons of copper ore which, when smelted, produced between 500 and 1,000 tons of metallic copper [figures from Jovanović (1985, p. 119)].

This intensive mining activity, it is now argued, resulted in the depletion of the oxide (and carbonate) copper ores of the Balkans by Late Eneolithic times,

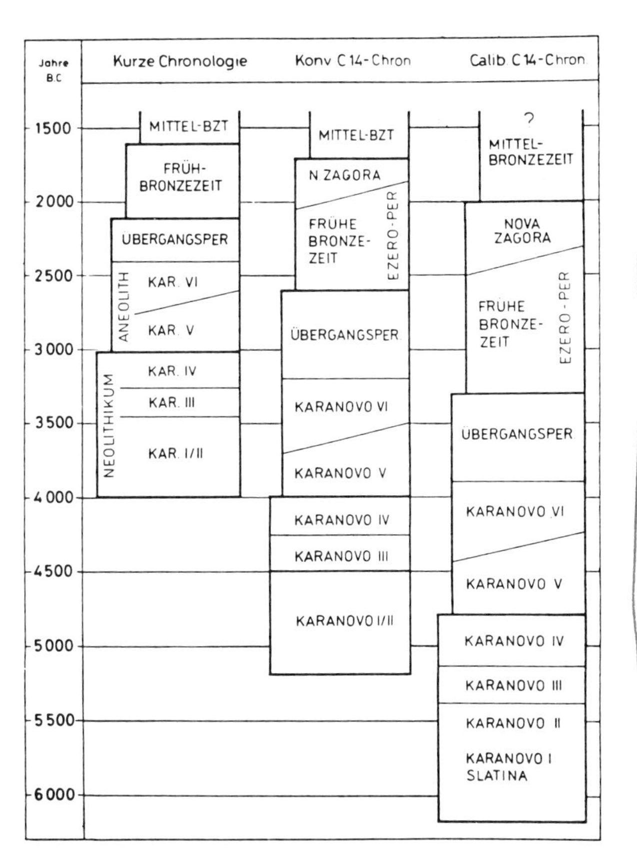

Figure 1.7 Chronology of Neolithic Europe [from H. Quitta, *Das Altertum* (1986), 32:116].

resulting in a great drop in metal production (Chernykh 1982; Todorova 1981). With this metal shortage came a period of experimentation and innovation resulting in the first production of arsenical copper (Jovanović 1985, pp. 118–119). The search for new sources of copper eventually led to the exploitation of the massive deposits of sulfide ores and a shift in the main centers of metallurgical development from the Danube Basin and the Carpathians to the Alps and the ore mountains of Czechoslovakia, both areas rich in copper sulfide ore deposits.

This means that arsenical copper came into use during the shift from using oxide copper ores to using sulfide ores. During the search for new sources of metal, following the depletion of the oxide deposits, the Eneolithic prospectors came upon some of the richly colored arsenic-bearing copper ores, which in turn led them to the far more extensive sulfide deposits. The addition of tin, a feature of the European Early Bronze Age, could then be seen as an attempt to turn the copper smelted from sulfide ores into a product at least as good as, if not better than, the arsenical copper of the transitional period.

If future research should confirm this hypothesis for the introduction of arsenical copper, followed by the shift to bronze, then this would raise the intriguing possibility that developments in western Asia followed a similar pattern. What is lacking for the ancient Near East is, of course, the evidence from the mining sites themselves. No fifth or fourth millennium copper mines have been discovered anywhere outside of southeastern Europe. The one possible exception is the area of Timna, in the southern Negev of Israel (figure 1.8), where Beno Rothenberg, the director of the Timna Project, thought copper mining had begun already in the fifth millennium B.C., with extensive operations by the fourth millennium [for survey, see Rothenberg (1983)]. I would argue, however, that there was no mining or smelting activity at Timna before the thirteenth century B.C. (Muhly 1984).

Nevertheless arsenical copper does come into use in western Asia during the late fifth and first half of the fourth millennia B.C., exactly as in the Balkans. The evidence for this is not as good as one would like it to be because of the lack of analytical data but, of the nineteen objects from the Susa A (or I) necropolis analyzed by Thierry Berthoud, six had at least 1.0% arsenic. The use of arsenical copper continues during the following Susa B–C period, with an increase in the number of objects made of arsenical copper along with an increase in the amount of arsenic present in each object. Of the eighteen objects analyzed, eleven were of arsenical copper, averaging 5.0% arsenic (Berthoud 1979, analyses 1180–1189, 1191–1193, 1195–1196, 1198, 1201, and 1203). Susa A dates to c. 4100–3900 B.C., with Susa B–C covering the rest of the first half of the fourth millennium B.C.

Roughly contemporary with the material from Susa B–C is the great collection of copper artifacts making up the Nahal Mishmar hoard, discovered in 1962 in a cave on the west bank of the Dead Sea (figure 1.9). Only 30 of the 423 copper artifacts making up this hoard have been analyzed, but 21 of these proved to be made of arsenical copper, averaging 5.6% arsenic (Bar-Adon 1980, pp. 235–243). Most of the artifacts from the Nahal Mishmar hoard must be classified as ritual or cult implements, however they might actually have been used. It is of special interest that, of the eight tools in the hoard that were analyzed, only one was made of arsenical copper. This means that the very objects, chiefly chisels, that could have best made use of the metallurgical properties of arsenical copper, were made instead of unalloyed copper. Perhaps arsenical copper was used at Nahal Mishmar not because it was a harder, more durable metal but because it would have facilitated the production of the intricate lost-wax castings that are the most distinctive objects in the hoard.

With the development of arsenical copper we are on the threshold of the Bronze Age. By the fourth millennium both Europe and western Asia had reached the point of a fully developed smelting, casting, and alloying technology. We are now in a position again to look beyond the ancient Near East, but this time to the East rather than toward the Mediterranean and Europe.

It must be emphasized that the chronological (and technological) comparisons made here among Europe, the Aegean, and southwest Asia are derived from the use of calibrated radiocarbon dates, the validity of which is by no means universally accepted within the discipline of prehistoric archaeology. The Hungarian scholar Janos Makkay has recently called attention to what he considers to be the "crisis of prehistoric chronology" (Makkay 1985a) brought on by what he regards as the totally unwarranted acceptance of calibrated radiocarbon dating.

For Makkay the ready acceptance of radiocarbon dating has been due, at least in part, to the strong support such dating supplied for (currently popular) antidiffusionist theories. If metallurgy developed in the Aegean earlier than in western Asia, then obviously European metallurgy could not have derived from oriental prototypes. Thus the chronology for Neolithic Europe based on radiocarbon dates seemed to make common cause with theories of independent invention and the local origins of metallurgical technology in the British Isles, in the western Mediterranean, in southeastern Europe, in the Aegean, and in the eastern Mediterranean.

This, according to Makkay, has raised such problems in the interpretation of European Neolithic cultural assemblages that he feels that "we have now reached the point where the deep, unbridgeable chasm between

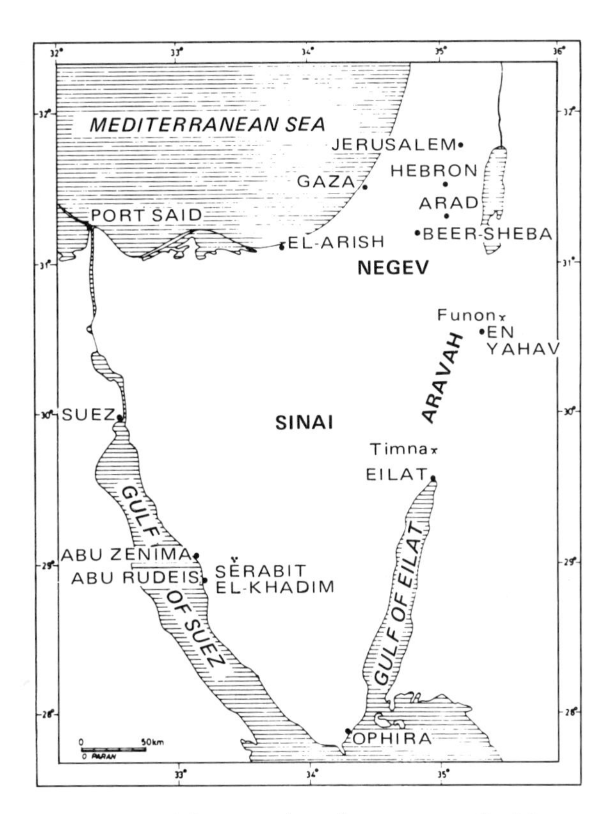

Figure 1.8 Mining and smelting sites in the Negev and Sinai. After I. Beit Arieh, *Tel Aviv* (1980), 7:46, fig. 1.

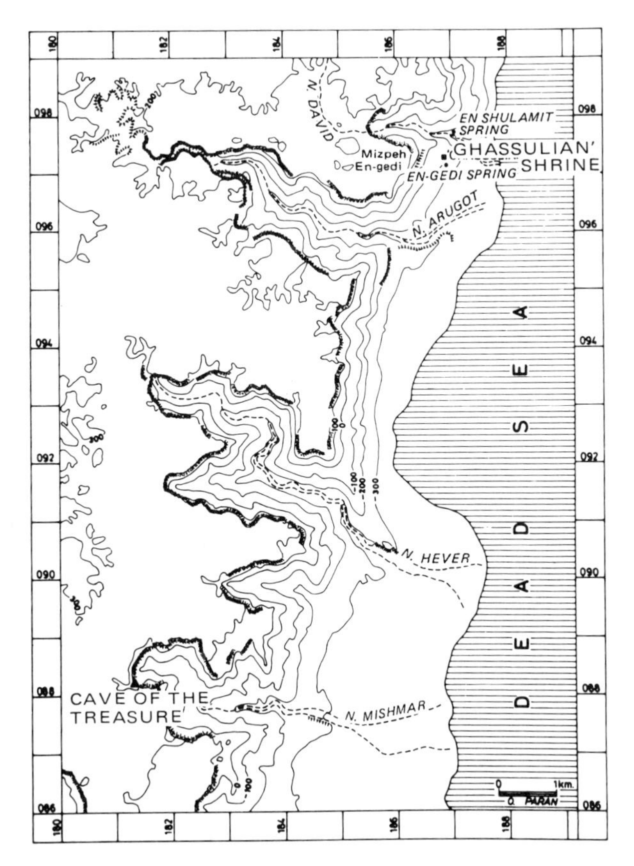

Figure 1.9 Location of the Cave of the Treasure, Nahal Mishmar, on the west bank of the Dead Sea (from D. Ussishkin, *Tel Aviv* (1980), 7:2, fig. 1).

diffusionism and antidiffusionism is most apparent, and in our view, also are the exaggerations and the failure of the current form of antidiffusionism" (Makkay 1985b, p. 4). For Makkay we have not actually done away with the traditional diffusionist map; no, now we just have the familiar black arrows pointing in the opposite direction. Such a map has already been proposed in a work which claims that whereas

once it would have been obvious that barbarian chiefs imitated the practice of civilized kings, now the arrows of diffusion seem to point the other way, for the royal tombs of Ur date to about 2800 B.C. while those in the Caucasus begin at a date closer to 3500 B.C. Were the kings buried at Ur party to a Kurgan dynasty? Was Gilgamesh himself . . . descended from a Kurgan chief? (Stover and Kraig 1978, p. 68, as quoted in Makkay 1985b, p. 4)

The best thing, of course, would be to do away with the black arrows all together. By now we should have advanced beyond the stage where cultural and archaeological developments are to be explained in terms of vast movements of peoples across the map regardless of how black the arrows or their orientation. The reason for Makkay's feeling of crisis is that he insists on seeing everything in terms of diffusion or antidiffusion. With material from fifth millennium Europe that he feels must somehow be related to similar material from third millennium Anatolia, he sees good reason for assuming a crisis situation.

It is not necessary to accept all of Makkay's arguments to see that things are not quite so straightforward as proponents of calibrated radiocarbon dating believe them to be. There are many problems still to be solved, including one of special interest for the development of metallurgical technology, that of the "missing fourth millennium B.C." in the Aegean.[2] What is needed is more real cooperation and exchange of ideas on all sides and, in particular, a breakdown of all uncompromising, hard-line positions. Some encouraging developments can be detected in the current recognition of "the general weaknesses of both purely endogenous and purely exogenous models of change" (Cherry 1986, p. 38).

In developing the theory of "peer polity interaction," Renfrew has in fact acknowledged the importance of the interchange, the diffusion of ideas, even if only between neighboring states roughly equal in their level of cultural attainment. As Renfrew now puts it:

Die isolierte Stadt *is a concept whose examination has indeed yielded useful insights and within which questions of the intensification of production and of the emergence of decision-making hierarchies in the face of increasing population density and other factors, can profitably be discussed. But the* form *of these control hierarchies and of the institutions by which intensification is achieved cannot so effectively be considered in isolation. (Renfrew 1986b, p. 1)*

To which I would add "origin" to the discussion of form. The reasons for this will, hopefully, become clearer by the end of this chapter.

Early Metallurgy in South and Southeast Asia

If we turn from the West to lands east of Mesopotamia, we find that throughout the vast area of south Asia, Southeast Asia, and east Asia our conception of the earliest phases in metallurgical technology has changed dramatically within the past ten years. In south Asia it has long been the case that there was no evidence for the use of metal before c. 3000 B.C., within the various phases of proto-Harappan cultures in India and Pakistan. The situation has now changed drastically thanks to material from the French excavations at Mehrgarh, in southern Baluchistan, under the direction of J. F. Jarrige.

Recent work at Mehrgarh has uncovered copper objects, principally beads, from graves dated to the transition from the Aceramic Neolithic to the Neolithic, a transition placed at c. 6000 B.C. (Jarrige 1984). This means that the first use of copper in Baluchistan came at about the same time as these developments took place in Mesopotamia and Iran. It is even more remarkable that at Mehrgarh the earliest copper artifacts appear in association with turquoise and lapis lazuli. Such associations suggest connections with Afghanistan and Iran and raise the possibility that the appearance of copper in Baluchistan was somehow related to the early use of copper at sites such as Ali Kosh and Yarim Tepe.

These elusive Mesopotamian connections seem to be maintained in the metallurgy of the Copper Hoard cultures of south Asia, now to be associated with Ochre Colored Pottery and dated to the early third millennium B.C. Some 50% of the Copper Hoard objects that have been analyzed were of arsenical copper, having at least 1.0% arsenic. Arsenical copper continued to be used in the following Harappan period, along with bronze (Agrawal 1982, pp. 235–236), but this seems to represent the eastern-most extension in the use of the alloy. According to the analytical evidence available at present, arsenical copper was not used in Southeast Asia or in east Asia.

There has been considerable interest during the past decade in the beginnings of metallurgy in Southeast Asia, especially in Thailand and in Vietnam. Archaeological work in Thailand during the 1960s and 1970s, principally at the sites of Non Nok Tha and Ban Chiang, generated claims for the use of tin and the production of bronze artifacts dating to c. 3500 B.C. (Gorman and Charoenwongsa 1976; Bayard 1980). It was possible then to claim that tin was used to alloy copper in Thailand earlier than anywhere else in the world (or at least as early).

This new view of the Bronze Age was greeted with great enthusiasm by some, with great skepticism by others. The ambiguities in the archaeological evidence were not helped by the confused series of radiocarbon dates from Non Nok Tha and the failure to publish the radiocarbon dates from Ban Chiang [for a summary of the controversy, see Muhly (1981)]. A review of the problem published in 1983, with contributions by several leading scholars in the field, demonstrated clearly just how confused the situation had become (Loofs-Wissowa 1983). There were charges and countercharges, but there was little that all participants could agree on. Donn Bayard did little to advance understanding in his 1979 paper at the Colloquy on South East Asia held in 1973 (Bayard 1979). A reviewer of this volume concluded that she belonged "with the people who have been taught to disbelieve until the case is proven" (Stargardt 1981, p. 334).

Further work by Joyce White on the material from Ban Chiang stored in the University Museum of the University of Pennsylvania and renewed excavation in Thailand by Charles Higham at sites such as Ban Na Di and Khok Phanom Di have provided additional support for the position that no metal finds from Thailand date before c. 2000 B.C. [White 1982; Higham 1984; see also Stech and Maddin (this volume)]. This means that Thailand can no longer be considered a pioneer or innovator in the development of bronze metallurgy.

It is equally true, however, that metallurgy in Thailand can no longer be seen as an industry borrowed from China. In the past scholars looked on metallurgical developments in Southeast Asia as if they were but a pale reflection of what was happening in the important metallurgical centers in the north, all located in China. The earliest metal artifacts in Southeast Asia were thought to be the Dông-sỏn drums, named after a village in northern Vietnam, and even they were of uncertain date. Many maintained, and some scholars still maintain, that the earliest Dông-sỏn drums are to be placed in the fourth or even third century B.C. (Sørensen 1979).

This dating may well turn out to be correct, but the Dông-sỏn drums can no longer be equated with the beginnings of metallurgy in Southeast Asia. Even though the high dates for Non Nok Tha and Ban Chiang have been drastically lowered, the revised chronology still places the beginnings of metallurgy in Thailand at the same date now used for China, roughly 2000 B.C. Furthermore, the same date seems to apply to Vietnamese metallurgy as well. A recent survey of work in Vietnam places the beginnings of the Bronze Age there (the Gò Bông Epoch) at c. 2000 B.C. [Davidson 1975, pp. 88–90; see also Hà Văn Tân (1980, pp. 125–127)]. From the site of Dôc Chùa, 25 km northeast of Ho Chi Minh City, comes a single radiocarbon date of 3145 ± 130 B.P. [ZK-422, as cited by Higham (1984, pp. 236, 256)]. When calibrated, this gives a date (according to the 1973 MASCA calibration curve) of 1500 ± 130 B.C. Excavations at Dôc Chùa produced sandstone molds for casting bronze artifacts (figure 1.10) and a material culture identical with that known from contemporary sites in northeast Thailand [for the Dôc Chùa date, see Hà Văn Tân (1980, pp. 135–136) and Bayard (1984, p. 313)].

It should also be pointed out that the development of bronze metallurgy in Thailand and Vietnam seemed very different from that of China. Northeastern Thailand and northern Vietnam developed a hammering and casting tradition that made use of simple stone molds and emphasized the production of basic tools and implements—axes, needles, fish hooks, woodworking and engraving tools, plowshares, and ornaments, such as bracelets and rings. Typologically and technically this industry seemed more akin to the metallurgy of western Asia and of Europe than to any metallurgical tradition known from China.

It must also be realized that the early development of bronze metallurgy in Thailand and Vietnam is not paralleled elsewhere in the area outside of China. In Korea, for example, the Early Bronze Age is placed at c. 700–300 B.C. The site of Songguk-ni, located in the hilly area of south-central Korea, has produced some of the earliest bronze finds from the country, dated c. 600–400 B.C. According to Richard Pearson, the leading Western scholar on the archaeology and prehistory of Korea, the latest interpretation would have it that "with the accumulation of finds from scientific excavations and radiocarbon dates, it appears that bronze technology began in Korea early in the first millennium B.C. and that bronze objects were actually manufactured on Korean soil" (Pearson 1982, pp. 25–26).

The evidence from the Malay peninsula is unsatisfactory. According to a survey, "The Origin of the

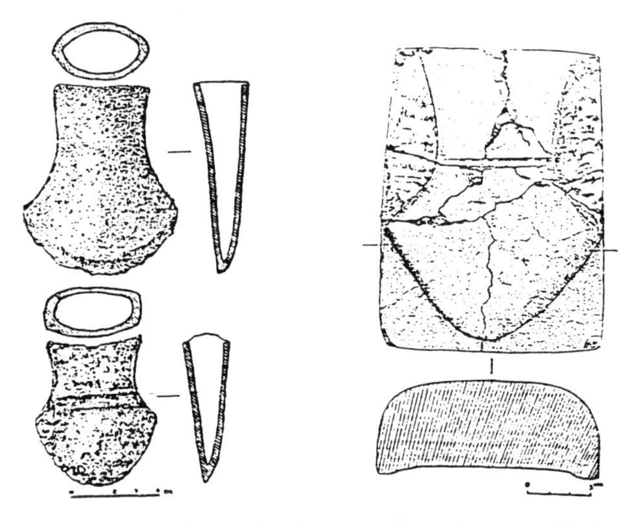

Figure 1.10 Bronze hand axes and sandstone mold for casting the same type of ax from the site of Dôc Chùa, northern Vietnam [from Hà Văn Tân, *Bulletin de l'École française d'Extrême-Orient* (1980), 68:153, fig. 10c, e].

Malayan Metal Age," published in 1962 but operating within what was, even for that time, an antiquated historical framework, "the first conclusions ... that an archaeologist reaches, is that there has never been an indigenous 'Bronze Age' in Malaya, for the simple reason that the country is virtually devoid of copper ores" (Loewenstein 1962, pp. 6–7). This conclusion seems to be supported by more up-to-date work in the area (Peacock 1979, p. 214).

Far more surprising is the late date for the beginnings of metallurgy in Japan. The production of pottery in Japan has a long history going back to c. 10,000 B.C. The pottery from the Fukui Cave site, Nagasaki Prefecture, is often called "the oldest pottery in the world" (Pearson 1978, p. 53). The production of pottery became highly developed by the Early Jomon Period (c. 5300–3600 B.C.), but it was only the much later Yayoi Period (300 B.C.–A.D. 300) that saw the development both of wet rice agriculture and bronze metallurgy (Ikawa-Smith 1980, pp. 141–142).

The earliest bronze finds, from Fukuoka Prefecture in North Kyushu, dating to Early Yayoi times (c. 300–100 B.C.), seem to be imports from Korea and from Han Dynasty China. It was only during the Middle Yayoi (c. 100 B.C.–A.D. 100) that local production of the characteristic bronze mirrors, bells, and weapons took place, making use of the local soft tuff for the manufacture of stone molds (Barnes 1981). We know from contemporary Chinese texts that during the Yayoi Period the Fukuoka Prefecture was known as the country of Nu to the Chinese of the Han Dynasty (Barnes 1981, p. 41). By Late Yayoi times (c. A.D. 100–300) Japan seems to have entered the Iron Age, a development that soon led to the Japanese annexation of the southeastern tip of the Korean peninsula, which aimed at controlling the iron ore deposits of that area, known to the Japanese as Mimana (Sansom 1958, pp. 16, 47).

Beginnings of Metallurgy in China

This brings us to the final and in many respects the most complex problem to be discussed here: the beginnings (and origins) of Chinese metallurgy. At the time of the "The Great Bronze Age of China" exhibition in the United States (1980), there was much talk about metal finds from China dating as early as the fourth and fifth millennia B.C. These finds include an amorphous piece of copper from Banpo in Xi'an, Shaanxi Province, said to have a high content of zinc and nickel, and a small disk from Jiangzhai, near Banpo, said to be of copper with 25% zinc. Both of these objects could be described as being made of brass (Chang 1980, p. 36).

It is quite impossible to accept the existence of brass artifacts from the time of the Yangshao Neolithic in

China (c. 6000–3000 B.C.). The objects in question are dated to c. 5000 B.C., and such a date is supported by four radiocarbon dates from Banpo itself (ZK-38, ZK-121, ZK-122, ZK-127). At question is not the date of the Yangshao Neolithic but the context of these two artifacts that are almost certainly to be interpreted as intrusive pieces of modern metal. The same probably holds for two brass rods from Sanlihe, Jiaoxian, Shandong Province, said to belong to the Longshan Culture, which, on the basis of a series of fourteen calibrated radiocarbon dates (Barnard 1983, p. 240), can be dated c. 3000–2400 B.C. (An Zhimin 1982–1983, pp. 53–58).

Although these artifacts have already found their way into popular works on Chinese archaeology, they cannot play a role in any serious discussion of early Chinese metallurgy. Much more interesting are the reports of two bronze knives from a third millennium B.C. context in China. The earliest of these is a bronze knife from Linjia, Dongxiang, Gansu Province, which belongs to the Majiayao Neolithic Culture, said to date to 2750 B.C. When calibrated, the two radiocarbon dates from this period (ZK-108, BK-77013) would fall at the end of the fourth millennium B.C. The Majiayao knife is now claimed to be "the earliest bronze artifact unearthed in China" (Zhu Shoukang 1982, p. 5 and figure 4). The knife was made of bronze with 6–10% tin and cast in an open mold (for a list of radiocarbon dates, see appendix).

The second bronze knife is from the site of Jiangjiaping, Yongdong, Gansu Province, and dates to the Machang Neolithic Culture, for which there is a series of radiocarbon dates (ZK-348, ZK-345, BK-75009, BK-75012, ZK-346) that, when calibrated, cover the period 2400–2000 B.C. (Zhu Shoukang 1982, p. 5). These finds are credible not only because of their archaeological context but also because of their geographical location; they come from the very region—Gansu Province—that has produced the metal finds that have transformed our understanding of the beginnings of copper and bronze metallurgy in China.

From the Qijia Culture—regarded as a Chalcolithic culture to be placed between its Neolithic predecessor, the Machang, and the Early Bronze Age Xia culture—come forty-three metal artifacts. Twenty-five of these have been subjected to chemical analysis. Roughly two-thirds (sixteen) were made of unalloyed copper, whereas the remaining were made of the tin, lead, and tin-lead alloys also known from the succeeding period. The two radiocarbon dates associated with the metal finds (ZK-15, ZK-23) suggest, when calibrated, a date of c. 2000 B.C. (Zhu Shoukang 1982, p. 5, table 2).

This is then the long-sought missing Copper Age, a Chalcolithic period during which Chinese smiths were making artifacts of unalloyed copper (Dōno 1975). Moreover, the Qijia artifacts are far removed from the bronze ritual vessels of the Shang Dynasty. During the time of the Qijia Culture, copper was used to fashion such tools as awls, knives, adzes, and sickles, the basic repertoire of the metalsmith in all other parts of the world. Moreover, this was a hammering and casting technology producing simple functional objects that were cast in stone molds and then cold-worked to final form. This is the tradition of metalworking said to be anathema to Chinese smiths and to basic Chinese concepts of the nature of metal and the relationship between metal and clay (Barnard 1965).

Whether or not riveted sheet metal work was also being produced at Qijia Culture sites, as strongly argued by some (Bylin-Althin 1946; Bagley 1977, pp. 197–200; Rawson 1980, pp. 27, 57) and stoutly denied by others, notably Noel Barnard [in numerous studies but especially in Barnard (1980–1981)], must remain problematic, pending discovery of at least fragments of the vessels themselves. On the basis of what has survived, it can no longer be maintained that metallurgy began in China in ways radically different from those known elsewhere. Nor were the Qijia developments an anomaly; the metallurgical technique now documented for the Qijia Culture carried over into the following period.

In 1976 the excavation of tombs in the vicinity of Huoshaogou, Yumen, Gansu Province, produced over 200 copper and bronze artifacts, consisting of axes, sickles, chisels, knives, daggers, awls, tubes, arrowheads, and spearheads (Zhu Shoukang 1982, p. 4). These objects soon came to be known as the Gansu metals. Here was the same use of metal for practical utilitarian tools and weapons, not for ceremonial wine vessels. Culturally the finds seemed to represent the material culture of the Xia Dynasty, the predecessor of the Shang. According to traditional chronology, the Xia Dynasty is to be dated 2205–1760 B.C. The relevant radiocarbon dates published by K. C. Chang (1980b, pp. 370–371), when calibrated, do seem to support the traditional dating.[3]

Of the over 200 metal artifacts discovered at Huoshaogou, 66 have been subjected to chemical analysis. Of these, roughly one-half (thirty) were made of unalloyed copper, the rest of tin, lead, or tin-lead alloys. This means that the basic copper alloys of the Shang Dynasty were already in use during the times of the Qijia Culture and the Xia Dynasty. Most of the Xia artifacts had been cast in molds, and some showed evidence of cold-working (An Zhimin 1982–1983, pp. 64–66).

The sequence of developments whereby the Gansu metals became the ceremonial and ritual bronze vessels of the Shang Dynasty is the subject of another paper, which I hope to publish in the near future. At present we must concentrate on the ways in which the beginnings of copper (and bronze) metallurgy in China were related to similar developments in Vietnam and Thailand and across Asia to the lands of ancient Mesopotamia and beyond.

First Use of Metal

I would like to return first to a point made at the beginning of this chapter, the extraordinary leap of ingenuity involved in grasping the possibility of turning rock into metal. How is it possible that such a profound discovery could have been made, in very much the same way, over and over again, at different times in different parts of the world. We are not talking about the way metallurgy *developed* in China, for in that sense Shang Dynasty bronzes represent Chinese metallurgy, a metallurgy that developed in special ways, because of conditions unique to China. What is at issue here is the genesis of the concept of working with metals and the initial development of that complex pyrotechnology necessary for the extraction of copper and tin from their minerals and ores.

The same point was made almost fifteen years ago by Cyril Stanley Smith, the scholar who has done more than anyone else to create both the field of archaeometallurgy and the modern discipline of theoretical research into the structure of materials. Smith made it clear that his conclusions represented very much his own point of view:

Personally, I believe that the idea of making metals from rocks is not easily arrived at and a second independent nucleation of metallurgy in China (or the New World either) seems highly improbable. This by no means denies that there is something uniquely Chinese in the technique and beauty of the superb bronzes of Shang and Chou China, but this seems to lie in the making of moulds and the decoration of them, not necessarily in the origin of metallurgy itself. (Smith 1973, p. 23)

A persuasive but dangerous argument; it could be seen as the first step on the path that led Heine-Geldern to find the origins of Peruvian metallurgy in the Caucasus (Heine-Geldern 1974, esp. p. 810). The difference again lies in the application of theory and the degree to which one is willing to push a basic idea.

The likelihood of the initial discoveries in metallurgical technology being unique ones was also emphasized over twenty years ago by the late Theodore Wertime in one of the most influential papers ever published in the field of archaeometallurgy, one that can rightly be said to have changed the course of scholarship in the field. Wertime argued that "one must doubt that the tangled web of discovery, comprehending the art of reducing oxide and then sulfide ores, the recognition of silver, lead, iron, tin, and possibly arsenic and antimony as distinctive new metallic substances, and the technique of alloying tin with bronze, could have been spun twice in human history" (Wertime 1964, p. 1266). This passage has recently been quoted, in order to refute it, in a new study on the finds from the Varna necropolis by Colin Renfrew (Renfrew 1986a).

Wertime and Renfrew debated the question of diffusion versus independent invention at the International Congress of Pre- and Protohistory held in Belgrade in 1971 (Wertime 1973; Renfrew 1973). Renfrew has now returned to the subject claiming, in reference to the Wertime passage just cited, that

this is an argument that need not be replayed here. It has effectively been demonstrated that metallurgy in the New World had separate origins and developments from those in the Old World. It is even likely that metallurgy in China developed independently from that in Western Asia. It has been argued on several occasions that there is a good case for the independent origin of metallurgy in southeast Europe. The same is true for Iberia. What interests us here is to understand more clearly how these independent innovations of a new commodity may have occurred. (Renfrew 1986a, p. 145)

Regarding the separate development of metallurgy in the New World, there would be no argument, but separate origin is another matter. It is not sound scholarship to use a hard-line stand on a controversial problem in order to refute a position taken by several eminent scholars in the history of technology. Few would argue with Renfrew's position that "the decisive innovation in the development of a new commodity is generally social rather than technical" (Renfrew 1986a, p. 146); the same point has been made by Cyril Smith on several occasions [especially Smith (1981, pp. 325–331)].

More interesting is Renfrew's observation that, with regard to the first use of metal, the real explanatory problem

is not to explain why people did not at once use this great technology. It is, on the contrary, to understand why they bothered at all. For until the technology develops a great deal further, as it can only do through intensive use, early copper metallurgy does not produce anything decisively useful at all. The artifacts that can be produced from native copper by an annealing process have very few properties to recommend them in comparison to well-chosen stones, and many that are lacking. (Renfrew 1986a, p. 146)

The faulty reference to annealing, only just noticed, insofar as early native copper metallurgy is concerned, in the artifacts from Çayönü discussed at the beginning of this chapter, makes clear that we have here the words of an archaeologist quite removed from the realm of materials science. Anyone who has worked with copper will recognize right away that the beads, pins, awls, and reamers that make up the corpus of early copper artifacts do not represent a random typological selection. They were made of copper in order to take advantage of the special mechanical properties of metal, notably that of ductility.

Right from the start, with the simple hammering of native copper, it was possible to do things with metal that could not be done with any stone. The evidence from Çayönü and from the numerous early metal-using sites in northern Mesopotamia indicates that both the decorative and the utilitarian qualities of metal were recognized from the start. Renfrew has caught only half of the picture when he argues that early metallurgy was practiced "because the products had novel properties that made them attractive to use as symbols and as personal adornments and ornaments, in a manner that by focusing attention, could attract or enhance prestige" (Renfrew 1986a, p. 146).

What do such purposes have to do with an awl from Çayönü or the awl from Tell Maghzaliyeh? Renfrew's hypothesis simply does not account for the ways in which metal actually was used during the period c. 7000–5000 B.C.

Renfrew's view of the beginnings of metallurgy in China is totally at variance with the evidence presented here. He maintains that "in the same way in China, evidence for a copper age or for the early development of bronze metallurgy may largely be lacking precisely because metal was used primarily for the production of prestige objects, notably bronze vessels, and stray finds of inconsequential objects are very rare" (Renfrew 1986a, p. 147). Such a position was defensible when we had only the evidence from looted bronzes of museum quality and that from the excavations at Anyang, but to present it in 1986 is to ignore all the archaeological work carried out in China since the end of the Cultural Revolution. The interesting question today is not *why* the beginnings of metallurgy in China were different from those in the West, because as we have seen they really were not so different, but *how* developments in China related to what was happening elsewhere in east Asia and Southeast Asia.

For relationships between Thailand and Vietnam on the one hand and China on the other, the vast, largely unexplored area of south China holds the key. We know almost nothing about the prehistory of this crucial area. In a recent study, devoted to many of the same issues discussed here, William Watson observes that "before the second half of the Shang reign the typical Shang bronzework is confined to the Zhengzhou region" (Watson 1985, p. 343). We have little idea what was going on in the south during the third and second millennia B.C.

Watson concludes that "only in the period from 2500 to 200 B.C. does the Chinese evidence for the use of metal accumulate to a significant degree, and pure copper, hammered and cast, is still more frequent than tin bronze" (Watson 1985, p. 332). The same holds true for the rest of the Old World.

Although Watson is surely wrong in accepting the fifth to fourth millennium B.C. date for the earliest metal artifacts in China (Watson 1985, pp. 329–330), his paper represents a welcome antidote to the heavily autochthonous flavor of much recent archaeological scholarship. His evaluation of the long-running controversy over the beginnings of bronze metallurgy in Southeast Asia deserves to be quoted in full: "This controversy, largely obscure to the outside world through the controversialists' appeal to unpublished material and inaccessible statements, turns on the vindication of the independence and originality, technical and artistic, of bronze-age culture present on China's south-west border and in the countries of the southern seas" (Watson 1985, p. 357). Yet, as Watson realizes, it is not just for chronological reasons that an understanding of the beginnings of metallurgy in Southeast Asia still eludes us.

What we lack most of all is any sense of the underlying cultural and political contexts associated with the emergence of bronze technology in Southeast Asia (Watson 1985, p. 357). In all other corners of the Bronze Age world—China, Mesopotamia, Anatolia, the Aegean, and central Europe—we find the introduction of bronze metallurgy associated with a complex of social, political, and economic developments that mark the "rise of the state." Only in Southeast Asia, especially in Thailand and Vietnam, do these developments seem to be missing, and explaining (or eliminating) this anomalous situation is one of the major challenges of archaeological and archaeometallurgical research during the next decade.

What work during the past decade has accomplished is the solid documentation of the long-missing formative state in the development of Chinese copper and bronze metallurgy. The beginnings of metallurgy in China have now been shown to be not so much different from what can be seen elsewhere throughout the ancient world. This neither proves nor disproves the independent invention of metallurgy in east Asia, but it does move the argument into a more comprehensible frame of reference and should eliminate once and for all the argument that Chinese metallurgical technology developed in ways totally different (and thus apart from) anything known in the West.

Appendix

List of Chinese radiocarbon dates cited in text, taken from N. Barnard, *Radiocarbon Dates and Their Significance in the Chinese Archaeological Scene* (Canberra: Department of Far Asian Studies, Australian National University, 1979).

ZK-15	3935 ± 105 B.P.
ZK-23	3645 ± 95 B.P.
ZK-38	6065 ± 110 B.P.
ZK-108	4525 ± 100 B.P.
ZK-121	5905 ± 105 B.P.
ZK-122	5840 ± 105 B.P.
ZK-127	5585 ± 105 B.P.
ZK-345	3865 ± 120 B.P.

ZK-346	3665 ± 80 B.P.
ZK-348	3970 ± 240 B.P.
BK-75009	3860 ± 90 B.P.
BK-75012	3750 ± 100 B.P.
BK-77013	4450 ± 90 B.P.

Addendum

The important article by Sun Shuyun and Han Rubin (1983–1985) lists 299 metal artifacts coming from pre-Shang and early Shang contexts, including 48 from the Longshan and Qijia cultures, over 200 from the Huoshaogou culture and 28 from Erlitou. Of these 299 artifacts only 30 have been subjected to detailed technical analysis (pp. 287–289, table 8). Of these, 11 were made of copper, 18 of bronze, and 1 of brass. The brass artifact in question is the one from Sanhile, Jiaoxian, Shandong Province (see p. 14), and the authors argue in some detail that "it was technically possible to produce brass at a very early date. It only required a place where copper and zinc ores both existed for brass to be made by primitive smelting (perhaps with repeated meltings)" (Sun Shuyun and Han Rubin 1983–1985, 270). This could indeed explain the isolated appearance of objects with high zinc contents.

In light of my comments on page 16, I would like to call attention to a conference paper by C. F. W. Higham (1988). To quote the opening line of Higham's abstract: "This paper addresses the issue of the social, economic, technological and chronological contexts within which Southeast Asian bronze working took root." I am sure that we all look forward to the publication of this paper, as well as the fifty-odd other papers delivered at the conference on Ancient Chinese and Southeast Asian Bronze Age Cultures, which was organized to pay tribute to the work of Noel Barnard, one of the foremost authorities on ancient Chinese bronzes.

Notes

An earlier version of this chapter was published in a special issue of the *Bulletin of the Metals Museum* (Sendai, Japan), vol. 11 (1986), which contains the Proceedings of the Symposium on the Early Metallurgy of Japan and the Surrounding Area (Sendai, 16–17 October 1986).

1. And the traffic in illicitly excavated Chinese bronzes still continues. See "Of Gansu pots and Tang tripods," *The Economist* (1 November 1986), p. 96.

2. The problems of the missing fourth millennium B.C. in the Aegean was the subject of a special symposium on the Final Neolithic, held as part of the International Symposium on Archaeometry held in Athens, Greece (Greek Atomic Energy Commission, Demokritos), 19–23 May 1986, organized by C. Doumas, A. Hourmouziades, and C. Renfrew.

3. There is a tendency in recent scholarship to argue that archaeology has confirmed the traditional account of China's prehistory. This is part of a general return to faith that can be seen in publications dealing with Mycenaean Greece, the Hittites, and the Trojan War and with the founding of the city of Rome and the "history" of Regal Rome. For Chinese mythological traditions connected with the Xia Dynasty, see Allan (1984). Archaeologically the Xia Dynasty is best seen as represented by the late Henan Longshan Culture and the first two phases at Erlitou (Zhao Zhiquan 1985).

References

Adams, W. Y., D. P. Van Gerven, and R. S. Levy. 1978. "The retreat from migrationism." *Annual Review of Anthropology* 7:483–532.

Agrawal, D. P. 1982. "The Indian Bronze Age cultures and their metal technology," in *Advances in World Archaeology*, F. Wendorf and A. E. Close, eds. (London: Academic Press), vol. 1, 213–264.

Allan, S. 1984. "The myth of the Xia Dynasty." *Journal of the Royal Asiatic Society* 1984:242–256.

Amiet, P. 1979. "Archaeological discontinuity and ethnic duality in Elam." *Antiquity* 53:195–204.

An Zhimin. 1979–1980. "The Neolithic archaeology of China. A brief survey of the last thirty years." *Early China* 5:35–45.

An Zhimin. 1982–1983. "Some problems concerning China's early copper and bronze artifacts." *Early China* 8:53–75.

Bagley, R. W. 1977. "P'an-lung-ch'eng: A Shang city in Hupei." *Artibus Asiae* 39:165–213.

Bar-Adon, P. 1980. *The Cave of the Treasure. The Finds from the Caves in Nahal Mishmar*. Jerusalem: Israel Exploration Society.

Barnard, N. 1965. "Review of Chêng Tê-k'un, Chou China (Cambridge, 1963)." *Monumenta Serica* 24:307–459.

Barnard, N. 1980–1981. "Wrought metal-working prior to Middle Shang (?): A problem in archaeological and art-historical research approaches." *Early China* 6:4–30.

Barnard, N. 1983. "Further evidence to support the hypothesis of indigenous origins of metallurgy in ancient China," in *The Origins of Chinese Civilization*, D. N. Keightley, ed. (Berkeley, Calif.: University of California Press), 237–277.

Barnes, G. 1981. "Early Japanese bronzemaking." *Archaeology* 34(3):38–46.

Bayard, D. T. 1975. "North China, South China, Southeast Asia or simply 'Far East'?" *Journal of the Hong Kong Archaeological Society* 6:71–79.

Bayard, D. T. 1979. "The chronology of prehistoric metallurgy in northeast Thailand: *Silābhūmi* or *Samrddhabhūmi*?" in *Early South East Asia*, R. B. Smith and W. Watson, eds. (Oxford: Oxford University Press), 15–32.

Bayard, D. T. 1980. "An early indigenous bronze technology in northeast Thailand: Its implications for the prehistory of East Asia," in *The Diffusion of Material Culture*, H. H. E. Loofs-Wissowa, ed. (Asian and Pacific Archaeology Series, no. 9) (Manoa, Hawaii: University of Hawaii), 191–214.

Bayard, D. T. 1984. "A checklist of Vietnamese radiocarbon dates," in *Southeast Asian Archaeology at the XV Pacific Science Conference*, D. Bayard, ed. (Dunedin, New Zealand: University of Otago Studies in Prehistoric Anthropology), vol. 16, 305–322.

Berthoud, T. 1979. *Étude par l'analyse et la modélisation de la filiation entre minerais de cuivre et objects archéologiques du Moyen Orient (IVe-IIIe millénnaire)*. Ph.D. dissertation, Université de Paris.

Bishop, C. W. 1942. *Origin of the Far Eastern Civilizations. A Brief Handbook*. Publication 3681. Washington, D.C.: Smithsonian Institution.

Bognár-Kutzián, I. 1976. "On the origins of early copper-processing in Europe," in *To Illustrate the Monuments: Festschrift Stuart Piggott*, J. V. S. Megaw, ed. (London: Thames and Hudson), 70–76.

Braidwood, R. J., H. Çambel, and W. Schirmer. 1981. "Beginnings of village-farming communities in southeastern Turkey: Çayönü Tepesi, 1978 and 1979." *Journal of Field Archaeology* 8: 249–258.

Bylin-Althin, M. 1946. "The sites of Ch'i Chia P'ing and Lo Han T'ang in Kansu." *Bulletin of the Museum of Far Eastern Antiquities, Stockholm* 18: 383–498.

Caldwell, J. R. 1968. "Pottery and cultural history on the Iranian Plateau." *Journal of Near Eastern Studies* 27: 178–183.

Çambel, H., and R. J. Braidwood. 1983. "Çayönü Tepesi: Schritte zu neuen Lebensweisen," in *Beiträge zur Altertumskunde Kleinasiens: Festschrift Kurt Bittel*, R. M. Boehmer and H. Hauptmann, eds. (Mainz: Philipp von Zabern), 155–166.

Cauvin, J. 1985. "Le Néolithique de Cafer Höyük (Turquie): Bilan provisoire après quatre campagnes (1979–1983)." *Cahiers de l'Euphrate* 4: 123–133.

Chang, K. C. 1977. *The Archaeology of Ancient China*, third edition. New Haven, Conn.: Yale University Press.

Chang, K. C. 1980a. "The Chinese Bronze Age: A modern synthesis," in *The Great Bronze Age of China*, Wen Fong, ed. (New York: Metropolitan Museum of Art), 35–50.

Chang, K. C. 1980b. *Shang Civilization*. New Haven: Yale University Press.

Chang, K. C. 1981. "In search of China's beginnings: New light on an old civilization." *American Scientist* 69(2): 148–160.

Chapman, J. C., and Tylecote, R. F. 1983. "Early copper in the Balkans." *Proceedings of the Prehistoric Society* 49: 373–376.

Chernykh, N. E. 1982. "Die ältesten Bergleute und Metallurger Europas." *Das Altertum* 28: 5–15.

Cherry, J. F. 1986. "Polities and palaces: Some problems in Minoan state formation," in *Peer Polity Interaction and Socio-Political Change*, C. Renfrew and J. F. Cherry, eds. (Cambridge: Cambridge University Press), 19–45.

Childe, V. G. 1944. "Archaeological ages as technological stages." *Journal of the Royal Anthropological Institute of Great Britain and Ireland* 74: 7–24.

Cooke, S., R. B. Nielsen, and B. V. Nielsen. 1978. "Slags and other metallurgical products," in *Excavations at Nichoria in Southwest Greece. Vol. I, Site, Environs and Techniques*, G. Rapp, Jr., and S. E. Aschenbrenner, eds. (Minneapolis: University of Minnesota Press), 182–224.

Creel, H. C. 1935. "On the origins of the manufacture and decoration of bronze in the Shang Period." *Monumenta Serica* 1: 39–69.

Davidson, J. H. C. S. 1975. "Recent archaeological activity in Viet-Nam." *Journal of the Hong Kong Archaeological Society* 6: 80–99.

Deshayes, J. 1972. "Dikili Tash and the origins of the Troadic Culture." *Archaeology* 25: 198–205.

Dōno, T. 1975. "On the Copper Age and its transitional period to the Bronze Age in ancient China," in *XIVth International Congress of the History of Science. Tokyo and Kyoto 1974* (Tokyo), vol. 3, 284–287.

Esin, U. 1969. *Kuantitatif Spektral Analiz Yardimiyla Anadolu'da Baslangicindan Asur Kolonileri Çağina Kadar Bakir ve Tunç Madenciliği*. Istanbul: Tas Matbaasci, Istanbul Universtesi Yarlinlar, Novaa.

Esin, U. 1976. "Die Anfänger der Metallverwendung und Bearbeitung in Anatolien (7500–2000 v. Chr.)," in *Les débuts de la métallurgie*, H. Müller-Karpe, ed. (Nice: UISPP), 209–239.

French, D. 1962. "Excavations at Can Hasan. First Preliminary Report, 1961." *Anatolian Studies* 12: 27–40.

Gale, N. H., and Z. A. Stos-Gale. 1982. "Bronze Age copper sources in the Mediterranean: A new approach." *Science* 216: 11–19.

Gorman, C., and P. Charoenwongsa. 1976. "Ban Chiang: A mosaic of impressions from the first two years." *Expedition* 18(4): 14–26.

Hà Văn Tân. 1980. "Nouvelles recherches préhistoriques et protohistoriques au Vietnam." *Bulletin de l'École française d'Extrême-Orient* 68: 113–154.

von Heine-Geldern, R. 1974. "American metallurgy and the Old World," in *Early Chinese Art and Its Possible Influences in the Pacific Basin*, N. Barnard, ed. (Taiwan: Interculture Arts Press), 787–822.

Higham, C. F. W. 1983. "The Ban Chiang culture in wider perspective." *Proceedings of the British Academy* 69(1983): 229–261.

Higham, C. F. W. 1988. "The social and chronological contexts of early bronze working in Southeast Asia." Paper presented at the conference Ancient Chinese and Southeast Asian Bronze Age Cultures, Kioloa, New South Wales, February 8–12.

Ho, Ping-ti. 1975. *The Cradle of the East.* Hong Kong and Chicago: The Chinese University of Hong Kong and the University of Chicago.

Ikawa-Smith, F. 1980. "Current issues in Japanese archaeology." *American Scientist* 68:134–145.

Jarrige, J. F. 1984. Personal communication, 21 May 1984.

de Jesus, P. S. 1980. *The Development of Prehistoric Mining and Metallurgy in Anatolia.* 2 parts. BAR International Series 74. Oxford: British Archaeological Reports.

Jovanović, B. 1985. "Smelting of copper in the Eneolithic Period of the Balkans," in *Furnaces and Smelting Technology in Antiquity,* P. T. Craddock and M. J. Hughes, eds. (London: British Museum), 117–121.

Kear, B. H. 1986. "Advanced metals." *Scientific American* 255(4):158–167.

Keightley, D. N. 1977. "Ping-ti Ho and the origins of Chinese civilization." *Harvard Journal of Asiatic Studies* 37: 381–411.

Lambert, N. 1973. "Les industries de la grotte néolithique de Kitsos." *Actes du VIIIe Congrès internationale des sciences préhistorique et protohistorique. Beograd 1971* (Belgrade), vol. 2, 502–515.

Loehr, M. 1974. "China: Prehistory and archaeology," in *Encyclopaedia Britannica, Macropaedia,* fifteenth edition (Chicago: William Benton), 297–301.

Loewenstein, J. 1962. "The origins of the Malayan metal age." *Journal of the Malayan Branch of the Royal Asiatic Society* 29(174) (1956):5–73.

Loofs-Wissowa, H. H. E. 1983. "The development and spread of metallurgy in Southeast Asia: A review of the present evidence." *Journal of Southeast Asian Studies* 14: 1–11 (with replies by D. Bayard and P. Charoenwongsa, pp. 12–17; W. G. Solheim, III, pp. 18–25, and answer by Loofs-Wissowa, pp. 26–31).

Majidzadeh, Y. 1979. "An early prehistoric coppersmith workshop at Tepe Ghabristan," in *Akten des VII. Intern. Kongress für Kunst und Archäologie* (München: Archäologische Mitteilungen aus Iran, Ergänzungband 6), 82–92.

Makkay, J. 1985a. "The crisis of prehistoric chronology." *Mitteilungen des archäologischen Instituts der ungarischen Akademie der Wissenschaft* 14:53–70.

Makkay, J. 1985b. "Diffusion, antidiffusionism and chronology: Some general remarks." *Acta Archaeologica Academiae Scientiarum Hungaricae* 37:3–12.

Moorey, P. R. S. 1982. "The archaeological evidence for metallurgy and related technologies in Mesopotamia, c. 5500–2100 B.C." *Iraq* 44:13–38.

Muhly, J. D. 1973. *Copper and Tin. The Distribution of Mineral Resources and the Nature of the Metals Trade in the Bronze Age.* Hamden, Conn.: Archon Books.

Muhly, J. D. 1981. "The origins of agriculture and technology—West or East Asia?" *Technology and Culture* 22:125–148.

Muhly, J. D. 1983. "Kupfer. B. Archäologisch," in *Reallexikon der Assyriologie und Vorderasiatischen Archäologie.* (Berlin: de Gruyter), vol. 6, 348–364.

Muhly, J. D. 1984. "Timna and King Solomon." *Bibliotheca Orientalis* 41:275–292.

Muhly, J. D. 1985. "Beyond technology: Aegean metallurgy in its historical context," in *Contributions to Aegean Archaeology: Studies in Honor of William A. McDonald,* N. C. Wilkie and W. D. E. Coulson, eds. (Minneapolis, Minn.: Center for Ancient Studies), 109–141.

Neuninger, H., R. Pittioni, and W. Siegl. 1964. "Frühkeramikzeitliche Kupfergewinnung in Anatolien." *Archaeologia Austriaca* 35:98–110.

Ottaway, B. 1973. "The earliest copper ornaments in northern Europe." *Proceedings of the Prehistoric Society* 39:294–331.

Patterson, C. C. 1971. "Native copper, silver and gold accessible to early metallurgists." *American Antiquity* 36: 286–321.

Peacock, B. A. V. 1979. "The later prehistory of the Malay Peninsula," in *Early South East Asia,* R. B. Smith and W. Watson, eds. (Oxford: Oxford University Press), 199–214.

Pearson, R. 1976. "Review of Ping-ti Ho, *The Cradle of the East* (Hong Kong and Chicago, 1975)." *Science* 193: 395–396.

Pearson, R. 1978. "Image and life: Fifty thousand years of Japanese prehistory." *Archaeology* 31(6):53–56.

Pearson, R. 1982. "The archaeological background to Korean prehistoric art." *Korean Culture* 3(4):18–29.

Quitta, H. 1986. "Radiocarbon und die Chronologie der Jungsteinzeit in Bulgarien." *Das Altertum* 32:113–117.

Rapp, G. 1982. "Native copper and the beginnings of smelting: Chemical studies," in *Early Metallurgy in Cyprus, 4000–500 B.C.,* J. D. Muhly, R. Maddin, and V. Karageorghis, eds. (Nicosia: Pierides Foundation), 33–40.

Rawson, J. 1980. *Ancient China: Art and Archaeology.* London: British Museum.

Renfrew, C. 1973. "Sitagroi and the independent invention of metallurgy in Europe," in *Actes du VIIIe Congrès internationale des sciences préhistorique et protohistorique. Beograd. 1971* (Belgrade), vol. 2, 473–481.

Renfrew, C. 1986a. "Varna and the emergence of wealth in prehistoric Europe," in *The Social Life of Things. Commodities in Cultural Perspective,* A. Appadurai, ed. (Cambridge: Cambridge University Press), 141–168.

Renfrew, C. 1986b. "Introduction: Peer polity interaction and socio-political change," in *Peer Polity Interaction and Socio-Political Change*, C. Renfrew and J. F. Cherry, eds. (Cambridge: Cambridge University Press), 1–18.

Rothenberg, B. 1983. "Explorations and archaeo-metallurgy in the Arabah Valley (Israel)." *Bulletin of the Metals Museum* (Sendai, Japan) 8 : 1–16.

Ryndina, N. V., and L. K. Yakhontova. 1985. "The earliest copper artifact from Mesopotamia." *Sovetskaya Arkheologiya* 1985(2) : 155–166. (In Russian, with English summary.)

Sansom, G. 1958. *A History of Japan to 1334*. Stanford, Calif.: Stanford University Press.

Schirmer, W. 1983. "Drei Bauten des Çayönü Tepesi," in *Beiträge zur Altertumskunde Kleinasiens: Festschrift Kurt Bittel*, R. M. Boehmer and H. Hauptmann, eds. (Mainz: Philipp von Zabern), 463–476.

Selimchanow, I. S. 1977. "Zur Frage einer Kupfer-Arsen Zeit." *Germania* 55 : 1–6.

Smith, C. S. 1973. "Bronze technology in the East," in *Changing Perspectives in the History of Science. Festschrift Joseph Needham*, M. Young and R. Young, eds. (London: Heinemann), 21–32.

Smith, C. S. 1981. "On art, invention and technology," in *A Search for Structure. Selected Essays on Science, Art, and History*, C. S. Smith, ed. (Cambridge, Mass.: MIT Press), 325–331.

Solecki, R. S. 1969. "A copper mineral pendant from northern Iraq." *Antiquity* 43 : 311–314.

Sørensen, P. 1979. "The Ongbah Cave and its fifth drum," in *Early South East Asia*, R. B. Smith and W. Watson, eds. (Oxford: Oxford University Press), 78–97.

Stargardt, J. 1981. "Review of *Early South East Asia*, eds. R. B. Smith and W. Watson (Oxford, 1979)." *Journal of Historical Geography* 7 : 331–334.

Stover, L. E., and B. Kraig. 1978. *Stonehenge: The Indo-European Heritage*. Chicago, Ill.: Nelson-Hall.

Sullivan, M. 1980. "Review of *The Great Bronze Age of China*, ed. Wen Fong." *New York Times Book Review* (31 August 1980), 10–11, 18.

Sun Shuyun and Han Rubin. 1983–1985. "A preliminary study of early Chinese copper and bronze artifacts." *Early China* 9/10 : 261–289.

Todorova, H. 1981. *Die kupferzeitlichen Axte und Beile in Bulgarien*. München: Prähistorische Bronzefunde, IX/14.

Trigger, B. G. 1986. "The role of technology in V. Gordon Childe's archaeology." *Norwegian Archaeological Review* 19 : 1–14.

Tylecote, R. F. 1976. *A History of Metallurgy*. London: The Metals Society.

Warren, P. 1980. "Problems of chronology in Crete and the Aegean in the third and earlier second millennium B.C." *American Journal of Archaeology* 84 : 487–499.

Watson, W. 1985. "An interpretation of opposites? Pre-Han bronze metallurgy in West China." *Proceedings of the British Academy* (1984), 70 : 327–358.

Wertime, T. A. 1964. "Man's first encounters with metallurgy." *Science* 146(3649) : 1257–1267.

Wertime, T. A. 1973. "How metallurgy began: A study in diffusion and multiple innovation," in *Actes du VIIIe Congrès internationale des sciences préhistorique et protohistorique. Beograd, 1971* (Belgrade), vol. 2, 481–492.

White, J. C. 1982. *Ban Chiang. Discovery of a Lost Bronze Age*. Philadelphia, Penn.: University Museum.

Willey, G. 1962. "The early great styles and the rise of the Pre-Columbian civilizations." *American Anthropologist* 64 : 1–14.

Yakar, J. 1984. "Regional and local schools of metalwork in Early Bronze Age Anatolia." *Anatolian Studies* 34 : 59–86.

Zhao Zhiquan. 1985. "On the ancient site of Erlitou." *Annali, Istituto Orientali di Napoli* 45 : 287–302.

Zhu Shoukang. 1982. "Early bronze in China." *Bulletin of the Metals Museum* (Sendai, Japan) 7 : 3–15.

2
On the Origins of Copper and Bronze Alloying

George Rapp, Jr.

The smelting of copper ores goes back perhaps six millennia. Over a considerable span of time ancient metalsmiths discovered the bounty of pyrotechnology applied to metalliferous rocks. Our knowledge of the origins of copper alloy metallurgy is primarily indirect, derived from analyses and interpretations of the composition and structure of artifacts, slags, and ores. Unfortunately archaeology has focused on the excavation of temples, graves, and habitation sites. Relatively few examples of Chalcolithic or Early Bronze Age metallurgical or mining sites are known, and these have not provided abundant evidence of ores or technology.

It is only since the time of John Dalton, a mere 200 years ago, that we have used the concept of the chemical elements in any scientific or technical sense. Although ancient languages, particularly those of the Near East, have separate words for distinct metals then in use, the development of alloy metallurgy must have come about without any clear notion by the artisans of elements and compounds. Metalsmiths must have been fully conscious of the efficacy of mixing different materials, that is, different ores, but they must have marveled at the results of smelting a variety of metallic and nonmetallic "stones."

Currently our earliest evidence comes from the Near East. China may not be a good area for such studies. Most Western scholars believe that China went more or less directly from stone age technology to a fairly sophisticated bronze technology, largely bypassing the use of native copper or unalloyed copper smelted from simple copper ore minerals (Barnard 1961). Perhaps recent research by our distinguished Chinese colleagues is changing this simple picture.

A variety of scenarios may be proposed to account for the initial discovery (repeated in various parts of the world) that copper metal can be separated from nonmetallic-looking minerals such as malachite and azurite. For example, bright green malachite and bright blue azurite could have been applied as decoration on the surface of pottery by Chalcolithic pyrotechnologists. If the pottery was then fired in a reducing atmosphere, copper beads would be formed. Malachite and azurite begin to decompose below 400°C (Simpson et al. 1964). It is likely that the earliest copper smelting was done well below the melting temperature of copper (Caldwell 1966).

Lead objects (which require smelting) and copper slag appear in the late seventh millennium B.C. in Turkey (Wheeler et al. 1979). Casting in the Near East may go back as far as 4000 B.C. There is an extensive literature on the origin of metallurgy in the cultural belt beginning in the Mediterranean and extending to India (Wertime 1973).

Parts of this region had a well-developed copper age, whereas other regions seemed to move rapidly into copper alloys (Renfrew 1967; Renfrew and Whitehouse 1974; Jovanović and Ottaway 1976; Selimkhanov 1962; Agrawal 1971). Mining in parts of Europe was well developed in the Neolithic (Jovanović 1980). Fortunately for early miners and metallurgists, the oxidized ores of copper lie above the primary sulfide ores.

From such beginnings it was a long road for most cultures to the ability to form alloys with precisely the casting and physical properties needed. The metal artifacts with the earliest dates are often ornaments rather than implements. It is possible that this is a result of the much broader range of compositions that would make satisfactory ornaments.

The earliest "alloys" contain appreciable quantities of arsenic and lead, sometimes nickel or antimony, and occasionally silver, bismuth, or tin. A primary question is whether the arsenic, lead, and tin were initially added deliberately or were the result of impure raw materials. If the early arsenical coppers were deliberate, as Charles (1967) and others believe, then what prompted the initial effort(s) in smelting "mixed" ores? Was it random experimentation with metallic-looking stones or the result of perceptive understanding of the smelting products of naturally complex ores?

McKerrell and Tylecote (1972) firmly believe that metalsmiths in the Scottish Early Bronze Age selected their ores to control the properties of the bronzes. They point to distinctly different compositions for halberd-rivet pairs, the rivets being of a softer alloy. However, it seems just as reasonable that the Scottish metalsmiths were simply perceptive enough to see that sometimes their products were softer and then used those for rivets. The wide range of all components in early alloys does not indicate careful selection of ores and careful control of compositions. Rather, the smiths chose wisely among their metal stocks when making an implement where certain physical properties were needed.

To expand our understanding of the origins of copper and bronze alloying, we need to continue systematic analyses of artifacts, slags, and ores, increase the use of metallographic and other physical investigations of early metals, and support and assist archaeologists in locating and excavating early mining and metallurgical sites. The Sumerian Committee in the 1920s was on the right track when they began systematically to analyze artifacts and local ores, but such investigations must be even more detailed and comprehensive. Here I focus on the chemistry of artifacts, ores, and technologies in an effort to increase our understanding of the origins of alloy metallurgy.

Analyses of Chalcolithic and Early Bronze Age Copper and Copper Alloys

For most of the region from Europe and the Mediterranean through the Near East to the Indian subcontinent, the first alloys were predominantly arsenic bronzes containing more than 1% arsenic. In the British Bronze Age the earliest bronzes were made with arsenic (McKerrell and Tylecote 1972). Renfrew (1967) states that tin bronze is rare in the Greek Aegean Early Bronze Age. In Crete in EB I–II, arsenic was more common than tin as an alloying element (Branigan 1974). In the eastern Mediterranean region tin bronzes predominate only at Troy. Eaton and McKerrell (1976) have summarized the data for the eastern Mediterranean and Near East. Agrawal (1971) has republished data from this broad region. In many areas tin bronze was not common until the Middle Bronze Age. Lead, nickel, antimony, and silver were found in addition to arsenic and tin in copper alloys in amounts greater than 1%. Lead was used frequently as an alloying element.

In the analyses of the earliest Chinese bronzes presented by Barnard (1961), it was lead rather than arsenic that served as the alloying element. Lead provided the castability desired by the Chinese metalsmiths.

Throughout much of the eastern Mediterranean and Near East it is difficult to get Chalcolithic and Early Bronze Age copper-based metal artifacts for analysis. Hence the Archaeometry Laboratory has undertaken a study of Chalcolithic material from India. Table 2.1 presents thirteen new analyses from the Chalcolithic of India. As can readily be seen, ten of the samples are unalloyed copper (less than 1% of any alloying element), whereas the remaining three (numbers 1, 6, and 10) are arsenic bronzes with 2.06%, 2.40%, and 4.84% arsenic, respectively.

In only one specimen (number 5) was tin present in an amount greater than 0.03%, and in this specimen it was less than 0.2%. Zinc is not abundant in these Indian Chalcolithic artifacts. In six of the ten artifacts not considered to be alloyed copper, the percentage of arsenic still ranges from 0.13% to 0.68%. Lead is present in amounts greater than 0.11% in six of the ten specimens. Nickel is present in concentrations above 0.11% in only three of the ten, and antimony is present in only one of the ten. Specimens 1 and 11 contain 2.5% and 2.6% iron, respectively. Specimen 11 showed an iron content of only 0.15% when reanalyzed after the ferromagnetic particles were removed.

Of the three arsenical coppers one was a weapon and two were ornaments. There is no correspondence between chemistry and function. These specimens are from the excavations of the Department of Archaeology, Deccan College, Pune, India.

Recognition of Smelted Copper

A number of archaeometallurgists have investigated or commented on the possibility of recognizing unalloyed smelted copper [for example, Coghlan (1962)]. In a recent study Maddin et al. (1980) concluded

Table 2.1 Analyses of Chalcolithic metal artifacts from India

Archaeology lab no.	Copper (%)	Tin (ppm)	Zinc (ppm)	Arsenic (ppm)	Lead (ppm)	Antimony (ppm)	Nickel (ppm)	Iron (ppm)	Deccan College sample description: Item	Locality	Deccan no.
1	88.65	7.74	296.0	20,650.0	5,340.0	2,460.0	5,950.0	25,710.0	copper ax	Ahar	2255c
2	99.91	10.4	< 47.9	32.0	6.70	351.0	235.0	< 47.6	copper ax	Navadatoli	5680
3	100.90	42.5	< 44.5	4,420.0	689.0	34.3	40.4	< 44.3	copper ax	Chandoli	–
4	99.11	135.0	< 49.6	6,810.0	365.0	42.7	179.0	402.0	copper ax	Jorwe	6
5	96.90	1,985.0	< 50.1	1,300.0	1,180.0	44.5	38.8	289.0	copper bangle	Nevasa	–
6	95.56	< 5.8	< 49.8	24,000.0	1,220.0	213.0	1,120.0	1,360.0	copper bangle	Kayatha	–
7	98.23	181.0	51.4	1,840.0	6,490.0	30.0	591.0	< 80.0	copper anklet	Inamgaon	1104-B
8	99.12	119.0	< 49.4	53.3	463.0	45.7	102.0	< 77.7	copper bangle	Inamgaon	1105-E
9	98.84	29.4	< 52.0	490.0	439.0	45.8	60.8	< 81.2	copper anklet	Inamgaon	1104-A
10	95.45	61.4	< 39.4	48,400.0	707.0	37.9	3,500.0	< 125.0	copper bangle	Inamgaon	1105-C
11	96.20	171.0	41.4	641.0	3,680.0	22.9	90.2	26,100.0	copper bangle	Inamgaon	2345
12	99.90	244.0	< 34.7	1,390.0	3,570.0	39.1	206.0	< 110.0	copper anklet	Inamgaon	1336
13	99.84	308.0	< 33.1	1,610.0	676.0	34.1	124.0	< 105.0	copper anklet	Inamgaon	1105-D

that no adequate criteria exist for distinguishing artifacts of native copper from those of worked and recrystallized smelted copper of high purity. They presented chemical and metallographic data. Adding to my earlier observations (Rapp 1982), I want to suggest that careful investigations of the chemistry of artifacts and related ores frequently can lead to distinguishing smelted copper from native copper.

Table 2.2 summarizes the results of the Archaeometry Laboratory analyses of over 1,000 native copper samples collected from geological deposits throughout the world. The reason that some elements have lower *N*'s (the number of samples analyzed) is that, from the beginning of our analytical program nearly two decades ago, we have regularly increased the number of elements sought.

One of the chief values of table 2.2 is that it presents the maximum concentrations found for the various trace elements in native copper. Thus it serves as a guide to limits of chemical discrimination for native coppers. The maximum value reported in table 2.2 can be taken as the maximum value likely to be found in artifacts made from native copper that has not been melted or chemically altered.

A special look at arsenic, antimony, cobalt, iron, nickel, silver, and zinc seems warranted. Of the 851 native coppers analyzed for arsenic, 19 (2.23%) contain over 1% arsenic and 44 (5.1%) contain over 0.3% arsenic. The mean arsenic content in native copper is 0.09%. These data may be somewhat skewed because of the large number of Lake Superior region samples in the database. Domeykite (Cu_3As) and algodonite (Cu_6As) are found in some native copper deposits from this region. However, domeykite and/or algodonite are also found in England, Germany, Iran, and elsewhere.

Antimony, on the other hand, is present only in very low concentrations in native copper. Not one specimen out of the 1,065 analyzed has over 0.01% antimony, and the mean value is only *4 ppm*. If copper ores of the tetrahedrite-tennantite (Sb–As) group were smelted, as suggested by McKerrell and Tylecote (1972) and others, the resulting alloy should contain much higher concentrations of antimony than are found in early copper-based metals. Likewise, the concentrations of cobalt in native copper are very low. Only one specimen in the 1,065 analyzed had over 0.01% cobalt.

Of the 1,062 native copper samples analyzed for iron, 17 (1.6%) had more than 1% iron. The maximum was 9.1%, and the mean was 0.097%. It appears that iron is not a good discriminator.

The data for nickel show that only 16 of the 851 samples analyzed had more than 0.10% nickel. Only one (containing 3.2%) had more than 0.63%, and the mean is only 0.01%. Therefore copper artifacts with

Table 2.2 Impurity elements in native copper[a]

Element	Mean (ppm)	Maximum (percent)	*N*
Ag	238	2.90	1,065
As	976	8.60	851
Au	0.18	0.0014	864
Cd	3.8	0.044	851
Ce	4.8	0.052	864
Co	14	0.38	1,065
Cr	9.4	0.11	1,065
Cs	11	0.091	851
Fe	970	5.00	1,062
Hf	0.96	0.0087	854
Hg	17	0.1200	1,065
In	13	0.12	864
Ir	0.12	0.0012	864
Mo	4.6	0.075	478
Ni	110	3.20	851
Pt	1.1	0.0097	467
Re	0.13	0.0034	478
Ru	3.4	0.20	851
Sb	4.4	0.089	1,065
Sc	0.43	0.0067	1,065
Se	9.4	0.31	1,065
Ta	0.09	0.0035	851
Te	4.0	0.11	851
Th	0.51	0.004	851
Zn	170	8.20	851
Zr	19	0.12	478

a. Analyses by neutron activation. Note that the mean is in ppm and the maximum is in percent.

more than 0.70% nickel must have been smelted. Some of the Harappan copper has more than 0.70% nickel (Agrawal 1971), as do some early Syrian, Sumerian, and Indian bronzes (Cheng and Schwitter 1957).

Of the 1,065 native coppers, 36 contain more than 0.05% silver and 5 contain more than 1%. Silver is a good discriminator element for distinguishing among native copper deposits, but with its wide range of concentrations it is likely to be less valuable in assigning artifacts to ore type.

The maximum percentage of zinc in native copper as shown in table 2.2 is somewhat misleading. Of 851 samples analyzed, 1 showed 8.2% zinc and 1 showed 1.9% zinc. However, these two samples were from the Franklin, New Jersey, zinc mine, and it is likely that the high values are from microscopic zinc minerals admixed in the native copper. For all other native copper localities only seventeen have over 0.1% zinc, the maximum being 0.58%.

Two decades ago Friedman et al. (1966) presented the results of an important investigation into the minor and trace elements found in copper that could be used to assign the copper to ore type: native copper, oxidized ore, or reduced ore (sulfide). They found that silver, arsenic, bismuth, iron, antimony, and lead were the most important metallic impurities in relating the metal to the original type of ore. They used a product function based on the concentrations of these elements as a discriminant function to assign metals to the ore type. For some reason this technique has not been used by archaeometallurgists and geoarchaeologists. The only exception of which I am aware is the study by Agrawal (1971).

In order to study the effectiveness of this technique, we subjected the analyses of the Chalcolithic samples from India shown in table 2.1 to discriminant analysis. The study is only partial because the analytical data contain only antimony, arsenic, iron, and lead. Friedman et al. (1966) indicate that adding silver and bismuth would increase the discriminating power.

Table 2.3 indicates the results of this study. Even though three of the thirteen results give ambiguous answers, it can be seen that native copper is an unlikely ore source. For the Chalcolithic of India, smelting provided the unalloyed copper.

The weakness in the method used by Friedman et al. (1966) and a general problem with artifact analyses is the lack of analysis for sulfur. With the addition of sulfur and the refinement of the other data this method should improve discrimination of ore type. (The large database of native copper analyses enables me to refine Friedman's probability tables on native copper.) It should be noted that sulfur can come into a furnace charge through slagging and fluxing components (for example, through jarosite admixed with gossan limonite) as well as by being part of the ore.

Successful use of this method should provide additional help in determining the antiquity of the technology of smelting sulfides. In my opinion, chemical determination of native copper and of ore type is achievable particularly as additional refinements in methodology are made.

Roles of Arsenic, Tin, Lead, Antimony, and Nickel

As indicated, analyses of Chalcolithic and Early Bronze Age copper-based artifacts show appreciable amounts of arsenic, tin, antimony, and lead, often greater than 1%. Because of lead's low melting point (327°C) and the ease of smelting galena (below 800°C), lead was an easy metal for early metalsmiths to use. Galena deposits are fairly available throughout the areas where metalsmithing originated. The chief question that concerns us is distinguishing lead added deliberately to improve the resulting alloy from lead that entered the metal because it was available as a component in the ore. Until more definitive data can be obtained, at this point it seems reasonable to use 1% as the demarcation between accidental and deliberate lead values.

Although antimony is present occasionally in early

Table 2.3 Discrimination of ore type for Chalcolithic copper artifacts from India

Artifact number	Antimony		Arsenic		Iron		Lead		Relative probability (percent)			Likely ore source
	ppm	unit	ppm	unit	ppm	unit	ppm	unit	I	II	III	
1	2,460.0	5	20,650.0	7	25,710.0	7	5,340.0	6	0.0	6.5	93.5	sulfide
2	351.0	3	32.0	1	47.6	2	6.7	1	30.4	12.7	56.9	sulfide
3	34.3	1	4,420.0	6	44.3	2	689.0	4	0.6	69.3	30.1	oxide
4	42.7	2	6,810.0	6	402.0	4	365.0	3	0.0	46.7	53.3	ambiguous
5	44.5	2	1,300.0	5	289.0	3	1,180.0	5	0.7	53.6	45.7	ambiguous
6	213.0	3	24,000.0	7	1,360.0	5	1,220.0	5	0.0	0.5	99.5	sulfide
7	30.0	1	1,840.0	5	80.0	2	6,490.0	6	33.7	28.2	38.1	ambiguous
8	45.7	2	53.3	2	77.7	2	463.0	4	1.5	78.5	20.0	oxide
9	45.8	2	490.0	4	81.2	2	439.0	4	0.3	70.6	29.1	oxide
10	37.9	1	48,400.0	8	125.0	3	707.0	4	0.9	61.6	37.6	oxide
11	22.9	1	641.0	4	26,100.0	7	3,680.0	5	0.2	36.5	63.3	sulfide
12	39.1	1	1,390.0	5	110.0	3	3,570.0	5	6.8	55.8	37.3	oxide
13	34.1	1	1,610.0	5	105.0	3	676.0	4	1.7	93.3	5.0	oxide

copper-based metals in amounts greater than 1% (thereby having an effect on the properties of the metal), it seems likely that antimony was always an accidental alloying element when it was a component of the sulfide ores. Nickel presents a greater problem. Cheng and Schwitter (1957) have shown that many Syrian, Sumerian, and Indian bronzes contain more than 1% nickel, and some contain more than 3%. They also found that the nickel composition was a function of the type of object, thus indicating deliberate choice. The question remains, How much of the choice was made before smelting and how much was based later on the working characteristics of the products of various smelts? Nickel is present in amounts greater than trace levels (>100 ppm) in many copper-based artifacts from most of the important metalworking regions in the Chalcolithic and Early Bronze Age.

Tin is a significant component in a few copper ores (for example, Cornwall, England), but except in rare cases it is unlikely that tin in amounts greater than 0.10% was accidentally incorporated in smelted copper. Tin percentages between 0.10% and perhaps 1.5% or 2.0% most likely result from reuse of scrap metal.

The most interesting questions involve the early arsenical coppers. Analyses from dozens of early metallurgical centers have shown that arsenic bronzes were fairly common in the Chalcolithic and Early Bronze Age. Coghlan (1972) and others have detailed the important property changes with the addition of varying amounts of arsenic to copper. The most intriguing question is how the metalsmith discovered the technology of alloying arsenic with copper. The use of arsenic was not only widespread, it was intensive. In a fourth millennium B.C. Near Eastern hoard of copper objects, twenty-four of thirty-six analyzed had an average arsenic content of 5.23% (Muhly 1977).

Unlike lead there is no evidence that arsenic was ever used as a metal (and native arsenic is *very* rare) or found as an ingot (such as copper and tin). Thus arsenic must have been alloyed with copper by the smelting of complex ores containing both copper and arsenic.

There seems to be a consensus among archaeometallurgists that the first copper minerals smelted were the hydroxycarbonates (malachite and azurite) and the oxides (cuprite and tenorite), yet the first alloys were arsenical. McKerrell and Tylecote (1972) and many others have suggested that the metallic-looking complex copper arsenic sulfide mineral tennantite is the most likely source of the arsenic. However, tennantite would come from the lower unoxidized zone (along with many metallic-looking copper sulfides).

What should be considered are arsenates from the oxidized zone of ore deposits. The following arsenates are all associated with malachite in the oxidized zone of copper deposits in Europe, the Near East (including the USSR), and elsewhere:

Olivenite	$Cu_2(AsO_4)(OH)$
Clinoclase	$Cu_3(AsO_4)(OH)_3$
Tyrolite	$CaCu_5(AsO_4)_2(CO_3)(OH)_4 \cdot 6H_2O$
Mimetite	$Pb_5(AsO_4)_3Cl$
Adamite	$Zn_2(AsO_4)(OH)$
Erythrite	$Co_3(AsO_4)_2 \cdot 8H_2O$
Annabergite	$Ni_3(AsO_4)_2 \cdot 8H_2O$

Use of the last four could account for the occasional high Pb, Zn, Co, or Ni contents of arsenic bronzes. And oxide copper ores rich in arsenic would lose less arsenic during smelting than the sulfide ores.

Ores of Copper, Tin, and Arsenic

Unfortunately the world's literature on the quantity, distribution, and accessibility from the surface of most copper ores, particularly native copper, is outdated, scattered, and difficult to assess. Cornwall (1956) has provided a good summary of the nature of native copper deposits, but we have insufficient data on the distribution of minor occurrences of this mineral. Determining the viability of ore sources will entail extensive on-site geological investigations. To understand the origin of alloy metallurgy, it is necessary to ascertain the type of ore involved if the complete technology is to be determined.

Copper ore deposits are widespread and exist in a great variety of mineralogies, complexities, and structures, but only a small number are credible as Chalcolithic and Early Bronze Age ore sources. Native copper undoubtedly supplied the initial copper but (outside of the Lake Superior region of North America) could not sustain the increasing demand. Until contrary evidence is offered, we can assume that the earliest smelting was done using oxide ores.

The beginnings of arsenic and copper alloying may well be coincidental, or nearly so, with the beginning of smelting complex sulfide ores. Although others have suggested the tetrahedrite-tennantite series, $(Cu,Fe)_{12}Sb_4S_{13}$–$(Cu,Fe)_{12}As_4S_{13}$, as the most likely ore, I believe that enargite (Cu_3AsS_4) is a more likely candidate. Tetrahedrite is much more common than tennantite, and early arsenical coppers usually do not have high concentrations of antimony. Enargite is often found in large quantities and is well known from deposits in Austria, Hungary, and at large deposits at Bor, Yugoslavia (Sillitoe 1983). All these copper ore minerals oxidize to malachite.

Although there are arsenical tin deposits at Cornwall, in Bolivia, and elsewhere, there is little evidence to indicate the early importance of this type of deposit. Interest in the mineral stannite (Cu_2FeSnS_4) may be revived with the recent discovery by Yener (1985) of stannite deposits near early metallurgical sites in southern Turkey.

Wertime (1973) and Agrawal (1971) have discussed the geographic relationships between Near Eastern ore deposits and early metallurgical sites. To date, nearly all such discussions have been too general. Detailed on-site investigations and detailed reviews of mining records of the mineralogy and structure of specific ore deposits will be necessary to determine ore sources for early copper alloys.

Conclusions and Recommendations

To expand our knowledge of the origins of alloy metallurgy, it is imperative that archaeological excavations be conducted at early mining and metallurgical sites. Ore minerals are nearly indestructible under most conditions of site burial so careful excavation should discover what ores were used, as the Chinese have done with their discovery of malachite in foundry areas (Barnard 1961). It will also be necessary to conduct further systematic analyses of Chalcolithic and Early Bronze Age copper-based artifacts to secure evidence of ore sources and smelting technology. These analyses should include the determination of sulfur content. I have not dealt here at all with the important contributions to be made through the study of early slags.

The Archaeometry Laboratory currently is continuing its program of artifact and ore analysis. We are pleased to contribute to the search for understanding of the development of civilization through the benefits of alloy metallurgy.

Acknowledgments

Since 1972 the neutron activation analyses have been supported in part by the Reactor Sharing Program under US Department of Energy contract E-(11-1)-2144 to the University of Wisconsin Reactor Facility. Richard Cashwell, director of the Reactor Facility, gave valuable assistance with many aspects of the neutron activation analyses. James Allert provided invaluable assistance with computer analyses of the data. Duane Long developed the procedures and did the atomic absorption analyses.

References

Agrawal, D. P. 1971. *The Copper Bronze Age in India.* New Delhi: Munshiram Manoharlal.

Barnard, N. 1961. *Bronze Casting and Bronze Alloys in Ancient China.* Canberra: The Australian National University and Monumenta Serica.

Branigan, K. 1974. *Aegean Metalwork of the Early and Middle Bronze Age.* Oxford: Oxford University Press.

Caldwell, J. R. 1966. *Tal-i-Iblis: The Kerman Range and the Beginnings of Smelting.* Illinois State Museum Preliminary Reports 7. Springfield, Ill.: Illinois State Museum.

Charles, J. A. 1967. "Early arsenical bronzes—A metallurgical view." *American Journal of Archaeology* 71:21–26.

Cheng, C. F., and C. M. Schwitter. 1957. "Nickel in ancient bronzes." *American Journal of Archaeology* 61:351–365.

Coghlan, H. H. 1962. "A note upon native copper: Its occurrence and properties." *Proceedings of the Prehistoric Society* 28:58–67.

Coghlan, H. H. 1972. "Some reflections on the prehistoric working of copper and bronze." *Archaeologia Austriaca* 52:93–104.

Cornwall, H. R. 1956. "A summary of ideas on the origin of native copper deposits." *Economic Geology* 51:615–631.

Eaton, E. R., and H. McKerrell. 1976. "Near Eastern alloying and some textual evidence for the early use of arsenical copper." *World Archaeology* 8:169–191.

Friedman, A. M., M. Conway, M. Kastner, J. Milsted, D. Metta, P. R. Fields, and E. Olson. 1966. "Copper artifacts: Correlation with source types of copper ores." *Science* 152:1504–1506.

Jovanović, B. 1980. "The origins of copper mining in Europe." *Scientific American* 242:152–167.

Jovanović, B., and B. S. Ottaway. 1976. "Copper mining and metallurgy in the Vinca Group." *Antiquity* 50:104–113.

Maddin, R., T. S. Wheeler, and J. D. Muhly. 1980. "Distinguishing artifacts made of native copper." *Journal of Archaeological Science* 7:211–225.

McKerrell, H., and R. F. Tylecote. 1972. "The working of copper arsenic alloys in the Early Bronze Age and the effect on the determination of provenance." *Proceedings of the Prehistoric Society* 38:209–218.

Muhly, J. D. 1977. "The copper ox-hide ingots and the Bronze Age metals trade." *Iraq* 39:73–82.

Rapp, G., Jr. 1982. "Native copper and the beginning of smelting: Chemical studies." in *Early Metallurgy in Cyprus, 4000–500 B.C.*, J. D. Muhly, R. Maddin, and V. Karageorghis, eds. (Nicosia, Cyprus: Pierides Foundation), 33–38.

Renfrew, C. 1967. "Cycladic metallurgy and the Aegean Early Bronze Age." *American Journal of Archaeology* 71:1–20.

Renfrew, C., and R. Whitehouse. 1974. "The Copper Age of peninsular Italy and the Aegean." *British School at Athens* 69:343–390.

Selimkhanov, I. R. 1962. "Spectral analysis of metal articles from archaeological monuments of the Caucasus." *Proceedings of the Prehistoric Society* 38:68–76.

Sillitoe, R. H. 1983. "Enargite-bearing massive sulfide deposits high in porphyry copper systems." *Economic Geology* 78:348–352.

Simpson, D. R., R. Fisher, and K. Libsch. 1964. "Thermal stability of azurite and malachite." *American Mineralogist* 49:1111–1114.

Wertime, T. A. 1973. "The beginnings of metallurgy: A new look." *Science* 182:875–886.

Wheeler, T. S., R. Maddin, and J. D. Muhly. 1979. "Ancient metallurgy: Materials and techniques." *Journal of Metals* 31:16–18.

Yener, K. A. 1985. "Tin in the Taurus Mountains: The Bolkardag mining survey." Paper presented at the Archaeological Institute of America meeting, 26–30 December 1985, Washington D.C.

3
Early Metallurgy in Mesopotamia

P. R. S. Moorey

My main purpose here is to provide a concise guide to the present evidence for the development of metalworking in Mesopotamia before 2000 B.C. in its social and economic context. Mesopotamia, more particularly the southern part (Sumer or Babylonia), has held an important place in general histories of early metallurgy for over fifty years. Woolley's excavations at Ur (1922–1934) revealed the remarkable achievements of Sumerian metalworkers about 2500 B.C. in gold (or electrum), silver, copper, and copper alloys. Their work, as illustrated by finds from the Royal Cemetery at Ur, in Cyril Smith's words, "reveals knowledge of virtually every type of metallurgical phenomenon except the hardening of steel that was exploited by technologists in the entire period up to the end of the nineteenth century A.D." (Smith 1981, p. 195).

It is not a coincidence that this achievement was contemporary with the appearance in Sumer, from sometime in the fourth millennium B.C., of the earliest literate urban civilization in the world. It is this conjunction of technological with social and political innovation, revealed both in the evidence of texts and artifacts, that makes this region and this early period of particular interest to students of ancient metallurgy.

Metal Sources and Workshops c. 8500–2000 B.C.

Mesopotamia, which had no local sources of metal (except iron), imported her metals already processed or semiprocessed from mining areas in Turkey, Iran, Afghanistan, and Oman. It is still extremely difficult to state with confidence which metals were coming from which mining area at which times into Mesopotamia. The situation certainly varied from time to time and from one part of Mesopotamia to another. With the exception of tin (perhaps from Afghanistan), both Turkey and Iran could have supplied all the main metals used; in addition, Oman offered copper and perhaps some lead.

Broadly, northern and central Mesopotamia (Akkad/Assyria) derived its metal from the mines of Turkey and Iran, whereas the south (Sumer/Babylonia) drew more regularly on Iran and Oman. When regular access to any particular mining area was blocked, it would have been possible to draw on alternative sources of supply.

In historic times texts indicate that metals were rigorously controlled by the bureaucracy and were regularly recycled. Consequently the actual amount of metal recovered through excavation at any period is no guide to the scale of contemporary use. Metal finds are rare in all periods on settlement sites. It is also exceptional, as in the Early Dynastic Period (c. 3000–2350 B.C.), for metal artifacts to be found in quantity in graves. Even when texts are available, knowledge of the trade in metals is wholly inadequate, because surviving texts made little reference to it. It is clear, however, that plundering and tribute taking in war made an important contribution to the supply of metals and to technology in most periods. For example, a famous Sumerian epic poem referring to the Early Dynastic Period, although known only from copies of the text made a thousand years later, describes how an expedition from Uruk into Iran seized not only precious metals but also the artisans skilled in working it with their tools, including molds for casting (Wilcke 1969, pp. 126–129).

Modern study of early metallurgy in Mesopotamia is virtually confined to the objects and to a few relevant texts. No metal workshops of the period have been excavated nor have any metal craftsman's tools been identified from Mesopotamia before 2200 B.C. A workshop of the later Ubaid Period (c. 4000 B.C.) has recently been excavated at Degirmentepe (Malatya) near the upper Euphrates in Turkey (Esin 1985, p. 188). Crucibles, molds, and pot bellows have not been recognized in Mesopotamian excavations before the late third millennium B.C.

The Emergence and Transformation of a Trinket Technology (c. 8500–3500 B.C.; Hassuna, Halaf, and Ubaid Periods)

The history of copper working in Mesopotamia is usually taken back to a pendant from the Shanidar Cave in the Zagros Mountains (Solecki 1969) in the ninth millennium B.C., but it may only illustrate the use of malachite as a semiprecious ornamental stone. Isolated rolled beads of cold- or hot-worked native copper from Tell Ramad, on the Euphrates River in modern Syria, and from Ali Kosh, in Khuzistan in modern Iran, in seventh millennium B.C. pre-pottery Neolithic village settlements provide evidence for the earliest lowland use of worked metal (Smith 1969; France-Lanord and de Contenson 1973). The repertoire of objects at this time has recently been extended by the discovery of an awl, said to be of cold-worked native copper, at Tell Magzallia in north Mesopotamia. Soviet analysts have argued that its impurity pattern indicates an origin in the Anarak ore field of Iran (Ryndina and Yakhontova 1984).

It had long been assumed that all the earliest copper objects from Mesopotamia were of native metal, but this conclusion is no longer accepted. Russian excavations on a series of sixth millennium B.C. village settlements around Yarim Tepe on the Sinjar Plain of northern Mesopotamia have indicated the presence of smelted copper at this time. At Yarim Tepe I pieces of malachite ore were scattered through all levels of the site, suggesting some degree of local smelting. This may be confirmed by the high levels of iron (1–10%) in the associated copper objects (generally rings, bracelets, and simple, flat blade tools). This has been explained by the use of an iron flux in smelting [see Merpert et al. (1977b, 1982)]. Some knowledge of smelting is further endorsed by the presence of a lead bracelet at Yarim Tepe I (Merpert et al. 1977a, p. 84, pl. IX). A lead bead from Jarmo in the Zagros Mountains may also belong to this early phase (Braidwood et al. 1983, p. 542, fig. 136:19). At this early stage both the metals and the knowledge of metalworking might have reached northern Mesopotamia from the highland mining regions of either Turkey or Iran.

By the sixth millennium B.C. melted native copper, perhaps already producing "arsenical coppers," and smelted copper were employed in northern Mesopotamia. Nothing is known at present about contemporary casting techniques. This early metalwork at Yarim Tepe I is associated with an equally unsuspected level of organized pottery production, represented by a large number of two-stage kilns situated in a clearly separate manufacturing area within the boundaries of the village. They remind us that from the outset developments in a number of fire-using mineral industries were aspects of a single technological phenomenon. For example, the production of the fine Hassuna/Halaf polychrome painted pottery and the appearance of alkaline glazed materials are contemporary innovations in the villages of northern Mesopotamia from the sixth into the fifth millennium B.C.

Evidence for metallurgy in Mesopotamia between about 5000 and 3500 B.C. is extremely poor, amounting to no more than a handful of copper ornaments and isolated pieces of lead. They appear to show no technological advances on the achievements of the pioneer metalworkers in the Sinjar Plain. There is no certain evidence for the use of gold, electrum, or silver before about 4000–3500 B.C. In two cemeteries of that period in the south, at Eridu (193 graves; Safar and Lloyd 1981) and at Ur (50 graves; Woolley 1956), there is no metalwork in the graves. At present we must assume, as the pattern of finds has now been consistent for over half a century, that this reflects a real situation in which even base metals served a minor, essentially decorative role in society.

This has surprised those who wish to project the remarkable achievements of later Sumerian metalsmiths far back into the prehistoric period; but, seen in its proper context, it is not surprising. Because excavation reports in Mesopotamia have always paid too little attention to stone and baked clay tools, it is not always appreciated how vital they were to farmers until well into the second millennium B.C. Stone and clay were the raw materials predominantly used by all prehistoric communities of farmers and craftsmen for their tool kits and for their weapons in war.

Therefore copper tools were not needed to increase the efficiency of food production or of carpentry in the relatively small, self-contained villages of Iraq before 3500 B.C. Indeed, at this stage stone tools were probably more efficient than copper and were easily made of accessible materials. There was no incentive to increase the potential supply of copper, even if there were the means, nor to improve the range and strength of copper tools. Metal was neither vital for subsistence nor yet valued as a prestige commodity. Brightly colored semiprecious stones, such as turquoise and lapis lazuli, or imitations of them in blue-glazed dark stones or faience served that purpose. Distance from the sources of such stones does not appear to have restricted their supply. Thus it is unlikely that distance would have seriously affected supplies of copper if the metal had been in demand. The Mesopotamian evidence suggests that there had to be major social and political changes in the settlements of the area before there could be technological advances in metallurgy and a marked expansion in the use of metals.

As evidence is rare everywhere in the Near East in the late fifth and early fourth millennia B.C., it is difficult to compare the situation in Mesopotamia with that in Turkey (de Jesus 1980) and Iran (Heskel and Lamberg-Karlovsky 1980; Moorey 1982b). But copper seems to have been more important closer to the mining areas, certainly by the end of this phase. A major settlement was established c. 4200–3800 B.C. at Susa in southwest Iran, with mud brick structures and burials containing over 2,000 bodies, accompanied by large and small flat axes, chisels, pins, and flat mirrors of copper (Amiet 1979). Although it is geographically part of Mesopotamia, Susa had more direct access to copper mines at Anarak and elsewhere in Iran (Berthoud et al. 1982). Indeed her significance may be explained by control of all external trade in Khuzistan (or Susiana). This concentration of worked copper is still unparalleled in Sumer and in northern Mesopotamia at this time. Deshayes (1960, vol. 1, pp. 408–409) long ago argued that at this stage Sumer was indebted to Iran, both for any metal she had and for related technical information. One particular technique, using two-piece molds for casting shaft-hole tools and weapons, certainly appears earlier in Iran than in Mesopotamia (Deshayes 1960, vol. 2, nos. 1841–1845, p. 95).

Metals in the Service of Temples and Palaces (c. 3500–2200 B.C.; Uruk IV–III, Early Dynastic I–III, Akkadian Periods)

About the middle of the fourth millennium B.C. objects from temples and graves at Tepe Gawra in northern Mesopotamia, at Telloh, and at Uruk in the south reveal remarkable changes in the status of metallurgy. The range of metals and techniques was already wide, and the level of skill was high. The quantity of both precious and base metals available had sharply increased. Understanding these changes is difficult. It is the climax of a period of innovation, not the complex processes leading up to it, that is revealed by the archaeological evidence.

The new role for metal coincides (1) with the first appearance of population concentrations in a small number of large fortified settlements with monumental mud brick buildings, (2) with clear evidence of social stratification and craft specialization, and (3) with written administrative records, especially in the south (Sumer). The city-states were competing against each other. No one city was able to monopolize export trade, primarily in agricultural produce and in textiles, or to control all trade routes into the region. The Sumerian city-states were linked by the Euphrates River with Syria and Turkey, by the gulf with Oman, and by land or sea with Iran and her eastern neighbors. These were all areas with important mines. Within Sumer and in her long-distance trade the availability of water transport by canal, river, or sea eased and encouraged the movement of quantities of metal, stone, and timber. Overland they were carried by donkey caravans, as the camel had not yet been domesticated throughout the Near East.

It was the demands of these growing urban communities in Mesopotamia and the commercial and industrial networks they created that stimulated the marked expansion of local metal industries. Luxury objects in precious metals were made for temples; fine jewelry was in demand to distinguish men and women of wealth and high status. However, it was copper, not silver or gold, that served as the primary means of exchange until the Akkadian Period, when silver replaced it (Lambert 1953; Limet 1972). Intensified craft production, particularly the manufacture of wheeled vehicles, involved the introduction of more metal tools: shaft-hole axes and adzes, saws, and chisels of all sizes (Piggott 1983, p. 21). But significantly no copper tools certainly for use in farming have been identified in Mesopotamia before the late third or early second millennium B.C. (Moorey 1971). Military competition encouraged the improvement of ax heads, daggers, and spearheads, as well as the development of simple copper armor. In Sumer sling stones were preferred to bows until the Akkadian Period, and until that time arrowheads and maces were predominantly of stone.

The sources of innovation in the late prehistoric period are obscure because relative chronologies are still uncertain and evidence of metalwork is rare. Berthoud et al. (1982) have recently argued, on the evidence of nearly 600 analyses of ores and artifacts of copper and copper alloys from Iran, Afghanistan, and Oman, that the innovative role long associated with the highland mining regions in Iran had lost its impact by the late fourth millennium B.C. [see also Moorey (1982b)]. After that, through the third millennium B.C., dynamic change in metallurgy was centered in the lowlands of Mesopotamia. However, not only the metalworkers of Iran but also those of Turkey (de Jesus

1980) and the makers of the Nahal Mishmar hoard from Israel (Bar-Adon 1980) must also be taken into account when considering the background of metallurgy in Mesopotamia, c. 3500–3000 B.C.

The emergence of polymetallism is a key feature of this period. It indicates fundamental changes in the primary processes of metal production in the highland zones producing metal. There is the first systematic exploitation of native gold and electrum. Silver is for the first time extracted from argentiferous lead ores, marked by a new, widespread use of silver metal (Prag 1978) and a sharp increase in manufactured lead in Mesopotamia (Moorey 1985, p. 122). Both texts and objects reveal the presence of iron. As the archaic Sumerian terminology for this metal is not yet fully understood (Vaiman 1982) and percentages of nickel are no longer regarded as a safe guide to meteoric iron (Piaskowski 1982), it is not certain whether this is both meteoric iron and iron smelted as a by-product of copper smelting (Cooke and Aschenbrenner 1975).

At the same time the complexity of copper metallurgy steadily increased. The secondary sulfide ores were being exploited in Oman and Turkey, perhaps also in Iran. The evolving techniques of smelting increasingly brought varied ores into contact with one another. More intensive smelting and casting stimulated experiments with alloys, although the proper separation of natural from manufactured copper alloys remains controversial among modern experts. Across the whole of the Near East in the second half of the fourth millennium B.C. copper alloying was widely known and intensively practiced, particularly alloys with arsenic and, to a lesser extent, lead. Arsenical copper was to remain the most common metal in Mesopotamia until well into the second millennium B.C.

Bronze (tin-copper alloy) emerges in Mesopotamia, including Susa, late in the fourth millennium B.C., or in the first half of the third, when it appears both in texts and objects. It was the needs of towns in Sumer and the Susa region that were to control the circulation of tin for centuries. If, as now seems likely, tin came from sources in Afghanistan and perhaps eastern Iran, it could have been shipped overland through Iran or by sea up the gulf through Bahrein (*Dilmun* ?). In the third millennium B.C. in Sumer the use of tin was sporadic. When the Sumerian Metals Project in Philadelphia is finished, it should be possible to quantify the use of tin more accurately; it seems to have always been a luxury metal. Already there is enough evidence to show that at this stage there was no controlled correlation between types of object and alloy: Sheet bronze vessels, for example, may contain as much tin as tools and weapons (Moorey 1985, pp. 51ff).

From about 3500 B.C. all available metals and alloys, except iron, were cast to make tools and weapons, ornaments, toilet articles, and statuary; all except iron were hammered into sheets for the production of jewelry (sometimes over cores of bitumen), decorative fittings, and vessels. Few if any vessels were cast in Mesopotamia before about 1000 B.C. They were formed out of sheet metal with chased or repoussé surface decoration. Cast fittings, usually only handles, were rivetted or soldered to the body. By at least the middle of the third millennium B.C. tin was used for soft soldering, with a low-melting-point alloy of silver, copper, and tin for hard soldering (brazing) (Craddock 1984; Roberts 1974).

Casting, notably lost-wax casting, first for small-scale and then for large-scale statuary, was the most important single technique to be conspicuously developed (c. 3500–2750 B.C.). Contrary to expectation, large-scale human statues are of copper long after the introduction of bronze, with little appreciable use of lead to facilitate the casting process (Frankfort 1939, no. 183; Strommenger 1985; Al-Fouadi 1976). Lead is, however, used for this reason in one of the earliest surviving lost-wax castings: a small lion pendant from Uruk IV (Heinrich 1936, pp. 25, 47, pl. 13a; Braun-Holzinger 1984, no. 1). The earliest Near Eastern objects cast by the lost-wax process, that is, before Uruk IV, are in the remarkable Nahal Mishmar hoard from Israel (Bar-Adon 1980), made c. 3700 B.C.

There are important distinctions in the available archaeological evidence for metalworking in Mesopotamia c. 3500–2200 B.C. In the late prehistoric period (before 3000 B.C.) only a few graves—at Tepe Gawra in the north, at Telloh and Susa in the south—contain metal objects. At Ur the "Jamdat Nasr" graves are now mainly dated in the early historic period (c. 3000–2650 B.C.). Otherwise the little evidence that there is comes from temples. In the early historic period (c. 3000–2200 B.C.) the evidence is mainly from graves, some richly equipped. Outside the temples in the Diyala region and at Tell al-Ubaid so far there is little metalwork from excavated temples. Almost no metalwork has been excavated from the large administrative buildings and houses of this period.

Temples

In Uruk IV–III scattered fragments and a few objects illustrate the use of gold (or electrum), silver, and copper alloys for temple furniture, for parts of statues largely made in stone or wood, and for small-scale lost-wax castings. In the Early Dynastic Period, cast stands for vessels, in animal or human shapes, and fine votive models such as the chariot from Tell Asmar, indicate a growing mastery of elaborate castings in base metal, some nearly life-size. They culminate in the Akkadian Period (c. 2350–2200 B.C.) in two masterpieces of metal casting: the life-size royal head from Nineveh and the fragment of a male statue inscribed for King Naram-Sin from Bassetki (Braun-Holzinger 1984, pls. 9–14).

The objects from Tell al-Ubaid, excavated over sixty years ago, remain a unique illustration of the base metal fittings of a Sumerian temple (Hall and Woolley 1929). They also indicate that an elaborate program of temple decoration was so expensive in cast metal that, where necessary, freestanding animals and parts of panels carved in relief were made of sheet copper hammered over roughly shaped wood and bitumen patterns and then rivetted in place; only the heads and some limbs were cast, both in copper and bronze.

Graves

Only at Tepe Gawra in north Mesopotamia have graves of the Late Prehistoric Period indicated the type of metalwork buried in the richer graves of the Late Prehistoric Period. One tiny object best illustrates craft achievement at this time. It is a small fitting (30 × 23 mm) in the shape of a wolf's head made of electrum sheet metal from tomb 114 in stratum X:

The neck is hollow and tubular, forming a socket which has two holes through its walls ... for insertion of tacks and dowels.... The entire head is a single piece of metal with the exception of the ears, the lower jaw and the teeth. The ears were attached by means of copper pins.... The lower jaw was carefully jointed into the rest of the head and held in position by an electrum pin.... Inside the head was filled with bitumen. (Tobler 1950, p. 92, pl. LIXb, CVIII, fig. 65).

At present the Early Dynastic Period (c. 3000–2350 B.C.) is the only time in Mesopotamia for which the evidence of metalwork in cemeteries illustrates the whole range of the social hierarchy. At Ur, where the royal and temple establishment of one of Sumer's most prosperous cities is buried (Woolley 1934), sixteen "royal tombs" contained great quantities of gold, electrum, silver, bronze, and copper objects of all kinds. There were hundreds more graves; at least 75% contained some base metal objects (usually weapons and vessels) and about 30% also had precious metal (usually jewelry and toilet articles). The range of shaping techniques is comprehensive, and all the major decorative techniques are present: chasing, cloisonné, filigree, granulation, various methods of gilding, and repoussé.

This may be contrasted with a small cemetery for a village at Tell al-Ubaid, near Ur. Here only 19% of the graves contain base metal objects and then usually only one simple object; only one grave out of ninety-four held precious metal (silver earrings). This contrast emphasizes the concentration of wealth at Ur and the marked variations of status in Sumerian society by this time.

Throughout Mesopotamia there was a steady increase in the base metal objects placed in graves between 3000 and 2500 B.C., usually pins, jewelry, toilet articles, and weapons. Only at such sites as Kish is the range increased to include decorated copper rein rings, tools (awls and saws), and a larger number of sheet metal vessels with cast supports (Algarze 1983–1984; Moorey 1978, pp. 61ff, 103ff) in graves also containing vehicles.

Conclusion

Before about 3500 B.C. only copper and lead played a minor role in the crafts of Mesopotamia [compare Heskel (1983)]. Precious metal was absent. At first, technical expertise was probably brought from the highland zone with the imported metal. The growth of a complex urban civilization c. 3500–3000 B.C. in Mesopotamia transformed the local metal industries, although in explaining this change, it is now difficult to separate cause from effect. Did changes in the mining regions trigger expansion of industries in the lowlands, or did expanding demand in the lowlands stimulate polymetallism in the mining areas? Whatever the prime cause, the range of metals available sharply increased as did experimenting with alloys. From about 3500 B.C. in the developing towns of Mesopotamia the demands of wealthy ruling groups stimulated innovation and experiment with techniques of forming and decoration that culminated in the material culture of the richest citizens of Ur c. 2650–2500 B.C. What those craftsmen achieved has rarely been surpassed since.

References

The two books marked with an asterisk contain full references for this subject; Braun-Holzinger (1984) is also fully illustrated.

Al-Fouadi, A.-H. 1976. "Bassetki statue with an old Akkadian royal inscription of Naram-Sin of Agade." *Sumer* 32:63–75.

Algarze, G. 1983–1984. "Private houses and graves at Ingharra: A reconsideration." *Mesopotamia* 18/19:135–155.

Amiet, P. 1979. "Archaeological discontinuity and ethic duality in Elam." *Antiquity* 53:195–204.

Bar-Adon, P. 1980. *The Cave of the Treasure: The Finds from Caves in Nahal Mishmar.* Jerusalem: Israel Exploration Society.

Braidwood, L. S., et al., eds. 1983. *Prehistoric Archaeology along the Zagros Flanks.* Oriental Institute Publication 105. Chicago: Oriental Institute.

Berthoud, T., S. Cleuziou, L. P. Hurtel, M. Menu, and C. Volfovsky. 1982. "Cuivres et alliages en Iran, Afghanistan, Oman, au cours des IVe et IIIe millénaires." *Paléorient* 8(2):39–54.

*Braun-Holzinger, E. A. 1984. "Figürliche Bronzen aus Mesopotamien". *Prähistorische Bronzefunde,* Abt. I, Band 4.

Cooke, S. R. B., and S. Aschenbrenner. 1975. "The occurrence of metallic iron in ancient copper." *Journal of Field Archaeology* 2:251–266.

Craddock, P. T. 1984. "Tin and tin solder in Sumer: Preliminary comments." *Museum Applied Science Center for Archaeology Journal* (University Museum, Philadelphia) 3(1): 7–9.

De Jesus, P. S. 1980. *The Development of Prehistoric Mining and Metallurgy in Anatolia*. BAR International Series 74. Oxford: British Archaeological Reports.

Deshayes, J. 1960. *Les outils de bronze, de l'Indus au Danube (IVe au IIe millénaire)*, 2 vols. Paris: Université de Paris.

Esin, U. 1985. "Degirmentepe (Malatya), 1984." *Anatolian Studies* 35: 188–189.

France-Lanord, A., and H. De Contenson. 1973. "Une pendologue en cuivre natif de Ramad." *Paléorient* 1: 109–115.

Frankfort, H. 1939. *Sculpture of the Third Millennium B.C. from Tell Asmar and Khafajah*. Oriental Institute Publication 44. Chicago: Oriental Institute.

Hall, H. R., and C. L. Woolley. [1929.] *Ur Excavations*. Vol. 1, *Al' Ubaid*. Philadelphia, Penn.: Museum of the University of Pennsylvania.

Heinrich, E. 1936. *Kleinfunde aus den archaischen Tempelschichten in Uruk*. Leipzig: Deutsche Forschungsgreminschaft.

Heskel, D. L. 1983. "A model for the adoption of metallurgy in the ancient Near East." *Current Anthropology* 24: 362–366.

Heskel, D. L., and C. C. Lamberg-Karlovsky. 1980. "An alternative sequence for the development of metallurgy: Tepe Yahya, Iran," in *The Coming of the Age of Iron*, T. A. Wertime and J. D. Muhly, eds. (New Haven, Conn.: Yale University Press), 229–266.

Lambert, M. 1953. "La periode présargonique." *Sumer* 9: 198–213.

Limet, H. 1972. "Les métaux a l'époque d'Agade." *Journal of the Economic and Social History of the Orient* 15: 3–34.

Merpert, N., R. M. Munchaev, and N. O. Bader. 1977a. "The investigations of Soviet expedition in Iraq, 1974." *Sumer* 33: 65–104.

Merpert, N., R. M. Munchaev, and N. O. Bader. 1977b. "The earliest metallurgy of Mesopotamia." *Sovetskaya Arkheologiya* 3: 154–165. (In Russian with French summary.)

Merpert, N., and R. M. Munchaev. 1982. "An den Anfängen der Geschichte Mesopotamiens." *Das Altertum* 28: 69–80.

Moorey, P. R. S. 1971. "The Loftus hoard of Old Babylonian tools from Tell Sifr in Iraq." *Iraq* 33: 61–86.

Moorey, P. R. S. 1978. *Kish Excavations 1923–33*. Oxford: Clarendon Press.

Moorey, P. R. S. 1982a. "The archaeological evidence for metallurgy and related technologies in Mesopotamia, c. 5500–2100 B.C." *Iraq* 44: 13–38.

Moorey, P. R. S. 1982b. "Archaeology and pre-Achaemenid metalworking in Iran: A fifteen year retrospective." *Iran* 20: 80–101.

*Moorey, P. R. S. 1985. *Materials and Manufacture in Ancient Mesopotamia. The Evidence of Archaeology and Art: Metals and Metalwork, Glazed Materials and Glass*. BAR International Series 237. Oxford: British Archaeological Reports.

Piaskowski, J. 1982. "A study of the origin of the ancient high-nickel iron generally regarded as meteoric," in *Early Technology*, T. A. Wertime and S. F. Wertime, eds. (Washington, D.C.: Smithsonian Institution), 237–243.

Piggott, S. 1983. *The Earliest Wheeled Transport: From the Atlantic Coast to the Caspian Sea*. Ithaca, N.Y.: Cornell University Press.

Prag, K. 1978. "Silver in the Levant in the fourth millennium B.C.," in *Archaeology in the Levant: Essays for Kathleen Kenyon*, P. R. S. Moorey and P. J. Parr, eds. (Warminster, England: Aris and Phillips), 36–45.

Roberts, P. M. 1974. "Early evolution of brazing." *Welding and Metal Fabrication* 42: 412–415.

Ryndina, N. V., and L. K. Yakhontova. 1984. "The earliest copper artifact from Mesopotamia." *Sovetskaya Arkheologiya* 2: 155–167. (In Russian with English summary.)

Safar, F., and S. Lloyd. 1981. *Eridu*. Baghdad: State Organization of Antiquities and Heritage.

Smith, C. S. 1969. "Analysis of the copper bead from Ali Kosh," in *Prehistory and Human Ecology of the Deh Luran Plain*, F. Hole, K. V. Flannery, and J. A. Neely, eds. (Ann Arbor, Mich.: Museum of Anthropology, University of Michigan), 427–428.

Smith, C. S. 1981. *A Search for Structure: Selected Essays on Science, Art and History*. Cambridge, Mass.: MIT Press.

Solecki, R. 1969. "A copper mineral pendant from northern Iraq." *Antiquity* 43: 311–314.

Strommenger, E. 1985. "Early metal figures from Assur and the technology of metal casting." *Sumer* 42: 114–115.

Tobler, A. J. 1950. *Excavations at Tepa Gawra II: Levels IX–XX*. Philadelphia, Penn.: University of Pennsylvania Press.

Vaiman, A. A. 1982. "Eisen in Sumer," in *Vorträge gehalten auf der 28 Rencontre Assyriologique Internationale in Wien (6–10 Juli, 1981)* (Horn, Austria: Archiv für Orientforschung Beiheft), vol. 19, 35–38.

Wilcke, C. 1969. *Das Lugalbandaepos*. Wiesbaden: Harrassowitz.

Woolley, C. L. 1934. *Ur Excavations II: The Royal Cemetery*. Philadelphia, Penn.: University Museum.

Woolley, C. L. 1956. *Ur Excavations IV: The Early Periods*. Philadelphia, Penn.: University Museum.

4
Early Copper Metallurgy in Oman

Andreas Hauptmann, Gerd Weisgerber, and Hans Gert Bachmann

Archaeometallurgical work on early copper production in Oman in the southeastern part of the Arabian Peninsula is connected with the most important question of whether the present Sultanate of Oman is identical with the Bronze Age copper land of Magan, which was a major supplier of copper for Sumer and Babylon in the third and second millennia B.C.

The archaeological and historical background to this question arose in the 1930s: During excavations in Ur, numerous cuneiform tablets dating to the third and second millennia B.C. were found. These written texts pointed out that intensive commerce, especially in copper and diorite, existed with Magan. The metal was imported in amounts of many tons, and the rock was sought for making statues.

The first attempts to identify the copper land of Magan in Oman and to find the connection to Ur and Sumer were undertaken by Peake (1928) and Desch (1929). They pointed out the huge slag heaps there (probably in the area of Lasail and Aarja) and compared analytical data of ores and slags with artifacts from Ur and Sumer. From this they stressed as characteristic the nickel contents of the ores and of the artifacts. More detailed investigations to find the connection between Magan (Oman) and Mesopotamia based on scientific evidence were undertaken by Berthoud (1979). He analyzed a limited number of ores and artifacts from numerous sites in southwest Asia. Using a multivariate method with thirty-one chemical elements, he distinguished several groups of ore deposits and artifacts. He found that the copper ores from Oman are comparable to artifacts from Susa D (2600–2400 B.C.). The method used by Berthoud contains some analytical and geological problems (Seeliger et al. 1985, p. 643). It seems, too, that too few objects were analyzed by Berthoud so that the final evidence for determining whether the copper came from Oman or the southern part of Iran is still a problem.

The German Mining Museum has carried out archaeometallurgical research following the prospecting work of the Oman Mining Company LLC from 1977 onward in order to prove the relics of early mining and metallurgy in the field. But, studying the numerous ancient smelting sites, we found that most of them date to the Early Islamic Period. Later, we discovered smelting sites from the third to second millennium B.C. and recognized the important role they had in this area. During this work it was possible not only to study the archaeology of this region but also to get data on the ores exploited, the early metallurgical procedures carried out, and the characterization of the metal and the artifacts that were produced in Oman.

Geology and Copper Ore Deposits of the Oman Mountain Range

The Oman mountain range consists mostly of basic and ultrabasic rocks, such as basalts, gabbros, granodiorites, and peridotites. They build up the so-called Samail Ophiolite, which is part of an ancient oceanic crust. The Samail Ophiolite Complex is the best conserved and most complete ophiolite complex known today. It extends from Ras Musandam in the north over 600 km down to the island of Masirah in the south.

One of the most impressive copper ore districts is located within this rock unit. During prospecting work in the last 25 years, more than 150 copper occurrences and mineralizations have been discovered. These deposits have been exploited since prehistoric times and have been the basis for 5,000 years of copper production within the present Sultanate of Oman. The ore deposits occur in nearly all lithological units of the ophiolite. Following recent investigations, all these deposits are believed to be of similar origin and are comparable in form and type with the famous copper deposits in Cyprus (Coleman et al. 1979).

For the purpose of archaeometallurgical research these deposits should be divided into two groups: the massive sulfide deposits and the veins.

The massive sulfide deposits, connected with volcanic rocks, are composed predominantly of pyrite with a copper content of 2–2.5%. Rich ores such as chalcopyrite ($CuFeS_2$) and bornite (Cu_5FeS_4) are rare in these deposits, and zones of secondary enrichment are not known (Iten, personal communication). At the surface these deposits show striking gossans. They were mostly exploited and worked partly by the ancient miners, although the activities discovered at a depth of 87.5 m were from early Islamic times. These deposits, of which only a few occur in the mountain range, are economically important today and are under exploitation, for example, in Lasail, Aarja, and Bayda near Sohar.

An overwhelming part of the mineralizations and ore deposits in Oman form veins that occur in gabbros and peridotitic rocks. They are generally controlled by tectonic faults. The main copper mineral under the surface is chalcopyrite, which is intergrown with small amounts of cobaltite, loellingite, and other Fe-Co-Ni-As minerals. The outcrops are mostly characterized by a distinctive and rich secondary copper mineralization with malachite (copper carbonate), brochantite (copper sulfate), and chrysocolla (copper silicate). These ores, which are of no economic value today, easily reach copper contents of up to 30% and more after hand picking. Most probably these ores were the decisive impulse for the earliest copper production in Oman.

The type and composition of the ore deposits provided sufficiently rich resources for the early copper production. The ores available were quite pure except for their iron content, although the geochemical pattern varies between the two types of deposits. For example, the high nickel content, which is stressed as typical for the Oman deposits in the previous archaeometallurgical literature, is not characteristic of all deposits. Nickel is low in the massive sulfide deposits (Coleman et al. 1979; Hauptmann 1985, p. 27, and unpublished data) but is clearly concentrated with cobalt and arsenic in veins situated in peridotitic rocks (Ni up to 0.6%, Co up to 0.12%, As up to 0.2%). The tin content correlates with the geological situation (ophiolitic rocks) and is generally below 10 ppm.

Periods of Copper Production in Oman

Based on the present state of knowledge, copper production started in Oman relatively early, during the Hafit Period in the first half of the third millennium B.C. (3000–2500 B.C.).

During excavations in Ras al-Hamra near Muscat, an Italian mission headed by M. Tosi found numerous artifacts (such as needles, chisels, fishhooks, and some pieces of raw copper) on top and in the uppermost layers of some shell middens close to the coastal strip of the Indian Ocean. These shell middens were the dumping grounds of fishing settlements related to the exploitation of marine resources. Radiocarbon dates place the beginning of these middens at about 5000 B.C. The copper instruments, however, most probably belong to the third millennium B.C. according to Iranian potsherds found in the same layer (Durante and Tosi 1977; Gnoli 1981). Analytical data show that these artifacts fit clearly into the trace element pattern of other metal finds from the Oman mountains. The artifacts from Ras al-Hamra correspond to relics of copper production in the interior of Oman, for example, the beginning of the smelting site of the metal-producing settlement of Maysar and the hoard of copper ingots from Al-Aqir in Wadi Bahla.

More extensive copper production, however, started during the Umm An Nar Period in the third to second millennium B.C. (2500–1900 B.C.), as shown by the numerous slag heaps all over the mountain. Important smelting sites next to Maysar and also larger ones with up to 4,000 tons of slag include Wadi Salh, As-sayab, Bilad, Muaidin, and Tawi Ubaylah (Hauptmann 1985; Weisgerber 1980, 1981, 1983). This period continued to the Wadi Suq Period (1900–1200 B.C.).

Another peak of copper production approaching an industrial scale can be recognized in the time 1200–600 B.C., in the Lizq Period. Later, especially in early Islamic times (A.D. 800–1000), copper was produced on a large scale. Archaeometallurgical surveys indicate that at these times copper ores were mined and smelted at nearly all the ore deposits in the mountains. But contrary to the earlier copper production, the largest sites are situated at the massive sulfide deposits. Little activity followed in the twelfth to nineteenth centuries, when a large part of the ancient slag heaps were recycled, most probably for domestic use.

The preliminary end of the copper production is provided by modern activities in Lasail, Aarja, and Bayda. These deposits are exploited by the Oman Mining Company LLC and are the only ones in the Arabian Peninsula.

Earliest Copper Production in Oman

Returning to the earliest periods, we should concentrate on the settlement of Maysar. During 1979–1982 the German Mining Museum carried out excavations there, as this was the only site with sufficient archaeological evidence.

In Maysar the copper production started in the Hafit Period, whereas the main activities are dated into the Umm an Nar Period. The distribution of finds show that the copper ores from mines situated nearby were transported to the settlement and smelted, apparently only on a small scale, in one area of the settlement. However, one of the characteristics of the site of Maysar is the large number of copper finds that date to the Umm an Nar Period. Among these are also an assortment of archaeometallurgical finds, such as ores, slags, furnace fragments, and pieces of matte. These finds provided the material for an extensive scientific investigation to reconstruct the smelting processes used at this time. For this reason we should consider the site of Maysar an excellent sample for studying early copper production in Oman (Hauptmann 1985).

Pyrometallurgical Techniques

Ores and Copper Products

In Bronze Age Maysar mostly relatively pure copper ores with about 30% copper and a low content of iron were used. Following the intensive intergrowth of the ores in the neighborhood of Maysar, oxide ores such as malachite and chrysocolla, sulfur-containing ores such as brochantite, and only partly sulfidic ores were smelted together. This fact is shown, too, by numerous pieces and inclusions of matte in the copper metal. Mineralogical and chemical investigations proved that the matte consists of nearly pure Cu_2S, sometimes with a few percent of iron. This is in contrast to the widespread occurrence of Cu-Fe sulfides in Oman. Because there is no evidence of roasting the ores before smelting in order to concentrate the copper, the existence of pure matte is a criterion that the sulfur content is mainly a result of the use of brochantite. Therefore matte and copper were produced in a one-step smelting process (figure 4.1a). As shown by the phase relations in the system $Cu–Cu_2S$ (Schlegel and Schüller

a

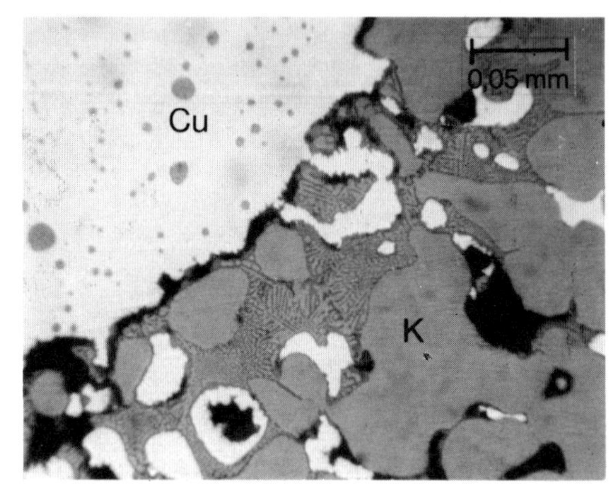

b

Figure 4.1 (a) Regulus of matte and metallic copper exsolved from a Bronze Age tap slag. Matte envelops the copper. Length of sample: 5 cm. (b) Myrmekitic intergrowth of cuprite (Cu_2O) and chalcocite (Cu_2S) between lamellae of chalcocite (K) and prills of copper (Cu). Inclusion in a Bronze Age dropping slag. Sample OM 214, Maysar 1. Reflected light, oil immersion.

1952), these were exsolved in the liquid state above 1100°C.

The precipitation of copper during smelting occurred not only by the reduction of the decomposed copper ore but also possibly by the principle of the roast reaction:

$$Cu_2S + \tfrac{3}{2}O_2 \rightarrow Cu_2O + SO_2,$$

$$Cu_2S + 2Cu_2O \rightarrow 6Cu + SO_2.$$

Even if there is no evidence for an extensive and deliberate use of this process, which is well known from the Mabuki copper smelting in Japan (Lewin and Hauptmann 1984), the chemical reaction was clearly observed microscopically in slags (figure 4.1b).

Some samples of more iron-rich mattes found in Maysar arise from the use of small amounts of chalcopyrite or bornite. However, these ores did not influence the precipitation of copper. As shown in the system Cu–Fe–S (Schlegel and Schüller 1952), even copper sulfides with an iron content of 10% coexist in the liquid state with pure metallic copper. Precipitation of metallic iron in copper during cooling is possible only if the sulfide melt contains more than 12% iron.

Therefore the complaints of a Sumerian broker, found on a cuneiform tablet, about the low quality of the copper delivered probably do not refer to its high content of (metallic) iron but perhaps to the sometimes imperfect separation of the metal from the coexisting matte. It has been shown that the copper ingots mostly contain between 0.2% and 1.2% sulfur, but sometimes up to 6% together with high values of iron. Optical investigations showed that these are caused by inclusions of matte in the metal. Thus the statement of Tylecote (1962) that the frequent appearence of iron in Bronze Age artifacts is due to the unintentional use of Cu-Fe sulfides, which occur as relicts in the copper, is confirmed.

We have no evidence in Maysar if and in what way the matte produced together with the metal was further treated. According to the relatively large amount of matte pieces, this intermediate product probably had not been refused.

The production of matte in this period is not surprising according to the form and type of mineralization. Here, the question of when humans started to use sulfur-containing ores (sulfide ores and sulfate ores!) is not of primary importance. The crucial point is to find out when humans first recognized matte as a high copper-containing material and when they were able to treat it as a metal.

There are other known examples for the early use of sulfur-containing ores, either by metal and matte inclusions in slags, finds of matte pieces, or the sulfur content of slags. For the Chalcolithic or Early Bronze Age in Timna, see Rothenberg (1978, p. 9) and Bachmann (1978, p. 23), and in Fenan see Hauptmann et al. (1985, p. 180).

Slags

So far the ores used and the by-products from this one-step process have been identified by studying the metal and sulfide objects and inclusions in the slag. But this does not provide us with a sufficient understanding of the whole process. Without investigating the slag as the waste material of the pyrotechnological activity, we are not informed, for example, of the kind of flux used or the chemophysical conditions under which the smelting process was carried out. Important parameters here are the temperature of firing, the gas atmosphere, especially the redox conditions in the furnace, and the viscosity of the (liquid) slag.

It has been pointed out by several scholars that the investigation of slags with chemical and mineralogical methods offers a powerful tool for the reconstruction of ancient smelting processes. Chemical analyses[1] show that the slags from Maysar have the composition of iron-rich slags well known in archaeometallurgy (figure 4.2 and table 4.1). They were produced in the Early Bronze Age (third millennium B.C.) as well as in the Umm an Nar Period (third to second millennium B.C.) with a similar bulk composition with respect to their SiO_2, Fe-oxide, MgO, and CaO contents. By comparing the chemical composition of these slags with the siliceous country rock of the ores, we can conclude that iron ores such as hematite and limonite were used as fluxes to reach compositions with advantageous chemophysical properties for the smelting operation.

Remarkable differences exist between the slags from the two periods with respect to their copper contents. Although slags from the later period contain mostly below 2% copper (some exceptions with up to 7%), slags from the Early Bronze Age sometimes show copper contents of up to 31% (Hauptmann 1985, p. 120), which points to inefficient copper production with low technical know-how. The separation of the metal was far from complete because of the smelting conditions with relatively low temperatures. Also the reduction of Cu^{2+} from decomposed ore was only partly reached.

The main component of the silicate slags from both periods is, according to the bulk chemistry, fayalite-rich olivine (figure 4.3).[2] It forms millimeter-size idiomorphic crystals and tiny skeletons in the matrix. Typical is a pronounced zoning of the crystals from Mg-rich compositions in the core to Fe- and Ca-rich compositions at the rim ($Fo_{60}Fa_{38}La_2 \rightarrow Fo_{18}Fa_{72}La_{10} \rightarrow Fo_3Fa_{35}La_{62}$), which forms a sharp rim around the crystal (see figure 4.5) consisting of two phases resulting from the solvus between fayalite and kirschsteinite (Hauptmann et al. 1984).

Further components are clinopyroxenes with composi-

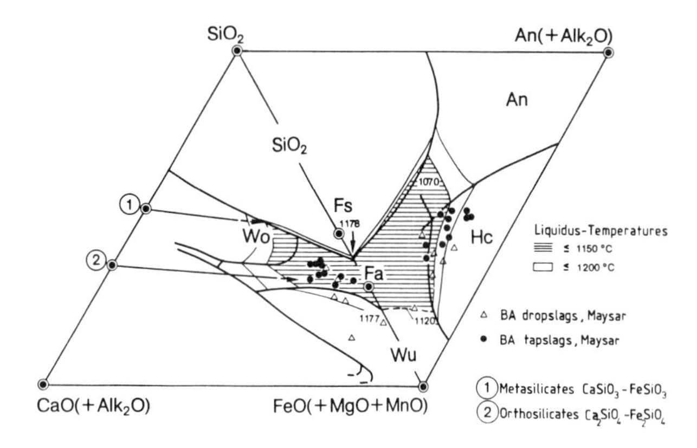

Figure 4.2 Projection of Bronze Age copper smelting slags into the systems $CaO(+Alk_2O)$–$FeO(+MgO + MnO)$–SiO_2 and anorthite (Alk_2O–SiO_2–$FeO(+MgO + MnO)$), which are modified parts of the quaternary system CaO–FeO–SiO_2–Al_2O_3. An, anorthite; Hc, hercynite; Fs, ferrosilite; Fa, fayalite; and Wo, wollastonite.

Table 4.1 Chemical composition, phase content and viscosity of several representative copper slags from the Early Bronze Age and the Umm an Nar Period found in Maysar 1

	Early Bronze Age		Umm an Nar Period		
Component	OM 114	OM 214	OM 48/5	OM 289/2	OM 350/2
SiO_2	28.5	10.6	28.9	28.6	26.4
TiO_2	0.20	n.d.	0.15	0.21	0.19
Al_2O_3	1.50	0.60	4.73	7.07	6.68
Fe_2O_3	22.3	47.0	11.0	8.40	16.2
FeO	23.4		33.2	34.1	27.3
MnO	0.70	1.40	0.20	0.32	0.31
MgO	3.60	3.0	11.5	7.14	10.7
CaO	7.60	9.70	5.73	10.0	6.05
Na_2O	0.61	0.14	0.61	0.61	0.54
K_2O	n.d.	n.d.	0.39	0.63	0.35
P_2O_5	0.02	2.0	0.08	0.11	0.08
S	0.70	1.60	0.27	0.22	0.34
Cu	7.30	22.0	2.21	1.72	3.78
Zn	n.d.	0.10	0.06	0.07	0.07
Ni	0.18	0.18	0.13	0.10	0.13
Total (wt. %)	96.69	99.32	99.16	99.30	99.12
Phase content[a]	Ol, Qz, Cpx Mt, G	Cu-Sp, Ol, Cpx, Mt, G	Ol, Cpx, G, Hc, Mt, Hm	Ol, Hc, Mt, Cpx, G	Ol, Hc, Mt, Wu (?), G
Viscosity (ln η) at 1200°C	–	–	1.71	1.85	1.83

a. Phase abbreviations: Ol, olivine; Qz, quartz; Cpx, clinopyroxene; Mt, magnetite; Wu, wüstite; Hc, hercynite, Cu-Sp, cuprospinel; G, glass.

Figure 4.3 Texture of Bronze Age copper slag. Main component is an iron-rich olivine (ol); magnetite (mt) is subordinate. The matrix consists of a fine-grained mixture of olivine, hedenbergite, anorthite, and glass. Sample OM 350/1, Maysar. SEM picture, secondary electron image.

tions between ferrosilite and hedenbergite ($Fe_2Si_2O_6$–$CaFeSi_2O_6$), crystallized after fayalite. Some of the clinopyroxene typically has a high Al content. The composition of the glassy matrix, which in fact changes from micro- to cryptocrystalline to true glass, varies between anorthitic and gehlenitic composition. Among the iron oxides mainly magnetite was determined. It occurs in two different forms: frequent zoned idiomorphic crystals with a Mg-rich core precipitated as the primary phase under relatively high oxidizing conditions and dendritic magnetite crystals found in the interstices between silicate phases throughout the slags and along chilled surfaces. The magnetite crystallization is due to the crossing of the equilibrium fayalite = magnetite + quartz during cooling and solidification of the iron-rich melt. In slags crystallized under high oxidizing conditions, hematite, delafossite ($CuFeO_2$), and cuprite are also seen.

Estimation of Liquidus Temperatures (Melting Temperatures)

The estimation of liquidus temperatures of slags is the basis for the estimation of one of the most important process parameters: the firing temperature in the furnace.

For this purpose several methods are discussed. The first is the projection of bulk analyses reduced to the main components into phase diagrams suitable in chemistry and the observed phase content. Here, especially two modified systems, $CaO-FeO-SiO_2$ and $SiO_2-CaAl_2Si_2O_8$ (anorthite)$-FeO$, from the quaternary system $CaO-Al_2O_3-SiO_2-FeO$ were used. These systems are known in archaeometallurgy (Tylecote and Boydell 1978; Morton and Wingrove 1969, 1972; Milton et al. 1976). However, this concept is suitable for only a limited range of slag compositions. For the overwhelming part of copper slags one word of warning is necessary. Because of the complex chemical composition of these slags, the simplified plotting in a system with four components can give only a wide range of liquidus temperatures that, in detail, may contain hazardous data (Kresten 1986). Furthermore, the systems are valid for lower redox conditions than those under which ancient copper slags are generally produced (Hauptmann 1985, p. 68; Hauptmann et al. 1985, p. 183; Keesmann et al., unpublished). For the Bronze Age slags from Maysar liquidus temperatures in the range 1110°–1230°C were found. Nevertheless, this concept should be regarded as a useful tool for a rough estimation, especially in cases when no analytical equipment is available.

The most important advantage of the use of phase diagrams is the possibility of judging the state of the art reached by the ancient metallurgists during charging. In figure 4.4 and in the complete phase diagrams given in Milton et al. (1976, p. 31) and Morton and Wingrove (1969, p. 1559), an area of relatively low liquidus temperatures is observed, and it is limited by a steep increase of the isotherms in the fields of SiO_2, hercynite, and wollastonite. Slag compositions inside these fields would require a drastic increase of temperatures to reach the liquidus.

The composition of the Bronze Age slags from Maysar (table 4.1) show a good range of tolerance toward the SiO_2 content. However, it is obvious that the Al_2O_3 content, which shifts the composition in the direction of hercynite, is too high in order to meet the ideal area with the lowest temperatures.

The second method of estimating liquidus temperatures is the study of metal-sulfide inclusions. This is a useful and advantageous method for simply composed copper ores, like those from Maysar, that have been smelted. A discussion of such phase diagrams as $Cu-Cu_2S-S$ and $Cu-Fe-S$ is possible directly and without drastic corrections. As, Ni, and other trace element contents have no decisive influence. Inclusions in slags from Maysar point to a minimum temperature of 1100°C.

A third method for liquidus temperature estimation is experimental study by differential thermoanalyses (DTA). In order to determine the solidus and liquidus temperatures (beginning of melting to the fully liquid state and reverse), some slag samples were studied by

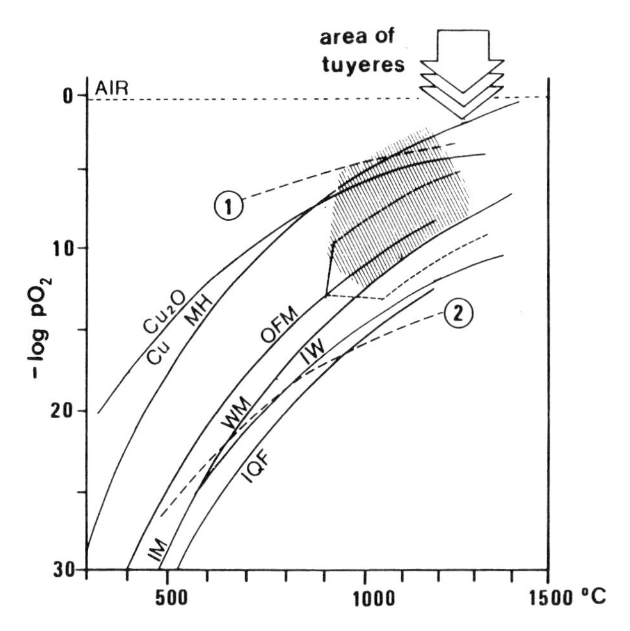

Figure 4.4 Eliminated p_{O_2} and temperature ranges of Bronze Age copper smelting slags from Maysar, shown in the p_{O_2}-T diagram of the systems Fe–Si–O, Cu–O, and C–O. IQF, iron + quartz – fayalite; IM, iron – magnetite; IW, iron–wüstite; WM, wüstite – magnetite; QFM, quartz + magnetite – fayalite; MH, magnetite – hematite. 1, $CO + CO_2 - CO_2$; 2, graphite $- CO + CO_2$. The highest oxygen pressure is near the tuyeres (arrow). For comparison, the stability range of iron slags is shown (Keesmann et al., unpublished). Data from Eugster and Wones (1962), Muan (1955), and Osborn and Muan (1960).

DTA. The liquidus temperatures lie in the range 1120°–1170°C. Some of the data deviate markedly from those of the projection in the system $CaO-Al_2O_3-SiO_2-FeO$.

Finally, liquidus temperatures can be estimated by investigations of furnace fragments. Tite et al. (unpublished) investigated several fragments of furnace linings from Maysar and found out that the fragments belonging to a smelting furnace were heated several hours at temperatures of 1150°–1200°C, whereas fragments of a crucible were heated only for a short time up to 1100°–1150°C.

The liquidus temperatures of the Bronze Age slags from Maysar, however, were only barely reached or were reached for only a short time. This is demonstrated by unreacted relicts of ores and fluxes as well as by the porphyritic texture of the slags. And it is also confirmed by the studies of the furnace fragments. The firing temperature in the furnaces apparently did not exceed the liquidus temperatures of the slags by more than about 50°C. Tapping has often been carried out with only partially liquid slags, preventing complete separation of slag, matte, and metal.

The Redox Conditions in the Furnace

The phases existing in the slags indicate disequilibria and deviations from a continuous smelting process. This could represent simply a variety of chemophysical conditions of different points in the smelting furnace [see the smelting experiments of Tylecote and Boydell (1978, pp. 32–38) and of Avery (1982, pp. 206–208)]. But because the furnace fragments found suggest a rather small reaction vessel with a diameter of only 0.4–0.5 m, the observed irregularities are caused more probably by a decrease in temperatures and/or an increase in oxygen in the furnace, which can result from irregular and changing handling of the bellows.

These deviations can be recognized by studying the texture of the iron-rich slags, which are heavily influenced by the oxygen content of the gas atmosphere, especially the CO/CO_2 ratio in the liquid state. Hereby, the Fe^{2+}/Fe^{3+} ratio is controlled, leading to a changing crystallization of iron silicates and iron oxides (Roeder and Osborn 1966). Using the buffer equilibria in the systems Fe–Si–O and Cu–O as functions of the oxygen pressure and temperature (p_{O2}/T diagram; see figure 4.4), the following three simplified interpretations are possible.

First, slags with Fe-rich olivine as the main component are crystallized in the stability range between the equilibria fayalite = magnetite + quartz and fayalite = iron + quartz. Most of the Bronze Age slags from Maysar belong to this type of slag, crystallized between $p_{O_2} \approx 10^{-7}$ to 10^{-10} atm (figure 4.5). They represent a rather strong reducing atmosphere.

Second, slags with a cotectic crystallization of magnetite and Fe-rich olivine roughly correspond to the buffer equilibrium fayalite = magnetite + quartz. This is due to relatively high oxidizing conditions. This texture is observed occasionally in slags from Maysar.

And third, slags with Fe-rich olivine and hematite (Fe_2O_3) compare with the equilibria fayalite = magnetite + quartz and magnetite = hematite. The buffer equilibrium magnetite = hematite cuts the curve of Cu/Cu_2O at about 1000°C. Thus these slags are crystallized under relatively high oxygen pressure. In fact, many slags from the Early Bronze Age, but also from the Umm an Nar Period in Maysar, show this phenomenon and contain high copper values in the form of Cu(II) oxides.

Compared with modern metallurgical procedures, this means that smelting not only was carried out according to the principles of a strong reducing process but also contained chemical reactions typical of the more oxidizing atmosphere of the reverberatory furnace. This agrees well with the observations from Czedik-Eysenberg (1958), who also proposed a similar smelting process and corresponding types of furnaces for the Bronze Age smelting of copper in the Alps.

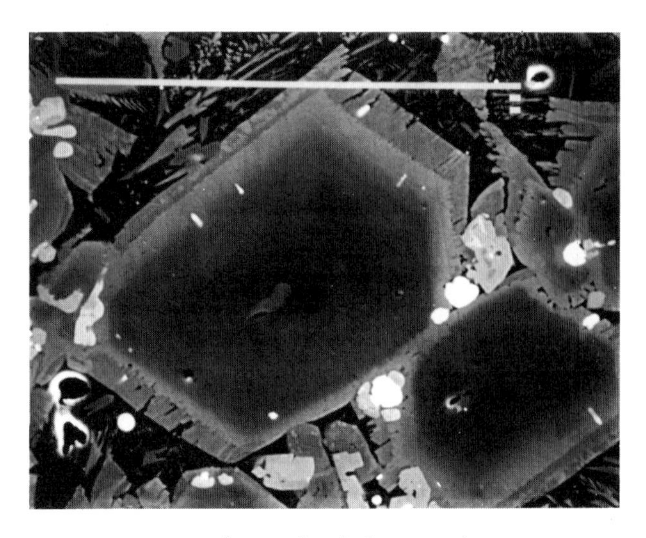

Figure 4.5 Zoned crystals of olivine with a magnesium-rich core and calcium-iron-rich rim (=kirschsteinite, $CaFeSiO_4$; Hauptmann et al. 1984). White, copper prills; gray, magnetite. Sample OM 350/1, Maysar. SEM picture, secondary electron image. Length of bar is 100 μm.

Viscosity

The separation of metal or matte from the slag depends essentially on its viscosity in the liquid state. For selected slag samples viscosities were calculated at 1200°C using the method of Bottinga and Weill (1972). Contrary to earlier calculations (Bachmann 1980), this method is worked out in more detail for silicate liquids. It considers the dual role of aluminium as a network former and as a network modifier. The method of Bottinga and Weill was confirmed by experimental studies and proved to be relatively accurate (Urbain et al. 1982). However, the problem with higher Fe^{3+} contents is not satisfactorily solved. The calculation, therefore, can give only an estimation of the viscosity.

The Bronze Age slags from Maysar show values below $\ln \eta = 2$. These values are not ideal: The viscosity of a modern copper slag should be around 5 poise in a temperature range of 1050°–1150°C. It has to be regarded, too, that the slags have been tapped in a viscous state: They started to crystallize within the furnace. Therefore the high copper contents even in the Bronze Age slags are not surprising.

The values of viscosity are independent of the morphological differences of the slags. Because the slags have no significant differences in chemical composition, this demonstrates that the phenomenology is caused by different techniques (tapping) rather than by physical propeıties, as postulated by Koucky and Steinberg (1982). The same observation was made by Craddock et al. (1985, p. 204), investigating silver slags from Rio Tinto.

The Metal Products

There is no doubt that the production of copper was one of the most important activities at Maysar: Numerous pieces of raw copper and objects such as needles, chisels, and axes from the Hafit and the Umm an Nar periods have been found. The most important find among these artifacts is without doubt a hoard consisting of twenty-two bun-shaped ingots or their fragments. The hoard found in Maysar 1, house 4, weighed more than 6 kg (figure 4.6). It is believed that this hoard was traded to Maysar as a caravansarylike settlement. A similarly impressive find is the copper hoard discovered recently in Al-Aqir (Bahla), which consists of sixteen complete ingots with a total weight of 19 kg.

In the Middle East and in the Indus Valley, bun-shaped copper ingots were found at several sites in India, Syria, Iraq, and Pakistan. We are, of course, not able to say that all these ingots come from Magan (Oman), but at least it is clear that bun-shaped ingots were the form in which copper was traded during the third to second millennium B.C. in this area.

The copper ingots from the Maysar hoard as well as some of the artifacts from Ras al-Hamra have been investigated. Chemical analyses[3] (table 4.2) of ingots together with pieces of raw copper show that these compare well with the trace element pattern of many copper ores from Oman. Although not enough analytical work has been done so far in order to define the whole variability of ores in the Oman mountain range, it seems that the copper hoard from Maysar could be representative of the Bronze Age copper of Oman.

The first remarkable characteristic is the high nickel content of the ingots, which is not amazing in relation to the ores (figure 4.7). The nickel content is between 0.1% and 0.5%. This fact has been pointed out already by Desch (1929) and Peake (1928). As shown by the investigations of Hauptmann (1985), the recovery of nickel in the metal from a pure copper ore is about 80%. If copper matte is produced at the same time as the metal, it is nearly free of nickel. About 20% of the original nickel content is substituted in the slag phases (spinel, olivine). A much stronger partitioning, however, can be observed during matte smelting using Cu-Fe sulfides: The nickel is always concentrated in the iron sulfide phases in the matte, which, of course, are slagged. In this case the nickel content of the metal is much lower than in the ore and the nickel content of the slag is higher.

Also, the arsenic content of most of the ingots and prills shows the same level as the nickel, but the variability is higher (figure 4.8). According to Tylecote (1980), the existence of copper objects with high arsenic contents could be an indication that only little refining, if any at all, was carried out in a crucible and that the objects were cast as soon as possible after the

Figure 4.6 The copper hoard from Maysar 1, house 4, location 31, contains 6 kg of metal in the form of bun-shaped ingots or their fragments. These ingots were the form in which the metal was traded.

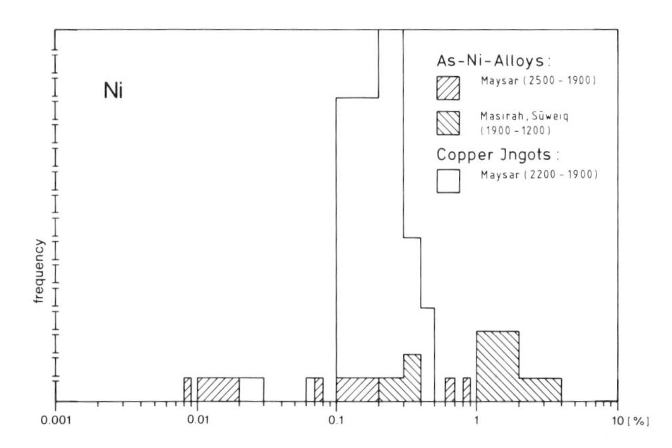

Figure 4.7 Nickel contents of copper ingots and artifacts from Oman.

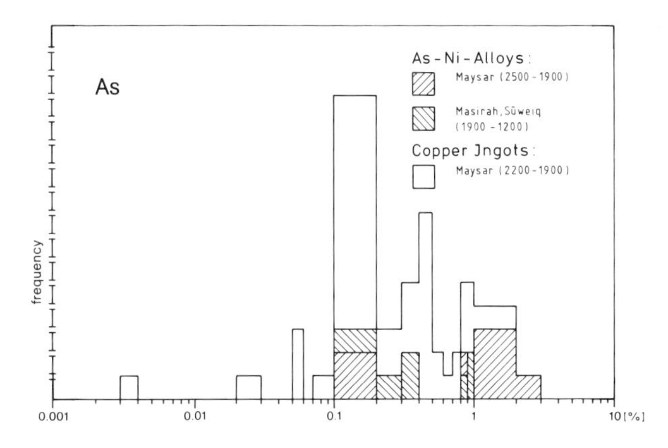

Figure 4.8 Arsenic contents of copper ingots and artifacts from Oman.

Table 4.2 Chemical composition of selected copper ingots and pieces of raw copper from Maysar and Ras al-Hamrah

Sample[a]	Cu (%)	Zn	Fe (%)	Co	Ni (%)	As (%)	Sb	Sn	Ag (%)	Pb	Cr	Cd	Co/Ni	Sb/As
DA 4840/A	87.0	230	0.31	1100	0.48	0.20	7	230	0.21	89	n.d.	34	0.24	0.004
DA 4840/B	97.0	360	0.51	530	0.29	0.14	61	300	0.22	47	200	33	0.19	0.044
DA 4840/C	96.2	4200	2.49	1200	0.28	0.82	570	n.d.	0.01	n.d.	n.d.	130	0.44	0.069
DA 4840/D	89.6	720	4.1	1300	0.22	0.95	29	46	0.02	24	n.d.	300	0.59	0.003
DA 4840/E	95.5	310	0.01	240	0.24	0.43	39	n.d.	n.d.	n.d.	n.d.	100	0.1	0.009
DA 4840/F_1	93.9	230	0.04	230	0.23	0.34	70	380	0.08	42	27	89	0.1	0.02
DA 4840/F_2	97.9	160	0.01	240	0.26	0.37	15	n.d.	0.29	42	n.d.	140	0.092	0.004
DA 4840/G	97.0	300	0.37	1400	0.28	0.06	51	230	0.10	64	n.d.	14	0.5	0.085
DA 4840/H	95.5	290	0.17	920	0.29	0.58	31	260	0.16	74	n.d.	150	0.32	0.005
DA 4840/I	98.2	73	0.01	240	0.24	0.41	18	n.d.	0.29	45	n.d.	65	0.10	0.004
DA 4008	96.4	230	0.14	73	0.24	0.14	52	440	0.01	28	45	78	0.030	0.037
DA 4028	98.6	290	0.80	320	0.31	0.17	n.d.	170	0.18	n.d.	n.d.	28	0.10	<0.001
DA 4030	97.6	260	0.14	340	0.46	0.41	44	320	0.09	53	n.d.	93	0.073	0.010
DA 4033	91.2	370	0.03	200	0.19	0.09	61	520	0.20	69	44	33	0.11	0.067
DA 4040	94.8	120	n.d.	n.d.	0.26	0.83	42	n.d.	0.06	1200	n.d.	120	<0.01	0.005
DA 4259	99.8	600	0.38	680	0.14	0.72	170	n.d.	0.01	24	n.d.	120	0.49	0.023
OM 46	93.4	610	0.24	1100	0.14	1.71	3300	520	0.03	7700	n.d.	230	0.78	0.19
OM 437	81.3	200	0.25	93	0.29	0.06	36	260	0.02	130	16	25	0.032	0.06
OM 441	94.8	210	0.10	74	0.43	0.81	200	n.d.	0.09	190	n.d.	130	0.017	0.02
OM 443	57.6	74	0.46	360	0.02	0.01	n.d.	310	0.01	8	5	4	1.8	<0.001
RAH O.N.	82.4	200	0.27	240	0.20	0.47	71	430	0.05	390	n.d.	130	0.12	0.015
RAH 5	80.3	170	0.02	22	0.07	0.68	52	250	0.07	48	2	170	0.031	0.008
RAH 10/k10	84.5	180	0.01	150	0.11	0.08	84	310	0.09	150	32	13	0.14	0.11

a. DA 4840/A–I and DA 4008–4259: Copper ingots, Maysar 1, house 4 (2200–1900 B.C.).
OM 46–443: Pieces of raw copper, Maysar 1 (2200–1900 B.C.).
RAH O.N. and RAH 5: Pieces of raw copper, Ras al-Hamra (third millennium B.C.).
RAH10/k10: Needle, Ras al-Hamra (third millennium B.C.).

metal was melted under charcoal. This corresponds to investigations on raw copper and copper ingots from Maysar by Hauptmann (1985), who suggested that the high sulfur and trace element contents and the complete absence of cuprite (Cu_2O) in some of the ingots excludes a refining of the raw copper produced before casting these ingots. The variability of the arsenic content may therefore represent a partly repeated casting, but it could also be related to the irregular smelting processes deduced from the investigation of slags.

In order to provide a more detailed comparison of the objects and the ores, it is necessary to consider those elements that show only a little partitioning during smelting. Seeliger et al. (1985) proposed to use the ratios of geochemically similar elements such as Co and Ni and Sb and As to characterize the ores and artifacts. The decisive point of this concept is that these element pairs should be concentrated or diminished in a similar way during the metallurgical process. In this way problems arising from the high variability of absolute concentrations of elements within the same ore deposit can be avoided. It is important to recognize that most of the ores and ingots have a Co/Ni ratio between 0.06 and 2 and a Sb/As ratio between 0.001 and 0.2 (figure 4.9) and are in good correspondence with each other. The Sb/As ratios of some ingots of <0.001 are due to Sb values below the detection limit at the lower level of $\mu g \cdot g^{-1}$.

The composition of the ores and ingots in the Co/Ni and Sb/As diagram mostly agree with the range proposed by Seeliger et al. (1985), which was based on less data. But it appears that the differentiation, especially of ores from Iran (Bardsir and Kermanshah), will be more difficult as more samples are available. Unfortunately this is one decisive point in the provenance studies of the Mesopotamian artifacts, because in the southern part of Iran numerous copper deposits occur in a geologically similar context (ophiolites) and are also possible metal sources of early Mesopotamia.

Some remarkable results were found during a comparison of the ores and ingots from Maysar and some artifacts from different localities in Oman. Table 4.3 and figures 4.7 and 4.8 show that the artifacts from the Hafit Period (3000–2500 B.C.) from Maysar contain the lowest nickel and arsenic values, partly below the average content of the ingots.

On the other hand, artifacts from the Umm an Nar Period (2500–1900 B.C.) from Maysar have arsenic contents of 1–2% that are not reached by either the ores or the ingots. In addition, two samples from this

Table 4.3 Chemical composition of artifacts from Maysar, Soweiq, and the island of Masirah[a]

Sample[b]	Cu (%)	Zn	Pb	Fe (%)	Co	Ni	As (%)	Sb	Sn	Ag	Cr	Cd	Co/Ni	Sb/As
DA 4029	81.6	82	60	0.06	10	770	1.67	n.d	350	43	31	280	0.013	<0.001
DA 4260	80.1	55	4600	0.25	20	0.17 (%)	0.87	2200	n.d.	34	n.d.	140	0.012	0.25
DA 4305	70.7	100	450	0.23	52	0.90 (%)	2.20	580	180	40	n.d.	340	<0.01	0.026
DA 4962	84.4	n.d.	1100	0.01	n.d.	0.69 (%)	1.46	1040	n.d.	1300	n.d.	220	<0.01	0.071
DA 4963a	83.9	30	34	0.03	10	120	0.19	n.d.	190	50	10	33	0.083	<0.001
DA 4963b	79.2	300	46	0.38	12	84	0.14	n.d.	n.d.	90	n.d.	28	0.14	<0.001
DA 5168	97.6	330	30	0.36	120	1.14 (%)	1.48	n.d.	n.d.	1300	n.d.	230	0.01	<0.001
DA 6516	99.4	110	38	0.24	860	0.22 (%)	0.10	3	200	61	n.d.	21	0.39	<0.003
DA 6518	94.0	280	350	0.51	2400	2.93 (%)	0.87	400	220	87	n.d.	140	0.082	0.046
DA 7358	97.3	70	57	0.15	1300	1.35 (%)	0.32	85	2600	90	n.d.	54	0.096	0.027
DA 7370	92.0	190	42	0.42	680	1.48 (%)	0.34	230	230	90	n.d.	60	0.046	0.068
DA 7397	80.6	180	450	0.45	290	3.83 (%)	1.00	400	1.22 (%)	54	n.d.	160	<0.01	0.40
DA 7347	46.5	260	1900	1.12	290	0.33 (%)	0.27	n.d.	n.d.	10	5	42	0.088	<0.001

a. Values given in $\mu g \cdot g^{-1}$ if not in %; n.d. = not detected.

b. DA 4029: Chisel head, Maysar 1, house 4, c. 2000 B.C. (oxidized material).
DA 4260: Ax, Maysar 1, house 31, c. 2000 B.C. (oxidized material).
DA 4305: Ax, Maysar 4, tomb 1, c. 2300 B.C. (oxidized material).
DA 4962: Chisel, Maysar 1, c. 2000 B.C. (oxidized material).
DA 4963a: Needle, Maysar 25, tomb 1, c. 2500 B.C. (oxidized material).
DA 4963b: Needle, Maysar 25, tomb 1, c. 2500 B.C. (oxidized material).
DA 5168: Ax, Masirah, site 38, c. 1500 B.C.
DA 6516: Pick, Masirah, site 38, c. 1500 B.C.
DA 6518: Spearhead, Masirah, site 38, c. 500 B.C.
DA 7538: Dagger, Masirah, site 38, c. 1500 B.C.
DA 7370: Knife, Masirah, site 38, c. 1500 B.C.
DA 7397: Spearhead, Suweig, c. 1500 B.C. (oxidized material).
DA 7347: Awl, Masirah, site 38, c. 1500 B.C. (oxidized material).

group (DA 4962, DA 4305) show considerably higher nickel contents.

The third group of artifacts come from Suweiq and the island Masirah and belong to the Wadi Suq Period. They contain up to 3.8% nickel, which clearly surpasses the values of the ores (max. 0.8%) and the ingots (max. 0.48%; figure 4.7).

The artifacts from the last two groups must therefore not be regarded as copper artifacts but rather as alloys of copper with arsenic or arsenic and nickel. From the chronological point of view this means that the first alloys used in Oman were composed of copper arsenic, which appeared during the Umm an Nar Period in the third to second millennium B.C. This fully corresponds with the first appearance of such alloys in the whole Middle East (Muhly 1980). In the subsequent Wadi Suq Period (1900–1200 B.C.) in Oman Cu-As-Ni alloys were used, and they are similar to those described by Craddock (1985) from Ras al-Khaimah. This type of alloy seems to be a peculiar feature of the southeastern part of the Arabian Peninsula.

Craddock could refer only to a few data about ore and copper from Oman, but his assumption on the unintentional production of these alloys caused by the composition of the ores seems to be correct.

As shown in figures 4.9 and 4.10, the Co/Ni and the Sb/As ratios of alloys from Masirah, Suweiq, and Ras al-Khaimah are in good correspondence to the copper ores from Oman. Even if the ores analyzed as yet (Berthoud 1979; Coleman et al. 1979; Hauptmann 1985) do not reach the nickel and arsenic values of these alloys, according to type and composition of the Oman copper deposits it cannot be excluded that suitable ores had once been available in larger amounts: In the ultrabasic rocks of the Samail Ophiolite Complex, Ni-Co-As-Fe minerals were mineralogically identified in tiny crystals in the copper ores. They were described from the Troodos Mountains (Panayiotou 1980) in Cyprus, which are geologically identical with the Oman mountain range.

In addition, Leese et al. (1985–1986, pp. 114ff) were faced with a similar problem when they analyzed ores and metal objects from Timna, Israel. Here, the arsenic content in the metal artifacts is much higher than in the ore. The explanation that the ore samples were not typical of those smelted in ancient time and that ores with a higher arsenic content were chosen in antiquity seems unlikely to Leese et al. An enrichment of As in the metal is noted also by Merkel (in press), who carried out smelting experiments with ores available today in Timna. This phenomenon, which is a serious problem for As-containing ores and copper-

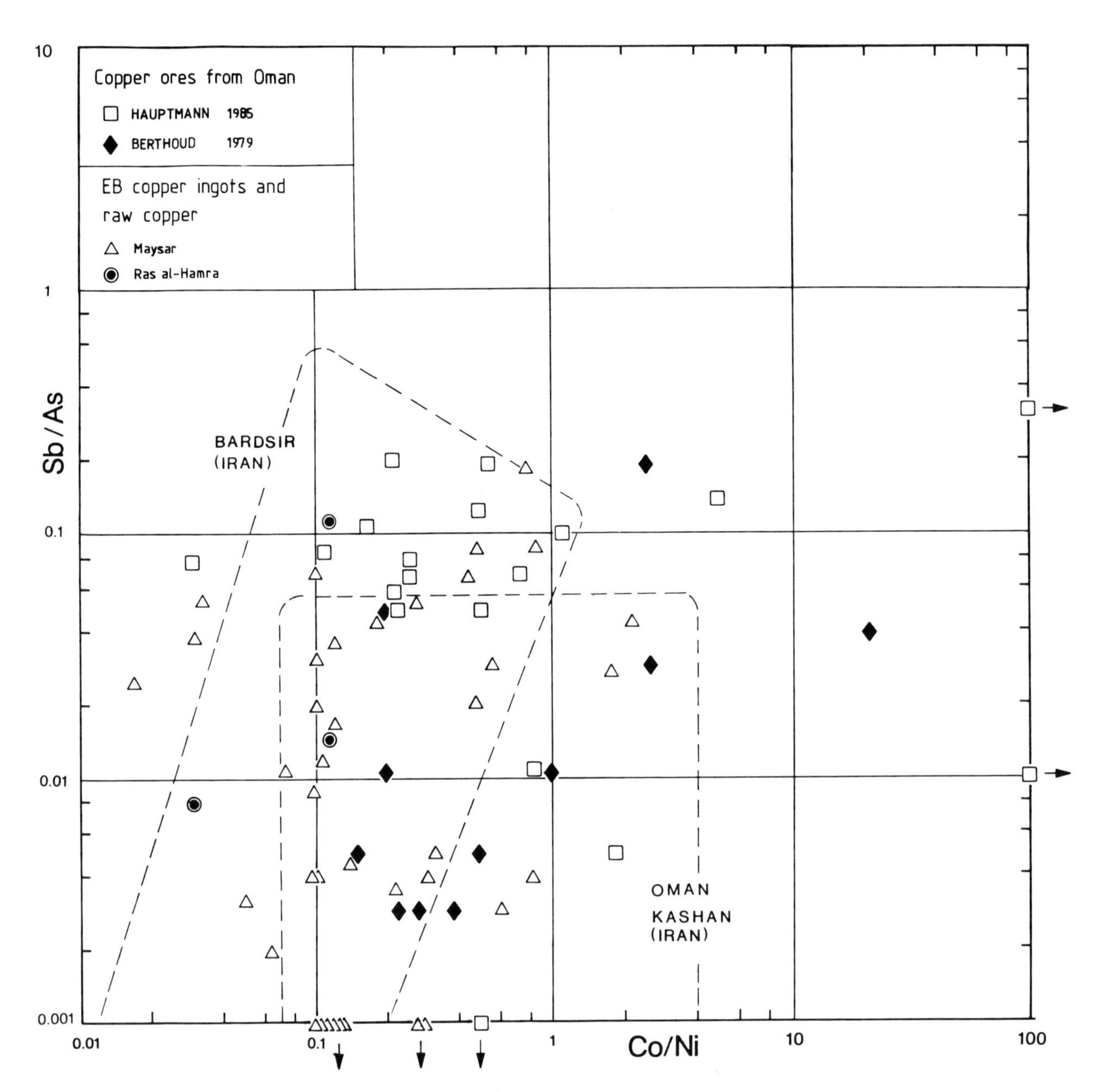

Figure 4.9 Sb/As and Co/Ni ratios in copper ores from several deposits in Oman in comparison to Bronze Age copper ingots and raw copper from Maysar and Ras al-Hamra. Horizontal arrows on the right-hand side mean that the points are outside of the area shown; vertical arrows at the bottom of the graph mean that the antimony content of the sample is below the detection limit. The diagram shows the range of compositions of the ores from Kashan and Bardsir (Iran) proposed by Seeliger et al. (1986), which was based on only a few data points. It is obvious that a differentiation between the ores from Oman and the southern part of Iran in this way is nearly impossible.

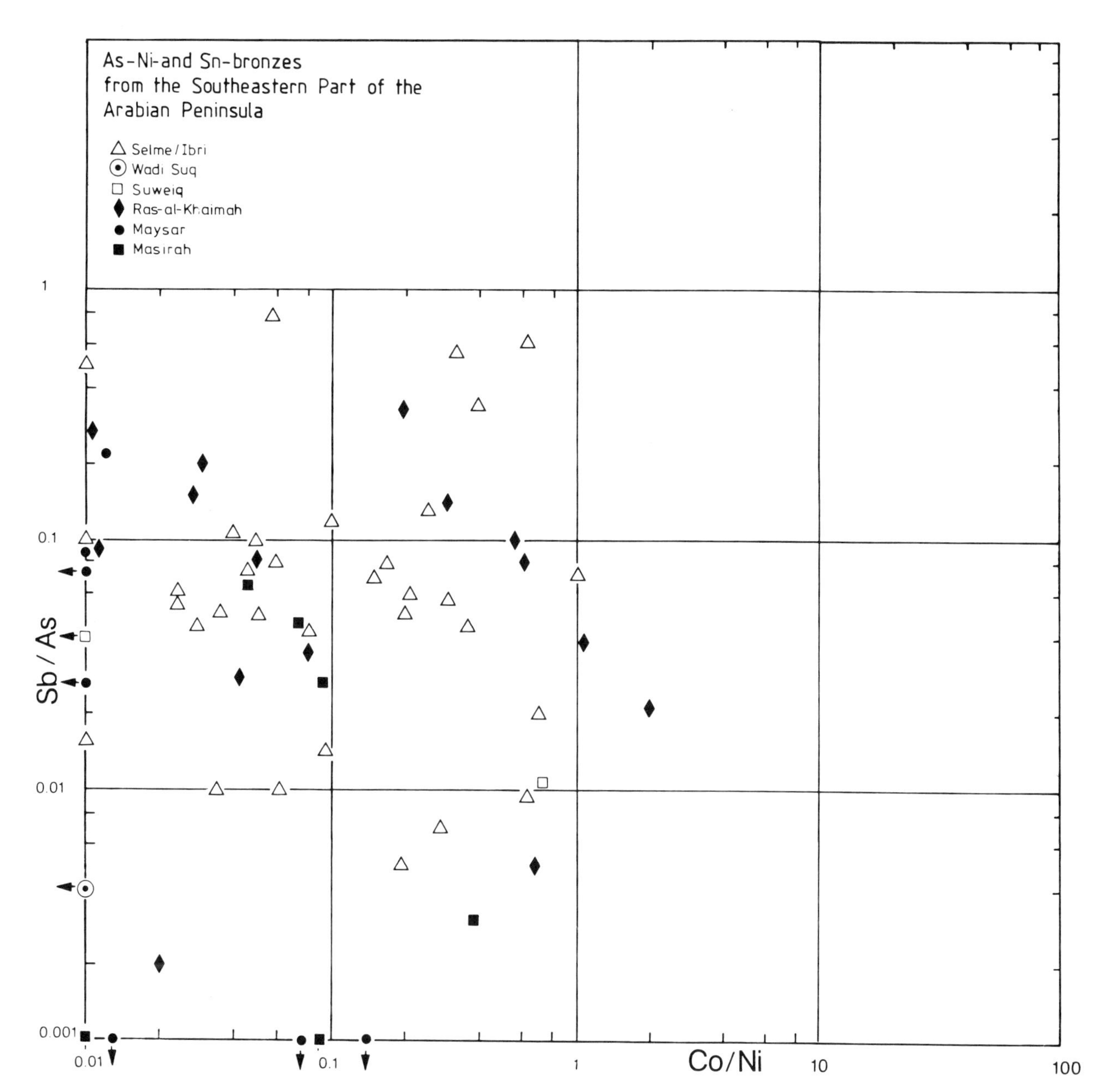

Figure 4.10 Sb/As and Co/Ni ratios in bronzes from several localities in Oman and from Ras al-Khaimah. The situation of artifacts from Maysar outside the diagram most probably is due to the intensive corrosion of the analyzed samples. The bronzes partly fit into the area of the ores and ingots as shown in figure 4.13 but clearly show a shift to lower Co/Ni ratios.

arsenic alloys, is reinforced. It is not understood so far but is clearly related to the smelting process. It seems possible, from looking at the nickel contents of ores and artifacts from Oman, that a similar enrichment of nickel occurs during smelting.

The fact that the artifacts from the Hafit and the Umm an Nar periods from Maysar differ by more than one order of magnitude from the ores and ingots discussed with respect to their Co/Ni and Sb/As ratios (figure 4.10) should not be surprising, because these samples were mostly corroded. The archaeological evidence together with the local situation of the copper deposits nearby point to production of the artifacts in the settlement of Maysar.

Following the question of how these alloys were made, the analytical data of artifacts that suggest a deliberate production (for example, with the addition of arsenic minerals to the copper) should not be regarded as isolated. The comparison to the analytical data of ores and the knowledge of the geological circumstances prove that the artifacts most probably were made using (unintentionally) As- and Ni-rich copper ores from deeper parts of the ore deposits.

An investigation of a comparable problem was carried out by Helmig (1986). He analyzed ore, matte, and copper (ingots, prills, artifacts) from the Bronze Age town of Shahr-i Sokhta (Sistan) (Tosi 1983). From a mineralogical point of view there is no doubt that the type of ore found in Shahr-i Sokhta in close context to slag and copper was the source of the metal produced. Analytical data show, however, marked discrepancies between the ore and the slag and metal, especially with respect to the arsenic and lead contents. Helmig stressed the necessity of a secure statistical basis with a sufficiently high number of objects in the discussion of provenance studies. Furthermore, Helmig demonstrated that arsenical copper with 1–3% As was produced unintentionally and only sporadically using somewhat As-rich ores. He could exclude a deliberate addition of (relatively pure) arsenic ores to the metal. With this point he also could confirm Lorenzen's (1965) and Tylecote et al.'s (1977) results, which pointed out that the recovery rate of arsenic in copper is quite high and that arsenical alloys can be obtained through the smelting of suitable ores.

The Appearance of Tin Bronze

One of the most significant events during the second millennium B.C. in Oman is the development of tin bronzes alongside the arsenical nickel alloys of copper (figure 4.11). The present state of knowledge suggests that they occurred together for about 600 years until true tin bronzes with the usual level of 10–12% tin appeared.

Important finds of mainly tin bronzes were made in Ibri (Selme) in the interior of the Oman mountain range and in Ras al-Khaimah close to the coast of the Persian Gulf.

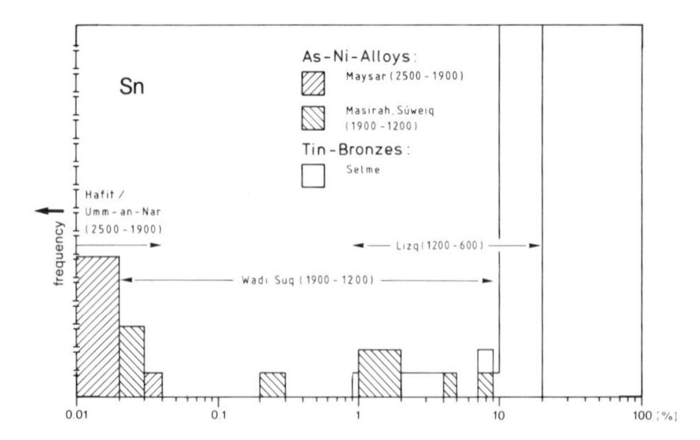

Figure 4.11 The development of tin bronzes in the southeastern part of the Arabian Peninsula from the Hafit and Umm an Nar periods (third and second millennia B.C.) to the seventh century B.C.

The Selme hoard was found in a tomb during agricultural work. It is thought to be the cache of an active tomb robber who plundered tombs mainly of the late second and early first millennia B.C. The treasure consists of some 600 items: bangles, vessels, and weapons of metal and numerous chlorite vessels. Two kinds of find can be distinguished: first, the objects that belong to the original burials in the Umm an Nar tombs (third millennium B.C.), and, second, most of the objects that were deposited in the tomb at a later date by a robber. Especially, the heavy bangles and the metal vessels belong to the first millennium B.C. (1200–600 B.C., Lizq Period). Some daggers date to the second millennium B.C. (Weisgerber 1980; Weisgerber and Yule, unpublished). Bronzes comparable to the types from Selme are published from Rumeilah (Al Ain) (Boucharlat and Lombard 1985).

With the discovery of the Selme hoard the knowledge of bronze production in the eastern part of the Arabian Peninsula has deepened. The objects show a high standard of craftsmanship as well as a considerable independence of the cultures in Mesopotamia and Iran. Although most of these bronzes seem to have been produced using local copper ores, the shape of some of the vessels, especially those from the first millennium B.C., reflects a certain Assyrian influence.

Most of the objects from Selme analyzed so far are tin bronzes with 9–13% tin[4] (table 4.4). The bangles belong to this group with one exception: sample DA 3737 has a tin content of only 1.69%. The composition of the daggers varies from 0.95% tin (DA 3634) to 13.5% (DA 3632). Compared with the geochemistry of the copper ores from Oman there is no doubt that even tin contents below 0.5% were intentionally added. Dagger DA 3634 was used as a ceremony instrument, according to our understanding. The same is true probably for ax DA 3618.

Table 4.4 Chemical composition of selected tin bronzes from Ibri, Selme, and other localities in Oman[a]

Sample[b]	Cu (%)	Sn (%)	Fe (%)	Zn	As (%)	Sb	Co	Ni (%)	Pb	Ag	Co/Ni	Sb/As
DA 3618	95.1	3.10	0.41	130	0.49	n.s.	69	0.02	210	n.s.	0.035	–
DA 3632	84.6	13.5	0.33	< 30	0.06	n.s.	130	0.01	130	n.s.	1.30	–
DA 3634	97.7	0.95	1.02	180	0.02	n.s.	320	0.06	630	n.s.	0.53	–
DA 3642	87.3	12.0	0.04	< 30	0.44	n.s.	10	0.01	36	n.s.	0.10	–
DA 3649	87.5	12.2	0.27	180	0.04	3	110	0.04	360	n.s.	0.28	0.007
DA 3721	85.0	8.40	0.03	32	0.29	300	< 10	0.02	6.35 (%)	< 16	< 0.05	0.10
DA 3731	88.2	11.3	0.14	40	0.09	94	< 10	0.29	62	20	< 0.01	0.10
DA 3734	84.7	11.6	0.22	70	0.12	64	130	0.37	210	66	0.035	0.053
DA 3735	87.2	11.5	0.23	100	0.13	10	430	0.26	< 1	18	0.17	0.008
DA 3737	95.7	1.69	0.17	60	0.04	17	70	0.09	7900	60	0.078	0.043
DA 3738	93.0	5.93	0.23	64	0.10	70	690	0.47	560	110	0.15	0.07
DA 3740	87.9	9.94	0.11	66	0.17	130	300	0.65	< 1	30	0.046	0.076
DA 3742	85.1	11.6	0.21	130	0.02	100	< 10	0.16	260	< 16	< 0.01	0.5
DA 3750	89.9	9.80	0.17	70	0.04	53	730	0.29	270	63	0.25	0.13
DA 3751	89.9	10.0	0.18	77	0.04	24	170	0.08	160	48	0.21	0.06
DA 5630	90.4	9.20	0.17	< 15	0.27	44	54	0.66	4200	80	< 0.01	0.016
DA 5632	87.2	11.8	0.17	< 15	0.23	14	190	0.82	< 1	65	0.023	0.006
DA 5635	90.3	9.80	0.20	< 15	0.27	320	< 10	0.01	440	95	< 0.1	0.12
DA 5638	89.1	9.50	0.17	< 15	0.14	72	730	0.36	550	37	0.2	0.051
DA 5640	81.4	9.36	0.06	380	0.05	5	320	0.52	160	20	0.062	0.010
DA 5641	89.3	8.20	0.24	< 15	0.30	30	170	0.49	330	58	0.035	0.010
DA 5642	89.0	9.20	0.07	< 15	0.24	110	110	0.38	270	56	0.029	0.046
DA 5943	88.5	8.76	0.27	1	0.21	280	940	0.02	680	100	4.7	0.13
DA 5018a	54.3	4.39	0.13	46	0.10	n.d.	16	0.20	17	14	< 0.0	< 0.001
DA 5018b	66.9	8.02	0.08	52	0.16	n.d.	14	0.22	9	14	< 0.0	< 0.001
DA 7397	80.6	1.22	0.45	180	1.0	400	290	3.83	450	54	0.076	0.04

a. Values given in $\mu g \cdot g^{-1}$ if not in %; n.s. = not sought; n.d. = not detected.

b. DA 3618: Ax, Ibri, Selme (1200–600 B.C.).
DA 3632, DA 3634, DA 3642, DA 5630, and DA 5632: Daggers, Ibri, Selme (1200–600 B.C.).
DA 3649–3751 and DA 5635–5642: Bangles, Ibri, Selme (1200–600 B.C.).
DA 5943: Vessel, Maysar 9 (500 B.C.).
DA 5018: Spearhead, Maysar 9 (1900–1200 B.C.).
DA 7397: Spearhead, Al-Hadhib, Suweiq (1900–1200 B.C.).

Bronzes with a low tin content were also found in Suweiq (a spearhead, DA 7397) and in Maysar 9 (a spearhead, DA 5018), which belong to the Wadi Suq Period.

Of special interest is the composition of artifact DA 3721. It contains 8.4% Sn and 6.35% Pb and is the only artifact that has such a high lead content. Regarding the composition of copper ores from Oman analyzed so far there is no reasonable explanation for where the metal could have come from. The only possibility of provenance is a small hydrothermal vein with sulfidic Cu-Pb ores near Bid Bid in the middle of the Oman mountain range. This occurrence deviates from the pattern of ore deposits described already, insofar as it is connected with dolomitic rocks stratigraphically below the ophiolite complex.

All other objects show trace element compositions well within the ranges of ores and ingots from Oman. Characteristic are, again, the high nickel contents—0.01–0.82% (figure 4.12). Also the arsenic contents vary in accordance with these (figure 4.13).

In Ras al-Khaimah, two different tomb structures were excavated (Donaldson 1984, 1985): a long-chambered tomb in which were found a pennannular ring with tapered ends of electrum, a ring fragment, an arrowhead, and a spearhead, and a ring-chambered tomb with thirty-five metal artifacts (nine arrowheads, one spearhead, two razor blades, and numerous other instruments).

Donaldson states that most of the metal finds from Ras al-Khaimah are chronologically nondiagnostic, but the arrowheads can be attributed to the period 1500–600 B.C. and certainly not later than 600 B.C. We propose that these could also belong to the Wadi Suq Period. We assume this not only because of archaeological characteristics but also because of their chemical composition, which would correspond better with the present picture of the Wadi Suq bronzes from Masirah and Suweiq insofar as they usually have tin contents below 1%. It should be remembered, however, that throughout the second millennium B.C. artifacts were made from copper alone or by alloying copper with tin or arsenic throughout the Near and Middle East (Craddock 1977; Branigan et al. 1976; McKerell 1977).

Some of the bronzes from Ras al-Khaimah show the same unusual and interesting features as the bronzes from Masirah and Suweiq: They contain up to 4.65% nickel and 1.5% arsenic (Craddock 1985). This corresponds mostly with the composition of the copper ingots from Maysar and suggests the production of these bronzes using ores from Oman.

Following the concept of Seeliger et al. (1985) we plotted the Co/Ni and Sb/As ratios of these bronzes in a diagram (figure 4.10). We found with a few exceptions a picture similar to that for the objects from Maysar with respect to the Sb/As ratio. In this point most of the objects from Selme and Ras al-Khaimah correspond to the ores and the ingots of copper. But the Sb/As ratio is shifted to lower values.

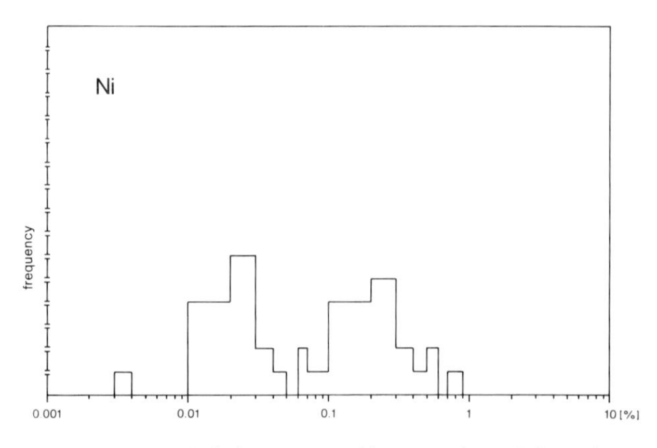

Figure 4.12 Nickel contents of bronzes from Selme (Ibri).

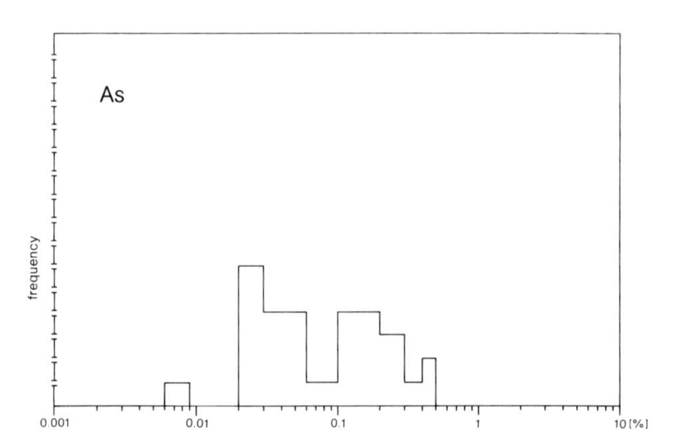

Figure 4.13 Arsenic contents of bronzes from Selme (Ibri).

Seeliger et al. (1985), investigating artifacts and ores from northwest Anatolia, stressed care in the interpretation of the Co/Ni and Sb/As ratios. Discussing provenance problems of artifacts from the Troas, they found discrepancies between ores and objects not only in the lead isotope composition but also—similar to the problem described here—in the Co/Ni ratios. They do not exclude that this could be a problem of the limited data available, but they also expect that the Co/Ni ratio could change in the metal during the smelting and alloying, as was assumed by Tylecote et al. (1977). The analyzed bronzes also show differences in the Sb/As ratio. This corresponds to the results of Zwicker (1980), who demonstrated that the As/Cu and Sb/Cu ratios change during smelting.

The interpretation of the Co/Ni and Sb/As ratios of the ores, ingots, and bronzes from the southeastern part of the Arabian Peninsula show that partitioning of these elements apparently is relatively low during smelting of the ores. But it seems that, during alloying, especially the Co/Ni ratios change to lower values.

The results of the bronze analyses show that true tin bronzes with more than 9% tin appeared in the southeastern part of the Arabian Peninsula after 1200 B.C.

That these were probably produced using local copper ores from Oman is indicated by the archaeological evidence and with a few exceptions by the trace element pattern.

The Chemical Composition of Ores and Artifacts and the Question of Magan

The archaeological and archaeometallurgical identification of the Sumerian copper land of Magan with the present Sultanate of Oman has been proved with a high degree of certainty. Several surveys in the Oman mountain range showed that the geographical situation of Magan (Oman) agrees fully with the text from the third millennium B.C. found in Sumer (Bibby 1977; Weisgerber 1980). Also the amount of copper produced in these periods, which is indicated by the huge amount of slag, is unique in the whole Middle East and supposes the localization of Magan in Oman; following calculations of Hauptmann (1985), during the third to second millennium B.C., 2,000–4,000 tons of copper were smelted.

The final evidence that copper produced in Oman was used in Mesopotamia has not yet been given by analytical data. The work presented shows the first of more detailed studies on the techniques of metallurgy used in the third and second millennia B.C. in Oman and analytical data on copper—and bronze—artifacts available. It also should point to some analytical questions in the interpretation of the process by which ore is converted to artifact.

Notes

The investigations were carried out under grants from the Stiftung Volkswagenwerk, Hannover, and the Foreign Office, Bonn, to which we are very much obliged. We thank J. Zallmanzig, Deutsches Bergbau-Museum, and G. Drews, formerly of Romisch Germanisches Zentralmuseum Mainz, for performing the analyses. The chemical and mineralogical preparation of the material investigated was carried out by A. Ludwig, Deutsches Bergbau-Museum, who has been a great help during the work. Our special thanks go to Ali Shanfari, director of the Department of Antiquities, Muscat, for allowing us to take samples for scientific investigation from the many artifacts found at several sites in Oman.

1. The slag analyses were performed using an X-ray fluorescence spectrometer (Philipps PW 1400) for the main components. Minor elements were determined by atomic absorption spectrometry using a Pye Unicam SP 2900. For the X-ray fluorescence analyses, the preparation of samples was carried out with the following method: 0.5 g powdered sample was mixed with 1.0 g $NaNO_3$, 7.0 g $Na_2B_4O_7$, and 0.5 g $NaBO_2 \cdot 4H_2O$ and heated for 10 min at 1150°C. The melt was poured on a red hot Pt-Au disk and quenched with air. The peaks were measured using the correction model of De Jongh. Further details are described in Hauptmann (1985, p. 118).

2. The determination of the mineralogical phase content was carried out by X-ray diffraction, microprobe analyses (ARL microprobe, type EMX), and optical microscopy. It was found to be convenient to work with polished thin sections because of the ability to study the sections under reflected and transmitted light.

3. The analyses were performed by G. Drews, Mainz, using an ICP-OES from ARL.

4. The analyses were performed by J. Zallmanzig, Bochum, using an atomic absorption spectrometer (Perkin Elmer type 3030). According to the concentration of the elements, the measurements were carried out with a flame or a graphite furnace (HGA 500).

References

Avery, D. H. 1982. "The iron bloomery," in *Early Pyrotechnology*, T. A. Wertime, and S. F. Wertime, eds. (Washington, D.C.: Smithsonian Press), 205–214.

Bachmann, H. G. 1978. "The phase composition of slags from Timna site 39," in *Archaeometallurgy. Vol. 1, Chalcolithic Copper Smelting: Excavations and Experiments*, B. Rothenberg, R. F. Tylecote, and P. J. Boydell, eds. (London: Institute of Archaeometallurgical Studies), 21–23.

Bachmann, H. G. 1980. "Early copper smelting techniques in Sinai and in the Negev as deduced from slag investigations," in *Scientific Studies in Early Mining and Extractive Metallurgy*, P. T. Craddock, ed. British Museum Occasional Paper 20 (London: British Museum), 103–135.

Berthoud, T. 1979. "Etude par l'analyse de traces et la modélisation de la filiation entre minerais de cuivre et objects archéologiques du Moyen-Orient." Ph.D. dissertation. Université de Paris.

Bibby, G. 1977. *Dilmun*. Hamburg: Rowohlt.

Bottinga, Y., and D. Weill. 1972. "The viscosity of magmatic silicate liquids: A model for calculation." *American Journal of Science* 272:438–475.

Boucharlat, R., and P. Lombard. 1985. "The oasis of Al-Ain in the Iron Age. Excavations at Rumeilah 1981–1983." *Archaeology in the United Arab Emirates* 4:44–62.

Branigan, K., H. McKerell, and R. F. Tylecote. 1976. "An examination of some Palestinian bronzes." *Journal of the Historical Metallurgy Society* 10(1):15–24.

Coleman, R. G., C. C. Huston, I. M. El-Boushi, K. M. Al-Hinai, and E. H. Bailey. 1979. "The Semail ophiolite and associated massive sulfide deposits, Sultanate of Oman," in *Evolution and Mineralization of the Arabian-Nubian Shield*, A. M. S. Al-Shanti, ed. (Oxford: Pergamon), vol. 2, 179–192.

Craddock, P. T. 1977. "The composition of copper alloys used by the Greek, Etruscan and Roman civilisation. 2. The archaic, classical and Hellenistic Greeks." *Journal of the Archaeological Society* 4:103–123.

Craddock, P. T. 1985. "The composition of the metal artifacts." in *Oriens Antiquus* 24:97–101.

Craddock, P. T., I. C. Freestone, N. H. Gale, N. D. Meeks, B. Rothenberg, and M. S. Tite. 1985. "The investigation of a small heap of silver smelting debris from Rio Tinto, Spain," in *Furnaces and Smelting Technology in Antiquity*, P. T. Craddock, ed. British Museum Occasional Paper 48 (London: British Museum), 199–218.

Czedik-Eysenberg, F. 1958. "Beiträge zur Metallurgie des Kupfers in der Urzeit." *Archäologia Austriaca* 3:1–18.

Desch, C. H. 1929. *Interim Report on Sumerian Copper*. London: British Association for the Advancement of Science, 437–441.

Donaldson, P. 1984. "Prehistoric tombs of Ras al-Khaimah." *Oriens Antiquus* 23:1–312.

Donaldson, P. 1985. "Prehistoric tombs of Ras al-Khaimah." *Oriens Antiquus* 24:86–142.

Durante, S., and M. Tosi. 1977. "The aceramic shell middens of Ras al-Hamra: A preliminary note." *Journal of Oman Studies* 3(2):137–162.

Eugster, H. P., and D. R. Wones. 1962. "Stability relations of the ferruginous biotite, annite." *Journal of Petrology* 3: 82–125.

Gnoli, G. 1981. "ISMEO activities. Oman." *East and West* 31:182–195.

Hauptmann, A. 1985. "5000 Jahre Kupfer in Oman. Bd. 1. Die Entwicklung der Kupfermetallurgie vom 3. Jahrtausend bis zur Neuzeit." *Der Anschnitt* 4:1–137.

Hauptmann, A., I. Keesmann, and B. Schulz-Dobrick. 1984. "Die Kristallisation von Fe-reichem Olivin in archäometallurgischen Schlacken." *Fortschrift für Mineralogie* 62(1):84–86.

Hauptmann, A., G. Weisgerber, and E. A. Knauf. 1985. "Archäometallurgische und bergbauarchäologische Untersuchungen im Gebiet von Fenan, Wadi Arabah (Jordanien)." *Der Anschnitt*, 37(5–6):163–195.

Helmig, D. 1986. "Versuche zur analytisch-chemischen Charakterisierung frühbronzezeitlicher Techniken der Kupferverhüttung in Shahr-i Sokhta/Iran." Diplomarbeit, Faculty of Chemistry, Ruhr-University, Bochum, 1–113.

Keesmann, I., H. G. Bachmann, and A. Hauptmann. Unpublished. "Parageneses and norm-calculation of iron-rich slags."

Koucky, F., and A. Steinberg, 1982. "Ancient mining and mineral dressing on Cyprus," in *Early Pyrotechnology*, T. A. Wertime and S. F. Wertime, eds. (Washington, D.C.: Smithsonian Press), 149–180.

Kresten, P. 1986. "Melting point and viscosities of ancient slags: A discussion." *Journal of the Historical Metallurgy Society* 20(1):43–45.

Leese, M. N., P. T. Craddock, I. C. Freestone, and B. Rothenberg. 1985–1986. "The composition of ores and metal objects from Timna, Israel," in *Wiener Berichte über Naturwissenschaft in der Kunst*, A. Vendl, B. Pichler, J. Weber, and G. Banik, eds. (Wien: Hochschule für Angewandte Kunst), vol. 2, no. 3, 90–120.

Lewin, B., and A. Hauptmann. 1984. "Kodo-Zuroku. Illustrierte Abhandlung über die Verhüttung des Kupfers in Japan aus dem Jahre 1801." *Veröffentlichungen aus dem Deutschen Bergbau-Museum* 29.

Lorenzen, W. 1965. *Helgoland und das früheste Kupfer des Nordens*. Ottendorf: Niederelbe-Verlag.

McKerell, H. 1977. "Nondispersive X-ray fluorescence applied to ancient metal in copper and tin bronze." *Journal of the European Study Group on Physical, Chemical, Biological and Mathematical Techniques Applied to Archaeology* 1:138–174.

Merkel, J. F. In press. "Ore benefication during Late Bronze/Early Iron Age in Timna, Israel."

Milton, D., E. J. Dwornik, R. B. Finkelman, and P. Toulmin, 1976. "Slag from an ancient copper smelter at Timna, Israel." *Journal of the Historical Metallurgy Society* 10(1):24–33.

Morton, G. R., and J. Wingrove. 1969. "Constitution of bloomery slags I. Roman." *Journal of the Iron and Steel Institute* 207:1556–1564.

Morton, G. R., and J. Wingrove. 1972. "Constitution of bloomery slags II. Medieval." *Journal of the Iron and Steel Institute* 210:478–488.

Muan, A. 1955. "Phase equilibria in the system $FeO–Fe_2O_3–SiO_2$." *Transactions of the American Institute of Mining and Metallurgy* 203:965–976.

Muhly, J. D. 1980. "The Bronze Age setting," in *The Coming of the Age of Iron*, T. A. Wertime, and J. D. Muhly, eds. (New Haven, Conn.: Yale University Press), 25–68.

Osborn, E. F., and A. Muan. 1964. "Phase equilibrium diagrams of oxide systems (1960)," in *Phase Diagrams for Ceramists*, E. M. Levin, C. R. Robbins, and H. F. McMurdie, eds. (Columbus, Ohio: American Ceramic Society), 60.

Panayiotou, A. 1980. "Cu-Ni-Co-Fe sulphide mineralization, Limassol Forest, Cyprus," in *Ophiolites*, A. Panayiotou, ed. (Nicosia: Ministry of Agriculture and Natural Resources and Geology Survey Department), 102–116.

Peake, H. 1928. "The copper mountain of Magan." *Antiquity* 2:452–457.

Roeder, P. L., and E. F. Osborn. 1966. "Experimental data for the system $MgO–FeO–Fe_2O_3–CaAl_2Si_2O_8–SiO_2$ and their petrological implications." *American Journal of Science* 264:82–125.

Rothenberg, B. 1978. "Excavations at Timna Site 39. A Chalcolithic copper smelting site and its metallurgy," in *Archaeometallurgy. Vol. 1, Chalcolithic Copper Smelting: Excavations and Experiments*, B. Rothenberg, R. F. Tylecote, and P. J. Boydell, eds. (London: Institute of Archaeometallurgical Studies), 1–15.

Schlegel, H., and A. Schüller. 1952. "Die Schmelz- und Kristallisationsgleichgewichte im System Cu-Fe-S und ihre Bedeutung für die Kupfergewinnung." *Freiberger Forschungshefte*, ser. B, 2:1–32.

Seeliger, T., E. Pernicka, G. A. Wagner, F. Begemann, S. Schmitt-Strecker, C. Eibner, Ö. Öztunali, I. Baranyi. 1985. "Archäometallurgische Untersuchungen in Nord- und Ostanatolien." *Jb. Röm. German. Zentralmuseum*, 32:597–659.

Tite, M., I. C. Freestone, and N. D. Meeks. Unpublished. "Report on the scientific examination of furnace fragments from Maysar, Oman."

Tosi, M. 1983. *The Prehistoric Sistan*. Rome: Instituto Italiano per il Medio ed Estremo Oriente.

Tylecote, R. F. 1962. *Metallurgy in Archaeology*. London: Edward Arnold Ltd.

Tylecote, R. F. 1980. "Summary of results of experimental work on early copper smelting," in *Aspects of Early Metallurgy*, W. A. Oddy, ed. British Museum Occasional Paper 17 (London: British Museum), 5–12.

Tylecote, R. F., and P. J. Boydell. 1978. "Experiments on copper smelting based on early furnaces found at Timna," in *Archaeometallurgy. Vol. 1, Chalcolithic Copper Smelting: Excavations and Experiments*, B. Rothenberg, R. F. Tylecote, and P. J. Boydell, eds. (London: Institute of Archaeometallorgical Studies), 25–49.

Tylecote, R. F., H. A. Ghaznavi, and P. J. Boydell. 1977. "Partitioning of trace elements between the ores, fluxes, slags and metal during the smelting of copper." *Journal of the Archaeological Society* 4:27–49.

Urbain, G., Y. Bottinga, P. Richet, 1982. "Viscosity of liquid silica, silicates and alumino-silicates." *Geochimica et Cosmochimica Acta* 46:1061–1072.

Weisgerber, G. 1980. "... Und Kupfer in Oman." *Der Anschnitt* 32(2–3):62–110.

Weisgerber, G. 1981. "Mehr als Kupfer in Oman." *Der Anschnitt* 33(5–6):174–263.

Weisgerber, G. 1983. "Copper production during the third millennium B.C. in Oman and the question of Makan." *Journal of Oman Studies* 6(2):269–276.

Weisgerber, G., and P. Yule. Unpublished. "The copper hoard from Ibri/Selme."

Zuicker, U. 1980. "Investigations on the extractive metallurgy of Cu/Sb/As ore and smelting products from Norsun-Tepe (Keban) on the Upper Euphrates (3500–2800 B.C.)," in *Aspects of Early Metallurgy*, W. A. Oddy, ed. British Museum Occasional Paper 17 (London: British Museum), 13–26.

5
Early Copper Mining and Smelting in Palestine

Gerd Weisgerber and Andreas Hauptmann

Fifty years ago the first remains came to light at Teleilat Ghassul of what later was to be called the Palestinian Copper Age (figure 5.1). The site is situated just at the end of the Jordan Valley north of the Dead Sea. Although the overwhelming majority of tools and weapons there are of flint, a remarkable number of copper objects were also discovered by Mallon et al. (1934, 1940). In those days few considered the question of the provenance of the metal.

Following World War II Perrot (1955) unearthed a kind of workshop in the artificially dug subterranean dwellings at Tell Abu Matar near Beersheba. Ore, slag, crucibles, and molds testify to smelting and melting there. Copper objects occurred in other dwellings of the settlement, including axes, a standard, one decorated and three plain globular maceheads, and some needles. Pottery finds date this lot of metal items to the period already defined at Teleilat Ghassul.

The excavations at Jericho (Level VIII) and Arad (Amiran 1978) also shed light on this period; copper objects were found at Arad.

The question of the origin of early Palestinian copper should have become dramatically important after the discovery of the sensational copper hoard in a cave at Nahal Mishmar (figure 5.1) in the Judean Desert west of the Dead Sea (Bar-Adon 1962, 1980). But despite the 400 copper items with a total weight of 140.2 kg (including the clay core in the maceheads) no concentrated efforts were undertaken to find the source for the metal. Given its lack of zinc, the copper of the high arsenical bronzes strangely enough was presumed to have been derived from Armenia or Azerbaijan (Key 1980). But only in 1973 did Amiran and her collegues first ask seriously about the provenance of Palestine's Chalcolithic copper. At that time only two copper-producing centers were known in the area in question: Sinai and Timna.

Copper Production in Sinai

Since the days of Petrie (1906) the Sinai Peninsula has been reported to have harbored remains of copper production. This is true but only for certain periods after c. 2500 B.C., for example, when the pharaohs sent expeditions. Many small copper mineralizations in the different periods were worked. The slag heaps from Bir Nasib point with their 100,000 tons to an important period of copper production (figure 5.2), but no Chalcolithic exploitation and production has been reported so far, despite the large number of smaller Chalcolithic and Early Bronze Age settlements (Beit-Arieh 1981). Even turquoise mining for the first period has been suggested (Beit-Arieh 1980).

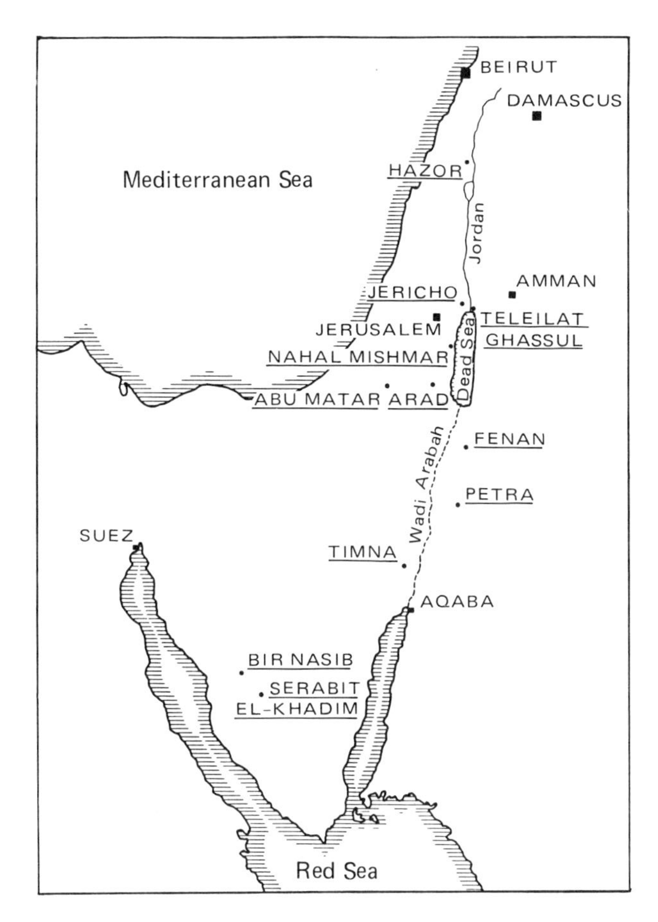

Figure 5.1 Map of sites mentioned in the text.

Figure 5.2 Bir Nasib in Sinai. The main heap with about 100,000 tons of slag near the well.

Copper Production in Timna

Beginning in the late 1950s Rothenberg became involved in field studies on the western side of the Wadi Arabah, the desert valley between the Dead and Red seas in the foothills below the plateau of the Negev Desert. Here, Frank had reported (1934) copper mines and smelting sites from his survey in the 1930s in Wadi el-Meneiyeh, which later was renamed Timna. Glueck had revisited the area (1945) but, regarding the subject of copper production, he contributed only a confused mass of wrong interpretations and thoughts, finally leading to his "Pittsburgh of Palestine" and making Solomon a wise copper king (Glueck 1959). Rothenberg's survey activities brought an astonishing number of sites to light—more than 300 in the Wadi Arabah and more than 400 in the Sinai Peninsula (Rothenberg 1970, 1972a). Many of them were connected with copper mining and smelting.

Chalcolithic Mining and Smelting

Below high cliffs surrounding the valley of Timna, situated in sandstone formations, copper ore nodules and impregnations occur below the surface. Erosion often exposes them in the embankments of the wadis. The main ore, malachite, is mixed with azurite, cuprite, paratacamite, and chalcocite. According to Rothenberg there exists at least one Chalcolithic smelting furnace (1973, 1978). In fact, the smelting furnace, site 39, (figure 5.3) cannot be dated (Muhly 1984), but the immediate proximity of a dwelling ruin (site 39A), where excavations have produced Chalcolithic flint material, ensures at least the presence of humans at the site in the Chalcolithic Period. Tiny pieces of crushed slag and a furnace with an open front correspond to what one encounters in the period (according to the new results) from Fenan. The evidence for Chalcolithic mining in Timna is rather scanty. Only a single grooved hammer made of basalt was collected beside a shallow opening of an unfinished adit.

Early Bronze Age Mining

Stone hammers of this period, mostly round or square with biconically bored shaft holes, were abundant (figure 5.4). They occurred mostly broken on the surface around shallow depressions, traces of deep mining. In the mines of Area T they were studied in detail. Their age becomes clear by the recent results from Fenan. In Timna finds other than flint were few.

The German Mining Museum investigated the Early Bronze Age mines of the third millennium B.C. in Area T. Irregular shafts sunk a few meters (2–4 m) in the white Nubian sandstone were observed; sometimes irregular steps of a sort remained, for example, at Area T, shaft 42 (figure 5.5 and 5.6). The exploitation areas starting from the bottom of the shafts were irregular in form (*Duckelbau*) and, corresponding to the stability of

Figure 5.3 Timna, site 39. Smelting furnace supposed to be Chalcolithic. Photograph taken some years after the excavation.

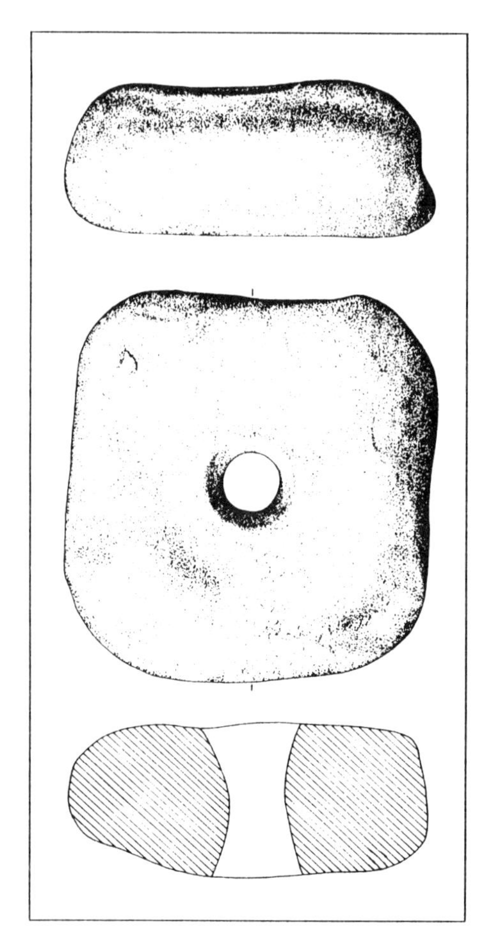

Figure 5.4 Timna, Area S, surface find. Early Bronze Age hammer stone with double conical bored hole for hafting.

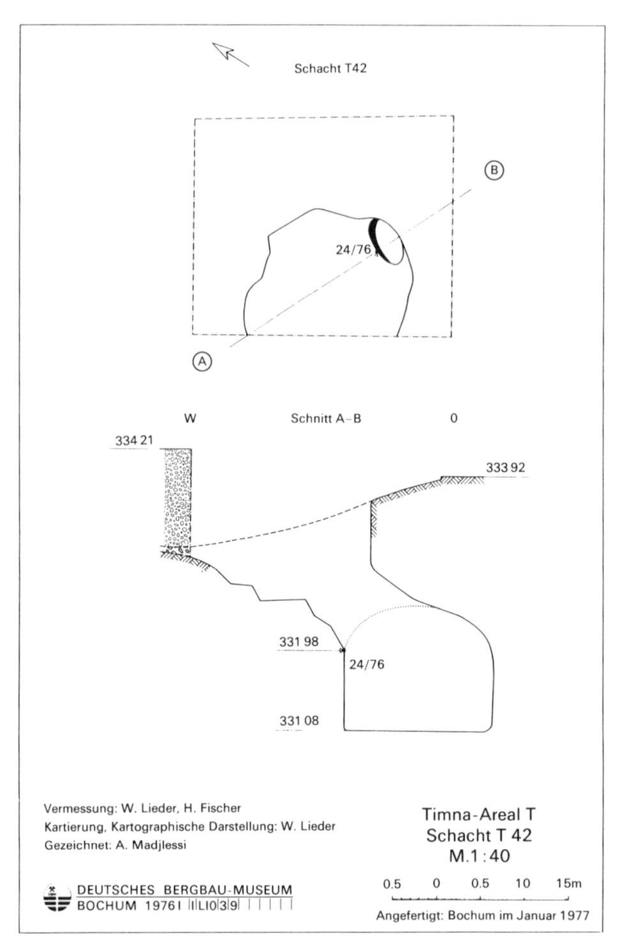

Figure 5.5 Timna, Area T. Shaft T 42 of an Early Bronze Age mine. Plan and section (Conrad and Rothenberg 1980).

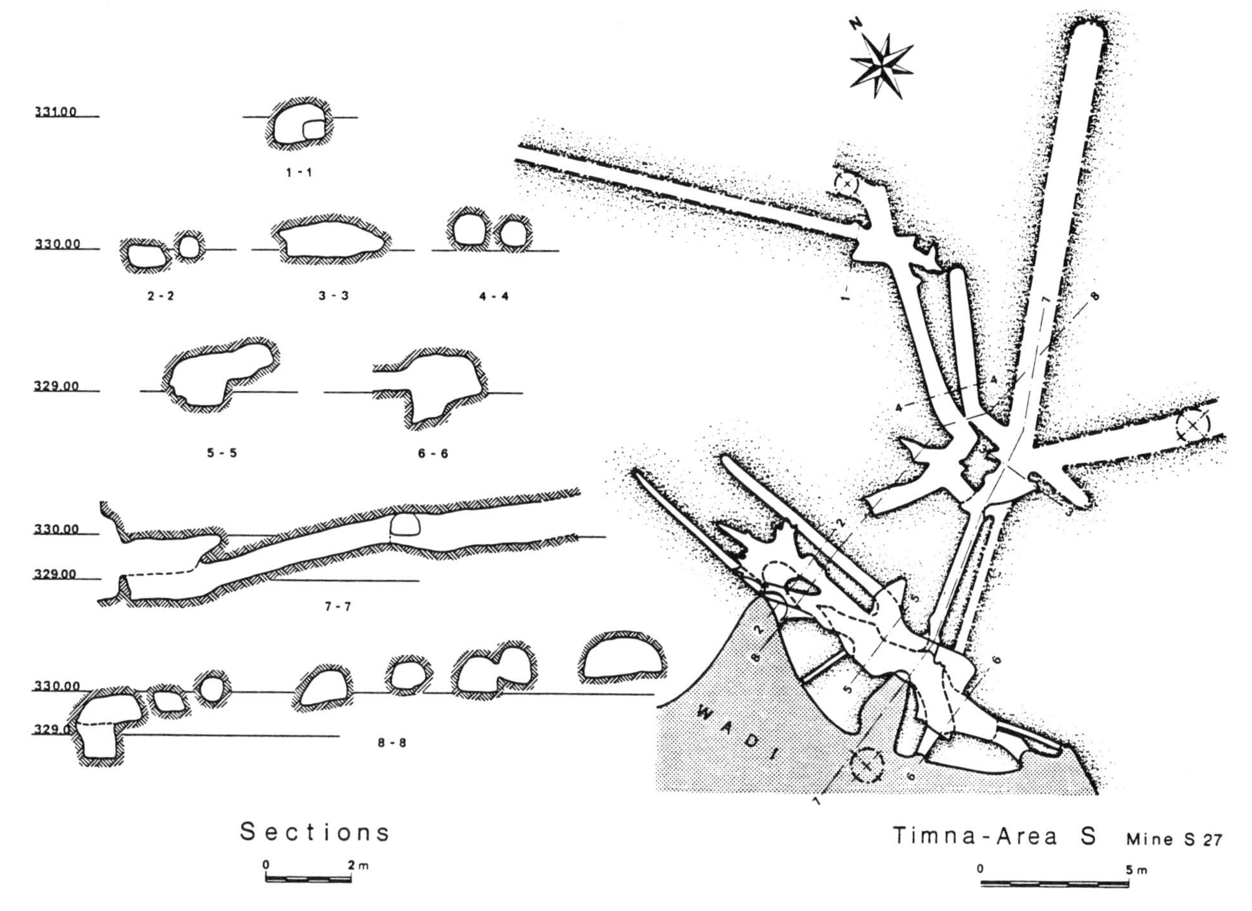

Figure 5.6 Timna, mine S 27. Ground plan of the Early and Late Bronze Age mine (Conrad and Rothenberg 1980).

the rock, could be hollowed to rather wide chambers. Shallow depressions from hammering the rock with a blunt tool remain as tool marks on several walls. As the different shafts were grouped closely together, these chambers could meet each other either by chance or by intention to facilitate ventilation. Mine T, with shafts 31, 40, 42, 43, and others, is a rather good example of the procedures I described in 1980 (Conrad et al. 1980, p. 48, plans 21–23). As is usual in the history of mines, mine T was later reopened and reworked. This happened during the second half of the second millennium B.C. The straight galleries, the deepening of the bottom of older galleries, and the chisel marks of the iron and wedge working technique (*Schlägel- und Eisen-Arbeit*), so clearly distinguishable from the older shallow depression marks (figures 5.7 and 5.8), result from these secondary activities. [The intensive mining operations of the Iron Age (Conrad and Rothenberg 1980; Rothenberg 1983) and the Roman Period (Rothenberg 1972b) are not treated here.]

Of the estimated 9,000 mines in the wider Timna area, at least a few hundred seem to originate from the Chalcolithic and Bronze Age.

Bronze Age Smelting

To judge from the smelting waste, the copper production of the area seems to have been rather restricted. Of the approximately 2,000 tons of slag, little belongs to the periods in question, thus leaving Timna of only minor importance with respect to mining in the Sinai Peninsula, once claimed to have yielded the copper of Chalcolithic and Early Bronze Age Arad (Amiran et al. 1973).

Copper Production in the Fenan Area

Curiously, the idea that Perrot had continously repeated has received only little attention. Correctly thinking that a "metallurgical industry can only originate and develop in the neighbourhood of metal ores and fuel" (Perrot 1955, p. 188), Perrot came to the conclusion that the "smiths of Abu Matar procured their ores from Trans-Jordan, from the copper lodes of the Punon Valley at a distance of more than 100 km from their furnaces" (Perrot 1957, p. 38; Fritz 1985). The Fenan (Punon) area (figure 5.9) remained forgotten despite contemporary geological and archaeological reports (Bender 1965; Kind 1965). Sinai and Timna were

Figure 5.7 Timna, mines S 28/2 and 3. The mines became visible in the embankment after erosion by the wadi to a depth of more than 5 m. The earliest drift on top shows flat depressions as tool marks, whereas the lower part is chisel marked. The wide lower cavity was cut through the bottom of the earlier drift.

Figure 5.8 Timna, mine T, view of the shaft. Early Bronze Age drift where the original bottom was cut by working to the dip.

claimed to have been the main copper sources of prehistoric Palestine.

Since 1983 a team from the German Mining Museum has intensively investigated the area around Fenan and the wadis el-Jariye and el-Ghuwebe (Bachmann and Hauptmann 1984; Hauptmann et al. 1985). The area is situated at the eastern border of the great rift valley Wadi Arabah that stretches from the Red Sea to the Dead Sea about three-quarters of the way from the Gulf of Aqaba northward. Here, at the foot of the Jordanian Plateau (reaching more than 1000 m), severe tectonic movements together with strong erosion have exposed two kinds of copper mineralization.

Widely distributed low-percentage secondary ores occur as diffuse irregular impregnations of nodules, lentoids, and carpetlike mineralizations or form up to fist-size rinds. The ores consist principally of relatively poor malachite ($Cu_2(CO_3)(OH)_2$) with brochantite ($Cu_4(SO_4)(OH)_6$), chalcocite (Cu_2S), and bornite (Cu_5FeS_4). Contents of copper around 15–25% or even more are obtainable only by a careful hand sorting of nut-size nodules. Mineralizations of this type were observed at the uppermost ends of Wadi Khalid and Wadi Ratiye, in Galb Ratiye, in Wadi Abiad, and at Umm el-Amad in the mountains south of Fenan. These sandstone ores were exploited in Chalcolithic and Roman times.

The second ore type is connected with the dolomite-limestone-shale formation as a crumbly, 1–3-m-thick stratigraphic sequence that gradually merges at the bottom into massive or bedded dolomite. Here, fissure mineralization and pockets of minerals such as chrysocolla ($Cu_2H_2Si_2O_5(OH)_4 \cdot nH_2O$) and malachite are evident and are closely intergrown with lenses of pyrolusite (MnO_2) and other manganese minerals (Bender 1965). These particularly characteristic intergrowths of copper and manganese minerals were exploited from the Early Bronze Age to the Late Iron Age. In contrast to the sandstone ore (type 1), this ore (type 2) occurs in a much more concentrated form. Measurements done with a portable X-ray fluorescence spectrometer have shown in the field that handpicked copper ores can easily contain up to 35% copper. Aside from the iron and manganese content this copper ore is relatively pure. With a few exceptions (Pb up to 2.8%), the trace elements Zn, Sn, Ni, Co, As, Sb, and Bi are below 1%, and the content of S is also less than 0.5% (Hauptmann et al. 1985).

Copper-manganese mineralizations are concentrated in the lower Wadi Khalid, in the most upper Wadi Abiad, in Wadi Dana, in Wadi el-Jariye, at el-Furn, and near Khirbet en Nahas (Ruins of Copper) and were recently discovered at Umm ed-Dhur, just north of Ain el-Fidan.

The copper production of the Fenan area spans the time from the Chalcolithic to the Islamic Middle Age and may be listed as follows:

Chalcolithic	4500–3100 B.C.
Early Bronze Age	3100–2100 B.C.
Middle Bronze Age	2100–1900 B.C.
Iron Age I	1200–1000 B.C.
Iron Age IIC	800–400 B.C.
Nabatean Period	first century A.D.
Roman Period	A.D. 200–400
Mameluke Period	thirteenth century A.D.

The Wadi Fenan area and especially the embankments of the mouth of Wadi Fidan to the Wadi Arabah were settled at least from the preceramic Neolithic onward (Raikes 1976, 1980). Recently a Neolithic (Yarmoukian) tell was found with a thick superimposed Chalcolithic layer. Fragments of crude pottery and of typical basalt bowls together with flint ax heads with a ground edge prove the dating. Atop the settlement mound is a large amount of smaller slag fragments that may derive from the same period; they indicate the copper production even at this major site, obviously the predecessor of the early settlement at the site of the later town of Fenan, where at least from the Early Bronze Age traces of nearly every period can be seen.

Chalcolithic Mining

The traces of Chalcolithic mining are abundant in the areas of the sandstone mineralization (figure 5.10). More than a hundred mines were dug into the slopes of Galb Ratiye and Wadi Abiad during this period. The adits reach up to about 10 m or more in the bedrock; their sections were intended to be semicircular. Mineral layers in the horizontal fissures were consequently exploited at the bottom of the adit. The mining tools were heavy stone hammers of quartz porphyry (figure 5.11). They were made of suitable elongated pebbles with a fine pecked groove in the middle to fix the handle. More irregular stones became notches for the same purpose. Other fine tools, namely picks, were made with an enormous investment of labor. A flat basalt pebble was flaked to a rhomboid form and then perforated by biconically drilling inward from both sides. Finally the pebble was roughed by pecking to a hammer pick with one pointed and one blunt end with triangular cross sections. Another kind of hammer was made by drilling biconical shaft holes in more or less irregular basalt pebbles. Fragments of these tools were collected on nearly all the surfaces of the mining dumps. Strangely enough, they often occur together with Roman pottery and lamp fragments. It is clear that this material was deposited on the dumps when the Romans cleared the prehistoric mines in order to rework them. Chalcolithic pottery was rare on the dumps; in contrast, flint occurs in abundance.

Near the mines of Galb Ratiye about fifteen ruins of small huts were found; they are attributed to the same period. These are the miners' working dwellings. Their permanent settlement lay on a high terrace to the south of and above Wadi Fidan. This site, first discovered by Raikes (1980, p. 55, Site E), yielded a large quantity of tiny ore pieces, some slags, typical Late Chalcolithic pottery and basalt vessels, and, more important, raw material—roughed-out, half-finished, and broken or discarded examples of all the miners' basalt implements mentioned so far. No doubt, the tools for the miners were produced at this site, called Wadi Fidan 4.

Chalcolithic Smelting

For the time being it is difficult to describe Chalcolithic smelting because the early smelting sites yielded little surface pottery. So it is not clear which ones started in the Chalcolithic and which ones are completely Early Bronze Age. On the surfaces of nearly all the Chalcolithic mining dumps, either in Galb Ratiye or in Wadi Abiad, small slag fragments were collected. They should come from test or productional smelting carried out in the immediate proximity of the mines. At any rate, this observation gives us an interesting idea of how the earliest mining and smelting was organized. During future surveys special attention should be given to these early smelting activities near the mines.

Bronze Age Mining

Based on the results of the 1984 season the beginning of the exploitation of type 2 mineralization was attributed to the Middle Bronze Age. In the lower Wadi Khalid several of the black mining dumps were situated immediately beside the wadi on the lower terrace. Sometimes the erosional inclination of the embankment and the dump forms one line. The mines were accessible by short shafts and followed the slightly inclined ore horizon to a depth of 15–20 m (figure 5.10). Especially the dumps of mines 6–10 delivered a bulk of Middle Bronze Age pottery. Other pottery of the second millennium B.C. could be collected in mines 42 and 43.

Mine 43 became accessible to the expedition because of the rather recent research of the Jordanian Natural Resources Authority, which penetrated the ore deposit by modern prospecting galleries, thus cutting the ancient mines and exposing them for mining and archaeological studies. Red burnished ware occurs, as does the whitish ware with comb motifs. In one special case a sherd of this ware was used as a lamp, as witnessed by the traces of soot on its edges. Its site is 22 m from the opening in the backfill of mine 43. Two other sherd lamps of the same kind and date were found in the Wadi Khalid mine 34.

The rather small external dumps of waste of these mines could easily but falsely be taken to represent the extension and importance of these activities. Most of the ore was still separated in the mines to keep the sterile host rock underground and to avoid unprofit-

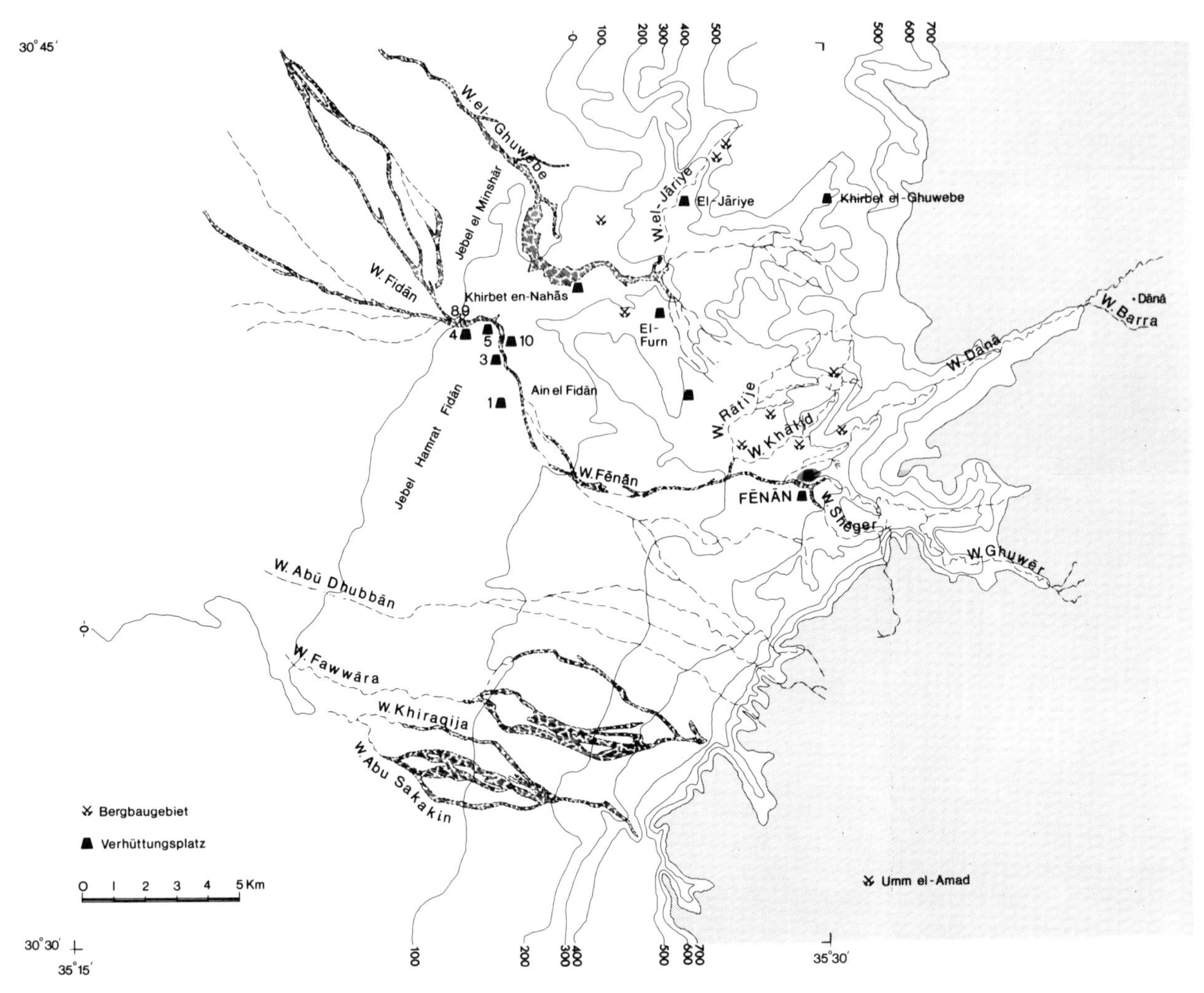

Figure 5.9 Fenan area with mining fields and smelters.

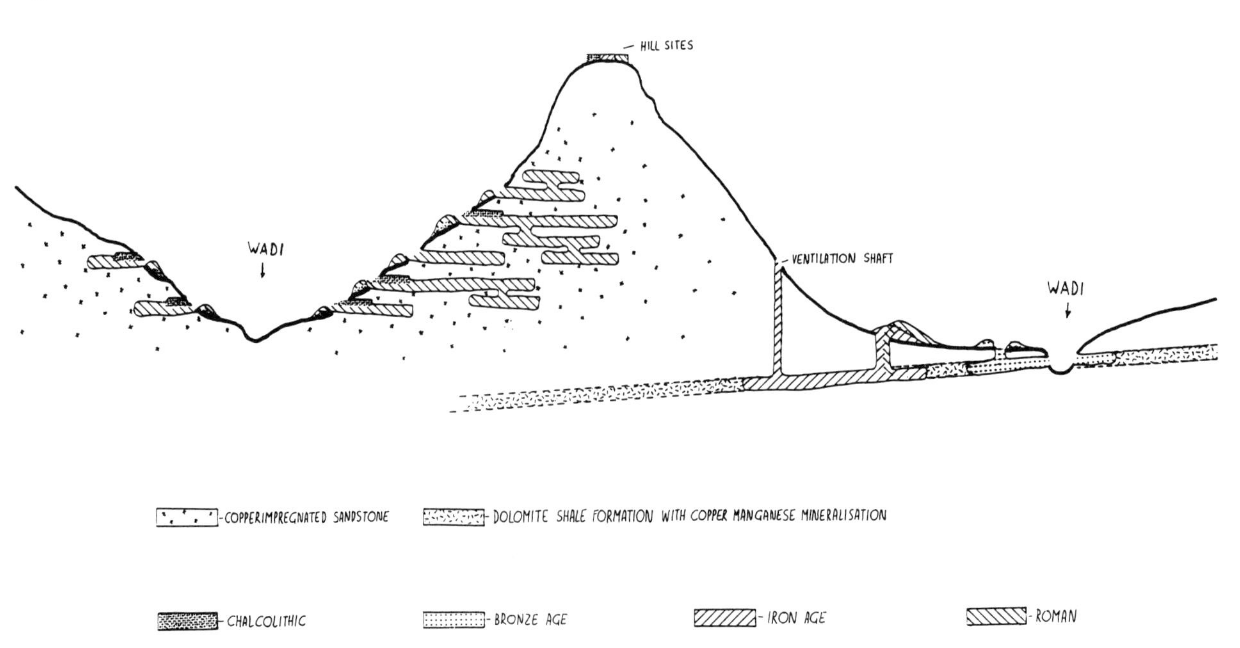

Figure 5.10 Fenan, schematic cross section from Wadi Abiad (left) to Wadi Khalid (right). The Chalcolithic and Roman mines exploited the copper ores in the sandstone completely neglected by workers in the Bronze and Iron ages. Their activities were concentrated in the ore of a shale formation.

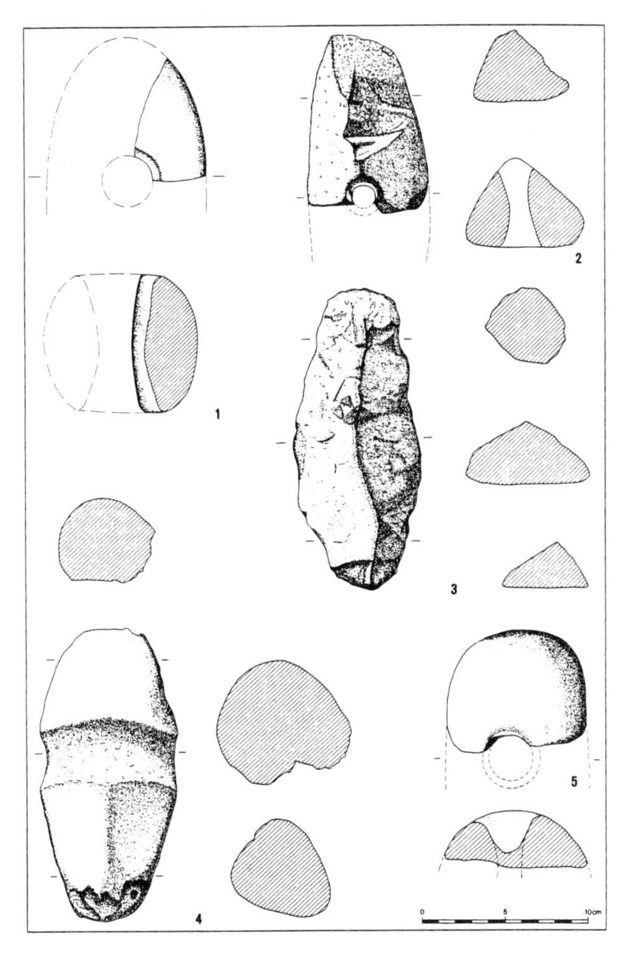

Figure 5.11 Fenan. Different types of Chalcolithic mining tools, some with biconical drilled holes and others with a nicely pecked groove for hafting.

able transportation. But mine 43 recently could be entered for 55 m without reaching the end!

The mines were built in a chamber-pillar construction. Pillars of the natural rock remained for security between the exploited chambers. Sometimes artificial pillars were built using wadi boulders that were brought in. The steriles were carefully backfilled up to the ceiling, thus preventing the accumulation of large waste piles on the surface. Thus the visible black tips give no true impression of the dimensions of the old workings! The subterranean height of these mines measures 0.80–1.50 m; laterally they extend at least 30 × 50 m and obviously could be connected with each other. The limits of natural ventilation probably effectively stopped activities when the air became too bad. So far, no artificial ventilation, for example, by ventilation shafts, can be attributed to these mines. (Later during the Iron Age the exploitation of this ore deposit could be restarted only by sinking new shafts plus special ventilation shafts near the cliff, not stopping at depths of 40 m.)

In 1986 in mine 2 in the lower Wadi Dana, which is excavated in the same ore deposit, undoubtedly Early Bronze Age sherds (rim of a hole mouth jar) were found in the subterranean backfill, pushing back the use of the type 2 ore deposits to the early third millennium B.C. This corresponds to the ore pieces collected on the smelting sites around the later town of Fenan, which have been attributed to the Early Bronze Age by a few archaeological finds and by thermoluminescence. Probably the change from Chalcolithic mining of the sandstone ores to the Early Bronze Age exploitation of the copper-manganese horizon was not as sharp as it was exposed here by purpose of clear definitions. Some kind of fluid transition seems possible. Nevertheless, the Chalcolithic mining tools of basalt described were never encountered in Wadi Khalid either inside or outside the mines exploiting the type 2 mineralization. At Khirbet en-Nehas, where underground mining cannot yet be studied and where at the surface traces of Iron Age activities predominate, two basalt hammers were collected on a dump, one of them grooved, one notched.

Bronze Age Smelting

At more than twenty sites copper ores were smelted in the wide Fenan area during the Early Bronze Age, as revealed by considerable heaps of slag. They are concentrated around Fenan itself (Fenan 8–16) but occur also 1 km to the east in Wadi Ghuwer (designated W. Fenan 2–5), to the west in Wadi Fidan, and on top of the windy hill Ras en-Naqab. Today small pieces of slag with a viscous structure are characteristic.

Among the slags, fragments of conical clay sticks are abundant (figure 5.12), as are small sandstone slabs slagged together and baked fragments of clay furnace linings. These furnaces were invariably dug into slopes and always occur in large batteries (figure 5.13). In 1986 archaeological work revealed these furnaces to consist of a backside supported by built stones or the natural bedrock, which had been relined many times; small side walls were made of stone and clay, and a semicircle bottom was dug out. Apparently no tuyeres were used for the smelting process; natural draft was enough. The conical clay sticks and the small sandstones were probably used to construct a grating at the front of the smelter to let air in. The team excavated twenty-five furnaces of this kind at Fenan 9. They stood side by side, sometimes in double row; the rear wall of each one often was relined, some more than twenty times (figure 5.14). As suggested by the fact that there are many more furnaces than the twenty-five excavated and judging by their frequent repairs, we estimate that at least 600 separate smelting processes would have taken place in Fenan 9. But thus far we do not know if the furnace really had to be relined for each smelting process. If this was necessary, however, this procedure must have been carried out more than a thousand times.

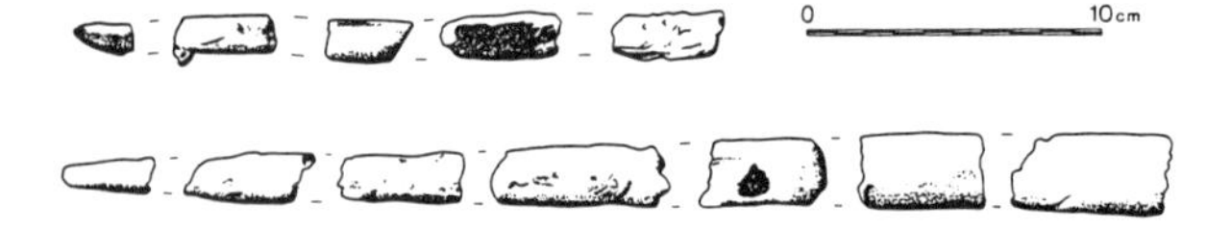

Figure 5.12 Fenan. Fragments of Early Bronze Age baked conical clay sticks.

Figure 5.13 Fenan, site 9. Situation of Early Bronze Age slag heaps.

Figure 5.14 Fenan, site 9. Ruin of Early Bronze Age furnaces before excavation as discovered in the ground. The relined backsides are clearly visible among slag fragments and stones.

Scientific Studies of the Fenan Slags

From the abundant archaeometallurgical finds of the early periods (slag, raw copper, furnace fragments, clay sticks), so far only slags have been analyzed using mineralogical and chemical methods.

Corresponding to the composition of the ores in the area investigated, with few exceptions (Mameluk, Turkish Period) only manganese silicate slags were produced; these are rare in ancient metallurgy. Similar slag is known only from Cyprus (Bachmann 1982) and Oman (Hauptmann et al. 1985) and, interesting enough, has been described for Timna in the tenth century B.C. Chemical analyses carried out by X-ray fluorescence and by wet chemical analyses revealed that the manganese-rich Jordanian slags contain 25–42 wt % MnO, 28–48% SiO_2, 1–13% CaO, 1–5% FeO, 0.5–4% MgO, and up to 6% Al_2O_3.

The following mineral phases were identified by X-ray diffraction and microprobe analyses. The main components are tephroite (Mn_2SiO_4) and rhodonite ($(Mn, Ca)SiO_3$), and, in the case of slag with a high degree of oxidation, hausmannite (Mn_3O_4). The slag usually shows a distinct tendency toward glassy solidification. These glassy slags are typically red because of abundant cuprite inclusions (Cu_2O).

The projection of the chemical composition in the ternary system CaO(+BaO)–Mn(+FeO + MgO)–SiO_2 and in the diagram MnO/CaO(+FeO) (figures 5.15 and 5.16) shows that the slags have different ranges of variation in dependence from the chronological periods. This is caused by the mixing of the available raw material in an attempt to standardize the composition, which was determined by the chemophysical characteristics of the corresponding liquid slag.

With the knowledge of the chemical composition of the slags, local variations can be understood and traced to the areas of exploitation. By this method it is possible to identify those parts of the deposit from which the ore for, say, the Chalcolithic and Early Bronze Age smelters came. These slags have a wider range of composition than Iron Age and Roman slags. Evidently in these epochs large quantities of dolomite-limestone-shale layers were exploited. Copper-manganese-rich ores embedded in limestone lenses were preferred for mining and smelting, as indicated by the high CaO and MnO content of the slag.

The crystallization of the Early Bronze Age manganese slags from Fenan apparently took place under relatively high oxidizing conditions. This is demonstrated by the common mineralogical paragenesis of tephroite and cuprite: The slag contains up to 15% Cu in the form of $Cu_2{}^+$! This paragenesis is unique and is not known in iron-rich slags. The tendency to attribute early copper smelting processes to more oxidizing conditions is also confirmed by numerous investigations on slags. In the case of Fenan, the manganese

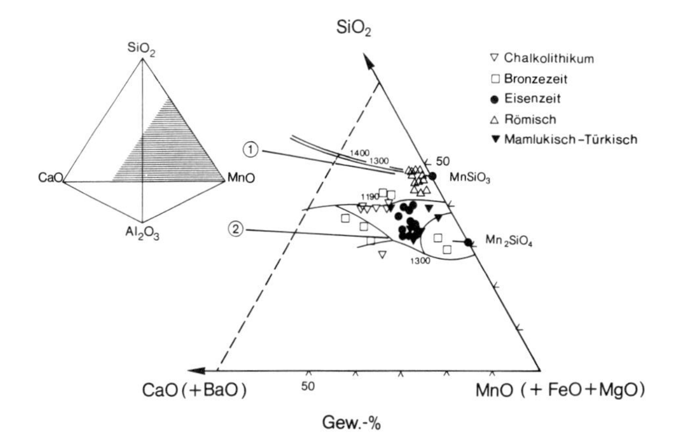

Figure 5.15 Chemical composition of copper slags from Jordan shown in the ternary system $CaO-(+BaO)-MnO(+FeO)-SiO_2$ (as part of the system $CaO-MnO-SiO_2-Al_2O_3$).

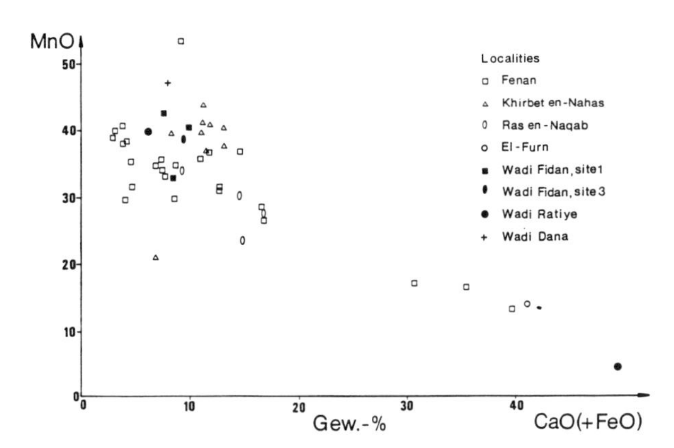

Figure 5.16 MnO/CaO(+FeO) contents of copper slags from Jordan. The values show a clear distinction between samples from the site of Fenan and of Khirbet en-Nehas.

oxide (MnO) that is the basis for the formation of slag plays a key role. It has a much larger temperature and oxygen pressure stability range than the "FeO" (wüstite). This means that the precipitation of metallic iron and manganese during the formation of Mn slags is prevented. This is a serious problem in iron-rich slags, where metallic iron often crystallizes if the smelting process is carried out under too strong reducing conditions (formation of iron sows). According to Bachmann (1982) and Steinberg and Koucky (1974) this is believed to be the reason for using manganese-containing fluxes in ancient metallurgy. On the other hand, the formation of Mn-silicate slags with tephroite as the main component is possible at temperatures around 1200°–1250°C, even under the influence of the oxygen of the free air. This is in strong contrast to iron-rich slags. Most important also is the fact that the crystallization of Mn oxides is suppressed in these slags, which has a decisive effect on the separation of the metal from the slag. This clearly shows that especially the last point was of importance for the smelting procedures. It also corresponds to the shallow construction of the furnaces used in the Early Bronze Age in Fenan. As outlined, these smelting furnaces had no kind of shaft construction that would have caused a stronger reducing gas atmosphere.

Summary

Slag, weighing 150,000–200,000 tons, in the Fenan area played a key role in the copper supply of the ancient Near East in general and of Palestine itself from the earliest times up to the Islamic Middle Ages. The most productive periods were the Iron Age and the Roman Period, but the Chalcolithic and Early Bronze Age (fourth and third millennia B.C.) also left considerable remains both in mining and smelting. The basis for these workings is a well-organized infrastructure. One temporary and three permanent settlements were discovered, and connections with copper production are evident. More than one hundred smaller Chalcolithic mines once were dug in the copper-bearing sandstone mountains around Fenan and even farther away in the high mountains. Early Bronze Age mining started with the exploitation of the copper-manganese horizon wherever it outcropped. But the main early activity in this kind of deposit took place in the Middle Bronze Age (early second millennium B.C.). Here the work was stopped because of the lack of advanced mining technology, which first became available in the Iron Age.

The remains of Early Bronze Age smelting mark the peaks and slopes of the hills around Fenan, partly as extremely long slag heaps. Taking the numerous furnaces at each of the more than twenty sites into account together with the multiple relining of their backsides, we estimate that thousands of smelting procedures must have taken place in order to enable the production of copper for export. Keeping this in mind, the 140-kg hoard from Nahal Mishmar is by no means surprising. It is a rare example of what could have been produced with highly developed craftsmanship and a sufficient supply of raw materials. For protohistoric Palestine the area of Fenan must be regarded as the main copper producer. Sinai and Timna at that time never could have been this important because of their limited ore base or their proximity to Egypt.

Considering the fact that the archaeological results presented here derive nearly exclusively from survey work, much more detailed information can be expected from future excavations both in the mines and at the smelting sites. Only excavations can produce a better chronological sequence of mining and smelting, of the techniques in use, and of the working and living circumstances of the people who did the work.

References

Amiran, R. 1978. *Early Arad. The Chalcolithic Settlement and Early Bronze Age City. First–Fifth Seasons of Excavations, 1961–1966.* Jerusalem: Israel Exploration Society.

Amiran, R., I. Beit-Arieh, and J. Glass. 1973. "The interrelationship between Arad and sites in southern Sinai in the Early Bronze Age II." *Israel Exploration Journal* 23: 193–197.

Bachmann, H. G. 1982. "Copper smelting slags from Cyprus: Review and classification of analytical data," in *Early Metallurgy in Cyprus, 4000–500 B.C.*, J. D. Muhly, R. Maddin, and V. Karageorghis, eds. (Nicosia: Pierides Foundation), 143–152.

Bachmann, H. G. 1983. "Frühe Metallurgie im Nahen und Mittleren Osten." *Chemie in unserer Zeit* 17(4): 120–128.

Bachmann, H. G., and A. Hauptmann. 1984. "Zur alten Kupfergewinnung in Fenan und Hirbet en-Nehas im Wadi Arabah in Südjordanien." *Der Anschnitt* 36: 110–123.

Bar-Adon, P. 1962. "The expedition to the Judean Desert, 1961: Expedition C—The cave of the treasure." *Israel Exploration Journal* 12: 215–226.

Bar-Adon, P. 1980. *The Cave of the Treasure. The Finds from the Cave of Nahal Mishmar.* Jerusalem: Israel Exploration Society.

Beit-Arieh, I. 1980. "A Chalcolithic site near Serabit el-Khadim." *Tel Aviv* 7: 45–64.

Beit-Arieh, I. 1981. "An Early Bronze Age II site near Sheikh `Awad in southern Sinai." *Tel Aviv* 8: 95–127.

Bender, F. 1965. "Zur Geologie der Kupfererzvorkommen am Ostrand des Wadi Arabah, Jordanien." *Geologisches Jahrbuch* 83: 181–208.

Bender, F. 1974. "Explanatory notes on the geological map of the Wadi Arabah, Jordan." *Geologisches Jahrbuch*, Reihe B, 10: 1–62.

Conrad, H. G., and B. Rothenberg, eds. 1980. *Antikes Kupfer im Timna-Tal. 4000 Jahre Bergbau und Verhüttung in der Arabah.* Bochum: Deutsches Bergbau-Museum.

Conrad, H. G., L. Fober, A. Hauptmann, W. Lieder, I. Ordentlich, and G. Weisgerber. 1980. "Beschreibung der untersuchten Grubenbaue," in *Antikes Kupfer in Timna-Tal*, H. G. Conrad, and B. Rothenberg, eds. (Bochum: Deutsches Bergbau-Museum).

Frank, F. 1934. "Aus der Arabah I." *Zeitschrift des Deutschen Palästina-Vereins* 57: 191–280.

Fritz, V. 1985. *Einführung in die Biblische Archäologie.* Darmstadt: Wissenschaftliche Buchgesellschaft.

Glueck, N. 1945. *The Other Side of the Jordan.* New Haven.

Glueck, N. 1959. *Rivers in the Desert.* New London: American School of Oriental Research.

Hauptmann, A. 1980. *5000 Jahre Kupfer in Oman.* Bd. 1, *Die Entwicklung der Kupfermetallurgie vom 3. Jahrtausend bis zur Neuzeit.* Bochum: Deutsches Bergbau–Museum.

Hauptmann, A., G. Weisgerber, and E. A. Knauf. 1985. "Archäometallurgische und bergbauarchäologische Untersuchungen im Gebiet von Fenan., Wadi Arabah (Jordanien)." *Der Anschnitt* 37: 163–195.

Key, C. A. 1980. "The trace-element composition of the copper and copper alloy artifacts of the Nahal Mishmar Hoard," *The Cave of the Treasure*, P. Bar-Adon, ed. (Jerusalem: Israel Exploration Society), 238–243.

Kind, H. D. 1965. "Antike Kupfergewinnung zwischen Rotem und Totem Meer." *Zeitschrift des Deutschen Palästina-Vereins* 81: 56–73.

Mallon, A., R. Koeppel, and R. Neuville. 1934. *Teleilat Ghassul I.* Rome: Institut Pontificii and Instituti Biblici.

Mallon, A., H. Senes, J. W. Murphy, and G. S. Maham. 1940. *Teleilat Ghassul II.* Rome: Institut Pontificii and Instituti Biblici.

Muhly, J. D. 1984. "Timna and King Solomon." *Bibliotheca Orientalis* 3/4: 275–292.

Perrot, J. 1955. "The excavations at Tell Abu Matar near Beersheba." *Israel Exploration Journal* 5: 17–40, 73–84, 167–189.

Perrot, J. 1957. "Les fouilles d'Abou Matar près de Beersheba." *Syria* 34: 1–38.

Petrie, W. M. F. 1906. *Researches in Sinai.* London: John Murray.

Raikes, T. 1976. "Ancient sites in the Wadi Arabah." Unpublished.

Raikes, T. 1980. "Notes on some Neolithic and later sites in Wadi Arabah and the Dead Sea Valley." *Levant* 12: 40–60.

Rothenberg, B. 1970. "An archaeological survey of south Sinai." *Palestine Exploration Quarterly* 102: 4–29.

Rothenberg, B. 1972a. "Sinai explorations 1967–1972." *Bulletin, Museum Haaretz Tel Aviv* 1972: 31–45.

Rothenberg, B. 1972b. *Timna, Valley of the Biblical Copper Mines.* New York: Stein and Day.

Rothenberg, B. 1973. *Timna. Das Tal der biblischen Kupferminen.* Bergisch Gladbach: Gustav Lübbe Verlag.

Rothenberg, B. 1978. "Chalcolithic copper smelting. Excavations at Timna Site 39," in *Archaeometallurgy. Vol. 1, Chalcolithic Copper Smelting: Excavations and Experiments*, B. Rothenberg, R. F. Tylecote, and P. J. Boydell, eds. (London: Institute of Archaeo-metallurgical Studies).

Rothenberg, B. 1983. "Explorations and archaeo-metallurgy in the Arabah Valley (Israel)." *Bulletin of the Metals Museum* 8: 1–16.

Steinberg, A., and F. Koucky. 1974. "Preliminary metallurgical research in the ancient Cypriot copper industry." *BASOR* 18: 149–178.

6
Tell edh-Dhiba'i and the Southern Near Eastern Metalworking Tradition

Christopher J. Davey

Research into ancient Near Eastern metallurgy is approaching the point where metallurgical traditions can be delineated. The remains of metalworkers' tools have been found from a wide range of sites throughout Iran, Mesopotamia, Anatolia, Cyprus, the Levant, and Egypt, and it is quite clear that there are a number of different metallurgical practices being used. But more important, there are a number of families of each type of implement being used for the same metallurgical process. It is these families that may form distinct traditions.

Here I investigate the metallurgical tradition that is represented by the metalworkers' collection found at Tell edh-Dhiba'i, a site now in the suburbs of modern Baghdad, Iraq (al-Gailani 1965; Davey 1983). The collection is dated to the Old Babylonian Period and may have been associated with a destruction of the city some time before 1750 B.C.

By the middle of the third millennium B.C., discoveries such as the Royal Cemetery at Ur reveal that metallurgy in Mesopotamia had reached a high level of sophistication (Smith 1981, p. 195). The evidence for the development of this industry has been documented by Moorey (1982 and this volume). Few metalworking implements come from third millennium B.C. Mesopotamia, but it is probable that the objects found at Tell edh-Dhiba'i represent the metallurgical tradition that developed in the area from the late fourth millennium and would therefore have been employed during the Early Dynastic and Akkadian periods. I propose that this tradition held sway from Mesopotamia to Egypt and that it should be called the "Southern Tradition" to distinguish it from that existing in Iran, Anatolia, the coast of the Levant, and the Mediterranean.

The Technology

Among the commonest objects in the Tell edh-Dhiba'i collection are the crucibles, and these objects by virtue of their shape are some of the most distinctive artifacts of the collection (figures 6.1 and 6.2). When the crucible is placed in its upright position, its shape is such that it will not hold a liquid. It was normally assumed that this problem was overcome by placing the crucible in the furnace in a laid back attitude and that, when the metal was to be poured, it was rolled forward. Such a method has been suggested for the Late Bronze Age crucible from Keos (Tylecote 1976, p. 18).

However, the Old Kingdom Egyptian tomb of Mereruka at Saqqara dating from the Fifth Dynasty, about 2400 B.C., has reliefs that depict this type of crucible in operation (figure 6.3). The reliefs show two crucibles of the Tell edh-Dhiba'i shape placed back to back in a furnace in an upright position. They also show that,

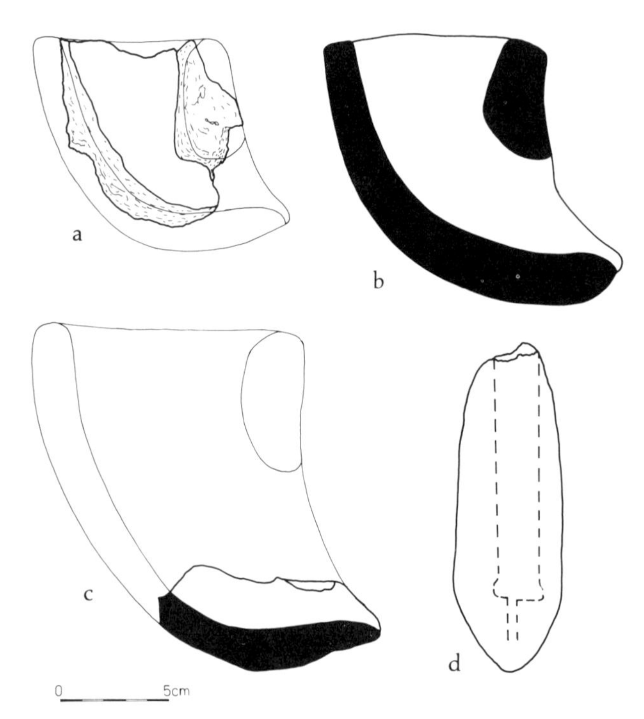

Figure 6.1 Drawings of (a) a reconstruction of the crucible from Tell el-Qitar, (b) a crucible from Tell edh-Dhiba'i (IM 65797), (c) a reconstruction of a crucible from Serabit el-Khadim, and (d) a possible bellows or blowpipe nozzle from Tell edh-Dhiba'i (Excavation 614/15).

Figure 6.2 A crucible from Tell edh-Dhiba'i. Scale: 1 cm divisions (IM 65798).

Figure 6.3 Metalworkers depicted in the reliefs of the Old Kingdom Tomb of Mereruka, Saqqara, Egypt (after Duell 1938, pl. 30).

at the time of pouring, the metal was released from the crucible by poking the front of it with a stick. It therefore seems that the crucible was used with a bot or plug placed over part of the entrance (figure 6.4). When it was time to pour the molten metal, the crucible was taken to the mold and the plug pushed aside.

The reliefs depict the use of this crucible in considerable detail, much of which may escape the casual observer. Laboratory attempts to use replicas of these crucibles have revealed the precision of the reliefs.

In the furnace scene, for example, the two crucibles have lids on top of them. It was found when using replicas of these crucibles that, to get the most rapid temperature increase within the crucible, it was necessary to place a lid over the top of it. This may seem quite logical, but it appears to have escaped the attention of most observers who have described this relief scene. It also confirms the suggestion that the crucibles were not laid back, because in that position they would not have been able to support a flat lid. However, nothing resembling a lid was found at Tell edh-Dhiba'i.

Another notable aspect of the tomb reliefs is the angle at which the draft is applied to the crucible. The people holding the blowpipes are pointing them directly down into the front of the crucible and over the top of the plug or bot. This angle is quite critical because it determines the temperature gradients within the crucible. In this instance it is important that the draft is applied directly to the position of the metal to establish a concentration of heat at that point.

The blowpipes have ceramic nozzles that would have had small apertures in them so that the velocity of the draft would have been increased. High-velocity draft rather than high-volume draft was vital for achieving a rapid rise in furnace temperature. A rapid temperature increase was important, as it would prevent the fusing of the plug to the crucible, which could result from extended periods of heating.

In the Tell edh-Dhiba'i collection, pot bellows (figure 6.5) are an addition not depicted in the Tomb of Mereruka. There does appear, however, to be a nozzle in the collection that could have been used as a blowpipe or maybe as a tuyere (figures 6.1 and 6.6). The identification of this object is not entirely certain, as there is an opening at only one end, although a small

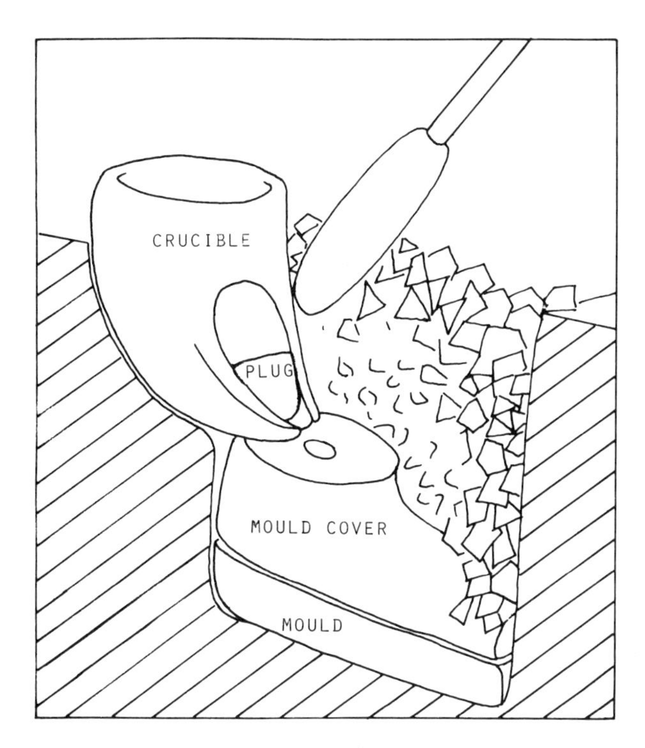

Figure 6.4 A reconstruction of a Tell edh-Dhiba'i furnace.

Figure 6.5 A pot bellows, diameter 0.4 m (Excavation 614/1).

Figure 6.6 The possible blowpipe or bellows nozzle. Scale: 1 cm divisions (Excavation 614/15).

hole nearly breaks through the other end (Davey 1983, p. 180). The object may be either a reject nozzle or possibly one that may have been repairable by removing some of the end. Whatever the case, an object of this shape with a small hole in the end would have provided a choked down draft, the velocity of which would have been increased. This arrangement is consistent with that described for Tell Zeror (Tylecote 1981, p. 116).

The technology used at Tell edh-Dhiba'i in the Old Babylonian Period had developed from that depicted in the Tomb of Mereruka. The use of pot bellows is one obvious example of this development, and this has been discussed in detail elsewhere (Davey 1979). Of more significance is the molding technology. The mold in the Mereruka reliefs is an open mold. Tell edh-Dhiba'i also had numerous such molds, but in addition there were mold covers, which reveal a more developed molding technology.

At Tell edh-Dhiba'i the covered molds could be placed at the bottom of the furnace, where they would be preheated, thus improving the molding process. In this situation the crucible need not be removed from the furnace to make the pour, and the metal would flow more freely in the mold, thus reducing the required superheating, the time for casting, and the consumption of charcoal.

Also at Tell edh-Dhiba'i were *cire perdue* molding, indicated by a broken lost-wax mold for a pin (figure 6.7), and sand casting, for which there was an ax head pattern and a core. The lost-wax mold was for a pin similar to those found on Mesopotamian sites from 3000 B.C. The mold therefore demonstrates the antiquity of the metallurgical tradition represented by the Tell edh-Dhiba'i assemblage. It also reveals the commonplaceness of the lost-wax molding method in that a simple pin was produced by what is generally regarded as a sophisticated procedure.

The ax head mold (figure 6.8) is a contrast to the lost-wax mold. It is also for an object common in third millennium B.C. Mesopotamia but, unlike the pin, which would have been extracted from the mold in its final shape, the ax head would have required a certain amount of hammering to complete its fabrication.

The mold assemblage from Tell edh-Dhiba'i reveals the use of a variety of molding techniques simultaneously. These techniques were used for different objects according to their appropriateness. It is noteworthy, however, that two-piece molds were not found at Tell edh-Dhiba'i.

The melting and casting processes at Tell edh-Dhiba'i were developed to conserve fuel, which in an arid environment would have been comparatively scarce. The procedure achieved a rapid melting of the metal in the crucible so that only a minimum of the charcoal placed in the crucible and around its entrance would have been used. The casting procedure within the furnace

Figure 6.7 The lost-wax mold. Scale: 1 cm divisions (Excavation 614/19).

Figure 6.8 The ax head pattern. Scale: 1 cm divisions (IM 65791).

also reduced the amount of necessary charcoal because it minimized the required superheating of the metal.

The crucibles were constructed from highly refractive clay so that, although they are comparatively weak, they are highly insulating. The crucible shape is one that provides the maximum strength for a vessel with a large hole in its side. The cylindrical top of the crucible in particular provides sufficient strength for it to be carried with its molten charge.

The technology of bronze casting was sophisticated at Tell edh-Dhiba'i. The crucible design, construction, and operation together with preheated molds optimized the efficiency of the melting operation with respect to fuel consumption. The tradition of metalworking is therefore one that was compatible with an arid environment.

Comparative Near Eastern Material

Until recently the Tell edh-Dhiba'i type of crucible does not appear to have been found elsewhere in the ancient Near East. The crucible found by Petrie in Sinai is of quite a different shape, having a comparatively smaller side opening and being capable of containing a liquid when in an upright position (Davey 1985).

One crucible fragment was found in 1984 at Tell el-Qitar in northern Syria at Tom McClellan's archaeological excavation (figures 6.1 and 6.9). This fragment is small, fragile, and not immediately recognizable as a crucible.

Other fragments have been found recently in the Sinai. There is a large number of fragments in the collection dug up from caves in Wadi Serâbît el-Khâdim by a Tel Aviv University expedition (Beit-Arieh 1985). The crucibles from the Sinai were all broken near their bases. Experiments conducted at The Royal Melbourne Institute of Technology with replicas of this type of crucible revealed the area near the base to be the weakest point of the crucible and a common breakage pattern.

At present it appears that the Tell edh-Dhiba'i type of crucible has not been found in Iran. Tal-i Iblis, where a large number of crucible fragments were found, provides no evidence of this shape (Dougherty and Caldwell 1966; Caldwell 1967, 1968). Nor, it seems, does Anatolia at places such as Kultepe (Ozguç 1955; de Jesus 1980). The crucible has not been found in the metallurgical collections from the coast of the Levant at such places as Byblos (Dunand 1939, 1954, 1958) or in Cyprus (Dikaios 1969; Branigan 1974).

The Tell edh-Dhiba'i crucible shape therefore seems to be, at this stage at least, limited to the more arid parts of the Near East, Mesopotamia, inland northern Syria, Sinai, and Egypt. It is of course in these parts that charcoal was less plentiful.

The traditions relating to bellows have already been described, noting that there are two particular means of operation: by the feet and by the hands. The hand-operated method has been observed with bellows from Anatolia, the coast of the Levant, and Cyprus. The foot-operated variety, however, has been shown to exist during the late third and the first half of the second millennia B.C. in Mesopotamia, the inland of the Levant, Sinai, and Egypt.

It is significant that the foot-operated pot bellows seem to come from areas where the crucibles of the Tell edh-Dhiba'i shape are also found. It is possible that foot-operated bellows, which would provide much greater draft than the hand-operated, were found to be most suitable in those parts of the Near East where metallurgy was constrained by the provision of fuel for the furnace. Whatever the case, the foot-operated pot bellows and the Tell edh-Dhiba'i crucibles appear to be associated with the same tradition of metalworking.

The most common mold at Tell edh-Dhiba'i is the open mold. This type of mold has often been found at sites in Mesopotamia (Mallowan 1947, p. 160; Speiser 1935, p. 104), northern Syria (Braidwood and Braidwood 1960, p. 450, fig. 350/1; Thureau-Dangin and Dunand 1936, p. 87, fig. 26, pl. 34; Woolley 1955,

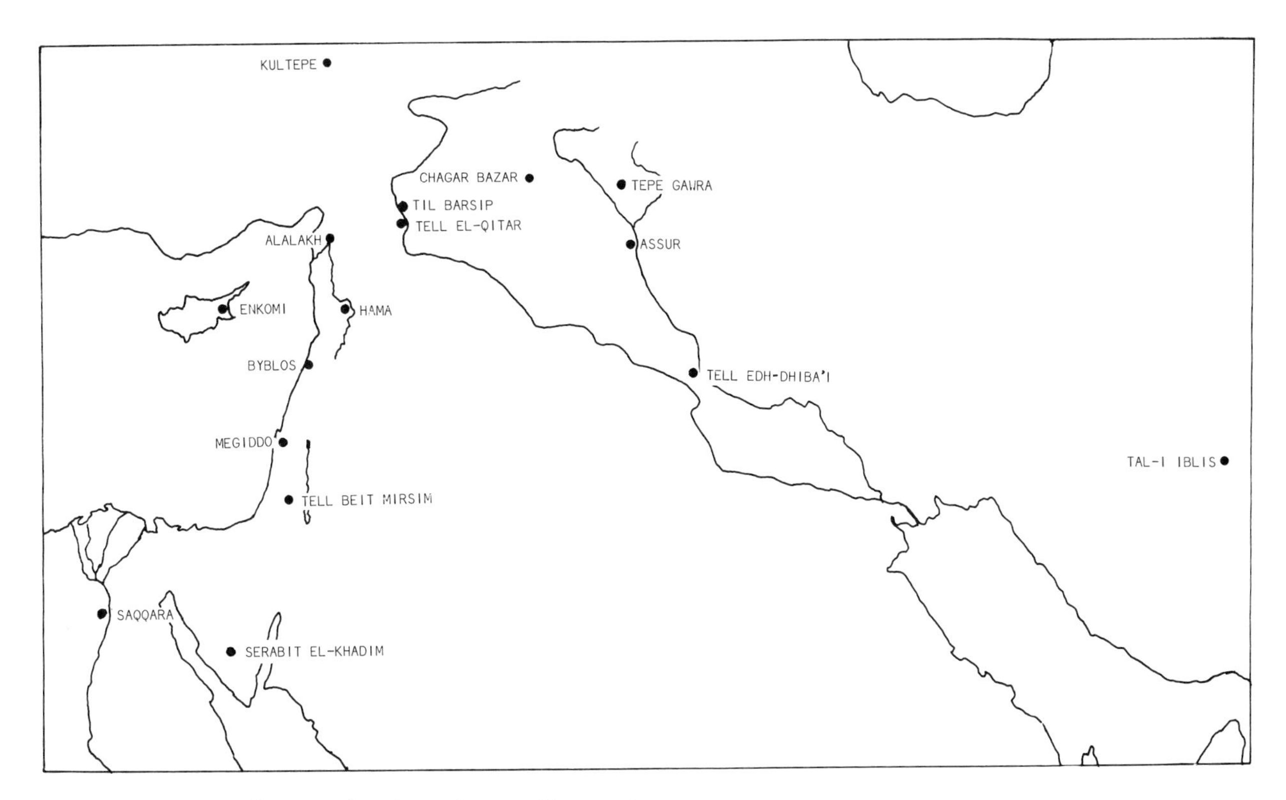

Figure 6.9 Map showing the sites where evidence of crucibles, pot bellows, and molds have been found.

p. 401, fig. 80b), and Palestine (Loud 1948, pp. 177, 185, pl. 269; Albright 1938, pp. 32, 53, pl. 43; Fugman 1958, p. 80, fig. 103, p. 95, fig. 117; Finet 1972, p. 66, pls. 13, 14). The recently published collection from Sinai has a large number of open molds made from stone (Beit-Arieh 1985). These molds are to be distinguished from the two-piece molds that are found in places such as Byblos (Dunand 1939, pl. 108, 1954, pl. 184) and Anatolia at Norsun Tepe. Open molds, of course, occur generally throughout the ancient world, but those from places such as Kultepe in Anatolia are often more finely made (Ozguç 1955, fig. 5).

The crucibles, bellows, and molds that compare to the Tell edh-Dhiba'i collection are found on sites on the plains of Mesopotamia and northern Syria, in Palestine, and Sinai from the first half of the second millennium B.C. The crucibles appear in Egyptian tomb reliefs of the Old Kingdom and the bellows in Egyptian tomb paintings of the New Kingdom. This metallurgical tradition is quite distinct from that found in the neighboring mountainous areas of Iran and Anatolia and the coast of the Levant, and it is suggested that the Southern Tradition is an appropriate name for it.

Discussion

It is clear that differences in metallurgical tradition do not necessarily relate to the type of metal that is being treated. During the Cappadocian trade between Assur and Kultepe, tin and copper were moving between the two locations, but the metallurgical traditions at Kultepe and at places in Mesopotamia such as Tell edh-Dhiba'i are remarkably different. Even where similar implements such as pot bellows existed, their mode of operation was quite different in the two areas.

The metal being used in the Mereruka scene is probably copper, whereas at Tell edh-Dhiba'i bronze was used. The same metalworking tradition therefore could be applied to different metals and alloys.

It is possible that the use of mold covers at Tell edh-Dhiba'i may have developed from contact with a metallurgical tradition that used two-piece molds that were preheated. It is also possible that the idea developed within Mesopotamia itself where *cire perdue* molds were used and probably preheated.

The molds found at Tell edh-Dhiba'i consist of open molds made from fired clay. The bellows were also made of clay. However, many of the open molds and bellows that have been referred to as being comparative to the Tell edh-Dhiba'i collection were made from stone and, in particular, limestone. This is especially true of those found in southern Palestine and in the Sinai. This occurrence reveals that within the same technological tradition there are two branches that have evolved to utilize most effectively the construction materials at hand.

It has been suggested that the metalworkers of the Sinai were itinerant and nomadic Semites (Beit-Arieh 1985, p. 115). However, stone molds and stone foot bellows of the early second millennium B.C. in Sinai and Palestine do not appear to be convenient tools for the use of nomadic coppersmiths. These are the tools of people who are settled and who are working in a copper or bronze trade that has a high-volume production.

The beginning of this tradition is probably in the fourth millennium B.C. or even earlier. Sites such as Buhen in Nubia are likely to have elements of this tradition within its metallurgical objects. We are not likely to find much of the nomadic traditions of the Near East, as nomads carry and use only the bare essentials. The pattern ax head at Tell edh-Dhiba'i, which was used for casting in sand, may be a vestige of an earlier itinerent tradition. The use of sand molding is no doubt the precursor to the open molds of clay and of stone that became so common on sites of the third and early second millennia B.C. in the Levant.

The open molds, therefore, probably point to the origin of this tradition, and that origin is no doubt with the nomadic people who may have been Semites. They were people who were used to using meager resources of charcoal and who, like the Bedouin of today in the area, cast into open sand or clay molds. As these people settled into communities of the Near East, this tradition developed, so they made their implements larger and heavier, out of stone and clay, which enabled more metal to be cast without the need of refashioning tools to do the work.

It is therefore proposed from this technological analysis of the metalworking tools found at Tell edh-Dhiba'i and comparable sites that between about 3000 and 1500 B.C. there was a single metallurgical tradition in Mesopotamia, northern inland Syria, Palestine, and Egypt and that this tradition originated with nomadic people from desert regions. The name "Southern Tradition" is suggested for the metallurgical tradition as a means of distinguishing it from others that are found in Iran, Anatolia, and the Mediterranean. This conclusion means that the beginnings of the metallurgy of Sumer, Egypt, and Babylonia are not to be sought in the surrounding mountains, but in the deserts, in places such as Oman, Fenan, and western Saudi Arabia, where people are known to have mined and engaged in metallurgy.

References

Albright, W. F. 1938. "Excavations at Tell Beit Mirsim." *Annual of the American Schools of Oriental Research* 2.

al-Gailani, L. 1965. "Tell edh-Dhiba'i." *Sumer* 21:33–40.

Beit-Arieh, I. 1985. "Serâbît el-Khâdim: New metallurgical and chronological aspects." *Levant* 17:89–116.

Braidwood, R. J., and L. S. Braidwood. 1960. *Excavations in the Plain of Antioch*. Chicago: University of Chicago Oriental Institute, vol. 1.

Branigan, K. 1974. *Aegean Metalwork in the Early and Middle Bronze Ages*. Oxford: Clarendon Press.

Caldwell, J. R. 1967. "Survey of excavations: Tal-i Iblis." *Iran* 5:144–146.

Caldwell, J. R. 1968. "Tal-i-Iblis." *Archaeologia Viva* 1:145–150.

Davey, C. J. 1979. "Some ancient Near Eastern pot bellows." *Levant* 11:101–111.

Davey, C. J. 1983. "The metalworkers' tools from Tell edh-Dhiba'i." *Bulletin of the Institute of Archaeology* 20:169–185.

Davey, C. J. 1985. "Crucibles in the Petrie Collection and hieroglyphic ideograms for metal." *Journal of Egyptian Archaeology* 71:142–148.

de Jesus, P. S. 1980. "The development of prehistoric mining and metallurgy in Anatolia." *Biblical Archaeology Review* S-74.

Dikaios, P. (1969) *Enkomi Excavations 1948–1958*. Maintz an Rhein: Philipp von Zabern.

Dougherty, R. C., and J. R. Caldwell. 1966. "Evidence of early pyrometallurgy in the Kerman Range in Iran." *Science* 153(3739):984–985.

Duell, P. 1938. *The Mastaba of Mereruka*. Chicago: University of Chicago Oriental Institute.

Dunand, M. 1939. *Fouilles de Byblos*. Paris: Geuthner, vol. 1.

Dunand, M. 1954. *Fouilles de Byblos*. Paris: Geuthner, vol. 2, part 1.

Dunand, M. 1958. *Fouilles de Byblos*. Paris: Geuthner, vol. 2, part 2.

Finet, A. 1972. "Apercu sur les fouilles Belges du Tell Kannas," *Annales Archeologique Arabes Syriennes* 22:63–74.

Fugman, E. 1958. *Hama fouilles et recherches 1931–1938*. Copenhagen: Wendt and Jensen, vol, 2, part 1.

Loud, G. 1948. *Megiddo*. Chicago: University of Chicago Oriental Institute, vol. 2.

Mallowan, M. E. L. 1947. "Excavations at Brak and Chagar Bazar." *Iraq* 9:1–87.

Moorey, P. R. S. 1982. "The archaeological evidence for metallurgy and related technologies in Mesopotamia." *Iraq* 44:13–38.

Ozguç, T. 1955. "Report on a workshop belonging to the late phase of the Colony Period (1b)." *Belleten* 19:77–80.

Smith, C. S. 1981. *A Search for Structure: Selected Essays on Science, Art and History*. Cambridge, Mass.: MIT Press.

Speiser, E. A. 1935. *Excavations at Tepe Gawra*. Philadelphia: Museum of the University of Pennsylvania, vol. 1.

Thureau-Dangin, F., and M. Dunand. 1936. *Til Barsip*. Paris: Geuthner.

Tylecote, R. F. 1976. *A History of Metallurgy*. London: The Metals Society.

Tylecote, R. F. 1981. "From pot bellows to tuyeres." *Levant* 13:107–118.

Woolley, L. 1955. *Alalakh*. Oxford: Oxford University Press.

7
Early Metallurgy in Yugoslavia

Borislav Jovanović

Systematic research of the early metallurgy and copper mining in the central and western Balkans, comprising the territory of modern Yugoslavia, began in earnest in 1968, when Rudna Glava was discovered. This Early Eneolithic (Chalcolithic) site is one of the earliest copper mines in this region (figures 7.1–7.3).

The first evidence of prehistoric mining in the central Balkans was published by M. M. Vasić in 1905. According to his evidence, prehistoric copper mining occurred at a large site near Bor, to which should be added evidence from the rich iron and copper ore deposits in Majdanpek. Both of these deposits are still actively exploited, and today they belong to the same mining zone as Rudna Glava; they are all part of the Timok eruptive basin in northeast Serbia (Vasić 1905, p. 597).

The traces of prehistoric mining in Jarmovac (southeastern Serbia) were pointed out by O. Davies when he published some typical mining tools of the times, a large maul with horizontally ground grooves (Davies 1937, p. 1).

The use of cinnabar as a red pigment in the period of the Late Neolithic [Kostolac Culture, Šuplja Stena mines, the Avala mountain in the vicinity of Belgrade (Milojčić 1943, p. 41)] was established along with the Early Eneolithic copper mining.

Records of the use of copper in the Early Eneolithic in Yugoslavia (Vinča Culture) originate from the eponymous deposits of Vinča, near Belgrade, also at Gomiolava, Farfos, Grivac, all in the Socialist Republic of Serbia. The remains of copper carbonate mineral smelting (primarily malachite) are confirmed from the deposits of the Vinča Culture at the Gornja Tuzla (Bosnia and Hercegovina). This is discussed more thoroughly later.

All of these data permitted a more precise knowledge of the oldest copper mining, hitherto completely undefined. The approximate quantity of copper oxide and carbonate minerals was estimated from the separate Eneolithic mines at Ai Bunar in southern Bulgaria (Černih 1978, pp. 72, 135). The quantity of copper extracted from these ores indicates the rapid development of the copper smelting and casting technology. Thus considerable quantities of high purity copper were processed in the beginning of the Early Eneolithic, that is, in the earliest phase of copper metallurgy.

A raw materials base for this quick development in the central and eastern Balkans existed in nine regions of northeastern Yugoslavia and in the south of Bulgaria. For Yugoslavia the confirmation exists in the Eneolithic mines at Rudna Glava, not far from the contemporary mining center at Majdanpek (eastern Serbia) (figure 7.1).

Continuing studies at Rudna Glava have been and are being conducted by the Institute of Archeology, Belgrade, and the Mining and Metallurgy Museum at

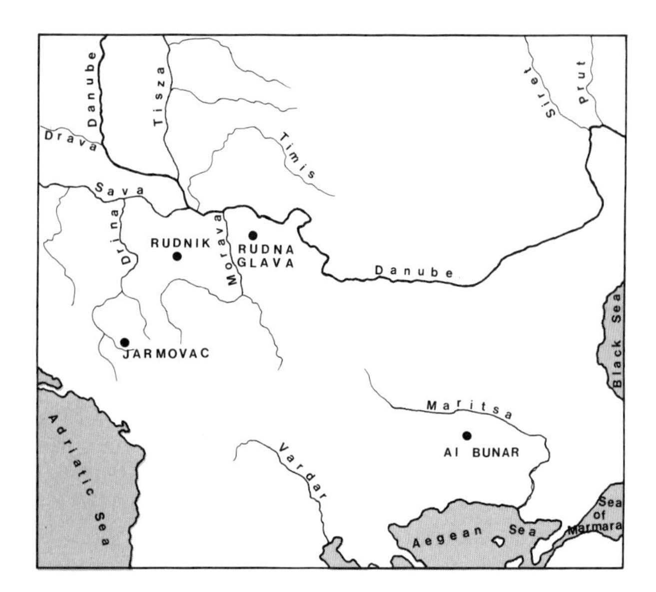

Figure 7.1 Geographical position of old mining monuments in the central and southern Balkans.

Bor. Forty shafts (pits) have so far been discovered; more precisely, these are exhausted natural veins. These Early Eneolithic pits are mainly situated along the northern profile of the contemporary magnetite strip mine (figures 7.2 and 7.3). Although the current exploitation of this mine has damaged or destroyed a part of the historic works, the open pit mining accounts for the actual discovery of the Early Eneolithic pits (Jovanović 1982, pp. 1–17).

The available data indicate the technology of the oldest copper mining in the Balkans established at Rudna Glava; these show the simplest, but at the same time the most efficient and easy to use, technological method for copper mineral extraction. The old miners of the Vinča Culture identified outcrops according to the color and composition of the ground. The main ore at Rudna Glava was chalcopyrite, apparently impregnated with magnetite along with other iron oxides (Bugarski and Janjić 1982, p. 120). The outcrops of the main vein are characterized by decomposition of the main ore body into a zone of secondary minerals such as malachite and azurite. The same process occurs with iron ores, and Rudna Glava contained an "iron hat" (gossan), making the deposit noticeable.

Vinča miners constructed a sort of access platform around the mining vein, removing the piled-up material and providing the approach to the trench. Further exploitation was carried out by simply cleaning the ore vein trench without changing its profile and dimensions. There existed smaller supporting dry walls of stone in places in order to prevent landslides. It is in one of these shafts (2n) that excavation of a small gallery was attempted in order to provide an easier approach to the vein, with its channel emerging through the loose conglomerate (Jovanović 1986, pp. 1–2). This may be the first example of a horizontal underground shaft in the prehistoric mining of southeast Europe.

This type of pit, exhausted vein channel, is therefore exclusively a natural creation. The size varied greatly, but the average depth was about 15–20 m with a diameter between 0.70 and 2 m. In most cases the pit is represented by winding narrow channels, sometimes linked at greater depths. Mining veins penetrated the strata, which consisted of crystallized calcite, crisscrossed with veins of quartz; hence most of the channels have rather solid walls (figures 7.4 and 7.5).

The depths that the ancient miners were able to reach depended on the inflow of fresh air, but the depletion of the secondary enriched minerals may also have played a role.

The ore was mined using the old fire-setting (heating and cooling) method on the rock. Antler tools were used to widen the fissures and cracks. The smaller blocks of ore were crushed by horizontal grinding using massive mauls. The ore was then pulled to the surface manually, as no traces of winches were discovered. It is probable that some sort of separation was conducted at the mining site.

The excavated evidence of Eneolithic mining tools, which includes stone mauls, confirms our reconstruction of the primary mining methods (figures 7.6 and 7.7). Mention should be made of the excavated pottery vessels and mining tool remains, which are of importance not only in dating the Early Eneolithic mines but also for a description of the technology (figures 7.8–7.11).

Stone mauls were found mostly in shafts, on access platforms, and as individual examples on the surface of the sites. There were group finds in shafts or at the bottoms (shaft 4a) as well as hoards. These artifacts are of particular value. The mauls consist of natural formations of igneous rock pebbles (gabbro, in most cases) taken from the detritus of the Šaška reka, which flows at the foot of Rudna Glava. The only adaptation by the miners was for them to grind a horizontal groove on the tool, to which was tied a rope or leather strip. The maul, used in this way, served as a kind of pendulum, enabling the worker to hammer in the narrow space of the shaft. By lengthening or shortening the rope or leather thong, the worker could easily reach the most convenient areas of the vein to break or crush the blocks of ore. As a rule, the ends of these mauls show damage from use; other examples had been damaged severely or destroyed. These were most likely discarded in the course of the work in the shafts. There are examples of some secondary use, as shown by grooves ground in damaged or broken tools.

More than 200 examples of mauls—pebbles with grooves—were found in the excavated areas of the Eneolithic mines (figures 7.12 and 7.13). Their quantity shows us that they were the main mining tools used in primary copper mining (Rothenberg and Blanco-

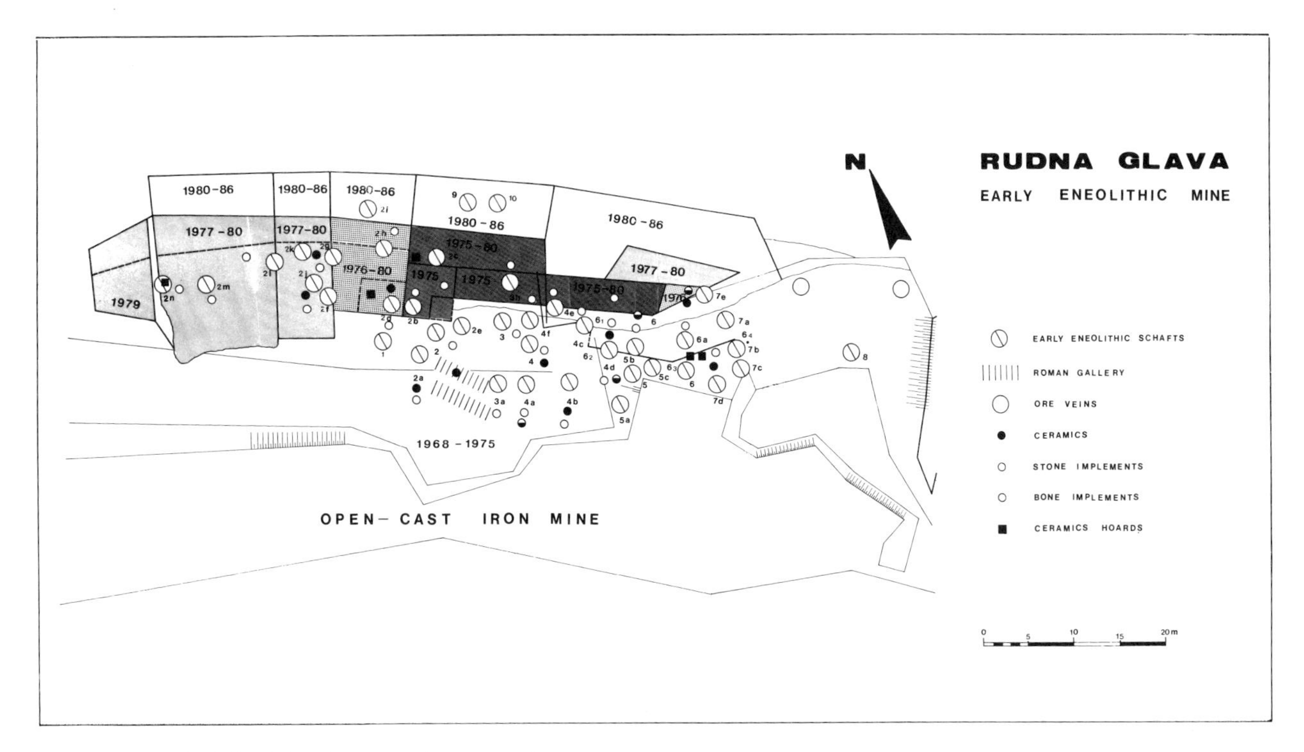

Figure 7.2 A draft of the Early Eneolithic pits disposition at Rudna Glava.

Freijeiro 1981). Although the natural form remained unchanged, it was possible to create a basic typological classification of the massive tools. This classification was made according to their basic purpose and their size and weight. This is the first attempt at classification of the oldest mining tools; although it is based mainly on the excavation of a single deposit, there was sufficient evidence from the mauls for specialization (Stanojević 1982, p. 53). This means that the selection was made in advance, with special attention paid to their future function. Heavier mauls were probably used for breaking ore blocks, smaller ones for crushing parts of the block, because different diameters of shafts required tool sets of corresponding size.

Antler tools were used as a sort of handpick because in all likelihood they were provided with a handle. Their number is noticeably smaller than that for the stone mauls—ten preserved or fragmented pieces were excavated at Rudna Glava. It is only logical that such a ratio of the chief tools depended mainly on the mechanical properties of the ore. At Rudna Glava, a mine that required breaking a homogeneous ore vein, maul pebbles were much more useful than were antler tools.

Mine hoards also help to confirm the technology described for the Early Eneolithic mining. Five hoards were uncovered from the mine area but there must have been many more. They had been carefully hidden, either placed in wider cracks between rocks (hoard 1 and 2), dug into loose material (hoards 3 and 4), or dug into a convenient place in the shaft itself (hoard 5) (figures 7.12 and 7.13). All of these hidden places, as a rule, contained large pottery vessels. Some were egg-shaped amphorae with two handles, and some were cylindrical pots with flat edges. There were some vessels of extraordinary quality, such as an amphora-ewer with one beltlike handle and one with a long funnel-shaped neck; both had well-polished black surfaces decorated by using a carving technique (Jovanović 1982, vol. 20, p. 1).

Fragments of pottery vessels were also found in shafts, on access platforms, or in the layers of accumulated material in the zones of old workings. All the vessels, with rare exceptions, belonged to the same categories: amphorae and large pottery.

The stratification of the Early Eneolithic mining areas at Rudna Glava is a certain indicator of the stepwise exploitation of the ore deposits. These are, however, technological steps because the technology at the sites situated far from each other could have been carried out simultaneously. The filling up of the abandoned pits with material containing pottery and stone and bone tools could have taken place in such a way only if the pits in the closest neighborhood had been exploited. On these occasions the discarded material was probably thrown out in the nearby emptied ore veins. This was more likely to happen if these pits—both the active and the abandoned ones—were separated by a definite height difference, for the mine was situated on the steep slopes of the mountain. Most of the pits were placed one above the other.

The stratification of the access platforms confirms

Figure 7.3 Rudna Glava. Northern profile of open pit magnetite mine at Rudna Glava. Damaged Early Eneolithic pits 4–7.

Figure 7.4 Rudna Glava. Entrance of the central trench of shaft 7.

Figure 7.5 Rudna Glava. Platform and entrance in shaft 2h.

this method of exploitation: The older pits were situated mainly on the lower slopes, and the newer ones were located higher up on the hill. The advantage of such a technical method, compared with the deposition of the discarded material, is apparent.

The inner stratification of single pits, especially the larger ones, shows a gradual sedimentation of the material. Massive mauls, pottery fragments, and pieces of ore lie in layers of grayish loose mass and crushed stone at various depths. They could have gotten there only if the pits had been gradually filled with matter from the higher pits. Otherwise, the pits (that is, the exhausted channels of the veins of ore) would have been filled with sterile earth washed down and deposited by atmospheric water.

The Rudna Glava site has been placed in relative chronology within the framework of the Balkans-Danube Neolithic and Eneolithic. If we judge from enclosed findings, which we consider to be safe in a typological and cultural sense (mainly by hoards and also by archaeological material from the pits), Rudna Glava is related to the change of the old to the new Vinča Culture. Exploitation was certainly conducted during the initial stage of the Vinča Culture. In absolute dates these oldest mining activities at Rudna Glava correspond to the middle and second half of the fourth millennium B.C. (according to the noncalibrated carbon 14 dates for the Vinča Culture) (Garašanin 1984, p. 65; Chapman 1981, p. 17; Tringham et al. 1980, p. 28).

The relative chronology of Rudna Glava may also be discerned through another type of pottery. These are the rectangular altars (or lamps) with shallow recesses ornamented with stylized deer heads and carved meanders. Altars of this shape are recognized as typical for the Vinča Culture and were represented at Rudna Glava in three specimens (figure 7.11). The most valuable one was the altar from hoard 3 (from the access platform of shaft 2h); for the typical elements of Early Eneolithic mining amphorae, altars, and stone mauls are found in the same closed entirety (Jovanović 1982, pp. 63–64).

Summary of Excavations at Rudna Glava

The oldest mining at the Rudna Glava massif has been only fragmentarily studied since 1968. At that time a considerable number of pits were destroyed with the opening of the current open pit magnetite mine. The old mines were situated alongside the inclined longitudinal fissure or a smaller cleft filled with a number of mining veins. The Early Neolithic pits were the

Figure 7.6 Rudna Glava. Cylindric maul-pebble with horizontally ground groove. Central platform.

Figure 7.7 Rudna Glava. Prismatic maul-pebble with horizontally ground groove. Western platform.

Figure 7.8 Rudna Glava. Pot from hoard 1.

Figure 7.9 Rudna Glava. Amphora with band-shaped handle and carved ornamentery from hoard 1.

Figure 7.10 Rudna Glava. Amphora from hoard 2.

Figure 7.11 Altar with deer heads. Found at the bottom of a destroyed shaft at a depth of 12 m.

Figure 7.12 Rudna Glava. Pottery vessels of hoard 3 in situ.

Figure 7.13 Rudna Glava. Pot and mauls of hoard 4 in situ.

most common alongside the richest mining veins. The Vinča miners did not know of horizontal galleries, although an initial pattern of this sort exists in pit 2n. The mining vein passed through loose earth, impregnated by poorer layers of the ore. By going deeper along the vein, the miners dug a single smaller horizontal gallery 3.5 m long and 0.80 m in diameter at a depth of 5.38 m. The hoard was hidden at the beginning of the gallery where it intersects a vertical pit entrance (at a depth of 2.60 m). The gallery does not follow the channel, which in the loose earth cannot be clearly formed. What Early Eneolithic miners did, in going in the main direction of the vein, was to widen the channel, making its fall less steep, so that they could reach the deeper ore. Shaft 2n did not reach the average depth of other Early Eneolithic mines (it reached only 5.38 m), and the other miners, because of the unsteady ground, abandoned further work. Judging by the ground structure and approximate horizontal line of the gallery, we think that a wooden platform wall was probably used, at least in its most primitive form (Jovanović 1986, pp. 1–2).

Completed studies on the Early Eneolithic pits at Rudna Glava give us reliable data on a wide scale of copper ore exploitation at the end of the Late Neolithic in the Balkan southeast European region. As I have already stated, only a small number of Eneolithic studies have been made. These have been conducted alongside the northern profile of the modern magnetite mine. Because of the appearance of the surrounding grounds and the stratification of the pits and access platforms studied, we can say that much bigger zones were involved in the mining activities in the Early Eneolithic Period. It is likely that the entire Rudna Glava massif covered in the past the gossan containing these works, whose disposition and character we cannot yet define precisely. It would not be a great surprise if we discovered copper ore being exploited during the later periods of prehistory.

The Early Eneolithic mining at Rudna Glava, reliably placed at the beginning of the late Vinča Culture, points to the use of old technological processes. Thus the use of the maul pebbles with horizontally ground grooves is known in the flint mining of the younger Neolithic, whereas the technique of fire setting (heating and cooling of rocks) was known before the use of metals. The mining tools found at Rudna Glava are also of local origin (figures 7.6 and 7.7). The same is true for the pottery vessels and altars. The bearers of this sort of mining are also known: They belong to the Vinča Culture, which occupied much of the south and central Balkans, as well as the southern part of the Pannonia Plain in the younger Neolithic and Early Eneolithic (Garašanin 1984, p. 62). Therefore it should again be stressed that Rudna Glava does not provide any evidence that mining techniques had been imported from Anatolian or east Mediterranean centers.

The earliest exploitation of the copper ore at Rudna Glava is extensive and occurred in a relatively short time. The results of all hitherto conducted analyses from Rudna Glava remain within the framework of the change from the old and late Vinča groups and the initial stages of the latter (Vinča Pločnik I, that is, Vinča C). It should be pointed out again that the completed research studies uncovered only a part of the Early Eneolithic works, which means that the quantity of the excavated copper minerals was much larger. Copper carbonate minerals, such as malachite and azurite (especially malachite), were used primarily. This by no means excludes other easily reduced copper sulfide minerals, such as brochantite (Bugarski and Janjić 1982, p. 120). Because these materials were the result of the secondary enrichment of the main body (chalcopyrite decomposition as the structural part of the magnetite ore), it was only natural that native copper was found at the beginning of the exploitation of Rudna Glava. Thus the Vinča miners began the exploitation of this exceptionally rich deposit driven by the surface evidence of native copper.

No evidence of the smelting of the ore has thus far been found in the nearby or removed surrounding regions of Rudna Glava. Moreover, in the mountainous regions of northeast Serbia there are no settlements of the Vinča Culture. The closest ones were situated in the valleys of the Mlava and Morava rivers to the west and the Timok River to the east. There is a lack of evidence of permanent camps of the Vinča miners at Rudna Glava itself. The hidden pottery closets, a characteristic of these deposits, may be the reason for the absence of permanent settlements. This forces us to conclude that the mining activities at Rudna Glava were seasonal in character. It seems that the excavated copper ore or, more precisely, the copper carbonate and oxides were transported to the neighboring plains, which were densely populated during the late Vinča Culture.

If these results of the Rudna Glava research studies give us reason to believe that this was an exceptional or even the only Early Eneolithic mine in the central Balkans, then that impression is wrong. Rudna Glava is just one of the few Early Eneolithic mines preserved by a fortuitous combination of circumstances. The evidence of M. M. Vasić, for example, published at the beginning of this century, mentions another prehistoric mine at Bor, an important modern copper production center in Yugoslavia. Similar data were published about prehistoric mines at the iron and copper ore deposits at Majdanpek, also a major modern mine. Descriptions of old mine tools and pits from the deposits at Majdanpek fully correspond to our conclusions about the exploitation at Rudna Glava (Schneiderhöhn 1941, p. 420). Therefore there is good reason to believe that the copper-bearing regions between Bor and Majdanpek in northeastern Serbia had already been exploited in the period of the primary copper metallurgy (Eneolithic and Chalcolithic periods). Many other deposits without economic value were also precious sources of metal in prehistory, but limited exploitation probably left no trace of activity.

There are two additional exploited copper deposits in the wider central Balkans: Rudnik and Jarmovac, near Priboj southwestern Serbia. The first deposit uncovered pits and a vast number of mauls with a horizontally ground groove (figure 7.14); at Jarmovac, which produced only one discovery of maul pebbles, prehistoric exploitation of this exceedingly large ore deposit is indicated. In both cases the question is whether the copper mines exploited carbonate and oxide ores. At Rudnik-Mali Sturac large quantities of flint are also represented. The jewelry made of this mineral is well known in the Vinča Culture (Milić 1972, p. 207; Jovanović 1982, p. 781).

The southern Balkans are also rich in copper ores used in the oldest metallurgy. The most significant mine of that period is Ai Bunar, which shows a wide range of exploitation. A few smaller open pit mines were examined there, some of which continued into pit works. According to excavated pottery copper carbonate and oxide minerals were used during the Late Eneolithic Period (Gumelnitza Culture, that is, the last quarter of the fourth and the beginning of the third millennia B.C.). However, there are indications that the mining activities at Ai Bunar had begun in the Early Eneolithic [the Marica Culture, that is, the second half of the fourth millennium B.C., which, in a chronological sense, corresponds to Rudna Glava (Černih 1978, p. 132; Todorova 1979, p. 41)].

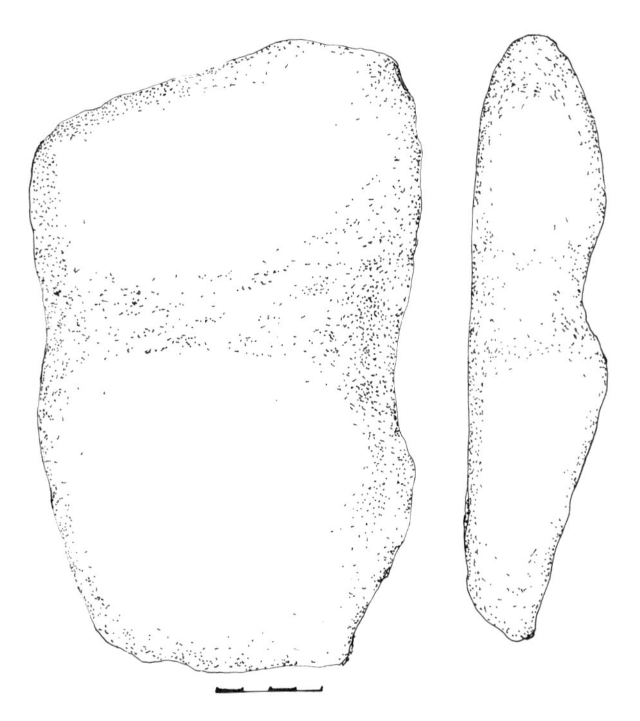

Figure 7.14 Rudnik. Prehistoric mine of Rudnik-Mali šturac. Maul pebble with horizontally ground groove. Surface evidence.

The small amount of data available, however, is sufficient to indicate rapid development of the copper industry of the central Balkans in the earliest stage of copper metallurgy. A high rate of primary mining production is unexpected because of the importance attached to native copper.

It has not been clarified yet at what places and in what ways such large quantities of copper were processed. At Ai Bunar, as at Rudna Glava, no traces of copper smelting have been found, just as we have no reliable evidence of smelting furnaces from the Early or Late Eneolithic. However, remains of copper smelting were found in the settlements of the Vinča Culture; these were the huts that I interpret as not containing any particular smelter or foundry.

Small pieces of copper oxide or of copper itself together with the remains of burning were found in the settlement of Gornja Tuzla in southeastern Bosnia (Čović 1984, p. 118). Similar mixtures of ash, tar, soot, malachite, and azurite were found in a pit at a younger Vinča Culture settlement at Fafos [in the vicinity of the Kosovska Mitrovica (Kosov)]. The settlement is situated in the neighborhood of the mountainous massif of Kapaonik, well known for its rich ore deposits (Jovanović and Ottaway 1976, p. 104). Numerous traces of copper ore processing were also found on the floors of the huts at the Vinča Culture settlement of Selevac, near Belgrade (wider region of the lower Pomoravlje) (Tringham et al. 1980, p. 28). Analyses suggest that the smelting of carbonate and oxide ores points to Rudna Glava or the basin of Bor-Majdanpek as a possible source (Glumac 1983, p. 137). Numerous fragments of copper oxide were also represented in the Vinča Culture settlements, and finds of the same material appear in the Vinča Culture settlement of Grivac in central Serbia (Jovanović and Ottaway 1976, p. 105).

Products of this early metallurgy are limited to smaller tools and jewelry. Massive copper bracelets with free ends are known from the necropolis of a younger Vinča Culture at Gomiolava in Posavina [western Vojvodina (Brukner 1980, p. 34)], as well as from hoard 3 from Pločnik in southern Serbia (Stalio 1964, p. 37). A group of hoards found at Pločnik (1–4) defines this place as a possible metallurgical center, although there is no evidence of the existence of a copper smelter or foundry. The hoards from Pločnik provide evidence for the development of a younger Vinča Culture metallurgy and the mastering of tool and weapon production—mainly chisels and axes. In a typological sense the oldest shaft-hole axes (hammers) follow the shape of the earlier shaft-hole stone axes. The developed form, a cross ax with oppositely placed edges but forming an original-type one-blade ax with a cylindrical hole, is a shape that has remained basically unchanged until now, although it evolved with different variants.

According to the available archaeological evidence, the technological circle of the early copper metallurgy of the Vinča Culture was completely closed. The first use of metals in the Balkans soon developed into production of jewelry, weapons, and tools. The first technological knowledge in mining, oxide and carbonate ore smelting, and the casting of single-sided and then double-sided molds were mastered.

The production of copper was not the final aim of Eneolithic mining and metallurgy. From the beginning of the use of metal, there were efforts to improve its properties, especially its hardness. As an example, consider the remains of the smelted galena at the bottom of a bowl from the Vinča settlement at Tuzla (Čović 1984, p. 112). It remains to be seen whether the first attempt at lead production was made. If so, this would be a step closer to leaded bronze production in the Late Neolithic (during the first half of the third millennium B.C.).

The composition of the earliest massive Eneolithic tools points to copper of high purity. The use of native copper is beyond any doubt, although this period did not last long (Maddin et al. 1980, p. 213). Otherwise, it is hard to believe in such extensive production of oxide and carbonate minerals. The copper extracted from these minerals must also have been of high purity with an insignificant amount of impurities that might have more accurately determined the corresponding ore deposits.

It is only later, in the Late Eneolithic, that massive weapons and tools were cast from arsenic-copper alloys or of leaded arsenic-copper alloys. Arsenical copper alloys predominate in that period (Jovanović 1985, p. 82), marking in this way the end of the exploitation of easily accessible deposits of copper oxides such as those at Rudna Glava and Ai Bunar. The further exploitation was limited to the sulfide ores of copper. A complex technological technique for extracting copper from these ores affected the production. A new technology for the smelting of sulfide ores had to be mastered, and this process took time, as judged by the sharp decline in the production of copper weapons. Some cultures of the Late Eneolithic, such as Baden and Kostolac, are known because of the paucity of excavated copper artifacts. There still were considerable quantities of pure copper in circulation, but these were from the smelting of the oxide and carbonate ores in the course of the Early Eneolithic. That is the reason why such copper tools of exceptional purity along with arsenical or leaded bronzes were found at the beginning of the Bronze Age (Sangmeister 1972, p. 103).

The beginning of the use of copper and gold in the Balkans is a consequence of the economic needs of the younger Neolithic culture. The only developed industry of stone tools and weapons of these cultures could no longer satisfy the needs of the growing population and the developing agriculture. The knowledge of metals and the production and technology of its prime production was of local origin.

According to the data available, the technological process for the production of copper tools and weapons during the whole Eneolithic Period followed this course. Contemporary mining oriented to oxide and carbonate ores provided the rapid inflow of considerable quantities of copper. Processing and smelting of malachite and azurite were performed in settlements, without forming any bigger metallurgical centers. The metal processing itself started with cold forging of native copper, but that stage was apparently short. The great majority of copper tools and weapons in the Early Eneolithic were produced by casting in open molds. The metallurgists of the Late Eneolithic, as shown by the examples from the Vučedou Culture, were already casting in double-sided molds (Durman 1983, p. 23).

After a sharp rise of early metallurgy, based on the exploitation of the rich deposits of the carbonate and oxide ores, the Balkans witnessed a general decline in production (Černih 1978, p. 274; Vulpe 1976, p. 152). The cause for such a state of affairs is of a technical nature, that is, a shift to the sulfide ores, once the carbonate and oxide ores had been exhausted. It is possible that many cultural and population movements inside the Carpathian Basin and Balkans during the Late Eneolithic were caused by the demand for the rich deposits of the copper in that region. Along with the use of sulfide ore there appear the first bronze alloys: arsenical and leaded bronzes. Only the complete mastering of the copper production from sulfide ores would cause the mass use of bronze, but this is considered to be a period of highly developed mining and metallurgy during the Middle and Late Bronze Age.

References

Brukner, B. 1980. "The settlement of the Vinča group at Gomolava (Neolithic and Early Eneolithic layer)." *Rad Vojvodjanskikh Muzeja* 26:5–35. (In Serbo-Croatian.)

Bugarski, P., and S. Janjić. 1982. "A turn to some former chemical and mineralogical investigations of mineral phenomena of Rudna Glava," in *Rudna Glava: The Oldest Copper Mining in the Central Balkans*, B. Jovanović, ed. (Bor-Beograd: Institute of Archaeology), 120–126. (In Serbo-Croatian.)

Chapman, J. 1981. *The Vinča Culture of Southeast Europe*, 2 vols. London.

Černih, N. E. 1978. *Mining and Metallurgy in Ancient Bulgaria*. Sofiya Bulgarian Academy of Science. (In Russian.)

Čović, B. 1984. "Prehistoric mining and metallurgy in Bosnia and Herzegovina: Condition and problems of research." *Godisnjak* 22:111–114. (In Serbo-Croatian.)

Davies, O. 1937. "Prehistoric copper mine at Jarmovac near Priboj na Limu." *Glasnik Zemaljskog Muzeja u Bosni i Hercegovini* 49(1):1–3.

Durman, A. 1983. *Metallurgy of the Vučedol Culture Complex.* Opuscula Archaeologica 8 (Zagreb). (In Serbo-Croatian.)

Garašanin, M. 1973. *The Prehistory of the Socialist Republic of Serbia,* 2 vols. Belgrade: Srpska Knjižèvna Zadruga. (In Serbo-Croatian.)

Garašanin, M. 1984. "The Vinča and Vinča Culture in the Neolithic of southeast Europe," in *The Vinča in Prehistory and Medieval Time* (Belgrade: Serbian Academy of Science and Art), 57–65. (In Serbo-Croatian.)

Glumac, P. 1983. "An archaeometallurgical study of the materials from Selevac." *Zbornik Narodnog Muzeja Beograd* 9:135–141. (In Serbo-Croatian.)

Jovanović, B., ed. 1982. *Rudna Glava: The Oldest Copper Mining in the Central Balkans.* Bor-Beograd: Institute of Archaeology. (In Serbo-Croatian.)

Jovanović, B. 1985. *The Bronze Age Metallurgy of the Central Balkans: The Eneolithic and Early Bronze Age of Some European Regions.* Krakow: Polish Academy of Science, 81–86. (In French.)

Jovanović, B. 1986. "New discoveries at Rudna Glava: The earliest shaft and gallery copper mine in Eastern Europe." *Newsletter, Institute for Archaeo-Metallurgical Studies* (London), 8:1–2.

Jovanović, B., and B. S. Ottaway. 1976. "Copper mining and metallurgy in the Vinča group." *Antiquity* 50:104–113.

Maddin, R., T. S. Wheeler, and J. D. Muhly. 1980. "Distinguishing artifacts made of native copper." *Journal of Archaeological Science* 7:211–225.

Milić, R. 1972, "Ore structure and factors controlling ore deposits in the multimetal deposit Rudnik." *Proceedings of the Seventh Geological Congress of Yugoslavia* (Zagreb), vol. 3, 207–222. (In Serbo-Croatian.)

Milojčić, V. 1943. "The prehistoric mine Šuplja Stena on the Avala mountain near Belgrade, Serbia." *Wiener Prähistorische Zeitschrift* 30:41–45. (In German.)

Rothenberg, B., and A. Blanco-Freijeiro. 1981. *Studies in Ancient Mining and Metallurgy in Southwest Spain.* Metal in History 1 (London).

Sangmeister, E. 1972. "Spectrum analysis of metal finds of the Mokrin necropolis," in *Mokrin: The Early Bronze Age Necropolis* (Beograd), vol. 2, 97–106. (In German.)

Schneiderhohn, H. 1941. *Lehrbook der Erzlagerstattenkunde, Vol. 1,* Jena. (In German.)

Stalio, B. 1964. "A new metal find from Pločnik near Prokuplja." *Zbornik Narodnog Muzeja Beograd* 4:35–41. (In Serbo-Croatian.)

Stanojević, Z. 1982. "Catalogue and classification of the archaeological finds from the Early Eneolithic mine at Rudna Glava," in *Rudna Glava: The Oldest Copper Mining in the Central Balkans,* B. Jovanović, ed. (Bor-Beograd: Institute of Archaeology), 20–59. (In Serbo-Croatian.)

Todorova, H. 1979. *The Eneolithic of Bulgaria.* Sofiya. (In Bulgarian.)

Tringham, R., D. Krstić, T. Kaiser, and B. Voytek. 1980. "The early agricultural site of Selevac, Yugoslavia." *Archaeology* 33(2):23–32.

Vasić, M. M. 1905. "Archaeological research in Serbia." *Srpski Knjizevni Glasnik* 15:520–687. (In Serbo-Croatian.)

Vasić, M. M. 1932. *The Prehistory of the Vinča,* Vol. 1. Belgrade. (In Serbo-Croatian.)

Vulpe, A. 1976. "Concerning the beginning of copper and bronze metallurgy in Rumania," in *The Introduction of Metallurgy* (Nice), 134–175. (In German.)

8
The Earliest Use of Metals in Hungary

J. Gömöri

The earliest copper objects found in Hungary are artifacts of the Middle Neolithic Szakálhát group (beginning of the third millennium B.C.) in the area of the Hungarian Great Plain [necklace with four copper beads (Hegedüs 1981, fig. 8)], and, in Transdanubia, copper awls (Makkay's excavations at Neszmély) and copper beads of the Zseliz culture (Bognár-Kutzián 1963).

The use of copper had been important from the beginning of the Late Neolithic at the Herpály tell settlement, as shown by the excavations of Kalicz and Raczky (1984).

The large-scale use of copper started in the Early and Middle Copper ages (about 2600 B.C.), in the Tiszapolgár and Bodrogkeresztur cultures (Kalicz 1985), and simultaneously gold appeared. Despite the occurrence of native copper around Rudabánya, Recsk, and Rozsnyú and in Transylvania, local copper production started only later (Bácskay 1985). The first copper and gold objects were imported from the Balkans, Near East, and Aegean cultures or were produced from collected native copper. Thus no mines have been found from this time in Hungary, only the fragments of melting pots. At the excavations of I. Ecsedy at Zók-Várhegy (Early Bronze Age, Zók-Vučedol Culture, 1900 B.C.) many crucibles, molds, and slags from copper metallurgy were found and analyzed (Ecsedy 1982).

The first traces of local copper melting are also connected to the Zók culture in Hungary, following the southward, eastward, and westward movements of the population. The first objects made of bronze were brought to the territory in the twentieth century B.C. (Kovács 1977). In the sixteenth to fifteenth century B.C. a significant bronze industry developed (Mozsolics 1967, 1973). In Lovasberény a molding workshop was excavated (Petres and Bándi 1969; Schubert 1981).

In Transdanubian areas the earliest iron object, a needle, was found in the Halstatt AB period cemetery at Sopron Krautäcker in a rich Urnfield Culture grave [Excavation by E. Jerem (1977, unpublished)]. In the present area of Hungary, as in other areas in central Europe, and in the nearby Illyrian, Italian, and Proto-Celtic settlement areas, iron objects appeared in greater quantities in the eighth century B.C. (Final Bronze Age, Halstatt B period) (Patek 1984; Pleiner and Bialeková 1982). The iron products of the southeast alpine Illyrian bloomeries can be found in significant quantities already in the seventh century B.C., and it can be supposed that the iron slags found sporadically in the important Pannonian centers, as in the hill forts of Sopron-Várhely (Burgstall; Patek 1976), Velem-Szent Vid, Szalacska-Nagybereki (Miske 1907; Darnay 1906), and Pécs-Jakabhegy (Maráz 1979), partly signal blacksmith activity from the Halstatt settlement period.

Bloomeries are, however, unknown from Transdanubia, even from the following Celtic La Tène period, despite the fact that the neighboring Burgenland in Austria (Bielenin 1977) is the site of several contemporaneous iron smelting furnaces (first century B.C.). Although iron smelting workshops were not found in eastern Hungary, the traces of iron making there are based on a metallographic investigation of the iron slag found in grave 52 of the Pre-Scythian cemetery of Mezäcsát (Patek 1984, p. 181, fig. 3). The population buried in Mezäcsát differs in their burial methods and in their objects of eastern origin from their surroundings and indicate new settlers in the Tisza area. Twenty-four iron objects were found in this cemetery, representing one-quarter of all iron objects found from the HB period between the Balkan Peninsula and Denmark (Pleiner 1981). Bracelets, pearls, knives, punches, a curb, and one piece of slag were found here. According to the investigations by Káldor and Tranta (1984), of the Department for Physical Metallurgy at the University for Heavy Industry, Miskolc, it can be stated that the slag is a piece of iron production slag with a primary crystalline phase of wüstite. The initial temperature of crystallization of wüstite in this sample was 1250°–1300°C. The crystallization of the eutectic between the dendrites took place at a lower temperature, at about 1200°C.

The earliest known iron-making furnace was found in Hungary in Sopron at the excavation of an Iron Age village (first century B.C.). The reheating furnace, with a basin diameter of 30 cm, had walls made of iron slag and fragments of tuyere bricks burnt from clay (Gömöri and Kisházi 1985). The activity of the Celtic blacksmiths is indicated not only by the iron slags of the settlements but also by blacksmithing tools and high-quality weapons and tools found in several hundred pieces in Celtic settlements and cemeteries (figure 8.1; Hunyady 1942, table LV). Anvils, pliers, and hammers obtained their final form in this age and were conserved through the following centuries.

In part of the Roman province of Pannonia, which lies presently in Hungary, no unambiguous traces of iron production were found. This province had to import iron from Noricum, Dalmatia, and Dacia, and from Celtic (Cotinus) Quad territories adjacent to the Roman Empire. Iron blooms were discovered at several places both from earlier and later Roman times (figure 8.2). The origin of the fourth- to fifth-century split iron blooms remain a mystery; they have average weights of 50 kg and are found in Late Roman Pannonian fortresses (for example, Fenékpuszta, Sopron) (Gömöri 1979, figs. 1–3; Rozsnoki 1979; Sági 1979). Long distance commerce brought prismlike wrought iron bars with weights of about 6 kg to local smithies (Heténypuszta; excavated by Tóth, Hungarian National Museum). In a stone house near the smithy of the Roman villa in Petöháza, excavated in 1985–1986, an iron bloom of a rectangular form was found. The villa of Petöháza was situated at the shore of the Ikva River on

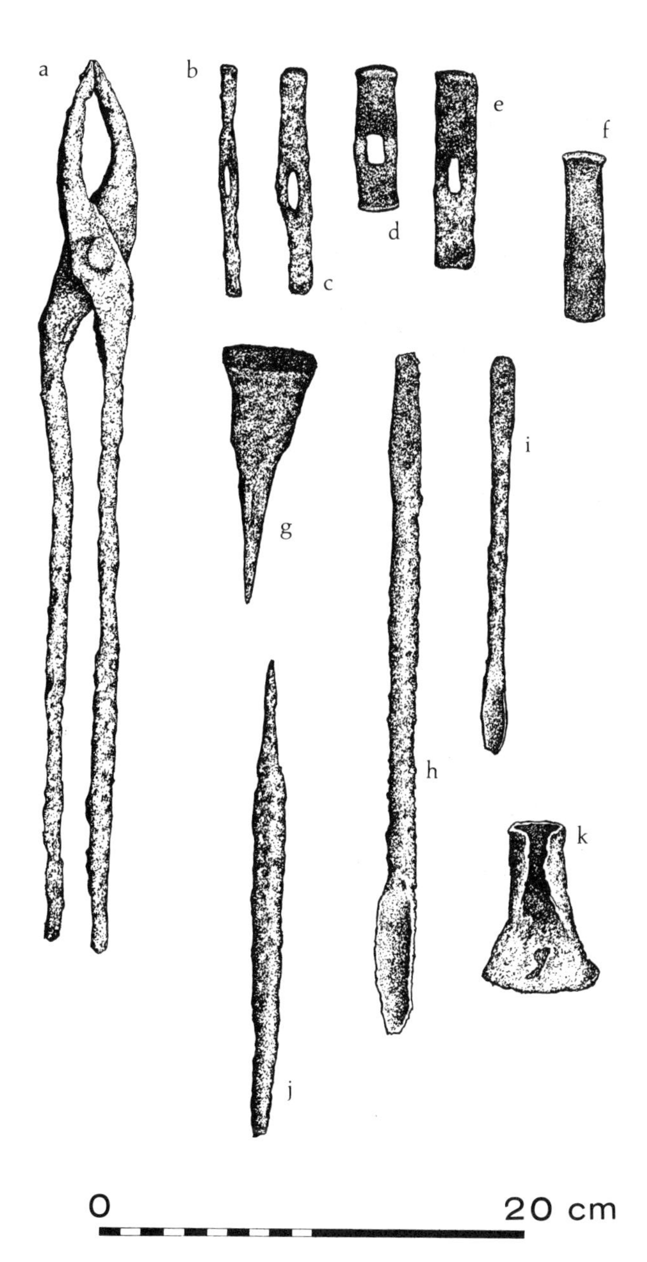

Figure 8.1 Some of the Celtic smithing and carpentry tools from Velem Szent-Vid [after Miske (1907)].

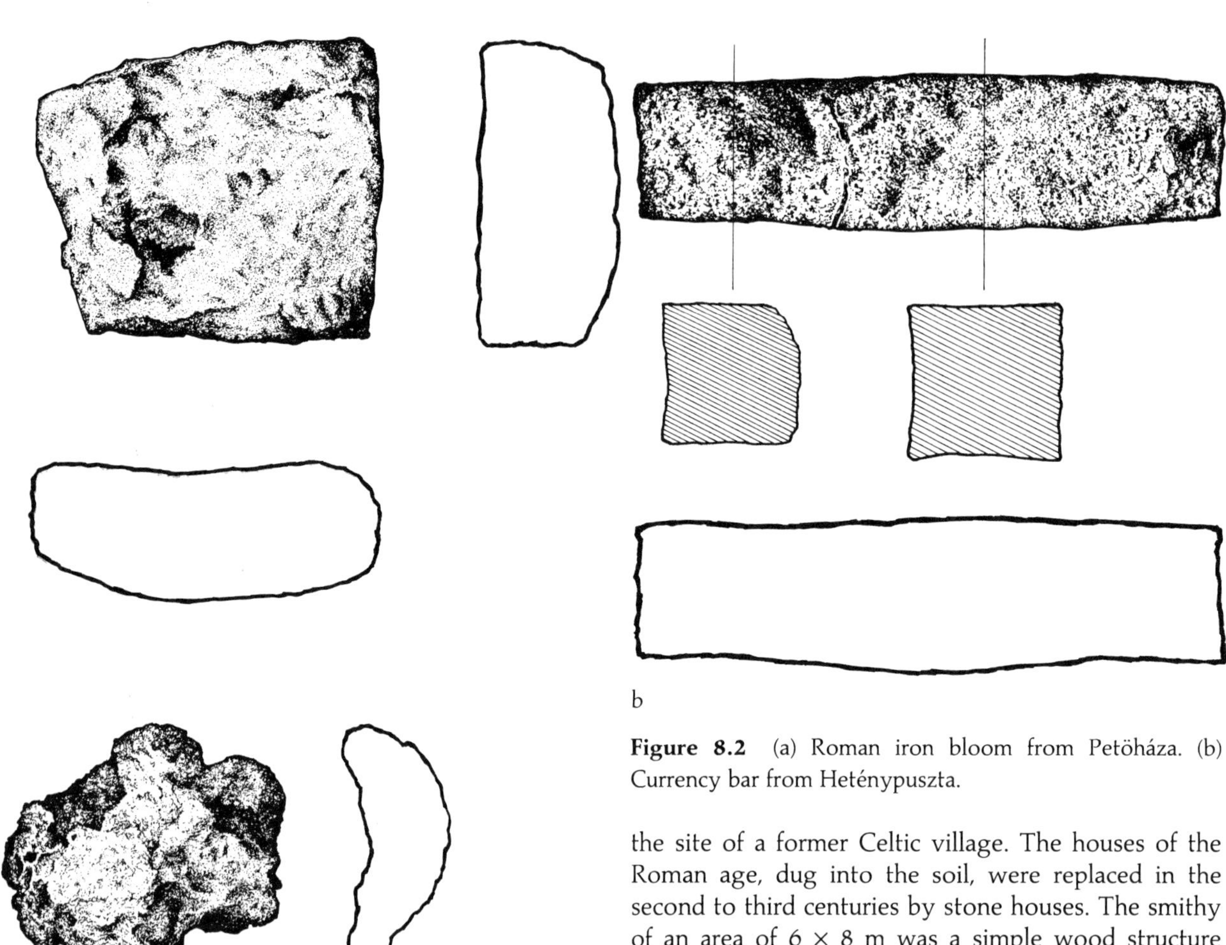

Figure 8.2 (a) Roman iron bloom from Petöháza. (b) Currency bar from Heténypuszta.

the site of a former Celtic village. The houses of the Roman age, dug into the soil, were replaced in the second to third centuries by stone houses. The smithy of an area of 6 × 8 m was a simple wood structure (figure 8.3) outside of the fence of the villa about 10 m from the last stone buildings. Only the postholes and the hearth remain from the workshop. The hearth is 20–30 cm in diameter. A sandstone slab of a height of about 25 cm high with traces of burning could be the protecting shield of the bellows. To the east of the hearth a circular pit (50 cm diameter and 40 cm deep) abuts the hearth. It was filled with slag mixed with charcoal and ash. The level of the workshop was indicated by pieces of iron slag, iron, and charcoal. A Roman coin (of Constantinus I; A.D. 306–337) was also found here. It proves that the workshop was in use in the fourth century A.D.

Heavy iron slags of diameters sometimes reaching 30 cm were scattered everywhere in the southern part of the villa. They hint at the presence of other workshops. At present it cannot be proven that the loose bog ore found near the workshop was processed there (Gömöri and Kisházi 1985, 324–327, fig. 1). The chemistry and metallography of the slags are being investigated now.

Some small-scale handicraftlike iron production cannot be excluded (figures 8.4–8.6). The built-in iron smelting furnace found near Scarbantia, Sopron, was also used for a handicraftlike iron production. It is dated by both thermoluminescence and archaeological evidence to the Roman age, to the third century A.D. (Benkö 1984a).

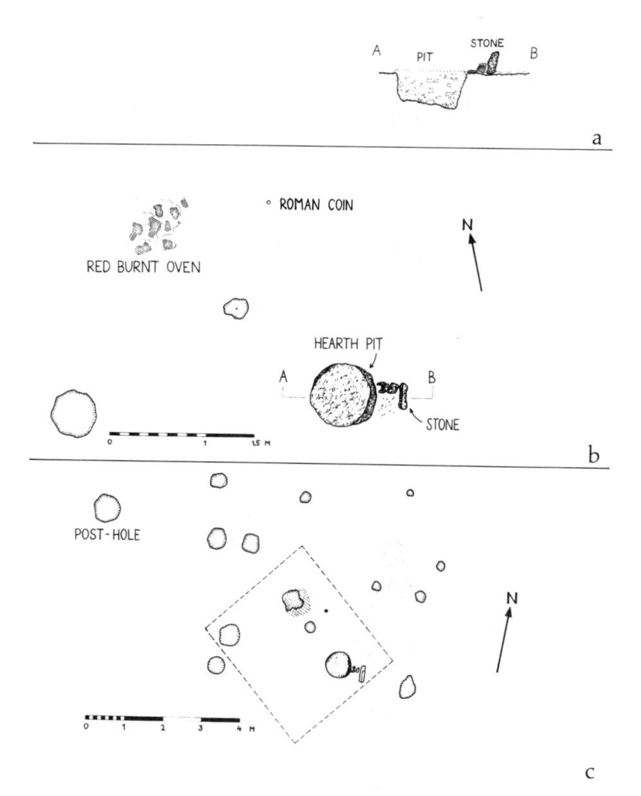

Figure 8.3 The smithy at Petőháza, fourth century A.D. (a) Section of the hearth showing stone for the smithing bellows. (b) Plan of the hearth. (c) Plan of the workshop with postholes.

In the Great Hungarian Plain, where in the Roman age Sarmatian, Jazyg, and Alan groups lived and spoke an Iranian language, archaeological traces of a well-developed pottery culture, that is, pottery kilns (Vörös 1984) and iron slags (Mágocs settlement, excavated by Vörös), were found at several settlements. In Tiszaföldvár even a bronze molding workshop was excavated (Vaday 1983). Iron slags originate from smithies, not from bloomeries.

It should be added, however, that in the southern part of the Great Plain, in Eperjes, blooms were produced in all likelihood from local bog ores (Cs. Bálint's excavations), as the site is far from the Transylvanian ore mines. The amorphous bloom with a weight of 8 kg gave the following results from metallographic and chemical investigations (investigations by Z. Remport and P. Kishazi) (smelting point, 1,300°C):

FeO = 29.30%, MnO = 2.50%, SiO_2 = 25.25%,
CaO = 11.82%, P_2O_5 = 0.731%.

Bloom iron: F (ferrite) = 95%, C = 0.05%, Si = 0.10%,
Mn = 0.57%, Cr = 0.13%, Al = 0.09%, V = 0%,
Ni = 0%, Cu = 0%, P = 0.076%.

Mineralogy: α-iron with magnetite, wüstite, quartz, larnite, bredigite (?), goethite, lepidocrocite, calcite, hercynite, mica, and plagioclase.

At the end of the migration period, in early medieval times, the local development of iron production and the smith industry received new impetus. At the time the Avar Empire united the whole Carpathian Basin under its rule. The Avars, or Zhuan-zhuans as Chinese sources call them (Laszlo 1955), came from the area of the Turkish Empire and reached from their ancient home around the Yenisei River, in the modern Touva autonomous republic, to the Carpathian Basin. Here they signed treaties with Byzantium and with the Langobards and conquered in A.D. 568 the Great Hungarian Plain, the settlement area of the defeated Gepids, later also Pannonia. It is interesting to note that the dug-in type of furnace used in their original home in Touva (Sunchugashev 1969) appeared later in the Carpathian Basin (figure 8.7). Seven iron-smelting furnaces excavated in a Pannonian Avar village belong, however, to a completely different type. The iron-smelting furnaces excavated in Tarjánpuszta-Vasasföld belong to the group of freestanding clay furnaces (Gömöri 1980).

It is interesting that the name of the site, Tarján, can be brought into connection with the Turk-Mongolian word *Tarquan*, tarcan being the name of a dignitary connected to iron working (Alföldi 1932). At the same time, Tarjan was the name of one of the tribes of the Hungarian people in the time of the conquest (tenth century). The site where the furnaces were found is called Vasas, and "Vasas" is the old Hungarian language word for miners and smelters. The Avar Empire itself was an "alloying crucible" of peoples. It is therefore difficult to say if this was the settlement of an Ugrian (ancient Hungarian) group of smelters at Tarjánpuszta. The possibility is, however, not excluded, as the runic script found in an Avar grave (Szarvas) was deciphered also in Hungarian language. Among others the inscription contains the word *vas* (iron), which reads as follows (Vékony 1985):

O1 = $\check{S}\beta = \beta^{\alpha}\check{S}\gamma$ = vas = iron.

It is likely that, in addition to Avar, Bulgaroturk, Ugrian, and Slavic parts of the population, some workshops conserving local traditions and technical processes were active in the Avar Empire, in the sixth- to eighth-century Carpathian Basin.

The changes of the supreme power in the ninth century have influenced significantly the development of iron smelting and processing in the Carpathian Basin, including the area of modern Hungary. In the northwestern part of the trisected Avaria, the Moravians established significant industrial centers, and traditions of the Avar bronze industry were revitalized. The typical Moravian-Slavic type of iron furnace, the Želehovice type (Pleiner and Bieleková 1982, p. 23, fig. 9), does not appear at the iron smelting sites connected to iron ore deposits in northern and western Hungary. No typical currency bars were found here (Pleiner and

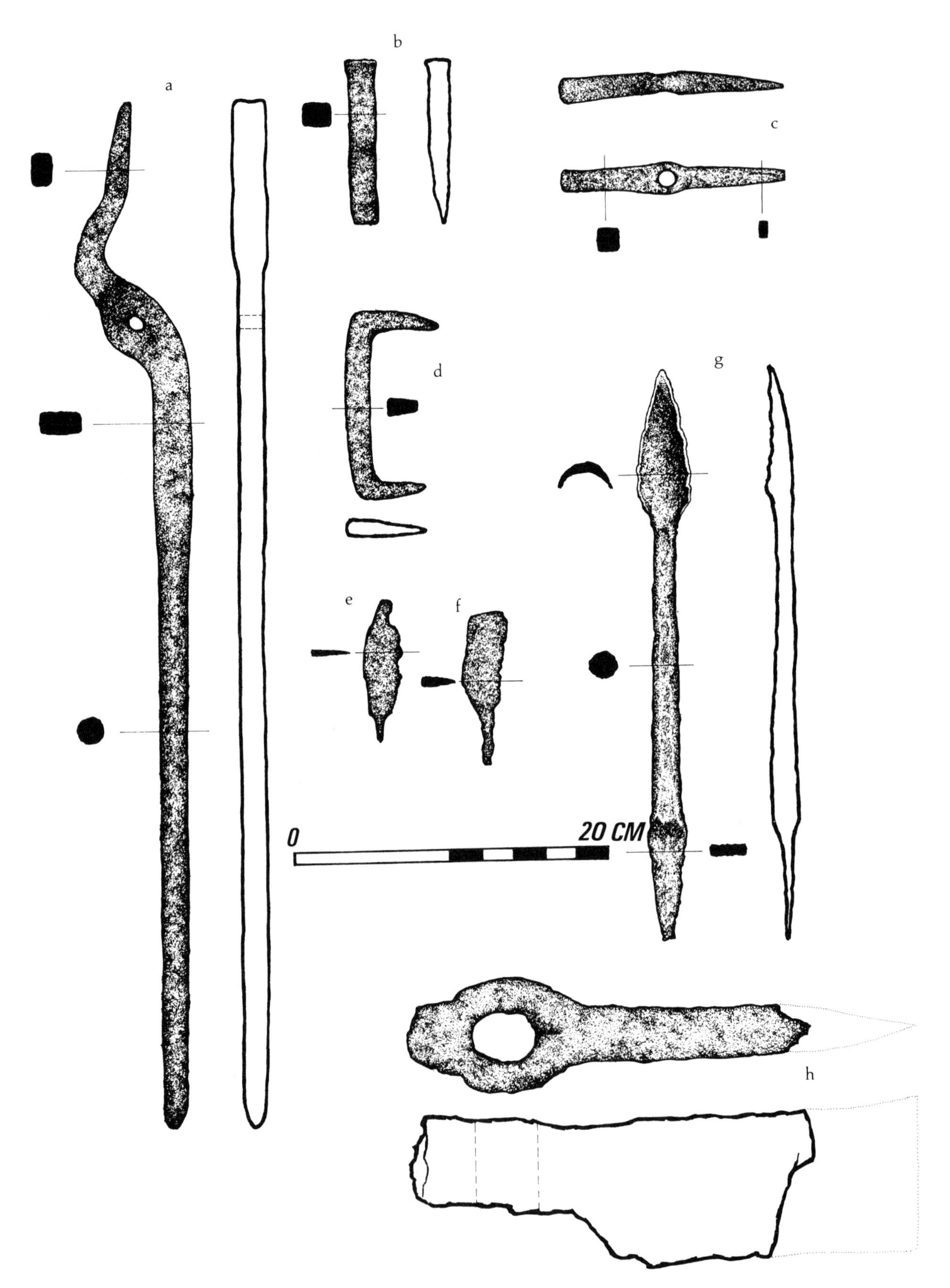

Figure 8.4 Some of the Roman smithing (a–c) and carpentry tools from the fourth-century workshop at Petöháza.

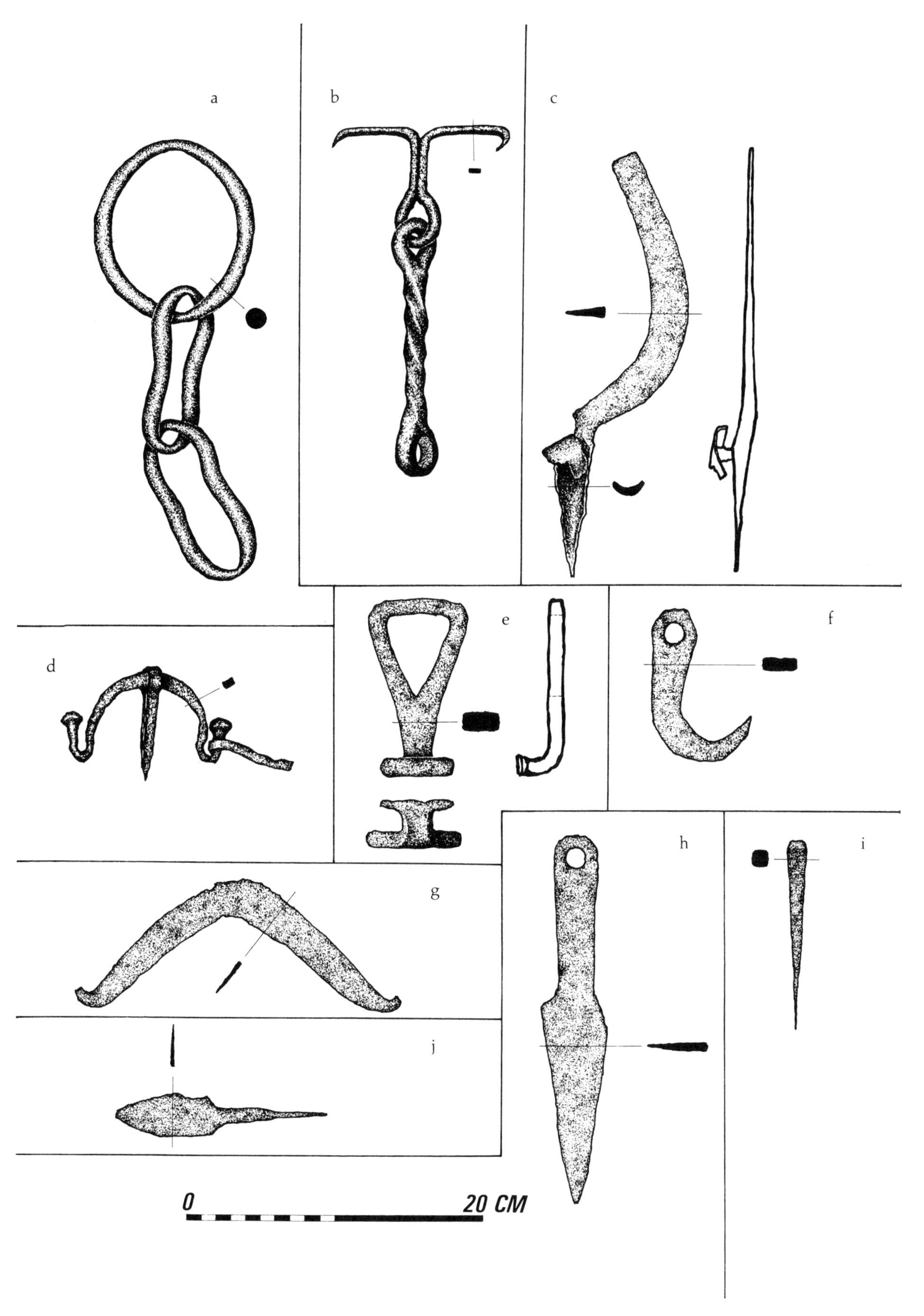

Figure 8.5 Various types of iron artifacts made in the Roman smithy at Petöháza.

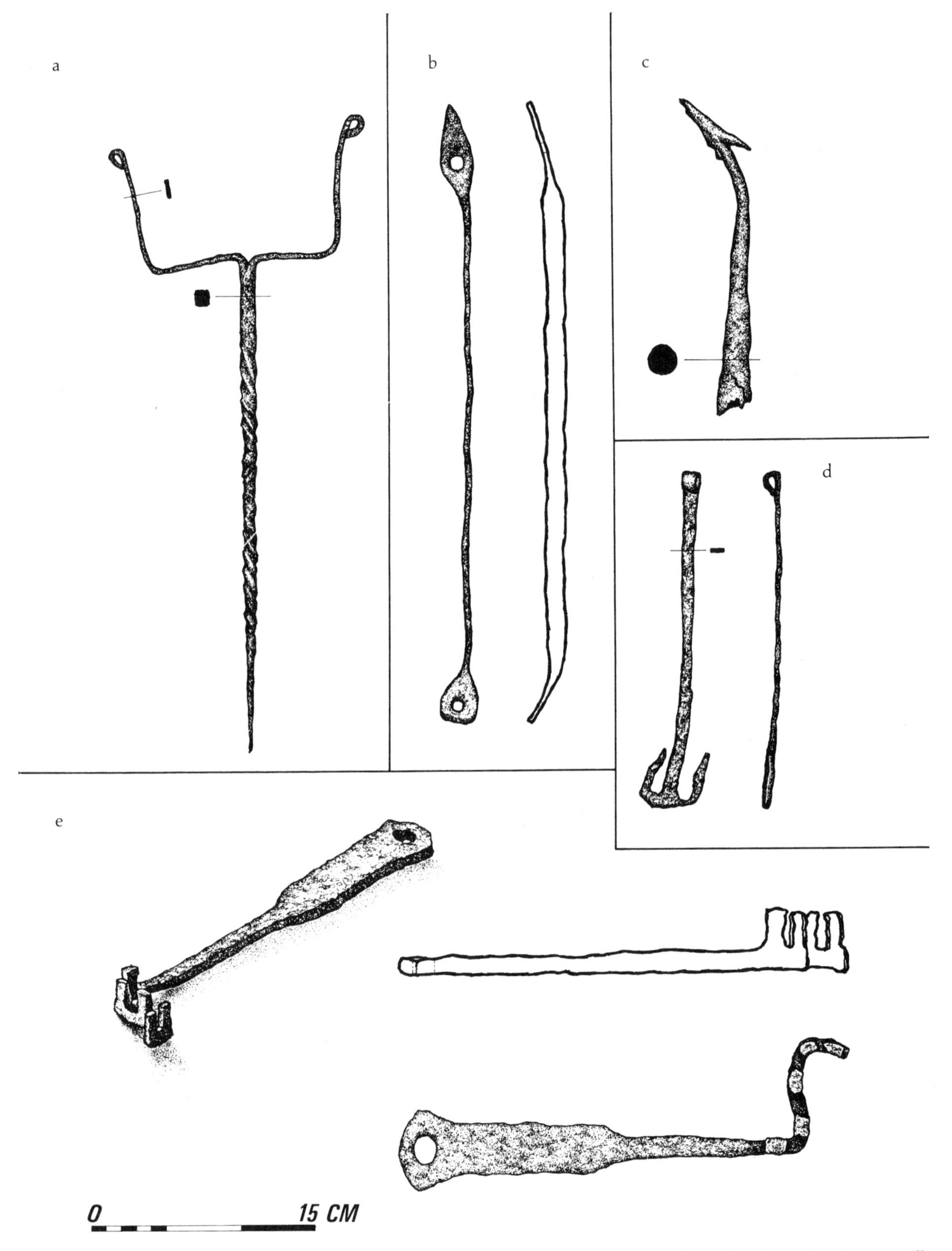

Figure 8.6 Iron artifacts found in the Roman villa at Petőháza.

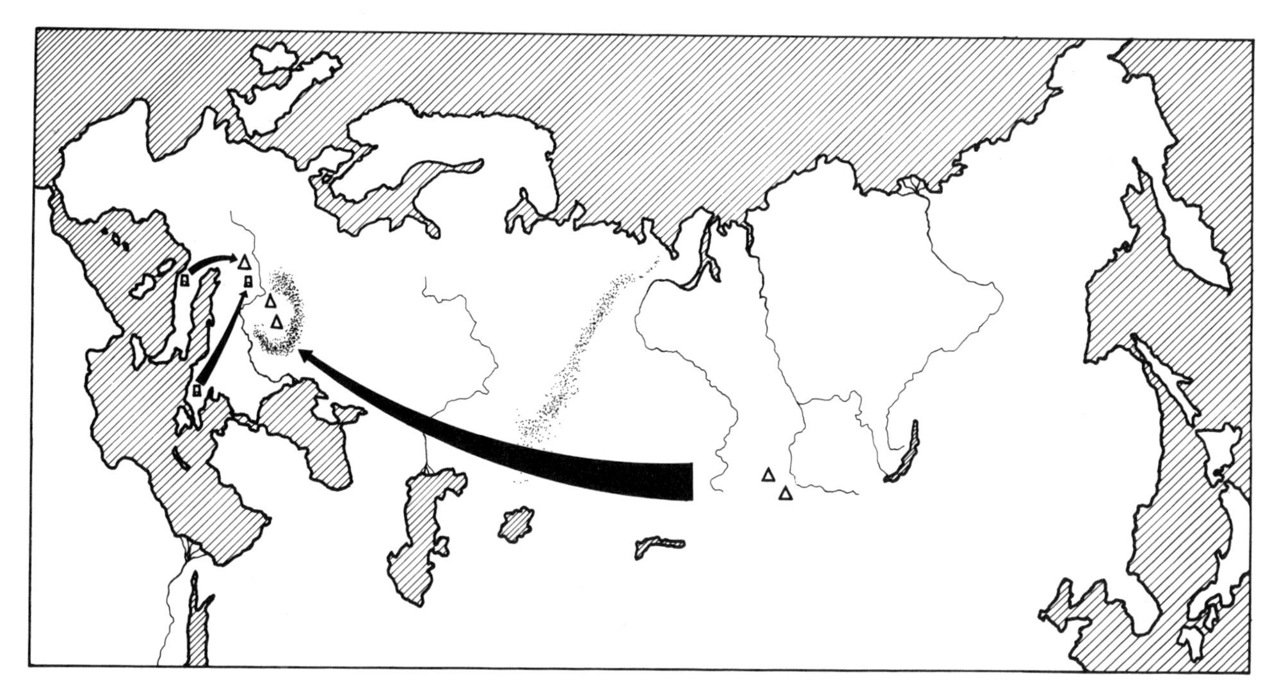

Figure 8.7 Map of the two types of early Hungarian iron smelting furnaces: Imola type (△), Nemesker type (⏢).

Bieleková 1982, p. 24, fig. 10). The plain areas close to Transylvania were under Bulgarian rule, and the already mentioned bloom from Eperjes may originate from this period. Pannonia under Frank rule gave refuge to vestiges of the Avar population, and in the settlement area of this Avar population the freestanding Nemeskér-type iron smelting furnaces were found (Gömöri and Kisházi 1985, p. 337).

A Slavic vassal of the Francs Pribina, who fled from the Moravian Nyitra, charged a fee in the marshy area west of Lake Balaton among lands of the Franco-Bavarian Church. Around Zalavár several small workshops (pit workshops) were found from this period (Valter 1981). In Zalavár a piece of smith's tong was also excavated (Sós 1963, table XC). Pribina asked, however, for his construction craftsmen from Bavaria, including smiths (Sós 1973). The workshops in Zalavár produced, among other things, spurs from a military outfit (Valter 1981, p. 127).

Interesting remains of this period are the pit workshop of great dimensions excavated in Várpalota by I. Bóna, together with the nearby reheating furnace (Valter 1981, fig. 5).

In tenth- to twelfth-century Hungary three types of iron smelting furnace can be distinguished: Nemeskér-type furnaces, Imola-type furnaces, and Vasvár-type furnaces.

The earliest type is the freestanding clay furnace occurring already in Avar times (seventh to ninth century) and in the time of Hungarian conquest (tenth century), the so-called Nemeskér-type known from modern Hungary in its western part. The characteristics are a basin of 30–40 cm diameter and a height of 70 cm (Gömöri 1980). The clay front wall includes a tuyere for double blowing. The tuyere panels, with an average diameter of 40 cm, were taken off after each smelting from the front opening of the furnace and were piled next to the furnace. They were found at the excavations, and they allowed conclusions on the diameter of the front opening to be drawn. Before the furnace a small slag outlet pit was found. The slag is thin and flows out from below the front wall of the furnace. In order to make the slag thinner, limestone was added to the furnace; lime was found at several furnaces (Iván, Tarjánpuszta).

The Nemeskér-type furnaces mostly occur at a brook on a plain area, at distances of 2–3 m from each other. The front openings of the furnaces face different directions, seemingly randomly.

Charcoal had been burnt on the spot before the construction of the furnaces, and the deforestation was created before the building of the furnaces. Charcoal-burning kilns were found in the lowest layer of the site of bloomeries (Iván, Nemeskér, Tömörd).

Dwelling houses dug into the soil were found in only one case (Tarjánpuszta) close to the furnaces and to some separate ovens for baking bread. A smithy with the appropriate tools was excavated close to the Nemeskér bloomeries (figure 8.8).

Imola-type furnaces are found everywhere in the country. They are due to the centralized organization of iron production in the tenth to eleventh century (Gömöri and Kisházi 1985, figs. 15–19). Their characteristics are a basin of an average diameter of 30 cm and a height of 70 cm. One or two furnaces side by side are dug into the wall of the workshop pit. The

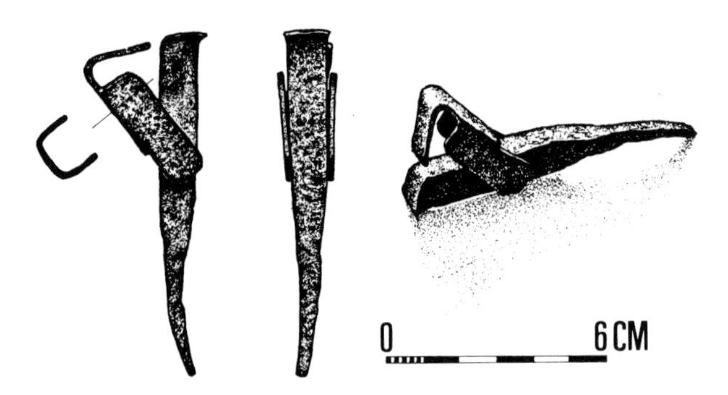

Figure 8.8 One of the smith's tools from the Nemesker smelting and smithing site (c. ninth to tenth century A.D.).

front opening, with a diameter of 15–20 cm, is coplanar with the wall of the pit; the throat opening, 10–15 cm in diameter, is at the level of the walking surface outside of the pit, and the shaft does not rise above the level of the walking surface. The basin is tilted inward, and there is no slag outlet from the furnace. In addition to spongy slag, pieces of flowing slag occur in only small samples at this type of furnace. Tuyeres with inner diameters of 2.5–3 cm are often found around the furnaces. No examples have been found that would indicate double blowing. Front panels have not been discovered. Therefore Vastagh (1972) supposed that these furnaces were operated with an open front. The inner walls of the furnaces were plastered with fireproof clay several times in many layers, thus hinting at a longer lasting operation.

Several workshops occur at a site, and they are often in a row along a brook (for example, in Trizs and in Imola) (Nováki 1969). At other sites the workshops are rather unordered, and younger ones are sometimes dug into the older ones (Szakony) (Gömöri 1983). No smithies or charcoal-burning kilns have been found in the vicinity of these bloomeries. The smiths living at this time were distributed in villages of the name *Vasverő* (Hungarian for "of smiths") and *Kovácsi* and the separate working places for smiths and smelters (the *vasas*) indicate a more developed division of labor. A dwelling house where arrowheads were also made, a rectangular pit house, was escavated near the iron smelting workshop in Sopron-Bánfalva, but even there the house is some tens of years later than the bloomery (Gömöri 1973).

Vasvár-type furnaces built into the wall of the workshop pit (eleventh to twelfth century) have so far been found only in western Hungary (Vasvár, Kőszegfalva, Olmód). It is typical to find two furnaces in a pit, with a characteristic basin diameter of 40–45 cm and a height of 50–60 cm. The furnaces are built of clay and have in the frontal part a slag outlet. Blowing was carried out from the side.

Nemeskér-type furnaces, which were also built of stone (Kányaszurdok, near Sopron), may have a provincial Roman origin; the origin of Imola-type furnaces is to be looked for in local tradition but Eastern, trans-Uralian influences are also possible (figure 8.7; Sunchugashev 1969). Vasvár-type furnaces have an alpine origin.

Fuel

Charcoal was produced for the tenth- to twelfth-century furnaces mostly in flat charcoal burning pits, 2–3 m in diameter. The charcoal studied was made mostly of oak (*Quercus robur, Quercus petrea*).

Ore

The raw material (Gömöri and Kisházi 1985) was obtained in eastern Hungary from the great siderite occurrence at Rudabánya and from smaller concretion-type ore deposits. The ore contained about 50–80% Fe_2O_3, with a rather high SiO_2 content.

In western Hungary the first discovered ancient ore field was surveyed in 1985. In a forest near the village of Kópháza some 250 conic dips were found with diameters of 3–6 m. The deposit is covered by Pleistocene terrace gravel; below it the fine muddy sand contains limonite concretions in a secondary position. The composition of the limonite is (from X-ray and derivatographic determinations by J. Ivancsics):

goethite (+limonite) 49 wt. %
quartz 38 wt. %
feldspars 4 wt. %
micas 4 wt. %
montmorillonite 5 wt. %

At an iron smelting furnace excavated about 200 m away from this mine are limonite concretions of the same type—brown, very fine grains, and a solid structure; Early Pannonian age, as in the mining pits. The ores found at the bloomery furnace were, however, according to investigations by Kisházi, partly roasted. The roasted ore differs from the fresh ore by its reddish color and its mineral composition. The original goethite amorphous limonite is transformed into hematite, after having been heated to about 350°C. The chlorite in the ore is transformed to vermiculite. Some samples contained some goethite and limonite, which hints at insufficient roasting.

The iron slags found at the iron smelting furnace of Kópháza contained the minerals fayalite ± pleonaste ± leucite ± wüstite ± magnetite, all originating from the smelting of the hardly melting part of the ore; quartz, plagioclase and micas are also found in the slag. The mineral composition shows that the furnace in Kópháza was used with a method that resulted in high iron loss at a low temperature for iron production.

The excavation was carried out after a geophysical survey using a proton magnetometer. The measurements made by J. Verő have shown a positive anomaly of about 10 nT at the site of the bloomery. The completely ruined clay furnace was dug into the soil; fragments of clay tuyeres prove artificial blowing. The inner diameter of the furnace was 30–40 cm, typical for the area.

Carbon 14 ages of charcoal samples from the iron ore mine and iron smelting furnace were determined by É. Csongor (Institute of Nuclear Research, Hungarian Academy of Sciences, Debrecen). The measurements give an age of A.D. 1190 $\pm$ 120 for the mine (Deb-490) and A.D. 990 $\pm$ 110 for the workshop pit of the iron smelting furnace (Deb-491). These dates are in agreement with the archaeological ages.

Similar iron ore mines were found in great number in the nearby territory of Austria (in Burgenland around Oberpullendorf) in the basins of the alpine boundary hills (Schmid 1977). The nearly continuous mining area was operated mostly by the Celtic population in the La Tène Period D (first century B.C.). A great part of the pits, however, is of an early medieval age. At the northwestern edge of the Alps, near Augsburg, mining pits were also excavated where a similar technique was used (Frei 1965–1966). The gist of this method is that barren rock from each pit is refilled into a neighboring earlier pit. Thus mining pits form rows or groups near each other. This method is economical from the viewpoint of treatment of the barren rock (Kaus 1981).

Physical Exploration of Bloomeries

Geophysical exploration measurements were carried out by J. Verő (Geodetic and Geophysical Research Institute, Hungarian Academy of Sciences, Sopron) in cooperation with the Ferenc Liszt Museum, which was responsible for the excavations. Geomagnetic measurements proved to be more effective than geoelectrics in detecting iron-making furnaces. Measurements were carried out in the last three years at sites of all three mentioned types of furnaces. Nemeskér-type furnaces were found, for example, in Dénesfa (Gömöri 1985; Verő 1985), Vasvár-type furnaces in Olmód, and Imola-type furnaces in Répcevis. In all cases the exact site of the bloomery was detected by the measurements. It is to be noted that the geomagnetic anomalies are not caused by iron slag; positive anomalies correlate with the burnt clay parts of the furnaces, mostly over autochthonous red or gray burnt clay. The working pits of the workshops can also be detected by this method, as the excavations in Répcevis (1985) and in Szakony (1982) have shown (Gömöri and Wallner 1984). Here the magnetic measurements enabled a direct excavation of the iron smelting workshops and of the ore roasting pit (figure 8.9), even though the only surface indication was a few pieces of scattered slag. Slag pieces were found, however, rather far away from the workshops because of transport by plowing.

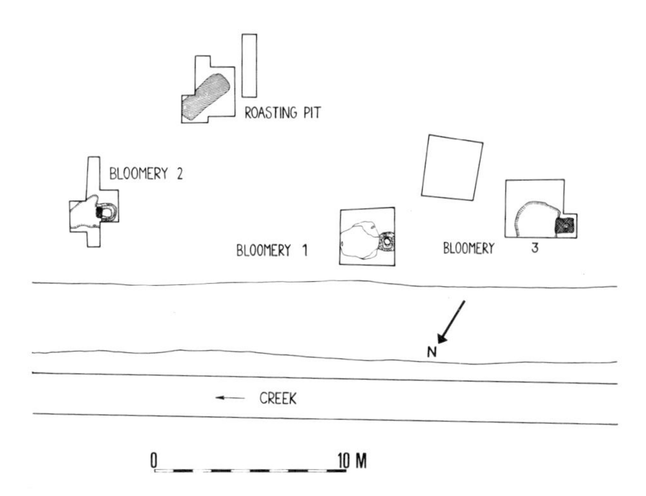

Figure 8.9 The Répcevis site where three Imola-type bloomery workshops and an ore roasting oven were excavated after high positive magnetic anomalies were found. Roasting pit, 19 $\pm$ 12 nT: bloomeries 30 $\pm$ 21 nT. Excavated by Gömöri; measurements by Verö (1985).

The present aim of the investigations is to explore the majority of the iron-making workshops in Hungary in order to determine the type, the applied technology, and the age of every site.

Physical Age Determination of Bloomeries

We consider the determination of the bloomery ages to be of the highest possible precision, as all three types of furnace originated in early medieval times. The archaeological age determination of the simple pottery found with the furnaces is difficult, as eighth-, ninth-, and tenth-century pottery is similar. Nevertheless, an eighth-century furnace could have been operated under Avar rule, that from the ninth century under Bavarian-Frank rule, and a tenth-century one under Hungarian rule, if the area of, for example, Transdanubia is considered. This means that the iron makers were from different ethnic groups and may indicate a different origin of the different furnace types and of the applied technologies.

At the same time, the continuity of the iron-maker groups, their survival, and the transfer of their technological traditions from one period to another is a realistic fact independent of any change in the supreme power.

That means that the tracing of the origins of the different iron-making technologies in early medieval Hungary is a difficult task. And it means also that we need as many sites of iron production as possible.

The solution of these problems should be promoted by the age determination of finds from different bloomery sites with different methods. The thermoluminescent age of the furnaces in Tarjánpuszta and Kànyaszurdok are, according to L. Benkő (Isotope Institute, Hungarian Academy of Sciences), A.D. 680–750 and A.D. 730–790, respectively (Benkő 1984b).

Archaeomagnetic measurements give an age for Tarjánpuszta around A.D. 650 (Tóth 1978–1979).

Successful archaeomagnetic age determinations were also carried out by P. Márton (Geophysical Institute, Loránd Eötvös University), who dated the Nemeskér-type Dénesfa furnace to the seventh century A.D. (Márton 1986). The Imola-type furnace of Szakony gave an age in the first half of the eleventh century A.D. (Márton 1984). These results are supported by the presence of freestanding furnaces in the Carpathian Basin in the Avar period. The Imola-type dug-in furnaces without front walls appear, however, only in the tenth to eleventh century in the area of modern Hungary, not considering the Roman dug-in furnace in Sopron.

Acknowledgments

I wish to thank all colleagues who took part in the exploration of the furnaces and who enabled me to use unpublished results and manuscripts in this review, namely, J. Iváncsics, E. Jerem, P. Kisházi, É. Csongor, J. Verő, Z. Remport, and, last but not least, Robert Maddin for reading and correcting the English manuscript.

References

Alföldi, A. 1932. "The origin of the dignity name 'tarchan'." *Magyar Nyelv* 28:205–220. (In Hungarian.)

Bácskay, E. 1985. "Prehistoric mining and utilization of some mineral raw materials in the Carpathian Basin and in the adjacent areas. Neogene mineral resources in the Carpathian Basin," in *Proceedings of the Eighth RCMNS Congress* (Budapest: Regional Committee on Mediterranean Neogene Stratigraphy), 559–565.

Benkő, L. 1984a. "TL dating of an iron smelting furnace found in the Roman cemetery of Scarbantia," in *Iparrégészet* (Industrial Archaeology), J. Gömöri, ed. (Veszprém: Hungarian Academy of Sciences), vol. 2, 140.

Benkő, L. 1984b. "TL dating of potteries, furnaces, and kilns," in *Iparrégészet* (Industrial Archaeology), J. Gömöri, ed. (Veszprém: Hungarian Academy of Sciences), 271.

Bielenin, K. 1977. "Einige Bemerkungen über das altertümliche Eisenhüttenwesen im Burgenland." *Wissenschaftliche Arbeiten aus dem Burgenland* 59:49–62.

Bognár- Kutzián, I. 1963. *The Copper Age Cemetery of Tiszapolgar, Basatanya*. Budapest, 333.

Darnay, K. 1906. "Celtic coinage and molding workshop at Szalacska, County Somogy." *Archaeológiai Értesitö* 26: 416–433. (In Hungarian.)

Ecsedy, I. 1982. "Excavations at Zók-Várhegy, 1977–1982. Preliminary report." *A Janos Pannonius Múzeum Évkönyve* 27:60–105.

Frei, H. 1965–1966. "Early iron mining in the northern Alps." *Jahresbericht der Bayerischen Bodendenkmalpflege* 6–7: 107–108. (In German.)

Gömöri, J. 1973. "Tenth century iron smelting workshop at Sopron." *Arrabona* 15:69–123. (In Hungarian.)

Gömöri, J. 1979. "Report on the archaeological research of the West-Hungarian iron ore field, II. The excavations at Kányaszurdok and the problem of the split iron blooms in Sopron." *Arrabona* 21:59–64. (In Hungarian.)

Gömöri, J. 1980. "Early medieval bloomeries from Tarjánpuszta and Nemeskér." *Acta Archaeologica Academiae Scientiarum Hungaricae* 32:317–343. (In German.)

Gömöri, J. 1983. "The bloomery workshops at Szakony." *Bányászati és Kohászati Lapok* 116:97–103. (In Hungarian.)

Gömöri, J. 1985. "The excavation of the Dénesfa iron smelting furnace." *Bányászati és Kohászati Lapok* 117:537–539. (In Hungarian.)

Gömöri, J., and P. Kisházi. 1985. "Iron ore utilization in the Carpathian Basin up to the Middle Ages with special regard to bloomeries in western Transdanubia," in *Neogene Mineral Resources in the Carpathian Basin*, J. Hála, ed. (Budapest: Hungarian Geological Society), 323–355.

Gömöri, J., and Á. Wallner. 1984. "Geophysical measurement at the excavation of the Árpád Period iron smelting furnaces in Szakony," in *Iparrégészet* (Industrial Archaeology), J. Gömöri, ed. (Veszprém: Hungarian Academy of Sciences), 227–242.

Hegedús, K. 1981. "Excavations at the Neolithic settlement of Csanytelek-Ujhalastó." *Archaeológiai Értesitö* 108:3–12. (In Hungarian.)

Hunyady, I. v. 1942. *Celts in the Carpathian Basin*. Budapest: Institute for Numismatics and Archaeology of the Hungarian Royal Pázmány Pèter University. (In Hungarian.)

Jerem, E. 1978–1979. "Sopron: Bécsi street, Krautacker site." *Mitteilungen des Archaologischen Instituts der Ungarischen Akademie der Wissenschaften* 8–9:218. (In German.)

Káldor, M., and F. Tranta. 1984. "Metallurgical study of iron objects found in Mezőcsát, Pre-Scythian cemetery, 8th century B.C.," in *Iparrégészet* (Industrial Archaeology), J. Gömöri, ed. (Veszprém: Hungarian Academy of Sciences), vol. 2, 192–193.

Kalicz, N. 1985. "On the chronological problems of the Neolithic and Copper Age in Hungary." *Mitteilungen des Archaologische Instituts der Ungarischen Akademie der Wissenschaften* 14:21–51.

Kalicz, N., and P. Raczky. 1984. "Preliminary report on the 1977–1982 excavations at the Neolithic and Bronze Age tell settlement of Berettyóújfalu, Herpály I. Neolithic." *Acta Archaeologica Academiae Scientiarum Hungaricae* 36:85–136.

Kaus, K. 1981. "Mining sites and production centers of Ferrum Noricum," in *Eisengewinnung und Verarbeitung in der Frühzeit* (Iron smelting and ironworking in early times), G. Sperl, ed. (Vienna: Montan-Verlag), 74–92. (In German.)

Kovács, T. 1977. *The Bronze Age in Hungary*. Budapest: Corvina. (In Hungarian.)

László, Gy. 1955. *Archaeological Studies on the History of the Avars Society*. Budapest. (In French.)

Maráz, B. 1979. "Preliminary report of the excavations in 1976–77 at Pécs-Jakabhegy." *Archaeologiai Ertesitö* 106:78–93. (In Hungarian.)

Márton, P. 1984. "The magnetic age of the Szakony medieval bloomery furnaces," in *Iparrégészet* (Industrial Archaeology), J. Gömöri, ed. (Veszprém: Hungarian Academy of Sciences), 243–248.

Márton, P. 1986. "The archaeomagnetic age of the Dénesfa furnace." *Bányászati és Kohászati Lapok* 118. (In Hungarian.)

Miske, K. 1907. *The Prehistoric Settlement at Velem-Szentvid*. Vienna: Konegen Károly. (In Hungarian.)

Mozsolics, A. 1967. *Bronze Finds from the Carpathian Basin*. Budapest: Akadémiai Kiadó. (In German.)

Mozsolics, A. 1973. *Bronze and Gold Finds from the Carpathian Basin*. Budapest: Akadémiai Kiadó. (In German.)

Nováki, Gy. 1966. "Remains of iron smelting in Western Hungary." *Wissenschaftliche Arbeiten aus dem Burgenland* 35: 136–198. (In German.)

Nováki, Gy. 1969. "Archaeological remains of iron metallurgy in northern Hungary from the 10th to 12th centuries." *Acta Archaeologica Scientiarum Hungaricae* 21(17):299–331. (In German.)

Patek, E. 1976. "The group of the Hallstatt culture near Sopron." *Archaeologiai Értesitö* 103:3–28. (In Hungarian.)

Patek, E. 1984. "The earliest iron articles in Hungary and their archaeological entourage," in *Iparrégészet* (Industrial Archaeology), J. Gömöri, ed. (Veszprém: Hungarian Academy of Sciences), vol. 2, 185–186.

Petres, É. F., and G. Bándi. 1969. "Excavations at Lovasberény-Mihályvár." *Archaeologiai Értesitö* 96:170–177. (In Hungarian with English summary.)

Pleiner, R. 1981. *The Path of Iron in Europe. The Earliest Iron in Europe*. Schaffhausen: Verlag Peter Meili, 115. (In German.)

Pleiner, R., and D. Bialeková. 1982. "The beginnings of metallurgy in the territory of Czechoslovakia." *Bulletin of the Metals Museum* (Japan) 7:19–20.

Rozsnoki, Zs. 1979. "Metallurgical study of iron blooms from the west of Hungary." *Arrabona* 21:87–106. (In Hungarian.)

Sági, K. 1979. "The historical background of the split iron blooms from Fenékpuszta, 5th century A.D." *Arrabona* 21: 113–115. (In Hungarian.)

Schmid, H. 1977. "Mining and geological conditions of prehistoric iron metallurgy in central Burgenland in the Basin of Oberpullendorf." *Wissenschaftliche Arbeiten aus dem Burgenland* 59:11–23. (In German.)

Schubert, E. 1981. "Early Bronze Age cultural connections in the Danube territory on the basis of metal analysis," in *The Early Bronze Age in the Carpathian Basin and Its Surroundings* (Budapest: Archaeological Institute of the Hungarian Academy of Sciences), 190–197. (In German.)

Sós, Á. 1963. *The Excavations of Géza Fehérs at Zalavár*. Budapest: Akadémiai Kiadó. (In German.)

Sós, Á. Cs. 1973. *Die slawische Bevölkerung Westungars im 9. Jahrundert*. Munich.

Sunchugashev. Ya. I. 1969. *Mining and Smelting of Metals in Ancient Touva*. Moscow: Akademia Nauk SSSR, 103–137. (In Russian.)

Tóth, J. 1978–1979. "Archaeomagnetic dating of the iron smelting furnace in Tarjánpuszta." *Arrabona* 19–20:163–167.

Vaday, A. 1983. "Tiszaföldvar: Brick works." *Régészeti Füzetek* 36:58.

Valter, I. 1981. "Árpád Period forge at Csatár," in *Iparrégészet* (Industrial Archaeology), J. Gömöri, ed. (Veszprém: Hungarian Academy of Sciences), vol. 1, 123–131.

Vastagh, G. 1972. "Metallurgical conclusions from iron finds from the 11th to 12th centuries." *Acta Archaeologica Academiae Scientiarum Hungaricae* 24:241–260. (In German.)

Vékony, G. 1985. "Runic scripts on objects of the Late Migration Period II." *Életünk* 27(2):154–155. (In Hungarian.)

Verö, J. 1985. "Geophysical measurements on the site of the Dénesfa iron smelting furnaces." *Bányászati és Kohászati Lapok* 117:540–541. (In Hungarian.)

Vörös, G. 1984. "Late Sarmatic pottery kiln in Sándorfalva, Eperjes," in *Iparrégészet*, (Industrial Archaeology), J. Gömöri, ed. (Veszprém: Hungarian Academy of Sciences), vol. 2, 147–154.

9
Early Metallurgy in Sardinia

Fulvia LoSchiavo

Early metallurgy in Sardinia can be divided into two periods, pre-Nuragic and Nuragic, the first running from the beginning of the third millennium to the middle of the second millennium B.C. and the second from that time to the Early Iron Age, around the fifth century B.C. (figure 9.1).

With the Nuragic civilization both cultural development and metallurgical technology reach their highest point. The technology occurs in consequence of the exploitation of local ores and comes to a level of evolution unparalleled in the western Mediterranean.

For this reason I devote most of my attention to Nuragic metallurgy, giving only a summary of the problems of the earlier periods. We need to remember that during the last ten years the topic has been slowly built up through careful analyses of archaeological data and extraordinary new discoveries; in the meantime the entire context of Sardinian archaeology has become much more definite, changing the common view from that of a backward and isolated land to a stepping stone in Mediterranean prehistory.

Pre-Nuragic Metallurgy

The Beginning of Metallurgy

The beginning of metallurgy took place in the framework of the "intercultural" connection created by the obsidian trade, a trade that probably began when the Early Neolithic was well settled (figure 9.2). Obsidian microliths of geometric shape are associated with cardial pottery not only in Sardinia but also in Corsica, which received obsidian and flint from Sardinia as early as the sixth millennium B.C. at Basi and Curacchiaghiu.

The levels referred to as the Middle Neolithic Bonu Ighinu Culture are rich in obsidian artifacts, and the production reaches its peak with the Late Neolithic Ozieri Culture, known to be the most important and characteristic of Sardinian prehistory. During the Ozieri Culture and following the Filigosa-Abealzu Culture of the Early Eneolithic, that is, at the end of the fourth and the beginning of the third millennia B.C., and at the same time in mainland Italy and in the Aeolian islands, the use of metals and the production of metal objects in Sardinia began. Copper and silver slags and artifacts are reported to have been found in the Ozieri level of Monte Majore Cave (Thiesi, Sassari) and in the huts of Su Coddu (Selargius, Cagliari); other artifacts include fragments of a flat copper dagger and of some pins with square cross section from the huts of Cuccuru Arrius (Cabras, Oristano), two silver rings from a tomb at Pranu Mutteddu (Goni, Cagliari), a spiral silver ring found in a rock-cut tomb at Filigosa (Macomer, Nuoro), a cylindrical silver spiral ring and a few simple rings in a dolmen at Corte Noa (Laconi, Nuoro), and four copper daggers, an awl, and spiral copper and silver rings in

DATE	PERIOD	CULTURES AND CIVILISATION OF SARDINIA
350000 120000	LOWER PALEOLITHIC	CLACTONIANO RIO ALTANA PERFUGAS
35000	MIDDLE PALEOLITHIC	
10000	UPPER PALEOLITHIC	GROTTA CORBEDDU OLIENA
6000	MESOLITHIC	
4000	EARLY NEOLITHIC	SU CARROPPU GROTTA VERDE
3500	MIDDLE NEOLITHIC	BONU IGHINU
2700	LATE NEOLITHIC	OZIERI
2500	EARLY ENEOLITHIC	FILIGOSA-ABEALZU
2000	MIDDLE ENEOLITHIC	MONTECLARO
1800	LATE ENEOLITHIC	BEAKER
1600	EARLY BRONZE AGE	BONNANARO
1300	MIDDLE BRONZE AGE	NURAGIC CIVILISATION
900	LATE BRONZE AGE	
750 535 238	IRON AGE	PHOENICIAN AND PUNIC CIVILISATION
b.C. A.D. 476	ROMAN AGE	ROMAN CIVILISATION

Figure 9.1 Chronology of ancient Sardinia (1986).

an "oven"-type rock-cut tomb at Serra Cannigas (Villagreca, Cagliari). Their association with Ozieri (Selargius, Cabras, Goni) and with the Filigosa-Abealzu (Macomer, Laconi, Villagreca) archaeological contexts are certain. The same can be said for a few metal objects and two small crucibles found in the excavation of the huge megalithic altar at Monte d'Accoddi (Sassari) (figure 9.3a). Those crucibles are identical with the ones discovered at Terrina IV (Aleria, Corsica) associated with slags, tuyeres, and an awl made of copper bearing traces of iron and magnesium oxides. The shapes and decoration of the pottery and the radiocarbon dating indicate a Chalcolithic culture about the end of the first half of the third millennium, both consistent with the Sardinian Filigosa-Abealzu Culture.

The copper and silver rings, bracelets, and copper daggers as well as the tuyeres found in the famous necropolis of the rock-cut tombs at Anghelu Ruju (Alghero, Sassari), on the other hand, were all found in collective burials from the Ozieri to the Bonnanaro Culture of the Early Bronze Age and cannot be related with certainty to the former period.

Monte Claro Metallurgy

The use of lead for pottery repairs is first documented in a beautiful bowl showing the most peculiar southern Monte Claro decoration; it was found in the Cuccuru Tiria cave at Iglesias (Cagliari), in the middle of the mining district of Sulcis-Iglesiente. Some awls are said to have come from that cave, but they are of widespread occurrence in Monte Claro as well as in Beaker and Early Bronze Age contexts.

An interesting type of tanged dagger appears during the Monte Claro phase, another most interesting and rich pre-Nuragic culture distinguished by a variety of regional aspects. These daggers were found in the tombs of Cresia Is Cuccurus (Monastir, Cagliari) of Via Basilicata (Cagliari) and of Serra Is Araus (S. Vero Milis, Oristano).

Another blade with a spatula shape was found in the huge fortification of Monte Baranta (Olmedo, Sassari) (figure 9.3b) and an unusual type of awl in the village of Biriai (Oliena, Nuoro) (figure 9.3c). As a whole and compared to the wide distribution in the Monte Claro Culture, the number of metal objects that can be related to the culture is not high, but the use of lead and the presence of a few characteristic shapes suggest local production, although no chemical or metallographic analyses have been made.

Statue-Menhirs Daggers

Elaborate double daggers (figure 9.4) are carved on the statue-menhirs of Laconi and Nurallao in the Sarcidano region of the province of Nuoro only a few miles from the copper mines at Funtana Raminosa. These are worth mentioning even though the chronology and

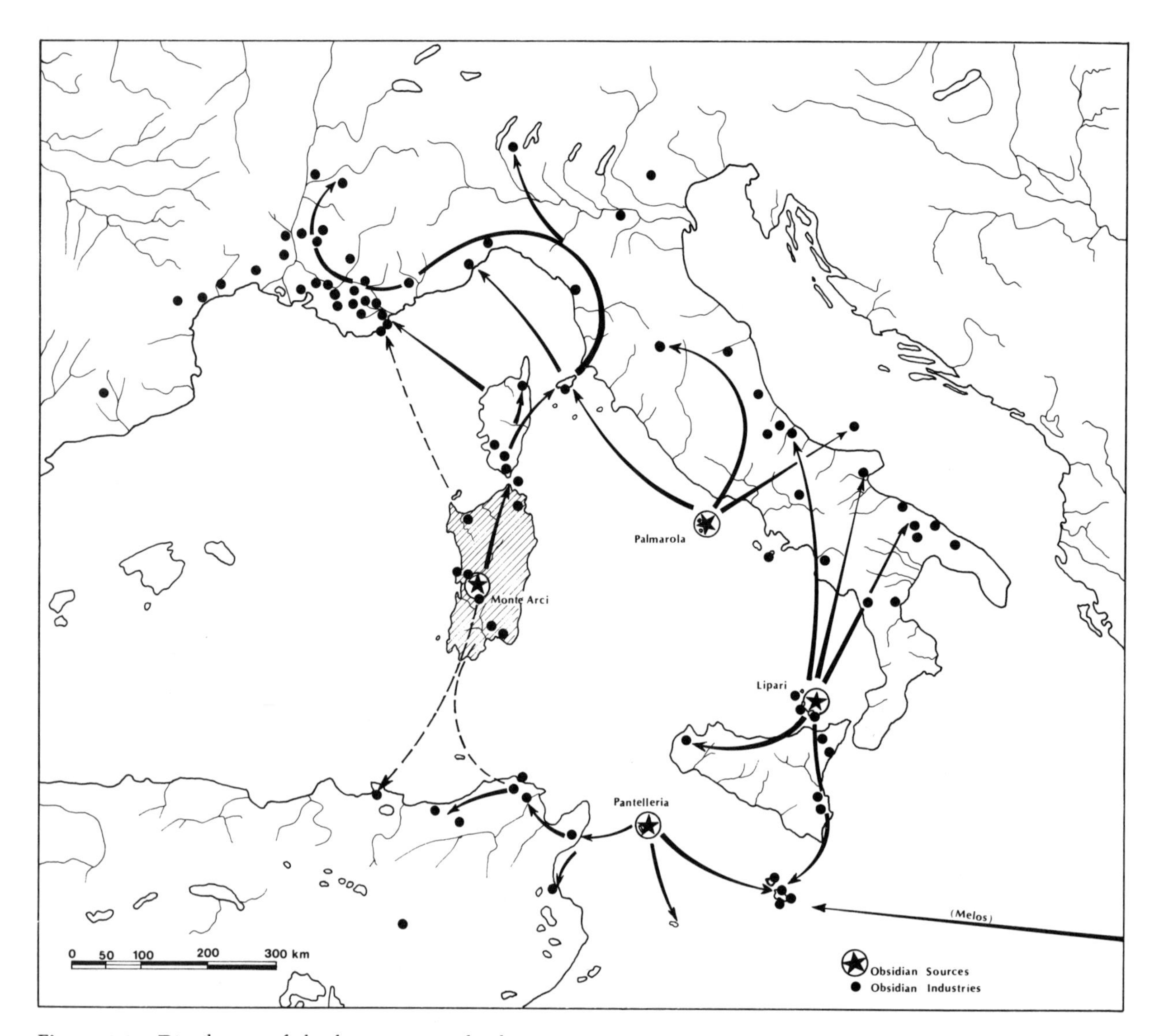

Figure 9.2 Distribution of obsidian sources and industries.

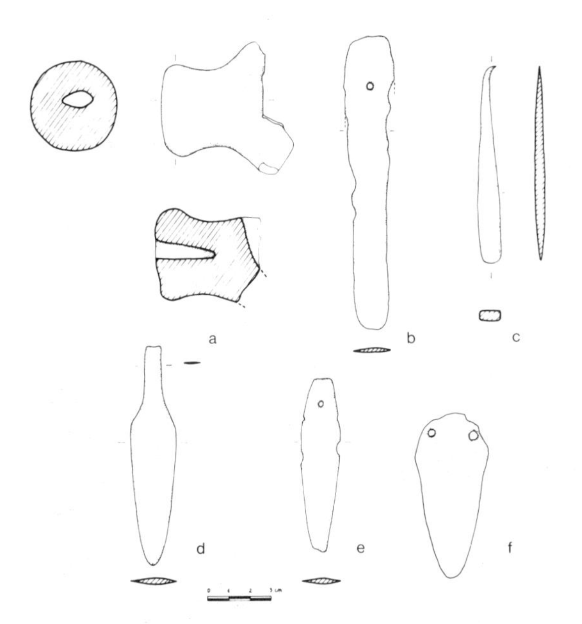

Figure 9.3 Pre-Nuragic metallurgy. (a) Fragment of a crucible from Monte d'Accoddi (Sassari); (b) blade from Monte Baranta (Olmedo, Sassari); (c) awl from Biriai (Oliena, Nuoro): (d and e) two small daggers from Frommosa Cave (Villanovatulo, Nuoro); (f) small dagger from Sa Turricula (Muros, Sassari).

Figure 9.4 Statue-menhirs with double daggers carved on the stone, from Laconi (Nuoro).

the cultural context of these monuments are still in discussion.

More than fifty statue-menhirs carved with eyes, brows, noses, an anthropomorphic figure on the breast, and a double dagger at the waist have been found, none of them in association with pottery or other cultural elements. But aniconic menhirs (standing stones without any carving) have been erected in the same region in close connection with chambered tombs with circular mounds and grave goods of pure Filigosa-Abealzu style. The shape of the dagger has been related to copper and flint Remedello daggers, but they have not been found in Sardinia [the only exception is a flint dagger from a tomb at Prano Mutteddu (Goni)]. For the time being the Laconi statue-menhirs are tentatively dated to about the second half of the third millennium B.C.

Beaker and Bonnanaro Metallurgy

Many small triangular daggers from Anghelu Ruju, Su Crucifissu Mannu (Porto Torres, Sassari), Serra Is Araus (S. Vero Milis, Oristano), S. Bartolomeo Cave (Cagliari), and Padru Jossu (Sanluri, Cagliari), two small flat axes from Su Crucifissu and S. Bartolomeo, a few rings and ring bracelets from Cuguttu (Alghero, Sassari), an arrowhead from Anghelu Ruju, and a silver sheet pendant from Padru Jossu are all the metal evidence found in association with the Beaker Culture. Their types generally repeat the shapes well known in the same period in western and central Europe.

There are daggers from Frommosa Cave I (Villanovatulo, Nuoro) connected to the Bonnanaro Culture of the Early Bronze Age (figure 9.3d, e) and from the village of Sa Turricula (Muros, Sassari) (figure 9.3f) and awls with two sharpened ends and a square section in the middle from Aiodda (Nurallao, Nuoro), Anghelu Ruju, etc.; here again this is an item well known in Early Bronze Age Europe.

The analysis (Junghans et al. 1960) of the Anghelu Ruju awls gives a composition close to the Remedello material: "mostly copper with variable contents of As, Sb, and Ag, which is fairly typical composition for the [Early Bronze Age]" (Tylecote et al. 1983, p. 66). It would be crucial to analyze *all* the metal objects from the Beaker and Bonnanaro periods in order to determine how many of them are actually made of arsenical copper.

The cultural connection of Sardinia with the Italian mainland, southern France, and the Iberian Peninsula (note the E1 Argar sword from southern Sardinia) are without doubt and, although only a few analyses have been made, the presence of arsenical copper may have a correspondence with the "exotic" shapes, suggesting an import of items rather than a flourishing local production.

Figure 9.5 Nuraghe Orrubiu (Orroli, Nuoro). The huge Nuragic complex with eighteen towers.

Nuragic Metallurgy

The Nuragic Civilization

The components of the Nuragic civilization taken singularly can be traced back in the prehistory of Sardinia; as a whole it appears as an extraordinarily rich and original culture characterized by elaborate architecture, plain pottery production, a taste for the carving of soft stones, and high technical skill in metallurgy.

The vocation of the land is predominantly pastoral with agricultural support in the mountains and hills of central and northeastern Sardinia and mostly agricultural in the southern and western plains of Campidano, Arborea, Logudoro-Meilogu, and Nurra. The political and social structure seems to be based on tribes and clans, apparently organized in regional "cantons." Archaeology has failed up to now to detect any rigid difference of classes, the existence of kings, or the predominance of one regional group over others.

All the rituals, including the ones connected with burial, bear a strong community character; no individual burial is known. What seems to distinguish a few huts in the village and a few chambers in the nuraghi is the presence of a round bench all along the walls that often can be found all around the semicircular structure in front of the "giants tombs" and always in the entrance hall of the sacred wells, suggesting the reunion of people sitting together for social and religious purposes.

The eminence of the Nuragic civilization among the other prehistoric communities consists in its mastery of building imposing and stout monuments with complex and elaborate plans and elegant refining.

More than 7000 nuraghi, no less than 400 villages, about 500 "giants tombs," and about 50 sacred wells are the numbers generally reported in the literature; they seem, however, to be underestimated. Moreover, the huge structure of the nuraghi, when a complex plan with one or two enclosures containing many additional towers are added to the first truncated conical basic shape, defies a solution by means of archaeological excavation. Nevertheless, many programs and studies are underway, and the results appear promising (figures 9.5 and 9.6).

Mycenaean pottery began to be found from 1976 to 1979, at first along the eastern coast and later in the Cagliari Gulf in the south. The clay analyses demonstrate provenance from Cyprus, Crete, Peloponnese, Rodi as early as Myc IIIB (1300–1200 B.C.). The last discovery is a surface find of an ivory head with a boar tusk helmet, comparable with examples from Philaki, Spata, Enkomi, Mycenae, etc. that can be dated

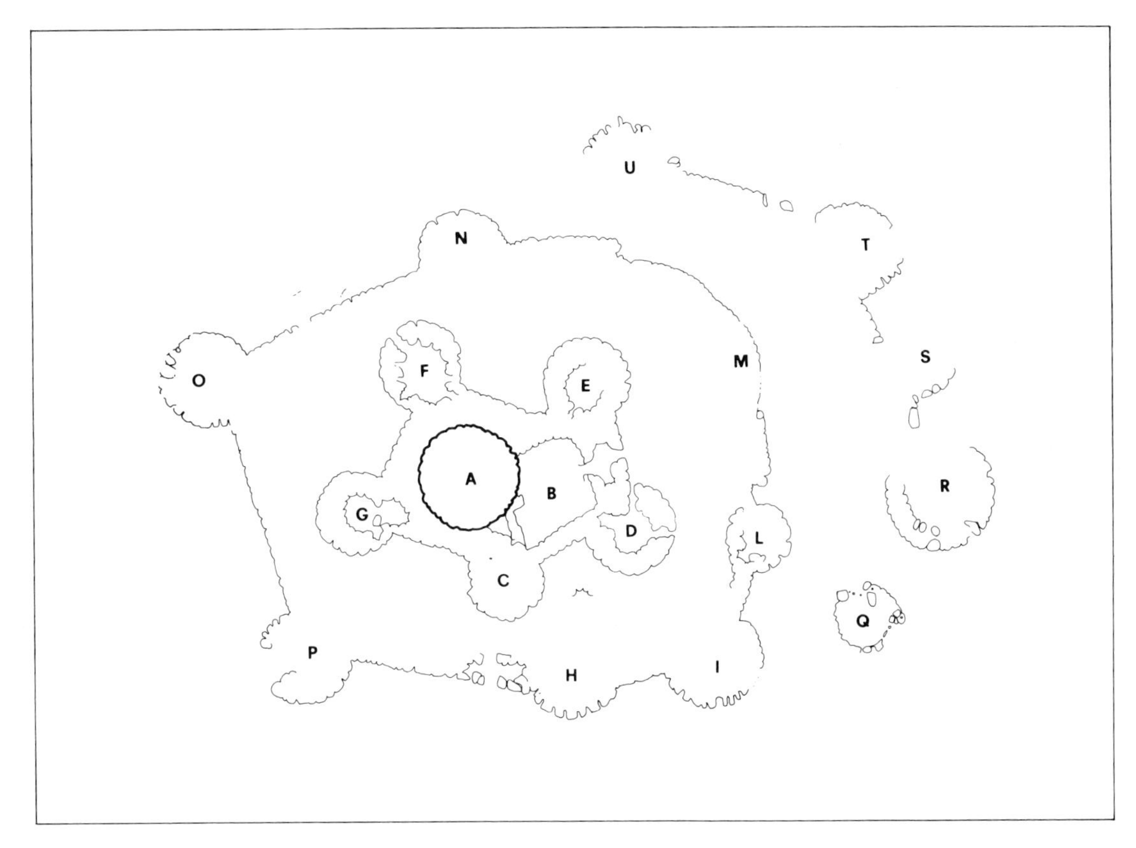

Figure 9.6 Plan of nuraghe Orrubiu (Orroli, Nuoro).

to Myc IIIA. This means an indisputable documentation of a connection of Nuragic Sardinia with the Mycenaean world as early as the fourteenth century B.C.

Leaving aside the problem of independent invention or transmission from the Aegean world to Sardinia of the technique of building up the tholos, mention must be made of the copper oxhide ingots.

Oxhide Ingots

The topic of the oxhide ingots (figure 9.7) is now under discussion largely because of the new discoveries that have opened the whole subject. These discoveries include a mold for an oxhide ingot at Ras Ibn Hani (Syria), a shipwreck with a cargo of copper and tin oxhide ingots from Ulu Burun (Kaş, Turkey), and finds of other fragments of oxhide ingots in archaeological contexts from Turkey (Bogazköy), Cyprus (Pila-Kokkinokremos), and Sardinia (Ittireddu, Villagrande Strisaili, Ozieri, Triei, Sardara, Soleminis, etc.).

The chronology of oxhide ingots seems to run from the end of MM III and LM I–II (sixteenth to fifteenth century B.C.; Ayia Triada, Zakro, Knossos, Tylissos, etc.) to the twelfth century. The chronological division of the different shapes as defined by Buchholz are no longer accepted; the chronology of the marks impressed or incised on them remains an open problem.

The hypothesis of one center of production of all oxhide ingots in the east and the west is generally dismissed; incontrovertible proof of provenance associating a group oxhide ingots with a particular ore body through chemical and metallurgical analyses has not been found; lead isotope analysis, claimed sometimes as capable of solving the problem, is still under discussion.

From an archaeological and historical viewpoint it seems highly probable that a few major copper mining centers existed. The metal was traded from the centers to the many sites where the oxhide ingots were produced; from there they were shipped or traded by land to all the sites where they have been found.

Western Sardinia is rich in copper ores, and many oxhide ingots, both complete and fragmentary, have been found (figure 9.8). It may be suggested that Sardinia was a partly autonomous center of mining production and internal distribution of this item. Certainly of external origin are the shape of the oxhide ingots and their Cypro-Minoan markings. In fact, either the Mycenaeans when they first came to Sardinia (Myc IIIA–IIIB from about the fourteenth century B.C. on) or the Cypriots of LC III (c. twelfth to eleventh century B.C.) could have been responsible for the introduction of the four-handled shape, perhaps in an attempt to build up a system similar to the one in use in the

Figure 9.7 Copper oxhide ingot from Serra Ilixi (Nuragus, Nuoro).

Aegean and in the Levant. The reason why it has not yet been possible to clarify this crucial problem is the absence in Sardinia of good chronological associations. Now it appears that this soon may be achieved thanks to new excavations and discoveries.

Metal Ores and Archaeological Evidence for Smelting

Wide regions of Sardinia contain metal ores: Sulcis-Iglesiente, Fluminese in the southwest, Sarrabus-Gerrei in the southeast, Barbagia, Sarcidano in the center and south, Baronia on the eastern coast, and Nurra in the northwest. The lead, zinc, iron, and copper ores are located almost exclusively in Paleozoic formations with the exception at Alghero of ores from the Mesozoic.

The most important copper mines are Calabona (Alghero), Canale Barisone (Torpé), Genna Scalas (Baunei), Funtana Raminosa (Gadoni), Rosas and Sa Marchesa (Narcao), Tiny (Iglesias), and Sa Duchessa (Domusnovas) (figure 9.9).

Iron mines of different genesis, although all in Paleozoic formations, are located at Monte Lapanu (Teulada), Punta di Candiazzus, Campo Pisano-Funtana Perda, (Iglesias), Antas, Enna Sa Spina, Su Libanu (Fluminimaggiore), Rio Piras, Salaponi-Nueddas (Gonnosfanadiga), Giancurru (Aritzo) in the northeastern region, and Nurra in the northwest.

Tin also occurs in hydrothermal ores between Fluminimaggiore and Gonnosfanadiga (Monte Mannu, "Canale Serci," Villacidro), where it is associated with chalcopyrite, sphalerite, and galena; S. Vittoria "Perdu Cara," Fluminimaggiore, has cassiterite embedded in quartz but, because of the scanty amount and the difficulty of extraction, these ores do not seem to have been exploited in antiquity; more research is needed, however.

The project actually underway, sponsored by the Soprintendenza Archaeologica di Sassari e Nuoro, Harvard University, and the University of Pennsylvania, examined the mines of Funtana Raminosa (Gadoni, Nuoro) generally thought to have been potential sources of Sardinian copper in ancient times and consisting mainly of chalcopyrite. Traces of activity predating the twentieth century, when the mines were again exploited, are now obliterated by tailings.

Because primary slags have been found in several nuraghi in the Nurallao district adjacent to the site of the mine and in the Gesturi district a few miles to the south, a temporary conclusion is that in ancient times some of the ores from Funtana Raminosa were transported over a high pass south of the mine to nearby areas where other materials necessary for smelting (kaolin for furnace lining, hematite for flux, charcoal, etc.) existed.

Confirmation of this hypothesis requires laboratory analyses of the slags as well as archaeological excavations of the nuraghi where the slags were found.

Archaeological Evidence for Melting: Foundries, Molds, and Crucibles

Almost every large nuraghe and Nuragic village seems to have had a local metallurgical activity: producing, melting, and repairing weapons, implements, and tools. Although up to now we have failed to find intact a smithing workshop, consistent traces of them are frequent.

Two crucibles have been found at Olmedo (1980) and at Ittireddu (1985) in the Nuraghe Funtana, where there have also been found oxhide ingots and a hoard containing not only oxhide ingots but also weapon fragments.

The number of stone molds uncovered is rising and is now greater than fifty, with a wide distribution on the island. They were constructed with soft stones from local deposits, such as steatite and chlorite; they are well refined and elaborate in shape. Monovalve molds, either simple or multiple, and bivalve molds are

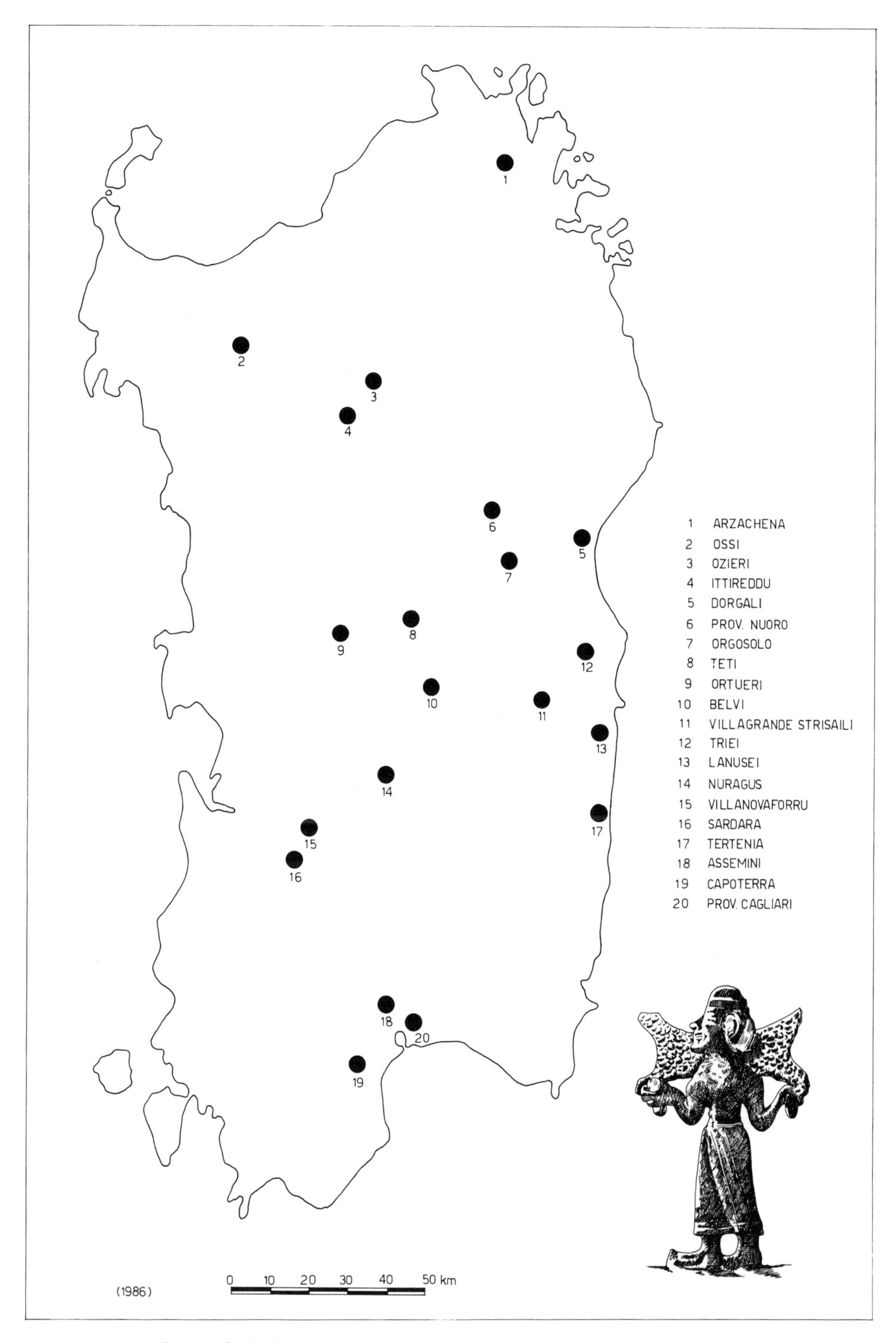

Figure 9.8 Distribution of oxhide ingots in Sardinia (1986).

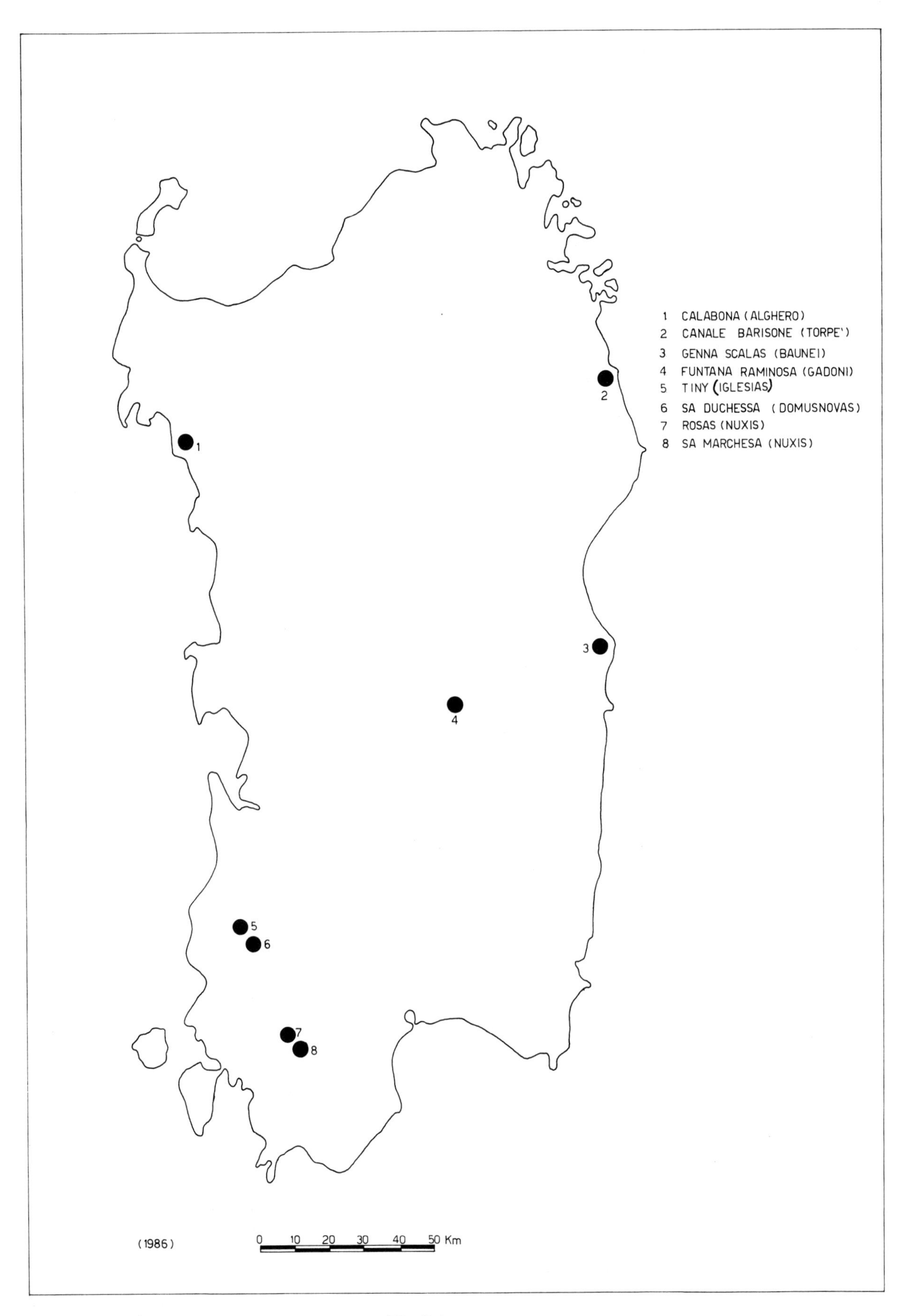

Figure 9.9 The most important copper mines of Sardinia.

known. In one interesting case (S. Luca, Ozieri), the stone of a simple monovalve mold with a carving of three sickles bears clear traces of having been used as a lid for another simple monovalve mold with the carving of three spatulae or daggers.

Fragments of clay molds are reported to have been recently discovered, but the reports are as yet unpublished. The ubiquitous existence of planoconvex ("bun") ingots found isolated or collected in hoards along with the frequent discovery of scrap hoards connected to workshops for remelting must be kept in mind.

The peak of Nuragic metallurgical activity, according to the chronology of the metal objects, took place in the Late Bronze Age and Early Iron Age (at least from the twelfth and eleventh to the eighth century B.C.).

Cypriot Influence on Local Bronze Work

It is now without doubt that there was a strong relationship between Cyprus and Sardinia, supported by much important evidence, including direct imitations chronologically near the models and later local derivations. The imports can mostly be dated to the twelfth century B.C. and consist of double axes, smithing tools (hammers, tongs, shovels for charcoal), and tripod stands. The direct imitations follow immediately and are perhaps contemporaneous with other imports in the eleventh century B.C.: double axes with vertical and converging blades, ax-adzes, large picks, shovels and other tools, mirrors, spiral handle attachments, tripod stands, and figurines of local production. Following that time, in the tenth to ninth century B.C., the external influence can be detected only as an original imprint of technical skill and of general taste.

It must be stressed that Sardinia was also the way through which Cypriot imports as well as Nuragic imitations and later derivations reached Italy, where in the Late Bronze Age a consistent exchange of objects is documented:

It has been argued ... that some Cypriot bronzeworkers also reached Sardinia during the late 2nd millennium B.C. and that, perhaps during the 11th and 10th centuries, bronzesmiths working in Sardinia were familiar with the skills of casting bronze by the investment or lost-wax technique. If the prototypes ... with the necessary skills were all present in Sardinia at the end of the 2nd millennium B.C., then it is at least possible that some Sardinian figurines were made during the Italian Late Bronze Age and long before the colonizing era. (LoSchiavo et al. 1985, p. 61)

Nuragic Bronze Figurines

The investment or lost-wax technique was known in the Aegean and Near East since EBA I, although it is difficult to date the older products because of the destruction of the clay molds. There is no need to state how rich the production of various objects and figurines was during the Bronze Age and particularly in the second millennium B.C. In Sardinia the time at which this technique was first applied is not yet certain, but there can be no doubt that it experienced a renaissance through the relationship with Cypriot metallurgy at the beginning of the Late Bronze Age, as can be demonstrated by the direct imitations and the later local derivations.

Thus, after a few imports of Oriental, Syro-Palestinian, and Cypriot figurines, there is an extraordinarily rich local production of figurines, beginning probably in the eleventh to tenth century B.C. and continuing during the Early Iron Age, at least until the eighth century B.C.

The technical skill and artistic level reached in the figurines make them a symbol of Nuragic Sardinia (figure 9.10). They represent warriors with swords and shields, archers with bows, quivers, and "directional feathers," the chief of the tribe with heavy cloak and wooden club, notables with gamma-hilted daggers in their crossbelts, seated women holding male figures,

Figure 9.10 Bronze figurine of a warrior of unknown provenance, Sardinia.

women covered with cloaks, and shepherds and common people offering piglets, lambs, lamb skins, jars, bread, etc. There is also a wide variety of animals, such as bulls, muflons, sheep, pigs, dogs, boars, deer, birds, and cocks. Animal figures are also produced as a decoration on small bronze boats, mostly at the prow.

Among the Nuragic figurines there are also models of nuraghi, stools, baskets, pithoi, wooden boxes, axes and double axes, daggers (mostly gamma-hilted), quivers, etc. These figurines are generally found as offerings in the sacred wells, sometimes in nuraghi and Nuragic villages, frequently in fragments in the hoards. The value and importance of the figurines were already high in ancient time; thus many of them have been found in Etruscan orientalizing tombs, evidently handed down and preserved for generations. The early datable export in the Cavalupo tomb at Vulci is of the second half of the ninth century B.C. at a period in which nothing similar existed in the Mediterranean (apart, as already said, from the Levantine Coast and Cyprus).

Iron Metallurgy

Recently we acquired evidence of an early introduction of iron in Sardinia, again in connection with the Cypriot influence on the metallurgy. This is a small fragment of a sheet of iron (3 cm) found in stratigraphic association with a wishbone handle of LC II (probably thirteenth century B.C.) in the lower level (stratum 4) of the upper room of tower c at Nuraghe Antigori (Sarroch, Cagliari), the Nuragic complex where most of the Mycenaean pottery has been found.

Another indication of an early date for the use of iron is an iron dagger with a steatite hilt found in a megalithic tomb with collective burials and amber beads of the "treasure of Tyrins" and Allumiere types.

Iron was used in the same period and in the same workshops where the bronze figurines and other common-use objects were made. There are iron elements in two bronze votive boats, an iron clasp in a bronze basin, a pin with an iron rod and a bronze head, and an attempt at casting an iron votive boat.

I suggest here that the beginning of the use of iron in Sardinia can be dated to the Late Bronze Age, where it was introduced from Cyprus, and that it is highly probable that Sardinia is responsible for the transmission of the technique to mainland Italy.

Conclusions

When I attended the Larnaca Symposium in June 1981, I tried for the first time to present an overview of Sardinian metallurgy and was compelled to borrow the conclusion from a recent publication on Cyprus by Stech-Wheeler, Muhly, and Maddin, adapting it to the western island (Stech-Wheeler et al. 1979, p. 142; LoSchiavo 1982, p. 278). In 1983, at the Tufts Colloquium in Boston, although I was tracing the archaeological background of Sardinian metallurgy, I was again forced to refer mostly to Cyprus, whose relationship to Sardinia was published in 1985 (LoSchiavo et al. 1985; LoSchiavo 1986).

During these years of uninterrupted study and research, many excavations and new fundamental discoveries have taken place that stress the importance of Sardinian metallurgy, of which the contemporary ancient world was well aware; the presence of the Mycenaean pottery from an early date confirms the historic reconstruction.

I can only hope that in the near future studies will move ahead in the same direction, helping to solve all the open problems and presenting a more precise picture of the metallurgical technology and history of Sardinia.

References

Because of the character of this chapter, the following references are mostly limited to publications in English. The most complete bibliography on the subject can be found in LoSchiavo et al. (1985) and in LoSchiavo (1986).

Balmuth, M. S., ed. 1986. *Studies in Sardinian Archaeology. Vol. 2, Sardinia in the Mediterranean.* Ann Arbor, Mich.: University of Michigan Press.

Balmuth, M. S., and R. J. Rowland, eds. 1984. *Studies in Sardinian Archaeology*, vol. 1. Ann Arbor, Mich.: University of Michigan Press.

Balmuth, M. S., and R. F. Tylecote. 1976. "Ancient copper and bronze in Sardinia: Excavations and analysis." *Journal of Field Archaeology* 3: 195–201.

Bass, G. F. 1967. "Cape Gelidonya: A Bronze Age shipwreck." *Transactions of the American Philosophical Society* 57(8): 3–177.

Becker, M. 1984. "Sardinian stone moulds: An interesting means of evaluating Bronze Age metallurgical technology," in *Studies in Sardinian Archaeology*, M. S. Balmuth and R. J. Rowlands, eds. (Ann Arbor, Mich.: University of Michigan Press), vol. 1, 163–208.

Catling, H. W. 1964. *Cypriot Bronzework in the Mycenaean World.* Oxford: Clarendon Press.

Catling, H. W. 1980. *Sardinian Craft and Culture from the Neolithic to the End of the Nuragic Period.* Austellung Karlsrule. (In German.)

Catling, H. W., ed. 1981. *Ichnussa from the Origins to the Classical Age: Sardinia.* Milan. (In Italian.)

Junghans, S., E. Sangmeister, and M. Schröder. 1960. "Metal analysis of Copper and Early Bronze Age finds from Europe." *Studien auf den Anfangen der Metallurgie* 1. (In German.)

LoSchiavo, F. 1981. "The economy and society of the Nuragic," in *Ichnussa from the Origins to the Classical Age*, H. W. Catling, ed. (Milan), 253–347. (In Italian.)

LoSchiavo, F. 1982. "Copper metallurgy in Sardinia during the Late Bronze Age: New prospects on its Aegean connections." *Early Metallurgy in Cyprus 4000–500 B.C., Acta of the International Archaeological Symposium*, J. D. Muhly, R. Maddin, and U. Karageoghis, eds. (Nicosia: Pierides Foundation), 271–282.

LoSchiavo, F. 1985. *Nuragic Sardinia in Its Mediterranean Setting: Some Recent Advances.* Occasional Paper 12. Edinburgh: University of Edinburgh, Department of Archaeology.

LoSchiavo, F. 1986. "Sardinian metallurgy: The archaeological background," in *Studies in Sardinian Archaeology*, M. S. Balmuth, ed. (Ann Arbor, Mich.: University of Michigan Press), vol. 2, 231–250.

LoSchiavo, F., E. Macnamara, and L. Vagnetti. 1985. "Late Cypriot imports to Italy and their influence on local bronzework." *Papers of the British School at Rome* 53:1–71.

LoSchiavo, F., R. Maddin, J. D. Muhly, and T. Stech. 1984. "Preliminary research on ancient metallurgy in Sardinia." *American Journal of Archaeology* 89:316–318.

Muhly, J. D., T. Stech-Wheeler, and R. Maddin. 1977. "The Cape Gelidonya shipwreck and the Bronze Age metal trade in the eastern Mediterranean." *Journal of Field Archaeology* 4:353–362.

Stech-Wheeler, T., R. Maddin, and J. D. Muhly. 1979. "Mediterranean trade in copper and tin in the Late Bronze Age." *Ann. Inst. It. Num.* 26:139–152.

Tylecote, R. F., M. S. Balmuth, and R. Massoli-Novelli. 1983. "Copper and bronze metallurgy in Sardinia." *Journal of the Historical Metallurgy Society* 17(2):63–78.

Zwicker, U., P. Virdis, and M. L. Ceruti. 1980. *Investigations on Copper Ore, Prehistoric Copper Slag and Copper Ingots from Sardinia.* British Museum Occasional Paper 20. London: British Museum, 135–153.

10
Early Nonferrous Metallurgy in Sweden

Gunborg O. Janzon

In Scandinavia early finds of metal influenced the theory of development and founded the chronology of the three ages—the Stone Age, the Bronze Age, and the Iron Age. This rough classification has never been satisfactory. For instance, the earliest metal objects found in south Scandinavia date back to the so-called Stone Age, about 3000 B.C. To this early period is also dated a find from Finland—a copper ring and typical Comb Ware pottery, phase II (3350–2800 B.C.; Taavitsainen 1982, p. 45).

There is today comprehensive literature focused above all on metal artifacts, their typology, function, socioeconomic significance, etc. But the question of when metallurgy, this new dynamic technology, made its first appearance and how it was practiced has been stressed little in Swedish and other Scandinavian archaeology for nonferrous metals. However, a basic and important work in this field has been made by Andreas Oldeberg, who published his work in 1942–1943 and later (1974–1976) finished the catalogue and description of early metal finds in Sweden.

In archaeology the term "archaeometallurgy" is often used to cover a range of metal- and iron-working activities, from ore prospecting and mining to smelting or melting of metal to produce the finished artifact. Each of these phases produces its own characteristic constructions, artifacts, and debris, but archaeological evidence of all these activities are seldom found in one area or at one site. The question, then, is which elements ought to be represented to allow the conclusion of metallurgical activities at one or more stages. I try to illustrate the problem with examples from Sweden and to some extent from the Scandinavian region.

First, it is not possible to claim metallurgy exclusively in an area from early finds of metal artifacts, but it is reasonable to assume that some metallurgical know-how reached an area with the first metal artifacts. The difficulty is knowing at what level the knowledge was acquired, say, c. 3000 B.C., the date for the well-known Bygholm find in west Denmark and for several stray finds of similar copper axes from south Scandinavia and Schleswig-Holstein (Randsborg 1978). Perhaps the metallurgical know-how at this time comprised only a general idea about metal derived from special rocks. How long did it take until the knowledge was more complete and metallurgy at all stages could be practiced? How did the society and its organization prepare and adjust to this new technology (Coles 1981)?

In what follows I suggest and try to define and exemplify three stages of metallurgy in the Swedish archaeological material: (1) prospecting of nonferrous ore deposits and mining, (2) metal production, and (3) manufacturing.

Prospecting of Ore Deposits and Mining

The first stage of metallurgy is the prospecting for ore deposits and their exploitation. When did these activities begin in Sweden?

Sweden is a mountainous country plentiful in raw materials for both ferrous and nonferrous metallurgy. A map made by U. Qvarfort (1980, p. 42) illustrates the main nonferrous deposits and mines in Sweden. In addition to the marked places there are many small deposits scattered throughout the country (figure 10.1).

In Fennoscandia there are no oxidized surface layers or secondary enrichment zones in the ore deposits because of the action of inland ice. Chalcopyrite is the dominant copper mineral. In northern Sweden there are also copper ores of the tetrahedrite type (Serning 1984, p. 4). Native copper is rare, but there are deposits in southern Norway, for example, Dalane, which have native copper with silver and high arsenic contents (Dons 1975, p. 51). In Dalane the copper occurs in sandstone, which makes it easily accessible. Native copper also exists in east Karelia, near the northwestern side of Lake Onega (Žuravlev 1975, map) (figure 10.2).

Raw material can also be alluvial in origin, for example, river gold and stream tin. There are possibilities that at least gold could have been collected from waterways in Sweden, in the provinces of Småland (Ädelfors, Emån) and Västerbotten (Skellefteälven) (figure 10.3).

The general opinion is that nonferrous ores were not prospected or worked in Sweden before A.D. 1200. The written sources first mention the copper deposits at Falun in central Sweden and later Åtvidaberg in southern Sweden (Serning 1984, p. 4; Sommarin 1910, pp. 141ff; Söderberg 1932).

This dating of early mining has recently been questioned. The results of a new method, developed by Qvarfort (Uppsala University) and dealing with analyses of sediments from Lake Tisken in Falun, suggest the opening of a mine by A.D. 700–800. The section analyzed and the dated sediments (carbon 14) show an increase in metal content, probably a result of mining. Similar results are also obtained from Garpenberg further south in the province of Dalarna (Qvarfort 1980, 1981, 1984) (figure 10.3).

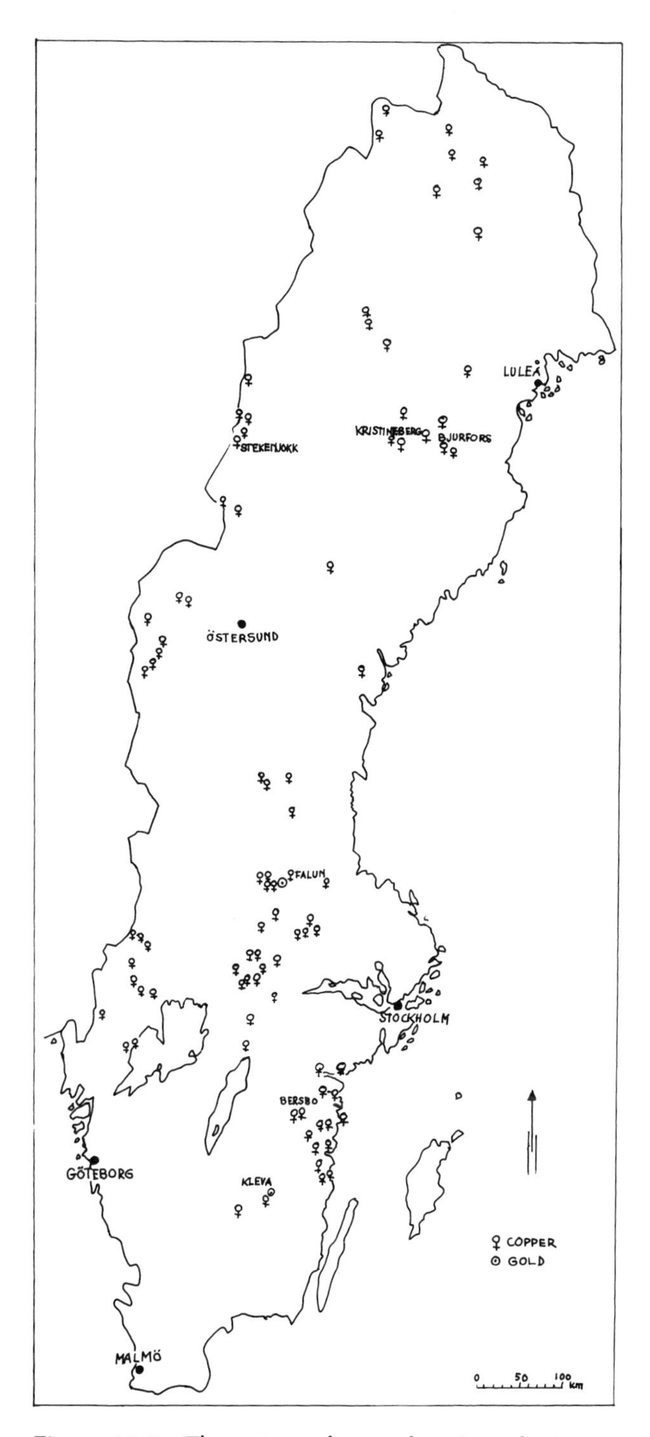

Figure 10.1 The main nonferrous deposits and mines in Sweden. After Qvarfort (1980). Kristineberg and Bjurfors indicate the Skellefte-field.

Archaeological Evidence

Are there any indications of early prospecting and mining activities in the archaeological material? I give two examples, the first from a group of graves.

In Sweden the first reliably dated copper artifacts are found in graves belonging to periods 3–4 of the Battle-Ax Culture (Malmer 1962, 1975). Consequently this gives an important time horizon in our material. The grave constructions and contents represent something quite new and different in the indigenous con-

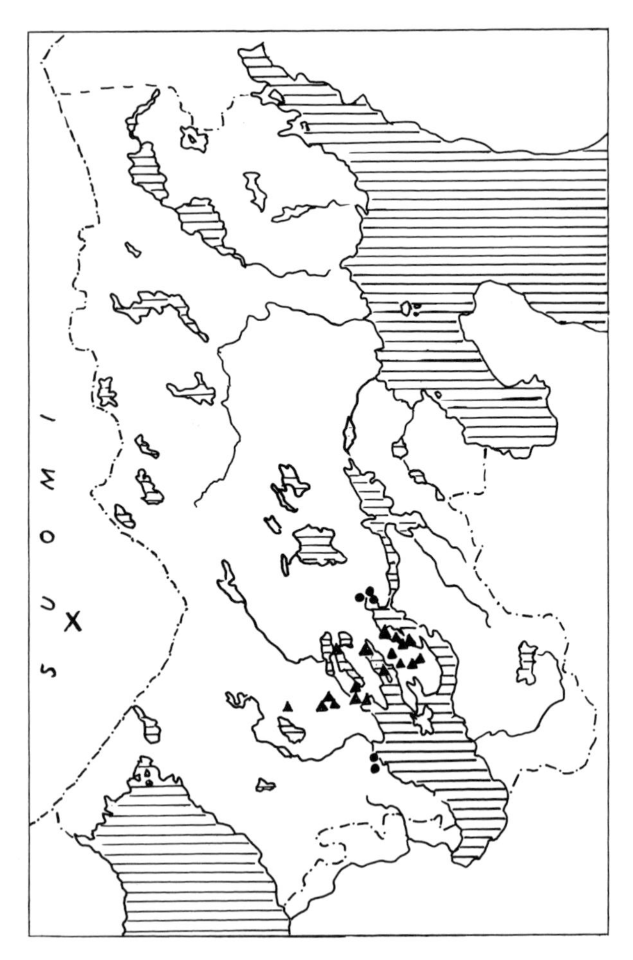

Figure 10.2 Deposits of native copper in east Karelia. ×, Polvijärvi Suovaara, the site where a copper ring and typical Comb Ware Pottery, phase II (3350–2800 B.C.) were found. Map after Žuravlev (1975) and Taavitsainen (1984).

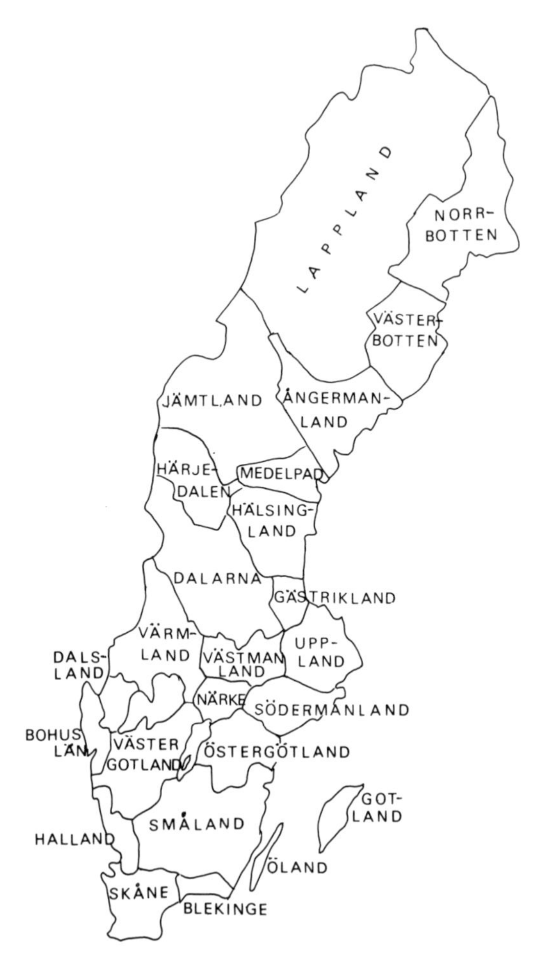

Figure 10.3 Provinces in Sweden.

text around 2500 B.C. Besides the tiny copper artifacts, the graves contain, for example, amber, flint artifacts, so-called battle-axes, and pottery.

The pottery (figure 10.4) introduced in this phase of the Battle-Ax Culture show, according to Birgitta Hulthén, a completely new ceramic technology. This is a typical example of discontinuity. Hulthén points out that other clay types were used and that chamotte (grog) was employed as a temper material. The vessels were modeled, not coil built, and were of globular shape with new decorations. Above all, a new firing practice was adopted—firing in a reducing atmosphere. This new technique shows all signs of perfection from the beginning, and no traces of intermediate transitional stages in the technical knowledge have been observed. The practice of the new technique is, however, only an interval, and the continued development shows a return to earlier methods (Hulthén 1976, pp. 120–121).

It is evident that the material from this group of graves (of approximately thirty graves, three contain the small copper artifacts) indicates contact with the European continent. My hypothesis is that this material represents foreign people with metallurgical knowledge who came individually or in small family groups to the northern area. Their aim was to prospect for minerals; they also probably practiced mining.

The prospecting may have even reached northern Sweden, where several large hoards of flint axes and a few potsherds indicate the presence of the prospectors in the same area as one of the richest deposits of sulfide ore, the Skellefte field (figure 10.1), where even gold is found. So-called battle-axes are also found high up in the northernmost part of Sweden.

My second example concerns a special kind of implement—the stone hammer. As it is often difficult to determine what is prehistoric mining or more recently worked pits, the archaeologist has to look for artifacts that can be connected with early activities. A tool of special interest in this case is the grooved stone hammer, which in Europe is often found in mining areas. It has been called the first specialized tool in connection with early mining (Jackson 1971; Jovanović

Figure 10.4 G- and H-pottery from graves of the Battle-Ax Culture. After Malmer (1962, 1975).

1980; Rothenberg and Bianco Freijeiro 1980; Schmid 1973). The groove to furnish the handle was probably more practical than a shaft hole, which would reduce the strength of the tool when hammering hard material.

In Sweden a number of these grooved stone hammers have been found. About 300 were catalogued and classified by Indreko (1956). To these should be added later registered finds. The number of grooved stone hammers in Swedish museums is now about 500 (Janzon 1984). As the map made by Indreko (1956, Abh. 22) illustrates (figure 10.5), these implements are found over the whole country. In the south of Sweden most of them are stray finds, but in the north they are mostly found on sites, often situated high in the mountains. The stratigraphy of those sites is generally mixed and with a relatively long chronology, from the Stone Age onward.

The stone hammer in some European areas is undoubtedly an indication of mining activity, but the artifact alone is not proof of mining, although the hammer is a necessary implement in most stages of metallurgy. In Sweden the grooved stone hammer obviously has been used for a long time and for different functions. Their varied size and weight, flattened or rounded percussion surfaces, sometimes with heavy blow marks, indicate the specialization. I give a few examples from different contexts.

Only one find of a stone hammer from a flint pit in Scania (Skåne), worked during Neolithic time in the third millennium B.C., is possibly connected with mining, specifically flint mining (Indreko 1956, pp. 34ff).

Several of the registered implements are found at sites in northern Sweden. One example is from Lappland, site 654 at the lake Malgomaj-Grundsjön (Vilhelmina parish, Svartviksudden, SHM 25960). There are registered two grooved stone hammers, one whetstone, one double mold of soapstone for a socketed ax, a thin sheet fragment of copper-based metal in a fireplace, etc. (figure 10.6).

Another example from the Vilhelmina parish is site 662 (Malgomaj-Varris, Lappvallen; SHM 25960), investigated 1957–1958 by Harald Hvarfner. The site has a long duration. There are registered fifteen grooved stone hammers, one dagger blade of bronze, one-half of a soapstone mold for a socketed ax, one double mold for a bronze blade, one fragment of a crucible (?), fired clay with asbestos, etc. (Janson and Hvarfner 1960, pp. 24ff; 116ff; Oldeberg 1974, p. 388) (figure 10.7).

In southern Sweden the excavation (1969–1971) of the Late Bronze Age complex at Hallunda in the neighborhood of Stockholm registered twenty-one grooved stone hammers in context with furnaces, finds of crucibles, molds of clay and soapstone, bronze bars, etc. About 185 hammer stones were also found at the site (Jaanusson 1972, pp. 48ff; 1981, p. 20; Jaanusson-

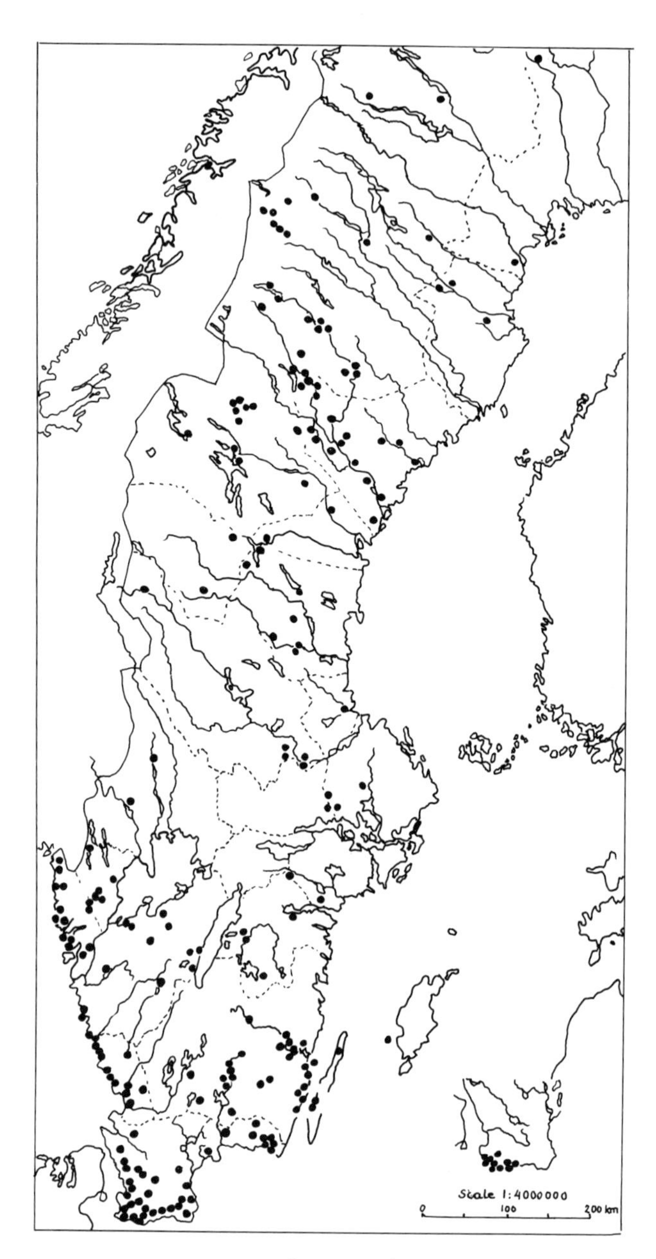

Figure 10.5 Registered grooved stone implements in Sweden. After Indreko (1956).

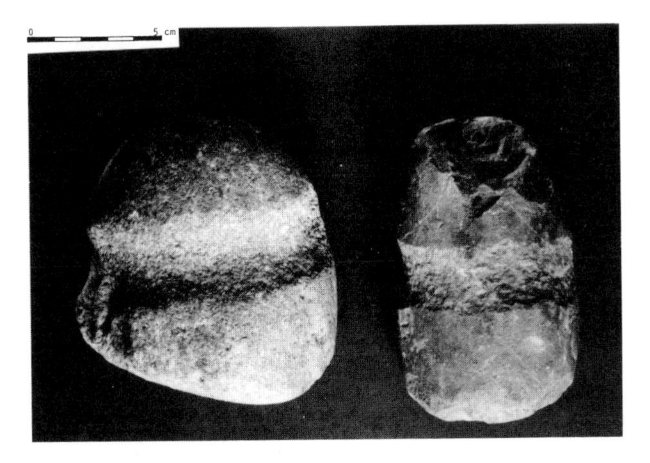

Figure 10.6 Two grooved stone hammers from site 654, Vilhelmina parish, Lappland. Photo Bengt A. Lunberg (1984).

Figure 10.7 Examples of grooved stone hammers from site 662, Vilhelmina parish, Lappland. Photo Bengt A. Lundberg (1984).

Figure 10.8 Three grooved stone hammers from Hallunda, Botkyrka parish, Södermanland. Photo Bengt A. Lundberg.

Vahlne 1975, fig. 7; Jaanusson et al. 1978, fig. 22) (figure 10.8).

Also from Scania (Skåne) is a find from the Bronze Age site of Hötofta 4:1, where a grooved stone hammer was found together with a fragment of a crucible, a bronze ring, etc. (Stjernquist 1969, pp. 125ff).

The given examples show that, except for Hallunda, only a few stone hammers are so far reliably dated and that those that are dated belong mainly to the Late Bronze Age and represent stage 3 of metallurgy, the manufacturing stage.

Smelting of Nonferrous Ores

The definition of the second stage, metal production, should include the presence of elements such as ore, charcoal, constructions for ore treatment and ore reduction (that is, roasting pits, furnace remains), products and debris from the activities, and, most important for interpretation, slag.

In Sweden we still do not have archaeological finds indicative of prehistoric smelting of nonferrous ores. This is not surprising; furnaces for copper smelting are scarce also on the European continent and on the British Isles. Maybe the picture will change in Sweden if we start to work with the hypothesis that our own local deposits have been used!

However, slag and metal do exist. There are many early sites with finds of slag, for example, in the northern mountainous ore-bearing area of Sweden. Survey and registration are going on, but few analyses have been made. We have recently started a project to change this lack of information (Grälls 1986).

Manufacturing and Alloys

The definition of the third stage, manufacturing, should operate with both the implements and constructions for manufacturing and the products and the alloys. When did manufacturing begin in the Scandinavian region?

The Danish archaeologist P. V. Glob suggested in 1952 that some metal objects were manufactured in southern Scandinavia by the Middle Neolithic time, about 3000–2300 B.C. So far, there is no proof for this interesting idea.

Flat axes of the Pile type, named after a hoard find in Scania, are interpreted as native products in Late Neolithic times, 2300–1800 B.C. (Stenberger 1964, p. 139). But hitherto no molds for flat axes or other traces of manufacturing from this time have been found in Sweden. The same is true for flanged axes, although the lack of molds is astonishing, considering the large number of axes.

It has been suggested that unsuccessful casting can be an indication of native production. Oldeberg mentions some early examples: one flanged ax from Öland

(Södra Möckleby parish; Cullberg 1968, p. 248, no. 725; Oldeberg 1974, no. 2053; Oldeberg 1976, p. 76) and a shaft-hole ax from a hoard find in western Sweden (Träslöv parish, Klastorp no. 5, Halland; Oldeberg 1942–1943, vol. 2, fig. 295; Oldeberg 1976, p. 78).

Workshops

There is today an increasing number of sites in southern Sweden that can be defined as workshops. The sites are identified both by manufacturing implements and by the products. Pyrotechnical traces are also present.

The earliest dated workshop is from western Sweden, from Grimeton parish, Halland. It was excavated in 1914–1918 by Georg Sarauw and is dated by bronze artifacts to the Bronze Age (Montelius, periods 2–5). The many crucibles of clay from the workshop are especially important. There are about twenty-five crucibles. (Three whole crucibles were found, seventeen were completely fused together, three were nearly complete, and further fragments represent another half of a crucible; Sarauw and Alin 1923, pp. 258ff). Other known sites are dated to the Middle and the Late Bronze Age (periods 4–6) (figure 10.9).

The characteristic location of the workshops is, generally, a relatively high position in the terrain, often on a slope (Weiler 1984, p. 67). The remains are often described as a circular mound with fire-cracked stones surrounded by a circle of bigger stones; the remains look similar to a grave.

Evidently there are also workshops in northern Sweden, but the often unstratified sites in this area are still not investigated with this classification in mind. A map published by Noel Broadbent (1982, p. 143; Bakka 1976) that marks the finds of Arctic bronze artifacts and molds gives an idea of the metallurgical activities in northern Fennoscandia during the Late Bronze Age (figure 10.10).

Hoards

Oldeberg registered (1942–1943) a series of finds from southern Sweden classified as "smith- and scrap-finds" (founders hoards). The special criteria for this category of hoard are that, together with a collection of whole and fragmented metal objects, they also contain metal bars, scrap from the casting process, and sometimes even pieces of ingot cakes. The hoard is often found near a big stone block and placed in a vessel or wrapped in textile. Like the workshops these hoards are mainly dated to the Late Bronze Age (Montelius periods 4–6). Oldeberg has suggested that the hoards belong to itinerant smiths or founders (Oldeberg 1942–1943, p. 190). I think that some of the masters moved from place to place.

From the metallurgical point of view the bars and pieces of ingots in some of the hoards are of great interest, for example, the Bräckan hoard found near the western shore of Lake Vänern (Oldeberg 1929).

Six rods and two pieces of one or two ingot cakes from the Bräckan find have been analyzed (Oldeberg 1942–1943, vol. 1, pp. 142, 208–209, nos. 263–271). The analyses are only semiquantitative. The rods show a marked similarity in composition, with a content of tin of about 8% and, with the exception of no. 264, low contents of antimony and arsenic. The two ingot pieces, on the other hand, have a low content of tin and a high content of silver; one of them has high contents of arsenic and antimony as well. The differences between the rods and the ingots might be explained by the loss of arsenic and antimony during casting and by the addition of tin. But the differences may just as well show that the material was collected for recasting from different sources (table 10.1).

The Bräckan find is from the province of Dalsland (figure 10.3). It is interesting to note that in this province there are numerous copper, silver, and lead deposits (Tegengren et al. 1924, p. 368). But we need more analyses of ore, ingots, bars, rods, and slags to discuss the difficult problem of the origin of the raw material of our early copper-based artifacts.

Alloys

In the Swedish material of early copper-based artifacts there are different proportions of metals and/or trace elements represented. The question is when we can speak about an intentional alloy.

Some quantitative analyses have been made from copper-based artifacts in Scandinavia. Cullberg's work (1968) "presents about 900 analyses of artifacts from Denmark, Norway, and Sweden—mainly flat and flanged axes. Later, Oldeberg (1976, pp. 120ff) added 125 analyses of Early Bronze Age artifacts from Sweden.

To illustrate the differences in the chemical composition I have grouped 341 of the analyzed Swedish artifacts, which contain 1–10% tin (Sn) and more than 1% of arsenic (As), antimony (Sb), silver (Ag), and nickel (Ni) [figure 10.11; compare Tylecote (1962, pp. 39ff)]. The diagram shows a group of thirty-seven copper-based artifacts containing less than 1% of other metals. The groups with $>1\%$ of arsenic, antimony, silver, or nickel contain few artifacts, less than twenty. The artifacts with tin have been divided into five groups. The first group, with 1.0–2.0% tin, is a small group with eighteen artifacts, but in the next group, 2.1–4.0% tin, there are fifty-seven artifacts.

It should be observed that the chemical composition can differ depending on where in the object the sample is taken. Therefore it would be recommended to take at least two samples from objects, especially if they had been cast in a vertically placed mold, for example, axes (Forshell 1984, p. 10).

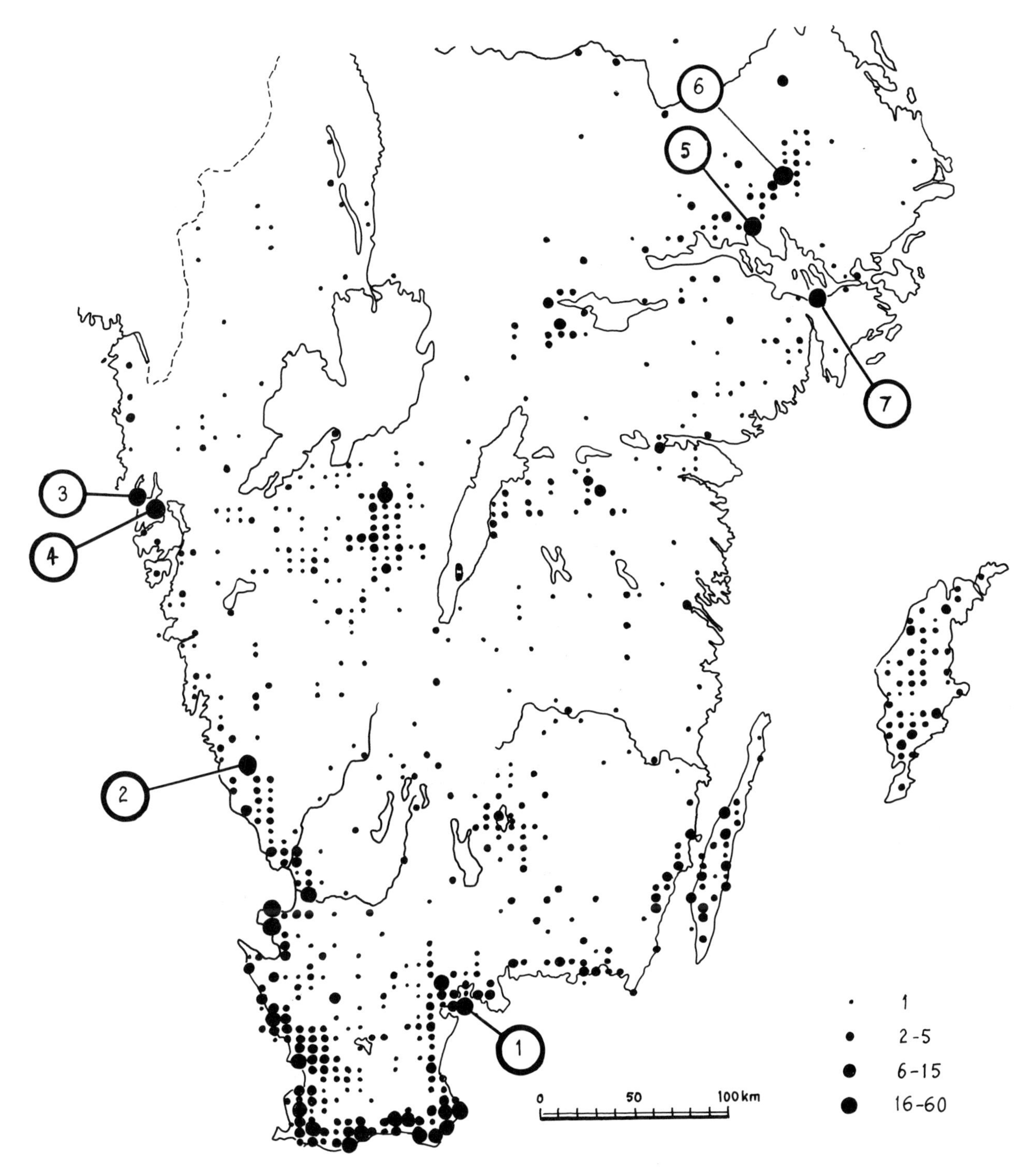

Figure 10.9 Registered finds of Early Bronze Age artifacts (after Oldeberg 1976, p. 92) and seven larger workshop finds (Montelius, period 2–5): (1) Bromölla, Ivetofta parish, Skåne (Petré 1961, p. 47); (2) Broåsen, Grimeton parish, Halland (Sarauw and Alin 1923, pp. 258ff); (3) Humlekärr, Lyse parish, Bohuslän (Särlvik et al. 1977, pp. 651ff); (4) Bokenäs, Bokenäs parish, Bohuslän (Niklasson 1948); (5) Skälby, Vårfrukyrka parish, Uppland (Oldeberg 1960); (6) Broby, Börje parish, Uppland (Schönbeck 1952); (7) Hallunda, Botkyrka parish, Södermanland (Jaanusson and Vahlne 1975).

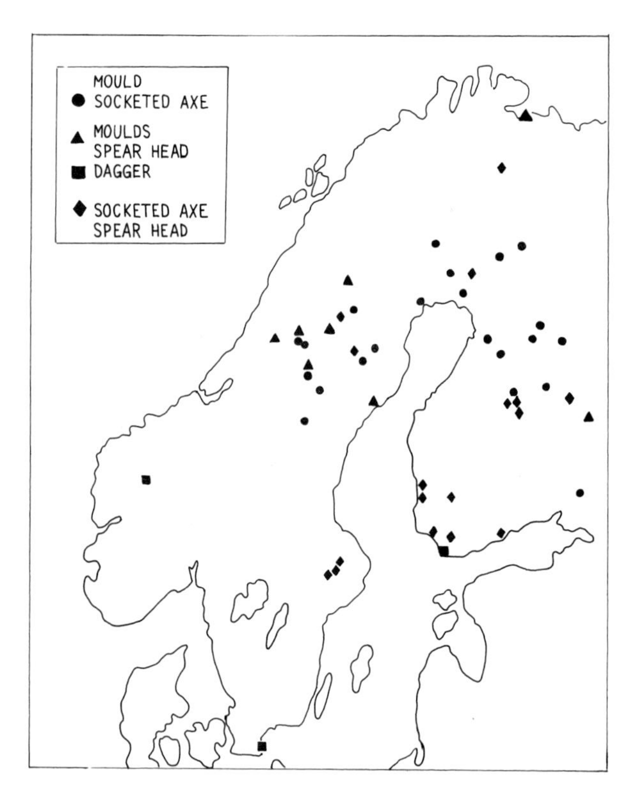

Figure 10.10 Finds of Arctic bronze artifacts and molds in Fennoscandia. After Broadbent (1982); compare Bakka (1976).

Table 10.1 Semiquantitative analyses of the Bräckan find, Jern parish, Dalsland[a]

Oldeberg (1974) artifact number	Site	Object	Sn (%)	Pb	Zn	Sb	As	V	Ni	Au	Ag
263	Bräckan	rod	8	—	—	tr[b]	tr	tr			+
264	"	"	8	—	—	+ +	tr	tr			+
265	"	"	8	—	—	tr	tr	tr			+
266	"	"	8	—	—	tr	tr	tr			+
267	"	"	8	—	—	tr	tr	tr			tr
268	"	"	8	—	—	tr	tr	tr			+
269	"	"	8	—	—	tr	tr	tr			+
270	"	ingot cake	+	—	—	tr	tr	tr			+ + +
271	"	"	+	—	—	+ +	+ +	tr			+ + +

a. After Oldeberg 1942–1943, vol. 2, nos. 263–271.
b. trace.

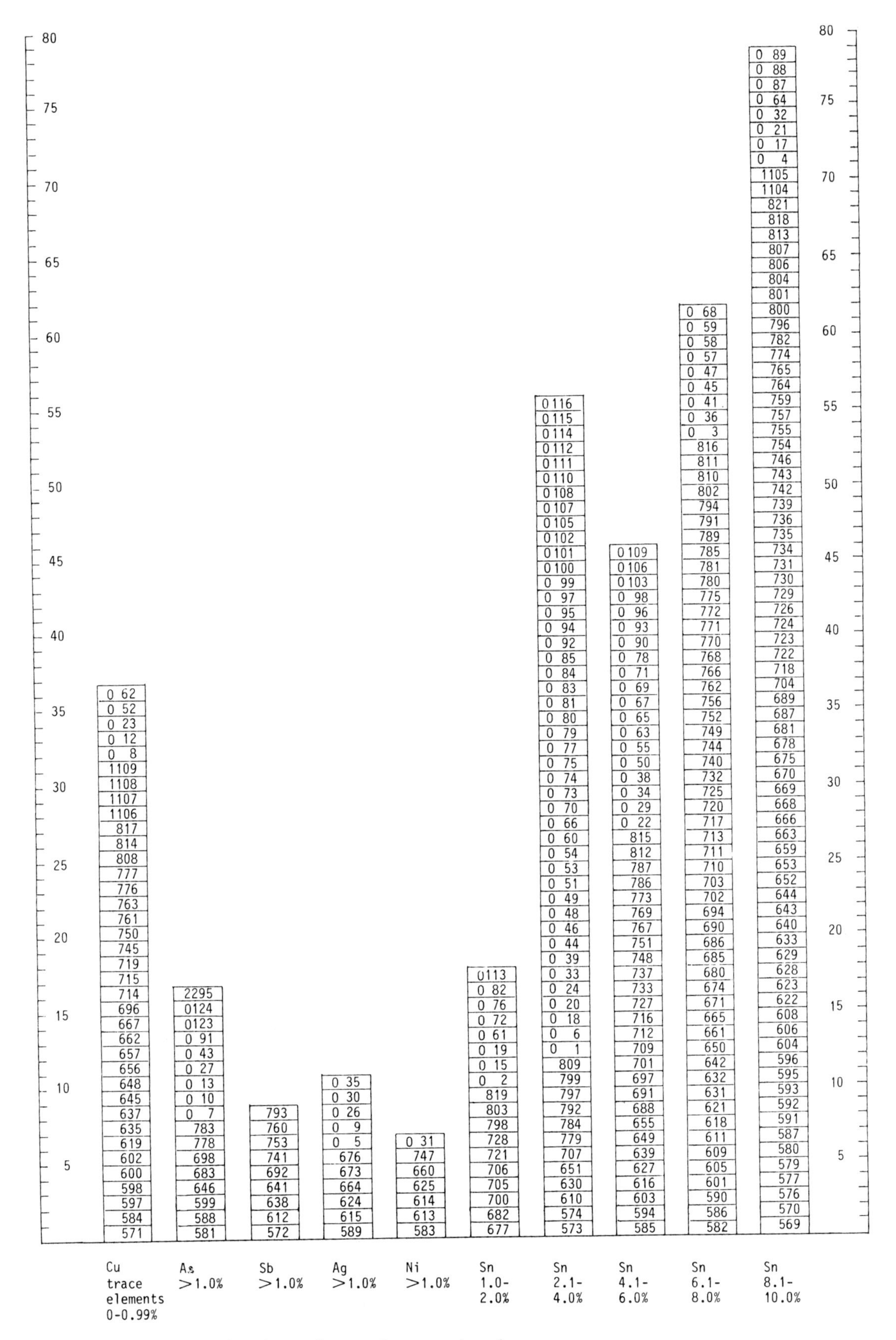

Figure 10.11 Grouped analyses of 341 early copper-based artifacts in Sweden, mainly flat and flanged axes (see figure 10.12). Number of analyses after Cullberg (1968), Oldeberg (1976) (numbers beginning with 0) and Oldeberg (1974) (column 2).

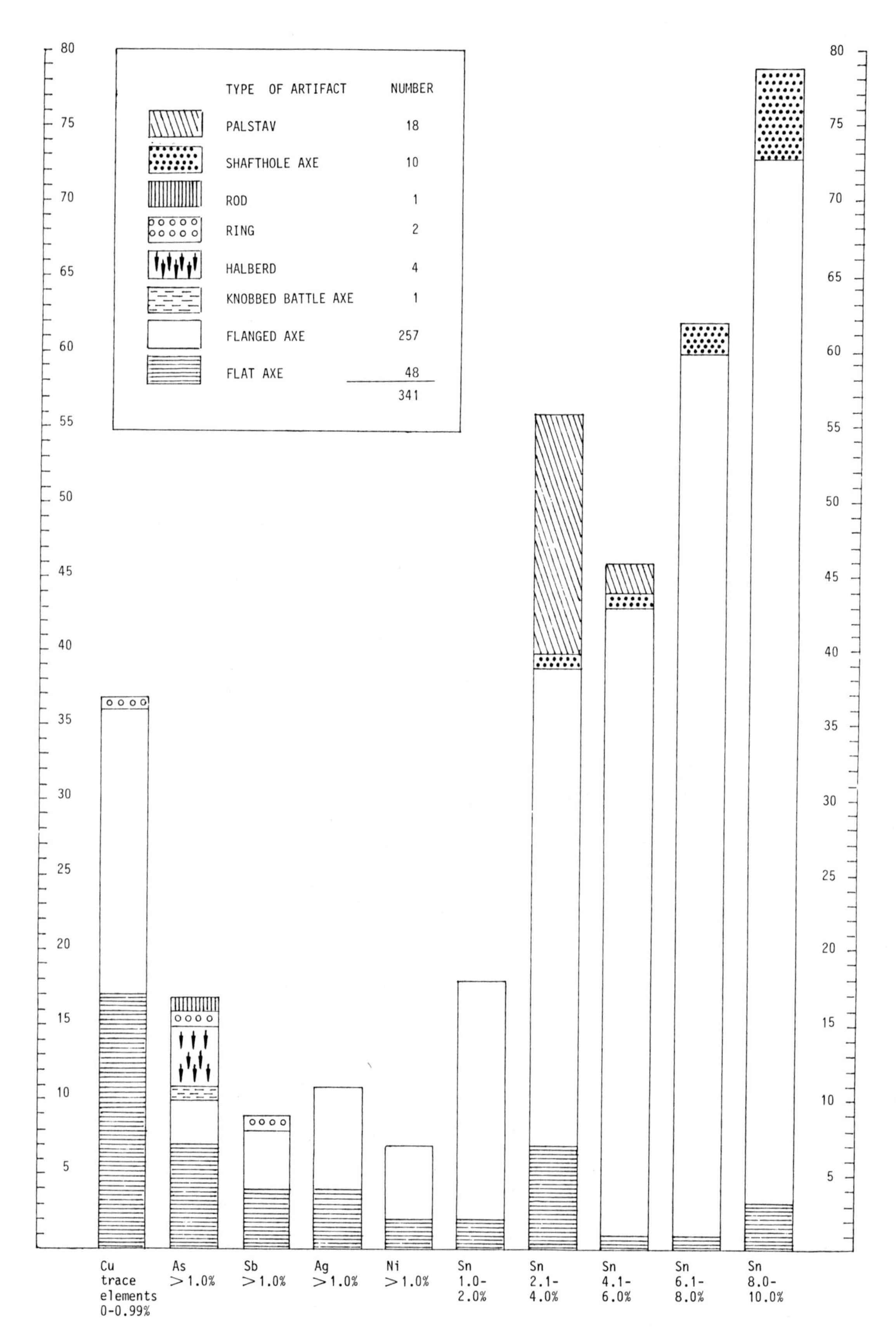

Figure 10.12 Eight types of artifacts represented in the ten groups of analyses in figure 10.11.

The eight types of artifact represented in the ten groups of analyses are shown in figure 10.12. Flat axes are most numerous in the "copper group", but they are present in all groups, and it is noteworthy that there are also flat axes containing at least 10% tin (Cullberg 1968, no. 658). The group with > 1% arsenic contains six types of artifacts, including a ring and a rod, probably raw material. Also the halberd blades are in the arsenic group. Most of the palstavs (sixteen) are in the group with 2.1–4.0% tin. Only two palstavs are in the next group.

The arrangement of the material in figures 10.11 and 10.12 shows how special types of artifacts, for example, halberds and palstavs, are tied to certain compositions of metal or alloy. Whether or not there are differences in the geographic distribution remains to be investigated.

Of special interest in the analyzed material are artifacts from the Early and the Middle Bronze ages, when the material still was not so much reused. This analysis may perhaps give us a chance to trace a certain ore with specific elements and/or specific composition (Coghlan and Case 1957, pp. 91–123; Tylecote 1962, p. 42).

Research

Since 1980 the Department of Archaeology, Stockholm University, has had a special section for archaeometallurgical studies (AMI). It is located in Håksberg in the heart of the mining district of central Sweden. Head of the institute is Inga Serning. Investigations and analyses made at the institute are published in a series of reports.

In Jernkontoret (The Swedish Ironmasters Association) the Historical Metallurgy Group also has a special comittee working with the problems of early nonferrous mining and metallurgy. Experts in modern industry have helped us with quite a few analyses.

The following projects are now in progress. In the Hallunda project Gunnel Vahlne studies the workshop material (furnaces, crucibles, molds, slag, etc.) from the Late Bronze Age site of Hallunda south of Stockholm (Vahlne 1974). Lena Forshell works with trace element patterns in medieval copper and their connection with mining and production (Forshell 1983, 1984, 1986). Ylva Roslund studies medieval lead and silver mines, and Ola Nilsson studies medieval lead and silver production (Nilsson 1986, pp. 113ff).

One project, of which I am head, works with the hypothesis that Sweden's own numerous sulfide ore deposits could have been used in prehistoric times. The site chosen for the investigation is Hellerö, located at the eastern coast in the province of Småland, southern Sweden, where we have one of the most intensive Bronze Age areas in the country. At the same site are represented culture layers, varied types of Bronze Age monuments, and copper deposits. We have today indications of the third manufacturing stage of metallurgy: slag and fragments of crucibles or molds. The fieldwork is still going on, and we hope eventually to get evidence also for mining and smelting activities at the site.

References

Bakka, E. 1976. "Arctic and Nordic in the Bronze Age of northern Scandinavia." *Det kgl. Norske Videnskabers Selskab. Miscellanea* 25: 1–58. (In Norwegian.)

Broadbent, N. 1982. *Skelleftea's History. Part 3, Prehistoric Development 6000 B.C. to A.D. 1000.* Uppsala: Almqvist & Wiksell. (In Swedish.)

Coghlan, H. H., and H. J. Case. 1957. "Early metallurgy of copper in Ireland and Britain." *Proceedings of the Prehistoric Society* 23: 91–123.

Coles, J. M. 1981. "Metallurgy and Bronze Age society," in *Studies on the Bronze Age: Festschrift for Wilhelm Albert von Brunn*, H. Lorenz, ed. (Mainz am Rhein: Verlag Philipp von Zabern).

Cullberg, C. 1968. "On artifact analysis: A study in the systematics and classification of a Scandinavian Early Bronze Age material with metal analysis and chronology as contributing factors." *Acta Archaeologica Lundensia*, ser. 4, no. 7.

Dons, J. A. 1975. "Telemark's geology: The province which has everything." *Telemark Bygd og By i Norge.* (In Norwegian.)

Forshell, H. 1983. *Outdoor Bronze Monuments: A Conservation Problem.* Armemuseum Report 1. Stockholm: Armemuseum.

Forshell, H. 1984. *Bronze Cannon Analysis: Alloy Composition Related to Corrosion Picture.* Armemuseum Report 2. Stockholm: Armemuseum. (In Swedish.)

Forshell, H. 1986. "Chemical analyses of ancient bronze alloys. Influence on corrosion picture of low or high Sn-Pb-Zn-Fe." Master's thesis, Department of Archaeology, Stockholm University.

Glob, P. V. 1952 *Danish Prehistoric Artifacts*, vol. 2. Copenhagen: Gyldendalske Boghandel. Nordisk Forlag. (In Danish.)

Grälls, A. 1986. "A study of slag deposits in the northern part of Norrland." Department of Archaeology, University of Uppsala. Unpublished. (In Swedish.)

Hulthén, B. 1976. "Technical investigations for evidence of continuity or discontinuity of ancient ceramic traditions." *Dissertationes Archaeologicae Gandenses* 16: 120–121.

Indreko, R. 1956. *Grooved Stone Tools.* Stockholm: Vitterhets Historie och Antikvitets Akademien Handlingar, Ant. ser. 4. (In German.)

Jaanusson, H. 1981. *Hallunda: A Study of Pottery from a Late Bronze Age Settlement in Central Sweden.* Study 1. Stockholm: The Museum of National Antiquities.

Jaanusson, H., and G. Vahlne. 1975. *Excavation 1969–71 of Settlement Remains at Hallunda in Botkyrka Parish, the Province of Södermanland.* Part 2. Report. Stockholm: Central Board of National Antiquities. (In Swedish.)

Jaanusson, H., L. Löfstrand, and G. Vahlne. 1978. *Late Bronze Age Graves and Settlement at Hallunda, Botkyrka Parish in the Province of Södermanland.* Report B. Stockholm: Central Board of National Antiquities. (In Swedish.)

Jackson, J. S. 1971. "Mining in Ireland." *Technology Ireland* 3(7):30–33.

Janson, S., and H. Hvarfner. 1960. *From Rivers and Lakes in Norrland.* Stockholm: Vitterhets Historie och Antkvitets. (In Swedish.)

Janzon, G. O. 1984. "Grooved stone hammers: Indication of early metallurgy?" *Jernkontorets Forskning,* ser. H, no. 32. (In Swedish.)

Jovanović, B. 1978. "The origins of metallurgy in southeast and central Europe and problems of the earliest copper mining," in *The Origins of Metallurgy in Atlantic Europe,* M. Ryan, ed. (Dublin: Stationery Office), 335–343.

Jovanović, B. 1980. "Primary copper mining and the production of copper," in *Scientific Studies in Early Mining and Extractive Metallurgy,* P. T. Craddock, ed. (London: British Museum), 31–40.

Lindahl, A. 1955. "The Stone Age double grave at Bergsvägen in Linköping I. The circumstances of the find and the grave construction." *Östergötslands och Linköpings Stads Museum,* 5–28. (In Swedish.)

Malmer, M. P. 1962. "Neolithic studies." *Acta Archaeologica Lundensia,* ser. 8, no. 2. (In German.)

Malmer, M. P. 1975. *The Battle Axe Culture in Sweden and Norway.* Lund: Liber Läromedel. (In Swedish.)

Niklasson, N. 1948. "A prehistoric settlement at Bokenäs's old church." *Göteborgs och Bohusläns Fornminnesförenings Tidskrift,* 45–79. (In Swedish.)

Nilsson, O. 1986a. "The Middle Age copper craft in Sweden." *Fjölner* 1:113–130. (In Swedish.)

Nilsson, O. 1986b. "The copper-smelting house at Kaspersbo in Garpenberg Parish," in *Prehistoric Iron Research 1980–83* (Jernkontorets Forskning, ser. H, no. 38). (In Swedish.)

Oldeberg, A. 1929. "The Bräckan find: A foundry find from the fourth period of the Bronze Age." *Rig* 12:37–48. (In Swedish.)

Oldeberg, A. 1942–1943. *Metal Technology in Prehistoric Time,* 2 vols. Leipzig: Kommissionsförlag Otto Harrassowitz. (In German.)

Oldeberg, A. 1960. "The Skälby find: A Late Bronze Age settlement." *Kungl. Vitterhets Historie och Antikvitets Akademien,* Ant. Ark. 5. (In Swedish.)

Oldeberg, A. 1974, 1976. *The Early Metal Age in Sweden,* 2 vols. Stockholm: Almqvist & Wiksell. (In Swedish.)

Petré, R. 1959. "A Bronze Age village in Bromölla," in *Skånes Hembygdsförenings Årsbok 1959* (Lund), 47–70. (In Swedish.)

Petré, R. 1961. "A Bronze Age mound at Nymölla, Ivetofta Parish, Scania." *Meddelanden från Lunds Historiska Museum,* 33–79. (In Swedish.)

Qvarfort, U. 1980a. "The cultivation influence in Lake Tisken and the opening of Falu copper mine." *Jernkontorets Forskning,* ser. H, no. 19, 1–13. (In Swedish.)

Qvarfort, U. 1980b. "The main sulfide ore deposits in Sweden," in *Malm Metall Föremål,* vol. 2, I. Serning, ed. (Stockholm: Stockholm University). (In Swedish.)

Qvarfort, U. 1981. "The beginning of sulfide ore mining in the Garpenberg and Öster Silvberg areas." *Jernkontorets Forskning,* ser. H, no. 20. (In Swedish.)

Qvarfort, U. 1984. "The influence of mining on Lake Tisken and Runn." *Bulletin of the Geology Institute of the University of Uppsala* 10.

Ramqvist, P. 1984. "Gene: On the origin, function, and development of sedentary Iron Age settlement in northern Sweden." *Archaeology and Environment* 1.

Randsborg, B. 1978. "Resource distribution and the function of copper in Early Neolithic Denmark," in *The Origins of Metallurgy in Atlantic Europe,* M. Ryan, ed. (Dublin: Stationary Office).

Rothenberg, B., and A. Blanco Freijeiro. 1980. "Ancient copper mining and smelting at Chinflon (Huelva, SW Spain)," in *Scientific Studies in Early Mining and Extractive Metallurgy,* P. T. Craddock, ed. (London: British Museum), 41–62.

Rothenberg, B., R. F. Tylecote, and P. S. Boydell. 1978. *Chalcolithic Copper Smelting. Archaeometallurgy.* IAMS Monograph 1. London: Institute for Archaeometallurgical Studies.

Sarauw, G., and J. Alin. 1923. *The Prehistoric Remains in the Götariver Area.* Göteborg: Göteborgs Jubileumspublikationer. (In Swedish.)

Särlvik, I., E. Weiler, and M. Jonsäter. 1977. *Excavations in Lyse Parish, Bohuslän.* Report B. Stockholm: Central Board of National Antiquities. (In Swedish.)

Schmid, E. 1973. "The extension of prehistoric Silex mines in Europe." *Der Anschnitt* 25(6):25–28.

Schönbeck, B. 1952. "Bronze Age houses in the province of Uppland." *Tor,* 23–45. (In Swedish.)

Serning, I. 1979–1980. "Ore—Metal—Artifact." Compendium in Archaeology, Stockholm University. (In Swedish.)

Serning, I. 1984. "The dawn of Swedish metallurgy." *Bulletin of the Metals Museum* (Japan) 9:3–19.

Söderberg, T. 1932. "Åtvidaberg in the Middle Ages." *Med Hammare och Fackla* 1932:4. (In Swedish.)

Sommarin, E. 1910. "The foundation of Sweden's oldest mine and its organization in the reign of King Magnus Ladulås." *Statsvetenkaplig Tidskrift* 1910:13. (In Swedish.)

Stenberger, M. 1964. *Prehistoric Sweden.* Uppsala: Almqvist & Wiksell. (In Swedish.)

Stjernquist, B. 1969. "Contributions to the study of Bronze Age settlements." *Acta Archaeologica Lundensia,* ser. 8, no. 8. (In German.)

Strömberg, M. 1975. "Studies of a grave field at Löderup." *Acta Archaeologica Lundensia,* ser. 8, no. 10. (In German.)

Taavitsainen, J.-P. 1982. "A copper ring from Suovaara in Polvijärvi, northern Karelia." *Fennoscandia Antiqua* 1:41–49.

Tegengren, F. R., et al. 1924. "Sweden's precious ores and mines." *Sveriges Geologiska Undersökning* (Stockholm), ser. CA, no. 17. (In Swedish.)

Tylecote, R. F. 1962. *Metallurgy in Archaeology: A Prehistory of Metallurgy in the British Isles.* London: Edward Arnold.

Tylecote, R. F. 1970. "The composition of metal artifacts: A guide to provenance?" *Antiquity* 44:19–25.

Vahlne, G. 1974. "The foundry finds from Hallunda (Södermanland) with stress on the interpretation of the molds." Department of Archaeology, Stockholm University. Unpublished. (In Swedish.)

Weiler, E. 1984. "The Bronze and Iron ages—1500 B.C. to A.D. 1000," in *From Flint Workshops to Industry* (Kungslv: Riksantikvarieambetet), 49–114. (In Swedish.)

Žuravlev, A. P. 1975. "On ancient copper metallurgy centers in Karelia." *Kratkiye Soobshcheniya* 142:31–38. (In Russian.)

11
Iron Production and Iron Trade in Northern Scandinavia

Gert Magnusson

In 1973 the Jämtland County Museum embarked on a minor research enterprise concerning low-technology ironwork in the county. For the first five years there was intensive fieldwork at Myssjö, Oviken, Rödön, Rätan, and Offerdal. During the same period a number of investigations were carried out, either in the form of rescue archaeology or at the instance of interested local communities or enterprises in Sveg, Ängersjö, and Klövsjö.

The fieldwork was financed by the National Labor Market Board (AMS). The AMS, together with grants from the County Council and the Dean Erik Andersson Memorial Fund, made possible, for the first time ever in Jämtland and, indeed, in Sweden, systematic and scheduled investigations. This proved to be a highly fruitful approach. A large volume of material was soon collected, and I describe here the results of several years' analysis. The inquiry has been conducted entirely on a regional basis, with the aim of illuminating the iron industry of Jämtland and Härjedalen from as many different angles as possible. Regional inquiries of this kind have subsequently been mounted in several other counties. Similar research has now been in progress for a number of years in the counties of Skaraborg, Örebro, and Västmanland. Similar programs are planned for the counties of Gävleborg, Södermanland, and Halland, and the Kalmar County authorities may be following suit (figure 11.1).

Methodologically, the work has proceeded retrospectively, using preindustrial conditions in Jämtland and Härjedalen to account for earlier conditions and to gauge the applicability of such a method to the material constituted by iron-manufacturing sites.

The inquiry has had a natural foundation in the findings of the Archaeological Remains Inventory. All the sites investigated were selected with reference to the qualities discernible from the inventory documents. This presented a number of problems in the early years because in a number of cases the sites chosen proved to have been completely destroyed. Since the 1975 field season, however, preparatory planning was entirely free from these problems, and each of the sites chosen was successfully investigated. In other words, the archaeological investigations revealed traces of slag heaps, roasted ore, and furnaces and an abundance of charcoal. Generally, the iron-manufacturing sites of Jämtland are among the absolutely best preserved in Europe. It is more the exception than the rule for a site to be so badly damaged that the work formerly done there cannot be reconstructed in detail. This presents antiquarian authorities with a major international duty of research and preservation.

With two exceptions—at the blast furnaces at Rönnöfors and Ljusenedal—iron production in the county of Jämtland employed what is known as the bloomery furnace process. In technical terms this pro-

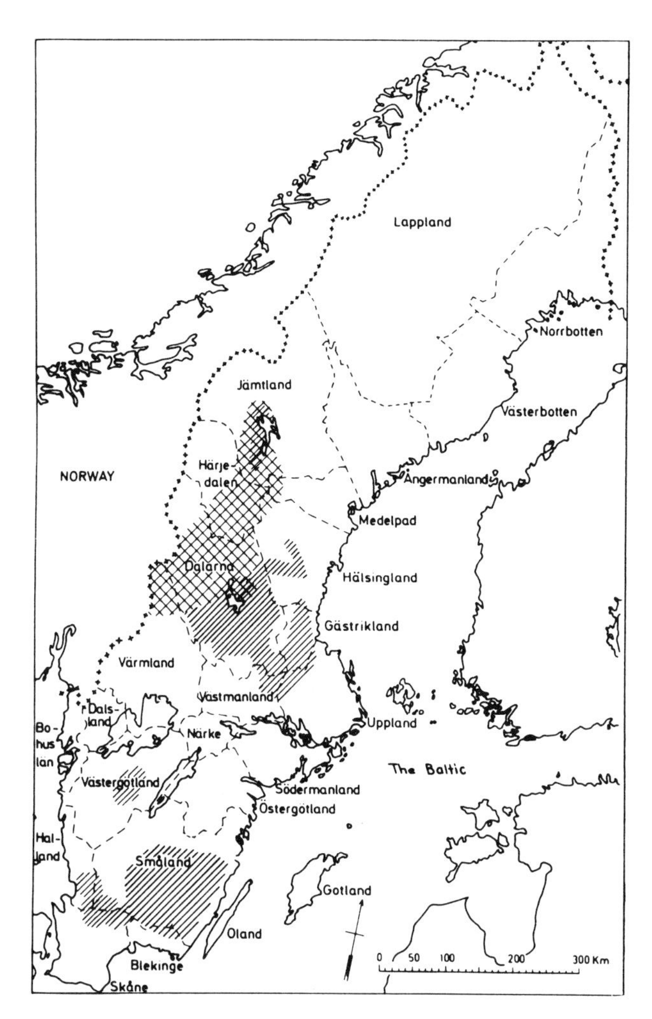

Figure 11.1 Areas with remains of primitive iron production. Cross-hatching indicates late type; single hatching, early type. After Hyenstrand (1979).

cess is distinguished from the blast furnace process by two main principles:

1. In a bloomery furnace, it is only the slag, not the iron, that is melted. In this way malleable iron is obtained directly. In the blast furnace both iron and slag are melted. The iron then carburizes into pig iron, which then has to be refined to make it valuable.
2. The bloomery furnace process is conducted intermittently, work in the furnace being interrupted when the iron is removed. The blast furnace operates continuously, at the same time as molten iron is teemed from it.

The bloomery furnace technique suited the kind of society existing in Jämtland during prehistoric times to the nineteenth century. It made an excellent sideline to agriculture and cattle farming. It did not require any heavy initial expenditure of labor or capital, and production could take place within the family, without any appreciable modifications of social organization. Work on an iron production site had to comply with a certain pattern. This was necessary in order for iron production to be feasible, and there were six important stages:

1. Collection of the raw material (in the form of ore, lake ore, bog ore, or other limonite precipitations) and drying of the ore.
2. Collecting fuel—firewood—for the process.
3. Roasting the ore to burn off organic material, crystallized water, and any sulfur.
4. Charcoal burning.
5. Reduction of the ore in a bloomery furnace.
6. Processing of the unmelted lump of iron, the bloom, to reduce the slag content still further. Hammering against an anvil stone.

This process implies that every iron production site must have a number of facilities for the different stages of the ironworking process, namely storage places for the accumulated ore, a roasting pit, a charcoal pit, a bloomery furnace, a slag heap, and an anvil stone where the last of the slag was hammered out. Adjoining some ironworking sites there may also be housing in the form of simple mud huts (figure 11.2). This depends more on the distance between prehistoric settlements and ironworking sites. Hut foundations have been discovered at ironworking sites in the parish of Berg.

The archaeological remains inventory has made it possible to deduce where ironworking was conducted on a significant scale. It should be pointed out, however, that few things are more difficult to catalogue than ironworking sites. Their location is virtually unpredictable. The only truly effective method is to track them down on foot, pacing every yard of the area. In Jämtland and Härjedalen this is an immense assignment and has not yet been undertaken. The inventory can be regarded as a good overview. Written sources and place names do not contradict the inventory findings in any way, but some areas appear to be underrepresented. This is above all the case with the Revsund *tingslag*, where, although the inventory does not mention any ironworking sites, we find sixteen farmers paying taxes in iron during the sixteenth century. Otherwise the inventory findings tally closely with the sixteenth-century cadastre (*jordebok*), but sixteenth-century iron dues are unlikely to be an infallible criterion of the location of ironworking sites.

In the inventory material there are five Jämtland parishes with a large number of ironworking sites. Above all we have Rätan, with eighty-six known sites, Oviken with seventy-four, Myssjö with fifty-three, and Offerdal with forty-six. In Härjedalen it is mainly the eastern parishes that have numerous known ironworking sites. The wealthy parishes in this respect are Sveg, with sixty-five sites, Ytterhogdal with forty-two, Alvros with about fifty-five, and Lillhärdal with thirty-four.

Altogether we have 436 known sites in Jämtland and 284 in Härjedalen (figure 11.3). It should be added

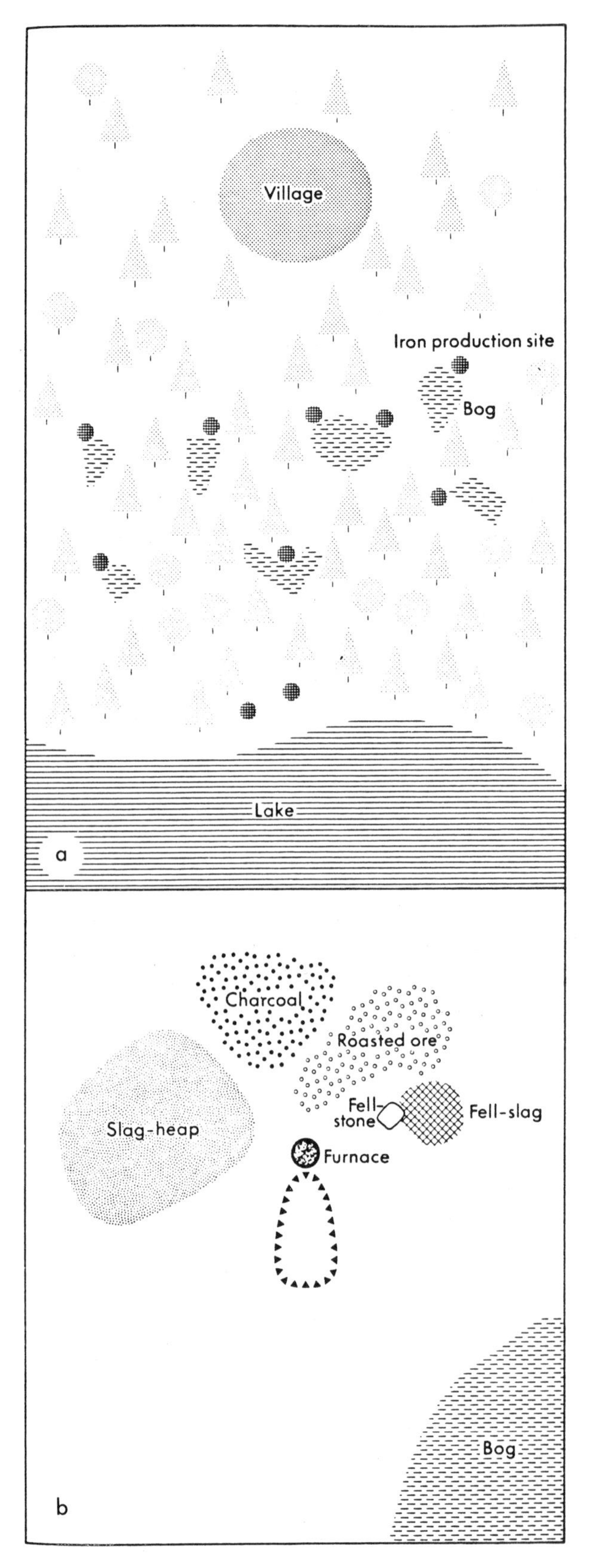

Figure 11.2 (a) Model of iron production sites and their relation to the permanent village. (b) Model of an ordinary iron production site.

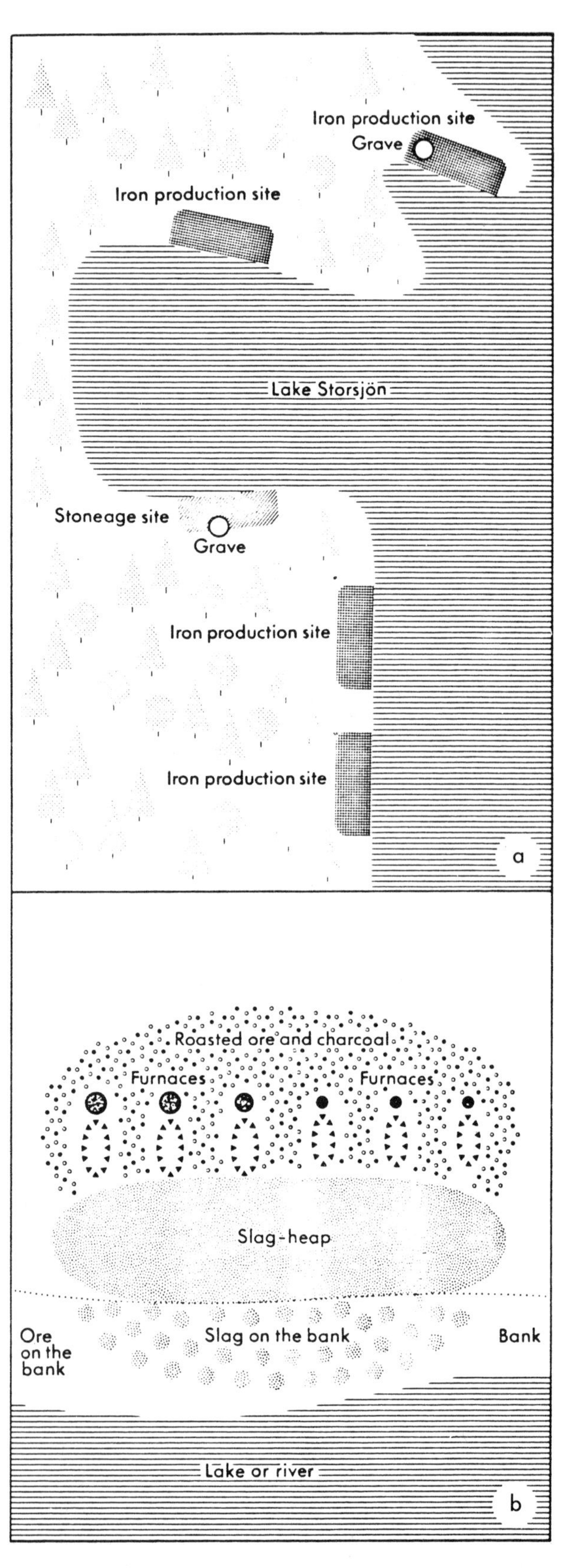

Figure 11.3 (a) Iron production sites at Josvedsviken, Rödöns Parish, Jämtland. (b) Iron production site in Rödöns Parish, Jämtland.

that these are minimum figures. The true number is difficult to gauge, but 1,000 sites or more would not be surprising. In other words, this is one of the largest ironworking areas in Sweden and perhaps—with regard to the number of sites preserved—in Europe.

The raw material used at the sites was mainly bog ore and limonite precipitations adjoining shale deposits around Storsjön, that is, a type of lake ore.

Roughly there are three types of ironworking site in Jämtland and Härjedalen. They are distinguished mainly by the type of furnace. One type is directly associated with the limonite precipitations around Storsjön. The other two are connected with bogs and fens. Most of the ironworking sites in Härjedalen come under this second category.

Three Types of Furnace

Three main types of furnace occur in Jämtland: (1) clay-lined pit furnaces, (2) clay-lined shaft furnaces, and (3) dry-stone walled pit furnaces. Altogether, using various criteria, one could distinguish seven different types, but five of these can be considered variations of type 3.

In principle, there was just one type of furnace—the clay-lined pit furnace—in use in Jämtland during prehistoric times. This appears to have been introduced during the Roman Iron Age. There is nothing to show that it was still being used after 1100. The design resembles similar furnaces in Tröndelagen (Norway) and in the Swedish provinces of Västergötland, Västmanland, and Närke, all of which are similarly dated.

The clay-lined shaft furnace occurs between 1100 and 1400. This type is rare and occurs at only a few sites in Jämtland. But the type is well known from other parts of Sweden. It is the commonest type of furnace in Västergötland and has its continental counterparts.

The third type is most common in both Jämtland and Härjedalen, and several hundred examples are known. The masonry of its walls ranges from dry stone to complete boulders. It can be dug into a sloping hillside or built above ground. It occurs between the thirteenth and mid-nineteenth centuries. Dry stone furnaces have been described by Mandelgren (1978).

Ages of the Sites

Dating the ironworking sites has been a problem in certain respects. Basically there is only one method available: carbon 14 analysis. This has many shortcomings and presents great difficulties. In fact, the carbon 14 method gives us only an indirect date for the operation of bloomery furnaces. What it really gives us is the death date of the tree that was used for charcoal in the furnaces. Consequently many specimens have to be studied and then subjected to a statistical procedure to reduce the effects of error. But a margin of error always remains. The ancient forests had plenty of old, dried trees that may have been put to use. There are records from the late eighteenth century suggesting that material of this kind was used, but this was at a time when forests began to be valued for their timber.

A test series of radiocarbon dates now exists for Jämtland and Härjedalen (figure 11.4); the series is based on about 100 specimens that were analyzed in Stockholm. The dates indicate two main horizons for the iron industry in Jämtland: a prehistoric horizon spanning from 200 B.C. to A.D. 1200, and a more recent horizon between about 1050 and 1850. These are wide ranges and require some discussion.

The older horizon is distinctive. It is shore bound and mainly confined to Storsjön. It also occurs on the Indalsälven River, on the Långan River at Lit, and on the Hårkan at Föllinge. There are also one or two sites in the mountains at Kall. These prehistoric sites are always characterized by a slag eroded out onto the river banks. In many cases excavations have shown parts of the slag heap to be incorporated in the embankment structure. Behind the slag heap are the bloomery furnaces, often in a line. Roasted ore, charcoal, and slag are scattered all over the area. There are ninety known sites in Jämtland, twenty of which (about 25%) have been carbon dated.

Dating of the more recent horizon is based on about 60 out of 600 dated sites (10%). The great majority of these sites are in forest areas, directly adjoining wetlands.

The average dating of the earlier horizon is between A.D. 300 and 600, with a certain overspill into the succeeding centuries. At present, therefore, one would venture to say that the iron industry in Jämtland began during the late Roman Iron Age and attained its apogee during the period of the great migrations. This was followed by a decline during the Vendel and Viking

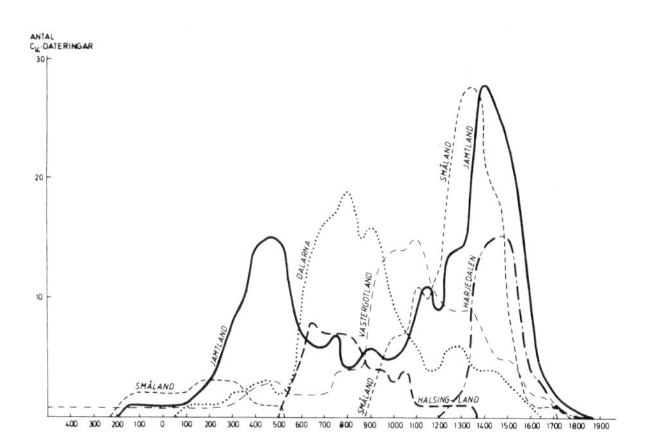

Figure 11.4 Radiocarbon dates from various parts of Sweden.

eras and then, in the High Middle Ages, by a new peak, this time employing a partly new furnace technology. The last peak occurs during the fifteenth and sixteenth centuries and is followed by a downturn until the industry disappears altogether in the nineteenth century. The fifteenth- and sixteenth-century peaks are clearly confirmed by written records from this period. The iron industry in Jämtland and Härjedalen would appear to have described an almost cyclical course ever since the Roman Iron Age. It is worth noting that the long-known early phase of Iron Age settlement at Storsjön coincides with this early phase of the iron industry.

The earliest ironworking phase may in fact have been quite heavily concentrated within a period of some 300 years, but it may also have been spread over, say, 900 years. This cannot be decided at present, but it does of course have an important bearing on the possible intensity of production. The same question may also apply to the fifteenth and sixteenth centuries.

Iron Production

Calculating the amount of iron produced in Jämtland in prehistoric times is both difficult and hazardous because there are so many different factors involved. An estimate of this kind, however, must necessarily be based on the quantity of slag remaining at the various ironworking sites.

Crucial factors governing the volume of output are furnace capacity and the level of skill (that is, the failure rate). Then again there is uncertainty regarding the size of the slag heaps—how much has been preserved, how much of each slag heap has been damaged, and so on. Even so, and despite the great uncertainty involved, I do not refrain from estimating the volume of outputs.

I have contented myself here with estimating the volume of the slag discharged on the shores of Storsjön and on the banks and shores of other rivers and lakes where these ironworking sites are found. The reference here is solely to shore-bound ironworking sites, that is, those dating from prehistoric times. I do not even attempt to gauge the number of slag heaps built into the embankments. Excavations have shown that these embankments could often incorporate quite considerable amounts of slag, but there are other sites that have been eroded away completely. After calculating the volume, I worked out an average for the specific weight of the slag. I then multiplied this by an average iron-to-slag ratio obtained from laboratory experiments and presented by a number of scholars of international repute.

In this way I have been able to estimate a minimum output volume for the prehistoric period. At least 2,821 tons of iron must have been produced in Jämtland. For the reasons already given, the true output was probably a good deal higher. If all this iron was forged into spade-shaped billets, the number of billets produced cannot have been less than 4,030,000. Of course, this is all a game with figures, but it conveys some idea of the magnitude of iron production. The following conjecture takes us a step further:

Time	Annual output	Number of billets per year
300 years	9 tons	13,428
800 years	3.5 tons	5,000

Thus the largest find of spade-shaped billets—126 billets at Månsta—equals roughly 0.9–2.5% of the presumed minimum annual output.

How large was Jämtland's iron output compared with other iron-producing areas? Jämtland is perhaps one of the more important ironworking areas in Scandinavia during certain periods of the Iron Age. Viewed internationally, however, it becomes a good deal less important (table 11.1). The comparison in the table has been undertaken in this way so as to prevent the duration of production periods from being decisive.

One is immediately struck by the great difference between the Roman Empire and northern Europe, the immense difference in size between the different societies. This, of course, is apparent in many other social terms, for example, population, towns and cities, and military organization.

In an attempt to construct models of the existence of ironworking in historical communities of Jämtland and Härjedalen (figures 11.5 and 11.6), a retrospective approach has been adopted in order to identify the social sphere within which ironworking existed there.

The eighteenth-century sites have almost exactly the same layout as the thirteenth- and fourteenth-century ones. Slag heap, furnaces, ore, and charcoal stores are identically sited throughout the period. The furnace technology is the same. Given these similarities, we may presume that ironworking filled a similar role in the agrarian economy of Jämtland and Härjedalen throughout the period. It was a more or less important sideline for an agrarian population whose social family pattern was much the same as in later centuries. Production was feasible within the family.

Table 11.1 Comparison of iron production in different areas

Country or province	Output (tons)	kg/day	Total
Jämtland	3.5–9.4	1–3	2,821
Dalarna	3	0.8	2,100
Italy, Populonia	1,200	3,200	575,000
Poland, H. Kreutzgebirge	10.8	3–4	5,400

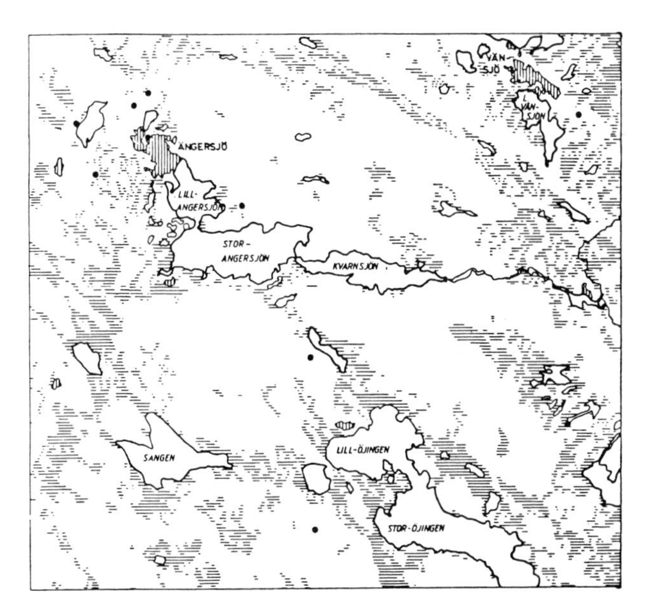

Figure 11.5 Iron production sites in the parish of Ängersjö in the county of Härjedalen. •, Iron production site; vertical hatching, village and cultivated area; horizontal hatching, bog; outlined areas, lakes.

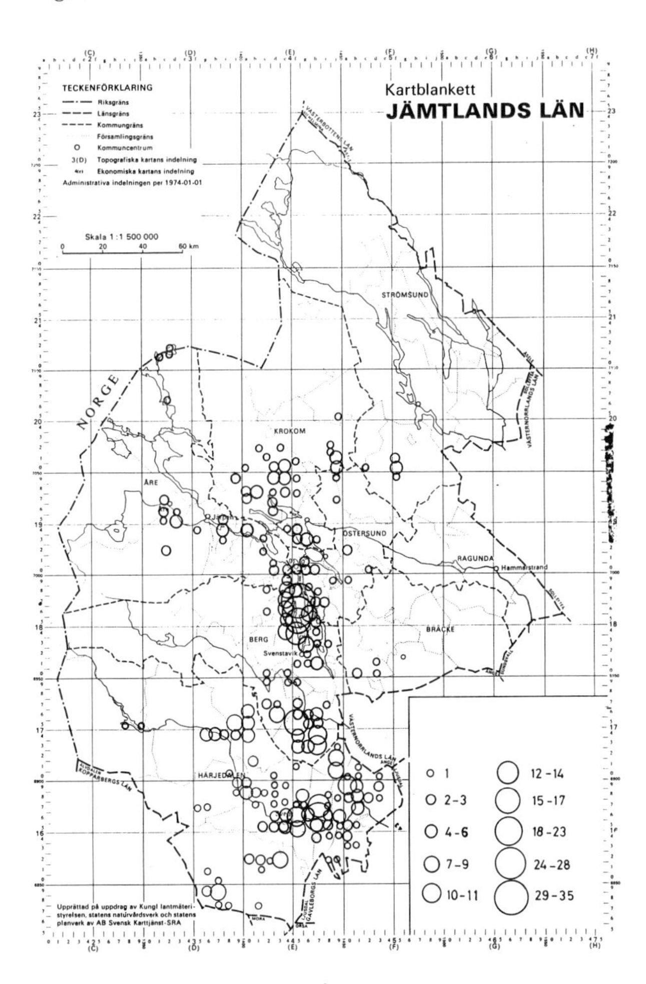

Figure 11.6 Distribution of iron production sites in Jämtland and Härjedalen.

The products could then be sold in various markets, above all in Norway. The importance of this sideline is reflected especially by the large volume of taxation the Danish crown derived from it in the mid-sixteenth century.

A completely different situation applies to the older shore-bound horizon. The furnaces no longer occur singly at the ironworking sites. Instead, there are whole batteries of them, up to ten or fifteen on a single site. The sites and workings convey the impression of large-scale operations conducted with skilled labor. Subject to certain reservations, it seems clear that this iron industry was at its height during the great migrations, the period from which a large number of rich burials are known to us. Those burials contrast starkly with the complete absence of Iron Age finds from the previous centuries, and the abundance probably reflects a widespread successful iron industry.

The Jämtland and Härjedalen iron industry existed for 1,700 years, using much the same technology throughout. With a few isolated exceptions, various types of pit furnace were used. Ironworking seems to have been subject to a trade cycle, with an initial boom during the Roman Iron Age and the great migrations (A.D. 200–600). Even so, that period has yielded relatively few finds, though some of the objects recovered are among the finest in Sweden.

The iron industry must have been relatively important to the people of Jämtland in prehistoric times. This has been considered previously, in view of the large number of spade-shaped billets. Judging by the output figures presented here, Jämtland must have been one of the biggest iron-producing areas in Scandinavia. Perhaps the construction of the hill fort should be viewed in this light. Investigations in recent years indicate that the fort was constructed during the Roman Iron Age or during the great migrations, the period when ironworking perhaps meant most to the people of Jämtland. It should perhaps be added that during the same period Medelpad experienced a substantial upturn, which a number of earlier scholars took to imply the existence of a kingdom in central Norrland during the great migrations. Medelpad has hardly any prehistoric ironworking sites at all. The billets found there resemble those from Jämtland in appearance and weight. Perhaps iron from Jämtland played a part in the creation of this conjectural kingdom. The idea has been broached by several scholars, from Nils Åberg (1953) to Baudou (1983), Selinge (1983), and Müller-Wille (1983), but their interpretations have invariably been based on the distribution of the spade-shaped billets.

This iron production probably formed part of a redistributive or market economy system in which the vast wilderness areas of Jämtland and Härjedalen were exploited from Medelpad, Hälsingland, or Tröndelagen. In the case of Jämtland this involved agricultural colonization and the establishment of a whole sequence of homesteads, one of which, Öneberget on the island of Frösön, was fortified. Öneberget was probably not so much a prehistoric fort in the true sense as a fortified homestead with a pivotal role in the economic and social system required for the more extensive exploitation of the inland wilderness expanses of southern Norrland. Hunting grew increasingly important during the late Iron Age and especially in the Middle Ages.

Examining the distribution of the spade-shaped iron billets (figure 11.7), one finds that they extend from the Mälaren Valley and Gotland in the southeast to Tröndelagen (Norway) in the west. Probably this pattern of distribution includes areas of both production and consumption.

On the strength of what has now been said, we can proceed to formulate a theory concerning the significance of the iron industry of southern Norrland. The rich and extensive forest areas of southern Norrland were exploited by individuals or by "trading companies" of various kinds, and in the process both fortified homesteads and ordinary farms were established, especially in the Storsjö area. The principal bases or homesteads of these companies were located in Medelpad or Tröndelagen and were directly or indirectly in touch with the continent of Europe. During the great migrations the iron industry came to play an important part, especially in relation to the many political crises occurring at this time. Those crises generated a heavy demand for iron for weapons, dramatically transforming metal supplies in the more densely populated areas and creating opportunities for the profitable exploitation of southern Norrland. Later, by the time considerable shifts occurred in the ensuing Vendel Period, with a large iron industry in Dalarna and Gästrikland, the earlier trading system had partly broken down. Ironworking moved into Jämtland and Härjedalen and at the same time was transformed from an industry supplying a big market to a sideline during an intensive phase of colonization, mostly during the Viking era and the early Middle Ages. Demand for iron increased during the Middle Ages, and ironworking became a profitable sideline for the agrarian population of Jämtland and Härjedalen. That sideline persisted and did not vanish altogether until the end of the eighteenth century.

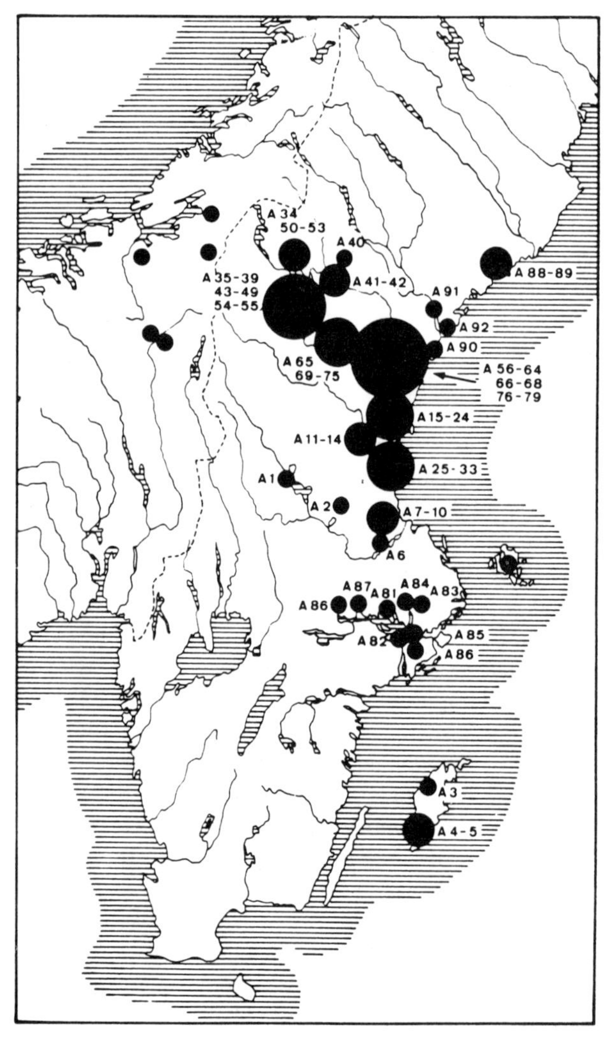

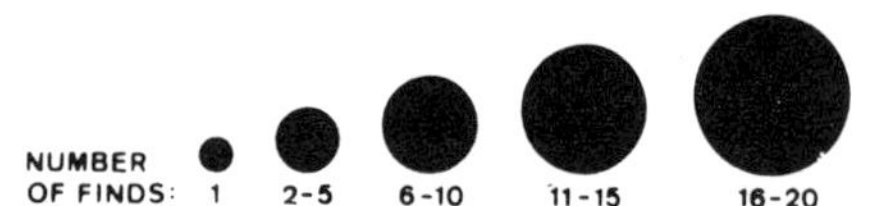

Figure 11.7 Distribution of spade-shaped currency bars found in Scandinavia.

References

Åberg, N. 1953. "The historical relationship between the time of the Great Migrations and a Kingdom." *KVHAA Handlingar* 82 (Stockholm). (In Swedish.)

Baudou, E. 1983. "Society and resource exploitation in southern Norrland in the early Iron Age." *Folk og Resurser i Nord.* (In Swedish.)

Hyenstrand, Å. 1977. "Foundries and the iron-manufacturing process: Some conclusions about the starting material." *Jernkontorets Forskning,* ser. H, no. 17. (In Swedish.)

Mandelgren, N. 1978. "A letter from a traveling artist." *Jamten Ostersund.* (In Swedish.)

Müller-Wille, M. 1983. "Der schmid im Spiegel archäologischer Quellen zur Aussage von Schmiedegräbern der Wikingzeit." *Das Handwerk in vorund frühgeschichtlicher Zeit.*

Selinge, K.-G. 1983. "A model for land exploitation within a fishing culture and earlier agrarian culture." *Folk og Resurser i Nord.* (In Swedish.)

12
Ancient Copper Mining and Smelting at Tonglushan, Daye

Zhou Baoquan, Hu Youyan, and Lu Benshan

An ancient underground mine with shafts and a gallery system was discovered at Tonglushan (Mt. Verdigris), Daye, in 1965 during the modern open cut working of the mine. Archaeological excavations between 1974 and 1985 have shown that the mining began at the end of the second millennium B.C. and continued until at least the early first century A.D. The excavations and an extensive program of research have provided us with a picture of ancient mining and metallurgical technology in the last millennium B.C.

General Description

Tonglushan is situated in Daye County, literally "grand smeltery," Hubei Province, on the southern bank of the Yangtze River, 140 km south of Wuhan.

Mt. Verdigris lies in the western end of a copper and iron mineral belt in the lower reaches of the Yangtze River. It was formed between the Sinian Period and the Triassic Era on the pre-Sinian base by seafloor and continental deposition. Later, because of the multispiralling crustal movements of Indochina and Yanshan, the area developed into a folded zone with a complicated geological structure. The active acidic magma movements in the Yanshan Period resulted in the formation of important copper and iron deposits in this area.

Mining and smelting at Tonglushan has a long history. According to a survey in the early 1950s, smelting slag in this region covers an area of 140,000 m^2, amounting to over 400,000 tons, and the residual copper in most slags is less than 0.7%. After the discovery of ancient underground working in 1965, from February 1974 to July 1985 archaeological excavations of seven mining sites and three smelting sites were carried out before modern operations began, in accordance with state requirements. Several hundred galleries and shafts with wooden pit props and a number of smelting furnaces were cleared and exposed. Stone, copper and iron mining of different periods, and pottery for daily use were recovered. The wooden and bamboo structures have survived because of the lack of oxygen and because they were "pickled" in the copper-salt-bearing water.

Pottery identification and carbon 14 dating of the wooden structure show that mining at Tonglushan started in the early Western Zhou Dynasty (eleventh century B.C.) and was active throughout the Spring and Autumn and Warring States periods until the late Western Han Period (A.D. 25).

The ancient mine has been classified as a national relic and was placed under state protection in February 1982. A museum was built to cover that part of the mine that was in operation in the Spring and Autumn Period (about 2,700 years ago), sacrificing an open cut pit, with a 400 tons per day capacity; more than

1.5 million tons of gravel and rock had already been removed.

Mining

At Tonglushan there are a number of ore bodies with large reserves. Most of the ore bodies outcrop or are only lightly covered. Native copper, malachite, and cuprite were formed by secondary enrichment in the fracture zone between the ore body and the surrounding rock, creating favorable conditions for ancient mining. Mining operation and archaeological excavation have revealed that all the exploited ore bodies had been mined before, and some of the early pits reach depths of 60 m.

Mine Prospecting

No one can be sure how the Tonglushan ore was first discovered, but the color of the outcrop certainly played a part. The County Annals of Daye (A.D. 18) read: "Tonglushan was so called since the hills were decorated, after a torrential rain, with verdigris flowers."

To follow the vein, a cut-and-trial procedure of prospecting and mining was employed by sinking holes and short drifting. Found widely among the relics of all periods are rectangular or boat-shaped trays similar to modern gold pans (figure 12.1). It is suggested that these vessels were used to estimate the richness of the ores by gravity analyses, a method still in use in some primitive gold mines today. The fact that areas with dense ancient pits are the places with high-grade ores shows the effectiveness of this method.

Shaft and Drift Digging and Support

Vertical shafts and horizontal and inclined drifts were employed. The cross-sectional area of shafts and drifts grew bigger with the expansion of productive forces. From the Western Zhou until the Spring and Autumn Period (eleventh to sixth century B.C.) net cross sections of shafts and drifts were smaller, the smallest being 50 × 50 cm^2; cross-sectional areas of shafts and drifts were roughly of the same size. The widest shafts had cross sections of 60 × 60 cm^2, and drifts 80 × 100 cm^2. In the Warring States Period, between the fifth and third centuries B.C., cross-sectional areas of shafts and drifts increased dramatically as bronze tools were replaced by iron ones for excavation. Cross-sectional areas of shafts and drifts were mostly around 100 × 130 cm^2 and 130 × 150 cm^2, respectively, and the largest section was 1,950 × 160 cm^2 (width × height).

Figure 12.1 Boat-shaped wooden vessel.

All tunneling took place at the contact bed of marble between the igneous rocks and the loose surrounding rocks. Shafts and drifts were supported by thick timbers. Between the tenth and sixth centuries B.C. square-set frames were used in shafts and drifts with equally spaced tenon frames (figure 12.2). Wooden sticks or planks and bamboo mats were used as lining sets and sheets. By the early fifth century B.C. close-set frames were used in shafts, but drifts were still supported by tenonned frames. By the late third century B.C., tenonned frames were completely replaced by square sets. Some of the drifts were supported by a new type of frame structure consisting of two forked column bars, a column arm, a support beam, and a sill piece (figure 12.3).

Mining Technique

Between the eleventh and eighth centuries B.C. mining was started by digging shafts from outcrops and followed at an appropriate depth by drifts. Blind galleries were excavated underneath drifts to follow the ore, forming a crisscross network of shafts, drifts, and galleries.

In the following three centuries horizontal drifts following a vein were cut in the rich ground about 20 m below the surface, until the vein of that particular level was exhausted; shafts were then sunk until the next vein was reached.

After the fifth century B.C. a kind of cut-and-fill method was used. Mine excavation was done by first digging shafts or inclined shafts to the depth of the base of the ore body and then excavating double-deck drifts, both with square setting, with the upper smaller one sitting on the lower larger one.

The ore was cut, picked, or dressed at the upper deck, and the lower deck was utilized for waste filling and as a roadway. This technique greatly improved the safety of mining and significantly reduced the transport of waste rock.

Hoisting, Water Draining, and Ventilation

In the beginning shafts and drifts were generally much shorter, and the delivery of implements and the hoisting of ores were done by hand. With the deepening of shafts and drifts and the expansion of production in the late Warring States Period, ores were hoisted with wooden winches. Two axles for winches have been found (figure 12.4).

In the eighth to sixth centuries B.C. underground water in the pits was drained by troughs along which water flowed to a sump to be hoisted in wooden buckets to the surface. Some shafts dated to the fifth to

Figure 12.2 Tenonned wooden frame.

Figure 12.3 Frame set.

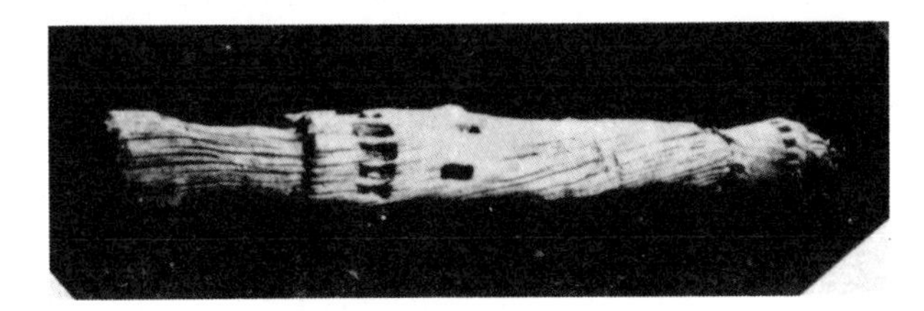

Figure 12.4 Wooden axle of a windlass.

third century B.C. reached as deep as 20 m below the water table, necessitating the building of a drainage system consisting of water sumps and drainage shafts. Numerous buckets were found.

No ventilation equipment has been discovered. It is suggested that natural ventilation was employed in the early stages. Ventilation equipment has been found, however, in pits of the late Warring States Period (fourth to third century B.C.). Some of the worked drifts were closed purposely. This suggests that the draft flow was directed to the working face.

Mining Implements

At the site of the Tonglushan ancient mine many implements have been unearthed, including tools made of bronze, iron, wood, bamboo, and stone. Tools for excavation included bronze hatchets, pickaxes, and chisels, iron hammers, drills, axes, and rakes, and wood shovels, spades, and hammers. Carrying implements were bamboo pans and bamboo and rattan baskets. Drainage facilities included wooden troughs, buckets, and scoops. Wooden winches and hooks, straw ropes, and stone counterpoises were used for hoisting. For ore dressing there were boat-shaped wooden trays, a gravity separator chute, stone anvils, and hammering stone blocks.

The implements so far unearthed at Tonglushan indicate that by the late Warring States Period (third century B.C.) bronze tools had been replaced by iron ones.

Of all the production tools unearthed, most are different from farm and handicraft implements. All the tools were specially designed for mining, of which the 3.5-kg bronze hatchet (figure 12.5), the 6-kg iron barrel-shaped hammers, and the 2.75-kg square iron chisels are examples (figure 12.6). This shows that mining activity had developed to a relatively high level.

Ore Dressing

Archaeological excavations and the ancient mine exposed during contemporary mining operation have revealed that ancient ore dressing was conducted by at least two methods. Ores were handpicked. It has been found that all smelting furnaces so far unearthed have platforms, stone anvils, and stone balls nearby for crushing.

Gravity separation was also used. A wooden trough hollowed out of a single block of wood was found in a pit from the fourth to third century B.C. The trough is 160 cm long, 36 cm wide at one end and 31 cm at the other, and 20 cm deep, with a wall thickness of 3 cm. There is an opening at the narrower end (figure 12.7). It is believed that this trough was a primitive gravity classifier washer. Dregs found in the vicinity of the

Figure 12.5 Bronze hatchet.

a

b

Figure 12.6 Iron hammer (a) and anvil (b).

Figure 12.7 Wooden channel.

trough were found to contain 12–20% copper, which provides evidence for the early use of this separation method.

Smelting Technique

Ores excavated and dressed were smelted into crude metal in the vicinity of mines.

Since 1976 three smelting sites and over ten furnaces in different states of preservation have been excavated. Figure 12.8 shows the remains of one furnace. A simulation study conducted on a reconstructed furnace demonstrated the functions of the various parts of this shaft furnace. Unearthed together with these furnaces are remains of furnace walls, slags, fire-resistant materials, ores, charcoal, and a few lumps of crude metal. The stratigraphic relation of this furnace at the time of excavation was clear-cut and gave a date of c. 800 B.C.

The furnace has the following structural features. It consists of a hollow bottom that supported the hearth and the burden and the furnace. The furnace reconstructed from remains has a height of about 1.5 m. The space underneath the furnace hearth was hollow, probably designed to keep dampness away in order to maintain the temperature in the furnace. Many shaft furnaces or cupulas of the Han Dynasty in Henan Province have hollow bottoms. In the late 1950s researchers built a shaft furnace of primitive design with its hearth in contact with the furnace bottom; it failed to produce liquid copper.

There are two tuyeres situated at the ends of the long axis of the furnace so as to maintain a uniform blast across the furnace section. No sign of a blower was found, however. On the front side of the furnace

Figure 12.8 Remains of a shaft furnace.

there is a door that could be partially closed to act as a tap hole for molten copper and slag or opened, if needed, for extracting the accretions, or burdens, with a rabble or other tool. The furnace wall was built of compacted refractory materials, and different parts used different combinations of materials, including red clay, kaolin, quartz sand, igneous rock chips, and iron ore dust. Near the furnaces are a feeding platform ground for crushing and screening materials and a puddling pit.

The smelting technology used oxide ores, mainly malachite, for the raw material. The ore was crushed, and that of higher grade was handpicked for smelting. From a material balance calculation of the slag, it can be deduced that iron ore was used as flux.

Over one hundred slag samples have been examined by chemical analyses. The slag composition seemed properly adjusted for reducing power and viscosity. The melting point was around 1200°C. The basicity $(CaO)/(SiO_2 + FeO)$ lies between 1.0 and 1.3, and the viscosity is 0.2 Pa·sec. The slags contain as little as 0.7% copper, and the crude metal contains more than 93% copper and about 5.4% iron.

13
Prehistoric Metallurgy in Southeast Asia: Some New Information from the Excavation of Ban Na Di

Charles Higham

Archaeological research in the valleys of the Red and Mekong rivers is over a century old, and the presence of a local metalworking tradition has been documented through the published material from excavations since 1879 (Noulet 1879). Samrong Sen is a settlement and probable cemetery site located in the valley of the Chinit River (figure 13.1). It is about 4.5 m deep. Although the stratigraphic details are not clear, Noulet (1879) and Mansuy (1902, 1923) have described not only bronze artifacts, including an ax, arrowhead, bracelet, bells, and a fishhook, but also part of a sandstone bivalve mold for casting a cutting implement (figure 13.2). Similar sandstone molds were discovered by Lévy during his examination of three small prehistoric settlements in the Mlu Prei area, near the headwaters of the Chinit and Sen rivers, about 160 km north of Samrong Sen. Lévy recovered specimens used in casting a sickle blade and an ax as well as several socketed axes and bangles (Lévy 1943; see figure 13.2). Lévy also recovered a number of iron artifacts in this area. Recognition of a vigorous metallurgical tradition along the coast of central Vietnam began in 1909 when Vinet encountered a group of urns containing cremated human remains at Thanh Duc, near the village of Sa Huynh (figure 13.1). The large and elegant vessels usually contain a variety of offerings. Much subsequent research has led to the recovery of iron slag, iron spearheads, knives, and sickles as well as bronze spearheads and bells (Trinh Can and Pham Van Kinh 1977). Some of the artifacts found in Sa Huynh contexts are paralleled in the rich assemblages of the Bac Bo Dong Son sites, which were first revealed at the site of Dong Son itself in the two decades preceding the Second World War. The range of bronze items is great, the best known being the great ceremonial drums.

Obviously, with the lack of radiometric dating techniques and the rarity of comparative material, early interpretations of the origin and nature of this widespread metallurgical tradition were speculative and usually relied on the then current diffusionist models. Excavations during the past two decades have greatly expanded available data, provided the basis of a chronological framework, and allowed some broader cultural implications of the use of metals to be isolated for discussion. This research has concentrated in Bac Bo, the Dong Nai valley, and northeast Thailand. More recently still, a similar metalworking tradition in the Chao Phraya valley has been revealed (Hanwong 1985; Natapintu 1985). Here I concentrate on the results of the excavation of Ban Na Di in 1980–1981, inasmuch as they have a bearing on the use of metals, and then set them in their wider Southeast Asian context. It is convenient at this juncture briefly to describe the general cultural framework employed. Bayard (1984)

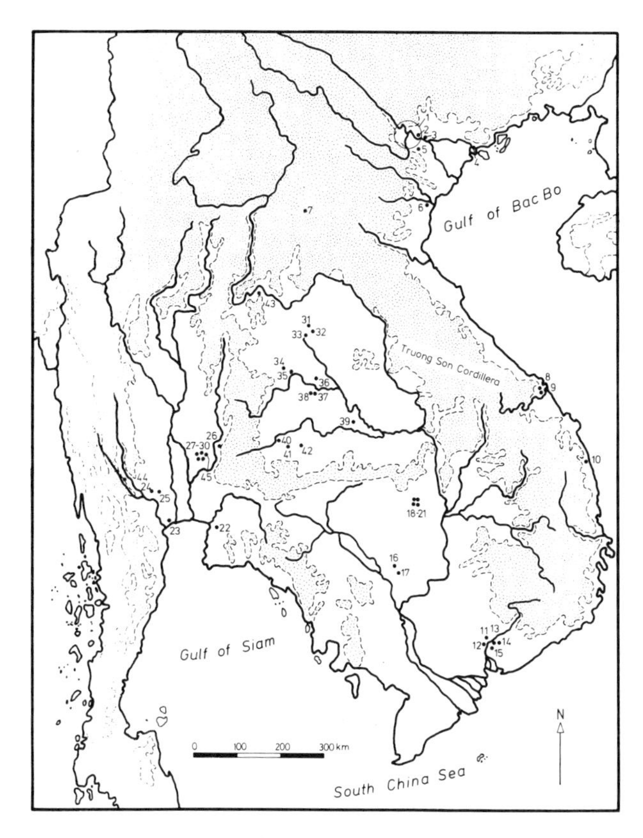

Figure 13.1 Map of mainland Southeast Asia showing the locations of the sites mentioned in the text. 1, Long Hoa, Phung Nguyen, Go Mun, Thanh Denh, Lang Ca, and Go Bong (for a map of this area, see figure 13.17). 2, Dong Dau. 3, Co Loa. 4, Viet Khe. 5, Chau Can. 6, Dong Son. 7, Ban Ang. 8, Tam My. 9, Binh Chau. 10, Sa Huynh. 11, Doc Chua. 12, Cu Lao Rua. 13, Cau Sat. 14, Hang Gon. 15, Phu Hoa. 16, Samrong Sen. 17, Long Prau. 18, O Pie Can. 19, O Nari. 20, O Yak. 21, Mlu Prei. 22, Ban Khok Rakaa. 23, Khok Phlap. 24, Ban Don Tha Phet. 25, Dermbang Nang Buat. 26, Khok Charoen. 27, Lopburi. 28, Phu Noi. 29, Ban Tha Kae. 30, Khao Wong Prajan. 31, Ban Chiang. 32, Ban That. 33, Ban Na Di. 34, Non Nok Tha. 35, Non Chai. 36, Muang Fa Daed. 37, Ban Chiang Hian. 38, Ban Kho Noi. 39, Non Dua. 40, Phimai. 41, Muang Phet. 42, Thamen Chai. 43, Phu Lon. 44, Ongbah. 45, Non Ma Kla.

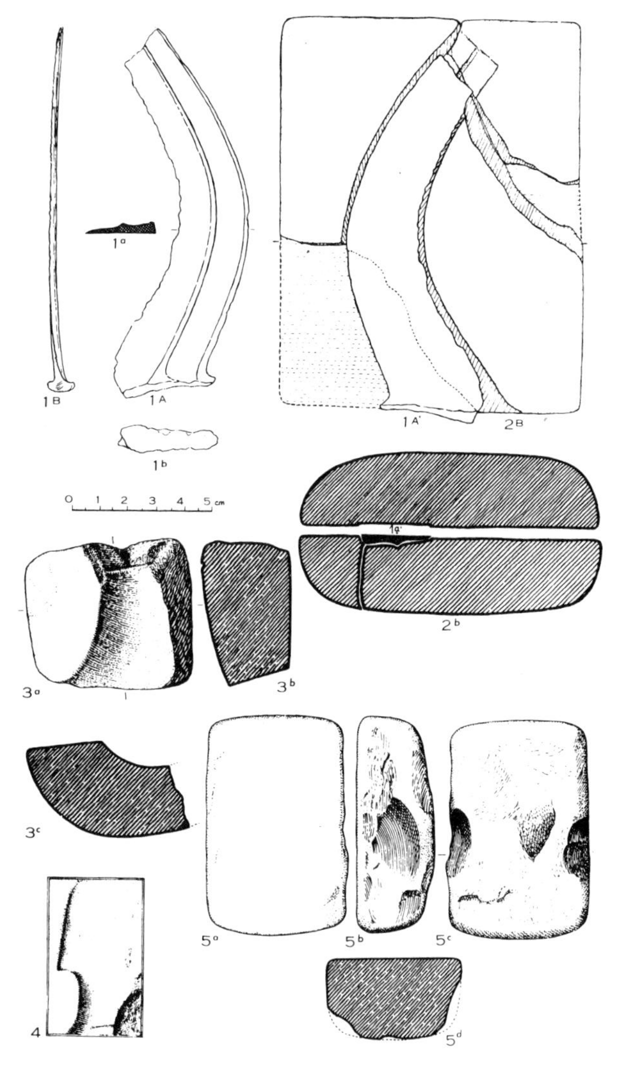

Figure 13.2 Examples of early published metal artifacts and molds from Samrong Sen (no. 4) and O Pie Can in the Mlu Prei region, northeast Cambodia. From Lévy (1943).

has proposed the employment of four general periods for later prehistory in northeast Thailand, and I follow these. The first, General Period A, covers the initial settlement of the interior river valleys of Southeast Asia by agriculturalists not yet conversant with metallurgy. General Period B involves the continued occupation within small village communities, but now incorporating bronze metallurgy. General Period C witnessed increasing centralization in association with ironworking. The dates of the general periods varies with the different regions.

Excavations at Ban Na Di

Like virtually all lowland settlements in northeast Thailand, Bac Bo, the Dong Nai valley, and central Thailand, Ban Na Di is located some distance from copper, tin, and lead ore deposits. Iron ore, particularly laterite, is more widely distributed (figure 13.3). The result of this imbalance between reasonable agricultural soils (as early bronze is invariably associated with rice cultivation) and ore deposits is that bronze in sites such as Ban Na Di is an exotic substance obtained through fairly extensive exchange networks.

Two areas at Ban Na Di were excavated (figure 13.4). The cultural deposits lie in a sandy loam matrix, but color, content, and texture vary sufficiently to allow the recognition of eight levels, numbered 1 through 8 from top to bottom. Only the basal five concern us here. Level 8 is an ashy soil deposit incorporating several flood lenses and charcoal concentrations. Several pits were excavated and found to contain organic refuse and potsherds. No burials were cut from this level, which is held to reflect occupation activities. Level 7 in part of excavated area was reserved for mortuary use, and this area is rich in superimposed inhumation burials. I subdivided these early mortuary remains into three subphases, described as mortuary phases 1a through 1c. The last burials in this phase were cut into Level 7 from basal Level 6. The latter level may not reflect occupation activity on the spot. This situation changed dramatically with Level 5, when the entire part of one of the excavated areas was given over to bronze casting, evidenced by a series of small clay furnaces ringed by crucible fragments, molds, charcoal, and flecks of bronze. This activity continued into Level 4, but at that time part of the site was reserved for jar burials containing the inhumed remains of infants in association with both bronze and iron artifacts. The radiocarbon dates from Ban Na Di suggest the following chronology:

Level 8: c. 1200–900 B.C.
Level 7: c. 900–500 B.C.
Level 6: c. 500–100 B.C.
Level 5: c. 100 B.C.–A.D. 200
Level 4: postdates A.D. 200

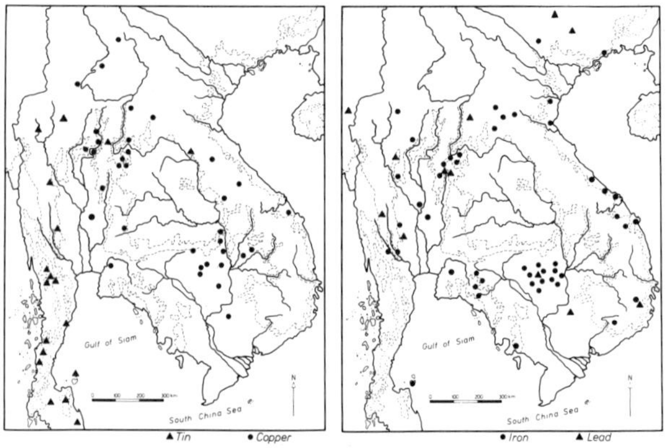

Figure 13.3 The principal deposits of copper, tin, lead, and iron in mainland Southeast Asia.

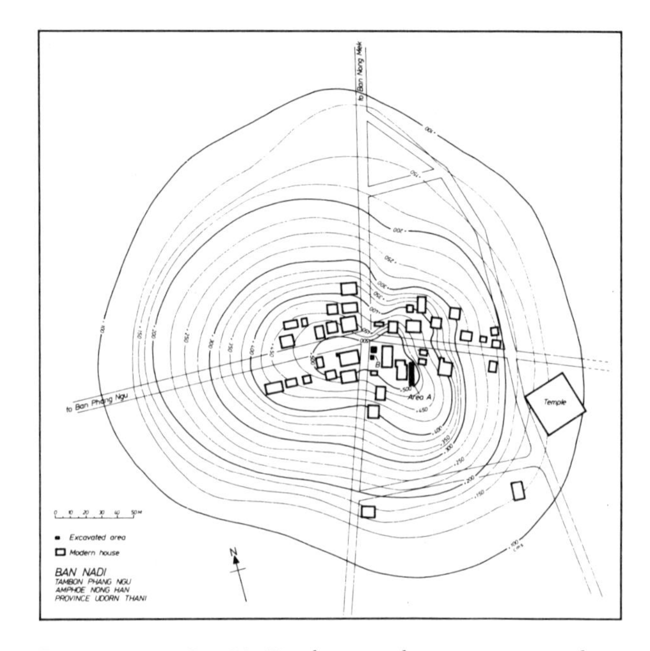

Figure 13.4 Ban Na Di, showing the areas excavated.

Levels 8 through 6 correspond with General Period B, and levels 5 through 3 with General Period C. The excavations furnished evidence for practically the entire sequence of bronze working activities involved in converting bronze ingots into artifacts.[1] I begin with the furnaces employed. Level 7, devoted mainly to burials, also incorporated a bronze working facility. It is composed of a series of clay blocks disposed around a central hollow full of large pieces of charcoal. There is a channel through the external wall that could have admitted the tuyeres. The furnace has a diameter of 0.60–0.75 m (figure 13.5) and is surrounded by two lenses of dense charcoal staining within which there are numerous fragments of crucibles and bronze. I interpret the charcoal staining as the remains of rake-out between successive uses of the furnace. This was the only such furnace in Level 7. Later in the sequence, however, in Level 5, we encountered eight such furnaces across the entire excavated area, which in this

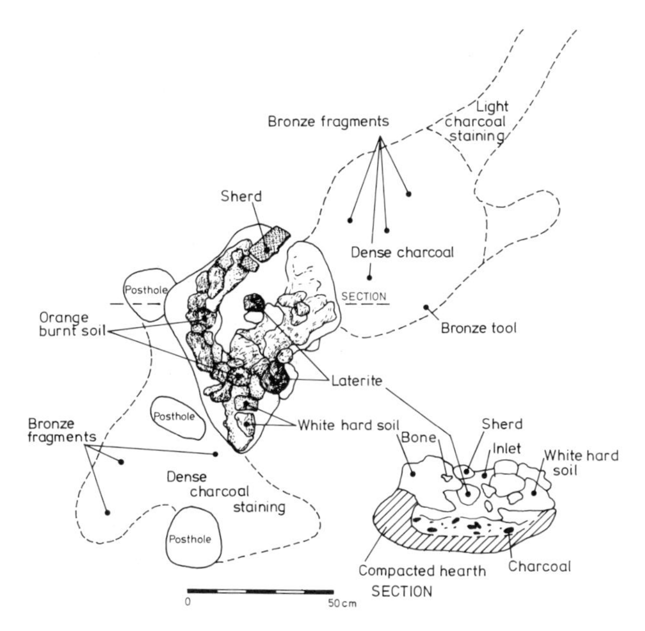

Figure 13.5 The bronze working furnace found at Ban Na Di, Level 7.

square covered 45 m². Although intended to concentrate heat sufficient to melt bronze, these furnaces differed from the one found in the earlier context. They are composed of loosely packed blocks of clay set in shallow pits: The clay has been subjected to considerable heat. The furnaces are small, a typical example measuring 0.4 m × 0.3 m, but perhaps because of reuse over time, one furnace attained a depth of 0.9 m. Each was ringed by a concentration of crucible fragments, flecks of bronze, and charcoal.

Molds were made of clay and stone (figures 13.6 and 13.7). There are two broken sandstone specimens that were originally components of bivalve molds. The first, which came from a Level 6 context, had been shaped to an oval cross section, 33 mm deep at its greatest extent. The inner surface was ground to a plane, within which was cut the outline of an ax blade. The second piece comes from Level 7 and contained on the interior surface two conduits, each leading to the form of projectile points. The upper surface of the mold is present, and there is a clear enlargement there to facilitate the pouring of molten metal into the mold when both halves were joined. There are several further examples of shaped sandstone that might have been molds spanning Levels 7 to 5.

Most of the clay molds were found in Level 5. The most complete was presumably never used, because it has survived with part of the section intact. The interior part designed to receive the metal is made of a thin layer of fine clay within which there is a decorative design. This inner section was coated with a layer of clay tempered with abundant rice chaff. The exterior of the mold is circular in cross section, with a diameter of 24 mm. It is probable that the template melted and flowed out of the mold when it was fired. There

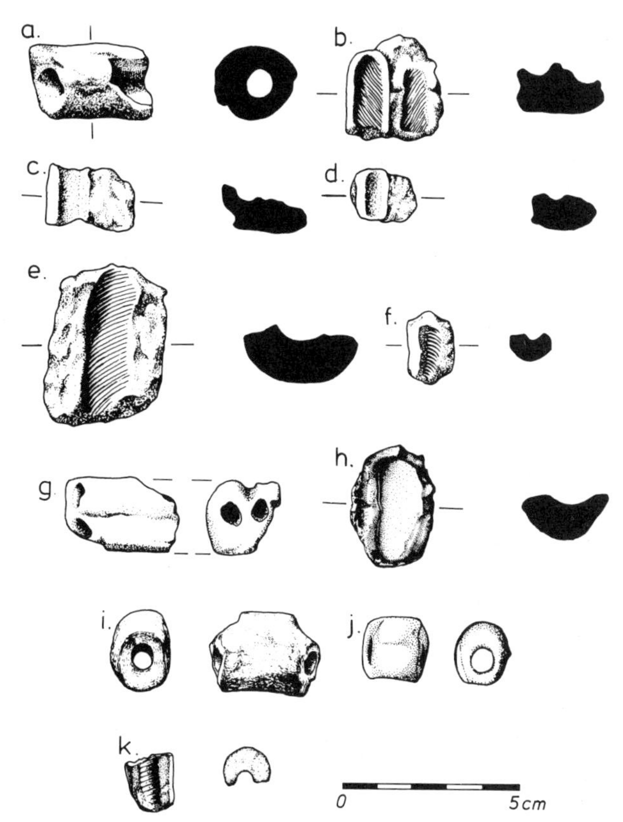

Figure 13.6 Clay molds from Ban Na Di, Levels 5 and 6.

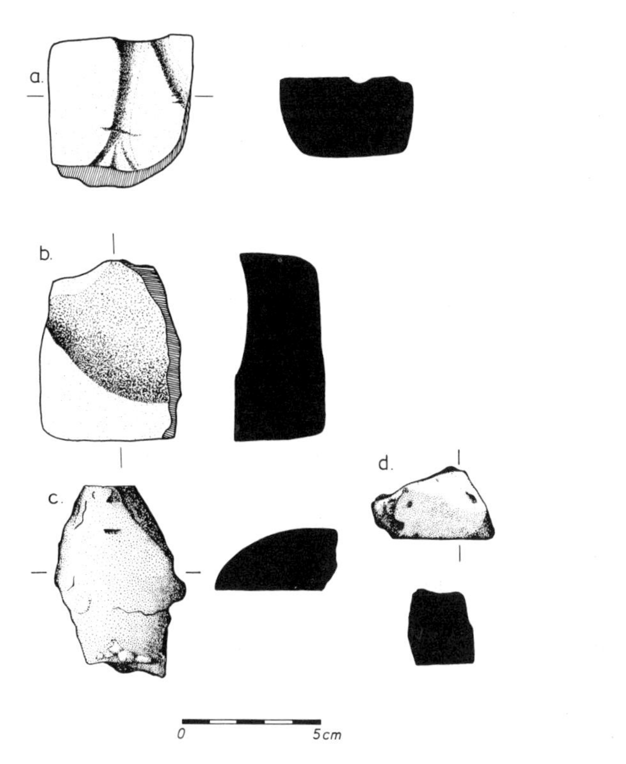

Figure 13.7 Sandstone molds for bronze casting from Ban Na Di, Level 6 (b) and Level 7 (a, c, d).

are several other fragments of clay mold, designed mainly for decorated bracelets and for bells. The issue of casting techniques highlights the significance attached to a piece of solidified metal from the pouring cone of a mold (see figure 13.8k). The metal is a tin-lead alloy in the ratio of 1:3. Rajpitak and Seeley (1984), in describing this object, have noted that its low melting point (between 180° and 270°C) would be advantageous if used to pour into a stone mold to create a template later covered in clay before casting in bronze. Under such conditions the stone mold would be far less likely to crack under thermal shock. That the occupants of Ban Na Di employed the lost-wax technique has been demonstrated beyond reasonable doubt by Pilditch (1984). While cleaning one of the bronze bracelets (from mortuary phase 1b, Level 7), Pilditch encountered a dark gray substance caught between the central clay core and the corroded metal. Two other bracelets on examination revealed the same material. Analysis by thin layer chromatography and gas liquid chromatography revealed that the material was a relatively unrefined form of insect wax.[2] Numerous fragments of crucible were recovered from the basal layer to Level 5. Two complete specimens were also found, both from Level 6 (figure 13.9). One had a capacity of 80 ml of molten metal, and the other was slightly smaller, with a capacity of about 75 ml. The crucible fragments from Levels 8–6 were made of a clay tempered liberally with rice chaff and bearing relatively thick adhesions of scoria. From Level 5 there was a change to sand temper and only a thin film of slag. Maddin and Weng (1984) have analyzed a specimen of scoria derived from a Level 8 crucible fragment. They concluded that the crucible had been used for melting rather than for smelting, with a preference for the melting of a preexisting alloy rather than a combination of separate copper and tin ingots.

It is evident that between 1200 B.C. and A.D. 200 the inhabitants of Ban Na Di melted bronze obtained through exchange and cast it using a variety of techniques into ornaments, weapons, and tools. It is now necessary to consider the nature of the implements and the alloys. Bangles were the most abundant artifact recovered from Levels 8–6 (figures 13.10 and 13.11). To judge from their presence in mortuary contexts, they were principally worn on the wrist, although one child was found buried with a solid bronze ring around each ankle. There is a range of ornamentation on the bangles, imparted by use of the lost-wax technique. Bronze beads from Levels 8–6 were by contrast rare. Only two were identified, each in the shape of a comma with a single hold for suspension. Two fishhooks were recovered from Level 7. There are at least three varieties of projectile point, probably used to tip arrows. One has a tang, another has a concave base, and the third, although tanged, has a broad cutting blade. All come from Level 7. The earliest bronze found in

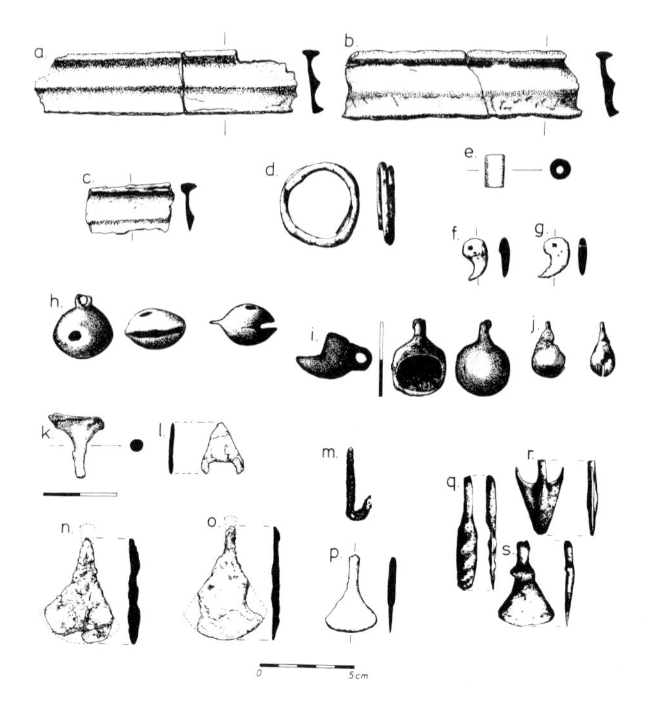

Figure 13.8 Bronze artifacts from Ban Na Di. (a–c) Bowl fragments, Levels 3–4. (d) Wound rod, burial 17, Level 6. (e) Tube, Level 3. (f, g) Beads, Level 7. (h, j) Bells, Levels 2–5. (k) Lead tin sprue, Level 4. (l) Arrowhead, Level 7. (m) Fishhook, Level 7. (n–s) Projectile points, Level 7.

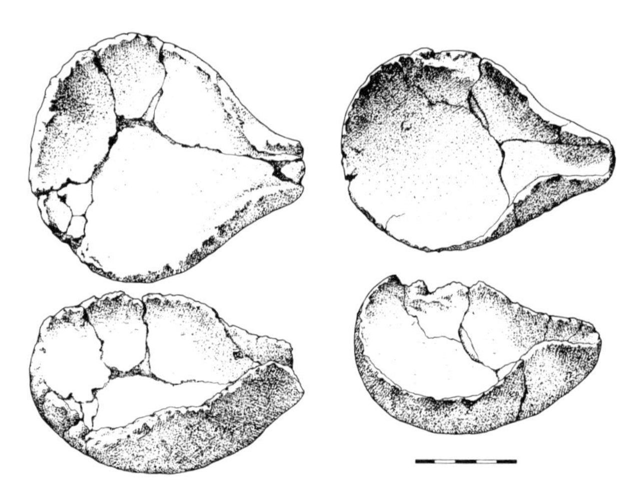

Figure 13.9 The two complete crucibles from Ban Na Di, Level 6.

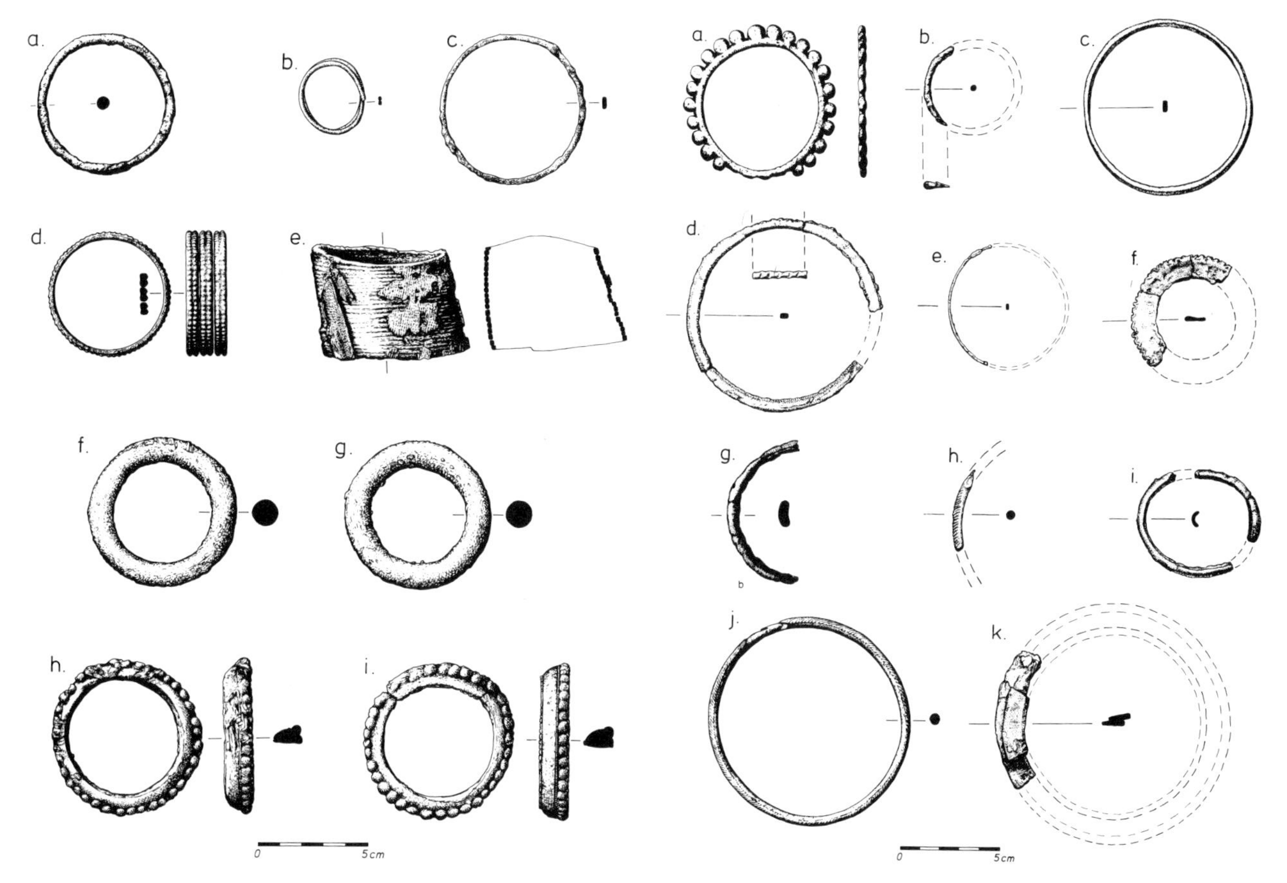

Figure 13.10 Bronze bracelets and a coil from Ban Na Di. (a) Mortuary phase 1. (b) Burial 5, mortuary phase 2. (c) Burial 3, mortuary phase 3. (d–i) Mortuary phase 1.

Figure 13.11 Bronze bracelets from Ban Na Di. (a) Mortuary phase 1. (b) Level 5. (c) Level 4. (d) Unprovenanced. (e) Level 4. (f) Level 7. (g) Level 5. (h) Level 5. (i) Level 6. (j) Mortuary phase 1. (k) Level 3.

mortuary contexts is a bronze "wire" used in joining sections of stone bracelets. The bronze implements from Levels 5 and 4 display a marked change. Rings small enough to fit the finger as well as decorated bells and bronze bowls were now cast.

Rajpitak and Seeley (1984) have analyzed the alloys, and Maddin and Weng (1984) have analyzed the composition of the wire. Rajpitak and Seeley revealed the presence of four major categories of alloy. The majority of objects were made from a low-tin bronze, which orginally contained between 2% and 14% tin. This alloy was used in all the arrowheads analyzed and many of the bracelets. Much rarer are three objects of leaded copper, one of which is a comma-shaped bead and the others, bracelet fragments. Twenty-five of the objects analyzed were made of a leaded tin bronze, all being rings or bracelets with the exception of three pieces of scrap. Finally, there is a high-tin bronze, used in the manufacture of three and possible as many as five of the objects subjected to analysis. This particular alloy, which involves a high fraction of tin (about 24%), was used to impart a goldlike finish and, because of its brittle nature, required particular metallurgical skill. Maddin and Weng, who examined the wire, are at pains to stress the extremely tentative nature of their findings at this early stage of consideration, and their results may well be modified by further analysis. On the basis of the cast structures they suggested that the wire was in fact cast in place. The alloy used was identified using the EDAX (energy dispersive analysis by X-ray) system, a technique that does not provide the exact composition. It does seem likely, however, that the alloy was a ternary mixture of copper, tin, and arsenic.

A scatter plot of the percentages of lead and tin indicates considerable variation in the quantities of these metals (figure 13.12). Because of the long time span involved, the percentage figures for the different metals were subjected to a factor analysis based on copper, tin, lead, and arsenic. Factor 1 in figure 13.13 reflects high values for copper, with both tin and lead being low and arsenic unimportant. Factor 2 reflects high values for lead and arsenic. Because most of the objects in question are bracelets or rings, it was decided to commence interpreting the results of the factor analysis from a chronological stance. Two subgroups were created, one composed of specimens from Levels 8–6 (that is, General Period B) and the other from Levels 5–3 (General Period C). This reflects the conversion of the area excavated from a cemetery to a bronze

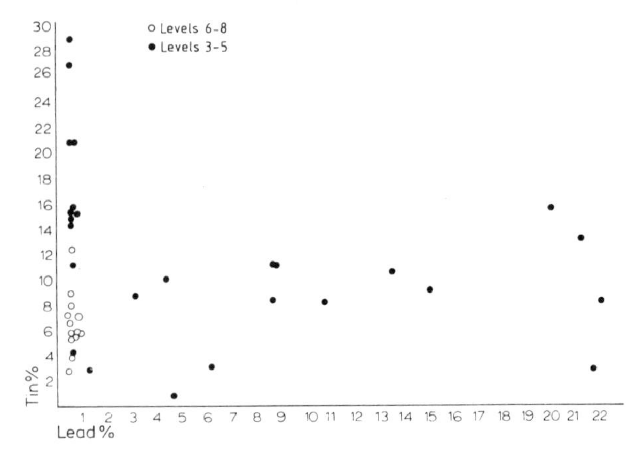

Figure 13.12 The relationship between the percentage values of tin and lead in the bronze artifacts from Ban Na Di.

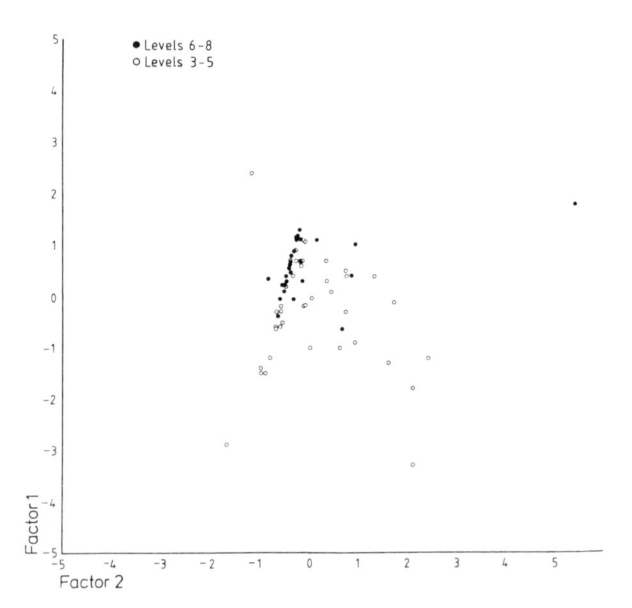

Figure 13.13 The results of a factor analysis based on the proportions of copper, tin, lead, and arsenic in the bronze artifacts from Ban Na Di.

working atelier. There was at the same time a major change in virtually all other aspects of material culture at the site. Consideration of figure 13.13 reveals that the earlier of the samples displays far less variability. They are almost all made from a relatively low-tin bronze. Only two specimens, a bead and a bracelet, have a significant lead content. The five projectile points analyzed are characterized by a low to very low tin content and the absence of lead. One bracelet is set apart by its high arsenic content. The later sample is much more varied. Although some of the specimens retain the low-tin bronze, there was a major development in the field of lead alloys. This is not, as it happens, correlated with a greater complexity of design; indeed, some of the later bracelets and rings are of simple shape. At the same time, at least four analyzed specimens display the extremely high tin content that has been identified at approximately the same juncture at other sites (Rajpitak and Seeley 1979). There are therefore strong grounds for concluding that there is a clear technological break with the advent of Level 5.

Although crucibles and bronze fragments are not unusual in occupation deposits at Ban Na Di, even from the basal level, bronze artifacts from mortuary contexts are decidedly rare. The only bronze found associated with the fifteen burials ascribed to mortuary phase 1a is that used in the repair of stone bangles. Three of the twenty-one burials in mortuary phase 1b have bronze bracelets: one burial, that of an adult female, has nineteen; another, also a female, has three; and a third disturbed burial of unknown sex, has one. Twenty-four burials were ascribed to mortuary phase 1c, and of these, three (two females and a male) had one or two bracelets. A coil of bronze was found with a very late male grave, and a child was found with two anklets of solid bronze. In other words, only about 15% of the burials recovered were provided with bronze grave goods. Nearly all the burials were, however, accompanied by pottery vessels and animal limb bones, and eighteen (30%) had jewelry made of probably exotic shell disk beads.

The Remains of Iron at Ban Na Di

Numerous fragments of iron were recovered at Ban Na Di. There was but one fragment in Level 7, nine in Level 6, and a dramatic increase to thirty-nine fragments in Level 5. This last level also provided the earliest specimen of iron slag. Only in the final burials of mortuary phase 1c does one encounter grave goods made of iron (figures 13.14 and 13.15). Burial 16 contained four iron circlets, which appear too large to be bracelets or anklets. They have a diameter of 114 mm, which is compatible with use as neck rings. This same burial also contained a piece of iron spear blade and a badly corroded bracelet with a square cross section. Burial 17 included a band of iron coiled to form a double ring and a nearly complete tanged knife. Iron

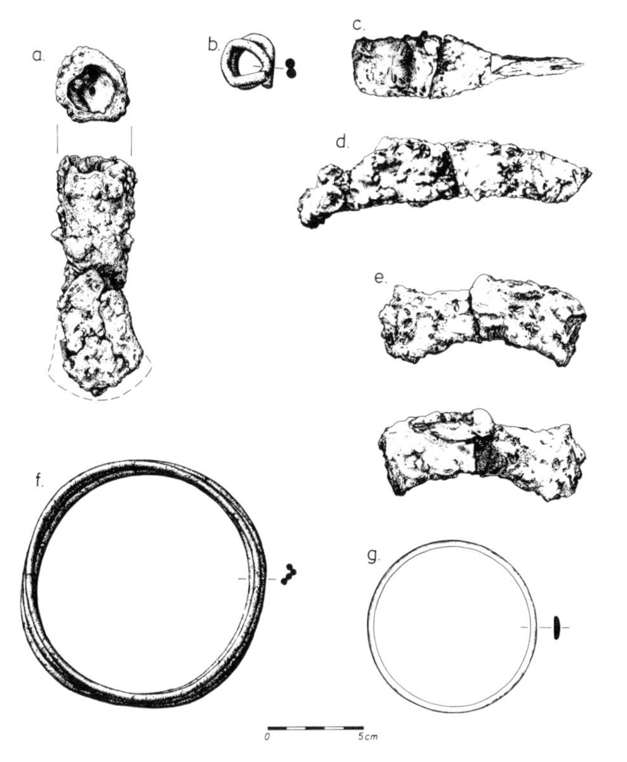

Figure 13.14 Iron artifacts from Ban Na Di. (a) Socketed hoe or digging stick, Level 4. (b) Iron coil, mortuary phase 1c. (c–e) Tanged knives, Level 4. (f) Four neck rings, mortuary phase 1c. (g) Bracelet, mortuary phase 1c.

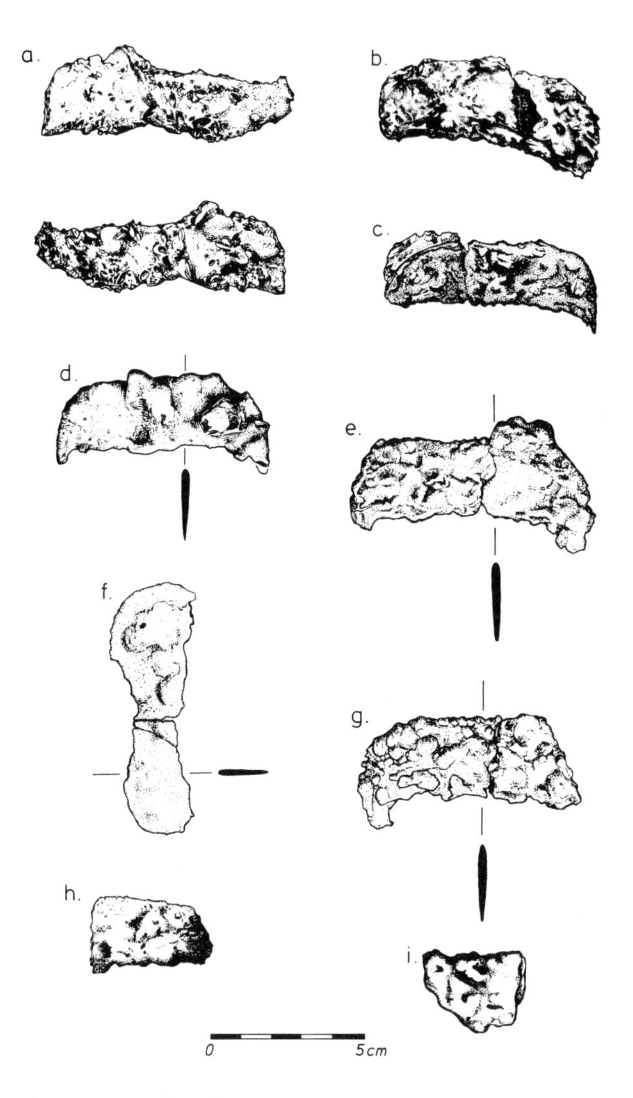

Figure 13.15 Iron knives from Ban Na Di. All artifacts are from Level 4 or 5.

goods are much more common in the five Level 4 lidded urns containing the remains of tiny infants. Burial 4 included an iron knifelike implement with hooked ends and the remains of a backing on the noncutting surface. It looks much like a hand-held sickle. A practically identical sickle was found in burial 8. Burial 5 yielded a socketed digging stick similar to those still used in agricultural pursuits today. In addition to the sickle, burial 8 also included a tanged knife, with the wooden handle still in place. Similar sickle blades, which resemble Han Period hand-held implements, are also found in occupation deposits in Levels 5–4. Unfortunately the material is so corroded that no detailed analysis of manufacturing techniques is yet available. It is, however, assumed here that they were forged. Local ores may have been smelted at the site, judging from the remains of iron slag in levels 5–3.

Metalworking at Ban Na Di: Summary

It has been possible to recognize two distinct traditions of metallurgy at Ban Na Di. The earlier, dated to c. 1500–100 B.C., involved the local casting of ornaments, projectile points, fishhooks and axes from a low-tin bronze. Ornaments were found sparingly in about 15% of the human burials. A rather special alloy was cast into the form of wire when used to repair rare and probably prestigious stone disks. The latest context in this early configuration incorporated two graves with iron implements. It was only in Level 5, however, that iron became at all common, and the first remains of slag indicate local smelting. At the same time, bronze alloys became much more variable, incorporating specimens with a high admixture of lead and some special pieces with high levels of tin. Bronze forms changed and included bells, bowls, rings, and different varieties of bracelet. All stages in the melting and casting of bronze on site are attested.

The Metallurgy of Ban Na Di in Its Wider Context

The occupation of Ban Na Di incorporated bronze working from initial settlement. Consequently no light has been thrown on the earliest contexts for metallurgy in northeast Thailand. Two sites in the same general area do exhibit a transition to bronze metallurgy. Ban Chiang is only about 24 km northeast of Ban Na Di, and excavations there in 1974–1975 revealed a long and important sequence. The bronze working tradition there is in most respects identical with that identified in Levels 8–6 at Ban Na Di (Wheeler and Maddin 1976), one distinction being the recovery of a socketed spearhead, a type that has not been identified at Ban Na Di. The crucibles are clearly of the same type, and a similar concentration of clay ringed by crucibles and casting spillage was recovered. Again, this tradition gave way to one involving iron-working and high-tin bronze. White's suggested date of c. 2000 B.C. for the initial presence of bronze there is clearly in harmony with the evidence from Ban Na Di (White 1982). The date of the transition to iron-working is a little more contentious, but White and I at least agree on the first millennium B.C., but I am rather more conservative. Non Nok Tha is a third site with clear affiliations in terms of bronze working with Ban Chiang and Ban Na Di. Again, sandstone bivalve molds and axes have been recovered over a horizon devoid of evidence for metallurgy. Now that the analysis of the Non Nok Tha data is complete (Solheim 1982–1983, p. 114), a detailed publication of the bronzes and their associated chronology is keenly anticipated. A major advance in our appreciation of the metalworking tradition in northeast Thailand has been the discovery and analysis of a copper mine and adjacent processing area at Phu Lon (Pigott 1984). This important site appears to have been in use toward the end of General Period B.

Early Metallurgy in the Chao Phraya Valley

Until recently evidence for a bronze industry comparable with that described for the earlier phase in northeast Thailand has been absent from the Chao Phraya basin. This situation has now been redressed by the excavations at Khok Phlap, Ban Khok Rakaa, and Ban Tha Kae. Khok Phlap is a low mound covering an area of about 1.5 ha (Daeng-iet 1978). No radiocarbon dates have yet been published, but the parallels with the General Period B sites in northeast Thailand are clear. Several burials were accompanied by whole pottery vessels containing seashells. Ornaments included bracelets made of turtle carapace, stone, shell, and bone as well as of bronze. One grave contained three anvils used in making pots and four complete vessels. The inhabitants of the site showed a predilection for colored stone beads and ear pendants, and their interest in bronze extended to barbed metal tips for their arrows.

There is no report of iron at Khok Phlap, but at a second relevant site, Ban Tha Kae, there is a clear succession from the use purely of bronze to a context including local ironworking. Ban Tha Kae is located a few kilometers north of Lopburi and covers an oval area measuring about 1000 m × 700 m. Excavations by the Thai Fine Arts Department in 1983 uncovered a sequence of considerable interest. The 2.5-m-deep stratigraphic layer was subdivided into three phases (Hanwong 1985). The lowest contained burials associated with bronze artifacts in the form of bracelets. Other offerings included shell beads and earrings as well as stone bracelets with a T-shaped cross section recalling those repaired with bronze from Ban Na Di. This same context revealed compelling evidence for the manufacture of shell bracelets. Using the marine tridacna shell as the source material, circular tabs were removed from the center of the blanks before final grinding to the desired shape. Stone axes were found in this early context, together with a pottery style incorporating the impression of curvilinear designs. Puthorn Pumaton, the excavator, then found that the succeeding middle period witnessed the first use of iron, together with blue and orange glass beads. The radiocarbon dating of the site or of similar exposures elsewhere has great priority.

The Lopburi area is clearly one of considerable relevance, and the Central Thailand Research Program is in the process of transforming our knowledge about it. Under the leadership of Surapol Natapintu, 105 prehistoric sites have been identified in two field seasons, and several have been excavated. Natapintu (1985) has excavated a small (3 m × 5 m) exposure at the site of Ban Phu Noi. Within it he recovered an assemblage of no fewer than thirty-two burials associated with marine shell bracelets and shell disk beads. Bronze was present at the site but only in the form of stray finds. Bracelets were made of seashell, turtle carapace, and ivory. These provide strong evidence for exchange networks linking interior sites with the coast.

Another major contribution has been made at Khao Wong Prajan, a hill only 15 km from Lopburi. There the Fine Arts Department, the University of Pennsylvania, and Pigott and Natapintu have found a considerable quantity of copper slag, seven separate copper smelting areas, and one prehistoric copper mine. Excavations in May 1985 revealed numerous clay molds for the casting of bracelets, axes, arrowheads, and spears, and clay crucibles. At one of the sites, Non Ma Kla, the excavation of a 4 m × 4 m square revealed a meter-deep stratum containing an arrowhead and an axhead of unalloyed copper and fragments of clay furnace lining and tuyeres. The excavator has suggested that this area of intensive copper working was contemporary with basal Ban Tha Kae, located only 5 km to the south. Among the surface finds at Khao Wong Prajan was a copper ingot, indicating the production of metal as well as of finished artifacts for exchange. When the radiocarbon dating for this industrial activity is available, it would be surprising if the time span was not within the period 1500–250 B.C.

During the course of the initial site survey of the Bang Pakong valley, many prehistoric sites were identified that will surely be found to relate to General Period B. Surface finds revealed bronze working and the import of glass and agate beads. One such site, Ban Khok Rakaa, has been subjected to trial excavations, and several burials were recovered in association with bronze artifacts. It is becoming increasingly evident that the adoption of bronze working in the Chao Phraya valley antedated the first use of iron in the area and that its availability incorporated just the same range of artifacts as have been recovered in similar contexts in northeast Thailand.

The Lower Mekong and Its Hinterland

Although the excavation of prehistoric sites in the lower Mekong area has a history of over a century, only during the last five years has the cultural sequence of the area assumed a structure of its own. In 1963 Malleret could describe the known prehistoric settlements in only the most general terms. Cu Lao Rua, better known as the Isle de la Tortue, was the best documented, having been described as early as 1888 by Cartaillhac. Excavations ensued in 1902 and 1937, and a local businessman assembled a collection of artifacts from the site. These included shouldered and quadrangular adzes, much pottery, and stone bracelets, pendants, and polishers. In 1937 a local villager showed Malleret a bronze ax that, he claimed, was found at this site. Rach Nui falls into the same category: It covers 120 m × 50 m, and four occupation layers were noted. These contained much pottery and quadrangular stone adzes, but local informants again claimed that a bronze ax had been found there.

Bronze metallurgy in this region was definitely attested at Hang Gon (Saurin 1963), although again the site has never been systematically excavated. Indeed, the material collected by Saurin was revealed by a bulldozer when clearing woodland. Hang Gon is one of many prehistoric sites in the Dong Nai valley. Surface sherds there cover an area of 350 m × 150 m. The site lies on a ridge that commands the junction of two streams. Its particular interest lies in the discovery of three sandstone molds, two for casting axes and the third for casting three ring-headed pins. Saurin stressed the typological similarities between the shape of the axes and specimens known from the Red River valley, Luang Prabang in northern Laos, and the Mlu Prei region in north central Cambodia. The site had a cultural stratigraphy of only 0.5–1.0 m, and neither iron nor glass beads were found among the material turned up by the bulldozer. A radiocarbon date of 2000 ± 250 B.C. has been obtained from an organic crust adhering to a potsherd. Little if any value attaches to such a sample because of its lack of any stratigraphic relationship to bronze.

The cultural sequence in this area and, in particular, the date of metallurgy there has, however, been clarified by recent Vietnamese investigations. Pham Quang Son (1978) has advanced a four-fold subdivision of later prehistory in the Dong Nai valley. The sequence begins with the site of Cau Sat, which is characterized by shouldered adzes and stone arm rings with a rectangular or trapezoidal cross section. The pottery from this site includes pedestaled bowls and tall jars with flat bases (Hoang Xuan Chinh and Nguyen Khac Su 1977). The second phase incorporates sites described by Fontaine (1972), of which Ben Do is the best known. Pham Van Kinh (1977) has described the many large shouldered adzes from this site. The shouldered form is much more abundant than the adzes with a quadrangular cross section and no shoulder. The third phase incorporates Cu Lao Rua. We have seen that there is a possibility of bronze being in use there. Certainly the large forms of shouldered stone adze fell away in popularity during this phase. It is likely that the sandstone molds of Hang Gon belong to this part of the Dong Nai sequence.

By phase 4 bronze was abundant and locally cast. In 1976 the important site of Doc Chua on the bank of the Be River was discovered, and excavations ensued then and during the following year. The finds documented a mature tradition of casting bronze (figure 13.16). No fewer than fifty mold fragments of clay or sandstone were found, and implements locally cast included tanged arrowheads, axes, small bells, socketed spearheads, harpoons, and chisels. The shape of the axes, chisels, arrowheads, and spearheads is remarkably similar to the examples from the General Period B contexts in northeast Thailand (Le Xuan Diem 1977). There are also general resemblances in some stone artifacts, such as the bracelets and adzes. There are few shouldered adzes at Doc Chua. The radiocarbon date of 1195 ± 130 B.C. derives from charcoal from a depth of 1.0 m at the site. This date is quite consistent with the bronze types found and suggests that bronze working was already established at a period equivalent to General Period B in northeast Thailand.

Indeed, there are hints that there was a network of bronze-using communities from the mouths of the Mekong up to the Ban Chiang area. A major link in this proposed chain is Samrong Sen, an occupation and burial site already alluded to, with a stratigraphy almost 6 m deep, situated on the banks of the Chinit River. The site was first investigated by Noulet (1879), but excavations adopting any semblance of stratigraphical control were undertaken much later. Mansuy (1902, 1923) recognized three layers, one of which attained a depth of 4.5 m and was said to incorporate numerous shell lenses.

Given the almost complete absence of stratigraphical control in a site clearly having a lengthy period of occupation, only the most general conclusions are possible. Among these is the clear relationships that exist between the shouldered and rectangular adzes of Samrong Sen and those from the lower Mekong sites. Armbands and beads of stone also link this site with others in the Mekong catchment. Pottery is less easily paralleled elsewhere. Much of the ware is plain, and

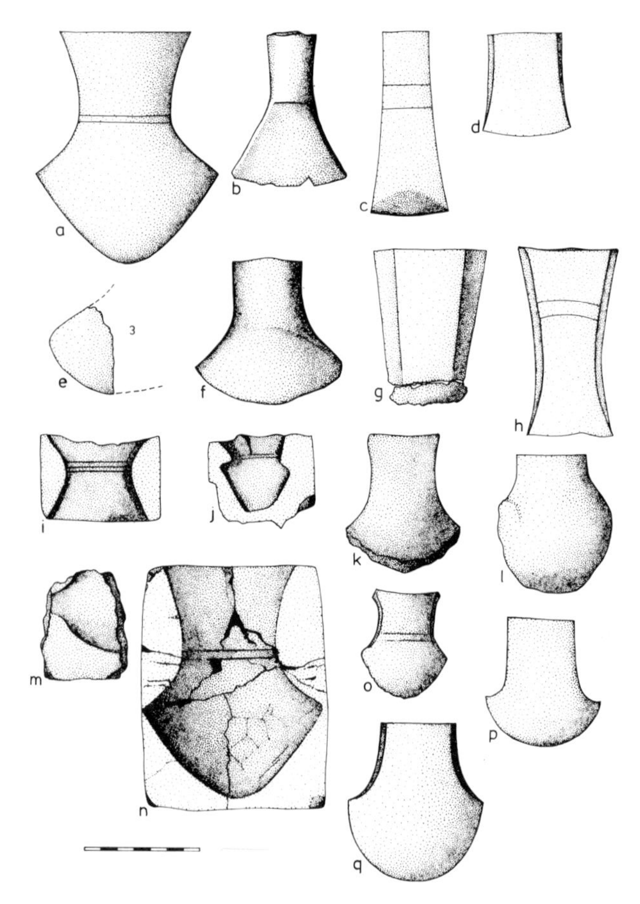

Figure 13.16 Bronze axes and molds from General Period B contexts in Southeast Asia. (a) Doc Chua. (b) Ban Chiang. (c) Dong Dau. (d) Go Mun. (e) Ban Na Di, from a mold. (f) Go Mun. (g) Non Nok Tha. (h) O Pie Can. (i–k) Doc Chua. (l) Samrong Sen. (m) Ban Na Di. (n, o) Doc Chua. (p) Go Mun. (q) Dong Dau.

the decoration is in general confined to impressed or incised geometric motifs. The complete vessels probably come from inhumation graves, although flooding is said to have redeposited the human bone in antiquity. A characteristic form has a broad pedestaled base supporting an open bowl. Some were decorated all over; others were left plain. Painting is quite absent. Although there are no precise parallels between these vessels and those from northeast Thai sites, Loofs-Wissowa has pointed out a remarkable similarity between two complete vessels from Samrong Sen and the earliest style recovered at Ban Chiang.

Mansuy recovered a fragment of bronze mold from Samrong Sen; unprovenanced bronze artifacts fall within the range of types documented in the lower Mekong, northeast Thailand, and coastal Vietnam. The stratigraphic association of the tanged arrowheads, socketed axheads, bells, arm rings, fishhooks, and chisels are unknown. There was a rich faunal assemblage at Samrong Sen, which from the vague and generalized reports available matches the findings at related sites in northeast Thailand. In addition to domestic cattle, pig, and dog bones there was a clear orientation toward water resources, documented by fishhooks, net weights, the bones of crocodile and water turtle, and numerous shellfish. Shellfish from two locations have been combined to provide one radiocarbon sample, which yielded a date of 1280 ± 120 B.C. The material dated comes from 1.0 m and 1.5 m below the surface and so at best dates only the later part of the cultural sequence. However, a date in the late second millennium B.C. would fit the typology of the bronze artifacts.

Samrong Sen must surely represent many similar sites in central Cambodia. Indeed, Mansuy was aware of a similar site, to judge from surface finds, at nearby Long Prao. It has never been excavated, but Lévy (1943) has examined three prehistoric sites located near the headwaters of the Sen and Chinit rivers (see figure 13.1). Surface remains from O Pie Can cover an area of about 1 ha. The site is located at the confluence of two streams, and the culture layer is 40 cm in depth. Lévy has reported finding pieces of sandstone molds on the surface and crucible fragments containing bronze and pieces of iron slag. The excavations also revealed molds for casting a sickle and an ax. The shape of the sickle mold is similar to molds from Go Mun contexts in Vietnam. The site also yielded many fragments of stone and clay bangles and clay stamps bearing deeply incised curvilinear designs. Reports from looters at nearby O Yak referred to human inhumation burials incorporating bronze bracelets, animal bone offerings, and brown-red glass beads. Burials were also noted in eroded material at O Pie Can. The surface of a third mound, O Nari, included much pottery and fragments of polished stone adzes and some bronze. As at Samrong Sen, the adzes were both shouldered and rectangular in shape. Stone beads and bracelets as well as round clay pellets were also recovered there.

Central Coastal Vietnam

The coastal plain that abuts the Truong Son cordillera between the Ca and Dong Noi valleys is in general narrow. Only where the Thu Bon, Con, and Ba rivers reach the sea are there enclaves of good agricultural soils. Archaeological research in these areas is in its infancy, but already it is clear that the Thu Bon valley was occupied in prehistory by a society proficient in metallurgy. The evidence for this finding comes from Binh Chau, where Ngo Si Hong (1980) has recently investigated occupation and burial remains. Three mounds have been excavated, the areas covering between 0.5 ha and 1.0 ha. Today the mounds command tracts of flat rice fields. Preliminary findings have revealed that the occupants inhumed the dead in association with several pottery vessels. Red, black, and white designs were painted on some burial vessels. The widespread form of split earring was rendered in fired clay at Binh Chau. Of particular note were the fragments of crucibles and molds found, in addition to bronze socketed axes and tanged arrowheads.

The Bac Bo Region

During General Periods A and B there are two aspects to the settlement in the Bac Bo region, one inland and the other coastal. The inland focus centers on the Middle Country, an area of rolling lowlands dissected by minor tributaries of the Red River (figure 13.17). It is studded with archaeological sites and much research has been devoted to them. The cultural sequence now emerging reveals close parallels with that 500 km to the southwest on the Khorat plateau, and there is simplicity in employing the same general cultural framework.

Three successive phases belong to General Periods A and B. They are, in chronological order, Phung Nguyen, Dong Dau, and Go Mun. The Phung Nguyen period has been subdivided on the basis of pottery typology, into three successive groups. Of the fifty-two sites excavated, only eleven have yielded bronze. Metal is recovered from the latest contexts of the Phung Nguyen phase and takes the form of corroded fragments and slag. As yet no artifacts have been recovered. Consequently, in terms of metallurgy at least, most Phung Nguyen phase sites appear to equate with General Period A in northeast Thailand. There are relatively few radiocarbon dates for this sequence. The Phung Nguyen phase appears to have developed into early Dong Dau by c. 1500 B.C. The date of the earliest Phung Nguyen contexts is not yet known but probably belongs to the third millennium B.C. As can be seen from figure 13.17, the distribution of Phung Nguyen, Dong Dau, and Go Mun sites lies in the same region, most sites being found above the confluence of the Red and Black rivers. They cover between 1.0 ha and 3.0 ha and are found on slightly elevated terrain near small stream confluences. The principal excavated site is at Phung Nguyen itself (Hoang Xuan Chinh and Nguyen Ngoc Bich 1978). This site covers 3 ha, of which 3960 m^2 have been excavated. The stratigraphy, as in most Phung Nguyen sites, is shallow, barely exceeding 1 m in depth. Excavations commenced in 1959 and were followed by two further campaigns. The earliest excavations were relatively restricted: Two excavation squares measuring 3 m × 9 m and 2 m × 7 m were opened, with six others covering 4.5 m × 4.5 m. The 1961 campaign was massive: seventeen 10 m × 10 m squares and five 20 m × 20 m squares. The final campaign in 1968 added a further small area excavation. Opening up such an immense area was possible because the cultural stratigraphy was only 0.5 m deep.

All this activity provided a large sample of material culture that is most interesting. No trace of bronze was recovered, but there was a substantial sample of

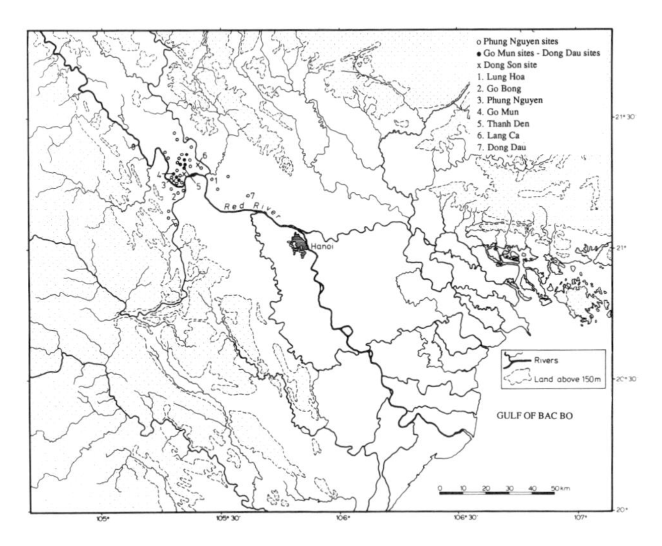

Figure 13.17 The distribution of General Period A, B, and C sites in Bac Bo.

pottery and stone artifacts. The stone adzes take a variety of forms. Rarest were the shouldered variety, of which only four were found. This compares with 777 examples of quadrangular form, some of which were sharpened consistently such that they are broader than they are long. When one adds the fragments of stone adze for which the shape is indeterminate, 1,138 adzes or adze fragments were recovered. There are also fifty-nine small stone chisels, some with cutting edges only 10 mm wide. Almost 200 grinding stones were found. They have grooves that resulted from sharpening stone adzes and chisels. Less abundant were stone projectile points. Three reach the dimensions of a spear point; the remainder are more likely tanged arrowheads.

The inhabitants of Phung Nguyen also fashioned stone rings, mostly in nephrite. The total sample of 540 specimens have been subdivided into eight types based on the shape of the cross section. Most are rectangular, but some are much more complex, having a range of ribs and flanges. Although some are large enough to rank as adult bracelets, others have small diameters and were designed either for children or perhaps for display as earrings. Stone beads were also made, most being tubular and measuring up to 1.3 cm in length.

There was also a vigorous tradition of working clay. There are clay pellets and bow pellets matching those found so often in all the General Period A–B sites described in Thailand. Clay net weights were also found, but most attention was accorded to the manufacture of clay vessels. It is on the basis of changing decorative styles that Vietnamese archaeologists have advanced their threefold subdivision of the Phung Nguyen sites. The earliest, called after the site of Go Bong, is characterized by incised parallel bands infilled with rows of impressions imparted with a pointed implement. The favored motif is in the shape of an S meander. The second phase incorporates a range of designs based according to Ha Van Tan (1980) on "geometric asymmetry." Again, incised bands infilled with dentate impressions alternate with bands left blank, forming a series of most attractive design fields. Extraordinarily this technique of decoration and the same asymmetry of motifs employed is matched in northeast Thai sites of approximately the same period. Watson (1983) has also drawn attention to the widespread distribution of this design technique.

One site, Lung Hoa, has yielded a sample of twelve burials in an excavated area of 365 m^2. The excavators believed that there was some evidence for differential wealth, although the sample is small. There were, for example, two burials associated with stone bracelets, beads, and earrings and pottery and polished adzes. Only the last two were found in association with the remaining ten graves. Clearly larger samples are necessary before this possibility can be tested.

Dong Dau is a most important site not least because of the unusually deep stratigraphic record (Ha Van Phung 1979). It is located to the east of most Phung Nguyen sites, within sight of the Red River (figure 13.17). It was recognized in 1961, and excavations undertaken in 1965 and 1967–1968 uncovered 550 m^2 to a depth of between 5 m and 6 m. The mound itself covers about 3 ha. Its basal cultural material can be described as a developed late Phung Nguyen horizon. It contains a sample of rice grains, which attests to rice cultivation as one of the subsistence activities of Phung Nguyen (Nguyen Xuan Hien 1980). Although the pottery of the Dong Dau site reveals Phung Nguyen origins at least in style and mode of decoration, there is also compelling evidence for a local and vigorous bronze industry. Trinh Sinh (1977) has stressed that many of the Phung Nguyen stone artifacts were replaced in bronze. The bivalve molds recovered by Ha Van Tan from a small (50 m^2) excavation in a Dong Dau context at Thanh Den are virtually identical with those from General Period B contexts on the Khorat plateau. The thirty or so fragments of stone and clay molds were designed for casting axes and fishhooks. The site has a cultural stratigraphy of only 1 m, and the three radiocarbon dates all match closely the dates obtained for General Period B in northeast Thailand (1280 ± 100, 1140 ± 70, and 550 ± 130 B.C.). Bronze was also employed to render socketed spearheads, arrowheads, and chisels. The Dong Dau ax began to take on the initial pediform shape so diagnostic of later decorated examples from the Dong Son phase.

Toward the end of the second millennium B.C. the Dong Dau developed into the Go Mun phase, named after the site of Go Mun. This site is located only 3 km northeast of Phung Nguyen, and indeed the twenty-five or so known Go Mun sites are located within the same general area as those of Phung Nguyen and Dong Dau. There have been four campaigns of excavations at Go Mun, commencing in 1961 and finishing a decade later. In all, 1500 m^2 have been excavated, the cultural stratigraphy being only 1 m deep. As at Phung Nguyen, stone adzes were mainly quadrangular. Of the eighty-seven recovered, eighty were quadrangular and only one was shouldered. Similar stone chisels to those from Phung Nguyen were also encountered. The inventory of bronze artifacts (figure 13.18) reveals that many forms in stone were copied in metal. Thirteen axes and seven chisels have been found. The lance or spearhead, arrowheads, and bracelets were likewise rendered in bronze. Fishhooks were the most abundant bronze artifact, followed by narrow projectile points. One sickle and the figure of a seated individual that was presumably cast by means of the lost-wax technique were recovered. Dong Dau pottery bears curvilinear and rectangular patterns that look like developed Phung Nguyen motifs, and it is in these that Ha Van Tan (1980) sees models for the decoration later found on the well-known Dong Son drums.

It is apparent from the excavation carried out at Go Mun and related sites that bronze working was increasing in intensity and in the range of artifacts cast. This phase, which lasted until about the seventh century B.C., brings us to the end of General Period B in the Bac Bo area. It has been intriguing to note the similar progress and chronology of bronze working in Bac Bo, northeast Thailand, and the lower Mekong valley. Detailed reports on biological finds and the rarity of burials, however, render the Bac Bo sequence less revealing than hoped for.

The Cultural Context of Early Bronze Working

Two general periods are in question. The later distinguished by the adoption of bronze working. It is now necessary to consider the significance of this widespread metallurgical tradition. Emphasis has been

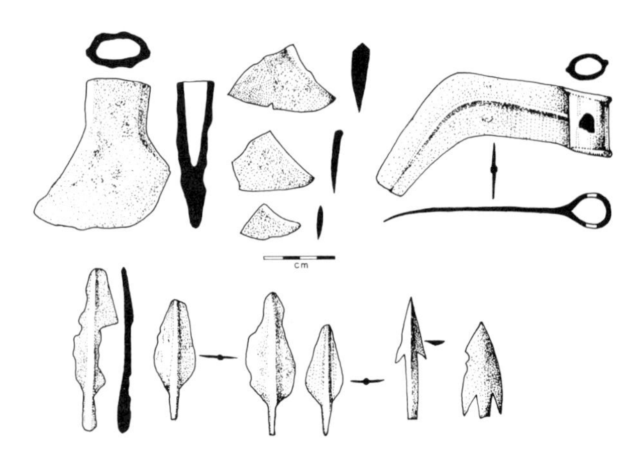

Figure 13.18 Bronze artifacts from Go Mun.

given to the issue of site size. As a general rule, it is assumed that there is some relationship between the size of a site and the number of people occupying it. Thus the modern village of Ban Na Di in Khon Kaen Province, northeast Thailand, has a population of 800–1,000 people in 23 ha of settled area. For Ban Chiang today, the figures are 4,500 people in 75 ha. During their site survey and analysis of the Central Valley of Mexico, Sanders et al. (1979) cited four different levels of population density. The lightest had 5–10 people per hectare, a figure rising to a high density at 50–100 people per hectare. Applying these ranges to prehistoric settlements, a risky but interesting exercise, would indicate a maximum of up to 500 people in a settlement covering 5 ha. The real figure may have been far less. One variable in our model for the settlement expansion into the inland terrain of Southeast Asia, then, is that during General Periods A and B, settlements did not exceed at the outside a population of 500 people. During General Period A they may have been much smaller.

During General Period A it is held that there was a need to acquire rare goods to fulfill rituals involved in the rites of passage. One such substance seems to have been marine shell jewelry. Stone axes too were important perhaps in ritual and economic activities. In a word, it is suggested that exchange for such goods was an adaptive mechanism to ensure that the social and technological necessities in settlement expansion were available. Let us take a closer look at the ritual and social factor. For this I rely on the evidence for mortuary behavior. Take, for example, the burials from Ban Na Di. It has been suggested that there were two groups of burials, and exotic or unusual artifacts have been shown to concentrate in one of them (Higham 1983). These objects include imported stone bracelets that were repaired when necessary with bronze wire. It is notable that the internal diameter of these bracelets is so small that they could only have been placed on the wrist when the owner was young. There are also marine shell bracelets and beads, cattle figurines, and in two late graves eight iron objects. Several graves also contained the remains of silken garments, a finding that suggests that the metal, stone, and shell objects were probably just the durable subset of goods exchanged during the currency of the first phase of burials at the site.

The circulation of such exotic goods has attracted considerable interest in the archaeological literature. Dalton, in particular, has emphasized their critical role in cementing the foundation and maintenance of alliances between autonomous communities (Dalton 1977). In the ethnographic cases he cited, he noted that the acquisition of such goods usually resides with lineage leaders who then dispose of them as part of the exchange system, thereby cementing war alliances and fostering the process of peacemaking. Success in obtaining valuables augments an individual leader's social ascendancy. Failure has the opposite effect. External relations in autonomous societies are horizontal and occur between settlements. Relative ascendancy turns on alliance structures, population size, and the disposal of valuables. Marriage and alliance, then, are intimately connected with transactions of food and treasures. Dalton views the economy, alliance networks, and social organization as mutually embedded.

The archaeological context of exotic goods at Ban Na Di is interesting. The goods are found concentrated in one part of the site rather than distributed evenly in both excavated parts. The origin of the shell beads and shell and stone bracelets is not known with precision but must have been remote. Clay figurines and exotic pots, perhaps with their perishable contents, may also have been perceived as valuables. There is then little doubt that the upper Songkhram area at least was linked with other contemporary communities through a widespread system of exchange, reaching out ultimately to the sea and the surrounding ore-bearing hills. This same situation has also been documented east of the Truong Son cordillera, among the stone and bronze-using communities of the Middle Country of Bac Bo. In the Chao Phraya, the middle Mekong, and the lower Mekong areas there was again a system incorporating the exchange of exotic goods. This situation presents a different connotation from that advanced by Wolters (1982). He suggested a prehistoric pattern of scattered and isolated settlements at the beginning of the Christian Era. This indicates little prospect of intersite or area contact. Naturally the alternative picture presented so far influences our appreciation of the processes invoked in what follows.

The juxtaposition of lowlands with upland massifs presents interesting parallels with the situation described for the Maya Lowlands by Voorhies (1973). She stressed the archaeological visibility of certain items from the upland, such as obsidian, when compared with that for lowland products. A similar situation obtains for the Southeast Asian theater, where the import of copper, tin, and lead illustrates clearly lowland-upland exchange. Again, the presence of marine shell reflects even wider contacts. There were doubtless many rarities or surpluses that originated within lowland and coastal contexts. Specialized ceramic vessels are one such product. Salt has a restricted distribution and was probably a crucial commodity then as now. We may also add silk as an archaeologically documented but perishable commodity. Certainly fabric in general is a strong candidate as a prestige item in exchange. The point at issue is that there was inequality of access to prestige goods. Under such conditions strategic location was advantageous in obtaining and maintaining prominence.

Within such an integrative exchange system, information and goods were open to transmission, with the lines of least resistance following river courses or coastal routes. Thus the knowledge and practice of bronze working from copper and tin could spread rapidly, with finished bronzes taking their place alongside stone and shell as objects signifying the status of the owner. The production and distribution of copper and tin surely added a further dimension to the existing exchange links. There were new routes to be opened and new opportunities to seize. Bronze goods were rare in mortuary contexts, and many of the products were clearly for display. This introduces a further point: The attainment of status was flexible rather than fixed. Thus, where a community and its leader commanded a strategic resource or distribution point, they could increase their prominence within the system. A breakdown in the source of supply or the opening of a new route would diminish their wealth and status. Likewise, group ownership of a rich tract of cultivable land would confer at least the possibility of realizing higher status if soil potential were realized. It is important to appreciate that the relative position of each autonomous settlement was given to fluctuation and therefore instability.

General Period C: The Development of Ironworking

Bac Bo: The Arrival of the Han

The Red River delta and the valleys of the Ma and Ca rivers have a less extreme climate than most of lowlying mainland Southeast Asia, largely because the dry season is tempered by moist winds that move across the gulf of Bac Bo. On reaching land, they form a low cloud cover often associated with drizzle. As Gourou (1955) has shown, this moist climate permits two crops of rice per annum on favorable soils. The Red River is given to sudden rises in level, a characteristic compensated for in the last few centuries by the construction of bunds. Such is the load of silt brought down by the Red River that the delta is still advancing rapidly. During the prehistoric period the delta area would have been subjected to widespread flooding, and indeed the Phung Nguyen sites are located in the more elevated middle country above the delta proper. After Phung Nguyen times there was a continuing process of expansion involving occupation of the upper and middle delta itself.

The sequence of Phung Nguyen, Dong Dau, and Go Mun phases described here witnessed a major transformation with the development of the culture of Dong Son. Paradoxically the site of Dong Son itself is peripheral to the main concentration of activity on the delta, being located on the southern bank of the Ma River. It was here that first Pajot and then Janse undertook excavations that revealed a rich prehistoric cemetery containing objects of bronze, iron, pottery, and semiprecious stones and objects of Chinese origin (Janse 1958). This was the first exposure to Europeans of a rich bronze- and iron-using society on mainland Southeast Asia, and its origins were sought in the context of prehistoric societies in eastern Europe, which exhibited similar proficiency in working metals. More recently, Vietnamese scholars have sought less exotic origins in stressing the continuity in terms of decorative motifs as between the Phung Nguyen pottery and the Dong Son bronzes. This has encouraged them to stress cultural continuity throughout the prehistoric occupation of the Red River valley.

Our appreciation of the Dong Son people is colored by literary allusions to the area that have survived in Chinese documents. The later part of the Dong Son phase was in fact contemporaneous with a major imperial Chinese expansion under the Han. Indeed, viewed from the Han capital, the Dong Son people were the most distant of several groups known as the southern barbarians. Although most of our information on Dong Son comes from burial sites, one major settlement probably dates, at least in part, to the period under review. This is Co Loa, located 15 km northwest of Hanoi on the floodplain of the Red River. There are three sets of ramparts at Co Loa, the two outer ones being moated and of an oval plan and the innermost one being rectangular. There is some literary evidence that Co Loa was established as a center during the third century B.C. This has been confirmed by recent archaeological excavations under the ramparts, which have revealed Dong Son style pottery under the middle rampart. In 1982 a complete Dong Son–style bronze drum was located near the central part of the site, containing over 100 socketed bronze plowshares.

One of the most critical points about Co Loa is its size. The outermost ramparts cover about 600 ha, 200 times the area of Phung Nguyen. No concentrated survey of settlement sites has been undertaken for the Dong Son phase, and most of our information is drawn from cemeteries and the contents of graves. This source of information is also restricted in the sense that many cemeteries are found in acidic soils and human bone has not survived. At Dong Son itself, for example, burials were recognized more from the disposition of artifacts than from the presence of human remains. Thus at Lang Ca, 314 burials but hardly any human bone were found. Even so, the excavators noted that a small group of graves were differentially rich, containing axes, daggers, situlae, and spearheads. Further evidence for rich burials has been found at Viet Khe and Chau Can (Luu Tran Tieu 1977). Both have yielded opulent boat burials. The richest at Viet Khe was interred in a hollowed tree trunk about 4.5 m in length and contains over 100 artifacts. No human remains were found, but there is little doubt that the bronze weapons, receptacles, and utensils were grave goods. This particular coffin also contained bronze bells, relatively small drums, and even a painted wooden box.

Wooden artifacts and human remains were found inside some of the eight boat burials found at Chau Can. These are smaller than the Viet Khe example and contain fewer grave goods: bronze axes and spearheads with their wooden hafts in place, earrings made of a tin-lead alloy, and a bamboo ladle. The suggested local origins of the Dong Son bronze industry has been supported by recent research at Dong Son itself. The first use of the cemetery involved burials accompanied by pottery vessels but only a few bronze axes, spearheads, and knives. These are said to equate with the Go Mun phase in the Red River valley, dating to c. 1000–500 B.C. The second phase saw a proliferation of bronze artifacts and extension of types to include daggers, swords, situlae, and drums. It is dated to 500–0 B.C. The third and last prehistoric phase has burials containing objects of Chinese origin, such as seals, coins, mirrors, and halberds. Ha Van Tan (1980) has suggested a date within the first century A.D. for these burials. Indeed, the site was later used for interments in the Han style following the incorporation of the delta region into the Han Empire during the first century A.D.

We can learn much about the people of Dong Son from these surviving bronzes. As has been demonstrated at the Dong Son cemetery, both the quantity and range of bronze artifacts increased greatly from c. 500 B.C. This intensification of production can be illustrated in two ways. The Co Loa drum, for example, weighs 72 kg and would have entailed the smelting of between 1 and 7 tons of copper ore (Nguyen Duy Hinh 1983). One burial from the Lang Ca cemetery contained the remains of a crucible and four clay molds for casting an ax, a spearhead, a dagger handle, and a bell. The crucible, although of the same shape as the earlier ones found in Dong Dau contexts, is much larger. Indeed, it could have held 12 kg of molten bronze (Vu Thi Ngoc Thu and Nguyen Duy Ty 1978). Many novel artifact forms were developed, and earlier forms reveal an interest in decoration. The drums, situlae, and rectangular ornamental plaques suggest an interest in ritual and ceremony, whereas the daggers, swords, and halberds reflect concern with personal weaponry. Nor was agriculture overlooked: The socketed bronze plowshares represent a profoundly important innovation in the field of agriculture, and the casting of bronze plowshares is a direct application of metallurgical skills to the intensification of agriculture.

The Dong Son metalworker was a master in the difficult field of bronze casting, but the status of local ironworking is not clear. During the Dong Son phase Chinese iron technology reached a high level of proficiency. Their methods involved the demanding system of iron casting rather than forging. It is known from documentary sources that iron objects were exported from China to the south during this period, and the presence of some bimetallic spears, with iron blades and cast-on bronze hilts, in Dong Son contexts shows that iron, if by now locally worked, remained rare. It is also instructive to note that the rich burials of Viet Khe and Chau Can do not contain any iron grave goods.

The status of iron during the Dong Son phase requires detailed analyses to determine whether the objects were cast or forged. Evidence for local iron smelting is also basic to an appreciation of iron and its importance to the local people. At present, the direct Chinese contact with Dong Son people is the most likely means whereby knowledge of iron reached Bac Bo. The specialized bronze workers there would doubtless have been interested in its properties. Iron did not threaten the centrol role of bronze in agriculture, war, or ceremony.

It is fortunate that the Dong Son bronze smiths decorated their drums with scenes drawn from the world around them. These allow a glimpse into the activities of the very lords described in Chinese documents. We can, for example, recognize the importance attached to elegant boats equipped with cabins and fighting platforms. They were crewed by paddlers and carried plumed warriors (figure 13.19). The spears, halberds, and arrows found in aristocratic graves are seen in action, either being fired or, in one case, chastising a captive. The Dong Son drums themselves are represented, mounted in sets of two or four on a platform. The houses were raised above ground on piles and had supported gable ends decorated with bird-head carvings identical with those seen on canoe prows.

The new excavations at Dong Son confirm a rapid change from the Go Mun to the Dong Son phase proper, with a major intensification in the quantity and range of bronze artifacts. These include for the first time swords and daggers as well as objects of ritual and ceremonial importance, such as drums, situlae, and

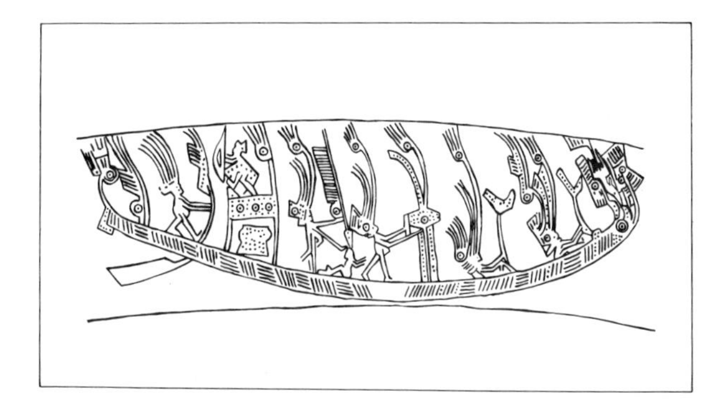

Figure 13.19 The representation of a boat from the Ngoc Lu drum. Bronze spearheads, ax halberds, and a drum are visible.

decorative body plaques. Intensification is also identifiable in subsistence and in exchange, particularly with Han China. This is again seen in the third and last brief period at Dong Son, when Chinese imports peaked. In A.D. 43 the warrior-aristocrats of Dong Son, who had survived many centuries of Chinese expansion, invasion, and subjugation, finally succumbed and were incorporated as a province of the Han Empire.

Although the culture of Dong Son is the best known expression of the transition to the centralized chiefdom in mainland Southeast Asia, it was by no means the only one. Indeed, the trend to centralization was widespread. Again, the foremost symbol of the new aristocracy, the Dong Son drum, is by no means confined to the valleys of the Red and Ma rivers.

The Chao Phraya Valley

There has been far less research in the Chao Phraya than in the Red River valley, but the few investigations undertaken reveal both similarities and differences. The Chao Phraya area was never subjected to the immediacy of foreign expansion, but it does command the eastern edge of the Three Pagodas pass, which was one of the routes providing Indian contact. Again, the Gulf of Siam permitted movement of people and exchange of goods by sea. The area was strategically placed to participate in the expansion of trade, which occurred during General Period C.

One of the most intriguing sites, Tham Ongbah, has yielded the remains of five drums of clear Dong Son affinities, hinting at direct contact with Bac Bo. Tham Ongbah is a massive cavern located in the upper reaches of the Khwae Yai River (see figure 13.1). Sørensen (1979) has recovered some information from the site, following its near destruction by looters. He found that the burial technique involved extended inhumation in wooden boat-shaped coffins. A radiocarbon determination of one such coffin gave an estimate of 230 ± 100 B.C. There were also several intact burials within the cave, of which ten were excavated. These lacked the wooden boat-shaped coffins, but both groups contained a similar assemblage of iron implements. Sørensen has suggested inferior social status rather than chronological change to account for such a disparity in relative wealth. The ten poorer burials have yielded several well-perserved iron implements, thereby affording insight into the uses of this metal presumably during the last century or two of the first millennium B.C. Burial 5 incorporated an iron hoe on the chest of the inhumed adult. Burial 6 included a tanged knife, possibly a spearhead, and five beads. Other burials yielded chisels and arrowheads. One contained seven iron objects, all placed near the ankles.

It is particularly unfortunate that we are unable to present a clear list of associations between funerary artifacts and the boat-shaped coffins because of looting. Sørensen, however, has recovered fragments of artifacts that hint at the wealth they once contained, and he has interviewed the looters' laborers to establish certain relationships between stray finds and the original contents of the coffins. On the basis of this admittedly unsatisfactory but necessary procedure, it is possible to point out several interesting features. The coffin burials included strings of beads around the waists and necks of the inhumed. They also included bronze and iron artifacts, including earrings, bracelets, and at least one bronze vessel. As has been mentioned, a set of drums was found near and probably associated with the coffins. These elaborate objects were decorated with zones of different motifs, including flying birds, humans, and geometric designs.

Contact of one sort or another between the occupants of the Chao Phraya and the Red River valleys is also indicated by a more recent find at Dermbang Nang Buat. Suchitta (1985) has reported a series of artifacts of clear Dong Son affinities, including a fragment of cast iron and the remains of ornamented drums and bowls. Iron casting is a technique developed in China and is alien to any Indian-derived method of ironworking. Whether this find reflects exchange or an actual movement of people from Bac Bo remains to be seen.

Ban Don Tha Phet is located on the western margins of the Chao Phraya lowlands. Its position on a low terrace commanding low-lying terrain matches the placement of the prehistoric settlements on the Khorat plateau. This cemetery site has been the object of several excavation seasons since its discovery in 1975 (You-Di 1978; Glover 1980, 1983; Glover et al. 1984), and it already plays a central role in our understanding of the late prehistoric period in central Thailand.

Unfortunately bone does not survive in the acidic soils there, but excavation has, as at Dong Son, revealed dispositions of artifacts with the occasional scrap of bone that are clearly the remains of human interments. When compared with General Period B burials, the grave goods are considerably more abundant and the range of materials is greater. Thus Glover (1983) has described four burials from the 1980–1981 season. The number of complete or fragmentary pots averaged twenty per burial, a far higher figure than in the General Period B burial assemblages. The burials were also associated with iron tools and weapons, bronze ornaments and bowls, and stone and glass beads.

The blades of socketed iron spearheads were often bent back to break or ritually kill the burial occupant before burial. One iron implement looks like a modern sickle blade; the application of iron technology is revealed by the socketed tips for either digging sticks or hoes, harpoons, and knives. The beads were manufactured from agate, carnelian, and glass. These are all exotic and recall specimens dated to the second and third centuries B.C. in India. There is a strong probability that the beads are Indian exports.

The burials at Ban Don Tha Phet have also furnished a series of thin-walled bronze bowls that exhibit interesting features (Rajpitak and Seeley 1979). They are decorated with incised motifs and occasionally representations of people. The alloy employed is a high-tin (19–21%) bronze, which is brittle and hard to work but imparts a goldlike color to the finished article. The number of such bowls at the site and the long tradition of established bronze working point alike to local and indeed specialized manufacture. A few similar bowls have been found in India. Two come from the site of Bhir, dating to the late first millennium B.C. The inferences are clear: that exchange linked Indian and Southeast Asian societies toward the end of the first millennium B.C. and that some individuals at Ban Don Tha Phet were interred with imported beads made of glass and semiprecious stone. It is unfortunate that the acidic soils at the site have ruled out the preservation of a large enough sample of charcoal for radiocarbon dating, but one determination made on the basis of the organic content of pottery indicated use of the site at least during the second to third century A.D.

The presence of moated settlements in the Chao Phraya lowlands has long been known, and some were clearly major centers of the Dvaravati civilization. Excavation at two of these sites, Sab Champa and Chansen, have provided evidence for initial occupation during the late prehistoric period. At Chansen Bronson found a basal cultural layer containing an inhumation burial associated with a socketed iron implement interpreted as a hoe or plowshare. This same context has yielded the remains of water buffalo bones. Veerapan (1979) has discovered inhumation graves at the moated site of Sab Champa, associated on this occasion with pottery vessels and molds used in bronze casting.

In many ways our knowledge of General Period C in the Chao Phraya lowlands remains sketchy and unsatisfactory. Excavations have not been undertaken in the context of regional settlement pattern studies, so no information on relative site sizes and possible presence of regional centers is available. Glover has stressed that we are not yet able to say whether the Ban Don Tha Phet cemetery was unusually rich or typical of other sites in the area as a whole. At Ban Tha Kae, however, there is some indication that the site ultimately covered a considerable area, probably in excess of 20 ha. The site itself has been destroyed, but early air photographs reveal a large mound. The question is, How much of the mound was occupied during General Period B? There is now no way of being sure.

An important step toward a fuller understanding of this period has recently been taken by Ho (1984). Within the context of her analysis of the material from the excavations of Khok Charoen, she undertook a series of site surveys. These suggested strongly that, after an initial occupation during what Ho calls the early metal age (equivalent to General Period C), there was a move toward centralization, wherein one site grew differentially large and, it is assumed, exercised political dominance over others in its area. Indeed, Ho was able to identify three distinct areas, each dominated by a large, ultimately moated site. The centers are about 30 km from each other. Ho has ascribed these sites to a "high metal age," which saw a marked increase in bronze working and the initial use of iron. This equates culturally and chronologically with General Period C. It is important to note how Ho's results from the Chao Phraya valley match those proposed by Welch in the Mun valley and by Higham and Kijngam in the Chi valley.

There are then some grounds for suggesting that there was a trend toward centralization during the mid to late first millennium B.C., but the chronological relationship between it and initial direct or indirect contact with exotic coastal traders remains to be determined. Contact with the Red River valley is also apparent, as evidenced in the drums from Tham Ongbah, the few bronzes in the Dong Son tradition at Ban Don Tha Phet, and the recently discovered objects of Dong Son type found at Dermbang Nang Buat (Suchitta 1985).

The Khorat Plateau

Toward the end of General Period B the occupants of the Khorat plateau lived in autonomous villages and followed a system of ranking as between the different descent groups. The expansion of settlement resulted from the fissioning of communities as they reached critical population thresholds thought to be in the region of about 300–500 people. This process of expansion was not constrained so long as there was sufficient land along stream or river margins with the necessary properties for the cultivation of rice and ample fishing, trapping, and stock maintenance, which characterized subsistence activities. Such enclaves of suitable terrain were not, however, limitless. The Khorat plateau includes substantial elevated terraces and hilly areas quite unsuitable for the cultivation of rice.

General Period B settlements are known in the favorable valleys that feed the Songkhram, Chi, and Mun rivers. Although pottery styles suggest tight regionality even to the village level, under appropriate analytical techniques settlements can be linked by the recognition of exchange in valuables, which reached into the surrounding hills for copper, tin, and stone and to the coast for trochus and cowrie shell ornaments.

It has been nearly forty years since Williams-Hunt (1950) published the results of his analysis of aerial photographs covering the Khorat plateau. These disclosed a considerable number of large mounds enclosed by moats and ramparted defenses clustered in the Mun-Chi drainage basin. Most sites are oval in plan, and some look as if they have been added to with time. Thus Muang Fa Daed is composed of three moated enclosures with a fourth feature thought to have been a reservoir nearby. Williams-Hunt also concluded that three sites were considerably larger than the rest, indicating a hierarchy of sites according to size. Similar oval moated sites are now known on the margins of the Bangkok plain and in northeastern Cambodia. These sites are intriguing not only for their distribution pattern, size, and date but also for their origins. It is this last point that I consider first.

Quaritch-Wales (1957) was attracted to these sites in the Mun valley, excavating small squares at Thamen Chai and Muang Phet. The sites yielded a cultural stratigraphy about 3 m and 2 m thick, respectively. The excavator noted the presence of iron down to the basal layers of Muang Phet and concluded on the basis of ceramic typology that the first settlement reflects at least influence from if not occupation by people from the Chao Phraya valley. Quaritch-Wales regarded the presence of iron as a result of contact with India and concluded that the sites were first occupied during the first millennium A.D. Moreover, Muang Phet has yielded a radiocarbon date of 140 ± 150 A.D. from a depth of 1 m from the surface.

A series of more recent settlement pattern studies and excavations favors a greater antiquity for some sites and innovation reflecting internal changes rather than incursions. The best documented settlement pattern analysis incorporating a moated site took place on the southern margins of the Chi River. The chosen study area incorporates the Chi floodplain, which reaches a maximum breadth of 7 km. Three tributary streams flow north across the floodplain and into the main river. They have smaller floodplains surrounded by low terrace terrain of varying width, which gives way to the slightly more elevated land of the middle terrace.

The prehistoric sites concentrate near the low terraces of the tributary streams and fringe the extensive tract comprising the Chi floodplain. One site in the latter area, Ban Chiang Hian, incorporates a double set of moats, a reservoir, and possibly the remains of ramparts. Three sites have been excavated, including Ban Chiang Hian itself. All three yield a distinctive red on buff painted ware in the lowest occupation layers. At Ban Kho Noi and Ban Chiang Hian, this was superseded in the mid first millennium B.C. by a plainer ware. At that juncture the excavators found the first evidence for iron and the water buffalo. The moated area of Ban Chiang Hian covers about 38 ha. It exceeds most significantly the size of all other prehistoric sites in the surveyed area and was clearly a special central site. As with all other such moated sites in the Khorat plateau, we cannot yet answer the question of the construction date. This is a critical issue because the existence of so large a site with such extensive earthworks is precisely the sort of evidence that is held to reflect the existence of centralized chiefdoms. There is, however, some evidence that suggests that such social groups were present in at least the Mun and Chi valleys by the period 300 B.C.–A.D. 300.

Let us begin by considering the site of Non Chai, which was excavated in 1978 by Pisit Charoenwongsa (Bayard et al. 1986). This most important site is located in the upper reaches of the Chi catchment. It is found on a small surviving tract of the old middle terrace but in such a location as to command low-lying alluvial soils, which are today classified as moderately suited to rice cultivation. The size of the site cannot be stated with precision because it has been removed for road fill. According to plans made after removal had commenced, it covered at least 18 ha. The excavators estimated an area of 38.5 ha. It was therefore considerably larger than any known General Period B site. Proximity to flooded areas is reflected in the considerable number of aquatic resources identified by Kijngam (1979). He described a considerable number of shellfish, fish, crabs, frogs, and water turtle from the middens there. At the same time the early settlers brought with them domestic water buffalo, cattle, dog, and pig. Like the occupants of Ban Chiang, they hunted extensively. The bones of deer, crocodile, rhinoceros, and many small mammals recur in the faunal spectrum (Kijngam 1979). The pottery from Non Chai is dominated by red-slipped and painted wares that echo later Ban Chiang styles (Rutnin 1979). In this context the radiocarbon dates confirm a relatively late prehistoric settlement and a rapid buildup of cultural material. Apart from one date of 3860 ± 240 B.C. for the lowest layer, the rest vary little throughout the cultural layers. Although the excavator has yet to comment on the context and status of the earliest date, an eroded section near the excavated area revealed clearly a thin band of occupation followed by a sterile soil buildup, then evidence for the continual occupation of the site. Thus some occupation may have occurred during the second millennium B.C., but the weight of evidence points to continuous occupation from toward the end of the first millennium B.C.

Charoenwongsa has suggested that phase I dates from about 400 B.C., phases II–III between 300 and 200 B.C., phase IV to 200–0 B.C., and phase V into the second century A.D. The radiocarbon dates reveal midpoints of 350 and 225 B.C. for samples derived from phase I, with 2σ ranges between 525 B.C. and A.D. 25. The material culture from Non Chai includes small amounts of iron slag from this phase, but iron slag only became common in phase IV. There are also 4 glass beads from contexts earlier than phase III at the site, and over 200 belong to phases IV and V. The surge in the number of beads probably dates to about 200–0 B.C. Clay molds for casting bronze bracelets and bells are likewise found in phase III–V contexts, although fragments from bronze and crucible fragments were identified in layers attributed to phases II–V.

The critical point about this well-dated site is that it covered at least 18 ha at some point during the period from 300 B.C. to c. A.D. 250, when it seems to have been abandoned. At present we do not know at what point in its sequence, if any, the site actually attained that area under continuous occupation. If a population figure of 50 people/ha is adopted for the area under continuous occupation, then the site would have harbored at its maximum extent about 1,000 people. If an estimate of nearly 40 ha for the site area is adopted, then the population could have been twice that figure. The pottery is distinctive in form, surface finish, and fabric. There are sharp differences between it and the contemporary wares at Ban Chiang Hian, although a few exotic sherds from the latter were found and probably represent imports.

The valley of the Lam Siao Yai is found about 60 km southeast of Ban Chiang Hian (see figure 13.1). Here again we encounter a large moated site. Non Dua is located so as to command an extensive deposit of rock salt as well as low terrace soils suited to rice cultivation. The salt exposure, known as Bo Phan Khan, is surrounded by evidence of industrial activity in the form of mounds and quantities of thick-walled pottery. At present the salt is obtained by removing the salty soil and passing water through it. The brine is then boiled in flat metal trays. Excavations in one of the mounds around the deposit exposed evidence for salt extraction to a depth of about 6 m, commencing, according to the radiocarbon date, in the first or second century A.D. Some examples of the crudely fashioned industrial wares were found during excavations within the moated site itself, which suggests that its occupants were concerned with the extraction of salt. The extent of the activity, measured in terms of the huge mounds that have accumulated around Bo Phan Khan, points to the production on a scale far greater than would have been necessary to satisfy local demand alone. Non Dua, the moated site, also yielded a deep stratigraphic sequence, and the initial phase of occupation has been assigned to the period 500–1 B.C. Some of the distinctively decorated rims and body sherds have been noted in phase II at Ban Chiang Hian, but otherwise, the pottery there was not matched in the Middle Chi valley.

Further information on the Khorat plateau settlement during General Phase C has been obtained by Welch. Working in a circumscribed survey area centered on the Khmer center of Phimai, he identified fifteen prehistoric sites. He found a concentration of them on recent terraces elevated above the floodplain. Excavations in two sites led Welch to propose a twofold subdivision of prehistoric settlement. The earlier, dated to 1000–500 B.C., is termed Tamyae and corresponds to General Period B. The following belongs to General Period C and is dated between 500 B.C. and A.D. 500. At least by the latter period some sites grew to a considerable size. Phimai itself, for example, is held to have covered 40 ha. There are allusions by Solheim (Welch 1984) to rich burials with imported grave goods and the use of iron at Phimai during this period. A particularly distinctive ceramic style was favored there, incorporating a lustrous black ware decorated with patterned burnishing lines.

The large moated sites in northeast Thailand are very much a feature of the Mun and Chi valleys. These sites are significantly larger than any other site in their surrounding territory. Non Chai as well as Phimai, Ban Chiang Hian, and Non Dua produced distinctive local styles of pottery, and there is some suggestion that some vessels were exchanged between centers or, at least, regions. Ironworking and bronze casting were undertaken, and glass beads are included among imported items. In no case has a Dong Son import been found in a stratigraphic context in one of the moated sites, but then excavations have been minimal. Objects of Dong Son origin, however, are known in northeast Thailand from surface or looted sites, and two ax halberds were found in a burial context at the site of Ban That (Kethutat 1976). Given the contemporaneity of these sites with Dong Son, it is noted with interest that no bronze plowshares have yet been found in the Khorat plateau. If we accept a figure of 50 people/ha as a reasonable population estimate for these prehistoric sites, then Non Chai would have accommodated at least 1,000 people, and Ban Chiang Hian, twice that number. The earthworks at Ban Chiang Hian are substantial and would have entailed much energy in their construction. Even if the moats and reservoir were a mere 1 m deep, then at a rate of 2 m^3 of soil moved by one person in a day, 500 well-fed adults would have taken a year to complete them (Chantaratiyakarn 1984). Again, the presence of a reservoir argues for the need to supply water to a large number of people during the dry season. The point is clear: that the moated sites represent a signal departure from the earlier system of village autonomy.

Moore (1986) has considered the distribution of these moated sites and noted that, although many cluster along the margins of the Mun floodplain, there are also several found on the surrounding older terraces. The latter are not obviously located so as to command good low-lying rice land but do possibly relate to the presence and local smelting of iron ore. The excavations at Ban Chiang and Ban Na Di, however, have revealed that a major cultural dislocation occurred at about 300 B.C. It will be recalled that Levels 6 and 7 at the latter site contained a cemetery, the latest two burials being accompanied by iron artifacts: a spearhead, circlets, and a knife. Level 5 represents a radical change in activity. It yielded the remains of a bronze working area, complete with crucibles, furnaces, and clay molds for casting bells and bracelets. The crucibles, although similar in shape and size to those in preceding layers, were now tempered with rice chaff rather than grog. The bronze alloys from Levels 8 to 6 inclusive contain 2–12% tin. That found in Level 5 was much more variable (see figure 13.13). Lead became a common additive, and a few pieces were made from the very-high-tin (over 20%) bronze that was used in the manufacture of the Ban Don Tha Phet bowls. The bronze industry during Levels 4 to 5 was concerned with the production of bracelets, bells, and bowls. Axes and spears, which characterized the General Period B repertoire, were no longer manufactured in bronze. It is considered highly likely that the metalworking at Non Chai represents an increased degree of specialization.

During the buildup of Level 4, part of the excavated area was used as a burial ground exclusively for infants. Their remains were found in lidded urns, associated with artifacts of bronze and iron and in one case with five blue glass beads. The bronzes are miniature bracelets, and the iron artifacts are knives and a socketed tip for a digging stick similar to those found at Ban Don Tha Phet. These were coated with the remains of rice, which survived through impregnation with iron oxide. The grave ritual clearly now distinguished the area used for burying infants from that used for adults, a distinct departure from the earlier Level 7–6 cemetery phase.

The range of artifacts found in Levels 5–4 also differs. There were no more clay figurines. Shell disk beads and shell bracelets were no longer exchanged, and most of the beads were made of orange or blue glass. Although a few iron artifacts were found in two Level 6 graves, Levels 5–4 yielded iron slag, which indicates local smelting of iron ore.

The Uplands of Laos

Hitherto our consideration of societies practicing metallurgy during General Periods B and C has been confined to the lowlands. Fifty years ago, however, Colani (1935) undertook fieldwork in the uplands comprising the northern Truong Son cordillera, specifically to enlarge our knowledge of a series of sites there characterized by large stone burial jars and freestanding stone slabs, or menhirs. Only with the recovery of glass and metal artifacts in the Khorat basin sites during the last few years has it been possible to appreciate the date and cultural affiliations of the upland groups (figure 13.20).

The most impressive and perhaps the most complete of these sites is known as Ban Ang. It is located at an altitude of just over 1,000 m, and occupies a prominent position on an extensive area known as the Plain of Jars. This site is dominated by a central hill within which Colani identified and excavated a pre-

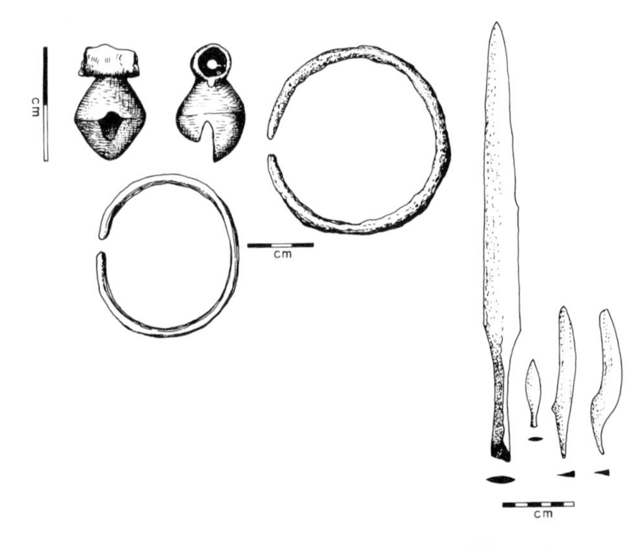

Figure 13.20 Some metal items recovered from the General Period C cemetery of Ban Ang, Laos.

historic crematorium. There were two groups of burial jars in its vicinity, one on a raised area that Colani ascribed to a ruling group. She found a richer set of grave goods there than in the other rather larger assemblage of jars. The burial offerings found at Ban Ang and indeed in the other related sites include glass and carnelian beads, cowrie shells of coastal origin, bronze helices, bells and bracelets, and iron knives, arrowheads, and spearheads. The presence of clay mold fragments reveals local bronze casting; iron slag is evidence for the local working of iron as well. Of course, some of the metal artifacts could have been imported.

The parallels with these artifacts are clearly with General Period C in the Khorat plateau sites, although one bronze figurine recalls Dong Son traditions of metalworking. A date in the region of 300 B.C.–A.D. 300 is consistent with the material found in and around the stone mortuary jars. Explaining the origins of the people in question and how they disposed the labor to move and shape such impressive stone funerary monuments is not so easy. Colani may well be correct in suggesting that these people were placed to control exchange routes connecting southern China, Bac Bo, and the burgeoning chiefdoms of the Khorat area. She also noted that the uplands in question are to this day a major source for salt, and indeed control of the salt trade may well have provided the resources for the import of exotic goods from some considerable distance.

Central Coastal Vietnam

Cremation is an unusual burial technique in prehistoric Southeast Asia, and again Colani may be correct in relating the occupants of upland Laos to the so-called Sa Huynh group. The discovery of the relevant sites along the littoral of central Vietnam occurred in 1909, when a French customs official, M. Vinet, encountered a collection of urns containing human cremated remains at Thanh Duc, near the village of Sa Huynh. This site was further examined by two amateur archaeologists in 1913, and the first professional consideration was given the following year by Henri Parmentier (Parmentier 1918). By then 120 jars had been unearthed over an area measuring 80 m × 50 m, and they were found to be grouped and to contain cremated human remains in association with a range of grave goods. This Sa Huynh burial practice stands out in Southeast Asia. Although it is true that jar burial occurred in northeast Thailand during General Periods A, B, and C, it was reserved for infants, and the body had never been cremated. Yet along the coastal tract almost from the mouth of the Mekong River to the southern boundaries of Bac Bo, this burial rite has been identified. The large and elegant burial vessels usually contain a range of offerings. Moreover, the charcoal associated with the cremated human remains and concentrations in the vicinity of the vessels are an unusually good source of radiocarbon samples.

Saurin (1963) has described an urn field at Hang Gon. The cemetery covered an area of about 100 m × 50 m, the lips of the burial urns being between 0.2 m and 1.0 m below the present ground surface. The urns were provided with covers but contained no bone, perhaps because of local acidic soil conditions. There was, however, much charcoal and burnt soil. A characteristic of this site was the practice of ritually breaking burial offerings. Smashed pottery vessels were placed inside the large urns. Polished stone adzes were damaged, bowls deformed, and even the sockets of iron axes broken. Small and durable items of jewelry, such as ear pendants and beads, however, remained intact.

Some beads of imported carnelian, agate, olivine, and zircon as well as blue and red glass beads and a solitary bead of gold were found at Hang Gon. There was also a distinctive stone ear pendant representing a two-headed animal of a type also found at Sa Huynh itself, in Bac Bo Dong Song contexts, and at Ban Don Tha Phet in the lower Chao Phraya area. Ironworking is documented in the form of slag as well as implements, the commonest of which are the ax and the sword.

There are three radiocarbon dates that confirm the contemporaneity of the Hang Gon urn field with the Dong Son sites further to the north. One sample from within a jar was dated to 240 ± 150 B.C., and a second from around the same vessel was 350 ± 150 B.C. The third sample comes from an organic residue in a vessel and is dated to 150 ± 150 B.C. Marginally earlier dates as well as the same type of two-headed animal ornament come from the nearby site of Phu Hoa (see figure 13.1; Fontaine 1972). The dates in question are 350 ± 140 and 540 ± 290 B.C.

Vietnamese prehistorians have recently concentrated their efforts on enlarging our understanding of Sa Huynh in the original area of discovery as well as slightly to the north in the Thu Bon valley. Ha Van Tan (1980) has concluded that the richest and most informative of the recently examined sites is Tam My. Trinh Can and Pham Van Kinh (1977) have recovered large funerary urns there. Again, the burial goods included iron artifacts, including spearheads, knives, and sickles, and bronze spearheads and bells. A two-headed animal ornament found there clearly matches those from Hang Gon and Phu Hoa. There is also a stone ear pendant that looks like an evolved form of those found in Go Mun contexts and at Khok Phlap in Thailand. The Sa Huynh sites confirm that by the end of the first millennium B.C. there was a string of communities occupying favorable coastal tracts along the shores of central Vietnam. By that juncture there was widespread use of iron and an exchange system incorporating exotic glass and stone jewelry.

The Adoption of Iron and Increasing Cultural Complexity

There is evidence in each of the lowland tracts and in upland Laos for major and contemporaneous cultural changes. Before the mid first millennium B.C. settlements covered about the same area and were, it is argued, autonomous. Thereafter prime centers developed and were the foci of craft specialization and intensified production. During the earlier period bronze working was adopted within the small village communities, which already had an established practice of exchange in exotic artifacts. These have been described under the heading of primitive valuables, and they underwrote the maintenance of economic, social, and political relationships between the participating communities. In a word, the exchange of valuables and the establishment of affine ties promoted alliance. Imbalance or fluctuations in access to valuables involved oscillations in rank. Within such a system much devolves on the leaders of dominant lineages. They had superior access to prestige valuables, could foster affine relationships, and played an entrepreneurial role in the transfer of goods.

The distribution of stone, marine shell, and metals—the principal surviving materials documenting such exchange—reflect goods crossing the landscape across a variety of routes and in different directions. Further research will surely illuminate the former existence of interlinked exchange networks with major concentrations in, for example, the Chao Phraya catchment, the Mekong valley, and the maritime tracts of Vietnam. It was, it is held, exactly along such nodes, particularly in the last two networks, that knowledge of bronze traveled. The system was also internally flexible with regard to permanence of rank. It is worth pausing to review such flexibility, as it will sharpen our perception of the transition to centrality.

The sites during General Period B were located in small stream valleys and along the margins of main river floodplains. The technique of rice cultivation probably involved the exploitation of enclaves of land subjected to only a limited degree of flooding following the onset of the monsoon. Climatic unpredictability involving draught in some years and major flooding in others, however, would doubtless have prejudiced predictable success in the rice harvest. The broad-ranging subsistence activities would have been a valuable buffer against the effects of climatic extremes. At the same time, however, intensifying production to permit participation in a prestige goods exchange system would have come more readily to those commanding the most extensive tracts of cultivable land. Indeed, for many river valleys good land and communications were the only major resource: There were no local deposits of ore, stone, or marine shell. There are then conditions under which certain communities might prosper unduly. Among these may be numbered access to circumscribed resource, be it copper, tin, or good land, control over a strategic position, such as a mountain pass or river crossing, or monopoly through the fortunes of geography over access to prestige exotic goods, such as glass beads, iron, or agate jewelry.

The transition to General Period C, with its fixing of rank underwritten by intensified production and exchange, represents the growth to prominence of one settlement and its occupants over others within its orbit. In Bac Bo we can isolate several variables that were intensified, although this area is a special case given its direct contact with Chinese expansion. In terms of agriculture, there was the advent of plowing and double cropping. As Goody has cogently shown, the application of animal traction to soil preparation greatly magnifies production and thereby makes it more feasible to concentrate a strategic surplus (Goody 1976). A rice surplus can be employed to attract and maintain followers, thereby concentrating people. It may also be converted into visible status objects: the great ceremonial drums, situlae, body plaques, and, in terms of prowess, daggers, swords, and ax halberds. There was a great proliferation of skill on the part of the Dong Son smith, and much more metal was mined, moved, and shaped. It is self-evident that the organization and purpose of all this activity, involving as it did the maintenance of permanent ateliers and control over far-flung ore sources, involved intensification. In this manner permanent ranking of an elite, craft specialization, and intensified agriculture form a triad of linked variables. Nor must it be forgotten that the Dong Son aristocrats controlled the maritime and riverine routes that exposed Bac Bo first to the import of Chinese goods and then to the admission of Chinese armies. Under a similar set of circumstances Haselgrove (1982) has applied a prestige goods control model to the initial contact between the expanding Roman Empire and the inhabitants of southern Britain, noting that, where a particular lineage or group is able to monopolize raw materials or new supplies of prestige goods, it should achieve a position of dominance over its former peers.

There is a consistent body of evidence that the use of iron and the import of glass and exotic stone beads took place along the coasts of central Vietnam, in the Chao Phraya area, and even on the inland Khorat plateau during the last few centuries of the first millennium B.C. In this last area, although shielded from Chinese contact by the Truong Son range and remote from coastal contact with Indian expansion, Non Chai still grew to cover at least 18 ha and during its occupancy (c. 300 B.C.–A.D. 200) witnessed the use of iron and import of exotic glass beads. At Ban Na Di, after the initial presence of a few iron objects in two graves, local iron smelting was initiated during the buildup of Level 5. Higham and Kijngam (1984) have suggested that the major changes noted at Ban Na Di and Ban Chiang, signaled by the advent of red-painted pottery and associated artifacts, reflects the expansion of groups from the fringes of the emergent Chi valley chiefdoms. The same phenomena of the growth of certain settlements, such as Ban Tha Kae, and employment of iron and imported jewelry have been isolated in the Chao Phraya valley.

These changes are seen here as consistent with the breakdown of the long-standing affine alliance and exchange system between independent communities. Occupants of such sites as Non Chai and Phimai, for example, had easy access to iron ore and commanded nodal positions in the upper reaches of the major Khorat plateau rivers. The Sa Huynh sites likewise were able to control coastal traffic. Any movement up the Chao Phraya River would involve passage through the territory of such growing centers as Ban Tha Kae. Several interacting variables, then, can be identified. Population growth and agricultural intensification are noted and were doubtless contributory factors. But the social change was critical in that it involved the growth of a few large central places as foci of population. There is some archaeological evidence that these centers controlled the large-scale production of salt, exchange of exotic artifacts, and certainly smelted iron ore. It is proposed that these seminal variables and their interaction reflect the flexibility and opportunism inherent in the system characterizing General Period B, wherein those controlling the best tracts of agricultural land and access to a new range of exotic prestige goods enjoyed considerable advantages. Intertwined in this model is the initial exposure to Indian and Chinese expansive forces.

Summary

Mainland Southeast Asia is richly endowed with copper, tin, lead, and iron ores. For over a century it has been known as an area that sustained prehistoric societies versed in metallurgy. The direction of recent research has involved the analysis of ore sources and processing, of casting techniques in areas removed from ores, of the implements made, and the social milieu within which metallurgists operated. Here I have been at pains to demonstrate that metallurgy was initially adopted by people with a long local period of occupation and that many of the early bronze artifacts copied forms previously rendered in stone. This occurred in Phung Nguyen–Dong Dau contexts in Bac Bo, and similar situations are alluded to in the Dong Nai valley, in northeast Thailand, and the Gulf of Siam. The same casting techniques and range of artifacts crosscut areas and groups of sites that reveal distinct regionality in other aspects of material culture. There is an apparent concentration of bronze casting activity along the coast of Vietnam, up the Mekong catchment, and perhaps slightly later in the basin of the Chao Phraya. We do not yet know anything about the origins of this metallurgical tradition. A judicious consideration of the available dating evidence points to a commencing date in the vicinity of 1500 B.C. It is argued that bronze artifacts took their place alongside artifacts made of exotic ceramics, stone, and shell, and doubtless more perishable items such as silk clothing as emblems of status among essentially autonomous but internally ranked communities.

There is a palpable dislocation in this tradition, which endured for at least 1,000 years, during the first millennium B.C. The evidence described for the site of Ban Na Di reveals the use of more variable bronze alloys and the beginning of ironworking. We are dealing with a large area, and it is not unreasonable to anticipate regionality in the date and manner of the adoption of iron. Thus it is possible that ironworking was developed locally, imported from China or India, or copied as a result of an initial acquaintance with the material through exchange contacts. Only detailed analyses on well-provenanced data will illuminate these issues. What we do know is that ironworking was soon adopted over much of Southeast Asia and that it was used in agriculture, as ornaments, and in conflict. Its use occurred as the old exchange systems and autonomous communities gave way to intensification in many fields and development of centralization. It is suggested that ironworking itself was one factor in these trends.

Notes

I am most grateful to the organizing committee of the Second International Symposium on the Beginning of the Use of Metals and Alloys for inviting me to participate and report on recent finds in mainland Southeast Asia. The excavation of Ban Na Di owed much to the efforts of my codirector, Amphan Kijngam, and the support of the Royal Thai Fine Arts Department and National Research Council. I am most grateful to Pisit Charoenwongsa for his guidance and interest in the project over many years. The Ford Foundation underwrote most of the costs incurred both in fieldwork and analysis of the data in Otago. To my specialist colleagues Robert Maddin, Y. Q. Weng, Nigel Seeley, Warankhana Rajpitak, and Jacqui Pilditch I owe a special debt.

1. I wish to acknowledge with gratitude the specialist analysis of the bronze by Warankhana Rajpitak, Nigel Seeley, Robert Maddin, Y. Q. Weng, and Jacqui Pilditch.

2. I am grateful to D. Avery, who drew my attention during the conference to the possibility that the wax in question was used in polishing the bracelet after casting. This alternative does not imply that the lost-wax technique was not employed at Ban Na Di.

References

Bayard, D. T. 1984. "A tentative regional phase chronology for northeast Thailand. Southeast Asian archaeology at the XV Pacific Science Congress." *University of Otago Studies in Prehistoric Anthropology* 16: 161–168.

Bayard, D. T., P. Charoenwongsa, and S. Rutnin. 1986. "Excavations at Non Chai, northeastern Thailand." *Asian Perspectives* 25(1): 13–62.

Chantaratiyakarn, P. 1984. "The Middle Chi research programme," in *Prehistoric Investigations in Northeast Thailand*, C. F. W. Higham and A. Kijngam, eds. BAR International Series 231. Oxford: British Archaeological Reports, pt. 2, pp. 565–643.

Colani, M. 1935. *Mégalithes de Haut-Laos.* Paris: Publications de l'Ecole Française d'Extrême Orient, nos. 25 and 26.

Daeng-iet, S. 1978. "Khok Phlap: A newly discovered prehistoric site." *Muang Boran* 4(4): 17–26. (In Thai.)

Dalton, G. 1977. "Aboriginal economics in stateless societies," in *Exchange Systems in Prehistory*, T. K. Earle and J. Ericson, eds. (London: Academic Press), 191–212.

Fontaine, H. 1972. "Deuxieme note sur le "néolithique" du bassin inferieur du Dong-Nai." *Archives Geologiques du Viet-Nam* 15: 123–129.

Glover, I. C. 1980. "Ban Don Ta Phet and its relevence to problems in the pre- and protohistory of Thailand." *Bulletin of the Indo-Pacific Prehistory Association* 2: 16–30.

Glover, I. C. 1983. "Excavations at Ban Don Ta Phet, Kanchanaburi Province, Thailand, 1980–1." *South-East Asian Studies Newsletter* 10: 1–4.

Glover, I. C., P. Charoenwongsa, B. Alvey, and N. Kamnounket. 1984. "The cemetery of Ban Don Ta Phet, Thailand: Results from the 1980–1 season," in *South Asian*

Archaeology 1981, B. Allchin and M. Sidell, eds. (Cambridge: Cambridge University Press), 319–330.

Goody, J. 1976. *Production and Reproduction.* London: Cambridge University Press.

Gourou, P. 1955. *The Peasants of the Tonkin Delta.* New Haven, Conn.: Human Relations Area Files.

Ha Van Phung. 1979. "In search of the relations between Go Mun and Dong Son cultures." *Khao Co Hoc* 29:43–61. (In Vietnamese.)

Ha Van Tan. 1980. "Nouvelles recherches préhistorique et protohistoriques au Viet Nam." *Bulletin de l'Ecole Française d'Extrême Orient* 68:113–154.

Hanwong, T. 1985. *Artifact Analysis from the Excavation of Ban Tha Kae, Amphoe Muang, Changwat Lopburi.* Master's thesis, Silpakorn University, Bangkok. (In Thai.)

Haselgrove, C. 1982. "Wealth, prestige and power: The dynamics of late Iron Age political centralisation in South-East England," in *Ranking, Resource and Exchange*, A. C. Renfrew and S. Shennan, eds. (Cambridge: Cambridge University Press), 79–88.

Higham, C. F. W. 1983. "The Ban Chiang culture in wider perspective. *Proceedings of the British Academy* 69:229–261.

Higham, C. F. W., and A. Kijngam. 1984. *Prehistoric Investigations in Northeast Thailand.* BAR International Series 231, 3 parts. Oxford: British Archaeological Reports.

Ho, C. M. 1984. *The Pottery of Kok Charoen and Its Farther Context.* Ph.D. dissertation, University of London.

Hoang Xuan Chinh, and Nguyen Khac Su. 1977. The late Neolithic site of Cau Sat (Dong Nai)." *Khao Co Hoc* 24:12–18. (In Vietnamese.)

Hoang Xuan Chinh, and Nguyen Ngoc Bich. 1978. *Excavations at the Archaeological Site of Phung Nguyen.* Hanoi: Nha Xuat Ban Khoa Xa Hoi. (In Vietnamese.)

Janse, O. R. T. 1958. *Archaeological Research in Indo-China. Vol. III. The Ancient Dwelling Site of Dong-So'n (Thanh-Hoa, Annam).* Cambridge, Mass.: Harvard University Press.

Kethutat, P. 1976. "Preliminary report on the test excavation at Ban That, Tambon Ban Ya, Amphoe Nong Han." *Journal of Anthropology* (Thamasat University, Bangkok), 1976:27–38. (In Thai.)

Kijngam, A. 1979. "The faunal spectrum from Non Chai." *Silapakon* 23(5):102–109. (In Thai.)

Le Xuan Diem. 1977. "Ancient molds for casting bronze artifacts from the Dong Nai basin." *Khao Co Hoc* 24:44–48. (In Vietnamese.)

Lévy, P. 1943. *Recherches préhistoriques dans la region de Mlu Prei.* Paris: Publications de l'Ecole Française d'Extrême Orient, no. 30.

Luu Tran Tieu. 1977. *Investigations at Chau Can.* Hanoi. (In Vietnamese.)

Maddin, R., and Y. Q. Weng. 1984. "The analysis of bronze 'wire'," in *Prehistoric Investigations in Northeast Thailand*, C. F. W. Higham and A. Kijngam, eds. BAR International Series 231. Oxford: British Archaeological Reports, pt. 1, pp. 112–116.

Mansuy, H. 1902. *Stations préhistoriques de Samrong-Seng et de Longprao (Cambodge).* Hanoi: F. H. Schneider.

Mansuy, H. 1923. "Contribution a l'étude de la préhistoire de l'Indochine. Résultats de nouvelles recherches effectuées dans le gisement préhistoriques de Samrong Sen (Cambodge)." *Memoires du Service Géologique de l'Indochine* 11(1):5–24.

Moore, E. 1986. *The Khorat Khmer.* Ph.D. dissertation, University of London.

Natapintu, S. 1985. *Prehistoric Investigations in the Lower Chao Phraya Valley, 1983–5.* Bangkok: Fine Arts Department. (In Thai.)

Ngo Si Hong. 1980. "Binh Chau (Nghia Binh). A newly discovered Bronze Age site on the central Vietnamese coast." *Khao Co Hoc* 33:68–74. (In Vietnamese.)

Nguyen Duy Hinh. 1983. "The birth of the first state in Viet Nam." *Otago University Studies in Prehistoric Anthropology* 16:183–187.

Nguyen Xuan Hien. 1980. "The vestiges of burned rice in Viet Nam." *Khao Co Hoc* 35:28–34. (In Vietnamese.)

Noulet, J. B. 1879. *L'âge de la pierre polie et du bronze au Cambodge d'après les découvertes de M. J. Moura.* Toulouse: E. Privat.

Parmentier, M. H. 1918. "Dépots de jarres à Sa-Huynh." *Bulletin de l'Ecole Française d'Extrême Orient* 24:325–343.

Pham Quang Son. 1978. "The first tentative links between the late Neolithic and early metal using cultures in the basin of the Dong Nai." *Khao Co Hoc* 25:35–40. (In Vietnamese.)

Pham Van Kinh. 1977. "Excavations at Ben Do (Ho Chih Minh City)." *Khao Co Hoc* 24:19–21. (In Vietnamese.)

Pigott, V. C. 1984. "The Thailand archaeometallurgy project 1984: Survey of base metal resource exploitation in Loei province, northeastern Thailand." *South-East Asian Studies Newsletter* 17:1–5.

Pilditch, J. 1984. "Sections on the material culture of Ban Na Di." in *Prehistoric Investigations in Northeast Thailand*, C. F. W. Higham and A. Kijngam, eds. BAR International Series 231. Oxford: British Archaeological Reports, pt. 1, pp. 57–222.

Quaritch-Wales, H. G. 1957. "An early Buddhist civilization in Eastern Siam." *Journal of the Siam Society* 65(1):42–60.

Rajpitak, W., and N. J. Seeley. 1979. "The bronze bowls from Ban Don Ta Phet: An enigma of prehistoric metallurgy." *World Archaeology* 11(1):26–31.

Rajpitak, W., and N. J. Seeley. 1984. "The bronze metallurgy," in *Prehistoric Investigations in Northeast Thailand*, C. F.

W. Higham and A. Kijngam, eds. BAR International Series 231. Oxford: British Archaeological Reports, pt. 1, pp. 102–112.

Rutnin, S. 1979. *A Pottery Sequence from Non Chai, Northeast Thailand.* Master's thesis, University of Otago.

Sanders, W. T., J. R. Parsons, and R. S. Santley. 1979. *The Basin of Mexico.* New York: Academic Press.

Saurin, E. 1963. "Station préhistorique à Hang-Gon près de Xuan Loc." *Bulletin de l'Ecole Française de l'Extrême Orient* 51:433–452.

Solheim, W. G., II. 1982–1983. "The dating of sites and phases in northeast Thailand." *Journal of the Hong Kong Archaeological Society* 10:112–116.

Sørensen, P. 1979. "The Ongbah cave and its fifth drum," in *Early South East Asia,* R. B. Smith and W. Watson, eds. (Oxford: Oxford University Press), 78–97.

Suchitta, P. 1985. "Early iron smelting technology in Thailand and its implications." *Research Conference on Early Southeast Asia* (Bangkok: Silapakon University), 25–40.

Trinh Can, and Pham Van Kinh. 1977. "Excavations of the urnfield of Tam My." *Khao Co Hoc* 24:49–57. (In Vietnamese.)

Trinh Sinh. 1977. "From the stone ring to the bronze ring." *Khao Co Hoc* 23:51–56. (In Vietnamese.)

Veerapan, M. 1979. "The excavation at Sab Champa," in *Colloquy on Early South East Asia,* R. B. Smith and W. Watson, eds. (Oxford: Oxford University Press), 337–341.

Voorhies, B. 1973. "Possible social factors in the economic system of the prehistoric Maya." *American Antiquity* 38: 486–489.

Vu Thi Ngoc Thu, and Nguyen Duy Ty. 1978. "A tool set for casting bronze from Lang Ca (Vinh Phu)." *Khao Co Hoc* 2:36–39. (In Vietnamese.)

Watson, W. 1983. "Pre-Han communication from West China to Thailand." Paper prepared for CISHAAN, Tokyo, September 1983.

Welch, D. 1984. "Settlement pattern as an indicator of socio-political complexity in the Phimai region, Thailand." *University of Otago Studies in Prehistoric Anthropology* 16:129–151.

Wheeler, T. S., and R. Maddin. 1976. "The techniques of the early Thai metalsmith." *Expedition* 18(4):38–47.

White, J. C. 1982. *Ban Chiang: Discovery of a Lost Bronze Age.* Philadelphia, Penn.: University of Pennsylvania Press.

Williams-Hunt, P. D. R. 1950. "Irregular earthworks in Eastern Siam: An air survey." *Antiquity* 24:30–37.

Wolters, O. W. 1982. *History, Culture and Region in Southeast Asian Perspective.* Singapore: Institute of South-east Asian Studies.

You-di C. 1978. "Nothing is new." *Muang Boran* 4(4) 15–16. (In Thai.)

14
Archaeological Investigations into Prehistoric Copper Production: The Thailand Archaeometallurgy Project 1984–1986

Vincent C. Pigott and
Surapol Natapintu

In the last two decades there have been important developments in the archaeology of Thailand. Of particular interest to archaeologists is the study of the emergence of metal use during the prehistoric period. The analyses of groups of excavated metal artifacts emphasized the complexity of prehistoric metalworking technology and raised questions concerning the sources of the ores and the nature of the processes by which such technology evolved in Thailand.

Especially important in this regard are the excavations of prehistoric village sites on the Khorat plateau in northeast Thailand, in particular at Non Nok Tha (Bayard 1971, 1979), at Ban Chiang (Gorman and Charoenwongsa 1976; White 1982, 1986), at Non Chai (Charoenwongsa and Bayard 1983), and at Ban Na Di (Higham and Kijngam 1984). This research has documented the significance and sophistication of copper-base metallurgy in the region from at least 2000 B.C. As a direct outcome of the Ban Chiang Project, a joint project was initiated in 1984 involving the Thai Fine Arts Department and The University Museum of the University of Pennsylvania. Primary support for this joint undertaking, the Thailand Archaeometallurgy Project (TAP), which has continued into 1988, has come from the National Geographic Society. The original mandate of the project was to investigate known ore sources in hopes of locating ancient mining and metalworking sites associated with them. The project thus far has developed in three stages: the first stage was devoted to a survey of ore sources for evidence of preindustrial activity; the second involved the excavation of a prehistoric copper mining complex, and the third, excavation at large-scale copper production sites.

Ore Source Survey and the Excavation of Phu Lon (TAP 1984–1985)

The initial 1984 research focused on base metal (Cu-Pb-Zn) resource exploitation in the Loei and western Nong Khai provinces in northeast Thailand. This mountainous area is adjacent to and along the western edge of the Khorat plateau. It is rich in mineral resources (Jacobson et al. 1969; Jantaranipa et al. 1981). With the guidance of Udom Theetiparivatra of the Loei office of the Department of Mineral Resources, we surveyed known ore deposits and documented ample evidence of preindustrial exploitation (Pigott 1984, 1985). The most important site surveyed was Phu Lon, a prehistoric copper mining complex on the southern bank of the Mekong River in Nong Khai Province (figure 14.1). The location of this mining site is directly associated with igneous rock formations of Permian-Triassic age. The ore deposit appears to be a typical contact metasomatic type consisting of a sulfide ore body with an indigenous oxide zone exposed at the

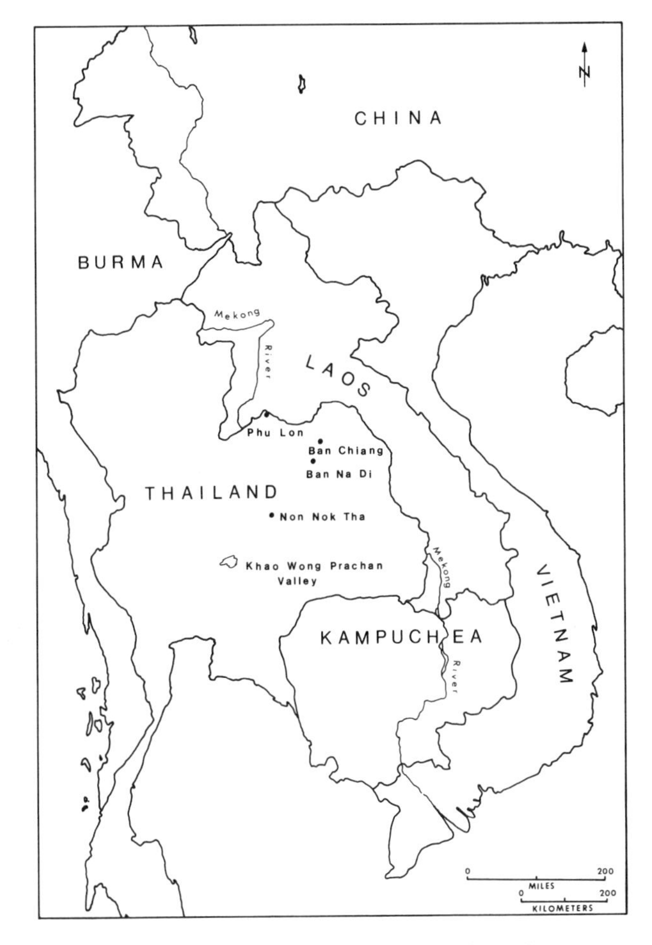

Figure 14.1 Map of Thailand and part of Southeast Asia with locations of major sites mentioned in the text.

surface. It is probably the result of emanations from a granodiorite intrusion into a limestone formation (Vernon and Theethiparivatra, personal communications). This geological activity may be associated with the collision of two land masses, namely the Indochina block and the Shan-Thai block, before the Jurassic, during the Triassic, or perhaps earlier (Buffetaut and Ingavat 1985, pp. 80–81). The result of this plate tectonic activity is Southeast Asia as we know it today.

The copper sulfide ore body at Phu Lon over the millennia weathered to its present gossan capped state. Prehistoric mining activity at the site appears to have been limited to the upper reaches of the ore body, where oxide-carbonate ores of copper, in particular malachite, were deposited in quartz veins disseminated in the iron-rich skarn host rock.

Phu Lon, or "Bald Mountain," may have received its name from the denuded nature of the mined out area at the site. Vegetation has never flourished in the mine tailings, which cascade down the mountainside from the area known as the Lower Flat (figure 14.2). The Lower Flat and the Peacock Flat were the focus of primary mining activity and presently consist of open areas of mining debris many meters in depth. They are situated in portions of the Phu Lon mountainside that have been hollowed out by mining. The pinnacle and adjoining steep cliff along the east side of the Peacock Flat form an imposing facade for the remaining portion of the mountain. Udom's rock across the Peacock and Lower flats is a massive outcrop of skarn that was never entirely mined; it sits as an outlying vestige of the former mountainside. The pinnacle is a monolith of skarn and, like Udom's rock, is a remnant of the topography that was isolated by mining activity around it. The Pottery Flat and Bunker Hill lie behind and to the east of the primary mining area. The Pottery Flat is a gently sloping wide area connected by a saddle to Bunker Hill. Bunker Hill is a small hilltop that on the north side drops off dramatically toward the Mekong River. Ban Noi is located in a flat open area quite close to the river.

The site of Phu Lon, excavated during the TAP 1985 season, may properly be described as a complex, as activities occurred in several locales and involved a variety of tasks conducted over an extended period, perhaps by various groups (Pigott et al. 1985). Our investigations focused on the primary mining areas of the Lower and Peacock flats and on a secondary mining area, Bunker Hill, and the adjacent area devoted to ore crushing and copper processing, the Pottery Flat. During the TAP 1985 season, Gerd Weisgerber, mining archaeologist at the Deutsches Bergbau-Museum in Bochum, Germany, studied the evidence for mining at Phu Lon and concluded that the earliest evidence for mining was preserved only at certain locations on the site, atop Udom's rock for example. Here shallow pits with rounded configurations are preserved; Weisgerber suggests that these pits were shaped with stone mauls during mining. Such mining mauls and ore-crushing hammers were readily available among the cobbles in the bed of the nearby Mekong River. Hundreds of these artifacts, mostly in the form of exhausted tools and fragments, litter the mining areas and dot the stratigraphic sections. These tools are commonly of hard intrusive rock, such as andesite, and were purposefully selected for the tasks of mining and ore crushing. Presumably, when metal tools became available, they too were used in ore extraction at Phu Lon. Some mine shafts at the site bear evidence of picking with tools that could well have been metal. Dating such activity is extremely difficult, as the shafts themselves rarely contain any datable materials.

An indication of the amount of human effort and time that must have gone into the extraction of copper at Phu Lon is found on the Lower Flat, where drilling by the Department of Mineral Resources determined that mining rubble is present to a depth of at least 10 m. Test trenches at several other locales on the Lower and Peacock flats revealed comparable stratigraphic situations. In one such case, at Udom's rock, layers of mining rubble alternate with layers of clean sand, suggesting some sort of periodicity in mining activity.

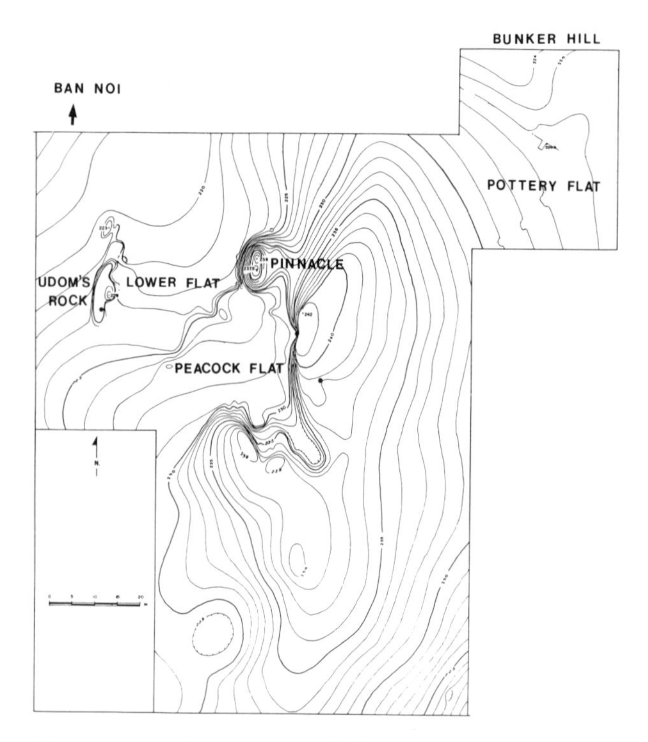

Figure 14.2 Contour map of the prehistoric copper mining complex at Phu Lon, Nong Khai Province, northeast Thailand.

It is important to note that, although ample evidence for mining and ore dressing and for some amount of metal processing was encountered at Phu Lon, only three finds of copper-base metal were made. One amorphous piece of metal came from the fill of a mine shaft at the base of the pinnacle, and another piece came from mining rubble in a test trench excavated along the east face of Udom's rock. Both were probably redeposited by erosion from elsewhere on the site. Proton-induced X-ray emission (PIXE) analysis of these two small fragments has determined that both are bronze in composition. They may represent some sort of casting splatter, as both have dendritic microstructures.

As mentioned previously, attempts to date mining activity at Phu Lon, as at most ancient mining sites, have proved difficult. A small pocket of charcoal from the otherwise sterile layers of clean sand and mining rubble at Udom's rock yielded a radiocarbon date in the first half of the first millennium B.C. As we will see, this date is in general agreement with other dates obtained from the Pottery Flat. Unfortunately, other than the mining mauls and a few prehistoric, cord-marked potsherds, no other cultural materials were obtained from excavation in the mining areas. The excavation of fill from the mine shaft at the base of the pinnacle did produce a ceramic pedestal vessel that is typologically akin to similar artifacts of the first millennium B.C. The vessel, although broken, appears to have been in situ and may have been placed in the mine shaft as some form of offering.

The Pottery Flat

The most important area at Phu Lon that we were able to excavate horizontally is Pottery Flat. Before excavation surface evidence in the form of cord-marked potsherds and indications of ore-crushing activity was detected. Excavation confirmed this to be a major ore-dressing location composed of a single stratum ranging in thickness from about 15 cm to 50 cm and covering an area over 100 m in length and 30 m in width. Ample charcoal from the Pottery Flat has provided a series of dates. Charcoal from the basal deposit of the ore-crushing stratum, presumably representing some of the initial activity in this locale, was dated to the first half of the second millennium B.C., and charcoal from the matrix of the stratum gave a range of dates in the first millennium B.C.

The matrix of the stratum consists primarily of crushed host rock, that is, skarn, from the ore deposit. Odd bits of malachite were found but in no significant amount. The malachite was the object of the ore-crushing activity at the Pottery Flat and would have been hand sorted from the crushed debris for subsequent smelting. On the surface of this stratum (as well as within it) were in situ ore-crushing hammers and anvils of selected intrusive rock. It is hoped that the study of the distribution of these tools across the stratum will enable us to identify specific loci of ore dressing by individuals.

Activities other than ore dressing were also undertaken at the Pottery Flat. Two fragments of different bivalve casting molds, one sandstone and one ceramic, were excavated. In addition, more than seventy small fragments of rice-chaff-tempered crucibles were recovered with a small amount of what is presumably smelting slag. The slag did not occur in a quantity that one might expect of a smelting locus. However, properly dressed malachite, freed of most of its gangue and low in impurities as such an ore is, would, when smelted, generate little slag. No features indicative of smelting furnaces were excavated, but the reduction of malachite could have been accomplished in a simple bowl furnace or in the crucibles themselves (Tylecote 1974). The large amount of charcoal present in the ore-crushing stratum is certainly related to the processing of the copper mined at Phu Lon.

Evidence of habitation activities is also present on the Pottery Flat in the form of a substantial concentration of cord-marked potsherds. Preliminary analyses by William Vernon [Museum Applied Science Center for Archaeology (MASCA)] have identified a minimum of three major ware types, all of which appear to have been locally made, as their mineral suites are typical of the Phu Lon area. A number of chipped stone adzes and the debris of their manufacture were mixed in the matrix of the ore-crushing stratum. In addition, a number of broken stone bracelet fragments occurred with indications that they too were being manufactured on the spot. The evidence for ore crushing, metalworking, and stone working, the large number of potsherds, and the thickness of the stratum all suggest that the activities of ore processing comingled with those of habitation over an extended period of time. Seasonal occupation of the Pottery Flat cannot as yet be ruled out.

Bunker Hill

Adjacent to the Pottery Flat is Bunker Hill, an area characterized by malachite-bearing quartz veins quite close to the surface. These were systematically exploited in prehistory and possibly into the historic period. Thousands of broken quartz crystals were excavated, among which were mixed some cord-marked potsherds and stone tools for mining and ore crushing.

Ban Noi

Between the Lower Flat and the Mekong River, near the modern village of Ban Noi, a second ore-crushing locus was identified that is much like that at the Pottery Flat, although not as extensive. At Ban Noi the deposit is over 1 m deep and is finely stratified. Unlike the Pottery Flat, these deposits appear to represent individual short periods of activity or perhaps even single ore-crushing events. The thickness of the deposit, however, suggests that the area was used on and off for a considerable period of time. A single radiocarbon date from this deposit places it in the first half of the first millennium B.C., comparable to other dated contexts elsewhere at Phu Lon. Of interest is the fact that the ceramics from Ban Noi appear to be distinct from those on the Pottery Flat.

Just as at the Pottery Flat, slag occurs at Ban Noi, suggesting that smelting took place there also. Crucible fragments also support the suggestion, if we presume that crucibles were being used for smelting. Small circular depressions in which crucibles might have rested were identified in the crushed rock matrix of the Ban Noi deposit. A single copper-base socketed ax was found in the basal layer of the Ban Noi deposit. This is the only identifiable metal artifact excavated at Phu Lon. It is currently undergoing metallographic and elemental analysis at MASCA.

Excavations at Non Pa Wai in the Khao Wong Prachan Valley (TAP 1986)

The third stage of TAP involved the excavations in 1986 of a group of sites where prehistoric copper ore processing as well as large-scale smelting took place. The sites, located in the Khao Wong Prachan valley in central Thailand, about 15 km northeast of the city of Lopburi, had recently been surveyed by Surapol Natapintu, director of the Central Thailand Archaeological Project (Natapintu 1987).

Non Pa Wai, or "Rattan Hill," was the first and most important of the three sites investigated. Topographic mapping by Andrew Weiss (MASCA) using a Topcon electronic distance measurer and theodolite demonstrated that the surface area of this site was over 5 ha (c. 50,000 m^2). The entire surface of the site is packed with copper smelting debris. This debris is so dense that the site cannot be cultivated. Its soil is distinct from the surroundings because of its presumed high ash and copper content from centuries of pyrotechnological activity.

The upper deposit of the site consists of gray ashy soil packed with the debris of copper production, including slag, crucible fragments, mold fragments (cup, conical, and bivalve), ore minerals, ore-crushing tools, potsherds, animal bone, and stone tool and bracelet fragments. The top meter of this deposit has been heavily disturbed by tree and shrub roots and by villagers searching for wild yams. The dense concentration of smelting and other debris on the surface of the site is due to strong prevailing winds eroding the loose soil matrix in which this debris is contained, thus deflating the surface of the mound. Below this disturbed top meter or so of deposit, the homogeneous gray ashy matrix continues, and stratigraphic units are difficult to ascertain. Artifacts are primarily fragmentary and appear discarded; they do not occur on working or living surfaces and have no recognizable associations. No in situ metalworking installations were identified.

Among the variety of artifacts found were thousands of thick-walled rice-chaff-tempered crucible fragments. It is our current opinion that such crucibles were used for smelting. As we have no indications that smelting furnaces, as fixed installations, were used in prehistoric Thailand, this suggestion is of particular interest here. Fragments of thick arc-shaped ceramic rims excavated at Non Pa Wai could represent the rims of bowl "furnaces," simple depressions dug into the surface of the mound that could have cradled the crucibles during the smelting process. It should be noted that the ores from the excavation at Non Pa Wai and from nearby Phu Ka sources consist of the copper silicate chrysocolla and weathered copper sulfide ores. The suitability of such ores to crucible smelting is an issue currently being addressed by Anna Bennett of the London Institute of Archaeology (Bennett 1986). It is clear from her investigations that the numerous planoconvex slag cakes that occur on the surface of Non Pa Wai and in the excavation are the direct result of pouring off onto the ground the contents of the large crucibles known from the site.

A second significant category of metalworking artifact found at Non Pa Wai is ceramic bivalve molds, often incised with geometric patterns on their exterior surfaces (figure 14.3). Such patterns may have allowed metalworkers to identify their molds and to match pairs of molds. Unfortunately many of the molds are fragmentary and worn, and identification of what was being molded is not always possible. Molds for socketed axes and arrow points are included in those that have been identified. In a number of instances the artifacts being molded would have been of an unusual thinness based on the depth of preserved mold impressions. Implement molds from the Khorat plateau sites are most frequently sandstone, whereas those known from Non Pa Wai and other sites in central Thailand are uniformly ceramic.

The most enigmatic finds were the staggering number of ceramic cup (figure 14.4) and conical molds (figure 14.5). These can range in interior diameter from as small as 1 cm to more than 10 cm. Their function is presumed to be for the casting of small ingots of copper. Evidence suggests that these molds may be used to cast copper poured directly from the large smelting crucibles after the bulk of the slag has been poured off, for we recovered variously shaped slags associated with these molds. For example, conical pieces of slag have a shape identical with the interior of the cup and/or conical molds. In addition, we excavated small L-shaped fragments of slag that also fit in the molds' interiors. Such slags may have been formed in the following manner. During casting, after pouring off the bulk of the slag in the smelting crucible, it would have been impossible in every instance to separate the remaining slag from the molten copper in the crucible. Thus some slag would have traveled from the crucible into the cup or conical mold receiving the molten copper. The ingots of copper cast in these cup and conical molds would be useful in themselves, as they could be easily counted and/or weighed, bagged, and transported. In addition, they could be quite easily melted down or hammered (with heating) into useful shapes.

Industrialized mass production of copper-base metal and artifacts is suggested by the huge size of the site of Non Pa Wai, the substantial volume of production debris, and the standardized nature of the artifacts used in metalworking (crucibles, molds, etc.). Moreover, the unprecedented scale of the copper smelting evidence suggests that bulk metal is being produced well beyond local needs. Finally, the depth and expanse of the gray ashy matrix that makes up most of the volume of the site suggest that the activities that created this debris were carried out over a substantial period by a sizable working population. Resolution of the chronological issues awaits the results of radiocarbon dating.

A hard, well-defined surface marks the transition between the gray ashy stratum that constitutes the bulk of the site and the shallow basal deposit in which a cluster of twelve burials was excavated. The burials,

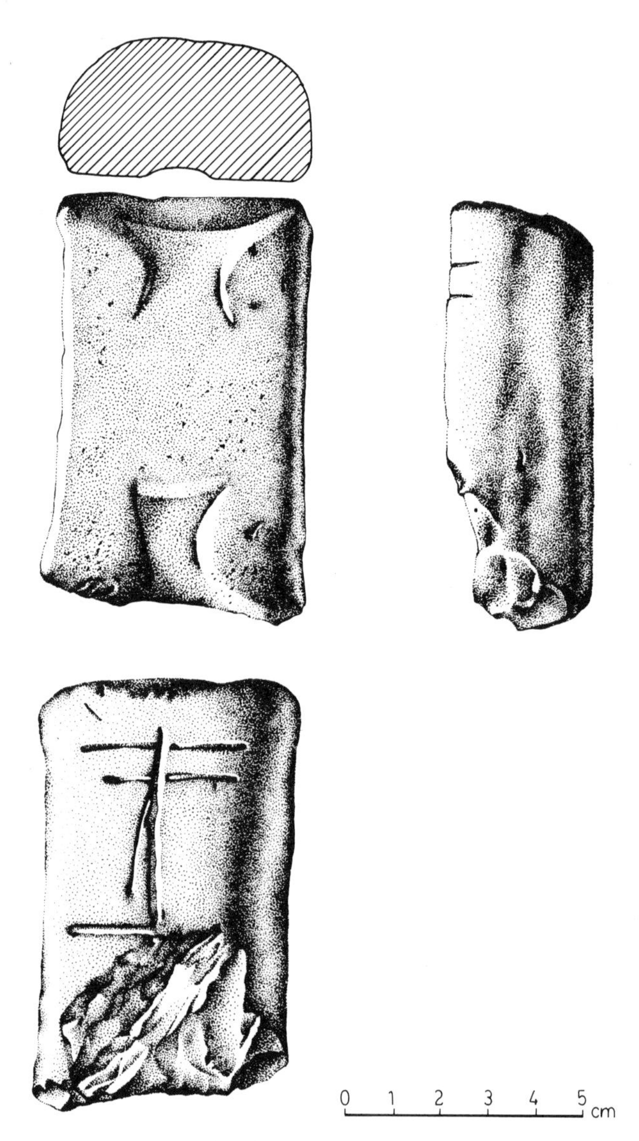

Figure 14.3 A bivalve casting mold from Non Pa Wai. The mold exterior has an incised geometric pattern, possibly a "maker's mark."

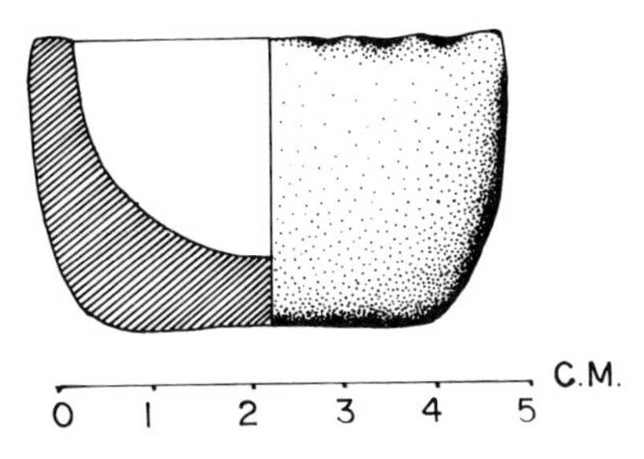

Figure 14.4 A cup mold from Non Pa Wai for the casting of small copper-based ingots.

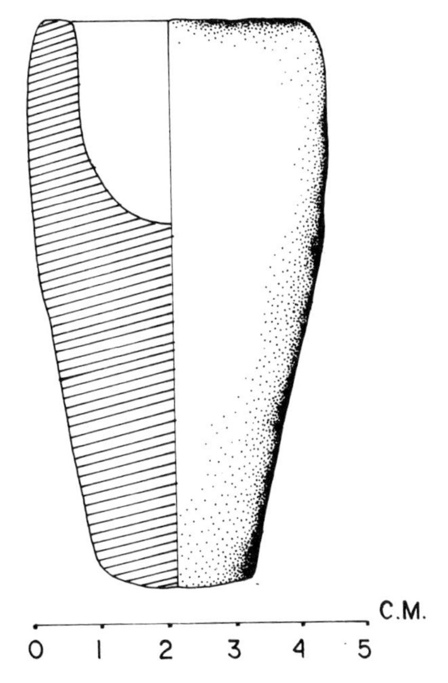

Figure 14.5 A conical mold from Non Pa Wai for the casting of small copper-base ingots.

although sealed by the well-defined surface, cut into several successive living surfaces contained in a dark, organically rich deposit no more than 40 cm deep. This basal deposit rested on virgin soil. The living surfaces were literally paved in occupation debris, including potsherds, snail shells of both forest and swamp varieties, animal bone, and chipped and polished stone adzes. The debris of copper production, mostly slag, crucibles, and ore fragments, did occur but not in quantities comparable to those in the upper deposit. No cup or conical molds were found in the basal deposit. The ceramics appear to be distinct from those of the upper deposit.

Among the burials grave goods were scarce but the occurrence of red ocher was quite common. One burial contained a socketed copper-base ax on top of which rested a circular disk of red ocher. The burial with the largest number of grave goods was that of a tall adult around whom were distributed five fragments comprising a complete pair of ceramic, bivalve, socketed ax molds. A mold fragment rested in each of the individual's hands. Other grave goods in this burial included two ceramic vessels and a grinding stone and palette for red ocher preparation. A group of small carved, circular shell ornaments were found near one ear. It would appear that this individual was a metalworker buried with the "tools of the trade."

On completion of the radiocarbon dating for the upper and basal deposits at Non Pa Wai, we will be in a position to discuss the duration of occupation and the potential for both cultural and technological continuity at this major site in the Khao Wong Prachan valley in central Thailand.

Excavations at Nil Kam Haeng (TAP 1986)

In the Khao Wong Prachan valley, within c. 2 km of Non Pa Wai and located at the base of the mountain Phu Ka, an important copper source, lies the site of Nil Kam Haeng. This site clearly rivals Non Pa Wai in its physical extent and volume of metalworking debris. Nil Kam Haeng has been damaged by bulldozing for a reservoir but appears to have been several hectares in size. Analysis of the crushed materials from the layers at the site will give us an indication of which ore sources may have been exploited.

A 6-m-long section of a bulldozer cut was cleaned, and test trenching into preserved lower strata was undertaken at this locus. From the surface of the site to virgin soil c. 5.50 m down, the strata consist of numerous thin layers of crushed ore host rock with some crushed slag. This layering of crushed host rock is reminiscent of that seen on the Pottery Flat and at Ban Noi at Phu Lon. It appears that large-scale hand crushing of ores and slags occurred in conjunction with copper smelting at Nil Kam Haeng. Smelting slags and crucible fragments were relatively common at the site, whereas cup and conical molds were not. Nevertheless, at Nil Kam Haeng intact slag disks that must have been cast in cup molds were excavated. The debris of habitation, including potsherds, animal bone, stone tools, and two burials, was also encountered.

As at Non Pa Wai, people appear to be living where they are conducting ore processing and metal production. When such activities occurred is not yet clear. However, one of two burials excavated contained a bimetallic bracelet composed of copper and iron. The presence of iron in the site in association with potsherds of prehistoric cord-marked ceramics argues for a first millennium B.C. date. Based on current understanding, this would make some of the occupation at Nil Kam Haeng contemporaneous with that of the upper deposit at Non Pa Wai, where the wares are thought to be later first millennium B.C. in date.

Concluding Remarks

Although the greater part of our analytical and interpretive tasks still lies ahead, it is important to emphasize that this research is of particular significance in that, within a single geographical area at two distinct copper-producing locales, we have investigated all the major steps in the process of prehistoric copper production—a technological continuum spanning mining, ore dressing, smelting, and casting. Furthermore, such evidence should allow us to go beyond technological analysis to identifying activity areas and the patterns of their use.

Although it is still unclear, it is possible that the ore deposits at Phu Lon and Phu Ka were sources of native copper, the manipulation of which has yet to be recorded in Southeast Asia. Moreover, the substantial scale of copper production at sites documented by the Thailand Archaeometallurgy Project thus far suggests that production was occurring for a far wider universe than the immediate vicinity of the sites themselves. The Mekong and its tributaries and other important river systems within and around Thailand must have served as conduits along which people and most probably metals traveled.

As we await the completion of radiocarbon dating for this project, our initial impression is that, at least during the first millennium B.C. if not earlier, copper production was intense, continuing well into the period when iron was readily available, in the later centuries of the first millennium B.C. Evidence from both Phu Lon and the Khao Wong Prachan valley sites supports this contention. Analysis and interpretation of materials from these sites, currently underway, should provide significant insights into the processes by which early copper production evolved in Southeast Asia.

References

Bayard, D. T. 1971. *Non Nok Tha: 1968 Excavation Procedure, Stratigraphy, and a Summary of the Evidence.* University of Otago Studies in Prehistoric Anthropology 4. Dunedin, New Zealand: University of Otago.

Bayard, D. T. 1979. "The chronology of prehistoric metallurgy in northeast Thailand: Silabhumi or Samrddhabumi?" in *Early Southeast Asia: Essays in Archaeology, History, and Historical Geography,* R. B. Smith and W. Watson, eds. (New York: Oxford University Press), 15–32.

Bennett, A. 1986. "Prehistoric copper smelting in central Thailand." Paper presented at the Meeting of the Association of Southeast Asian Archaeologists in Western Europe, London, September 1986.

Buffetaut, E., and R. Ingavat. 1985. "The Mesozoic vertebrates of Thailand." *Scientific American* 253(2):80–87.

Charoenwongsa, P., and D. T. Bayard. 1983. "Non Chai: New dates on metal working and trade from northeastern Thailand." *Current Anthropology* 24(4):521–523.

Gorman, C. F., and P. Charoenwongsa. 1976. "Ban Chiang: A mosaic of impressions from the first two years." *Expedition* 18(4):14–26.

Higham, C. F. W., and A. Kijngam. 1984. *Prehistoric Investigations in Northeastern Thailand.* BAR International Series 231. Oxford: British Archaeological Reports.

Jacobson, H. S., H. T. Pierson, T. Danusawad, T. Japakesetr, B. Inthupreti, C. Siriatauamaougkol, S. Prapassourukul, and N. Phophan. 1969. *Mineral Investigations in Northeastern Thailand.* Geological Survey Professional Paper 618. Washington, D.C.: US Government Printing Office.

Jantaranipa, W., R. Vongpromek, T. Sukko, and J. Preammanee. 1981. *Application of Enhanced Landsat Imagery to Mineral Resources of Loei Province, Northeastern Thailand.* Economic Geology Bulletin 30. Bangkok: Department of Mineral Resources.

Natapintu, S. 1987. "Current research on prehistoric copper-based metallurgy in Thailand." *SPAFA Digest* 8(1):27–35.

Pigott, V. C. 1984. "The Thailand Archaeometallurgy Project 1984: Survey of base metal resource exploitation in Loei Province, northeastern Thailand." *Southeast Asian Studies Newsletter* 17:1–5.

Pigott, V. C. 1985. "Pre-industrial mineral exploitation and metal production in Thailand." *MASCA Journal* 3(5):170–174.

Pigott, V. C., S. Natapintu, and U. Theetiparivatra. 1985. "The Thailand Archaeometallurgy Project 1984–5: Research in the development of prehistoric metal use in northeast Thailand." Paper presented at the Research Conference on Early Southeast Asia, Bangkok and Nakhon Pathom, April 8–13, 1985.

Tylecote, R. F. 1974. "Can copper be smelted in a crucible?" *Journal of Historical Metallurgy* 8(1):54.

White, J. C. 1982. *Ban Chiang: Discovery of a Lost Bronze Age.* Philadelphia: The University Museum, University of Pennsylvania, and the Smithsonian Institution Traveling Exhibition Service.

White, J. C. 1986. *A Revision of the Chronology of Ban Chiang and Its Implications for the Prehistory of Northeast Thailand.* Ph.D. Dissertation, Department of Anthropology, University of Pennsylvania.

15
Reflections on Early Metallurgy in Southeast Asia

Tamara Stech and
Robert Maddin

We entered the controversial field of Southeast Asian metallurgy over ten years ago, when we became involved in the analysis of metals from the prehistoric village site of Ban Chiang in northeastern Thailand. Since the preliminary publication of this material (Stech-Wheeler and Maddin 1976), further metallographic studies have been performed[1] and a series of elemental analyses has been obtained by proton-induced X-ray emission (PIXE).[2] It therefore seems reasonable to go back to Ban Chiang at the outset and present the new information; then we examine it in broader context.

Ban Chiang

The Early Period

Of course, the basic problem in discussing Southeast Asian metallurgy is that of chronology [see, for example, Higham (1983, pp. 1–7)]. We are not independently qualified to pass judgment on the various chronologies proposed for Non Nok Tha, Ban Chiang, and Ban Na Di, but we do feel that White's basic outline for Ban Chiang is reasonable (White 1982, fig. 18 on p. 20). The fundamental question in this inquiry is when metal first occurred at Ban Chiang; White now believes that this happened c. 1700 B.C. Therefore the early period in terms of metal usage is placed conservatively at c. 1700–1000 B.C.

At this point in the study of the Ban Chiang artifacts it is not possible to determine the total quantity of metal that can be assigned to each period. Artifacts range in size from droplets to axes, according to White, so a simple count would not adequately reflect the nature of production or usage.

At present, metallographic studies of six Early Period artifacts have been completed and elemental analysis has been performed on five of these. The three bracelets were left in the as-cast condition, which means they were not worked (figures 15.1 and 15.2), and tin is present in all of them in amounts ranging from 5.5% to 12.4% (see table 15.1). It should be noted that all are internally corroded, and therefore the levels of tin now detectable may not reflect the original composition. The high tin readings could result from an enhancement of the tin value by the corrosion process or from an aspect of the PIXE instrumentation or both. An adze, left in the as-cast condition (BC 694/1203; figure 15.3) is also so corroded that the presence of 18.4% Sn is questionable; indeed the microstructure is metallurgically incompatible with that amount of tin; that is, other metallurgical phases should be visible in the microstructure.[3] An adze with this amount of tin would not be functional in any operation involving impact because it would be highly brittle. The same comment concerning the metallurgical microstructure would also apply to an unidentified fragment (BC 679/1071). Spearhead BCES 762/2834, originally touted by us as

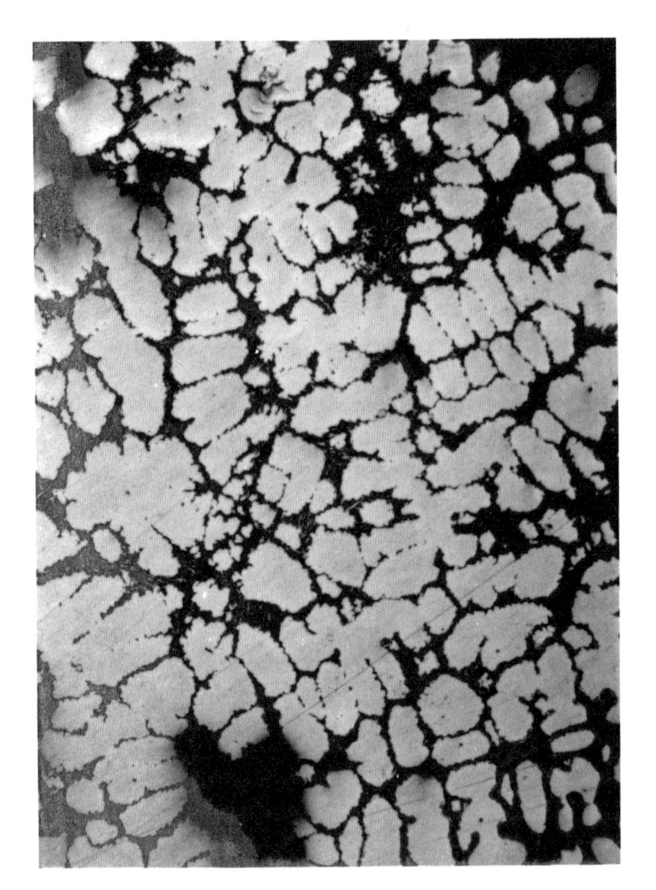

Figure 15.1 Bracelet BC 693/1203 shows a well-defined cast structure and a large grain size. × 100.

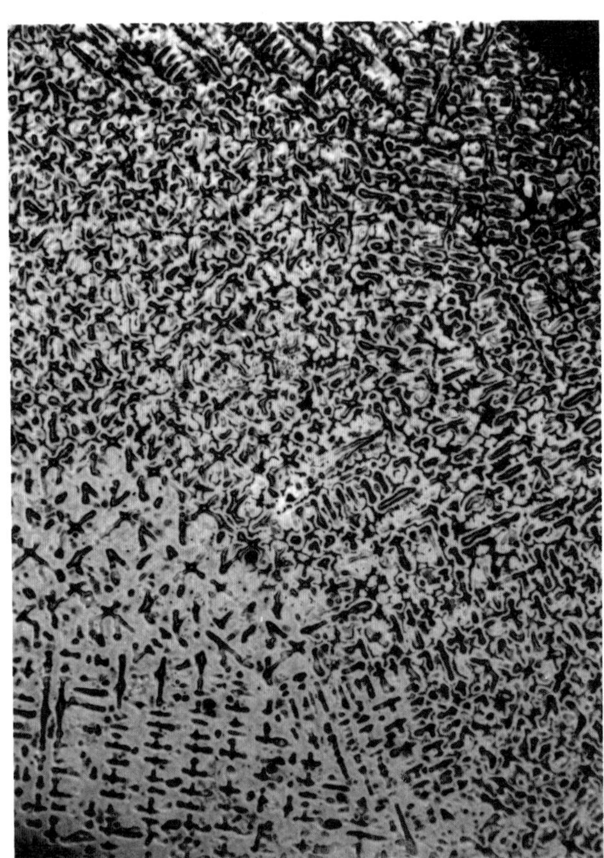

Figure 15.2 Bracelet BC 722 shows a dendritic pattern characteristic of a cast structure. × 100.

Table 15.1 Elemental analysis (PIXE)[a]

Sample	Cu	Sn	As	Ni	Pb	Fe	Ag	S	Si	Sb
Early Period										
BC 693/1203	86.7	12.4	0.095	<0.011	0.107	0.049	0.052	0.019	0.01	0.032
BCES 596/1984	89.6	9.5	0.114	0.029	0.044	0.017	0.117	0.19	0.12	0.055
BC 694/1203	80.4	18.6	0.094	<0.011	0.122	0.03	0.145	0.066	0.066	0.047
BC 679/1071	79.7	18.4	0.093	<0.011	0.121	0.08	0.047	0.101	0.066	0.047
BCES 762/2834										
Socket	90.0	9.17	0.101	<0.012	0.084	0.153	0.029	0.039	0.02	0.026
Blade	91.7	7.73	0.105	<0.012	0.081	0.111	0.018	0.048	0.016	0.035
Middle Period										
BCES 490/1286	92.9	6.2	0.057	<0.01	0.079	0.016	0.027	0.40	0.133	0.023
BCES 491/1286	86.0	13.3	0.163	<0.01	0.026	0.019	0.011	0.375	0.126	0.018
BCES 591/1981	84.4	12.1	0.135	<0.01	1.46	<0.018	0.149	0.267	0.096	0.397
BCES 395A/1115	90.1	5.5	0.119	<0.012	4.03	0.041	0.051	<0.006	0.016	0.054
BCES 616/2097	75.3	14.1	0.04	<0.009	9.93	0.055	0.052	<0.013	0.072	0.011
BCES 617/2097	84.5	13.9	0.141	<0.01	0.054	0.031	0.017	0.296	0.101	0.045
BCES 609/2069	69.6	16.1	0.071	0.011	13.0	0.069	0.05	<0.02	0.149	0.064
BC 708/1594	93.6	5.6	0.92	<0.011	0.036	0.282	0.035	0.36	0.098	0.086
BC 2188/530	88.3	10.0	0.313	<0.01	0.336	0.282	0.035	0.36	0.098	0.086
BCES 480/1367	85.2	10.6	0.056	0.016	1.27	<0.02	0.045	0.227	0.06	0.04
Late Period										
BC 2160/276	68.8	12.3	5.63	0.029	12.5	0.032	0.23	<0.007	0.031	0.297
BC 604/492	61.2	18.8	0.376	0.017	18.3	0.02	0.289	0.018	0.234	0.118
BCES 288	89.4	6.8	0.274	0.019	3.11	0.034	0.104	<0.007	0.019	0.169
BC 2156/322	77.2	10.1	0.125	<0.010	12.0	0.042	0.09	<0.011	0.036	0.086
BC 2161/781	74.0	24.8	0.205	<0.008	0.04	0.06	0.151	0.025	0.257	0.152
	80.7	17.9	0.154	<0.009	0.032	0.03	0.119	0.235	0.088	0.098

a. All values given as percentages.

Figure 15.3 Adze BC 694/1203 is heavily corroded and shows a cast structure. × 100.

"the earliest piece excavated at Ban Chiang dated to 3600 B.C." (Stech-Wheeler and Maddin 1976, p. 41), was cast as a unit (figure 15.4 bottom and figure 15.5 right), presumably in a bivalve mold, worked, and annealed. The micrographs show that the recrystallization of the grains strained by cold-working was not complete. The spearhead has 9.17% tin, according to the PIXE analysis. In 1976 we reported that this spearhead contained 1.3% tin but noted that the real value could be as much as three times that number because the method of detection (optical emission spectroscopy) was only semiquantitative with respect to tin. In addition, the caveats applied to the PIXE analysis should be noted—possible enhancement of the tin value by corrosion or an aspect of the PIXE instrumentation or both. The fact remains that, because of the extent of the internal corrosion, we will never obtain elemental analyses that show the original composition.

What does this limited sample tell us about the metals used in the Early Period at Ban Chiang? First, that casting was competently handled, alloying of copper with tin was known, and working and annealing were practiced. These basic techniques of copper working seem to have appeared full blown at Ban Chiang.

Figure 15.4 Spearhead BCES 762/2834 (bottom).

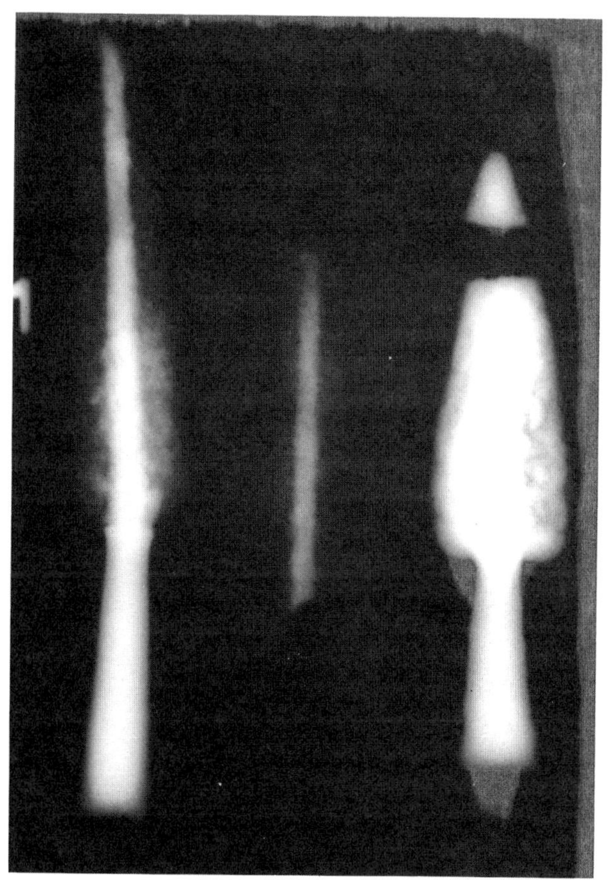

Figure 15.5 Radiograph of spearhead BCES 762/2834 (right).

The Middle Period

Bronze ornaments, primarily bracelets and anklets, were frequent in Middle Period graves, with crucibles appearing early in the period. This span is dated c. 1000–300 B.C. and is the time when iron enters the record at Ban Chiang, at least by the middle of the period, although a few small fragments are present in the earlier portions of the Middle Period, according to White (personal communication).

Eight bracelets and anklets (BCES 490/1286, BCES 491/1286, BCES 591/1981, BCES 395/1115, BCES 616/2097, BCES 617/2097, BCES 609/2069, BC 708/1594; figures 15.6–15.10) and two other artifacts of bronze (BC 2188/530, BCES 480/1367) have been studied metallographically. All were left in what is defined metallurgically as the as-cast condition, a technological indication that is probably prejudiced by the preponderance of bracelets in the sample but one that demonstrates the skills of early Thai metalworkers in achieving the desired product. Tin contents range from 5.5% to 16.1%, but again all the samples are corroded so these values may not reflect the original compositions. In contrast to the bronzes of the Early Period, however, are the levels of lead—five of ten artifacts analyzed contain more than 1% (1.27%, 1.46%, 4.03%, 9.93%, 13.0%). As we discuss later, the addition of lead seems to be deliberate.

Although the bronze industry is stable and accomplished, innovation occurs in the making of bimetallic—bronze and iron—and solely iron artifacts. The collection consists of iron bangles and two bimetallic spear points. One of the spear points (BCES 548/1582) has been studied. Its iron blade is extensively corroded, so it has yielded little information, but it appears to have been made from terrestrial iron because it contains only a minute amount of nickel. Nickel in excess of 4% and a metallographic structure characteristic of meteoritic iron are needed to prove a nonterrestrial origin. The blade must have been forged into shape and the bronze socket cast into it. The socket is bronze (determined qualitatively) with large discrete globules of lead distributed throughout at least the area of the sample.

The Late Period (c. 300 B.C.–A.D. 200)

White notes that, although no iron ornaments occurred in the excavated burials dating to the Late Period, iron was used for tools and weapons (White 1982, p. 45). Bronze bangles continue and bells appear, but the major innovation was a variant of bronze technology—a high-tin alloy with unusual manufacturing requirements.

Six artifacts of normal bronze were studied—four bracelets or bangles (BC 2160/276, BC 604/492, BCES 288, BC 2156/322; figure 15.11) and two fragments, only one of which (BC 2161/781) was analyzed. Five were left in the as-cast condition, as proved by their

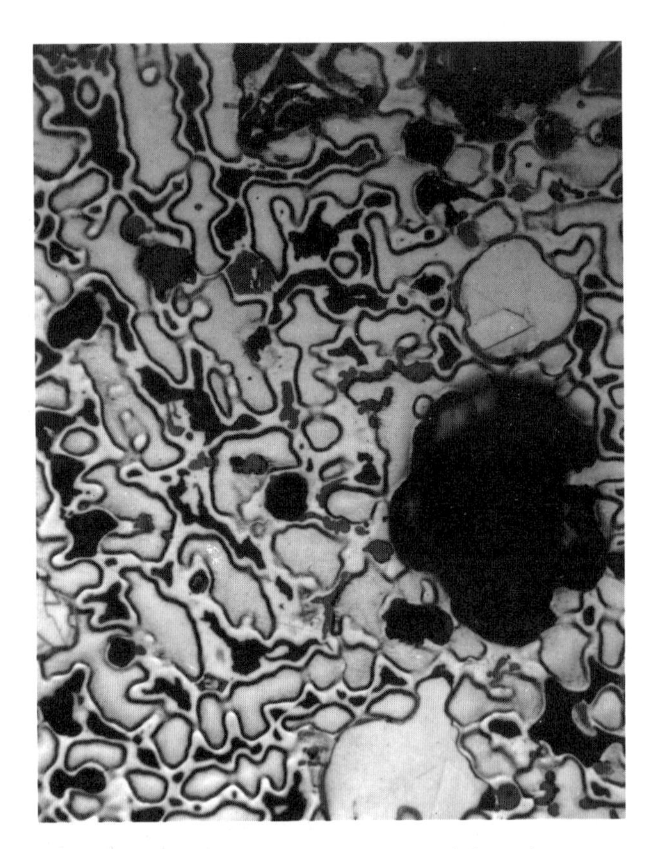

Figure 15.6 Bracelet BCES 490/1286 left in the as-cast condition. × 200.

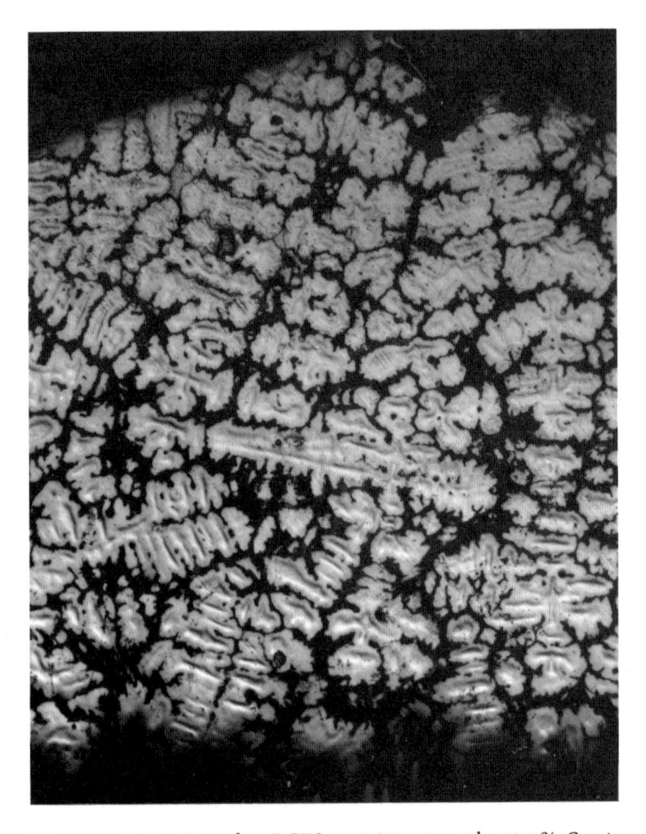

Figure 15.7 Bracelet BCES 591/1981, with 12.1% Sn, is not as heavily corroded as BCES 491/1286, and it shows a well-defined dendritic pattern. × 100.

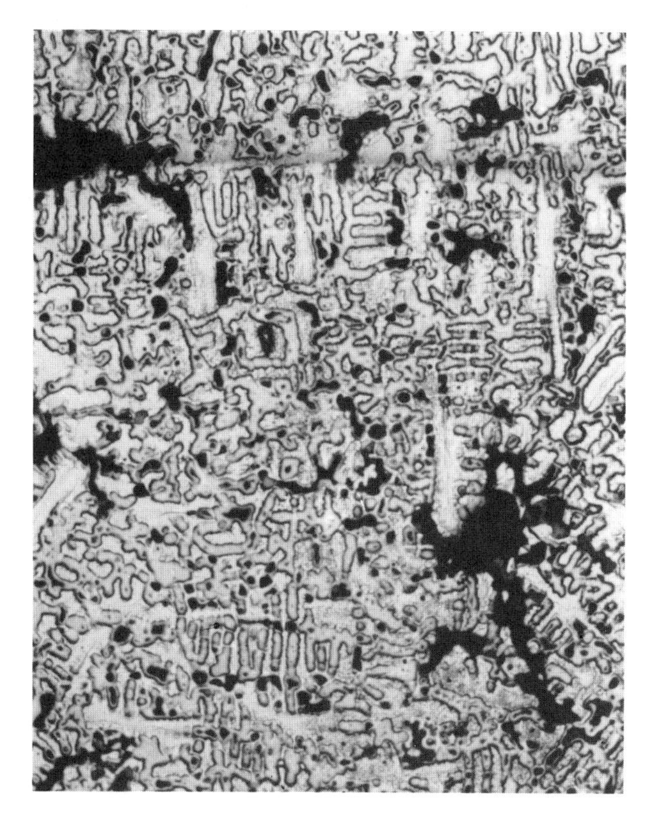

Figure 15.8 Bracelet/anklet BCES 395/1115, with 5.5% Sn and 4.03% Pb, shows a two-phase structure and an as-cast structure. The black areas are due to corrosion. × 100.

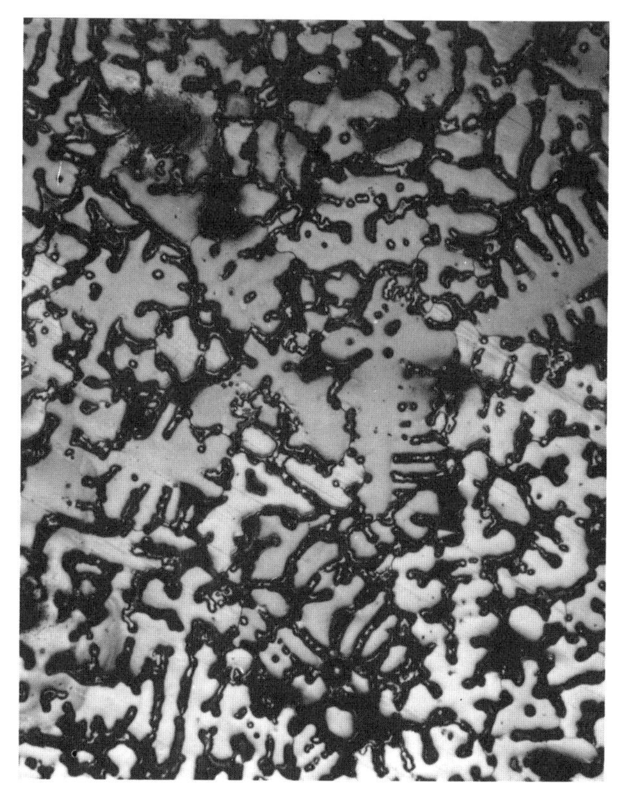

Figure 15.10 Bracelet/anklet BCES 708/1594, a low-tin bronze (5.6%), was left in the as-cast condition. × 100.

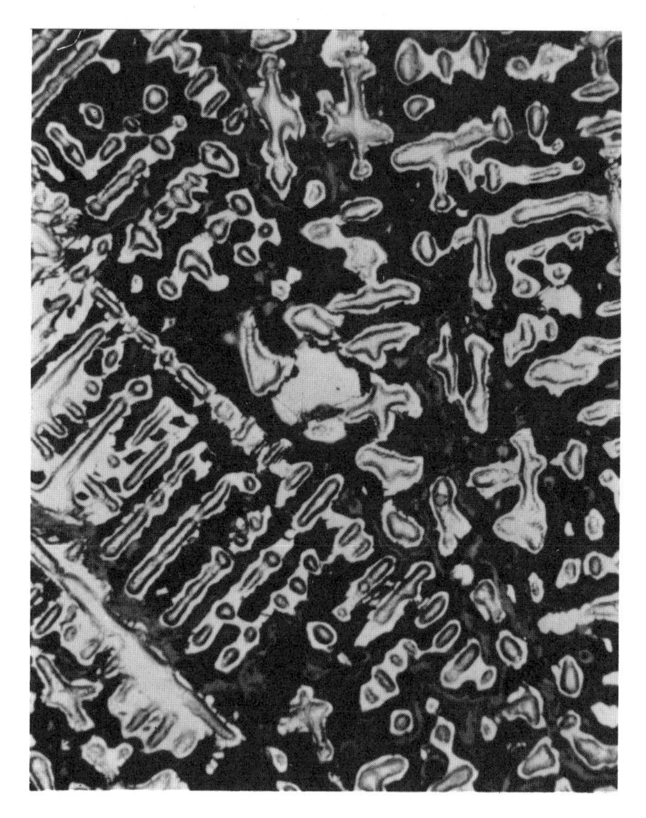

Figure 15.9 Bracelet/anklet BCES 617/2097, with a chemical content similar to BCES 395/2097 but with little Pb (0.054%), shows a sharp dendritic pattern. The black areas are the interdendritic sections that have corroded. × 200.

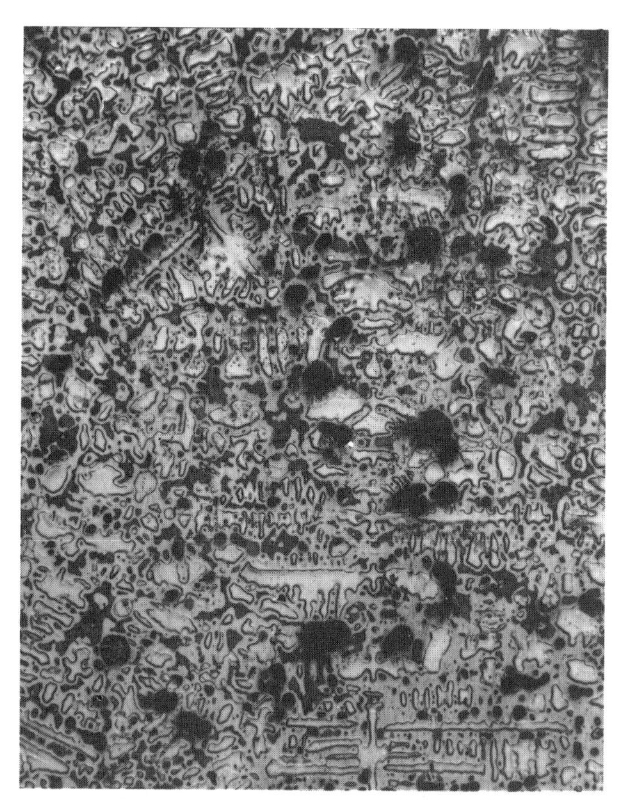

Figure 15.11 Artifact, not analyzed, shows a two-phase structure with a dendritic pattern characteristic of the as-cast condition. × 100.

dendritic structures. The sixth is extremely corroded, so it is difficult to determine the original structure; it might have been worked. Of the four for which we have elemental analyses, the tin contents range from 6.8% to 24.8%. Those quantities of tin above 11% seem suspicious, however, because the microstructures are not compatible with the amounts. The analyses must be exaggerated, for the reasons given before.

The high-tin bronze from Ban Chiang from the grave of a five-year-old child (BC 918/10; figures 15.12 and 15.13) that we have examined is a necklace made up of hundreds of thin rods, some straight and some helical. We discussed the metallurgy of this necklace extensively in 1976 (see figure 15.13). Since then we have obtained an elemental analysis of one section that is undoubtedly too high at 68.7% tin. Subsequent studies have shown that alloy to be represented at Don Klang, near Non Nok Tha, in levels dating to the first century A.D.

An oddity is a brass ring, which may be the accidental result of smelting ore with zinc minerals included perhaps in the attempt to add lead.

Two iron tools and an unidentified iron piece were also analyzed. One tool (749/2669) shows a significant carbon content along the edge (figure 15.14), which *may* suggest an attempt at case carburization—allowing the finished artifact to remain in the forge long enough to render the edges steel and hence making those edges much stronger. There is no further evidence of manipulation of "steel" in terms of quenching and tempering. The other two artifacts have essentially ferritic structures with no evidence for extensive contact with carbon (see figure 15.15). One artifact (1205/580) contains slag stringers, which demonstrate the direction of forging.

Thus, to summarize our scientific evidence on Ban Chiang, bronze working started in the first half of the second millennium B.C., with some slight evidence for a stage of experimentation immediately preceding the first accomplished attempts at production. This competent bronze industry, characterized by considerable skill in casting intricate shapes, persisted into the first centuries A.D. About a thousand years after the first bronze appears in the burial sequence at Ban Chiang, iron was introduced. This iron seems to have been forged from the products of smelted ores. There is at present little evidence that intentional steel was made at any time during the ancient occupation of Ban Chiang, but the sample studied is small. In addition, there is no way of determining metallurgically whether the "ferritic" iron examined was not obtained by the decarburization of a high-carbon iron, that is, cast iron. A further testament to the skills of the bronze workers is the use after 300 B.C. of a high-tin alloy, which is at best difficult to work, for ornamental purposes.

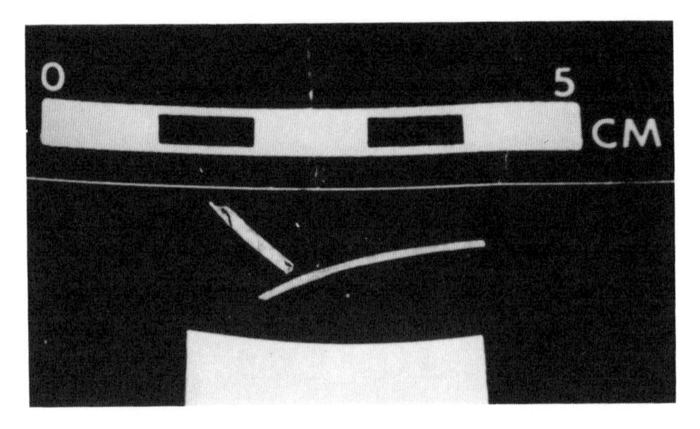

Figure 15.12 Pieces of high-tin bronze that made up "Bianca's" necklace.

Figure 15.13 High-tin bronze (BC 918/10 with as much as 24% Sn) shows the beta-Sn martensite structure resulting from quenching the bronze from above 520°C. × 100.

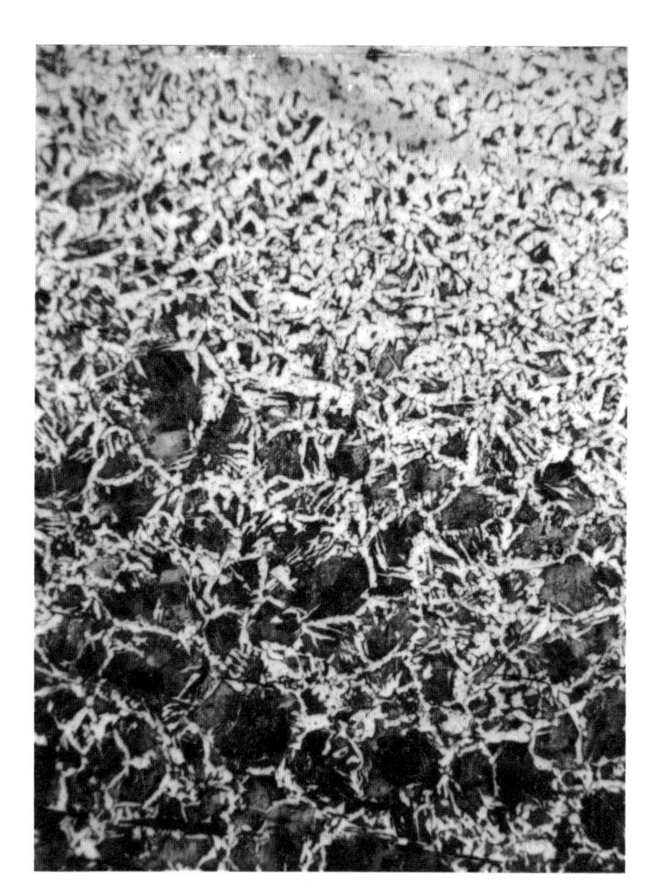

Figure 15.14 Iron tool BCES 749/2669 shows a significant amount of carburization along its edge. × 100.

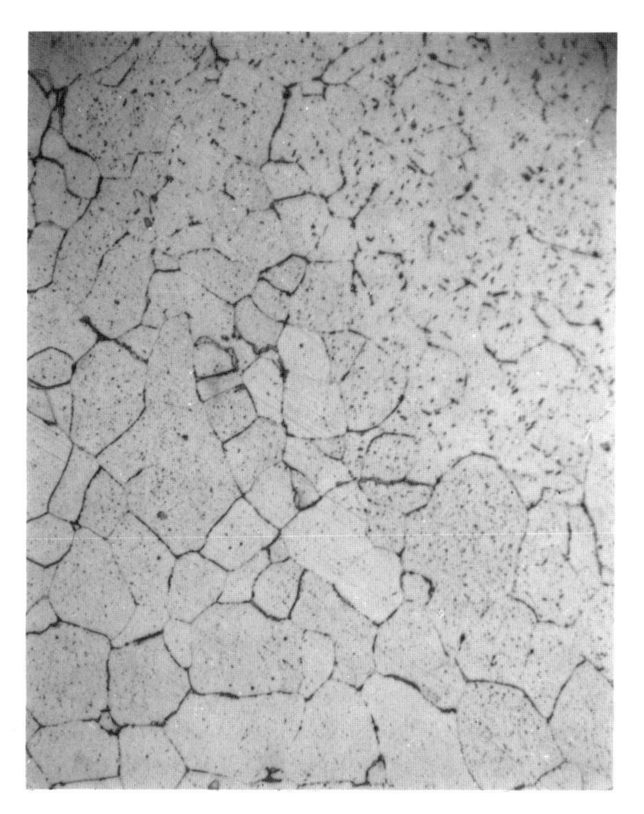

Figure 15.15 Ferritic structure showing no evidence for any carburization. × 500.

The Origins of Bronze Making in Thailand

It is clear that at this time we cannot say if the Thai bronze industry was independently invented or derivative. On the basis of the published evidence, the earliest bronze is claimed for Non Nok Tha, at some time during the third millennium B.C. (calibrated dates) (Bayard 1984, p. 165). This evidence would then predate that at Ban Chiang, a site that was also inhabited since the fourth millennium B.C. In the absence of the final report on Non Nok Tha, we cannot assess at the evidence in relation to other excavated sites. It is certainly not impossible that different sites started using bronze at different times, but the sample is so variable that we cannot tell. In the broader context of Southeast Asian bronze production, it is important to note that bronze metallurgy can be documented in Vietnam in the second millennium B.C. (Diep 1978; Higham 1983, p. 6), a date more consistent with that proposed for Ban Chiang than for Non Nok Tha. Higham (1983, note 4) questions that Non Nok Tha was inhabited before the second millennium B.C., so perhaps we can generally place the earliest bronze usage in the early second millennium B.C. The firm point is that present evidence shows that bronze did appear full blown, with little or no evidence for a period of experimentation.

Given, however, the geological situation—deposits of copper in northeastern Thailand and of tin not too distant in Cambodia and Laos—it might be reasonable to assume that experimentation took place in the source areas and that the abundance of the resources made only a (relatively) short period of experimentation necessary. Our view of Southeast Asian metallurgy has been strongly affected by the roughly evolutionary sequence of metallurgical development in Southwest Asia—native copper usage[4] preceding copper smelting, arsenical alloys of copper followed by those with tin. The area of Southwest Asia in which, so far as we now know, copper metallurgy advanced most rapidly is the resource-poor Mesopotamia. The introduction of tin as an alloying element may relate more to its extreme rarity in Southwest Asia rather than to the inexorable march of technological progress [see most recently, Stech and Pigott (1986)]. We simply do not have a model to deal with the development of bronze alloying in an area rich in the necessary resources. Given the availability of the appropriate resources, there is no reason why early Thai metalworkers could not have worked out the technology in their own milieu, although the low-fired pottery does not suggest great pyrotechnological skill.

Although Southeast Asia is well known as the site of the world's richest tin deposits, little geological research has been done from the point of view of prehistoric metalworkers. Modern economic viability of ore deposits is not the same as ancient viability, so reconstruction of the locations of deposits potentially useful to ancient miners is at present difficult.

Bronson (1985b) has compiled evidence for copper and tin deposits. Copper deposits cluster in three subregions: the northern and central Philippines, the northern mainland, and Sumatra and Java. Only the Philippines and the northern mainland can be described as moderately copper rich. The rest of Southeast Asia is more or less deficient in copper deposits, including the Mekong delta and Cambodia and a broad belt that extends from the Malay Peninsula eastward to the Maluccas. Even if we had no historical information on the subject, we could be justified in concluding that much of the region has always been short of smelted copper (Bronson 1985). Although, as elsewhere, there must be small deposits that are not economically viable and therefore not the subject of modern geological research, the implication is that the northern mainland—Burma, Thailand, and a few parts of Laos and Cambodia [the last deduced from Bronson (1985b, fig. 2), although this is in conflict with his text]—would have been the most likely areas in which the alloying of tin and copper could have first taken place. The Philippines only began to make metal in the closing centuries of the first millennium B.C., and the metal of choice was iron, a development that came about because of Chinese influence.

On tin, Bronson says:

Estimates of reserves are traditionally unreliable in the tin industry, but all authorities agree that Southeast Asia with Yunnan contains more than half of the world's tin.... The northern mainland and southern China contain a fairly large number of scattered deposits. All the rest of the tin is concentrated into a single belt that runs from the eastern side of central Burma down through the Kra isthmus and the Malay Peninsula to the islands of Singkep, Bangka and Belitung. Adjacent parts of Sumatra have a few small deposits and the whole of the Philippines have none at all. (Bronson 1985b, p. 30)

Thus southern Burma, northern Thailand, and northern Thailand, and northern Cambodia contain both elements necessary to make bronze. The archaeological evidence for northeastern Thai bronze production, although it could be skewed by chance, may be better than it appears, because the broad general concurrence of archaeology and geology is striking. Because Vietnam is not one of the geologically favored areas, the presence of bronze there in second millennium B.C. contexts may indicate that exchange of the raw materials and communication of technological knowledge took place rapidly.

The broad general resolution of the chronological issue, Bayard and Solheim's preliminary statements notwithstanding, leads to the striking realization that a technological continuum seems to have existed across Southeast Asia, starting in the case of metals with bronze working in the second millennium B.C. and continuing into the first millennium B.C. with leaded bronzes, iron, and high-tin alloys of copper. In a forthcoming article on metal articles from Samrong Sen, Cambodia, now in the Peabody Museum, Harvard University, Robert Murowchick cites the similarity of techniques, crucibles, and bivalve molds and adduces that "this close correlation between the Samrong Sen material and that from Ban Chiang, Non Nok Tha, Mlu Prei and other sites suggests that the metallurgical know-how was disseminated along the various trade routes joining Thailand, Cambodia and Vietnam." The area in which the expertise was developed cannot be located precisely, if there was indeed a single location, but odds are that it was one with both copper and tin deposits.

Leaded Bronzes

As Murowchick points out, leaded bronzes are a phenomenon of the first millennium B.C. Seeley and Rajpitak (1984, pp. 106–107) detected lead in Ban Na Di bronzes in 32% of the artifacts. Again, the earliest occurrence of leaded bronze on a regular basis would seem to be at Non Nok Tha, where it was made by Middle Period 3, some time in the third millennium. The large number of bronze nodules, which Bayard identifies as casting spillage, are, however, virtually free from lead, indicating that the lead was added after the initial melting of the bronze (Bayard 1981, pp. 697–698). Following the common wisdom, Seeley and Rajpitak and Bayard attribute the presence of lead to the improved fluidity it imparts to the casting of the alloy, thus enhancing the smith's ability to make complicated decorative items. White has noted that the shapes of bracelets in the Middle Period at Ban Chiang are, in fact, considerably more elaborate than those of the Early Period. The Ban Na Di sample consists largely of bracelets and rings, as does that from Ban Chiang. Seeley and Rajpitak wonder whether the presence of lead has chronological implications (1984, p. 107), which might be indicated by the sequence at Ban Chiang, where leaded bronzes occur first in the Middle Period and never in the Early Period sample. The evidence from Non Nok Tha, if the dating is anywhere near correct, would, however, vitiate this tentative conclusion.

Four of the five artifacts from Samrong Sen analyzed by Murowchick contain lead (6.82–26.47%) but, because of their uncertain dating, they merely confirm the general trend rather than contribute to the debate. It is again clear that we cannot even begin to resolve such questions until the final reports on Ban Chiang and Non Nok Tha are available.

For information on sources of lead, we turn again to Bronson:

Lead and zinc, which tend to occur together in the same deposits, are distributed rather differently from copper.... The northern Mainland is abundantly furnished with lead-zinc ores, and part of that subregion—Burma—is exceptionally rich. The two metals are also much more likely than is copper to be present in other parts of Southeast Asia.... Only Cambodia and the Mekong Delta are actually lead and zinc deficient. The rest of the region has at least moderate quantities of the relevant ores. (Bronson 1985b, p. 25)

Because zinc volatilizes or passes off as a gas (ZnO) at temperatures where the lead is still liquid, most of it would have disappeared during smelting, resulting in a fairly pure lead as the end product. The analyses from Ban Chiang (table 15.1), Ban Na Di (Seeley and Rajpitak 1984, p. 108 and table 3-21), and Samrong Sen (Murowchick, forthcoming) show that the leaded bronzes contain low levels of zinc. Lead-zinc, copper, and tin deposits occur in relative proximity in central Burma, northern Thailand, and northern Cambodia. The intimacy of association of the three mineral types in most places is not known; that is, it is not known if in other places in the region the lead-zinc ores actually co-occur with those of lead and/or tin.

High-Tin Bronzes

The few high-tin bronzes known from Late Period Ban Chiang and the nearby site of Don Klang are paralleled at Ban Na Di in three certain and two probable specimens. Most of the known artifacts made of alloys of this type are ornaments; the bowls of Ban Don Ta Phet (Rajpitak and Seeley 1979) are an interesting exception, probably made for a special decorative purpose. The consensus is that high-tin bronze was made because it was lighter in color and could be rendered shinier than normal bronze (Seeley and Rajpitak 1984, p. 107; Stech-Wheeler and Maddin 1976, p. 43). The difficulty of fashioning must have contributed to the apparent infrequency of use and therefore the value.

Containing about 24% tin, high-tin bronze presents a striking phenomenon in a region where silver and gold are rare to nonexistent in the archaeological record. Its occurrence may reflect the desire to convey status through adornment, with the time and care needed to produce ornaments factors determining value, perhaps almost as much as the appearance. Thus the frequency of tin in Southeast Asia may be the technological factor that enabled the cultural need for display to be fulfilled.

Iron

Iron appears at Ban Chiang in the middle of the Middle Period, by 500 B.C. or earlier in the first millennium. The conclusion reached on the basis of studying a few artifacts, as described earlier, was that they are made of forged bloomery iron, not meteoritic and probably not decarburized cast iron, and that no deliberate attempt was made to convert them into steel.[5]

In assessing how iron use at Ban Chiang fits into the general scheme for Thailand and the rest of mainland Southeast Asia, we are once again indebted to Bronson. His review of the available radiocarbon dates for early iron in Thailand leads him to conclude that this phenomenon can be dated to the first half of the first millennium B.C., which is well before any intensive contacts with India and China (Bronson 1985a, pp. 205–208; Pigott and Marder 1984).[6] This date is roughly in line with that of the flourishing ironworking technologies in many other parts of the world, including southwest Asia, the eastern Mediterranean, and Europe.

The technological arguments adduced by Pigott and Marder (1984, pp. 278–281) support the contention that Thai ironworking could have developed out of the millennium-old bronze working tradition. In their tentative reconstruction iron could have been encountered first under certain conditions in the copper smelting furnace, if iron oxide contained in copper ore (for example, chalcopyrite) or a hematite flux was used in smelting the siliceous ore. Although detailed evidence on the mineral form in which all Southeast Asian copper minerals occur is difficult to assemble, Murowchick does note the presence of chalcopyrite in both Thailand and Cambodia. There is at present no reason to suppose that iron smelting was not an indigenous Southeast Asian development.

Ban Don Ta Phet dates to the last centuries of the first millennium B.C. but before the strong Indian and Chinese influences on Southeast Asia. Most of the thirteen artifacts analyzed by Bennett were made of low-carbon steel and were air cooled and sometimes edge-hardened by hammering. Bennett concludes:

The smelting of small sized limonite pellets found in the area would have produced the fairly homogeneous low carbon steel. That this had been extensively forged was shown by the small number of slag inclusions remaining.... The tools, and the rest of the weapons, were made by simple techniques and no attention seemed to have been paid to the degree of decarburization during forging. There was no evidence of the use of sophisticated quenching and tempering treatments. However, the implements appeared to be well adapted to their function. They were of a hardness which would have prevented them from becoming blunt too quickly and would have enabled them to be easily sharpened. The working edges of some tools appeared to have been hardened during use. (Bennett 1982, secs. 9.2 and 9.3)

The second major metallurgical study analyzed among others four artifacts excavated at Don Klang (Pigott and Marder 1984, pp. 283–289). Two of these are made of low-carbon, possibly air-cooled steel, and a third artifact has a higher carbon content but shows no evidence of having been quenched; these three date between 300 B.C. and A.D. 210 (calibrated carbon 14 dates). The fourth artifact dates earlier, c. 390–395 B.C. (calibrated carbon 14 date); it is so completely carburized that Pigott and Marder believe it to have been intentionally steeled and cooled in an accelerated fashion, perhaps by quenching. Of the four additional artifacts reported, one appears to have been deliberately carburized, whereas the others are low-carbon steels that probably became steeled in the forge. All were apparently cooled in an accelerated fashion.

The evidence, then, is somewhat variable in details, that is, as to whether good steels could be made, but it is clear that low-carbon steel could be produced fairly regularly and that it would have probably been adequate for most of the requirements of agrarian people in the environment of Southeast Asia—digging, harvesting, and processing of roots, grains, and fibers and of animal products such as skins.

Final Remarks

It is most productive, particularly given the incomplete state of our knowledge, to try to understand Southeast Asian metallurgy in its own context rather than to compare it to technologies half a world away. For example, apparently lacking in Southeast Asia are the steps that led to the bronze industry in southwest Asia; but the raw materials are not lacking, which is interesting in relation to several arsenical copper artifacts from Ban Na Di (Seeley and Rajpitak 1984, p. 107), one containing 3.56% arsenic and 4.85% tin, and from Ban Chiang (no. 2160/276; see table 15.1). Although they fall late in the first millennium B.C. at both sites and thus cannot occupy a place in any evolutionary sequence, they do indicate that arsenic minerals were available, but they seem to have been ignored or not recognized. From his review of the literature, Bronson concluded that arsenic must have been present in the copper deposits of Southeast Asia, although only nineteenth-century smelting of copper-arsenic ores in northern Luzon can be cited in support (Bronson 1985b, p. 33). The rare presence of arsenic in late bronze artifacts in Thailand was probably not perceptible to the metalworker because the amounts are too low to have had much effect on the physical appearance or obvious properties.

In ironworking also the industry appears to remain largely stable so far as it has been traced, perhaps because the product responded sufficiently to the relatively constant needs of its users. Lack of change in some part of the cultural and ecological system that would have occasioned changes in the products needed to sustain the system meant that ferrous technology and its tools were and remained adequate.

In this context we should cite the fact that early iron in southwest Asia was by no means always fully manipulated, that is, carburized, quenched, and tempered. By comparison, current evidence indicates that the eastern Mediterranean littoral was the site of such developments [see, for example, Stech-Wheeler et al. (1981), Davis et al. (1985), and Maddin (1982)], whereas the Assyrians of inland southwest Asia, who are the documented consumers of tons of iron, were using virtually pure iron (Curtis et al. 1979) as were their neighbors in northwestern Iran (Pigott 1981). The reasons for the differences in approaches to ironworking techniques must lie to a certain extent in the nature of raw materials available, to environmental conditions prevailing in each area, and to the individual cultural requirements of each iron-using group.

In this vein one might speculate along the lines of the "peaceful Bronze Age" of Thailand (White 1982), which would have given way to a "peaceful Iron Age" until perhaps Chinese and Indian influences exerted different pressures on local development. To harvest rice, dig roots, and chop bamboo, low-carbon steels would have been adequate. In Southeast Asia the indications of differentiated social structure are not as sharp as they are in southwest Asia and the eastern Mediterranean in the early part of the first millennium B.C. In Thailand Higham sees a "major discontinuity" between 400 and 200 B.C. (Higham 1983, p. 16), perhaps occasioned by an externally generated disruption of the "peaceful" cultural continuum. In the eastern Mediterranean a "major discontinuity" occurred c. 1200 B.C., one that seems to have involved considerable hostilities and movements of peoples, with much of the activity centering around areas that are certainly not as agriculturally sympathetic as Thailand. A concomitant catalyst for relocation in marginal or undeveloped agricultural zones might have been the development of what we would characterize as a more sophisticated ferrous technology (that is, one producing carburized, quenched, and tempered steel) that responded to changes and uncertainties in all aspects of life, including the major one of producing food.

The explanation of why Neo-Assyrians and their Iranian neighbors did not participate in or share the developments of the Mediterranean littoral is even more complicated. One possibility is that the technology did not come about because the adversity of various aspects of life did not exist. Neo-Assyrians used bronze regularly; the mention of great quantities of iron in their texts may reflect stockpiling of a material that was culturally valuable in that it conferred status on the new empire. Pigott (1981) has suggested that Iranian iron usage was in emulation of Assyrian usage. We might extend that hypothesis to say that Neo-Assyrians used iron in imitation of their neighbors to the west, without the specific knowledge of what made iron essential to the peoples of the Mediterranean littoral.

Therefore the skills of the Thai ironworkers are not to be denigrated but rather praised for the successful development early in its history of an iron that filled their needs. We know too little about the sequel to the "major discontinuity" in Thailand to comment on the response of technological systems to changed circumstances, but we can speculate that the ironworking skills demonstrated by the existing metallographic studies were appropriate in their context.

Rather than look for evolutionary patterns and inventors, we might disagree with Higham and Kijngam's statement (1984, p. 1) that "the indigenous development of bronze and ironworking in inland Southeast Asia in a context of unchanging, stable, small-scale communities" is an interpretation that "if validated, would run counter to any notion of even the most general regularities of culture process." Technological systems, such as metallurgy, should be interpreted as integral parts of the cultural context in which they occur rather than as entities possessing their own dynamic, independent of context, which is uniform through time and space. Bronze metallurgy in southwest Asia flourished at a time of urban nucleation when some elements of the population had a strong cultural need to demonstrate status. Metals, as imported materials often brought from considerable distances, partially fulfilled that need in their role as exotic commodities. In Southeast Asia metallurgy developed in the context of an agrarian society that did not, before perhaps the last centuries B.C., experience discontinuities on a scale with those occurring in southwest Asia with some relative frequency starting by at least the late fourth millennium B.C. Southeast Asians also had much easier access to all metals than did their counterparts in the southwest. Metallurgy did not change the cultural dynamics in either end of Asia but was a dependent subsystem in both areas, as it was elsewhere. Southeast Asian metallurgy can be understood only when it is viewed as an industry that developed in a fairly uniform and peculiarly regional manner, based on an abundance of local natural resources. This feature makes Southeast Asia distinct from Old World culture areas on which our current models of metallurgical development are posited and requires that we make new models appropriate to a different environmental and cultural setting.

Notes

An earlier version of this chapter appeared in the *Bulletin of the Metals Museum* (1986), 11:43–56.

1. This chapter could not have been written without the help of Joyce White and Vincent Pigott, to both of whom go many thanks. We are grateful to Surapol Natapintu (see his 1982 thesis) and Christine Abiera for undertaking the metallographic studies under the direction of R. Maddin and Vincent Pigott.

2. The PIXE analyses were performed by S. Fleming, MASCA, University Museum, University of Pennsylvania, and C. Swann, Bartol Research Foundation, University of Delaware. For the technique, see most recently Fleming and Swann (1985).

3. As the tin content rises above about 11%, the beta phase begins to appear. A bronze with 24.8% tin, if quenched from above 520°C shows a preponderance of the beta phase. No beta phase or, for that matter, no decomposition products of the eutectoid beta phase were observed in the artifacts.

4. Recent fieldwork by the Thailand Archaeometallurgy Project, codirected by V. Pigott and S. Natapintu, is providing detailed evidence of just what types of copper ore were being exploited in prehistoric Thailand. Weathered copper sulfide ore deposits, with strong suggestions of the presence of native copper, and chalcopyrite have been recorded (Pigott and Natapintu, personal communication).

5. Differentiating decarburized cast iron from the normal bloomery iron on the basis of their microstructures is not at all certain, particularly when the decarburization occurs extensively. Chinese metallurgists studying early first millennium iron have concluded (Ko Tsun, personal communication) that the "bloomery" product was in fact decarburized from cast iron.

6. K. C. Chang (personal communication) believes that, although there may not have been *intensive* contact with south China, there was sporadic contact earlier than the first half of the first millennium B.C.

References

Bayard, D. 1981. "Temporal distribution and alloy variation in early bronzes from Non Nok Tha." *Current Anthropology* 22:697–699.

Bayard, D. 1984. "A tentative regional phase chronology for northeast Thailand," in *Southeast Asian Archaeology at the XV Pacific Sciences Conference* (University of Otago Studies in Prehistoric Anthropology), vol. 16.

Bennett, A. 1982. *Metallurgical Analysis of Iron Artifacts from Ban Don Ta Phet, Thailand.* Master's thesis, University of London.

Bronson, B. 1985a. "Patterns in the early Southeast Asian metals trade. Notes on the history of iron in Thailand." *Journal of the Siam Society* 73:205–225.

Bronson, B. 1985b. "Patterns in the early Southeast Asian metals trade." Unpublished.

Curtis, J. E., T. Stech-Wheeler, J. D. Muhly, and R. Maddin. 1979. "Neo-Assyrian ironworking technology." *Proceedings of the American Philosophical Society* 123(6):369–390.

Davis, D., R. Maddin, J. D. Muhly, and T. Stech. 1985. "A steel pick from Mt. Adir in Palestine." *Journal of Near Eastern Studies* 44:41–51.

Diep Dinh Hoa. 1978. "On the metal artifacts from the beginnings of the Bronze Age in Viet Nam." *Khao Co Hoc* 2:10–20. (In Vietnamese.)

Fleming, S. J., and C. P. Swann. 1985. "The application of PIXE spectrometry to bronze analysis: Practical considerations." *MASCA Journal*, Archaeometallurgy Supp., 3(5): 142–149.

Higham, C. F. W. 1983. "The Ban Chiang culture in wider perspective." *Proceedings of the British Academy* 69:229–261.

Higham, C. F. W., and A. Kijngam. 1984. *Prehistoric Investigations in Northeast Thailand.* BAR International Series 231. Oxford: British Archaeological Reports, pt. 1.

Maddin, R. 1982. "Early iron technology in Cyprus," in *Early Metallurgy in Cyprus*, J. D. Muhly, R. Maddin, and V. Karageorghis eds. (Nicosia: Pierides Foundation), 303–312.

Murowchick, R. Forthcoming. "An analysis of bronzes attributed to Samrong Sen, Cambodia."

Natapintu, S. 1982. *The Development of Prehistoric Bronze Metallurgy in Thailand.* Master's thesis, Department of Anthropology, University of Pennsylvania.

Pigott, V. C. 1981. *The Adoption of Iron in Western Iran in the Early First Millennium* B.C.*: An Archaeometallurgical Study.* Ph.D. Dissertation, Department of Anthropology, University of Pennsylvania.

Pigott, V. C., and A. R. Marder. 1984. "Prehistoric iron in northeast Thailand," in *Southeast Asian Archaeology at the XV Pacific Science Conference.* (University of Otago Studies in Prehistoric Anthropology), vol. 16, 278–301.

Rajpitak, W., and N. Seeley. 1979. "The bronze bowls from Ban Don Ta Phet: An enigma of prehistoric metallurgy." *World Archaeology* 11:26–31.

Rajpitak, W., and N. Seeley. 1984. "The bronze technology," in *Prehistoric Investigations in Northeast Thailand*, C. F. W. Higham and A. Kijngam, eds. BAR International Series 231. Oxford: British Archaeological Reports, pt. 1, pp. 102–112.

Stech, T., and V. Pigott. 1986. "The metals trade in Southwest Asia in the third millennium B.C." *Iraq* 48:39–64.

Stech-Wheeler, T., and R. Maddin. 1976. "The techniques of the early Thai metalsmith." *Expedition* 18(4):38–47.

Stech-Wheeler, T., J. D. Muhly, K. R. Maxwell-Hyslop, and R. Maddin. 1981. "Iron at Taanach and early iron metallurgy in the eastern Mediterranean." *American Journal of Archaeology* 85:245–268.

White, J. C. 1982. *Ban Chiang: Discovery of a Lost Bronze Age.* Philadelphia, Penn.: University Museum, University of Pennsylvania.

16
Early East Asian Metallurgy: The Southern Tradition

Joyce C. White

The development of metallurgy can be addressed from several points of view, many of which are represented in this volume. One approach to the topic is summarized in the subtitle of a recent synthesis for a lay audience entitled *Out of the Fiery Furnace: The Impact of Metals on the History of Mankind*, by Robert Raymond (1984). Although a valid point of view not only for writers for the public but also for scholars, the book has latent biases and limitations. The underlying historical inquiry focuses on where and when certain developments that in hindsight have proved significant to the use of metals today came about. Data not fitting into some sort of a progression to today's use of metals may be treated covertly or overtly as less important or peripheral.

An inquiry on metallurgical development from the viewpoint of an anthropologist might focus on the hows and whys or what archaeologists like to call the "processes" of metallurgical development. The hows and whys of any expression of metallurgical use, and even nonuse, would be examined in the contexts of specific cultures. In other words, the anthropologist might ask, What was the impact of humankind on the history of metals?

Southeast Asia is a relative newcomer in the discussion of the beginnings of metallurgy. Here I try to place into anthropological perspective the general significance of the early metallurgy that has been found recently in Southeast Asia to the study of early metallurgy as a whole. In this discussion the phrase "Southeast Asia" will refer to mainland Southeast Asia, including the southern portion of China. The data with which I have firsthand familiarity are from northeast Thailand, particularly the site of Ban Chiang.

Over the past few years the chronology of Southeast Asian metals has undergone detailed reevaluation (Bayard 1984; Higham 1984; Higham and Kijngam 1984; White 1986). The current consensus on the dating of bronze and iron at least for northeast Thailand is that bronze appears around 2000 B.C., give or take a couple hundred years. Iron appears in the first millennium B.C. with some disagreement as to whether it appears before or after 500 B.C. These current best estimates are not to my knowledge seriously out of phase with the limited Southeast Asian evidence outside of northeast Thailand, primarily Vietnam. Readers interested in a detailed discussion of the chronology for the controversial site of Ban Chiang can see White (1986).

It should be emphasized that current chronological understanding is based on excavations of only a few sites and on minimal data on metals. Refinements, revisions, and amplifications should be expected in the future as archaeological research in Southeast Asia expands. I challenge those working in other areas to examine their chronological data with the detail and rigor that we who have worked in northeast Thailand have been forced to do. In particular, the stratigraphic relationship between dated samples and metal artifacts needs to be explicitly demonstrated, and individual site sequences should be supported with regional evidence.

How does the revised chronology for metals in northeast Thailand affect the significance of early metallurgy in Southeast Asia for the study of early metallurgy in general? Back when the date for the appearance of bronze in northeast Thailand was suggested to be in the fourth millennium B.C., some speculated that Southeast Asia was the source for Chinese and even Near Eastern bronze (Solheim 1970; Muhly 1976). Obviously the current dating does not support any claims that Southeast Asian bronze is earlier than Near Eastern. Nor is there any firm archaeological basis to claim that Southeast Asia was the primary source for the early bronze of northern China. Does this put us back to the picture two decades ago of Southeast Asia as a retarded cultural backwater that passively received technological advances from outside, invented nothing on its own initiative, and had no influence on any other region? If Southeast Asian bronze was not earlier than bronze in one of the major Asian urban civilizations, does it have any significance to the understanding of the development of metallurgy?

To broach this topic, let us examine some of the details that are emerging on the early Southeast Asian metals and their cultural context. Two points can be made on the general significance of the dating of metals in Southeast Asia as we now understand it. First, I think it is fair to say that the bronze-producing period in Southeast Asia is approximately contemporary with that of northern China. At present there may be a clearer case for third millennium B.C. copper-based artifacts in China (Ko 1986) than there is for Southeast Asia. The third millennium copper and copper alloy objects from China are described as smallish flat cast items, such as knives and awls. These third millennium copper-based objects have come to attention only fairly recently as archaeology in China has expanded in areal and temporal scope.

In northeast Thailand, so far the oldest metal artifacts are relatively sophisticated socketed bronze tools cast in bivalve molds. Two of the main objects of concern are the tool from burial 90 of the 1968 excavation at Non Nok Tha and the spear point from burial 76 of the 1975 (BCES) excavation of Ban Chiang. [Photographs of both can be found in White (1982).] Given that the sophistication of these items is unlikely to have appeared spontaneously, it seems probable that in the future bronze items that were cast using simpler techniques or even artifacts of hammered copper will be found somewhere in Southeast Asia with dates in the third millennium. That a cultural horizon with such items has not been identified to date is not surprising, considering that southern China and Southeast Asia probably are less well covered in terms of archaeological research relative to northern China. For vast portions of the southern region nothing is known of the prehistoric life for the time period of concern. Considering that our current understanding is based on so few sites and such a tiny proportion of the region and given the known antiquity and relative sophistication of the earliest known metal artifacts of the south, it seems reasonable to propose for the time being that in both the south and the north of east Asia we have approximately contemporary metallurgical traditions. Therefore the Southeast Asian metal tradition need not be viewed as delayed or retarded but rather as not significantly out of step with the metallurgical tradition to the north.

This point is important because up until recently the appearance of metals in Southeast Asia was considered to be much later, in the realm of 500 B.C. (Clark 1971, p. 238). Thus recent excavations have shown that bronze was present on the order of 1,500 years before this estimate. The revision in date for the appearance of bronze has considerable significance for the general interpretation of Southeast Asian prehistory in that formerly it was thought that metal technology appeared at the same time as the Dongson decorative motifs—perhaps both brought simultaneously by western barbarians according to Heine-Geldern's interpretation [summarized in van Heekeren (1958)]. We now know that bronze technology was locally established for a considerable time period before the classic Dongson civilization. It is the metallurgy before the classic Dongson civilization that is my main interest and the main focus of the following discussion.

The early metallurgy found in southern east Asia is beginning to show distinctive patterning, even though the research is in the preliminary stages. Metallographic studies show that certain attributes that characterize the bronze tradition as a whole are present from even the earliest bronze artifacts recovered thus far, such as the socketed bronze spear point from BCES burial 76, the lowest metal grave good recovered from Ban Chiang during the Penn/FAD excavations. This socketed spear point (found in phase III from the Early Period) is older than 1500 B.C. and may be as old as 2000 B.C. [See White (1986) for a detailed discussion of the dating.] Stech-Wheeler and Maddin (1976) have reported that this implement was cast in a bivalve mold, is composed of a deliberate tin bronze, and shows evidence of cold-working and annealing along the edge.

The socketed tool from Non Nok Tha burial 90 was initially reported as composed of copper, but subsequent examination by Maddin has revealed that this corroded object probably included tin and probably was worked along the edge. A socketed ax from Non Nok Tha burial 69 that may be broadly contemporary with the Ban Chiang spear point from burial 76 shows similar characteristics. That is, it was cast in a bivalve mold from a tin-bronze alloy, and it has evidence of post-casting working along the edge (Smith 1973).

Metallurgists claim that these early objects show a competent and sophisticated array of basic bronze casting and working technology (Smith 1973, pp. 28, 32; Stech-Wheeler and Maddin 1976, p. 49). The combination of techniques employed in making the second millennium socketed objects from northeast Thailand would not be expected in an incipient stage of metals experimentation or an amateurish attempt to copy without comprehension a trade item that happened to find its way into the area. Because the same types and techniques and the same types of crucibles and molds show up over broad spatial and temporal ranges in Southeast Asia (Murowchick, this volume), I would argue that not only does bronze appear c. 2000 B.C. but also that a technological tradition has appeared by that time.

What other kinds of objects are found among the early bronze artifacts? From a slightly later burial phase dated to the mid-second millennium B.C. at Ban Chiang were found some plain rings around the ankles of a child (BCES burial 38 from Early Period IV). They were cast bronze with no indications of post-casting working. Bracelets and anklets were by far the most common type of bronze artifact recovered from Ban Chiang. In almost all analyzed cases they were cast tin bronze with no evidence of cold-hammering or annealing. Seeley and Rajpitak (1984, p. 106) note similar results for the bangles from Ban Na Di, a site 20 km from Ban Chiang excavated by Higham and Kijngam (1984). They comment that "decorative objects were fabricated by casting . . . and other fabrication techniques such as cold working and joining were strenuously avoided." With rare exceptions, similar results have been found for bangles from Non Nok Tha and Ban Chiang (Smith 1973, p. 29; Stech and Maddin, this volume).

Currently bangles followed by socketed implements dominate the inventory of early bronze artifacts recovered from prehistoric northeast Thai sites. It should be noted, however, that the major sites excavated have been cemeteries. Bronze artifacts in good condition generally come from graves, and this of course biases the picture. A few other types of bronze artifacts have been found in levels below the appearance of iron. Small arrowheads and small fish-hooks have been found at Ban Tong, Ban Na Di, and other sites. The edge of one arrowhead from Ban Na Di showed evidence of annealing and working (Seeley and Rajpitak 1984, p. 110). At Ban Na Di were reported comma pendants and so-called wire, which held together portions of a stone bracelet. Maddin and Weng (1984) report that the wire was cast directly into place.

To summarize briefly the current picture of the bronze artifacts from northeast Thailand from the time period before the appearance of iron, most artifacts are either items of personal adornment or implements. The majority of identifiable objects recovered are bangles (bracelets, anklets, and rings) of cast bronze usually with no post-casting modification. Socketed implements, the majority of which are adzes or axes, are common. Spear points seem to be less common. Small tanged implements, including small arrowheads, and small fishhooks are occasionally found, but these may be underrepresented in terms of frequency of recovery because of the bias of the sample toward grave goods. With few exceptions, items appear to be cast bronze alloys with some proportion of tin. The edges of the implements, including axes, spear points, and arrowheads, are likely to show evidence of annealing and working, which suggests that the implements were functional and not just ceremonial replicas.

When wrought iron appeared in the Ban Chiang sequence during the first millennium B.C. in the Middle Period, it is striking that the earliest iron objects were bangles and socketed spear points. Wrought iron bangles were found in burial 26 (BCES) of Middle Period phase VII, and a wrought iron spear point with a cast-on bronze socket was found in burial 24 of the same phase. The earliest iron objects from Ban Na Di have been interpreted as neck rings. Thus the early iron artifacts imply cultural continuity in terms of what the ancient society considered an appropriate use for metal, despite the appearance of a new and in many ways rather different metal technology.

With the appearance of iron, bronze technology did not disappear. Bronze bangles continue in the Middle Period and Late Period. If socketed bronze implements continued, none were recovered at Ban Chiang (except as the sockets on the iron blades). In the Late Period (after c. 300 B.C. until the early to mid-first millennium A.D.) iron implements of various sorts were found, although the typological range has not been addressed. A high-tin bronze (with tin content over 20%), which is found in thin wirelike necklaces, appears during the Ban Chiang Late Period. High-tin bronze has been found in necklaces, bangles, and other fragments at Ban Chiang, Ban Na Di, and Don Klang in northeast Thailand and in bowls at Ban Don Tha Phet in central Thailand, all at about the same general time period. In sum, although metal artifacts show some technological and typological development during the prehistoric time period, metals continue to be used in the region primarily for personal adornments and implements.

Evidence for where and how the metal items were made is beginning to come to light. Melting and casting took place at the village sites on the lowland interior. Small crucibles and clay casting hearths were found at both Ban Chiang and Ban Na Di (Higham, this volume). Fragments of sandstone bivalve molds came from these latter two sites, and complete sets of bivalve molds were found at Non Nok Tha. The sandstone molds were in general used in the casting of implements. As has also been suggested for Vietnam (Davidson 1979, p. 105), Higham and Kijngam (1984, p. 83) have suggested on the basis of a lead-tin casting sprue that the stone molds may have been used in a lost-lead casting process. The casting of bracelets most likely employed the lost-wax technique (Smith 1973, p. 29; Stech-Wheeler and Maddin 1976, p. 43; Seeley and Rajpitak 1984, p. 109). At Ban Na Di clay mold fragments from bracelets were found, and one bronze bracelet even had remains of insect wax (Higham and Kijngam 1984, pp. 81, 124).

Current evidence does not indicate that ore sources were in close proximity to sites of the Ban Chiang cultural tradition or that smelting took place in sites such as Ban Na Di or Ban Chiang. The question of where the copper ores came from and where the smelting took place is being addressed by the Thailand Archaeometallurgy Project, directed by Vincent Pigott and Surapol Natapintu (Pigott 1985; Pigott and Natapintu, this volume), and by Anna Bennett (1986). Pigott and Natapintu have conducted excavations at potential copper ore sources along the Mekong River in northeast Thailand and near Lopburi in central Thailand. At both sites they found evidence of second millennium human presence at the ore locales. A project from the Institute of Archaeology is investigating potential tin sources in western Thailand (Coote 1986). Although less evidence is available on the ore sources and manufacturing techniques for iron, these investigations are opening a new era of research into the technological side of the prehistoric metal-using period of the region.

Technological investigations are also proceeding in the laboratory. In addition to the presence of tin in the earliest bronzes in northeast Thailand, other elemental constituents have been identified, although the meaning of the elemental variation is not yet fully understood. Arsenic appears in second millennium B.C. metals but usually only at levels consistent with being an impurity. Arsenic may appear at higher levels in what seem to be first millennium slags in central Thailand. Hence, if arsenical copper ores were exploited for their special properties during the prehistoric time period, thus far the evidence does not indicate that this exploitation preceded the use of tin alloys, in contrast to the metallurgical sequence in the Near East.

Lead, which positively affects casting properties but negatively affects mechanical properties of a copper alloy, is another element that may have significant variation. Seeley and Rajpitak (1984, p. 109) note the absence of lead in Ban Na Di arrowheads, despite its presence in some of the decorative objects. They conclude that the choice of the binary tin-copper alloy "was firmly based on a knowledge of its properties, and in critical applications the most appropriate formulation was selected" (Seeley and Rajpitak 1984, p. 109). On the other hand, Seeley and Rajpitak (1984, pp. 119–120) notice an increase in leaded alloys in their sample corresponding to the Late Period, but they claim that this increase does not correspond to greater complexity of design. In contrast, Stech and Maddin (this volume) note an increase in lead in Ban Chiang Middle Period bronzes. This might relate to greater complexity of form in Middle Period bangles (White 1982, p. 40). As for the Non Nok Tha bronzes, Pittioni (1970) suggests that lead was in some cases deliberately added, but Selimkhanov (1979) concludes that in his sample lead is present only at levels indicating impurity. In sum, with respect to lead we can suggest at this time that the variation in content may indicate some degree of experimentation or purposeful technological flexibility. A larger sample of bronzes from more sites may elucidate whether significant variation in lead content can be more consistently related to technological considerations.

Elemental analyses are not the only area where potentially interesting variation seems to be occurring. There is considerable intra- and interregional variation in the bronze artifacts found at individual sites. For example, within northeast Thailand Non Nok Tha is relatively rich in socketed axes and bivalve sandstone molds. Ban Chiang produced only one socketed adze and one socketed bronze spear point and a fragment or two of a sandstone mold for bivalve cast tools. Ban Na Di produced no socketed implements. Some sites seem to lack metals in deposits that are contemporary with other sites that have metals. Thus for a while it was thought that central Thailand had little or no second millennium B.C. bronze on the basis of two sites with little or no bronze—Ban Kao and Khok Charoen. Recent excavations by Natapintu, Pigott, and others will probably change that conclusion.

Various reasons could account for the site to site variability in the presence of metals, including both sampling and cultural factors such as trade networks or social values. Archaeological sampling error could be involved, for it should be remembered that especially in the earliest phases bronze artifacts are rare. Another factor may have to do with the role of the bronze items in the prehistoric society. Throughout much of the prehistoric sequence in Thailand, current archaeological evidence indicates that society was village based with limited ranking. Higham has proposed that in this context the metal artifacts functioned as primitive valuables (Higham 1984, p. 248). Whether or not an individual village placed bronze in graves could in part depend on how an individual village fit into the trade network for prestige items. Bronze is not the only artifact that has an uneven distribution. For example, Ban Na Di graves produced numerous bracelets made from trochus, a shell with marine origins. Not a single example of a trochus bangle was excavated by the Penn/FAD project from a Ban Chiang grave.

On a larger geographical scale, even though early bronzes from elsewhere in Southeast Asia seem generally of the same basic technological tradition as described for northeast Thailand (Murowchick, this volume), interesting differences and variations are emerging. For example, at their central Thai site, Natapintu and Pigott found a new typological range of artifact (judging from the molds) that, although they still seem to be implements, are smaller and thinner than the relatively substantial socketed tools of the northeast. Sandstone is not the only material used for molds in the casting of implements. Clay ax molds have been found at both sites excavated by the Thailand Archaeometallurgy Project in Thailand and in Vietnam. During the first millennium B.C. northern Vietnam shows many developments in its metallurgy that are distinctive from those of northeast Thailand and that culminate in the classic Dongson Period. Perhaps within Southeast Asia we can begin to discuss regional subtraditions.

So what is all this adding up to in terms of the meaning of early metals in Southeast Asia for the study of early metals in general? I have proposed (White 1982, p. 48) that, based on the antiquity and the emerging typological and technological configuration of metallurgy in this region as discussed in general terms, we can talk about an indigenous "Southeast Asian metallurgical province" [following the use of the phrase by Chernykh (1980)]. Although there is site to site and subregion to subregion variation, in terms of typology and technology the configuration of the prehistoric metals in Southeast Asia shows both an internal coherence and, taken as a whole, considerable distinction from other major areas of early metallurgy. The internal variation within the region and over time suggests that metal technology was neither stagnant nor imitative. Instead it was probably adapted to local resources and responsive to temporal, geographic, and cultural conditions within the region.

Scholars specializing in the Near East or China—areas with metals of great typological richness, ornateness, and technological virtuosity—may consider the Southeast Asian metals to be rather prosaic. Alternatively one may view them as an appropriate technology—appropriate to their nonurban, village-based role. This is not the place to document all the differences between the second millennium B.C. societies in China and Southeast Asia. It is sufficient to note that in second millennium Southeast Asia we have nothing to compare with the evidence in China for the degree of social hierarchy, the level of systematic militarization, and the prominence of human sacrifice that is patently evident during the Shang Period. That the second millennium Southeast Asian metals are morphologically simple makes them no less appropriate to their context. Their formal simplicity does not lessen their technological integrity.

This leads to some final points about the significance of the study of early Southeast Asian metals. In my opinion, within the Asian context and the time period of concern, I try not to use the term "independent," which many other anthropologists and I distinguish from the term "indigenous." Cultural phenomena that develop independently of each other have come to imply societies that are cut off, isolated, and have no direct or indirect interaction with other groups. Thus, for example, the development of agriculture and civilization in Mesoamerica is considered independent of those developments in the Old World. In terms of scientific reasoning they can be treated as separate experiments.

There exists little indisputable evidence, such as trade items, to prove that Southeast Asians were receiving cultural contact from China or the Near East during the fourth, third, or early second millennium B.C. However, there are too many gaps in the archaeological record to accept that argument as proof that the region was cut off or "independent." Archaeologists will probably not fill in much of that unknown area in our lifetimes.

When archaeologists are in the position where they cannot prove or disprove something (which is much of the time), they find themselves having to develop the best working assumptions or estimates. What is the best working assumption for the relationships among the cultures of Asia during the third millennium B.C.? There is mounting indirect evidence—linguistic, botanical, zoological—that Asian people were interacting during this time period. We may have a difficult time characterizing this interaction because discussions based on the migration and diffusion are now perceived as at worst wrong and at best simplistic. Nonetheless, although independent invention may be an appropriate concept when discussing the emergence of New World versus Old World agriculture, it has come to the point of naiveté to use it in third millennium Asia. I would argue that the best working assumption for Southeast Asia during the third millennium B.C. is that the region was not totally cut off from the rest of Asia but interacted at some level, perhaps sporadically, minimally, or indirectly, with other societies on the continent. When more archaeological research is conducted in Burma, south China, central Asia, and the coastal regions of Southeast Asia, we may be able to specify the interaction.

In the meantime, in place of independent invention I urge us to substitute the phrase "indigenous innovation" as a more suitable concept for the current discussion. In arguing that the Southeast Asian metallurgical province is an example of indigenous innovation, one can point to numerous differences between the metallurgical traditions found in Southeast Asia and in other areas of early metallurgy. Only a few differences will be mentioned. In terms of manufacturing techniques, lost wax does not appear in northern China until the sixth century B.C., and hammering and annealing are extremely rare in early Chinese bronze implements. In terms of typology, the use to which the early Southeast Asian and Chinese societies put metals was quite different. Bangles, which are so characteristic of northeast Thailand, are virtually nonexistent in second millennium China. Of course, the bronze vessels and their piece mold method of manufacture are absent in second millennium Southeast Asia. Socketed axes are common in Southeast Asia, whereas tanged implements are common in northern China. Comparisons might be made between Southeast Asia and the Near East, illustrating that, although there are a few points of similarity between the two areas, such as the presence of annealing in both, there are also vast differences in the range of techniques used and the inventory and morphology of objects made.

Some of the differences can be attributed to vastly different social contexts, in that metallurgy developed in both China and the Near East in urban, more militarized societies. Nevertheless the metallurgy of second millennium Southeast Asia cannot be explained simply as a watered-down version of either of these other two traditions. Taken as a typological and technological whole, prehistoric Southeast Asian metallurgy displays a unique configuration of traits that seems to impress the specialists who have looked at the objects with the competence and appropriateness of the technology. The typological range would also seem to be appropriate for a village-based society. It is the appropriate selection and development of techniques and artifact types over time that makes the Southeast Asian metallurgical province an example of indigenous innovation.

If the Southeast Asian metallurgical province is an example of indigenous innovation, as I propose, what is the contribution of this nonurban, competent, but relatively unelaborate technology to the understanding of the beginning of the use of metals and alloys? I would like to return for a moment to Robert Raymond (whose book, although written by a layperson, will be read by more people than the number who read our technical articles). Raymond suggested that Southeast Asian metallurgy might be an example of a "false start" (1984, p. 43). The problem with this viewpoint is that it implies a sort of "manifest destiny" or "technological imperative" in the development of metals technology.

I propose that prehistoric Southeast Asian metallurgy was not a false start but rather an *alternative pathway*. If we are interested not only in the wheres and the whens but also in the hows and the whys, if we are interested in the processes of metallurgical development, we need to be able to compare differing examples in order to separate what is intrinsic to the process from what is specific to the individual example. The Southeast Asian metallurgical province shows among other things that complex social structure involving slave or highly organized coerced labor is not necessary to the systematic exploitation of metals. It may be necessary for *large-scale* exploitation of metals for societies with a high demand for large quantities of objects, such as weapons to outfit a standing army, or for large objects, such as huge ceremonial cauldrons, but it is not necessary for systematic exploitation of metals per se. Nor does the regular exploitation of metals necessarily lead in rapid order to greater social complexity or militarization. Urbanization and militarization eventually come to Southeast Asia but considerably later than the appearance of bronze. Although the metal technology did develop further in conjunction with the social development at least in northern Vietnam, the roots of that development are better sought in social factors. And most likely, Southeast Asia is not the only case where humankind had an impact on the development of metallurgy.

References

Bayard, D. 1984. "A tentative regional phase chronology for northeast Thailand," in *Southeast Asian Archaeology at the XV Pacific Science Congress* (University of Otago Studies in Prehistoric Anthropology). Dunedin, New Zealand: University of Otago, vol. 16, 161–168.

Bennett, A. 1986. "Prehistoric copper smelting in central Thailand." Paper presented at the Meeting of the Association of Southeast Asian Archaeologists in Western Europe, London, September 1986.

Chernykh, E. N. 1980. "Metallurgical provinces of the 5th–2nd millennia in Eastern Europe in relation to the process of Indo-Europeanization." *Journal of Indo-European Studies* 8:317–336.

Clark, G. 1971. *World Prehistory: A New Outline*. Cambridge: Cambridge University Press.

Coote, V. 1986. "Ancient base metal and tin mining in western Thailand." Paper presented at the Meeting of the Association of Southeast Asian Archaeologists in Western Europe, London, September 1986.

Davidson, J. H. C. S. 1979. "Archaeology in northern Viet-Nam since 1954," in *Early South East Asia: Essays in Archaeology, History, and Historical Geography*, R. Smith and W. Watson, eds. (New York: Oxford University Press), 98–124.

Higham, C. F. W. 1984. "The Ban Chiang culture in wider prespective." *Proceedings of the British Academy, London* 69: 229–261.

Higham, C. F. W., and A. Kijngam. 1984. *Prehistoric Investigations in Northeastern Thailand*. BAR International Series 231. Oxford: British Archaeological Reports.

Ko, T. 1986. "The development of metals technology in ancient China." Paper presented at the Symposium "The Beginning of the Use of Metals and Alloys II," Zhengzhou, China, October 1986.

Maddin, R., and Y. Q. Weng. 1984. "The analysis of the bronze wire," in *Prehistoric Investigations in Northeastern Thailand*, C. Higham and A. Kijngam, eds. BAR International Series 231. Oxford: British Archaeological Reports, pp. 112–114.

Muhly, J. D. 1976. "University Museum-Thai Fine Arts Department Northeast Thailand Archaeological Project: Introduction." *Expedition* 18(4):11–13.

Pigott, V. C. 1985. "Pre-industrial mineral exploitation and metal production in Thailand." *MASCA Journal* 3:170–175.

Pittioni, R. 1970. "Spectro-analytical research on bronze from northeastern Thailand." *Asian Perspectives* 13:158–162.

Raymond, R. 1984. *Out of the Fiery Furnace: The Impact of Metals on the History of Mankind*. University Park, Penn.: Pennsylvania State University Press.

Seeley, N., and W. Rajpitak. 1984. "The bronze technology," in *Prehistoric Investigations in Northeastern Thailand*, C. Higham and A. Kijngam, eds. BAR International Series 231. Oxford: British Archaeological Reports, pp. 102–112.

Selimkhanov, I. R. 1979. "The chemical characteristics of some metal finds from Non Nok Tha," in *Early South East Asia: Essays in Archaeology, History, and Historical Geography*, R. Smith and W. Watson, eds. (New York: Oxford University Press), 33–38.

Smith, C. S. 1973. "Bronze technology in the East," in *Changing Perspectives in the History of Science: Essays in Honor of Joseph Needham*, M. Teich and R. Young, eds. (London: Heinemann), 21–32.

Solheim, W. G., II. 1970. "Northern Thailand, Southeast Asia and world prehistory." *Asian Perspectives* 13:145–162.

Stech-Wheeler, T., and R. Maddin. 1976. "The techniques of the early Thai metalsmith." *Expedition* 18(4):38–47.

van Heekeren, H. R. 1958. *The Bronze-Iron Age of Indonesia*. The Hague: Martinus Nijhoff.

White, J. C. 1982. *Ban Chiang: Discovery of a Lost Bronze Age*. Philadelphia: The University Museum, University of Pennsylvania, and the Smithsonian Institution Traveling Exhibition Service.

White, J. C. 1986. *A Revision of the Chronology of Ban Chiang and Its Implications for the Prehistory of Northeast Thailand*. Ph.D. dissertation, University of Pennsylvania.

17
The Development of Early Bronze Metallurgy in Vietnam and Kampuchea: A Reexamination of Recent Work

Robert E. Murowchick

Europe and the Near East have witnessed a long and fruitful history of archaeometallurgical studies, but Southeast Asia only relatively recently has enjoyed the benefits of such study. Recent excavations have finally allowed archaeologists to cast aside the burdensome misconception long attached to the region—that it had long served only as a geographic funnel through which the cultural developments of India and China spread. Although the best known work has taken place in Thailand, during the past two decades impressive advances have been made in the study of the development of bronze metallurgy in Vietnam. Here I briefly review the current Vietnamese evidence tracing bronze metallurgy through the second and first millennia B.C. and present lab analyses of selected Late Bronze Age Vietnamese bronzes from the Peabody Museum at Harvard. I also examine the state of archaeometallurgical research in Kampuchea, as major finds in southern Vietnam and the recent analysis of selected Kampuchean bronzes in the Peabody Museum collection suggest that the often maligned Kampuchean bronze finds made decades ago warrant a vigorous reevaluation (figure 17.1).

Pajot's discovery of the Dongson bronze culture in 1924 initiated a new period in the archaeology of Southeast Asia. Unfortunately he did not put into

Figure 17.1 Map of mainland Southeast Asia showing the sites mentioned in the text.

practice the careful archaeological methods conscientiously applied by many of his contemporaries working in Europe and the Near East, and his excavations were marred by a lack of stratigraphic control, a total disruption of layers, and the intermingling of objects having different temporal and spatial provenance. Indeed, his Vietnamese workers later recalled that this excavation was little different than "digging for sweet potatoes" (Long 1975, p. 84). Dismayed by Pajot's earlier work, Olov Janse excavated Dongson, Han, and later Chinese tombs and reinforced the prevailing opinions of the day that China was the obvious origin of the Vietnamese bronze culture. Fortunately the Dongson site was not unique, and in recent years a series of important finds has enabled Vietnamese archaeologists to establish a preliminary chronology for the development of metallurgy in Vietnam preceding the Dongson Period, now clearly understood as a Terminal Bronze Age/Early Iron Age culture. This chronology has already been clearly summarized by Chinh and Tien (1980); I retain their use of three separate centers (north, central, and southern Vietnam) in the brief review given in what follows.

Northern Vietnam

In the Red River delta of northern Vietnam, the Late Neolithic–Bronze Age sequence spans four basic periods: (1) late Phung Nguyen or Go Bong (Late Neolithic/Early Bronze Age, c. 2000–1500 B.C.), (2) Dong Dau (Middle Bronze Age, c. 1500–1000 B.C.), (3) Go Mun (Late Bronze Age, c. 1000–500 B.C.), and (4) Dongson (Terminal Bronze Age/Early Iron Age, c. 600–258 B.C.).

Late Phung Nguyen or Go Bong Period (c. 2000–1500 B.C.)

First discovered in 1959 at the site of Phung Nguyen in Vinh Phu Province, the so-called Phung Nguyen culture complex is now represented by more than 100 sites, of which some 300 sites of the Go Bong Period are concentrated along the Red and Black rivers, including Dau Duong, Phung Nguyen, Lung Hoa, Dong Dau (lower levels) in Vinh Phu; Tu Son at Ha Bac; Van Dien (lower levels) at Hanoi; and Chua Gio and Hoang Ngo at Ha Tay. A radiocarbon date of 3330 ± 100 B.P. (Klein midpoint: 1650 B.C.) has been obtained from the lower levels at Dong Dau, whereas Trang Kenh (Hai Phong Province) has yielded a date of 3405 ± 100 B.P. (Klein midpoint: 1760 B.C.) (Tan 1980, p. 126; Bayard 1984, p. 309). It is interesting to note that this concentration of Phung Nguyen sites corresponds exactly to the traditional location of the first "kingdom" of Vietnamese legend, that of the Vanlang under the Hung kings.

Although no complete bronze implements dating to this period have been found, several late Phung Nguyen sites indicate that bronze metallurgy was already being practiced, at least on a limited basis within a Late Neolithic context dominated by a highly developed lithic technology. No reports of copper artifacts from this period have been made; although we cannot exclude out of hand the possibility that copper artifacts might be found in the future, at present the earliest metallurgical remains of the late Phung Nguyen Period show that copper, tin, and lead alloys were already known. Excavations at the site of Go Bong (Vinh Phu) in 1965 and 1967 yielded corroded bronze nodules (Hoa 1978, p. 12; Thich and Tan 1970, p. 129), and 1974 excavations at Bai Tu factory, Tuong Giang village (Ha Bac), yielded similar bronze droplets, although the stratigraphy at this site is not clear. A bracelet fragment from Bai Tu and a small ball from Chua Lai proved to be of lead (Hoa 1978, p. 12; Huyen 1984, p. 175) and possibly belong to the transitional period between Go Bong and Dong Dau.

Other possible indications of metallurgy are the presence of tubular beads of opal and other semiprecious stone found at the sites of Phung Nguyen and Go Bong; various Vietnamese archaeologists [cited in Davidson (1979, p. 100)] have suggested that only a tiny metal drill could have been used to produce the holes. A small lead-bronze awl and bronze wire fragments were found in 1974–1976 among the remains of the Hoa Loc culture in the coastal region of Thanh Hoa just to the south of the Phung Nguyen center. The dating of this culture is still being debated; it has been compared to the site of Ma Dong (Hanoi Province), which produced similar pottery and has been radiocarbon dated to 4145 ± 60 B.P. (Klein midpoint: 2738 B.C.) (Huyen 1984, p. 175; Bayard 1984, p. 309), yet no metallic artifacts have been found at this site.

Dong Dau Period (Middle Bronze Age, c. 1500–1000 B.C.)

Sites from the Dong Dau Period are distributed over approximately the same area as those of the late Phung Nguyen. Indeed, many sites contain Phung Nguyen remains in the lower levels with Dong Dau remains above. Carbonized wood from the Dong Dau levels of Vuon Chuoi (Ha Son Binh Province) yielded a radiocarbon date of 3070 ± 100 B.P. (Klein midpoint: 1343 B.C.) (Tan 1980, p. 127; Bayard 1984, p. 309), whereas the type site of Dong Dau produced a date of 1377 ± 100 B.C. (Long 1975, p. 55). Dong Dau pottery types share many features with the Go Bong material, suggesting a direct development from the preceding period. Although stone artifacts (polished stone axes, rectangular adzes, bracelets, and earrings) are still dominant, bronze is now much more common than in late Phung Nguyen sites. Dong Dau sites have yielded bronze arrowheads, fishhooks, spearheads, and socketed axes in a variety of shapes. Of particular interest is the appearance of the asymmetric pediform ax that becomes quite common at later Dongson sites. Metallographic analyses of some of these bronze artifacts show that they are alloys of roughly 80% copper and 20% tin [Tan (1980, p. 128), citing Phi et al. (1970)]; other analyses on seven Dong Dau bronze artifacts revealed a tin content ranging from 11.5% to 19.0%, with a spearhead and arrowhead having the latter unusually high proportions of tin [Selimkhanov (1984, p. 275), citing Ty (1981, p. 35)]. Clay and sandstone bivalve molds for the casting of axes, chisels, and arrowheads have also been reported from sites of this period.

Go Mun Period (Late Bronze Age, c. 1000–500 B.C.)

Sites of the Go Mun Period are concentrated south of Vinh Phu along the Red and Black rivers and are also found further south in the Ma River delta in Thanh Hoa Province. For the first time bronze implements actually outnumber those of stone. For example, at the site of Vinh Quang (Ha Son Binh Province), radiocarbon dated to the eleventh century B.C., 62% of the tools are bronze (Tan 1980, p. 128). These include a variety of tall socketed spearheads with a pronounced central rib, virtually identical with the common forms found in subsequent Dongson levels. The socketed axes and adzes also take a variety of shapes, some bearing simple thread relief decoration near the socket (Long 1975, figs. 117–130). Finely cast arrowheads with barbs and tangs, fishhooks, bracelets, small cast figurines of domesticated animals, and sickles with a medial rib and transverse socket have also been found.

Dongson Period (Terminal Bronze Age/Early Iron Age, c. 600–258 B.C.)

Many Dongson sites yielding a rich diversity of bronze goods have been excavated in northern Vietnam since the 1920s. Although this bronze culture was long held to have developed only under Chinese influence, Vietnamese archaeologists now strongly believe that Dongson metallurgy was the culmination of an unbroken line of development in the Red and Ma river deltas of northern Vietnam from the Phung Nguyen through the Go Mun periods (Tan 1980). Numerous radiocarbon dates securely place the Dongson Period in the mid to late first millennium B.C. The rich diversity of bronze goods from major sites such as Viet Khe and Lang Ca include such tools as knives, chisels, plowshares, and axes, vessels, and musical instruments, including the highly decorated bronze drums and bells. Weapons such as daggers, spearheads, and swords are much more common at Dongson sites than at Bronze Age sites in central or southern Vietnam. It is quite possible that this is related to increasing contact with the contemporaneous Dian culture of Yunnan (which shares many cultural traits with Dongson) and other northern groups who are well documented as having engaged in warfare for the capture of booty, slaves, and sacrificial victims. By the late first millennium B.C., cast iron was widely used at Dongson sites (Huyen 1984, p. 176); these include a large number of iron tools and weapons as well as a few bimetallic weapons.

Laboratory analyses are providing further information about Dongson metallurgy. Seven Dongson bronzes selected from the collection of the Peabody Museum of Archaeology at Harvard University (figures 17.2–17.8) were analyzed by Hedy Yuen and me in 1984–1985. All of these pieces were excavated at Thanh Hoa or purchased from villagers living near the site by Olov Janse. A wedge-shaped sample of bronze was removed from each artifact using a rotary Dremel Moto-tool; the wedge was mounted in epoxy, then ground, polished, and etched according to standard metallographic practice. Further samples from each artifact were analyzed by J. F. Merkel on a Perkin-Elmer 3030 atomic absorption spectrometer, following the guidelines suggested by Hughes et al. (1976) and Matterson et al. (1981). The results are given in table 17.1.

As will be detailed in a separate report on these Dongson bronzes, the etching of these pieces revealed a number of interesting features. When etched with potassium dichromate, the spear samples clearly showed dendritic structures, indicating that they were cast; a subsequent etch of ferric nitrate + hydrochloric acid + water revealed annealing twins and heavy strain lines (figures 17.9 and 17.10), indicating that the blades were cold-worked, probably to increase their hardness. Dagger 7692, a beautiful specimen 20.3 cm tall with an upturned curl flanking the blade at each side of the top of its hollow handle, was cast in one piece, the blade being hammered and annealed (figure 17.11) to increase its hardness. Strain lines from cold-working are also seen (figure 17.12) in samples taken from one of the other two dagger blade fragments. The slightly later Han bowl from Man-thon, measuring 9.7 cm in diameter and 3.8 cm tall, was expected to show that its very thin walls (1.35–1.70 mm) were produced by hammering; quite to the contrary, the potassium dichromate etch revealed that the vessel was cast, the beautiful cored dendritic structure showing no evidence of post-casting treatment (figure 17.13).

A most interesting discovery of clay bivalve molds from the third century B.C. Dongson necropolis at Lang Ca provides clear evidence of how small bronzes such as these might have been cast. Among the 314 graves excavated at this site in 1976 and 1977, 1 grave has been interpreted as that of a metalworker because it contained pottery molds for casting a square-heeled ax, a spearhead, the handle of a short sword, and a bell, as well as a pottery crucible for melting and pouring bronze (figures 17.14–17.18).

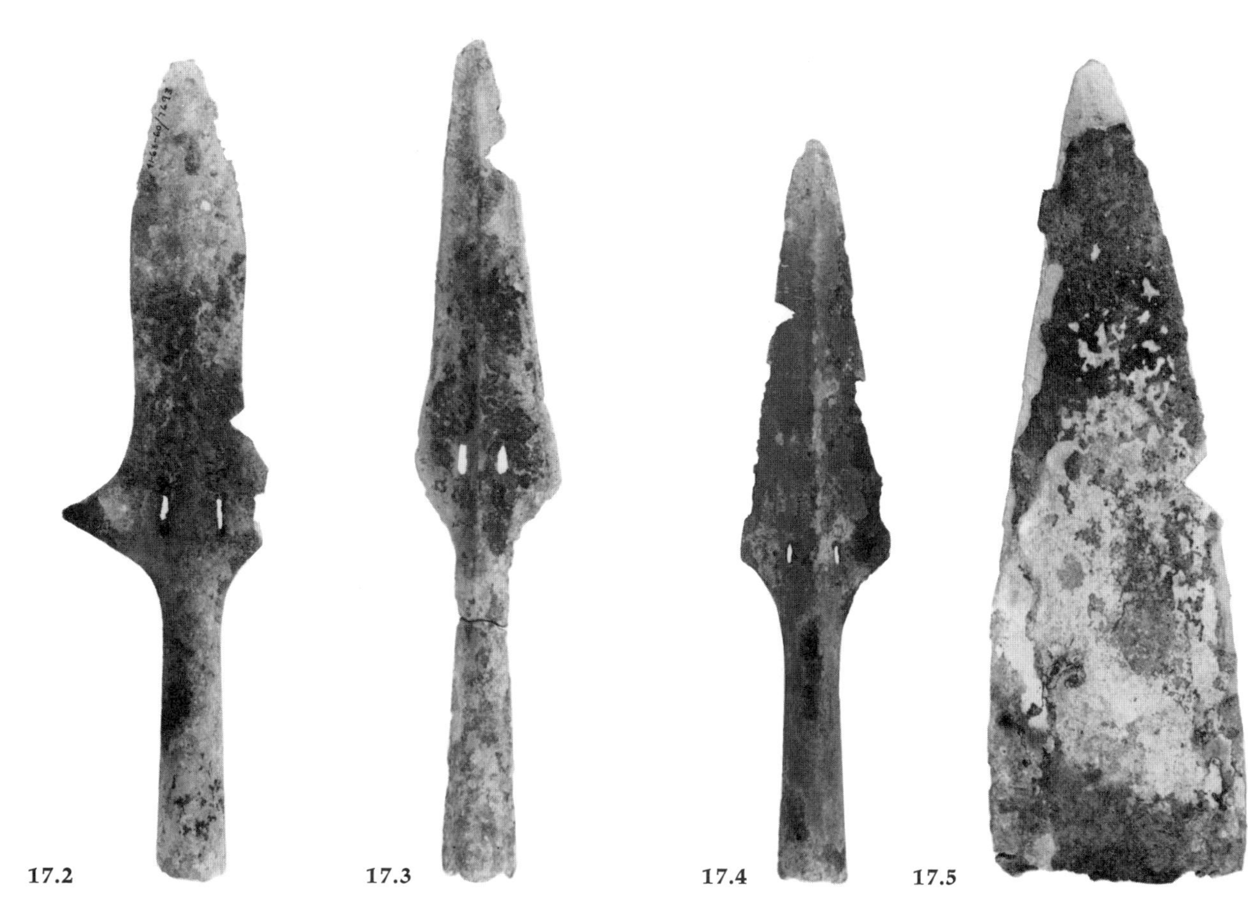

Figure 17.2 Bronze socketed spearhead (Peabody Museum 41-63-60/7693) excavated by Janse at Dongson. Max. length: 26.7 cm. Sample removed from blade. Photograph courtesy of the Peabody Museum of Archaeology and Ethnology.

Figure 17.3 Bronze socketed spearhead (Peabody Museum 41-63-60/7712) purchased by Janse at Dongson. Max. length: 9.6 cm. Sample removed from blade. Photograph courtesy of the Peabody Museum of Archaeology and Ethnology.

Figure 17.4 Bronze socketed spearhead (Peabody Museum 41-63-60/7711) purchased by Janse at Dongson. Max. length: 21.8 cm. Sample removed from blade. Photograph courtesy of the Peabody Museum of Archaeology and Ethnology.

Figure 17.5 Broken bronze dagger blade (Peabody Museum 41-63-60/7715) purchased by Janse at Dongson. Max. length: 11.9 cm. Sample removed from blade. Photograph courtesy of the Peabody Museum of Archaeology and Ethnology.

17.6 17.7

Figure 17.6 Broken bronze dagger blade (Peabody Museum 41-63-60/7714) purchased by Janse at Dongson. Max. length: 20.8 cm. Sample removed from blade. Photograph courtesy of the Peabody Museum of Archaeology and Ethnology.

Figure 17.7 Bronze dagger (Peabody Museum 41-63-60/7692) excavated by Janse at Dongson. Max. length: 20.3 cm. Sample removed from blade. Photograph courtesy of the Peabody Museum of Archaeology and Ethnology.

Figure 17.8 Bottom view of bronze bowl (Peabody Museum 41-63-60/7885) excavated from Han tomb 1B at Man-thon. Diameter at rim: 9.7 cm. Height: 3.8 cm. Thickness: 1.35–1.70 mm. Sample removed from rim. Photograph courtesy of the Peabody Museum of Archaeology and Ethnology.

Table 17.1 Atomic absorption analysis of selected Dongson bronzes[a]

Specimen[b]	Cu	Sn	Pb	As	Zn	Fe	Ni	Ag	Total
7692 Dagger	78.31	5.80	12.27	0.16	0.002	0.027	0.034	0.010	96.61
7693 Spearhead	77.53	8.20	12.85	0.070	0.007	0.032	0.035	n.d.	98.72
7711 Spearhead	74.17	3.16	21.31	0.065	0.007	0.76	0.055	0.050	99.58
7712 Spearhead	85.96	2.79	3.85	0.35	n.d.	0.019	0.080	0.12	93.17
7714 Dagger blade	81.12	11.10	4.44	0.34	0.005	0.59	0.054	n.d.	97.65
7715 Dagger blade	74.22	8.58	13.78	0.19	0.012	0.14	0.028	n.d.	96.95
7885 Bowl	79.61	4.45	12.91	0.67	0.023	0.023	0.081	0.10	97.87

a. All values are given in percent (n.d. = not detected). Supporting SEM studies were performed by Maurice Dumais of the McKay Laboratory, Harvard University.

b. The Peabody Museum accession numbers for these pieces carry the prefix 41-63-60/(specimen number). The artifacts are from the following locations: 7692 and 7693, excavated at Thanh-hoa Dongson; 7711, 7712, 7714, and 7715, purchased from villagers at Dongson; 7885, excavated from the Han tomb 1B at Thanh-hoa Man-thon.

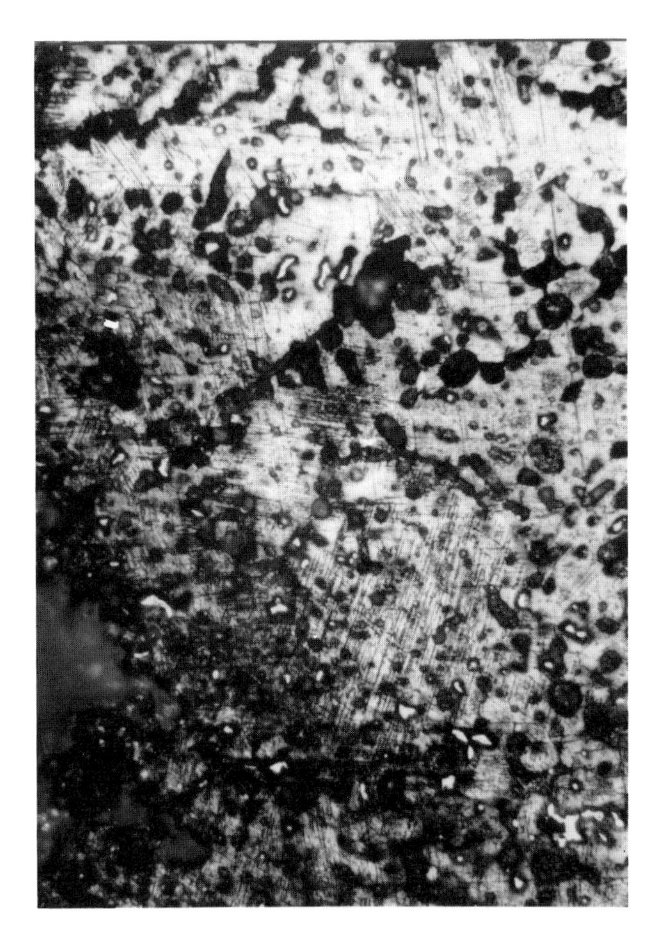

Figure 17.9 Polished section from blade of Dongson spearhead 7693. Ferric nitrate + hydrochloric acid + water etch. × 256. Photograph courtesy of the Peabody Museum of Archaeology and Ethnology.

Figure 17.10 Polished section from blade of Dongson spearhead 7711. Ferric nitrate + hydrochloric acid + water etch. × 256. Photograph courtesy of the Peabody Museum of Archaeology and Ethnology.

A careful examination of the molds has revealed details of their preparation (Thu and Ty 1978). A suitably sized piece of baked clay was created to form the foundation of the mold. The casting face of the mold was then covered with a layer of soft fine clay, wrapping around the edges of the baked clay to ensure proper adhesion. A bronze ax (or model) was then pressed into the clay and air dried, after which more soft clay was added to create the opposite face of the bivalve mold. Finally, another block of baked clay was added to this assembly to provide the support for the second half of the mold. After being air dried, the ax could be lifted out and the mold baked. A pouring channel was then cut into the mold and the halves carefully reassembled and bound; the mold was then ready to receive molten bronze. The mold for casting the handle of a short sword is slightly different, containing two distinct layers of clay, an "inner lining" and an "outer shell," which according to Thu and Ty (1978, p. 37) enabled the mold to withstand high temperatures and quickly vent escaping gases, although it is not clear why some molds lack the two-layer lining.

The spearhead mold clearly shows that the blade was cast in something close to its final shape; the flared "wings" on the blade were not created by hammering, although they were probably cold-worked to increase their strength. Unfortunately this mold is incomplete and does not give any indication of the method used to create the two slots so commonly found at the base of the blade of Dongson spearheads.

The Lang Ca find also revealed another interesting feature: The short sword mold represents only the handle. If the blade was also of bronze, the weapon would probably have been cast in one pouring using two separate bivalve molds (one-piece cast bronze short swords are commonly found in southwest China, although I am not familiar with any mold finds for these). This arrangement would necessitate a careful fitting of the handle mold to the blade mold to ensure proper alignment and to minimize leakage of the molten bronze between the mold parts. It is also possible that the Lang Ca mold was used to cast a bimetallic weapon with a bronze handle and an iron blade.

Like the molds, the fragmentary crucible is made of baked paddy clay. It originally took the form of an oblong sphere 16 cm in height, the rim pinched slightly to create a pouring spout. The walls are 3 cm thick, and the inner surface is still coated with a blue-

Figure 17.11 Polished section from blade of Dongson dagger 7692. Ferric nitrate + hydrochloric acid + water etch. × 256. Photograph courtesy of the Peabody Museum of Archaeology and Ethnology.

Figure 17.12 Polished section from blade of Dongson dagger 7715. Ferric nitrate + hydrochloric acid + water etch. × 256. Photograph courtesy of the Peabody Museum of Archaeology and Ethnology.

green layer of corroded bronze and other sediment. It has been estimated that a crucible of this size could contain 12 kg of molten bronze.

Central Coastal Vietnam: Sa Huynh Culture

The developmental sequence of metallurgy in the central coast is not as clear as that of the north, but tantalizing evidence for long-distance trade with the other regions of Vietnam and with the islands of Southeast Asia makes the archaeological research in this area important to an understanding of the relationships between southern and northern Vietnamese metallurgy. The dominant culture in this central region has traditionally been referred to as Sa Huynh, and its development can be divided into three periods (Chinh and Tien 1980, p. 59). The lower levels contain an abundance of stone tools, including a variety of axes, hoes, and chisels, and a peculiar type of four-point stone earring. The artifacts are occasionally found at distant sites. No evidence of metallurgy has yet been found in this early period of the Sa Huynh culture.

The middle period of Sa Huynh is named after the recently excavated coastal site of Binh Chau, although some scholars feel that Binh Chau represents a separate culture, as many ceramic features distinguish it from purely Sa Huynh sites (Hong 1980, p. 74). Binh Chau has yielded a number of broken pottery crucibles with inward-turning rims as well as fragments of pottery casting molds. Finely cast bronze implements include socketed axes and winged arrowheads. No radiocarbon dates are yet available from this site, and heavy disturbance of the stratigraphy from construction and combat during the Vietnam War has cast doubt on some artifact associations. The excavators, however, date Binh Chau to the late second or early first millennium B.C.; they believe that the high level of bronze casting technology at this site is fully comparable to that from Go Mun and Doc Chua (Hong 1980, p. 74).

The final period of Sa Huynh is represented by a variety of sites in the Nghia Binh, Dong Nai, and Da Nang regions. The use of bronze drastically declines, whereas iron axes, sickles, hoes, spades, spears, knives, and swords become common. Most of the limited bronze finds are bracelets (Chinh and Tien 1980, p. 60).

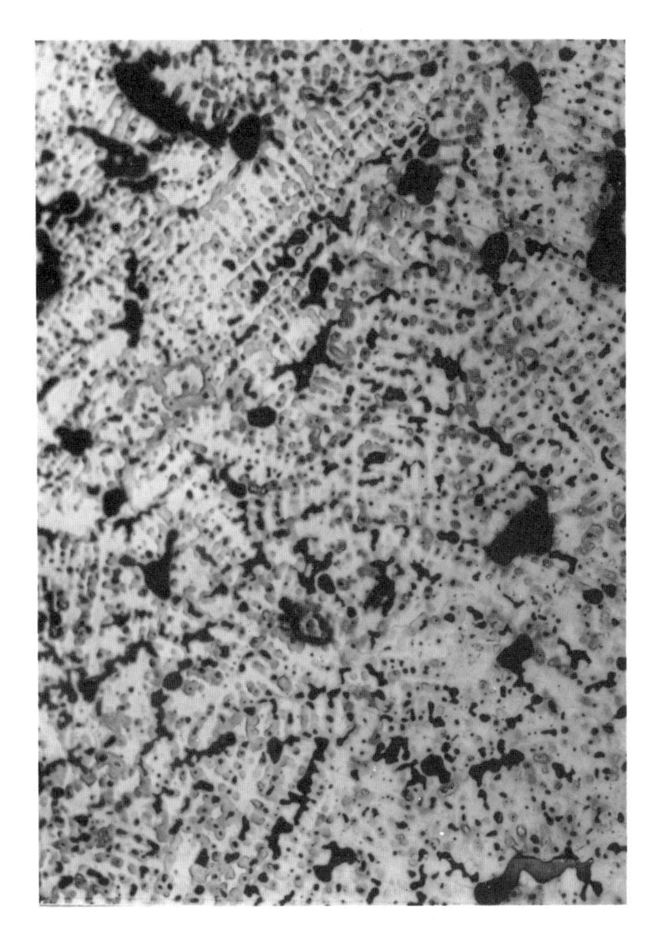

Figure 17.13 Polished section from rim of Han bowl 7885 from Man-thon. Potassium dichromate etch. × 128. Photograph courtesy of the Peabody Museum of Archaeology and Ethnology.

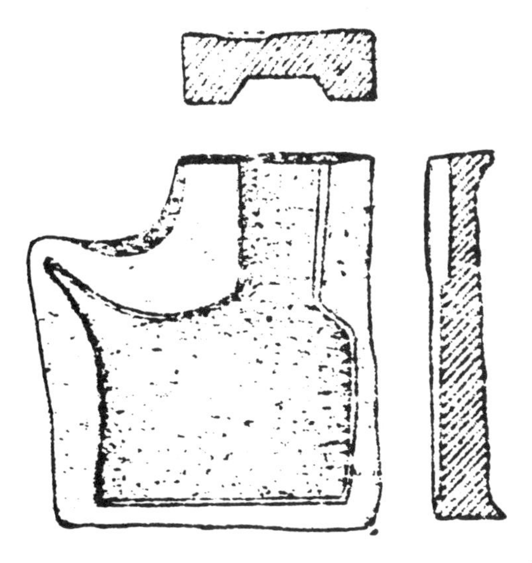

Figure 17.14 Terra-cotta mold for casting a heeled ax, from Lang Ca, northern Vietnam. Scale 1/3 [after Thu and Ty (1978, pp. 36–39, fig. 1)].

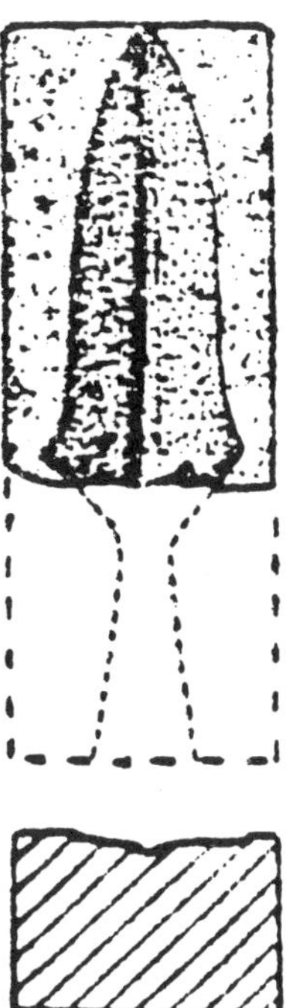

Figure 17.15 Terra-cotta mold for casting a spearhead, from Lang Ca. Scale 1/3 [after Thu and Ty (1978, pp. 36–39, fig. 4)].

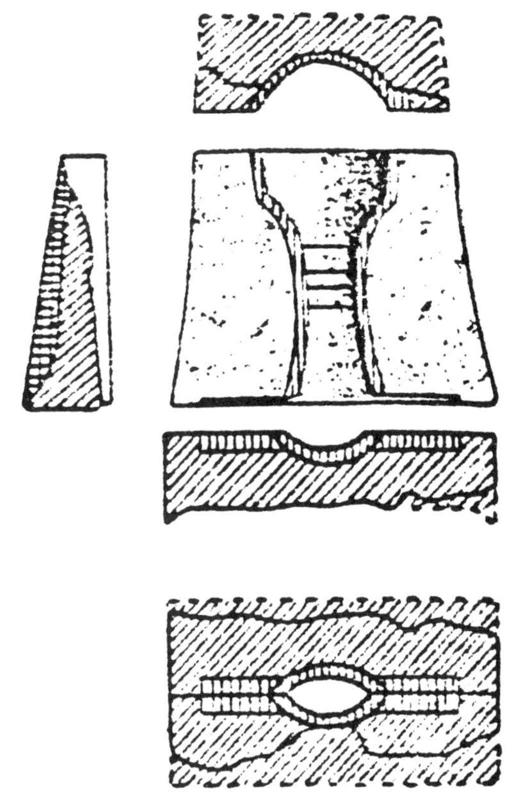

Figure 17.16 Terra-cotta mold for casting a dagger handle, from Lang Ca. Scale 1/3 [after Thu and Ty (1978, pp. 36–39, fig. 2)].

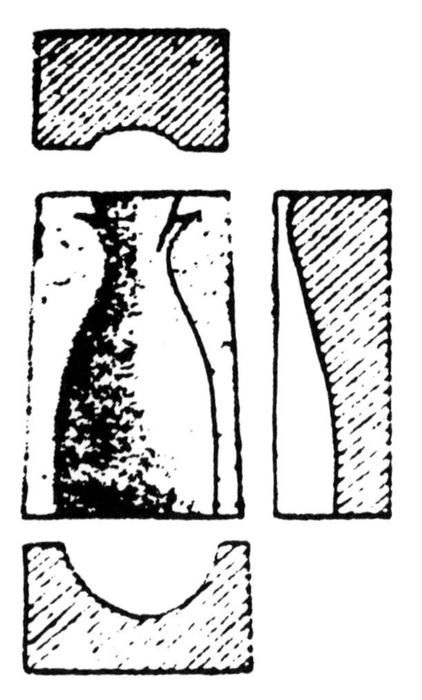

Figure 17.17 Terra-cotta mold for casting a bell, from Lang Ca. Scale 1/3 [after Thu and Ty (1978, pp. 36–39, fig. 3)].

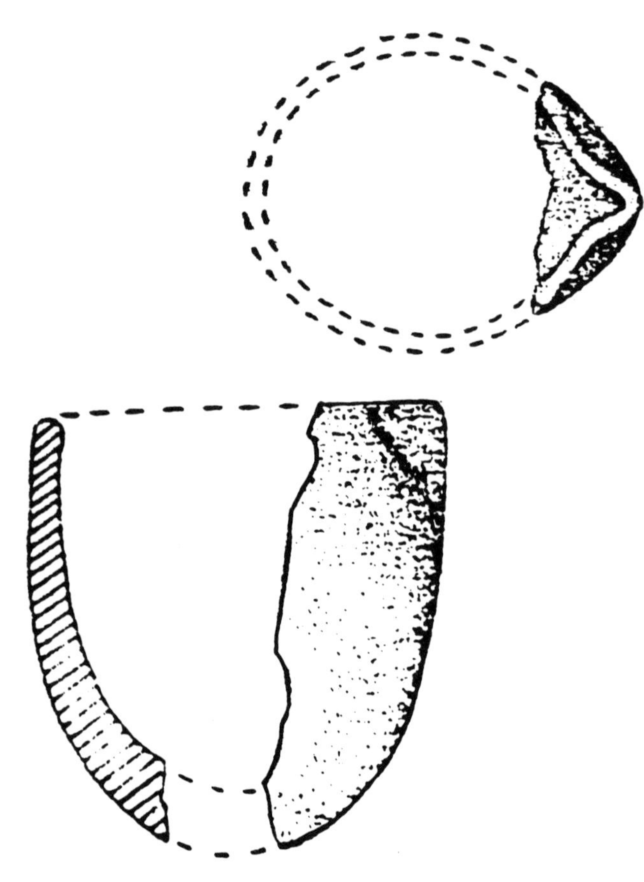

Figure 17.18 Pottery crucible from Lang Ca. Scale 1/3 [after Thu and Ty (1978, pp. 36–39, fig. 5)].

Southern Vietnam: The Dong Nai Basin

A number of stray mold finds have been made in this region during the past century, but only recently have enough major sites been excavated that the extent of bronze casting in this area has been realized and a tentative reconstruction made (Tan 1980, p. 135, citing Son 1978). The earliest two phases, placed in the Late Neolithic, are named after type sites at Cau Sat and Ben Do; although stone tools are both abundant and well made, no traces of metallurgy have yet been found. In the third phase, represented by the site of Cu Lao Rua and the lower levels of Doc Chua, the stone tools and pottery of the earlier phases are intermixed with only a few finds of bronze axes and casting molds for axes with flared blades.

The final phase in this area is represented by the main site of Doc Chua, located near the Dong Nai River some 36 km northwest of Ho Chi Minh City (formerly Saigon). Excavated in 1976 and 1977, this site has yielded a stunning assortment of metallurgical evidence in addition to stone axes, adzes, chisels, hoes, and bracelets (Con 1977; Ty 1977). A single radiocarbon date of 3145 ± 130 B.P. (Klein midpoint: 1420 B.C.) places this material in the mid to late second millennium B.C. (Tan 1980, p. 136; Bayard 1984, p. 310); a radiocarbon date of 3950 ± 250 B.P. (Klein midpoint: 2448 B.C.) from the nearby site of Nui Gom (Hang Gon), which has also yielded a few sandstone ax molds (Saurin 1963), seems too early for bronze associations in this region and must await further dates for confirmation.

Bronze finds from Doc Chua include numerous socketed spearheads and axes (figures 17.19 and 17.20) that lack decoration except for occasional axes bearing single or paired raised lines near the socket. Small bells and halberds (Viet *gua*; Chinese *ge*) and a small bronze two-headed animal figurine of uncertain use (Long 1977) have also been found.

Particularly striking are the more than fifty sandstone casting molds found at the site (figures 17.21–17.23). Because the molds are in various stages of completion, the mold-making process has been tentatively reconstructed (Diem 1977). The most rudimentary molds are roughly shaped trapezoidal blocks of sandstone with rectangular cross sections; others have one face ground flat. The shaping of the blocks was accomplished by cutting, chiseling, and polishing. After this preliminary preparation, the shape of the object to be cast was somehow imprinted onto the flat face of the mold. Several of the Doc Chua molds retain faint traces of yellow clay outlines, whereas others suggest that a model of the object to be cast was held against the mold and exposed to a smoky fire, staining the outline of the model onto the mold surface. The shape was then chiseled out, a delicate process subject to errors that forced the Doc Chua mold makers to discard a

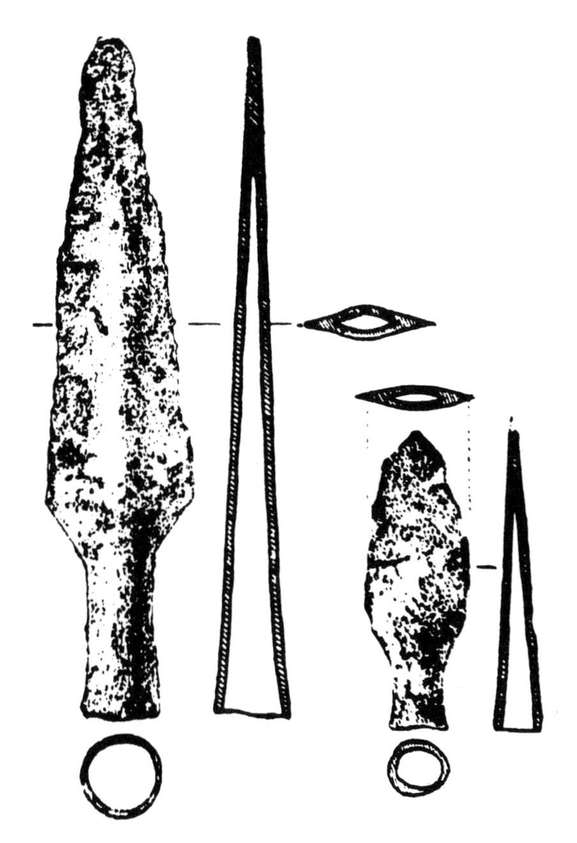

Figure 17.19 Bronze socketed spearheads from Doc Chua. Scale 1/2 [after Con (1977, pp. 29–32, fig. 4)].

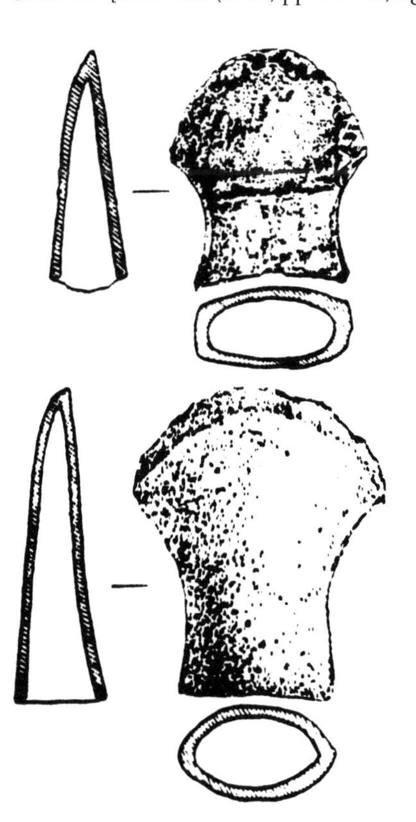

Figure 17.20 Bronze socketed axes from Doc Chua. Scale 1/2 [after Con (1977, pp. 29–32, fig. 3)].

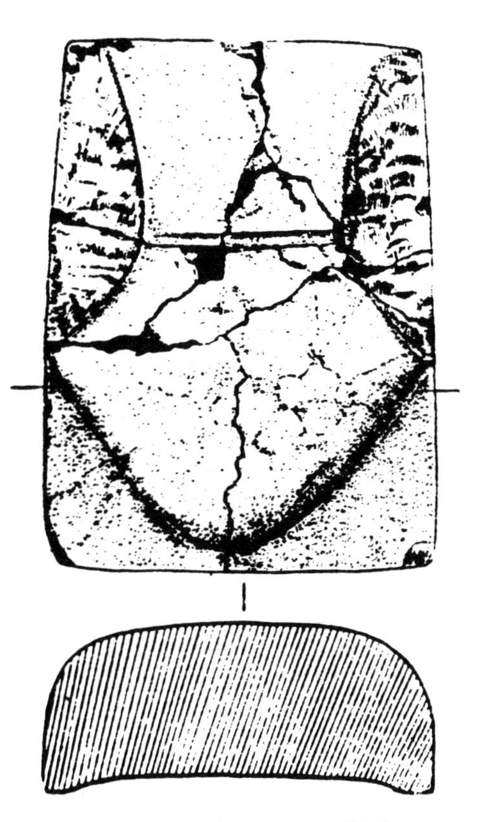

Figure 17.21 Sandstone ax mold from Doc Chua. Scale 1/2 [after Ty (1977, pp. 33–40, fig. 5)].

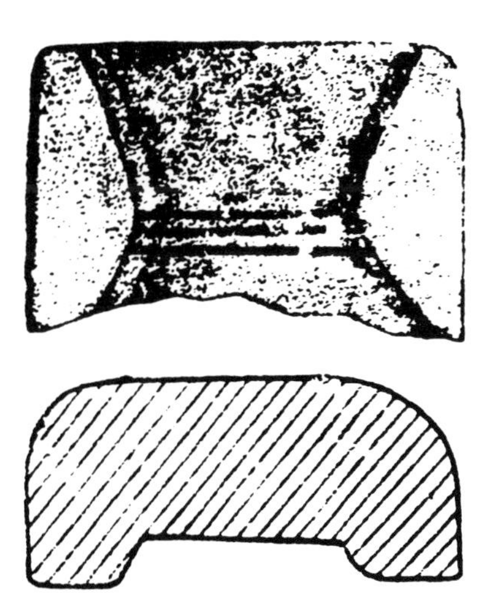

Figure 17.22 Sandstone ax mold from Doc Chua. Scale 1/2 [after Diem (1977, pp. 44–48, fig. 6)].

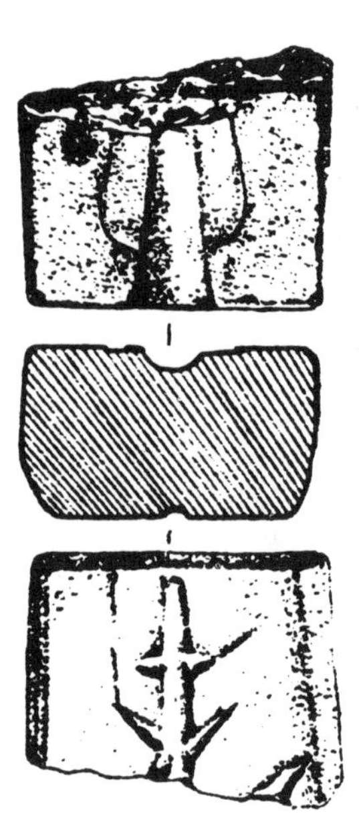

Figure 17.23 Sandstone mold for casting spear and barbed harpoon, from Doc Chua. Scale 1/2 [after Ty (1977, pp. 33–40, fig. 6)].

number of partially finished molds. Finally, the inner surfaces of the mold were carefully polished; a few of the molds received simple decoration at this stage, usually consisting of one or two raised ribs running around the body of the ax between the socket and the blade. Many of the molds show a variety of incised lines on their outside edges, undoubtedly meant to facilitate the careful alignment of the mold halves before pouring the molten bronze.

It has recently been claimed by some archaeologists [for example, Higham (1984, p. 140)] that sandstone molds could not withstand the heat of molten bronze and were used to produce a cast model of the object in lead, which melts at a much lower temperature. These lead models were then used to create pottery molds that, when baked, were durable enough to contain molten bronze without cracking. Although the existence of sprue-shaped objects made of a lead-tin alloy have been reported (Higham, personal communication), I find this argument unconvincing; pottery molds were certainly used for bronze casting in Southeast Asia and to an even greater extent in Shang and Zhou China, but sandstone is readily available thoughout Southeast Asia and, in addition to those sites mentioned, was the material of choice at a variety of other sites in Thailand (White 1982; Solheim 1967), Laos (Saurin 1951), southern Vietnam (Saurin 1963), and even at the late Shang site of Wucheng in Jiangxi Province, China (Jiangxi Provincial Museum, 1975, 1978; Jiangxi Museum of History, 1979). Carefully planned casting experiments using sandstone molds should clarify the matter considerably.

Kampuchea

Based on the lack of recent reports, it seems that archaeological investigations in Kampuchea have come to a virtual halt during the past decade of civil strife there. However, a number of bronze finds made earlier in this century at several Kampuchean sites suggest that bronze was being locally melted and cast there. Although these finds have been of limited value because they were either poorly excavated or simply purchased from local inhabitants, the case for Kampuchean bronze casting in the second millennium B.C. has been strengthened by the variety of bronze tools and weapons and by the abundance of casting molds in the Dong Nai basin. Because both regions are poor in the ores needed to produce bronze, it is likely that they were tied into a metal trade network along the Mekong River.

The site of Samrong Sen, one of the first prehistoric sites investigated in Southeast Asia, has produced a number of bronze artifacts since the nineteenth century, although none can be shown to have actually been excavated from the site (Worman 1949; Noulet 1877; Moura 1883; Vitout 1914; Finot 1928; Chantre 1891). Shells from the Samrong Sen mound have recently provided a radiocarbon date of 3230 ± 120 B.P. (1400–1160 B.C.) (Carbonnel 1979, p. 223), although this is of limited value until the stratigraphy of the site is better understood and controlled excavation provide datable material clearly associated with metallurgical remains.

The vast mound at the site was first excavated archaeologically by Henri Mansuy in 1901 and 1902; no metal artifacts were excavated, but Mansuy did purchase several bronzes said to come from the site by the modern inhabitants of the mound (Mansuy 1902). Mansuy's excavations continued in 1923, during which he acquired two bronze bells, a chisel, a fishhook, and a simple ring from the local villagers (Mansuy 1923). He did excavate one-half of a sandstone bivalve ax mold (figure 17.24) quite similar to a mold found at the site some forty years earlier (Pavie 1904) and similar to the numerous stone molds from Doc Chua.

The most recent collection of bronze artifacts from Samrong Sen was made during a site visit by Olov Janse, who was conducting excavations of Dongson and Han tombs in northern Vietnam from 1938 to 1940. These included bronze bracelets, socketed spearheads and axes, and a bell acquired from villagers at the site, as well as a pottery crucible (Janse 1936, 1941, 1951). As this expedition was partially sponsored by the Harvard-Yenching Institute, a sample of Janse's finds was given to Harvard at the close of his work. Like the Dongson bronzes mentioned earlier, I selected several specimens from Samrong Sen for metallurgical analysis (figures 17.25–17.29). The atomic absorption results are summarized in table 17.2; a more complete

Figure 17.24 Sandstone bivalve ax mold excavated by Mansuy at Samrong Sen. Approximate length 11 cm [after Mansuy (1923, pl. IX: 1)].

Figure 17.25 Bronze spearhead (Peabody Museum 41-63-60/8055) attributed to Samrong Sen. Length 11.8 cm, weight 50.8 g. Sample was removed from right edge of the blade. Photograph courtesy of the Peabody Museum of Archaeology and Ethnology.

report of these analyses is made elsewhere (Murowchick, n.d.).

Internal corrosion in all the specimens supports their alleged antiquity. An ammonium peroxide etch of the specimens clearly reveals that all the pieces were cast and that hammering and annealing were also practiced here by ancient metalsmiths. The socketed spearhead is of a shape commonly found throughout Southeast Asia, although the blade shows evidence of having been cleaned, recut, and sharpened in modern times, perhaps in an effort to increase its appeal to modern collectors. Although annealing twins were obvious, an optical examination of the specimen removed from the blade failed to show the expected strain markings that would have resulted from a final cold-working of the blade, features clearly noted on the blades of Dongson spearheads. This is probably because the current edge from which the sample was removed is not the original edge of the blade.

The two axes from Samrong Sen are different. Ax 8059 is a very heavy (93.3 g) socketed ax with a spindle-shaped body beveled along the top edge, decorated near the socket with two parallel ribs similar to those found on several axes and ax molds from Doc Chua. Its curved blade was apparently cast in something close to its final shape, as the mold marks from the bivalve mold run all the way to the ends of the blade. In contrast to this, the thin walls make the other ax (8060) relatively light (50.8 g), bearing in mind the fact that half of one face is missing. This ax is trapezoidal in section with two flat edges and two faces, one face being slightly larger than the other. Although lacking any decoration, each face includes a small triangular hole, a feature commonly found on Southeast Asian and southern Chinese axes and hoes that probably facilitated hafting the ax to a wooden handle. No mold marks are evident, but beautiful dendrites indicate this ax was cast, probably in a bivalve mold of clay or sandstone (figure 17.30).

Figure 17.26 Bronze socketed ax (Peabody Museum 41-63-60/8059) attributed to Samrong Sen. Maximum length 7.5 cm, weight 93.3 g. Sample removed from rim of socket. Photograph courtesy of the Peabody Museum of Archaeology and Ethnology.

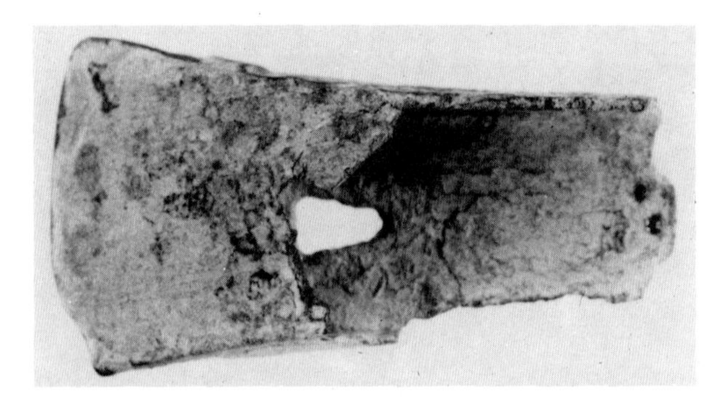

Figure 17.27 Bronze socketed ax (Peabody Museum 41-63-60/8060) attributed to Samrong Sen. Length 7.7 cm, weight 50.8 g. Sample was removed from the broken edge next to triangular hole. Photograph courtesy of the Peabody Museum of Archaeology and Ethnology.

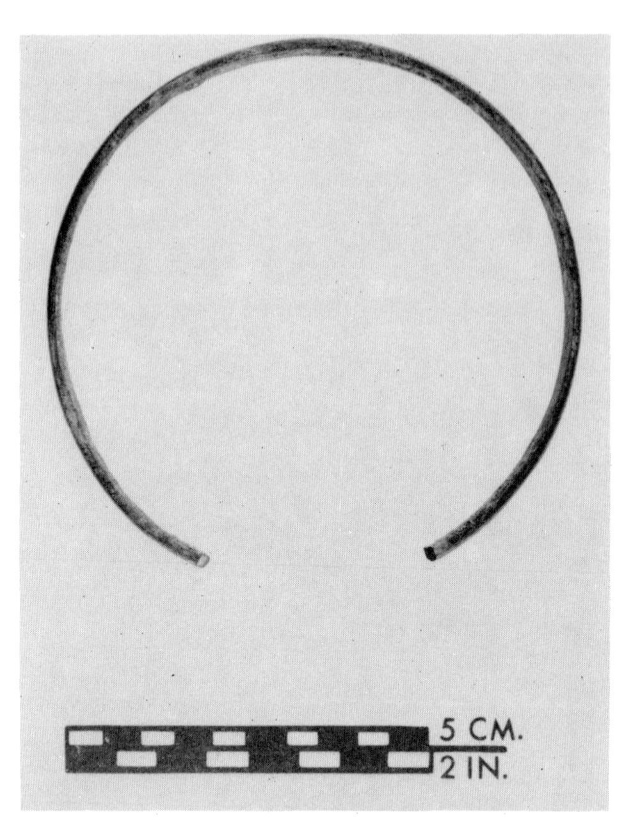

Figure 17.28 Bronze bracelet (Peabody Museum 41-63-60/8063a) attributed to Samrong Sen. Outer diameter 7.5 cm. Sample was removed from broken end of bracelet.

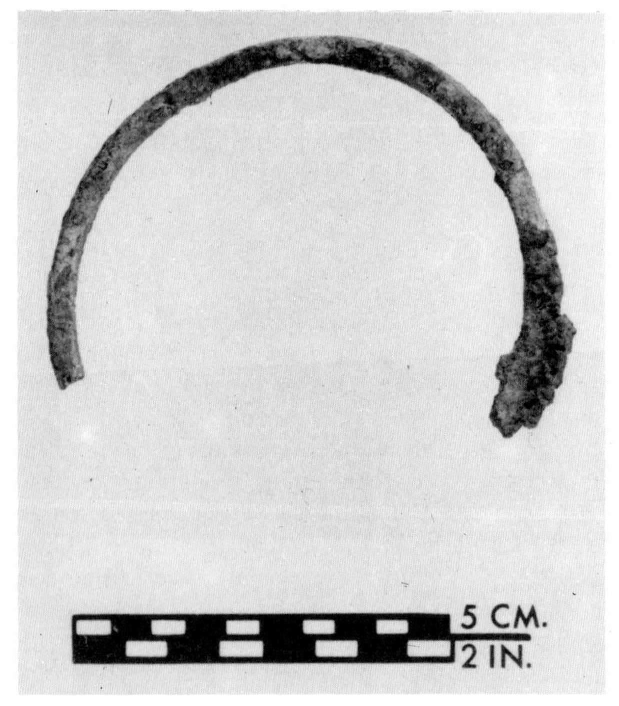

Figure 17.29 Bronze bracelet (Peabody Museum 41-63-60/8063b) attributed to Samrong Sen. Outer diameter 7.0 cm. Sample was removed from broken end of bracelet. Photograph courtesy of the Peabody Museum of Archaeology and Ethnology.

Table 17.2 Atomic absorption analysis of five bronzes attributed to Samrong Sen, Kampuchea[a]

Specimen[b]	Cu	Sn	Pb	As	Zn	Fe	Ni	Ag	Total
8055 Spearhead	87.84	9.88	0.95	0.83	0.096	0.028	0.005	0.10	99.73
8059 Ax	76.95	6.95	13.22	n.d.	0.012	0.098	0.006	0.05	97.28
8060 Ax	75.55	6.70	11.74	0.19	0.013	0.009	0.029	n.d.	94.32
8063a Bracelet	79.09	6.75	6.82	0.02	0.004	0.004	0.008	0.15	92.85
8063b Bracelet	71.83	0.97	26.47[c]	n.d.	0.002	n.d.	0.008	0.05	99.33

a. All values are given in percent (n.d. = not detected). SEM studies were performed by Maurice Dumais of the McKay Laboratory, Harvard University.

b. All Peabody Museum accession numbers for these pieces begin with the prefix 41-63-60/(specimen number).

c. The unusually high lead value for this artifact is high even for Southeast Asian bronzes and might be due to sampling from a lead pool.

The pottery crucible (figure 17.31) acquired by Janse, who wrongly identified it as a pottery spoon (Janse 1951, p. 78), is similar to those found at the early Bronze Age sites of Non Nok Tha and Ban Chiang in northeast Thailand (Bayard 1980; White 1982). Details of the crucible, including the existence of casting residue on its inner surface, are not noted in any of the descriptions of the piece.

Molds and bronze artifacts excavated by Lévy in a series of sites near Mlu Prei, some 70 km northeast of Samrong Sen, provide further evidence for local bronze casting. Lévy's finds include fragments of a sandstone mold showing the socketed end of an ax as well as a complete sandstone mold for the casting of a bronze sickle, an actual example of which was also found and analyzed (89.2% Cu, 10.06% Sn, and traces of Pb) (Levy 1943, p. 38). Other finds include a rounded sandstone block believed by Lévy to be an unfinished mold (figure 17.32) and fragments of a pottery crucible with bronze dross adhering to the interior pits. Bronze bracelets and axes (figure 17.33), stone and pottery bracelets, stone and shell beads, and abundant pottery all show distinct similarities to the material from Samrong Sen. Iron objects attest to a later occupation of the site, the terminal date of which Lévy estimated to be the sixth to seventh century A.D. Although we must be cautious in accepting Lévy's dates and interpretations because many of the objects he found were eroded out of the archaeological beds, his bronze finds are certainly not unusual when compared with those of Samrong Sen and the Dong Nai basin.

Geological surveys of Vietnam and Kampuchea have shown that the distribution of ores needed for bronze metallurgy is uneven (Workman 1972; United Nations Industrial Development Organization, 1972). Northern Vietnam is rich in copper, tin, and lead ores easily accessible to the ancient metalworkers of the area. It is possible that trade along the Red River made the abundant ore sources of Yunnan available to the northern Vietnamese bronze centers, but this trade was certainly not a necessity for the development of bronze in the north. The situation in the south, however, was quite different: Although tin sources are known to exist at the head of the Dong Nai River, southern Vietnam is notably lacking in copper and lead ores. The same situation is found in Kampuchea: Lead, tin, and zinc indications have been located west of Phnom Penh, but virtually no exploitable copper sources have been found near Samrong Sen or Mlu Prei.

This being the situation, the increasing evidence of local remelting and casting of bronze in southern Vietnam and Kampuchea suggests that a vibrant trade in crude metals (rather than the ores themselves) was taking place along the Mekong River from the north. A number of recent archaeological developments in Thailand have yielded supporting evidence of this ancient metals trade. Recent excavations at the site of Phu Lon in Loei Province, part of the ongoing research of the Thailand Archaeometallurgy Project jointly organized by the University Museum of the University of Pennsylvania and the Thai Department of Fine Arts, has yielded a prehistoric copper mining complex on the banks of the Mekong River (Pigott et al. 1985). The site, dated to the second half of the first millennium B.C., has yielded evidence of the ancient mining of malachite and native copper. The excavators believe that the huge amount of mining debris may indicate that the site was being worked much earlier than the radiocarbon dates indicate. Further work at this site will certainly clarify the situation. It is certainly possible that similar ancient mining activity was taking place on the Laotian side of the Mekong, where tin, lead, and copper sources are also found, although the paucity of archaeological work here makes such a claim difficult to support at this time.

Further evidence of a prehistoric trade in metals comes in the form of actual ingots recently looted from two Ban Chiang culture sites located near the site of Ban Chiang itself. As is the case with most looted antiquities, details surrounding the find are sketchy. At least twenty-five circular ingots were found, ranging in size from 4 cm to 5 cm in diameter and 1–2 cm thick and weighing between 60 g and 130 g. They

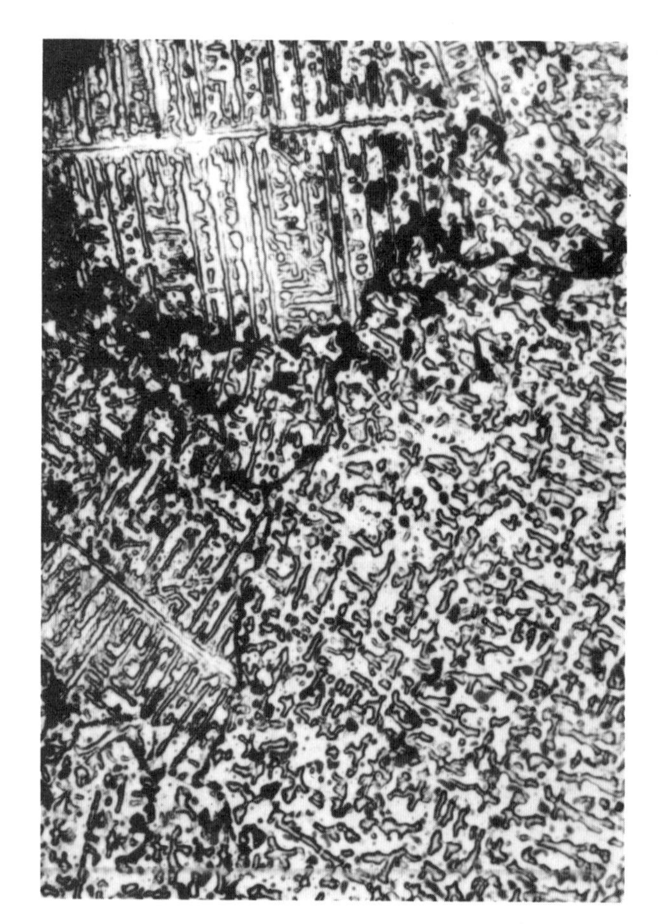

Figure 17.30 Photomicrograph of section of ax 8060, ferric chloride etch. × 128. Photograph courtesy of the Peabody Museum of Archaeology and Ethnology.

were apparently cast in roughly carved depressions. Although they have not yet been analyzed, their green corrosion suggests that they are made of copper or bronze. Three of the ingots are black and may prove to be pure tin. The looters reported that the ingots from the two sites were found grouped together "as if they had been in bags" (Sumitr Pitiphat, personal communication). Indeed, a cursory examination of the pieces shows textile impressions in the corrosion of several of the pieces. It is certainly possible that metal ingots such as these served as a form of currency in ancient Southeast Asia; with the large number of Ban Chiang sites awaiting proper archaeological excavation, perhaps further ingots will be found in situ so that more information about their production and use might be known.

Finally, recent work by the Thai Fine Arts Department in the area of Lopburi, some 140 km north of Bangkok, has yielded a rich assortment of pottery molds, casting debris, and slag. The molds include a large number of small conical shapes in fairly standardized sizes; these might have been used for the casting of standardized ingots, perhaps for trade (Surapol Natapintu, personal communication). The terrain between central Thailand and Kampuchea or southern

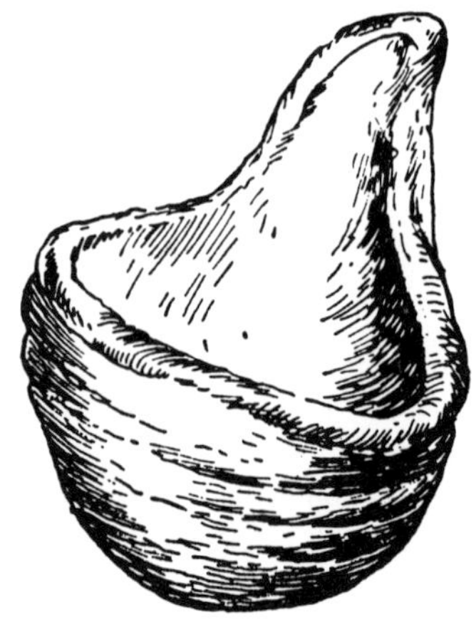

Figure 17.31 Pottery crucible from Samrong Sen [after Janse (1951, p. 78, fig. 52)].

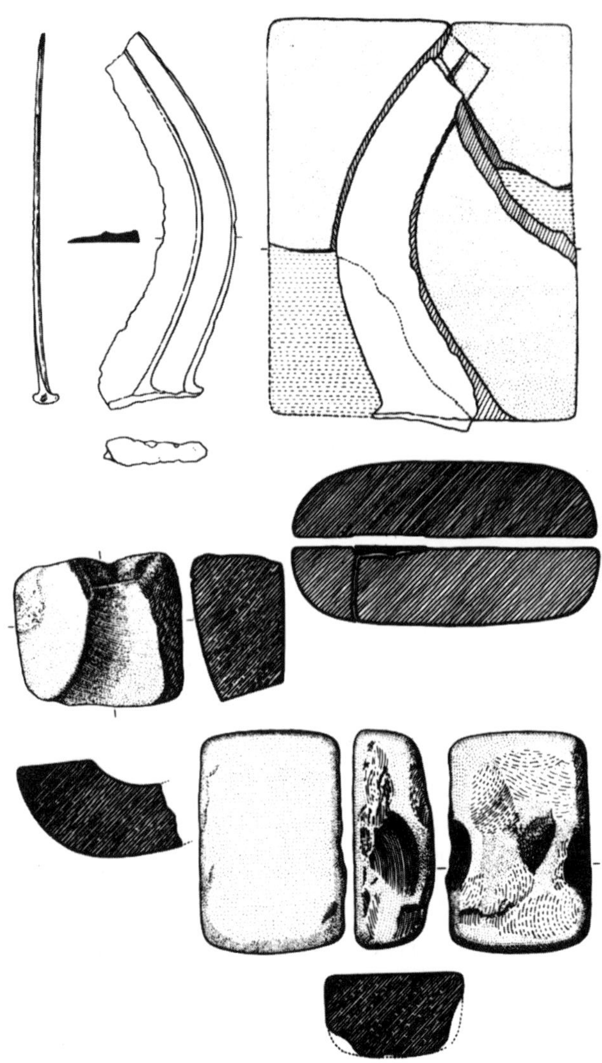

Figure 17.32 Metallurgical remains from Mlu Prei [after Lévy (1943, pl. XXII)]. (Top left) Bronze sickle (length 12 cm). (Top right) Sandstone bivalve sickle mold. (Middle left) Fragmentary sandstone ax mold (width 6 cm). (Lower right) Rounded sandstone block believed to be an unfinished mold (length 8 cm).

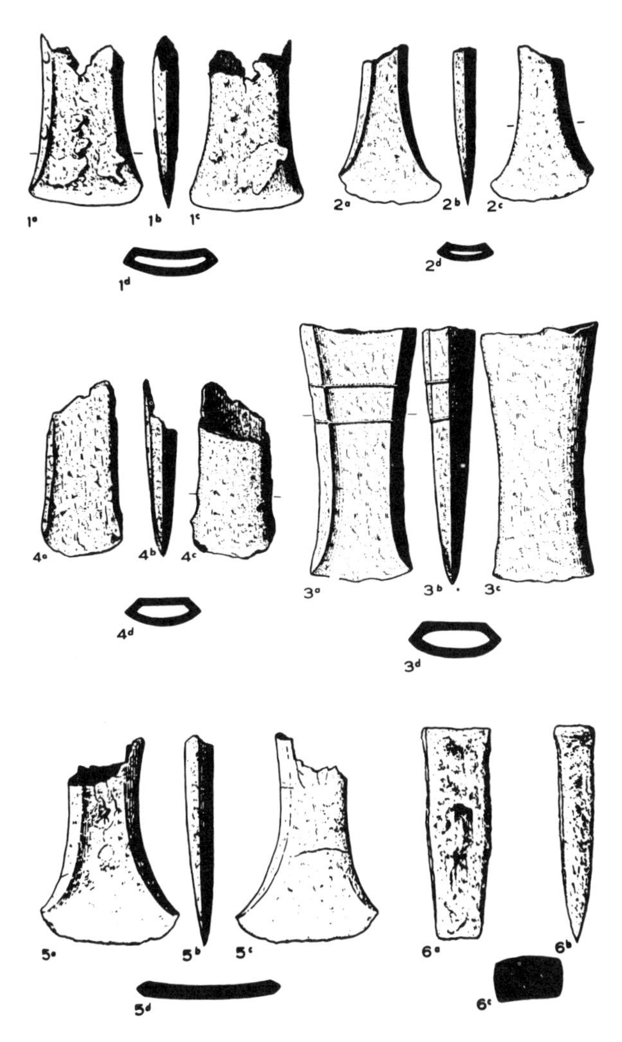

Figure 17.33 Bronze bracelets and axes excavated near Mlu Prei, Kampuchea [after Lévy (1943, pl. 24, 23)].

Vietnam presents no great barriers to such a trade over land. However, it is also believed that the southern coast of Thailand reached much farther inland than the present shoreline; during the second millennium B.C. the Lopburi area might have been only 10 or 20 km from the coast, suggesting that maritime trade with the rest of Southeast Asia was a possible route for the transmission of metals to resource-poor regions. It should be noted that the four-point earrings and animal-headed earrings typical of the Sa Huynh culture have been found at Suphan Buri (near Lopburi), as well as in the Dong Nai basin of southern Vietnam, central Vietnam, and at Dongson sites in the north. The date, nature, and extent of this metals trade with Kampuchea and Vietnam must await further archaeological and geological work in the region, but the similarities and increasing quantities of ancient metallurgical finds (molds, ingots, and the bronzes themselves)—and the absence of exploitable ores in the south—strongly suggest that such a trade must have flourished during the second millennium B.C.

Despite recent advances, major questions remain unanswered. Did the development of metallurgy in Southeast Asia necessarily involve a period of copper tool use (as seen in other parts of the ancient world), or was such a stage unnecessary in this region of abundant tin sources? What was the nature of influence between the Neolithic and Bronze Age cultures of Yunnan, which apparently enjoyed its own indigenous development of metallurgy while engaged in a variety of contacts with contemporaneous cultures of Sichuan and Gansu [see Wang (1980), Li and Xiao (1980), Cao (1981), Zhang (1981), and Zhang et al. 1981)]? Were the metallurgical developments in northern Vietnam influenced at all by the early metallurgy of northeast Thailand? What was the nature of the relationships among the cultures of north, central, and southern Vietnam? And if coastal trade was involved, might it not also have involved the coastal regions of the Gulf of Thailand, giving further access to areas with abundant tin and copper sources?

Vast areas of Southeast Asia remain archaeologically unexplored. Our understanding of the development of bronze metallurgy in Vietnam has greatly increased during the past few decades, but at present archaeological evidence provides little more than a preliminary chronology for this development. Admittedly the present evidence for early bronze metallurgy in Kampuchea is weak if it must stand alone, but it is strengthened by virtue of its similarity and geographic proximity to well-excavated finds from the Dong Nai basin. Lab analyses are helping to reveal valuable details from those finds already made, but thorough surveys and reexcavations of those sites already known are needed to establish firm dates for these finds. We have been tantalized by past discoveries and are greatly encouraged by the energy of current work in Vietnam. As the pace of research steadily increases, the richness of the archaeological material from Vietnam and Kampuchea will undoubtedly make great contributions to our understanding of bronze metallurgy throughout mainland Southeast Asia.

Acknowledgments

This material is based on work supported under a National Science Foundation (NSF) Graduate Fellowship. I would also like to thank K. C. Chang (Department of Anthropology, Harvard) for permission to study the specimens from the Peabody Museum and for valuable encouragement throughout the project; Robert Maddin [Center for Archaeological Research and Development (CARD), Harvard] for technical guidance and support; John Merkel (CARD) for sample preparation and AAS analysis; Sumitr Pitiphat and Pratomarik Ketutat (Thammasat University), Pisit Charoenwongsa and Surapol Natapintu (Archaeology Division, Thai Fine Arts Department); Seehawat Naen-na and Virasit Choo-sangthong (Khon Kaen National Museum) for their hospitality and enlightening discussions in Thailand; Garth Bawden (Peabody Museum) for research funds; and Tue Nguyen (MIT) and Ta Van Tai (Harvard) for Vietnamese translation assistance.

References

Bayard, D. 1980. "An early indigenous bronze technology in northeast Thailand: Its implications for the prehistory of east Asia," in *The Diffusion of Material Culture*, H. H. E. Loofs-Wissowa, ed. (Manoa, Haw.: Social Science Research Institute, University of Hawaii). Asian and Pacific Archaeology Series 9, 191–208.

Bayard, D. 1984. "A checklist of Vietnamese radiocarbon dates." *University of Otago Studies in Prehistoric Anthropology* 16:305–322.

Cao, Xianmin. 1981. "The casting technology of Yunnan bronzes," in *Collected Essays on Yunnan Bronzes* (Beijing: Wenwu Chubanshe), 203–209. (In Chinese.)

Carbonnel, J. P. 1979. "Recent data on the Cambodian Neolithic: The problem of cultural continuity in southern Indochina," in *Early South East Asia*, R. B. Smith and W. Watson, eds. (Oxford: Oxford University Press), 223–226.

Chantre, E. 1891. "Objets antiques en pierre et en bronze du Cambodge." *Bulletin de la Société d'Anthropologie de Lyon* 10:44–46.

Hoang Xuan Chinh and Bui Van Tien. 1980. "The Dongson culture and cultural centers in the Metal Age in Vietnam." *Asian Perspectives* 23(1):55–65.

Dao Linh Con. 1977. "The first season of excavation at Doc Chua (Song Be)." *Khao Co Hoc* 4:29–32. (In Vietnamese.)

Davidson, J. H. C. S. 1979. "Archaeology in northern Viet-Nam since 1954," in *Early South East Asia*, R. B. Smith and W. Watson, eds. (Oxford: Oxford University Press), 98–124.

Le Xuan Diem. 1977. "Ancient molds for casting bronze artifacts from the Dong Nai basin." *Khao Co Hoc* 4:44–48. (In Vietnamese.)

Finot, L. 1928. "Ludovic Jammes, Préhistorien." *Bulletin de l'Ecole Française d'Extrême Orient* 28:473–479.

Higham, C. F. W. 1984. "Prehistoric rice cultivation in Southeast Asia." *Scientific American* (April), 138–146.

Diep Dinh Hoa. 1978. "On the metal artifacts from the beginning of the Bronze Age in Vietnam." *Khao Co Hoc* 2:10–20. (In Vietnamese.)

Ngo Sy Hong. 1980. "Binh Chau (Nghia Binh): A newly discovered Bronze Age site on the central coast of Vietnam." *Khao Co Hoc* 1:68–74. (In Vietnamese.)

Hughes, M. J., M. R. Cowell, and P. T. Craddock. 1976. "Atomic absorption techniques in archaeology." *Archaeometry* 18:19–37.

Pham Minh Huyen. 1984. "Various phases of the development of primitive metallurgy in Viet Nam," in *Southeast Asian Archaeology at the XV Pacific Science Congress on the Origins of Agriculture, Metallurgy, and the State of Mainland Southeast Asia* (Dunedin, New Zealand: University of Otago), 173–182.

Janse, O. R. 1936. "Rapport préliminarie d'une mission archéologique en Indochine, auprès de l'Ecole française d' Extrême Orient." *Revue des Arts Asiatiques* 9(3–4) and 10(1).

Janse, O. R. 1941. "An archaeological expedition to Indo-China and the Philippines." *Harvard Journal of Asiatic Studies* 6:247–268.

Janse, O. R. 1951. *Archaeological Research in Indochina*, vol. 2. Harvard-Yenching Monograph Series 10. Cambridge, Mass.: Harvard University Press.

Jiangxi Provincial Museum. 1975. "Brief report on the excavation of the Shang Dynasty remains at Wucheng, Qingjiang, Jiangxi." *Wenwu* 7:51–71. (In Chinese.)

Jiangxi Provincial Museum. 1978. "Important results from the fourth excavation of the Shang Dynasty remains at Wucheng, Qingjiang, Jiangxi." *Wenwu Ziliao Congkan* 2: 1–13. (In Chinese.)

Jiangxi Museum of History. 1979. "Thirty years of Jiangxi archaeology," in *Thirty Years of Archaeological Work, 1949–1979* (Beijing: Wenwu Chubanshe), 240–251. (In Chinese.)

Lévy, P. 1943. *Recherches préhistoriques dans la region de Mlu Prei, Cambodge.* Hanoi: Publications de l'Ecole Française d'Extrême Orient.

Li Kunsheng and Xiao Qiu. 1980. "Discussions on the Neolithic cultures of Yunnan." *Wenwu Jikan* 2:133–142. (In Chinese.)

Nguyen Phuc Long. 1975. "Les nouvelles recherches archéologiques au Vietnam." *Arts Asiatiques* 31 (special edition).

Vu The Long. 1977. "About a zoomorphic bronze figurine recently discovered at Doc Chua." *Khao Co Hoc* 4:41–43. (In Vietnamese.)

Mansuy, H. 1902. *Stations préhistoriques de Somron-Seng et de Longprao, Cambodge.* Hanoi: F. Schneider.

Mansuy, H. 1923. "Résultats de nouvelles recherches effectuées dans le gisement préhistorique de Somrong Sen (Cambodge)." *Mémoires du Service Géologique de l'Indochine* 10(1).

Matterson, K. J., C. L. Peers, and K. N. Shettle. 1981. "Appendix VI: Analytical Data," in *Studies in Ancient Mining and Metallurgy in South-West Spain,* B. Rothenberg and A. Blanco-Freijeiro, eds. (London: Institute for Archaeo-Metallurgical Studies), 290–312.

Moura, J. 1883. *Le royaume du Cambodge.* Paris: Ernest Leroux.

Murowchick, R. E. 1985. "An examination of some early bronzes from Samreng Sen, Kampuchea." Report of metallographic studies of selected bronzes in the Peabody Museum of Archaeology, Harvard University. Unpublished.

Noulet, J. B. 1877. "L'âge de la pierre dans l'Indo-Chine." *Matériaux pour l'Historie Primitive et Naturelle de l'Homme,* ser. 2(8).

Pavie, A. 1904. *Mission Pavie Indo-Chine, 1879–1895. Etudes diverses. Tome III, Recherches sur l'histoire naturelle de l'Indochine Orientale.* Paris: Ernest Leroux.

Pham Hung Phi, Nguyen Khac Tung, and Hoang Xuan Chinh. 1970. "Spectrographic analysis of archaeological samples from Dong Dau." *Khao Co Hoc* 7–8:130–132. (In Vietnamese.)

Pigott, V. C., S. Natapintu, and U. Theetiparivatra. 1985. "Current research of the Thailand Archaeometallurgical Project: The development of prehistoric metal technology in northeast Thailand." *Research Conference on Early Southeast Asia,* 41–51.

Saurin, E. 1951. "Sur un moule de hache trouvé à Nhommalat (Laos)." *Bulletin de l'Ecole Française d'Extrême Orient* 45: 71–74.

Saurin, E. 1963. "Station préhistorique à Hàng Gòn près Xuan-Loc (Sud Viet-Nam)." *Bulletin de l'Ecole Française d'Extrême Orient* 51:433–452.

Selimkhanov, I. R. 1984. "On the question of Southeast Asian tin and its use in the bronze metallurgy of ancient Azerbaijan," in *Southeast Asian Archaeology at the XV Pacific Science Conference on the Origins of Agriculture, Metallurgy, and the State of Mainland Southeast Asia* (Dunedin, New Zealand: University of Otago), 270–277.

Solheim, W. G., II. 1967. "Molds for bronze casting found in northeastern Thailand." *Journal of the Siam Society* 55(1): 87–91.

Pham Quang Son. 1978. "First attempt at discerning the evolution of late Neolithic–early Bronze Age cultures in the Dong Nai basin." *Khao Co Hoc* 1:35–40. (In Vietnamese.)

Ha Van Tan. 1980. "Nouvelles recherches préhistoriques et protohistoriques au Vietnam." *Bulletin de l'Ecole Française d'Extrême Orient* 68:113–154.

Phan Van Thich, and Ha Van Tan. 1970. "A Determination of the amount of lead found in copper remains dating from the Bronze Age and early Iron Age." *Khao Co Hoc* 7–8: 126–129. (In Vietnamese.)

Vu Thi Ngoc Thu and Nguyen Duy Ty. 1978. "A tool set for casting bronze from Lang Ca." *Khao Co Hoc* 2:36–39. (In Vietnamese.)

Nguyen Duy Ty. 1977. "Further excavations at Doc Chua." *Khao Co Hoc* 4:33–40. (In Vietnamese.)

Nguyen Duy Ty. 1981. "The appearance of ancient metallurgy in Viet-Nam." *Papers Presented by Vietnamese Scientists at the XIV Pacific Science Congress, Khabarovsk, USSR,* 33–36.

United Nations Industrial Development Organization. 1972. *Non-Ferrous Metals (Copper, Aluminum, Tin, Lead, and Zinc): A Survey of Their Production and Potential in the Developing Countries.* New York: United Nations.

Vitout, F. 1914. "Note sur quelques objects préhistoriques de Samreng Sen, Cambodge." *L'Anthropologie* 25(1–2): 114–120.

Wang Ningsheng. 1980. *The Archaeology of Yunnan.* Kunming: Yunnan Renmin Chubanshe. (In Chinese.)

White, J. C. 1982. *Ban Chiang: Discovery of a Lost Bronze Age.* Philadelphia, Penn.: University Museum and the Smithsonian Institution Traveling Exhibition Service.

Workman, D. R. 1972. *Mineral Resources of the Lower Mekong Basin and Adjacent Areas of Khmer Republic, Laos, Thailand, and Republic of Vietnam.* Mineral Resources Development Series 39. New York: United Nations.

Worman, E. C., Jr. 1949. "Somreng Sen and the reconstruction of prehistory in Indo-China." *Southwestern Journal of Anthropology* 5:318–329.

Zhang Zengqi. 1981. "A Preliminary examination of the bronze cultures of western Yunnan," in *Collected Essays on Yunnan Bronzes* (Beijing: Wenwu Chubanshe), 92–107. (In Chinese.)

Zhang Zengqi, Sun Taichu, and Wang Dadao. 1981. "A general survey of Yunnan bronzes," in *Yunnan Bronzes,* Yunnan Provincial Museum, ed. (Beijing: Wenwu Chubanshe), 206–210. (In Chinese.)

18
The Ancient Heat Treatment of White Cast Iron

William Rostoker

In China as early as 400 B.C. a variety of farm implements were produced in large number as white cast iron castings (Rostoker et al. 1983). The full implements served as the plow, hoe, adze, or spade. The casting provided a hard and wear-resistant tip for the tool. The attachment of the metal casting to the wooden shaft or handle was accomplished in one of two ways: Either the casting was designed as a thin-walled sleeve that permitted a slipperlike fit to the wooden component, or the casting had a through-hole or socket by which a handle was pinned or wedged into position, much as the modern ax or hammer (Needham 1958).

White cast iron (often abbreviated to "white iron") is a good material for such applications. In its as-cast condition, the metal is extremely hard and possesses unusually good abrasive wear resistance. Where the ground contained a high content of sand and gravel, the white iron tip to the implement would have increased its useful life many times over contemporary metals such as iron, unhardened steel, and bronze. The use of white cast iron led to a major increase in the national acreage of arable land. Because food production was the dominant factor controlling the growth and development of societies in preindustrial times, the widespread availability of cheap, wear-resistant farming tools was one of the great milestones in Chinese history.

The Field Museum of Natural History in Chicago possesses a number of these ancient implements. My colleagues and I have been permitted to take small samples for metallurgical evaluation. These examinations have revealed that the cast iron had been heat treated to develop what in modern terms is called "malleable iron." By this is meant that the massive carbides in the cast metal have been thermally decomposed to produce in the microstructure irregularly shaped graphite nodules in a matrix of iron or steel.

The main body of the metal has been transformed to the malleable iron structure, but the surface to a depth of about 1 mm was free of graphite nodules and resembled a pure iron or a carbon steel. This kind of surface structure is familiar as the result of heat treating in a decarburizing atmosphere. In modern Britain such heat treatment has been used for many years and is called "white-heart malleableizing." In the United States, the same heat treatment without the decarburization is called "black-heart malleableizing." Unlike white and gray cast irons, malleableized cast iron possesses some ductility. However, it is not forgeable like low-carbon irons and steels. Before the twentieth century the term "malleable" was used in a broader sense to connote forgeability and formability and was applicable in that sense to any metal.

The use of white iron for farm implements was an intelligent and useful decision, but the heat treatment to produce a decarburized, softer, and much less wear-resistant surface is puzzling. We might dwell on the question of why this heat treatment was deliberately introduced into the production process.

One reason might have been to engender greater toughness in edges and points. As cast, white iron is brittle, although it has reasonable strength and is certainly not fragile (Rostoker 1987). However, if the soil was stony, a plow tip might have chipped and dulled too quickly. A fully malleableized structure would have much greater resistance to spalling, but it would have lost most of its wear-resistance advantage. However, a partially malleableized structure might have provided an optimum combination of toughness and wear resistance. The decarburized surface, even when hardened by cold hammering, would not have adequate wear resistance. We must believe that it was an adventitious condition that was tolerable because, after the decarburized surface was worn off, the body of the metal provided an abrupt reduction in the rate of wear.

The decarburization might have been put to good use in another context. If the white iron was cast in thin plates, for which there is evidence, the cast iron could have been completely converted to steel plates capable of hardening by quenching or by cold-working, thereby developing a good serviceable edge. Percy (1864, pp. 804–805) recounted that cutlery was produced during the first half of the nineteenth century in England by "pack decarburization" of blades precast to shape as white cast iron. I have suggested elsewhere (Rostoker et al. 1985) that plates of steel so produced and heat treated were the basis for the "hundred refined" swords famous in Chinese history.

The decarburization process involves a reaction at elevated temperatures between carbon dioxide (CO_2) and the carbon dissolved in austenitic iron, $[C]_{Fe}$ as follows:

$$CO_2 + [C]_{Fe} \rightarrow 2CO.$$

The reaction proceeds from left to right if the ratio of CO_2 to CO is regulated appropriately as described in figure 18.1. Either a continuous flow of diluted CO_2 or a secondary reaction in a closed system that converts the carbon monoxide (CO) back to CO_2 is necessary. When the latter is done in sealed crucibles, the process is called "pack decarburizing." The white iron pieces are packed in iron oxides: Any of Fe_2O_3 (hematite), Fe_3O_4 (magnetite), or FeO (wüstite) will serve according to published formulations. Using wüstite or mill scale as the example, the reverse reaction is therefore described as

$$FeO + CO \rightarrow Fe + CO_2,$$

and the CO_2 returns to take more carbon out of the casting. In modern times the process is conducted at 950°–1050°C.

Equilibrium diagram for the system Fe–O–C: % of CO_2 of $[CO + CO_2]$ versus temperature ($CO + CO_2 = 100\%$).

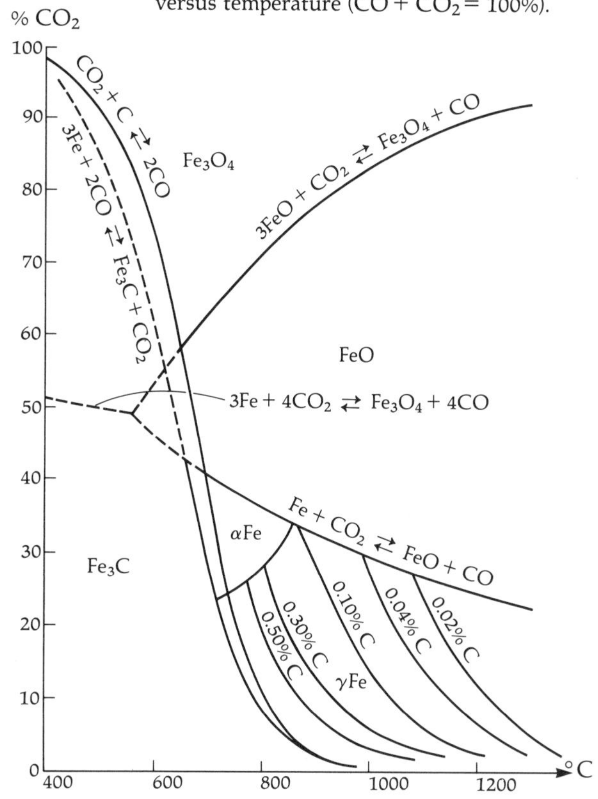

Figure 18.1 Conditions of decarburization and oxidation of iron as a function of gas atmospheres defined by the proportions of carbon dioxide and carbon monoxide. It can be seen that by changing the temperature or the carbon content of the iron, either carburization or decarburization can be made to occur.

The process is actually not so simple because the white iron can oxidize at the same time as it is being decarburized:

$$CO_2 + Fe \rightarrow FeO + CO.$$

As shown in figure 18.1, this can proceed if the CO_2/CO ratio exceeds about 2.0 (greater than about 35% CO_2). With two competing reactions the oxidation could be faster than decarburization, and the decarburization would never be apparent after the fact.

Some experiments have been performed in our laboratories to illustrate the conditions governing malleableizing, decarburizing, and oxidizing as concurrent and conflicting reactions. White iron was cast in the form of 25-mm-diameter bars. Their composition is 3.35% C, 0.78% Si, and 0.27% S. The cast structure is predominantly iron carbide eutectic (ledeburite). The sulfur was added to ensure the solidification to white iron at any rate of cooling. It also conforms to compositions of Chinese cast irons after the advent of the use of coal for smelting c. eleventh century A.D.

(Rostoker et al. 1984). Samples were cut as disks of about 6 mm in thickness.

Results of Experiments

Packed in mixtures of hematite or mill scale with carbon, the specimens oxidized so rapidly that no decarburized zone could be seen. The competitive oxidation rate was slowed down adequately by packing in a neutral granular material such as sand or iron powder. To provide a small amount of carbon dioxide as a starter, the specimen was preoxidized to form a film that was less than 0.5 mm thick. Under those conditions the loss of thickness of the specimen was no more than the oxide layer. The results of experiments under these conditions are as follows:

1. At up to 700°C neither surface decarburization nor malleableizing occurs.
2. At 800°C a decarburized zone about 2 mm thick can be realized in five days. This is illustrated in figure 18.2. Below the decarburized surface layer graphite nodules develop as the product of progressive decomposition of the eutectic iron carbide.
3. At 900° and 1000°C the decarburized zone is thicker but is not free of graphite nodules. Evidently the malleableizing rate is rapid and is approximately that of the decarburization rate. This is illustrated in figure 18.3. The malleabilization rate could have been substantially reduced if the silicon content was smaller than in our sample.

Summary

The decarburized surface layer discovered on the Chinese implement artifacts has been reproduced. We infer that the heat treatment was conducted at about 800°–900°C for several days. It could have been done by submerging the castings in a deep bed of hot ashes kept hot by a slow fire above.

It would also have been possible with this same technology to decarburize 6-mm-thick plates of white iron completely to a carbon steel. Plates so produced could have been forge- (pile-) welded together to produce stock from which to fabricate the famous thirty and one-hundred refined swords described in the contemporary Chinese archaeological literature (see "Iron and Steel Making Techniques," 1978; Rostoker et al. 1985).

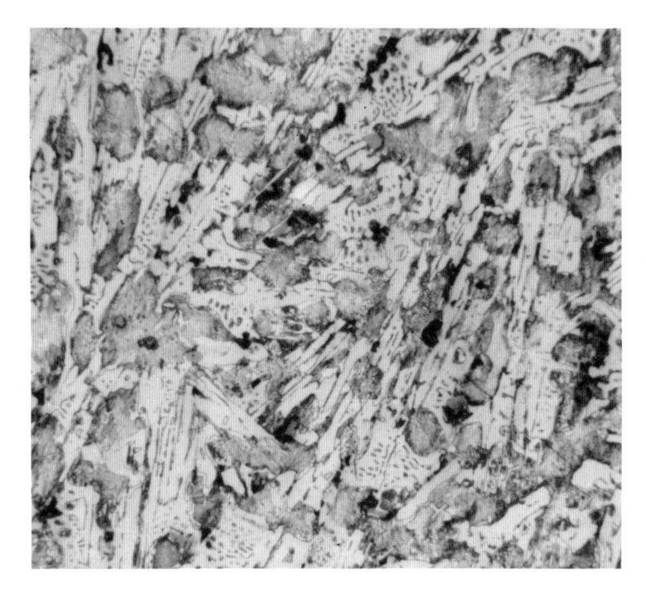

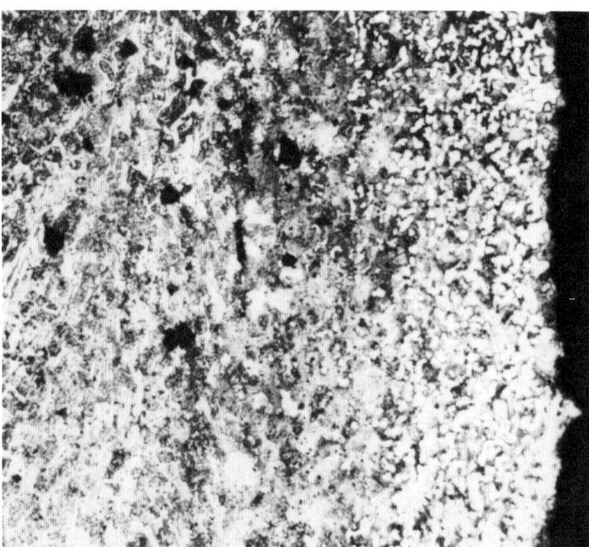

Figure 18.2 Micrographs showing the initial white iron structure (top; × 200) and the results of annealing at 800°C for five days (bottom; × 100). A carbon steel surface layer, the result of decarburization, is seen at the surface, and the interior shows graphite nodules, which are the product of the malleableizing transformation of the massive carbides in the as-cast structure.

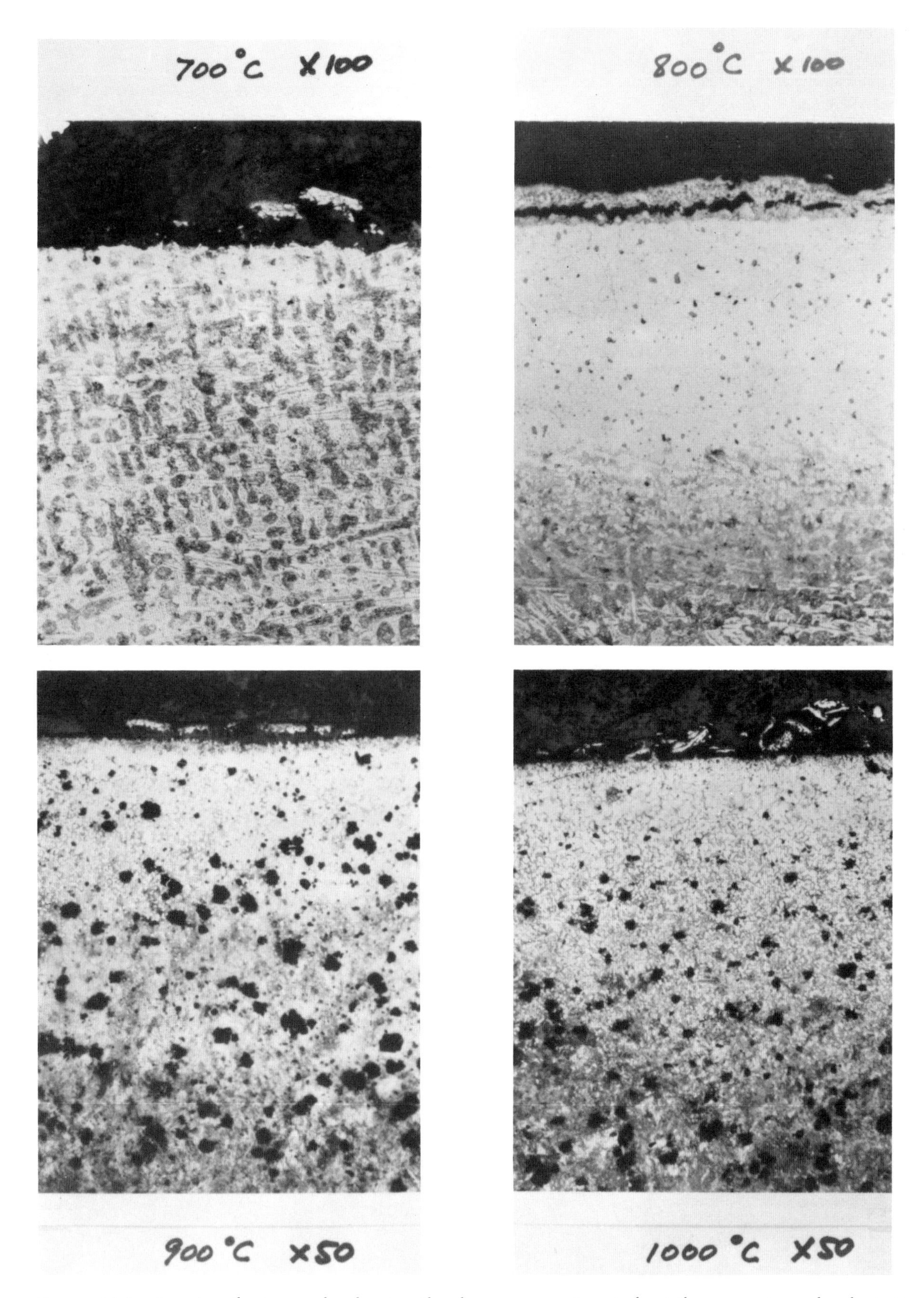

Figure 18.3 A series of micrographs showing the decarburization zones at the surface as the result of annealing at each of a series of increasing temperatures. What may be seen is that at higher temperatures malleabilization is progressing at about the same rate as decarburization so that the decarburization zones contain graphite nodules that appear to be quite stable.

References

"The iron and steel making techniques of the Han Dynasty in Honan." 1978. *Kaogu Xuebao* 1:1–26.

Needham, J. 1958. *The Development of Iron and Steel Technology in China*. Cambridge: The Newcomer Society.

Percy, J. 1864. *Metallurgy: Iron and Steel*. London: John Murray. Reprinted as *De Archaeologische Pers Nederland, Eindhoven* 2(3).

Rostoker, W. 1987. "White cast iron as a tool and weapon material." *Archeomaterials* 1(2):145–148.

Rostoker, W., B. Bronson, and J. Dvorak. 1984. "The cast iron bells of China." *Technology and Culture* 25(4):750–767.

Rostoker, W., B. Bronson, J. Dvorak, and G. Shen. 1983. "Casting farm implements, comparable tools and hardware in ancient China." *World Archaeology* 15(2):196–210.

Rostoker, W., M. B. Notis, J. R. Dvorak, and B. Bronson. 1985. "Some insights on the "Hundred Refined" steel of ancient China." *MASCA Journal* 3(4):99–103.

19 Social and Historical Implications of Early Chinese Iron Technology

S. J. Taylor and
C. A. Shell

Early iron technology in China and its impact have been studied by many Chinese, Japanese, and Western scholars. Although there are notable exceptions, much of this work has been based on incomplete knowledge of recent finds and of metallurgical science. Technical studies of iron artifacts are mainly in Chinese or difficult to access, and some of the better known earlier results have since been questioned or refuted. Moreover, historians of metallurgy, and archaeologists and historians have tended to work independently of one another, resulting in a rarely bridged gap between scientific analyses and historical reconstructions of the ancient iron industry. As a result the literature on this major technological innovation, which occurs in a period of profound social, political, and economic change [the Warring States Period (475–221 B.C.) and the Qin (221–207 B.C.) and Han (206 B.C.–A.D. 220) dynasties], is full of misunderstandings and contradictions.

We intend to summarize and discuss some of the important metallurgical and historical issues on the subject in the light of the technical background, particularly the various microstructure analyses that have been carried out on ancient Chinese iron objects. We also intend to add to this relatively small corpus of analyses by presenting the interim results of a preliminary examination by C. A. Shell of two pieces from the Museum of Archaeology and Anthropology of Cambridge University. The technical information gained from these analyses should at least clarify, if not settle, some of the issues in question.

Preliminary Remarks on Iron

Some Questions of Terminology

A cross-cutting of two systems of classifying ferrous metals, by production method and carbon content, is a common source of confusion in archaeological literature on iron. As we are particularly interested in the mechanical properties of iron, which are in the first instance determined by carbon content, we have chosen a terminology based on that criterion. We use the terms "wrought iron," "cast iron," and "steel" to indicate low-, high-, and medium-carbon iron, respectively. Additional terms indicate the method of production. These are "direct process" ("forged") and "indirect process" for wrought iron. Because cast iron is almost always produced by casting liquid metal, it seems unnecessary and confusing to use the term "cast cast iron." Therefore the method of production in this case is assumed to be casting unless otherwise indicated.

Untreated cast iron ("pig iron") is distinguished from "cast-iron" (with a hyphen), by which we mean malleableized cast iron. A further source of confusion is the common use of "cast iron" as a general term covering both pig iron and cast-iron; this is perhaps clearer in Chinese: They call cast iron "raw iron" and cast-iron "cast iron."

With regard to steel, it is important to distinguish in ancient artifacts areas within heterogeneous wrought or cast iron objects that accidentally fall within the steel range of carbon content from deliberately produced steel.

The Requirements of Iron Metallurgy

There are a number of basic requirements in iron metallurgy, whatever the methods used. These include ore, water, fuel, transport, a reasonable level of furnace technology, and labor.

Although they involve different fuel and furnace requirements, the production of cast and wrought iron in antiquity potentially differs most in the labor force and degree of central organization needed. There is a good deal of debate over the extent to which forced or free and skilled or unskilled labor were used in ancient mining, smelting, and metalworking (Barnard 1979–1980, p. 124; Franklin 1983, p. 287; Wilbur 1943). It is clear at least that production of wrought iron tends to be small scale and labor intensive, involving a small, skilled, and usually free work force (although it can be carried out in workshops dependent on or controlled by a social elite). Casting, on the other hand, is most efficient if centralized, in which case a larger proportion of unskilled and possibly forced labor can be used (although modern observations of nonindustrial iron smelting and casting show them to be generally on a relatively small scale) (Hommel 1937, p. 28).

Major Issues in the Development of Iron Metallurgy in China and Its Impact

Metallurgical Issues

Now that questions of terminology and requirements of iron metallurgy have been presented, it is possible to discuss the technical issues surrounding the early use of iron in China. Two general issues—the nature of the changeover from bronze to iron and diffusion versus independent invention—were discussed more in the earlier literature on the topic (Read 1937, pp. 403–405; Wertime 1961, p. 51) and in recent years have been (not entirely deservedly) somewhat ignored. On the other hand, numerous recent finds have allowed a fairly exact sequence of technological developments to be established (Hua 1982; Huang 1957).

The Changeover from Bronze to Iron Two misunderstandings regarding the changeover from bronze to iron plague archaeological literature. (Although these may be obvious to those trained in metallurgy, they are so common in archaeological literature as to bear being raised again here.) They are that iron is "better" than bronze and that the difficulty in smelting iron lies in the high temperatures required. In fact, it is only steel that is consistently stronger than bronze, although malleable cast-iron might approach bronze in quality, and in the Mediterranean and southwest Asia the change from bronze to iron as the commonly used metal seems to have been forced by a tin shortage (figure 19.1; Maddin et al. 1977, pp. 123, 126).

Furthermore, a usable bloom, or molten iron, if it contains a large amount of carbon or other impurities, can be produced at temperatures close to or even below those needed to melt copper or gold. In any case Neolithic potters in China could produce temperatures of up to 1200°C and a good reducing atmosphere in their kilns (Barnard 1961, p. 65). The difficulty with iron lay not in obtaining high temperatures but in developing the new techniques necessary to hot-forge the bloom or—in China, as there was little or no early casting in the West—to decarburize the brittle cast iron to produce a usable material. It is thus not surprising that, in the West and to a lesser extent in China, a long period of use during the Bronze Age of small amounts of meteoric iron and possibly of accidentally produced fragments of soft bloomery iron preceded the Iron Age proper (Wainwright 1936, p. 23).

The changeover was by no means complete, and there is evidence for the coeval use of the two metals in China at sites such as Yanxiadu (a Warring States site with an assemblage mainly of weapons, 65.8% of which were iron and 32.5% bronze, the latter being used only for specific weapons, such as crossbow

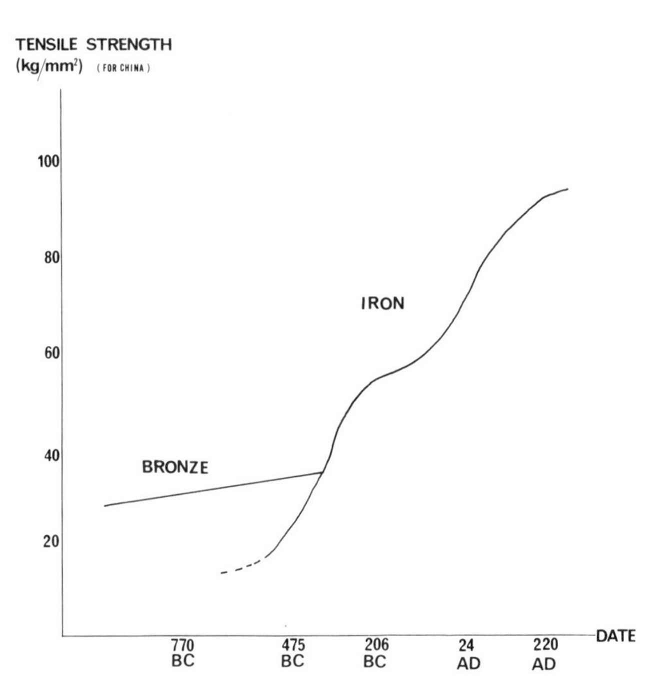

Figure 19.1 Comparative strengths of bronze and iron artifacts. After Hua (1982, fig. 5).

bolts) and in Europe in the Halstatt culture ("Excavation report of tomb 44," 1975, p. 230; Forbes 1964, p. 220). No doubt the quality of objects made of both metals fluctuated greatly, and in China the advantage of cast iron may have been cheap mass production as much as technical superiority (Cheng 1963, p. 248). A number of other factors—taste, superstition, local resources, etc.—probably also affected the ratio between what the Chinese termed the "lovely metal" (bronze) and the "ugly metal" (iron), as they did in Greece (Read 1937, p. 401; Keightley 1976, p. 31).

Two factors distinguish the appearance of iron in China from other areas. Some scholars suggest that bronze was hardly used for tools and implements in northern China (though Zhang Wan Zhong stated at BUMA II that finds from the Jin state capital site of Houma include a large number of bronze tool molds), in which case the changeover and the comparison of quality would have been between iron and stone, bone, or wood, rather than iron and bronze (Cheng 1963; Wagner, personal communication). Once introduced, iron seems to have been used quite soon on a large scale for tools, implements and weapons in China. This contrasts with many other parts of the world (for example, Egypt), where iron was used for small, rare luxury items for a long time after its introduction (Coghlan 1956, p. 66).

How the first iron was actually produced has been a matter of great speculation and disagreement. Because this event no doubt occurred in many different places in different ways and at different times, it is unlikely to have left much trace in the archaeological record, and further discussion seems largely fruitless. Most scholars agree that the discovery was probably accidental (with the notable exception of H. H. Coghlan), and two possible explanations of how it came about are through the smelting under unusually high temperatures or reducing conditions of high-iron copper ore or of iron ore mistaken for copper ore or by accidental reduction of the iron-bearing red ocher used to decorate Neolithic pottery (Coghlan 1956, p. 46; Wagner, personal communication).

Most early discussions of why iron came into use center on technical questions, but a newer approach emphasizes the importance of social and economic conditions in the cultures where the technology developed, because, as was seen, there is no technological reason why iron would be "inevitably" adopted on a large scale (and in fact it was not in many cases). Ursula Franklin (1983, p. 285), for example, in a discussion of bronze metallurgy concentrates on the societal and organizational requirements behind this complex two-step process (extraction and smelting of ores). She suggests that, for China at least, the amount of planning and organization and the large pool of labor, either forced or free, necessary to undertake the difficult and dangerous work involved in early metallurgy were more important prerequisites than were technical considerations. Thus she views bronze casting primarily as an indicator of a certain level of social organization, much as writers on state formation view technical achievements such as monumental architecture.

This approach is especially useful for China, where early casting of both bronze and iron required a large-scale centralized industry to be truly efficient and where iron production seems to have "taken off" so quickly. Although many Chinese authors follow a similar approach, Barnard (1979–1980, p. 124) rejects it as "organizational determinism." There is disagreement in the literature about the extent to which a large pool of forced labor was actually used and whether iron production caused or was caused by sociopolitical and economic changes (Barnard 1979–1980, p. 124; Thorp 1980–1981; Wilbur 1943). For the moment it seems that the archaeological and textual evidence is insufficient to resolve these debates, and they remain largely on a theoretical level.

Independent Invention versus Diffusion In the context of the "organizational approach," the question of independent invention versus diffusion becomes largely irrelevant. Moreover, many previous discussions on this topic have been based more on nationalism, ethnocentrism, or blind belief in diffusion than on solid archaeological evidence.

The question is linked to that of the order of discovery of forged and cast iron. Casting was certainly a Chinese invention, as it appears nowhere else on any scale until many centuries later (Needham 1964, p. 47). However some writers feel that forging iron preceded or was contemporaneous with casting and was imported into China from the north or northwest (Tylecote 1983, p. 436). Although recent early iron finds include both cast and forged objects ("Newly discovered late Spring and Autumn artifacts," 1978), and the Scythian and Sauromatian cultures were spreading iron metallurgy to the Urals, the Black Sea, and the Caspian in the sixth century B.C.—about the time iron appears in China—there is no archaeological evidence of borrowing in either direction (Watson 1971, p. 80). The comparatively sudden development of iron technology in China can be seen as much as a result of preceding advanced levels of bronze casting, pottery, and kiln technology as of cultural borrowing (Needham 1964, p. 47). In any case the isolated and unique nature of the developed industry in China are its most important and interesting features, whatever its origins.

Sequence of Technological Developments In recent years an increasing number of finds of iron artifacts and foundry sites have shed some light on the questions of when, why, and how iron began to replace bronze in many of its functions and on the question of the technical and social conditions preceding this change.

Hitherto these had been discussed in something of a vacuum. The finds have also allowed the establishment of a fairly exact chronology of technological development. Considerations of space do not allow the discussion of this chronology in any detail here. Nonetheless, it is worth summarizing a few points as background for the discussion that follows.

The date of the appearance of iron smelting (as opposed to chance working of meteoric iron) in China has been widely discussed on the basis of both textual and archaeological evidence. Most scholars suggest a date ranging from the fifth to the seventh century B.C. (Blunden and Elvin 1983, p. 68; Hua 1982, p. 5). Huang Zhanyue, in an article published in 1976, gives a "state of the art" resumé of this question. Rejecting all the textual evidence and much of the archaeological evidence as unreliable regarding dates, he finally cites what he considers the earliest reliable evidence: several knives or scrapers, several spades, a lump of iron, and a rod from five sites (Longdongpo, Chengqiao, Shiziling, Qichu tomb, and Changde Deshan) dated to the second half of the Spring and Autumn Period (c. 666–475 B.C.). [Since then three new finds of similar date—a tripod cooking pot, a steel sword or dagger, and a scraper—have been found at Yangjiashan ("Newly discovered late Spring and Autumn iron artifacts," (1978).] Familiarity with bronze casting would mean that a relatively short period of experimentation would be needed to produce these pieces, some of which are fairly small and simple. On the other hand, the steel artifact suggests that the earliest iron preceded these finds by at least 100 years, giving it a date of about 600 $\pm$ 50 B.C., if one accepts the dating of these sites (which is open to serious question) [Wagner, personal communication; at the conference Huang Zhanyue and Ke Jun (Ko Tsun) described recent, apparently earlier (Spring and Autumn Period) finds.]

Huang Zhanyue's report, if not conclusive, at least substantially clears up the long-standing debate surrounding the order in which forged and cast iron appeared. The presence of both forged and cast objects in this earliest group of artifacts supports the growing consensus among Chinese and Western scholars that both processes developed either at the same time or else so close together as to leave no perceptible gap in the archaeological record (Hua 1982, p. 10; Wagner 1983, p. 28; Watson 1971, p. 84). Generally, however, cast objects do seem to predominate, and the earliest foundry sites (all first century B.C. or later) have more blast furnaces than bloomery furnaces (Institute of Archaeology, 1961, p. 78; Wagner 1983, p. 29). Moreover, at this time or possibly as early as the Warring States Period, the indirect method of making wrought iron was developed, so that, although wrought iron production increases in the Han, the bloomery method virtually disappears (Wagner 1983, p. 29).

A tempting explanation of a more or less simultaneous development of forged and cast iron is the use of a crucible process, producing both a bloom and cast iron, similar to that observed by Read in China in the 1930s (Read 1937, p. 405). However, archaeological evidence for this technique is meager: The Western Han foundry site of Tieshenggou, Henan Province, has eighteen smelting furnaces and only one crucible furnace (Wang 1982, p. 125). It is also possible that the two methods developed separately or that the first forged iron was the result of experiments on blooms produced accidentally from a failed melt.

A number of other metallurgical processes developed at an early date in China. As Hua Jueming (1982, pp. 8–10) points out, malleableizing was unlikely to have been an accidental discovery but rather the result of the technological necessity of improving on cast iron and a demand for more efficient cheap tools. A partially decarburized ax, found at Luoyang and dated to the fifth century B.C., shows a structure thought to be the progenitor of malleable cast-iron, and by the fourth century a large number of malleable cast-iron objects occur (Li 1975, p. 21). By the mid–Western Han this technique had reached a high level, and incompletely decarburized objects occur less and less (Hua 1982, p. 8).

The earliest steel comes from a grave at Yangjiashan, the date of which is uncertain (seventh to third century B.C.) but suggested to be fifth century by the excavators ("Newly discovered late Spring and Autumn iron artifacts," 1978, p. 46). Other sources suggest slightly later dates (Needham 1964, p. 47), which may be more prudent given the uncertain dating of Yangjiashan. [But at the conference recent apparently earlier finds were described by Huang Zhanyue and Ke Jun (Ko Tsun).] Certainly by the Han, steel was fairly common and quench-hardening well understood (Tylecote 1976, p. 48).

By the first century B.C. spherical graphite cast-iron, produced in certain high-carbon irons by long annealing and quick cooling in iron molds, was in existence. The earliest find is a pick head from Tieshenggou. Quality comparable with modern spherical graphite cast-iron was obtained in this relatively rare structure (Guan and Hua 1981, p. 1).

Other important technological developments include quenching, from the third century B.C. [a quenched steel sword from Yanxiadu dates to this time ("Excavation report of tomb 44," 1975, p. 48)], and the welding together of sheets of metal of different carbon contents to produce a mixed structure, by the Western Han or earlier (Hua et al. 1960, p. 77; Watson 1971, p. 84). A number of changes in furnace design and operation reflect the leap in technological level and the major increase in production generally felt to have taken place from the late Warring States Period through to the Han (Wang 1982, p. 127). Bellows may have been used as early as the Warring States, and the highly efficient double-acting piston bellows were in use by the second century B.C. (Needham 1964, p. 47). By the first century A.D. at the latest these were being run by water power (Needham 1964, p. 47).

During this period of technological advance metal casting molds appear, although they do not entirely replace the earlier clay ones, and double-faced and multiple stacking molds are used increasingly (Archaeological Team 1962, p. 39; Hua 1983, pp. 121, 127). These new forms of mold allow much more efficient large-scale casting. A trend toward standardization of form in products such as axle bearings can be seen both in the decrees of Emperor Qin Shi Huangdi (reigned 221–210 B.C.) and in the finds from Mianchi in Henan (Li 1976, p. 61). This suggests industrial scale production for export to other areas within China rather than for local consumption.

Historical and Anthropological Issues

What Iron Was Used for and the Impact of Its Use The preceding discussion is not merely of metallurgical interest. Because different types of iron are generally held to be more or less well suited to different uses [wrought iron or ideally steel for most weapons and objects to survive sharp blows; cast-iron for ornaments, pots, and less than ideally tools and implements (Read 1937, p. 405; Trousedale 1975, p. 65); and raw cast iron virtually useless (Wagner 1982, p. 4)], this sequence affects what sort of objects were being made of iron. Hence it determines where the introduction of the metal had the most impact.

Despite this, early finds of untreated white cast iron objects, such as a tripod cooking pot and a spade edge, have been made in China (Huang 1976, p. 68; "Newly discovered late Spring and Autumn iron artifacts," 1978, p. 44). Observations in the 1930s of cast iron plow shares (albeit stronger than the very small, fragile Warring States implements) in use and lasting up to two years (Hommel 1937, p. 42) belie suggestions that such implements would be little better than stone or wooden ones and that only steel gives ferrous tools worth the trouble of producing (Maddin et al. 1977, p. 125). It should be remembered that nontechnological factors, such as price, would also affect the type of metal that was used. Broadly, however, Chinese finds do seem to fit the pattern described: Weapons are virtually all of wrought iron and tools and implements mainly of cast iron or cast-iron (Hua et al. 1960, p. 73). With the early introduction of steel and various types of malleable cast-iron, there develops a sophisticated matching of these different materials to the uses to which they are best suited, a practice that was well established by the Han (Wang 1982, p. 128).

Given these considerations, what then was iron primarily used for? A standard pattern for the introduction of iron in the West involves early use for ornaments and objects of value, followed by use in weaponry (in conjunction with bronze) and then in tools and implements with the virtual disappearance of bronze weapons (Forbes 1964, p. 215). Writers such as Loehr (1956, p. 79) have tried to see this pattern in China, and certainly Warring States tombs like those at Zhongzhou Lu, Luoyang, contain few iron objects other than belt hooks, often inlaid with gold (Huang 1976, p. 66). The small group of objects said to date to the Spring and Autumn Period suggests a broader range, however; weapons, vessels, and tools all occur. A similarly broad range in Warring States finds suggests that all these uses developed more or less simultaneously (Huang 1976, pp. 66–68). However, the proportions of these different types of object are unequal, with agricultural implements and to a lesser extent craft tools dominating strongly.[1] Hua Jueming (1982, fig. 1) shows that, of fifty-two iron objects analyzed metallurgically in China between 1949 and 1978, forty-four fall in these two categories, and of these, twenty-six are agricultural implements. The products of foundry sites such as Guxing and Yingzhen (both Han) show similar ratios (Cheng 1978; "Examination of Han cast iron smelting techniques," 1978, p. 47). It is not really until the Eastern Han (A.D. 25–220) that iron predominates in the full range of agricultural and craft tools, household items, and weapons, virtually replacing bronze for all but ceremonial uses (Wang 1982, pp. 122–123).

It seems likely, therefore, that the major impact of early iron use was on agricultural production, water management, and other large-scale public works, although it may have been some time before iron technology was sufficiently advanced to become an effective innovation. Opinions on whether this impact had broader repercussions fall into two schools of thought. Ch'ao Ting-chi and the authors of *Shang Zhou Kaogu (Archaeology of the Shang and Zhou Dynasties)* believe that the introduction of iron was a major factor, if not *the* major factor, in the changes in land ownership, trade and social structure, and the increase in population, which occurred in China in the Warring States Period and the early Han (Ch'ao 1936, p. 63; Thorpe 1980–1981). Others, on the contrary, hold a view similar to Boserup's theory that it is an increase in population that causes technical innovation and believe that iron became widespread only because the new social and economic conditions of the time demanded it (Hsu 1978, p. 260).

The first major irrigation projects (in the sixth to fourth century B.C.) do coincide with the introduction of iron, and the growing class of wealthy Warring States merchants included many iron founders (Ch'ao 1936, pp. 66–68; Huan 1967, p. xix), but it seems excessive to attribute a role of prime mover to iron tools. A more complex situation, in which social and economic conditions provided the initial incentive for the leap in quantity and quality of iron production that occurred in the late Warring States and early Han, seems more likely. This leap may have in turn influenced these conditions in a form of mutually amplifying change. However, little evidence for any of these views has been adduced, and Hsu Cho-yun's assertion that no satisfactory theory exists relating improvement in production with the changes in social institutions still seems justified in this case (Hsu 1965, pp. 130–131).

Geographical Variations in Iron Production and the Effect of the Han Iron Monopoly An important topic that the preceding discussion did not touch on is the variation in distribution and use of iron in various parts of China. These variations and those of the distribution of mines, smelters, and foundries on a local scale and the extent to which the state iron monopoly of the Han affected them are important issues in assessing the extent of the impact of iron metallurgy.

Traditionally it was thought that iron appeared first in the north rather than in the south of China (Eberhard 1968, p. 374). (It should be noted that, for ancient China, anywhere south of the Yangtze River is termed "south China," the far south at that time being inhabited by a non-Chinese people.) However, several of the earliest finds of iron objects are from Changsha, within the area of the ancient southern state of Chu (Huang 1976, pp. 67–68; see figure 19.2). Moreover, both ancient texts and modern geological evidence point to this area as the one with the most abundant and prominent ores in China (Cressey 1934, p. 117; Swann 1950, p. 42). The states of Wu, Yan, Yue, and Shu and Ba in Sichuan also seem to have been producing iron from an early date (Hsu 1978, pp. 120–121). In any case the archaeological evidence, as compiled in distribution maps by Barnard and Sato (1975) and by Huang Zhanyue (1957) suggests that by the mid–Warring States iron production was widespread although concentrated mainly in the central states area (Shanxi, Shaanxi, Hebei, Henan, and Shandong) and in the south (Hunan-Jiangxi area). The small but widespread ore bodies of China, the difficulty of transport in ancient times, particularly in the north China plain, and the literary evidence support this picture of a widespread, localized development (Cressey 1934, p. 109; Wilbur 1943, p. 69; Huan 1967, p. 33).

It is unclear whether this local production was carried out on a large scale, with mine, smelter, and foundry or forge close together, or on a small scale by local smiths and cast iron merchants, using raw materials mined and smelted elsewhere. These two forms of production have quite different requriements and social consequences. The former would be most efficiently carried out close to the ore bodies (generally far from cities) and would require large groups of forced or mobile free labor and good transportation and high initial investment. It would probably lead to greater standardization of products and, if not government controlled, to the rise of a class of wealthy iron industrialists. The latter, on the other hand, would require a small group of skilled metalsmiths working close to their market (although a large labor force might still be required for mining). It would lead to a flexible, locally varied production and the growth of a class of free artisans.

The literary and archaeological evidence on this topic before the Han is conflicting. The *Tong Yue* (59 B.C.) and a section of the *Zhou Li* (Wilbur 1943, pp. 60–61; Hsu 1978, p. 121) refer to what seems to be household production of various items, and the early Iron Age copper smelting site of Tonglushan shows no evidence of casting at the same location ("Preliminary appraisal of iron and bronze tools," 1975, pp. 21–22). On the other hand, certain states, such as Yan, already had government iron offices before the Han (Cheng 1963, p. 45). Mencius speaks of farmers being incapable of producing their own iron tools (Lei 1980, p. 9), and many texts record the appearance of a class of rich iron merchants (for example, the *Yan Tie Lun* and the *Han Shu*).

By the time of the establishment of the government iron monopoly in 117 B.C., however, the trend was clearly to large-scale centralized production. Although the Emperor Wudi's (reigned 141–87 B.C.) monopoly system, established to fill the Imperial Treasury's coffers, never worked completely and was fairly rapidly dismantled (Yü 1967, p. 17), it was certainly a sign of the times. With the exception of Nanyang, excavated Han ironworks are large scale (for example, 3,000 m^2 in area) and tend to combine the entire production process from mining to the finished product (Wang 1982, p. 125). Goods were increasingly standardized and multiple molds were used for efficient large-scale casting. The 419 government iron offices employed some 100,000 corveé, conscript, and convict laborers in 48 B.C. (Barnard 1961, pp. 31–36). Some of these offices, in areas without much ore, cast iron from ores mined and smelted elsewhere (Swann 1950, p. 277), and no doubt local smiths or iron casters existed to repair the farmer's broken plowshare. However, these small-scale operations were overshadowed by the larger ironworks, whether government run or private.

Experimental Work

In the preceding discussion the mechanical properties of various forms of iron and those forms actually in use for different purposes in ancient China have often been referred to as important factors. In the final section of this essay we intend to look more closely at the type of iron contained in two allegedly Han Dynasty objects using microstructural analysis. We compare them with results from similar analyses. These results may well shed further light on the issues discussed.

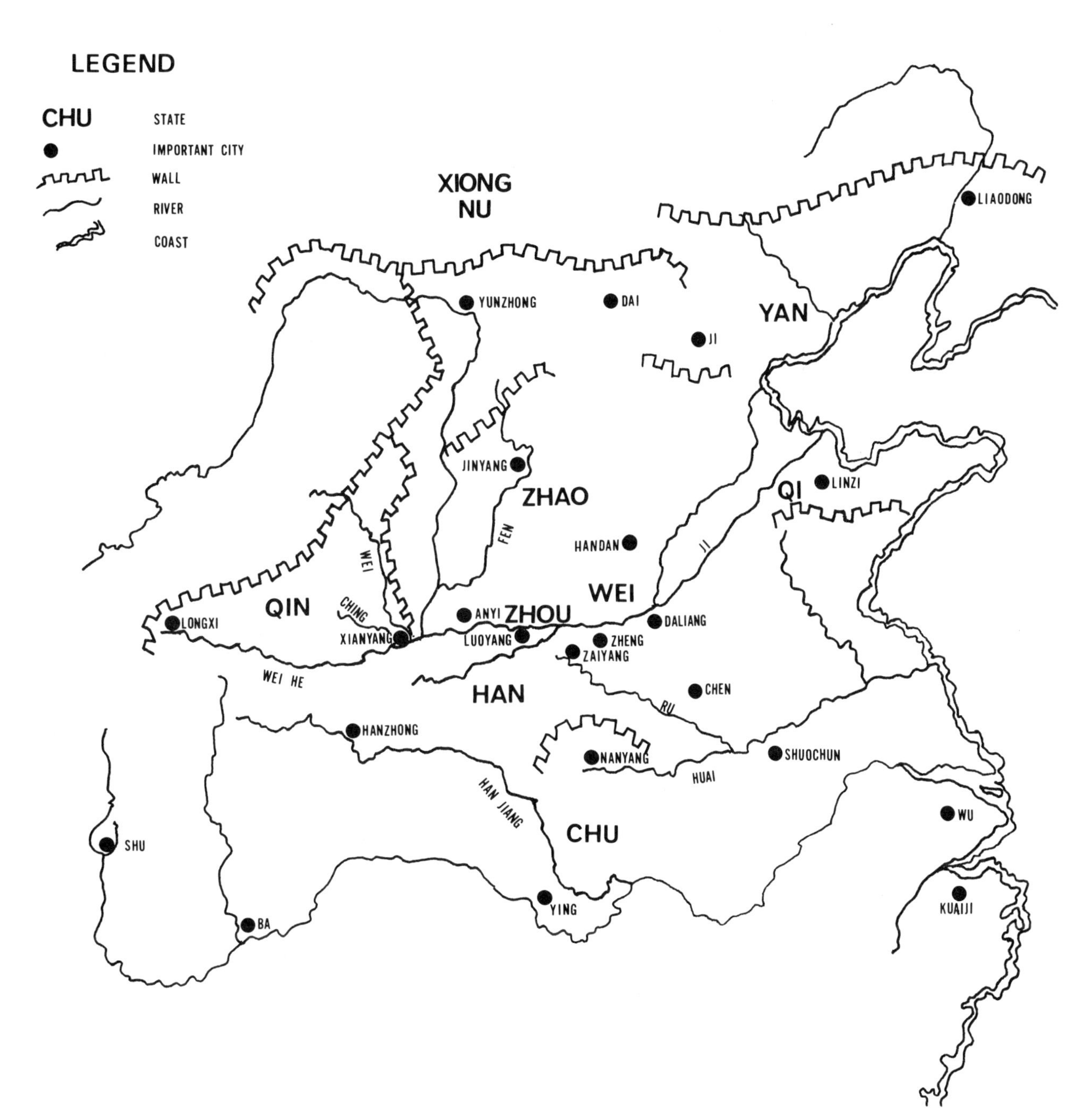

Figure 19.2 Map of China during the Warring States Period, 475–221 B.C. After Cheng (1963).

The Artifacts

The two objects examined by microstructural analysis are catalogued as numbers 27.839 and 33.16 in the collection of the Museum of Archaeology and Anthropology of Cambridge University. The first is a socketed hoe blade, said to be from a Han tomb, and the second is described as a *ji* lance, supposedly found near Luoyang and of Han date. Although neither piece is securely dated, as they were acquired through dealers, the hoe blade is of a form typical of Han and earlier agricultural implements and its small size and fragility suggest a pre-Han date (Watson 1971, pp. 81–82; Wagner, personal communication). Its only unusual features are the scrolls and the character "千" (one thousand) cast in relief on its surface. These suggest that the hoe may possibly have been for ceremonial use, for instance, the digging of a tomb. (It has definitely been used, because the working edge shows signs of grinding and dips toward the middle, indicating that it has been sharpened, possibly several times.) Interment of tools used in tomb construction in or around the tomb is thought to have been common practice in ancient China (Lei 1980, p. 8). Although no other example of this character of an iron object has come to our notice, similar scrolled inscriptions were found on iron artifacts from the Guxing site, and it has been suggested that the inscription is either a place or person's name or falls within the quite common group of auspicious or quality-vaunting inscriptions found on early tools (Hua and Wagner, personal communications).

The "lance," which could be more properly called a halberd, is a rather unusual form. A basically similar piece was found at the Han graves at Mancheng and similar although differently hafted forms occur in the Zhou (Institute of Archaeology 1980, pl. 69.1; Zhou 1957, pl. 22). Thus the Han dating, although not secure, seems a reasonable terminus ad quem for both pieces.

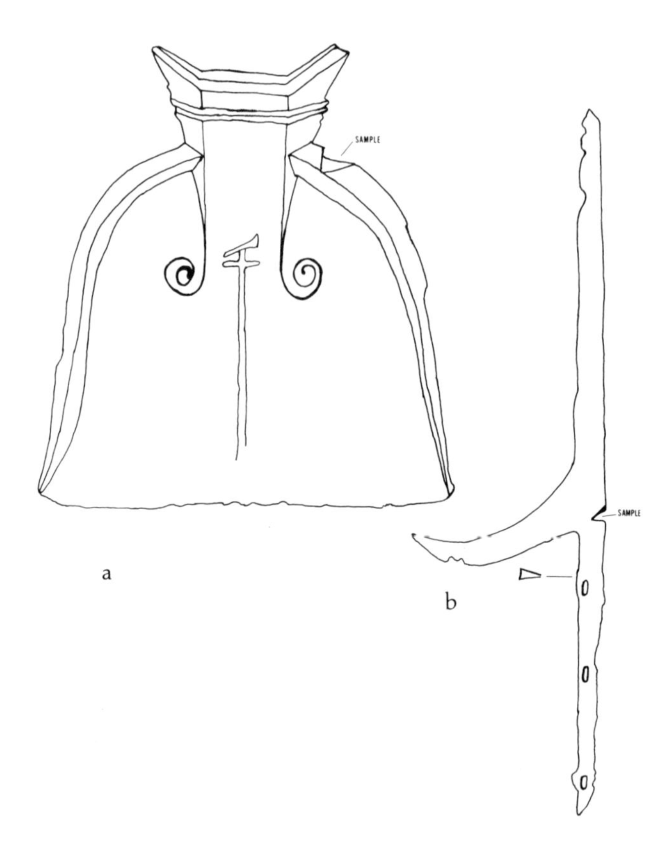

Figure 19.3 The artifacts. (a) Hoe, scale 1 : 2.5 (b) Lance, scale 1 : 7.5.

Results

Samples were taken from the two objects at the points shown in figure 19.3. A surface of about 8 mm × 10 mm, originally at a 30° angle to the shoulder edge, from the hoe, and one of about 5 mm × 16 mm, originally at a 60° angle to the upper edge, from the lance were polished and etched with 2% nital (2% nitric acid in ethanol). A preliminary examination of these showed the hoe to have been made of white cast iron that had been heat treated to cause graphitization (formation of nodules of graphite through precipitation of carbon). The massive cementite in the central region of the sample had a hardness value of 600 HV (using a 100 g or 0.981 $\bar{N}$ load), whereas the ferrite in this region gave values of 280–400 HV. However, a narrow outer layer had been heavily decarburized, probably to hypoeutectoid composition (<0.8% C). As a result, the border between the central region and this decarburized layer, a mixture of ferrite and pearlite, gave hardness values of 280–400 HV, whereas in the outer layer the ferrite gave values as low as 185–210 HV (185 HV in the outermost and hence most decarburized area). Given that about half a millimeter of the outer surface of the decarburized layer has been removed by corrosion and cleaning, it is possible that this area may have been pure ferrite and hence extremely soft. Any crack in the hoe would have had to be propagated from the outside. (For micrographs of the samples, see figures 19.4 and 19.5.)

The lance appears to have been made of martensitic—and hence quenched—steel. The fairly large amounts of slag, even distribution of carbon, and slightly higher carbon content toward the outside edge of the sample all suggest that it was made from bloomery iron heavily carburized to produce a eutectoid (about 0.8% C) or hypereutectoid (>0.8% C) steel, after which it was quenched and annealed. Hardness values for the sample vary greatly, with a range of 145–362 HV (using a 100 g or 0.981 $\bar{N}$ load) and an average of 303 ± 33 HV (1σ range). There is a possible although not statistically significant increase in hardness toward the cutting edge.[2]

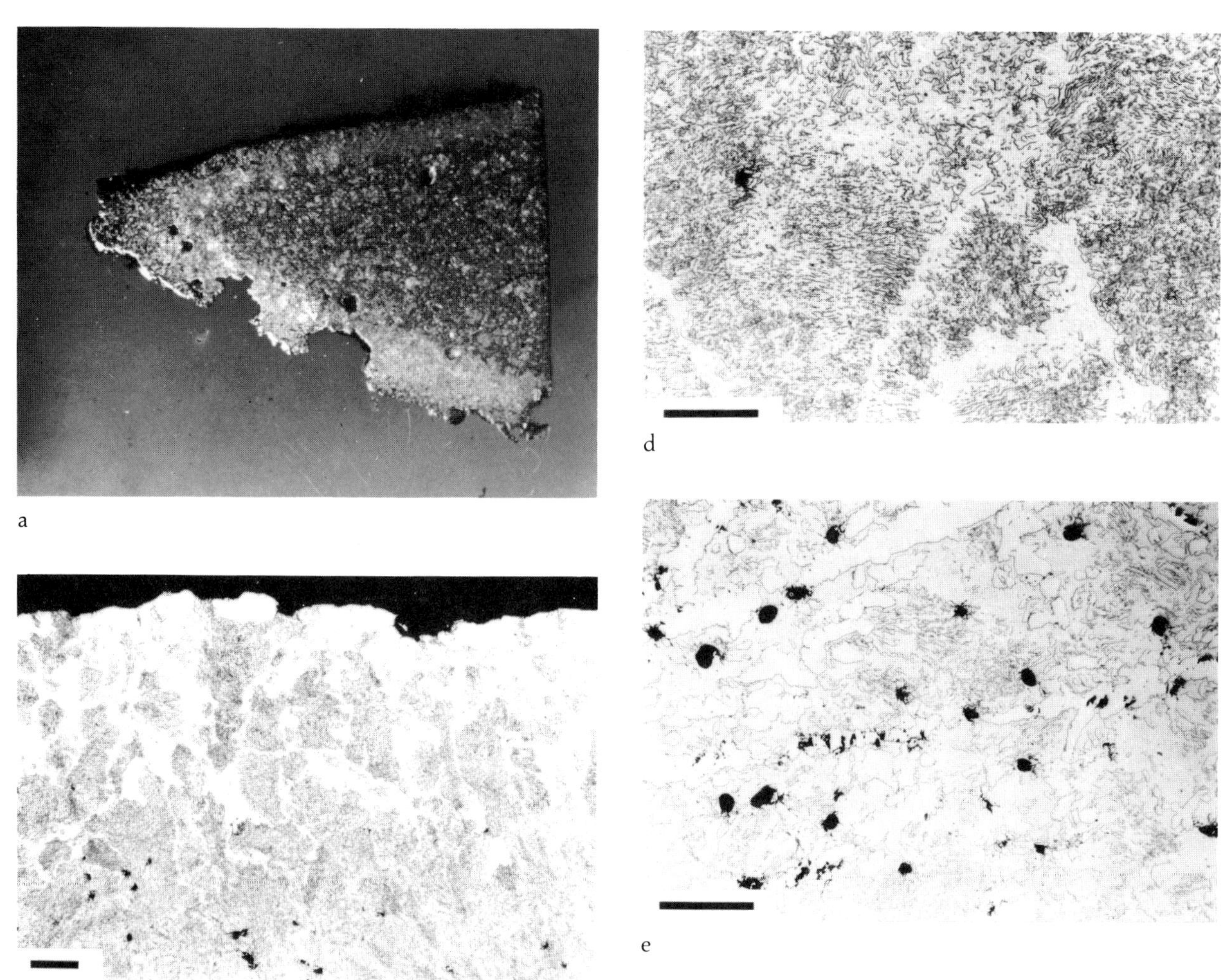

Figure 19.4 (a) Sample from the hoe (× 6). (b) Decarburized surface layer of ferrite and pearlite above central region of the hoe with massive cementite and graphite (scale bar: 0.1 mm). (c) Graphite in groundmass of massive cementite and pearlite (cavity from casting porosity; scale bar: 0.1 mm). (d) Partly spheroidized cementite in pearlite of decarburized surface of the hoe (scale bar: 0.05 mm). (e) Massive cementite and varied graphite morphology in high-carbon central region of the hoe (scale bar: 0.1 mm). Photograhs © 1986 C. A. Shell. Reprinted with permission.

a

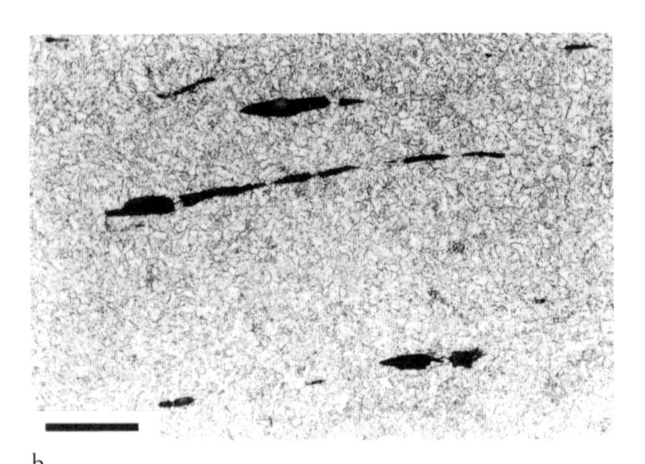

b

c

Figure 19.5 (a) Sample from the lance (× 7). (b) Elongated slag inclusions from high-temperature forging; fine spheroidized cementite in heavily annealed martensite (scale bar: 0.05 mm). (c) Fine cementite outlining fine ferrite and possibly prior austenite structure (scale bar: 0.02 mm). Photographs © 1986 C. A. Shell. Reprinted with permission.

Analysis and Comparison of Results

Tables 19.1 through 19.6 summarize the results of microstructure analyses of over 175 artifacts, divided into two groups, carried out in China and the West. We have tried to cover the body of available analyses as completely as possible, but some have been reported in too little detail to be incorporated into the tables and many others remain unpublished; moreover, not all the analytical results are definite or reliable. Nevertheless, the sample should be large enough to permit some generalization, so long as the figures are taken as rough indicators rather than absolute values.

One area in which the results do not seem to mirror entirely accurately the situation for Chinese iron artifacts in general is in the ratios of different objects produced over time. (Oddly enough the smaller sample of fifty-two artifacts seems to produce more typical results in terms of dominance of agricultural implements.) However, given the small number of artifacts that have been analyzed relative to the total number of iron finds and the extent to which selection of artifacts to be analyzed is nonrandom, this bias is not surprising. In any case it is clear from the results that the categories of agricultural implements and tools, when combined, outnumber iron weapons substantially before the Han, after which the ratios become much closer.

The ratios of different types of ferrous metals clearly reflect the dominance of cast iron and cast-iron over forged iron, wrought iron, and steel in China. This is particularly noticeable in the group of fifty-two artifacts, all of which were originally cast, although some were then decarburized to steel. Both white cast iron and malleable cast-iron are fairly common. Given the highly sophisticated forms of malleable cast-iron, approaching steel in quality, being produced by the Han, this is hardly surprising. It may seem more so in the case of cast iron, generally dismissed as of poor quality for useful tools. The appeal of white cast iron may in part have been its cheapness and its advantages, however minimal, over stone and wooden tools, but the structure of the hoe from the Cambridge collection suggests that tools of this material may not have been of such poor quality after all.

The hoe has a narrow outer decarburized zone, a common feature in many of the objects examined. Because the edges of the hoe are thin, even a relatively short and shallow decarburization could have produced the small zone at the edge that is hypoeutectoid all the way through. This soft band would allow the edge of the tool to be resharpened and make it less likely to break if it hit a rock, particularly in those tools that are edge reinforcements or "shoes" for wooden tools. One of the reasons why Warring States tools were so thin may have been to aid the production of this type

Table 19.1 Distribution of different artifact types by period (group of 127 artifacts)

Category	Fifth century B.C. or earlier	Warring States	Western Han	Eastern Han	Han (unspecified) or later	?	Total
Agricultural implements	2	10	2	1	10	3	28
Tools	9	5	5	2	8	1	30
Weapons	2	8	6	6	7	6	35
Other	2	4	1	1	8	9	25
Unidentified	2	1	0	0	2	4	9
Total	17	28	14	10	35	23	127

Sources: "Report on the microstructure of specimens" (1957), Coghlan (1956), Franklin (1974), Guan and Hua (1981), Henger (1975), Hua et al. (1960), Institute of Archaeology (1980), Lei (1980), Li (1975, 1979), Sun (1960), Tylecote (1976, 1983), Wagner (1982, 1983), Watson (1971), Xia (1973), Yang (1960), Yang and Zheng (1973), "An eastern Zhou tomb" (1965), "Preliminary report of a metallographic examination" (1975), "Preliminary appraisal of the iron and bronze tools" (1975), and "Newly discovered late Spring and Autumn iron artifacts" (1978).

Table 19.2 Distribution of different artifact types by period (group of 52 artifacts)[a]

Category	Early Warring States	Warring States	Western Han	Eastern Han	Han or Wei	Total
Agricultural implements	1	5	10	5	5	26
Tools	1	4	4	2	7	18
Weapons	0	1	1	0	0	2
Other	0	0	2	4	0	6
Total	2	10	17	11	12	52

a. Analyses by Hua Jeuming (Hua 1979).

Table 19.3 Distribution of different structures by period (group of 127 artifacts)

Category	Fifth century B.C. or earlier	Warring States	Western Han	Eastern Han	Han (unspecified) or later	?	Total
Cast iron							
White	6	5	3	1	4	7	26
Gray	0	0	0	0	3	1	4
Mixed	0	0	0	0	4	2	6
Malleable cast-iron							
Incompletely malleabilized	1	3	2	1	0	0	7
Black-heart	0	1	0	0	2	2	5
White-heart	1	0	0	0	0	1	2
Unspecified	1	3	0	1	5	0	10
Spherical graphite	0	0	1	0	3	0	4
Steel							
Forged	1	4	4	3	9	2	23
Cast	0	1	0	0	2	0	3
Unspecified	0	0	1	1	0	1	3
Wrought iron							
Unspecified	1	3	1	1	0	4	10
Direct process	5	3	0	0	1	2	11
Indirect process	0	2 (?)	0	0	2	0	4
Uncertain							
Steel or malleable cast-iron	0	2	1	2	0	1	5
Other	2	1[a]	1[a]	0	0	0	4
Total	17	28	14	10	35	23	127

Sources: See table 19.1.
a. A mixture of cast iron with wrought iron or steel.

Table 19.4 Distribution of different structures by period (group of 52 artifacts)

Category	Early Warring States	Warring States	Western Han	Eastern Han	Han or Wei	Total
Malleable cast-iron						
Black-heart	1	2	6	1	2	12
Incompletely decarburized white-heart	1	7	2	1	0	11
White-heart	0	1	2	1	1	5
Spherical graphite	0	0	1	1	1	3
Steel from cast iron	0	0	6	5	8	19
Other (raw material of decarburized cast iron)[a]	0	0	0	2	0	2
Total	2	10	17	11	12	52

Source: Hua (1979).
a. Intended for further working. The two objects were "boards" of metal, the raw material for finished implements.

Table 19.5 Distribution of different structures by artifact type (group of 127 artifacts)

Category	Agricultural implements	Tools	Weapons	Household items	Other	?	Total
Cast iron							
White	2	9	0	4	10	1	26
Gray	1	0	0	1	2	0	4
Mixed	0	1	0	0	1	4	6
Malleable cast-iron							
Incompletely malleabilized	2	3	2	0	0	0	7
Black-heart	5	0	0	0	0	0	5
White-heart	1	0	0	1	0	0	2
Unspecified	7	0	0	1	2	0	10
Spherical graphite	1	3	0	0	0	0	4
Steel							
Unspecified	0	0	2	0	0	1	3
From wrought iron	0	4	16	0	1	2	23
From cast iron	2	0	1	0	0	0	3
Wrought iron							
Unspecified	1	2	7	0	0	0	10
Direct process	2	5	3	0	0	1	11
Indirect process	0	0	3	0	1	0	4
Uncertain	4[a]	3	1	0	1	0	9
Total							
By casting	23	16	6	7	16	5	73
By forging	2	9	19	0	1	3	34
Bloomery iron unspecified	3	5	10	0	1	1	20

Source: See table 19.1.
a. A mixture of cast iron with wrought iron or steel.

Table 19.6 Distribution of different structures by artifact type (group of 52 artifacts)

Category	Agricultural implements	Tools	Weapons	Other	Total
Malleable cast-iron					
Black-heart	10	1	1	0	12
Incompletely decarburized white-heart	6	5	0	0	11
White heart	3	2	0	0	5
Spherical graphite	2	1	0	0	3
Steel from cast iron	5	9	1	4	19
Other (raw material of decarburized cast iron)	0	0	0	2	2
Total	26	17	2	6	52

Source: Hua (1979).

of completely decarburized band [or of malleableized structures (Hua 1982, p. 14)].

Wrought iron and steel are about equally common, although wrought iron predominates before the Warring States and steel after that period. Early wrought iron seems to be mainly forged bloomery iron, with the indirect process of production from cast iron appearing possibly during the Warring States and definitely by the Han. This tallies with Li Zhong's statement (1975, p. 2) that before the third century B.C., bloomery iron was the raw material for wrought iron and steel. So, although the domination of casting was strong, especially by the Han, a separate, parallel tradition of forging does seem to have existed from the beginnings of the industry. An important early (Warring States) development is that of quenching steel, as indicated by the appearance of martensitic structures. This technique was used particularly for blades.

Of the objects not identified in the original analysis as being cast iron, wrought iron, or steel, it is worth noting that two are mixtures of cast iron and either wrought iron or steel. This mixture is often termed "softened" or "irrigated" iron.

One of the most interesting sets of results is that of the division of the different types of object on the basis of the structures associated with them (tables 19.5 and 19.6). They show substantial differences in the materials used to make agricultural implements, tools, and weapons consistent with patterns suggested by Trousedale (1975, p. 65). The first group is made almost entirely of cast iron and cast-iron or (rarely) steel made from cast iron. The few wrought iron implements, however, are forged from bloomery iron. The tools are more mixed, though cast- and cast iron still predominate. Wrought (mainly forged bloomery) iron tools are reasonably common, but steel remains rare. Weapons show a completely different pattern, with virtually no cast iron and a roughly even balance between wrought iron (produced by a variety of methods) and steel (produced primarily from forged wrought iron). The miscellaneous category, consisting mainly of casting molds and household objects, is once again mainly cast iron and cast-iron, with wrought iron and steel rare (particularly steel if one excludes two unidentified fragments from the total).

Broadly, then, the objects analyzed can be divided into two groups. The first of these, a wide range of tools and implements that were common and designed for frequent use, was produced mainly by casting, with a few forged objects—mainly tools whose simple shape or need for shock resistance made cast iron of little use. This does not necessarily mean that the overriding consideration in the production of these tools was cheapness at the expense of quality. Quite apart from the high-quality malleable cast-iron being produced at an early date, white cast iron, as was the case with the described hoe, could be rendered fairly serviceable. The following advantages conferred by casting were probably far more important: the large scale of production it allowed, possibilities of standardization and greater centralized control of tool types, and the ease with which complex, particularly socketed, shapes could be produced.

The second group consists of iron weapons. Although simple forged iron weapons can be produced, the method is generally more labor intensive than casting, especially if weapons of any quality are required. Moreover, many of the weapons analyzed are of the luxury metal steel ("luxury" because of its high quality and comparative rarity among all the objects). This suggests that weapons themselves were closer to being a luxury good, as technical considerations seem to dominate over economic ones in their production. [It might be argued that they have to, because a soldier's life depends on his weapon. However, a quick look at the varying quality of ancient iron weapons produced in Europe shows that quite poor weapons were often used by fighters who had little choice in the matter (Forbes 1964, p. 203)—hence the mystique that surrounded a few presumably high-quality swords in popular legends.]

A case in point is the lance from the Cambridge collection. It would have been quicker to cast a blank for it and decarburize this to steel. However, this would preclude much further working, as steel is difficult to hot-work, and would make the tough quenched martensitic structure difficult to obtain. By using the slower, more labor-intensive method of forging the lance from bloomery wrought iron and then carburizing and quenching it, a thin hammered cutting edge and a tough martensitic structure were obtained.

The importance of labor-intensive methods like this, which require an individual smith's skill, and the great variability in the methods used to produce the weapons analyzed suggest a much-smaller-scale localized production in comparison with that of implements or tools. Such a mode of production would probably have been less subject to government or merchant monopoly control. Fewer of these luxury items could be produced than common tools, and the complex methods used to produce them may have taken slightly longer to develop (although this could be due more to the slowness with which technological innovations would spread in a decentralized industry than to the level of complexity of the techniques used, which was after all similar to that of some techniques used on cast iron, such as malleabilization). Thus the impact of iron weapons production was probably initially slighter and later in date than that of iron tool production.

Conclusions

The results confirm and expand many of the conclusions drawn earlier. They portray an industry in which forging techniques and casting techniques developed more or less simultaneously. [Given Barnard's (1980–1981) convincing arguments that there is no evidence for early wrought bronze in China, this points either to possible, although as yet unproven, import of the technique from the outside or to a slightly earlier start for forging to allow more time for experimentation.] Yet it is casting that predominates, especially from the Han onward, and gives the industry its unique character. The range of products expands from white cast iron, early malleableized cast-iron, possibly steel, and forged wrought iron around the fifth century B.C. to a wide selection of products of various sophisticated techniques by the Han.

Although the issue of the extent to which bronze tools were produced remains unresolved, it is difficult to draw firm conclusions about the differences in use between bronze and iron. Nevertheless, the importance of ceremonial vessels and weapons in the bronze assemblage and the economic limits on producing a large number of bronze tools suggest that the large absolute and relative numbers of common agricultural and artisans' tools in the early iron assemblage was a new phenomenon. Thus for the common people the Warring States Period saw their gradual entry into the Metal Age for the first time. This change was no doubt a significant factor in the agricultural revolution and ambitious water-control projects that were well under way by the Han, but it had sociopolitical implications as well.

That the nobility had controlled bronze production during the Shang and Zhou dynasties was probably neither here nor there for the vast majority of peasants. Metal had become a part of their daily life and was being produced—as the analyses show—virtually entirely by casting, a method highly susceptible to centralized control, whether governmental or private. Standardization, introduction of more efficient tool types, and economic exploitation all could and did arise from this type of production. The debate on the iron and salt monopolies in the *Yan Tie Lun* (*Discourses on Salt and Iron*) shows this clearly (Huan 1967). The literati complain on behalf of the people of the poor quality of government-produced tools and suggest that the monopoly exists only to refill the Imperial Treasury. The legalists respond by pointing to the exploitative practices of the private iron merchants in the past and the government's concern with modernizing and improving agriculture and water management (the latter through vast public works using corveé labor). Thus the common people seem to have paid for their iron tools by providing yet another opportunity for government control over them and economic exploitation of them by the emerging "middle class" of merchants and landowners.

In the early stages of iron production iron weapons seem to have been rare relative to tools and implements. By the Warring States their number was increasing, and by the Han weapons had become an important component of the iron assemblage. The weapons analyzed are predominently of forged wrought iron or steel, which suggests a different mode of production from the one described for tools. This gives yet another blow to the theory that the state of Qin owed its victory over the other Warring States to cheap mass-produced swords of either cast-iron or simple forged iron because the majority of weapons analyzed use sophisticated techniques and are of high quality, often of steel. The priority was clearly not economical mass production. Although iron use may have had only a minor influence on the internal conflicts of the Warring States—at least for the state of Qin—its importance in the Han campaigns against the "northern barbarians" is demonstrated by government edicts forbidding the export of iron weapons or tools beyond the Great Wall (Yü 1967, pp. 117–120). Despite this, the mode of weapons production was not one that lent itself to centralized control, although it would not be impossible to have government smithies, or to large-scale production. Possibly common soldiers were armed with arrows and spears, the heads of which can be efficiently made of cast-iron, whereas the production of high-quality weapons for officers remained a relatively small-scale luxury industry more similar to bronze weapon production and substantially different from contemporary production of tools and implements for the commoner.

The number of microstructure analyses of ancient Chinese iron now stands at a respectable total. There is room for much more work of this kind, however, particularly on securely dated objects from the early stages of iron production. Equally important is the need to bring together the scattered, often inaccessible or unpublished analytical reports. This chapter has, we hope, shown the usefulness of just such an exercise both in confirming the general outline of the history of iron metallurgy in China, as provided by scholars such as Huang Zhanyue, Li Zhong, and Needham, and of updating and expanding their conclusions. The development of iron technology in China is not only an important topic in the history of science but also one that can shed a good deal of light on the economic and sociopolitical history of that country, and as such it is to be hoped that more attention will be devoted to it in the future.

Notes

This chapter is partly based on S. Taylor's master's thesis, submitted in 1984 to Cambridge University. We wish to thank D. Phillipson, curator of the Museum of Archaeology and Anthropology, Cambridge, for allowing experimental work to be carried out on the two pieces from the museum collection. We would also like to thank D. Wagner, of the East Asian History of Science Library, Cambridge, and G. L. Barnes, of the Archaeology Department, Cambridge University, for their invaluable help and advice. Any remaining errors or omissions are entirely the authors'.

1. Not only do tools and implements dominate over weapons in the iron assemblage, but perhaps more telling, the ratio of iron to bronze in tools and implements is much higher than in weapons (Huang 1957, p. 104; Watson 1961, p. 141).

2. Analysis of these two samples is still underway, and full results will be published by C. A. Shell in due course.

References

Archaeological Team, Bureau of Culture, Henan Province. 1962. *Tieshenggou, Gong County*. Beijing: Wenwu Chubanshe. (In Chinese.)

Barnard, N. 1961. *Bronze Casting and Bronze Alloys in Ancient China*. Tokyo: Monumenta Serica and Australian National University.

Barnard, N. 1979–1980. "Some observations on metal-winning and the 'societal requirements' of early metal." *Early China* 5 : 124.

Barnard, N. 1980–1981. "Wrought metal-working prior to the middle Shang (?): A problem in archaeological and art-historical research approaches." *Early China* 6 : 4–30.

Barnard, N., and T. Sato. 1975. *Metallurgical Remains of Ancient China*. Tokyo: Nichiosha.

Blunden, C., and M. Elvin. 1983. *Cultural Atlas of China*. Oxford: Phaidon.

Ch'ao Ting-chi. 1936. *Key Economic Areas in Chinese History*. London: Allen & Unwin.

Cheng Te-K'un 1963. *Archaeology in China. Vol. 3, Chou China*. Cambridge and Toronto: Heffer and University of Toronto Press.

Cheng S. P. 1978. "An iron and steel works of 2000 years ago." *China Reconstructs* 27(1):32–34.

Coghlan, H. H. 1956. *Notes on Prehistoric and Early Iron in the Old World*. Oxford: Oxford University Press.

Cressey, G. B. 1934. *China's Geographic Foundation*. New York: McGraw Hill.

"An eastern Zhou tomb at Chengqiao, Luhe county, Jiangsu." 1965. *Kaogu* 3:105–115. (In Chinese.)

Eberhard, W. 1968. *The Local Cultures of South and East China*, second edition, A. Eberhard, trans. Leiden: E. J. Brill.

"An examination of Han cast iron smelting techniques on the basis of the ancient remains at Yingzhen." 1978. *Wenwu* 2:44–47. (In Chinese.)

"Excavation report of tomb 44, Yanxiadu, Hebei Province." 1975. *Kaogu* 4:228–240. (In Chinese.)

Forbes, R. J. 1964. *Studies in Ancient Technology*, vol. 9. Leiden: E. J. Brill.

Franklin, U. M. 1974. "Technical studies on two samples of early Chinese iron." Offprint in the East Asian History of Science Library, Cambridge University.

Franklin, U. M. 1983. "On bronze and other metals in early China," in *The Origins of Chinese Civilization*, D. N. Keightley, ed. (Berkeley, Calif.: University of California Press), 279–296.

Guan, H. Y., and Hua Jueming. 1981. "Research on Han-Wei spherical graphite cast iron." Manuscript in the East Asian History of Science Library, Cambridge University.

Henger, G. W. 1970. "The metallography and chemical analysis of iron-base samples dating from antiquity to modern times." *Bulletin of the Historical Metallurgy Group* 4(2):45–52.

Hommel, R. P. 1937. *China at Work*. New York: John Day.

Hsu Cho-yun. 1965. *Ancient China in Transition*. Stanford, Calif.: Stanford University Press.

Hsu Cho-yun. 1978. "Agricultural intensification and marketing agrarianism in the Han Dynasty," in *Ancient China: Studies in Early Civilization*, D. T. Roy and T. H. Hsien, eds. (Hong Kong: Chinese University Press).

Hua Jueming. c. 1982. "Ancient bronze smelting, casting and early iron casting in China." Offprint in the East Asian History of Science Library, Cambridge University.

Hua Jueming. 1983. "The mass production of iron castings in ancient China." *Scientific American* 248(1):120–128.

Hua Jueming, Liu S. Z., and Yang G. 1960. "Preliminary report of a metallographic examination of some Warring States and Han iron objects." *Kaogu* 1:73–87. (In Chinese.)

Huan, K. 1967. *Discourses on Salt and Iron: A Debate on State Control of Commerce and Industry*, second edition, E. M. Gale, trans. Taipei: Ch'eng-Wen.

Huang Zhanyue. 1957. "Iron implements of the period of the Warring States and Han Dynasty found in recent years." *Kaogo Xuebao* 3:93–108. (In Chinese.)

Huang Zhanyue. 1976. "Some questions concerning the beginnings of iron smelting and the use of iron implements in China." *Wenwu* 8:62–69. (In Chinese.)

Institute of Archaeology, Academia Sinica. 1961. *A Collection on the Archaeology of the New China*. Beijing: Wenwu Chubanshe. (In Chinese.)

Institute of Archaeology, Academia Sinica. 1980. *Excavation Report of the Han Tomb at Mancheng*. Beijing: Wenwu Chubanshe. (In Chinese.)

Keightley, D. 1976. "Where have all the swords gone? Reflections on the unification of China." *Early China* 2:31.

Lei C. Y. 1975. "A survey of iron agricultural implements of the Warring States Period," D. B. Wagner, trans. Offprint in the East Asian History of Science Library, Cambridge University.

Li Zhong. 1975. "Inquiry into the development of iron and steel smelting techniques at the beginning of the feudal period in China." *Kaogu Xuebao* 2:1–22. (In Chinese.)

Li Zhong. 1976. "Examination of the level of achievement of ancient Chinese smelting techniques on the basis of the Mianchi implements." *Wenwu* 8:59–61. (In Chinese.)

Li Zhong. 1979. "Studies on the iron blade of a Shang Dynasty Yueh-axe unearthed at Kao-ch'eng, Hopei, China." *Ars Orientalis* 11:259–290.

Loehr, M. 1956. *Chinese Bronze Age Weapons*. Ann Arbor, Mich.: University of Michigan Press.

Maddin, R., J. D. Muhly, and T. S. Wheeler. 1977. "How the Iron Age began." *Scientific American* 237(4):122–131.

Needham, J. 1964. *The Development of Iron and Steel Technology in China*. Cambridge: Heffers.

"Newly discovered late Spring and Autumn iron artifacts and a steel sword from Changsha." 1978. *Wenwu* 10:44–48. (In Chinese.)

"Preliminary appraisal of the iron and bronze tools excavated at the ancient mining site of Tonglushan." 1975. *Wenwu* 2:19–25. (In Chinese.)

"Preliminary report of a metallographic examination of the iron implements from tomb 44 at Yanxiadu." 1975. *Kaogu* 4:241–243. (In Chinese.)

Read, T. T. 1937. "Chinese iron: A puzzle." *Harvard Journal of Asiatic Studies* 2:398–407.

"Report on the microstructure of specimens of Chinese casting from the Royal Ontario Museum." 1957. Manuscript in the East Asian History of Science Library, Cambridge University.

Sun T. L. 1960. "Metallographic examination of some iron implements excavated at Hui county." *Kaogu Xuebao* 2: 125–140. (In Chinese.)

Swann, N. 1950. *Food and Money in Ancient China: The Earliest Economic History of China to A.D. 25 (Han Shu 24).* Princeton, N.J.: Princeton University Press.

Thorp, R. 1980–1981. "The Chinese Bronze Age from a Marxist perspective." *Early China* 6:97–102.

Trousedale, W. 1975. *The Long Sword and Scabbard Slide in Asia.* Smithsonian Contributions to Anthropology 17. Washington, D.C.: Smithsonian Institution.

Tylecote, R. F. 1976. *A History of Metallurgy.* London: The Metals Society.

Tylecote, R. F. 1983. "Ancient metallurgy in China." *The Metallurgist and Materials Technologist* (September), 435–439.

Wagner, D. B. 1982. "Theoretical considerations on iron casting." Offprint in the East Asian History of Science Library, Cambridge University.

Wagner, D. B. 1983. "Notes on the traditional iron industry of southern Henan." Offprint in the East Asian History of Science Library, Cambridge University.

Wainwright, G. A. 1936. "The coming of iron." *Antiquity* 10:5–24.

Wang Z. S. 1982. *Han Civilization,* K. C. Chang and collaborators, trans. New Haven, Conn.: Yale University Press.

Watson, W. 1961. *China before the Han Dynasty.* London: Thames and Hudson.

Watson, W. 1971. *Cultural Frontiers in Ancient East Asia: The Rhind Lectures 1965–1966.* Edinburgh: Edinburgh University Press.

Wertime, T. A. 1961. *The Coming of the Age of Steel.* Leiden: E. J. Brill.

Wilbur, C. M. 1943. "Industrial slavery in China during the former Han Dynasty." *Journal of Economic History* 3(1): 56–69.

Xia N. 1973. "The Shang period remains at Xicun, Gaocheng, Hebei Province." *Kaogu* 5:266–271. (In Chinese.)

Yang G. and Zhang Z. G. 1973. "An examination of Warring States casting techniques on the basis of the pottery molds from Houma and iron molds from Xinglong." *Wenwu* 6:62–65. (In Chinese.)

Yu Ying-shih 1967. *Trade and Expansion in Han China.* Berkeley, Calif.: University of California Press.

Zhou W. Z. 1957. *A History of Chinese Weapons.* Beijing: San Lian Bookstore Press. (In Chinese.)

20
Lost Wax and Lost Textile: An Unusual Ancient Technique for Casting Gold Belt Plaques

Emma C. Bunker

Many spectacular gold ornaments looted from unidentified burial mounds in southern Siberia were presented to Catherine, wife of Peter I of Russia, to honor the birth of their son in 1715 (Bunker et al. 1970, pp. 109–110). These ornaments, now housed in the Hermitage Museum in Leningrad, are not all of the same provenance. One distinctive group of openwork plaques among these treasures depicts scenes of animals and humans from the oral mythic and epic traditions of the ancient horse riding tribes who inhabited the grasslands beyond the Great Wall (Rudenko 1962, pls. I.4, 5, II.5, 9, IV.2, 3, 5, V.1–3, VI.3, 4, VII.1, 7, VIII.1–3, IX.1, 2, 6–8, XII.4, 5, XXIII.28, 29). So far, all attempts to place these plaques successfully in time, space, and history have failed, although a good deal of romantic speculative prose, including my own, has been expended (Salmony 1947, 1948, 1949, 1952; Haskins 1959, 1961, 1962; Bunker 1978). Now, new observations concerning their art historical relationships and the manner in which they were manufactured suggest a Far Eastern origin for these plaques rather than the Central Asian provenance currently promoted by Russian scholars (Sarianidi 1985, p. 54).

The distinguishing feature of this group of belt plaques is that the reverse of each has the appearance of a coarse, loosely woven fabric that is usually erroneously described as an imprint "of the coarse cloth with which the mold was covered during casting" (British Museum 1976, p. 17). Actually, on close examination, the back of each plaque displays a positive textile relief that duplicates in casting the original textile that must have reinforced the model (figure 20.1). A gold hook buckle with the textile appearance on the reverse in the Metropolitan Museum of Art in New York reveals undercutting in the decoration that could only have been accomplished with the use of a model made of wax (figure 20.2). Therefore the casting technique employed can be referred to as "lost wax and lost textile" (Bunker and Ternbach 1970; Hunt 1980, p. 73). In other words, the wax and the textile were both replaced by the metal induced into the mold during the casting process. A brief study of more belt plaques with textile relief backs in European and American collections allows for a better understanding of this technique.

Primarily, the lost-wax and lost-textile casting method was employed for producing plaques of gold and silver (Hearn 1987, fig. 78). Only a few bronze plaques, most of which were mercury gilded, have the textile relief on the reverse (Karlbeck 1966, pl. 32.1, 2; Bunker et al. 1970, fig. 119). The fabric reinforcement allowed for a thinner model that would reduce the amount of metal expended and diminish the weight in the finished product (figure 20.3). Bronze plaques that were not cast from a fabric-reinforced model but are decorated with designs that derive from those found on lost-wax

Figure 20.1 Reverse of gold hook buckle, third to second century B.C., 7.8 cm in diameter.

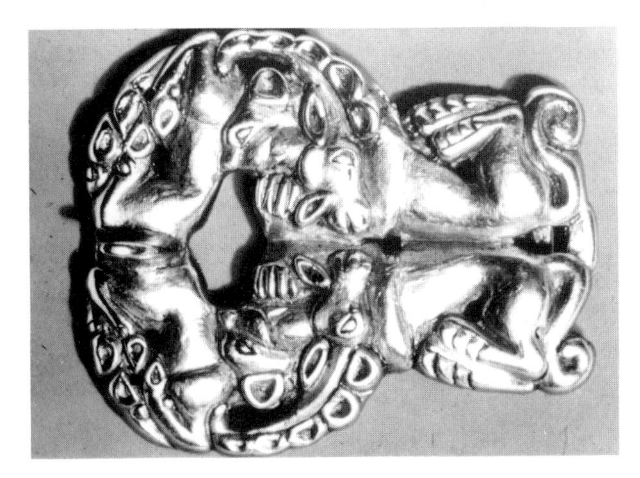

Figure 20.2 Gold hook buckle, third to second century B.C., 7.8 cm in length.

and lost-textile examples are much thicker and heavier by comparison with their earlier prototypes (figure 20.4).

What inspired the unprecedented addition of a piece of textile into the lost-wax casting process? During the late Warring States and early Han periods, one of the methods for producing the wide assortment of lacquer objects was one in which the "pieces were lacquered over a simple wooden core, or reinforced with lacquer-impregnated fabric applied to the core" (Kuwayama 1982, introduction). Judging from X-rays of a few Han examples in the Asian Art Museum of San Francisco, this fabric appears to be a loosely woven coarse hemp or similar cloth (Richard Barden, Asian Art Museum of San Francisco, Conservation Department, personal communication). It is tempting to see a connection between this lacquer technique and the lost-wax and lost-textile method of casting that was utilized from the third to first century B.C. to produce belt ornaments for the Yuezhi and Xiongnu tribes on China's northern borders.

The plaques with the fabric appearance on the backs were probably made in the following manner (Richard Kimball, master goldsmith and jewelry scholar, personal communication). A model for each plaque was made out of fabric-reinforced wax. The fabric on the edges of the model and any openwork edges were trimmed with a knife, evidenced by the slightly ragged appearance of the edges seen on the backs of the plaques. The resulting thin fabric-reinforced wax model was invested in a clay mold. Thus, when the wax and textile were burned out and replaced by molten metal, the resulting plaque carried a reproduction of the model's reinforcing fabric on its reverse. The attachment loops on the reverse of each plaque were then cast onto the finished plaque.

Many plaques, which have the textile appearance on the back or are stylistically related to plaques that do, have been found in northern China and beyond the Great Wall in Inner Mongolia and Ningxia in recent decades with material that can be dated to between the third and first centuries B.C. A bronze buckle without the fabric appearance that is of the same type of hook buckle as the Metropolitan Museum example already discussed was recently excavated from a cemetery belonging to either the Yuezhi or the Xiongnu at Xigou, Zhungeer Banner, western Inner Mongolia near the Great Wall, along with two gold plaques that are inscribed with third century B.C. Chinese characters (Yih Ju Meng Relics Work Station, 1980, p. 8; Li 1985, pp. 333–334). Two gilt bronze plaques of recumbent yaks with the textile appearance (Stockholm Museum) belong to the material found at Shouxian in the Huai valley (Karlbeck 1966, pl. 31.1, 2). One Shouxian plaque is the same as a plaque belonging to the Shaanxi Provincial Museum in Xi'an.

The plaques with the textile relief on the back exhibit a full range of styles and subject matter designed to fulfill the mythic and totemic requirements of the various tribes that were ultimately united into a vast steppe empire by the Xiongnu early in the Western Han Period. China had supplied the pastoral herdsmen to the north with luxuries for centuries in trade for horses, furs, jade, carpets, and rugs. Lacquer and silk abound in the tombs of Pazyryk in southern Siberia, along with a fragment of a late Warring States mirror (Rudenko 1970, p. 115). In many instances these plaques exhibit telltale characteristics of style or motif, which suggests a close awareness of things Chinese. The sword worn by the rider on the famous Hunting Plaque in Leningrad is a Chinese weapon with a third century B.C. trapezoidal shape suspended by a typical Chinese scabbard slide (figure 20.5). The flipped up hind hooves represent an independently invented Chinese version of the "flying gallop," also seen on a painted Changsha lacquered box (Jaffe 1983; Sullivan 1962, pl. 60), which does not derive from Central Asia or further west. Even the tree forms are Chinese and belong to the same species as those on a lien from

Figure 20.3 Belt plaque, third to second century B.C. Gold with turquoise inlay. Hermitage, Leningrad.

Figure 20.4 Bronze belt plaque, second to first century B.C., 10.5 cm in length. Los Angeles County Museum of Art, Los Angeles.

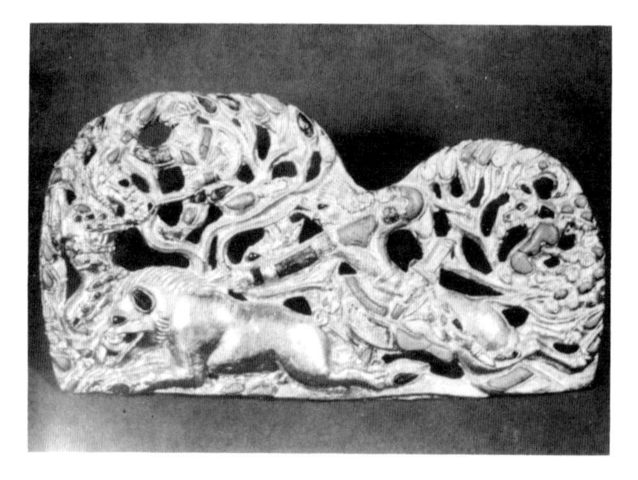

Figure 20.5 Hook buckle, third to second century B.C. Gold with turquoise inlay, 19 cm in length. Hermitage, Leningrad.

Changsha (Sullivan 1962, pl. 7). The use of rich inlay and small pictorial scenes in relief abound on the *boshanlu* of the Han Period. Further art historical comments could be considered but are beyond the scope of a metallurgy conference.

The discovery of a gold plaque from Verkhne-Udinsk near Lake Baikal (Bunker et al. 1970, fig. 18) and a pair among the finds at Tillyatepe in Afghanistan that have the fabric appearance have prompted the Russians to claim a Central Asian source for these Peter the Great gold plaques, a claim that cannot be substantiated (Sarianidi 1985, pp. 42, 246). The gold plaque from Lake Baikal (figure 20.6), known to be a Xiongnu summer campground, is an earlier version of a bronze plaque found at Xichagou in Liaoning (Sun Shoudau 1960, p. 33); it was associated with *wushu* coins not minted before 118 B.C. (figure 20.7) in an area that the Xiongnu had conquered. Each Tillyatepe buckle depicts a figure with Chinese features driving a chariot drawn by winged felines that are similar to Han Period engravings found at Yinan in Shandong province (Yinan 1956, pp. 97–98). One openwork gold plaque with a textile appearance on the back (figure 20.8) in the Peter I treasure is decorated with creatures that are related to Chinese dragons found on numerous Warring States and Han jades (Lawton 1982, p. 144). This plaque is also inlaid with glass paste, which, according to Li (1985, p. 341), often replaced jade for inlay during the late Warring States and Han periods. An almost identical version of this plaque in bronze without the textile relief on the back was recently excavated by Wu En with recognizable Xiongnu ceramics and *wushu* coins not minted before 118 B.C. in Ningxia, another area into which the Xiongnu had expanded (Wu, personal communication; Museum of Ningxia Hui, 1987). This plaque and other pictorial plaques were found near the waists of the dead, both male and female (figure 20.9).

The evidence for the manufacture of the plaques cast with lost wax and lost textile points to a Far Eastern source for the technique. Whether they were cast in one workshop or several, or even by itinerant Chinese artisans, is as yet unclear, but it is clear from the variety of thread thicknesses seen on the backs that the technique was not standardized. A jade plaque that copies a Xiongnu plaque excavated in Ivolga (Watt 1980, fig. 6; Davidova 1971) was certainly Chinese made, as the Xiongnu did not know how to carve jade (figure 20.10). Cruder bronze versions or stylistically related examples of the gold and silver plaques with the textile relief on the backs occur in burials all over the eastern Eurasian steppe lands from Liaoning to southern Siberia during the heyday of Xiongnu power. Consequently a Siberian provenance is perfectly acceptable for ornaments made for the Xiongnu elite. Further research on existing plaques and future

Figure 20.6 Gold belt plaque found at Verkhne-Udinsk, third to second century B.C. Hermitage, Leningrad.

Figure 20.8 Belt plaque, second century B.C. Gold with glass paste inlays. Hermitage, Leningrad.

Figure 20.7 Bronze belt plaques excavated at Xichagou in Liaoning Province, second to first century B.C. Liaoning Provincial Museum. [After Sun Shoudao (1960, p. 33)].

Figure 20.9 Bronze belt plaque, second to first century B.C. [after Museum of Ningxia Hui (1987, no. 1)].

Figure 20.10 Jade belt plaque, third to second century B.C., 16.4 cm in length. Private Collection, Hong Kong.

archaeological discoveries should provide more evidence for the circumstances surrounding their manufacture in the years to come.

Acknowledgments

I would like to thank Lore Holmes of the Arthur M. Sackler Foundation, W. T. Chase of the Freer Gallery, and Pieter Meyers of the Los Angeles County Museum of Art for their help and advice for this project. The conclusions and any errors are my own. I am also indebted to my colleague Wu En, of the Institute of Archaeology of the Chinese Academy of Social Sciences, for his generosity in sharing his unpublished material from his recent excavations in Ningxia with me on my visit to Beijing.

References

British Museum. 1976. *Frozen Tombs*. London.

Bunker, E. C. 1978. "The anecdotal plaques of the eastern steppe regions," in *Arts of the Eurasian Steppe Lands*, P. Denwood, ed. (London: University of London), 121–142.

Bunker, E. C., and J. Ternbach. 1970. "A variation of the 'lost wax' process." *Expedition* 12(3): 41–43.

Bunker, E. C., C. B. Chatwin, and A. R. Farkas. 1970. *"Animal Style" Art from East to West*. New York: The Asia Society.

Davidova, A. V. 1971. "On the question of the Hun artistic bronzes." *Sovyetskaya Arkheologiya* 1: 93–105. (In Russian.)

Haskins, J. 1959. "Sarmatian gold collected by Peter the Great." *Artibus Asiae* 22(1/2): 64–78.

Haskins, J. 1961. "Targhyn the Hero, Ag-zhunus The Beautiful and Peter's Siberian gold." *Ars Orientalis* 4: 155–157.

Hearn, M. K. 1987. *Ancient Chinese Art, The Earnest Collection in the Metropolitan Museum*. New York: The Metropolitan Museum.

Hunt, L. B. 1980. "The long history of lost wax casting." *Gold Bulletin* 13(2): 63–79.

Jaffe, I. B. 1983. "The flying gallop: East and West." *The Art Bulletin* (June), 183–200.

Karlbeck, O. 1966. "Selected objects from ancient Shou-chou." *Museum of Far Eastern Antiquities (Stockholm) Bulletin* 38: 83–192.

Kuwayama, G. 1982. *Far Eastern Lacquer*. Los Angeles, Calif.: Los Angeles County Museum of Art.

Li Xueqin. 1985. *Eastern Zhou and Qin Civilizations*, K. C. Chang, trans. New Haven, Conn.: Yale University Press.

Lawton, T. 1982. *Chinese Art of the Warring States Period: Change and Continuity, 480–222 B.C.* Washington, D.C.: Freer Gallery of Art, Smithsonian Institution.

Museum of Ningxia Hui Autonomous Regions and Tongxin District Cultural Objects Management Institute, and Ningxia Archaeology Section, Institute of Archaeology, Chinese Academy of Sciences. 1987. *Kaogu* 1: 33–37. (In Chinese.)

Report on the Excavation of the Ancient Tomb with Pictorial Stones in Yinan. 1956. Beijing: Bureau of Cultural Relics, Ministry of Culture. (In Chinese).

Rudenko, S. I. 1970. *The Frozen Tombs of Siberia*. Berkeley, Calif.: University of California Press. Originally published in Russian in 1962.

Salmony, A. 1947. "Sarmatian gold collected by Peter the Great." *Gazettes des Beaux Arts*, ser. 6, 31: 5–14.

Salmony, A. 1948. "Sarmatian gold collected by Peter the Great." *Gazettes des Beaux Arts*, ser. 6, 33: 321–326.

Salmony, A. 1949. "Sarmatian gold collected by Peter the Great." *Gazettes des Beaux Arts*, ser. 6, 35:5–10.

Salmony, A. 1952. "Sarmatian gold collected by Peter the Great." *Gazettes des Beaux Arts*, ser. 6, 40:85–92.

Sarianidi, V. 1985. *The Golden Hoard of Bactria*. Leningrad and New York: Aurora Art Publishers and Abrams.

Sullivan, M. 1962. *The Birth of Landscape Painting in China*. Berkeley, Calif.: University of California Press.

Sun Shoudao. 1960. "Discoveries from a group of ancient tombs of the Xiongnu culture at Xichagou." *Wenwu* 8–9: 25–35. (In Chinese.)

Watt, J. C. Y. 1980. *Chinese Jades from Han to Ch'ing*. New York: The Asia Society and Weatherhill.

Yih Ju Meng Relics Work Station and the Inner Mongolia Cultural Relics Work Team. 1980. "The Xiongnu tombs at Xibouban." *Wenwu* 7:1–10. (In Chinese.)

21
A Bronze-Casting Revolution in the Western Zhou and Its Impact on Provincial Industries

Jessica Rawson

Here I describe relationships between Chinese metropolitan bronze casting centers and those of the southern provinces during the Western Zhou Period (c. 1050–771 B.C.).

In Henan and Shaanxi ritual vessels were cast in complex ceramic piece molds, a specific technology not exploited in quite the same form anywhere else in the ancient world. When ritual vessel casting was undertaken using identical techniques outside these metropolitan areas, in southern China for example, the technology must have been borrowed. Quite early Shang vessels reproducing metropolitan examples are found in both central-southern and southeastern China, but a continuous industry did not develop from this base. Instead southern China seems to have been receptive to metropolitan techniques for rather limited periods and to have lost interest in the intervening stages.

When contact between the metropolis and the provinces was close, ritual vessels were accurately reproduced; when that contact was interrupted, the south deviated from northern practices, making their own highly inventive vessels. At times some areas seem to have given up making vessels altogether. A revolution in the repertory of ritual vessels employed by the Western Zhou is particularly helpful in locating moments of contact between the center and the periphery. Some southern regions copied metropolitan vessels before the ritual changed and were subsequently unaffected by the revolution; other areas followed the sequence of developments at the capital more closely.

I first examine the nature of the ritual vessel change and its probable date and then describe differing responses to the changes in central-southern and southeastern China.

Vessel Sets from Metropolitan Areas

The changes in vessel repertory can be best described in terms of the sets of vessels used for ritual sacrifices to the ancestors. A vessel set is here defined as a regularly recurring group of different vessel types required by a family or an individual of a given period to perform the accepted ritual food and wine offerings to their ancestors. This account ignores the complications introduced to a set by the exceptionally high rank of an individual, such as Fu Hao, consort of the Shang king Wu Ding, whose tomb contained a number of bronzes far in excess of the sets now to be considered.[1] The areas of China here treated as metropolitan in the Western Zhou Period include Shaanxi, parts of Shanxi, and most but not all of Henan, Hebei, and Shandong provinces.

Tomb groups provide useful information on vessel sets. At any one point in the Western Zhou Period, tombs in the metropolitan area seem to have contained a standard series of vessels. Such burials were probably equipped with vessels used in daily life and required by the occupant after death so as to continue making offerings to his or her ancestors; alternatively the vessels may have been part of the paraphernalia employed at the burial ritual. Unfortunately for some periods, especially the late Western Zhou, excavated burials are few, and information has to be gleaned from hoards containing vessels that share inscriptions indicating that they were made by a particular individual.

Early Western Zhou sets are represented by the contents of a tomb excavated at Shaanxi Jingyang Gaojiabao (figure 21.1).[2] The vessels comprised two damaged *ding*, a pair of *gui* on square bases (one now damaged), a *yan*, two *jue*, a *gu* (damaged), a *zhi* (damaged), a cylindrical *zun*, two unmatched *you*, a *pan*, and a *he*. The square bases on which the *gui* were cast and the decoration of several vessels with coiled dragons are features of early Western Zhou casting. Vessel sets of the next stage, the middle Western Zhou, reveal a change not so much in vessel type as in style. Smooth rounded containers decorated with bird motifs were fashionable. Vessels from tomb M19 at Shaanxi Fufeng Qijiacun represent the period with two *ding*, two *gui*, two *jue*, a *zun*, a *you*, a *zhi*, a *yan*, a *pan*, and a *he* (figure 21.2).[3] All these vessels are recognizably descended from late Shang and early Zhou vessel types.

Typical early and middle Western Zhou bronze ritual vessel sets, therefore, comprised the following: food vessels (pairs of *ding*, pairs of *gui* with S-shaped profiles, and a single *yan*), wine vessels (pairs of *jue*, one or two *gu*, and a single *zhi*, cylindrical *zun*, and one or two *you*), and water vessels (single *he* and *pan*).

These vessels shapes and this particular grouping cannot be traced much later than the second half of the middle Western Zhou. Rather suddenly many of these vessel shapes and this combination of types disappeared, being replaced by a quite different assemblage. This change can be documented in and its approximate date established from vessels in a large hoard found at Shaanxi Fufeng Zhuangbai (figure 21.3). The hoard contained vessels belonging to several successive generations of one family. The vessels are identified by the names of the individuals who cast them: Zhe, Feng, Shi Qiang, and Wei Bo Xing (Xing). Zhe is thought to have lived at the end of the early Western Zhou, Feng and Shi Qiang in the middle Western Zhou, and Wei Bo Xing probably lived at the end of the middle or beginning of the late Western Zhou. The vessels of the first two generations, Zhe and Feng, parallel the Gaojiabao and Qijiacun tomb groups closely (figure 21.1 and 21.2). The Shi Qiang vessels probably fall slightly later. The Xing vessels introduce new types; among the new shapes are *dou* with openwork foot rings, *gui* decorated with vertical ribbing used in large sets (there are eight Xing *gui*), large *hu* in pairs, and sets of bells known as *zhong*. *Jue* in the Xing set are a lingering remnant of the Shang inheritance.[4]

Jue were, however, no longer present in the late Western Zhou sets, such as the vessels found in a hoard at Shaanxi Fufeng Shaochencun (figure 21.4).[5] The hoard was composed of five *ding*, eight *gui*, two

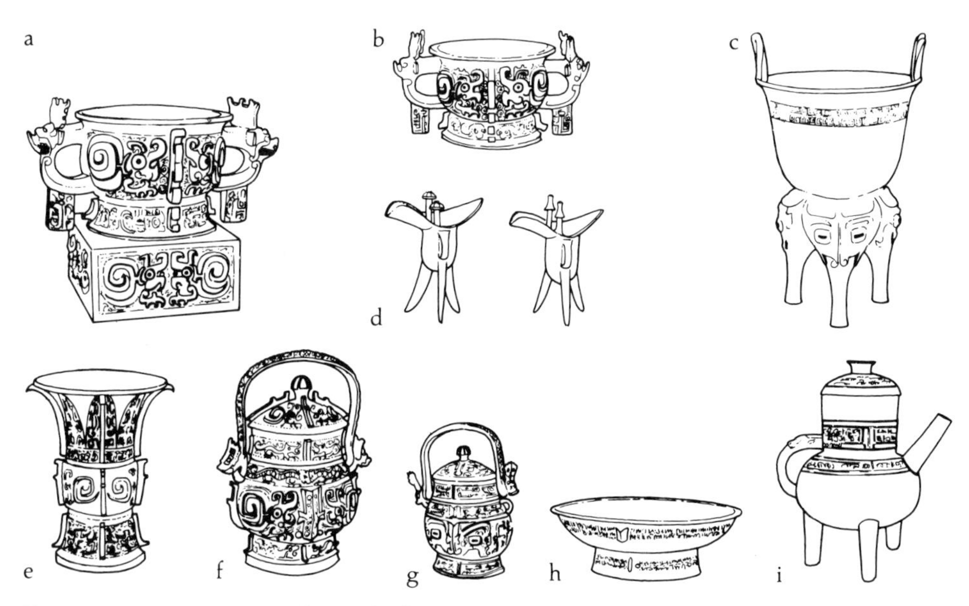

Figure 21.1 Drawings of vessels found in a tomb at Shaanxi Jingyang Gaojiabao: (a, b) two *gui* on square bases (one base damaged), (c) a *yan*, (d) two *jue*, (e) a cylindrical *zun*, (f, g) two *you*, (h) a *pan*, (i) a *he*. Two *ding*, a *gu*, and a *zhi* were also found but were too damaged to be restored. All artifacts from the early Western Zhou. After Research Institute of Shaanxi Province et al. (1984, vol. 4, nos. 136–145).

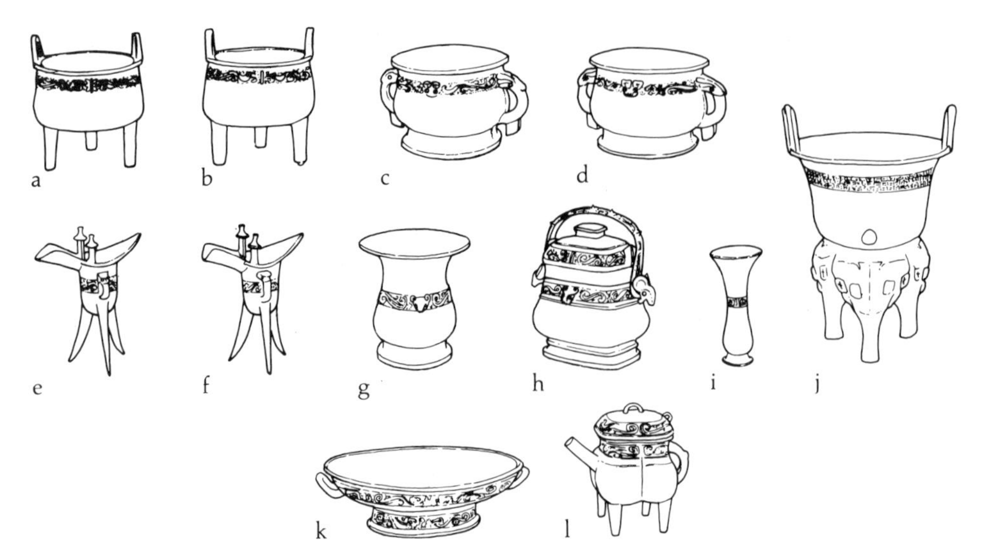

Figure 21.2 Drawings of vessels from Shaanxi Fufeng Qijiacun tomb M19: (a, b) two *ding*, (c, d) two *gui*, (e, f) two *jue*, (g) a *zun*, (h) a *you*, (i) a *zhi*, (j) a *yan*, (k) a *pan*, (l) a *he*. Middle Western Zhou. After Research Institute of Shaanxi Province et al. (1980, vol. 3, nos. 15–26).

hu, an *yi*, a *pan*, and two wine ladles. Only four of the *ding*, five of the *gui*, and the two *hu* share inscriptions of a type that indicate that they are contemporary. However, in terms of their functions the remaining vessels seem to have contributed to the set.

The main features of the set are a large number of *ding* and *gui*, which were probably part of a series of vessels indicating rank; a pair of large *hu*; an *yi* replacing the lobed *he*; and the disappearance of all the wine cup types inherited from the Shang. In addition, several of the vessels, especially the *gui* and *hu*, take shapes not used in the early and middle Western Zhou.

Principal Features of the Change

The change thus described can be summarized as follows:

1. Ancient wine cups inherited from the Shang were abandoned, including *jue*, *jiao*, *gu*, and *zhi*, as were the wine containers *you* and cylindrical *zun*, which had been popular with the Western Zhou. This change alone must have entailed a revolution in ritual.
2. New vessels were introduced: large *hu*, *dou*, *xu*, *fu*, and *yi*.[6]
3. New forms of old vessels were introduced, including *gui* with vertical ribbing (this vessel type seems to revive a much earlier form); *gui* with horizontal grooves in place of *gui* with S-shaped profiles; *li* decorated with striations and with flat rims in place of *li* with vertical necks; and a new form of ewer called a *ying*. These last three vessels were based on ceramic prototypes.
4. A new type of bell, *zhong*, with a tubular handle, was introduced from the south.
5. Identical *ding* and *gui* in sets of nine *ding* and eight *gui* (or seven *ding* and six *gui*, etc.) were employed to indicate rank.

One of the most extraordinary features of the change is the use of several new vessel types based on ceramic forms and the replacement of well-established types by these new alternatives. It is necessary to emphasize how unusual it is for a cheap material, ceramic, to be copied in an expensive one, bronze. Commonly goods made first in costly materials are afterward reproduced in cheaper ones, so that commodities originally too expensive for most aspiring customers later become widely available. The reverse process is always less usual. It had happened when bronze casting was first developed in Henan, for at that time the principal models available to casters were Neolithic ceramics. It happened again when bronze casting spread to areas where it had been previously little known. The new shapes, the *li*, *gui*, and *ying* (figure 21.5), were perhaps developed in western China, west of the major Zhou capital near Xi'an. It is possible that bronze casting was less familiar in this area than at the capital and that it was therefore natural to copy familiar ceramic forms in bronze.

The development of these shapes in the first place is not as surprising as their rather sudden ascendancy at the end of the middle Western Zhou. The metropolitan bronze casters of middle and late Western Zhou were the inheritors of an ancient tradition and of a repertory of well-established shapes; they did not need to look to ceramics for want of other and presumably more prestigious models. The complete predominance attained by the new shapes in the late Western Zhou suggests deliberate policy and some centrally made decision to discard the older vessels and adopt the new.[7] The completeness of the change sets a great divide across the Western Zhou. This divide is particularly useful in helping to identify moments of contact between metropolitan areas and southern regions not under direct Zhou control.

Figure 21.3 Drawings of vessels cast for four generations of a single clan. (Top row) The Zhe vessels, early Western Zhou: a *fang yi*, a *zun*, a *gong*, a *jia*. (Second row) The Feng vessels, middle Western Zhou: a *you*, a *zun*, a *jue*. (Third row) The Shi Qiang vessels, second half of middle Western Zhou: a *pan*, a *jue*. (Fourth and fifth rows) The Wei Bo Xing vessels, late Western Zhou: a *gui*, three *jue*, a *xu*, a *li*, a *pen*, two bells, a *dou*, two *hu* (in the case of identical vessels, only one of each type has been shown). After Research Institute of Shaanxi Province et al. (1980, vol. 2, nos. 14–20, 24, 25, 27, 29, 31, 33, 41, 44, 47, 49, 51, 54, 55).

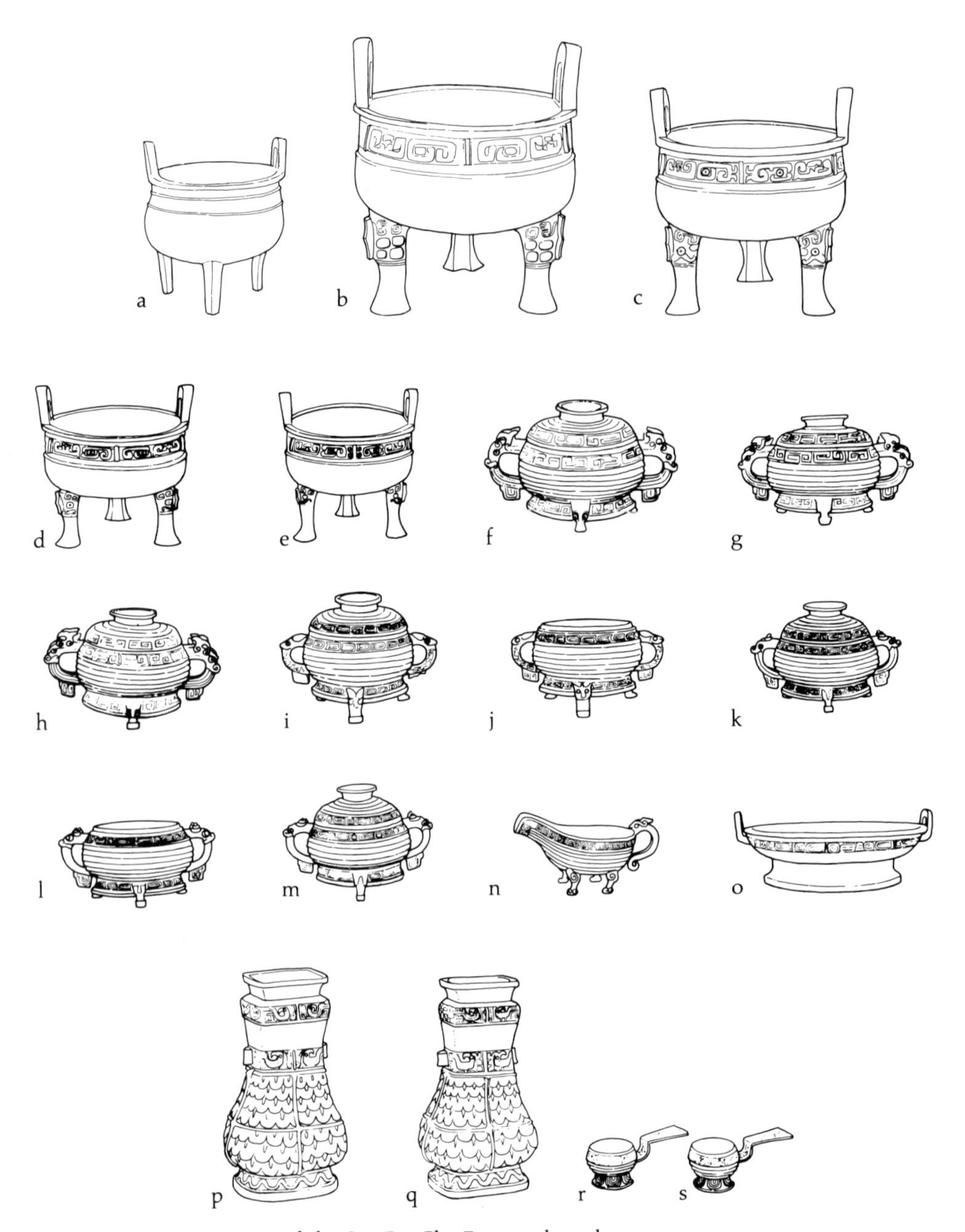

Figure 21.4 Drawings of the San Bo Che Fu vessels and others found with them at Shaanxi Fufeng Shaochencun: (a) one early *ding*, (b–e) four San Bo Che Fu *ding*, (f–j) five San Che Fu *gui*, (k–m) three X Shu Shan Fu *gui*, (n) an *yi*, (o) a *pan*, (p, q) two San Che Fu *hu*, and (r, s) two ladles. Late Western Zhou. After Research Institute of Shaanxi Province et al. (1980, vol. 3, nos. 113–129).

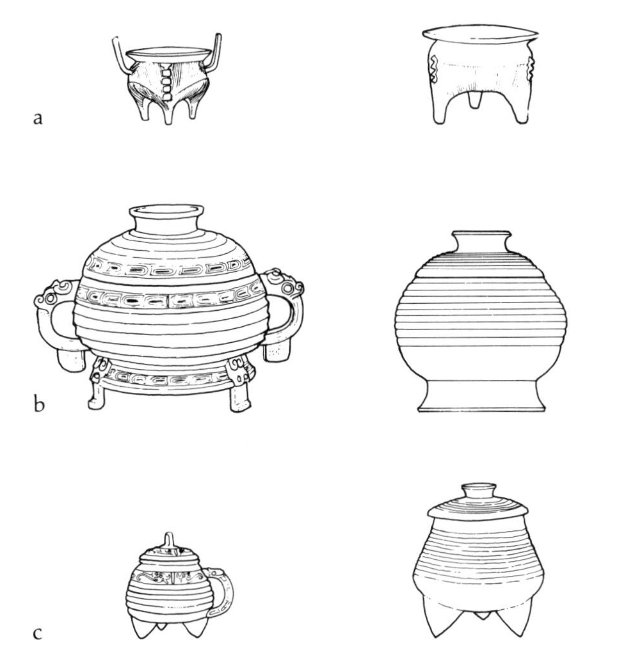

Figure 21.5 Drawings of three bronze vessels in the Sackler Collections, a *li*, a *gui*, and a *ying*, to illustrate their dependence on ceramic prototypes. From left to right: (a) a bronze *li* compared with a ceramic *li* from Shaanxi Chang'an Puducun M1, after Shi Xingbang (1954, fig. 8); (b) a bronze *gui* compared with a ceramic *gui* from Beijing Fangshan Liulihe M51, after Chinese Archaeological Research Institute (1974, fig. 15.7); (c) a bronze *ying* compared with a ceramic vessel from Shaanxi Chang'an Puducun M2, after Cultural Relics Committee of Shaanxi Province (1957, pl. 5.4).

Southern Central China: Southern Henan and Hubei Provinces

Shang bronze casting had been extremely influential in Hubei and Hunan provinces: Early Shang bronzes from those areas were closely dependent on Henan prototypes; later local Shang Period bronzes departed much further from metropolitan Shang examples.[8] Early Western Zhou bronzes from Hubei, by contrast, illustrate a renewal of close links between this peripheral area and the capitals by now established near Chang'an in Shaanxi Province and near Luoyang in Henan. Tombs at Hubei Huangpi Lutaishan contained standard early Western Zhou ritual vessels. Individual tombs in this group were supplied with such vessels as a single *ding*, two *jue*, a *zhi*, and a *zun*. These tombs are not an isolated find; other conventional Western Zhou bronzes have been found in the area.[9]

Interestingly this contact seems to have ceased during the middle Western Zhou. There is no evidence as yet of the import or manufacture of middle Western Zhou bronzes, such as those from Shaanxi Fufeng Qijiacun M19 (figure 21.2) or of the use of bird motifs characteristic of that date.

At the end of the Western Zhou contact was again renewed. Possibly in connection with the fall of the capital in the west to the invading Quan Rong, members of the Zhou aristocracy moved south. Bronzes found in a hoard at Hubei Jingshan Songhequ copy northern vessels types. Here are the matching sets of *gui* and *li* (taking the place of *ding*), with two large *hu*, *dou*, *yi*, and *pan* (figure 21.6).[10]

These vessel shapes could not have been directly evolved either from the early Western Zhou bronzes employed in the area or from a hypothetical middle Western Zhou stage about which we know nothing at present. This hoard is evidence of contacts between Hubei and the metropolitan Western Zhou area at the end of the Western Zhou Period. Late Western Zhou vessels or indeed bronze casters must have reached Hubei from the Xi'an area, stimulating casting of bronzes that exactly matched late Western Zhou ritual vessel sets. The quality of the casting is good, suggesting that the techniques developed in the metropolitan areas were transplanted to southern Henan and Hubei.

The Southeast: The Provinces of Anhui and Jiangsu

A quite different sequence of contacts between centers in Shaanxi and Henan and the southeastern provinces of Jiangsu and Anhui can be documented. In the Shang period the pattern of contacts followed the stages illustrated in the Hubei and Hunan provinces. Early Shang Period vessels were faithfully copied; later Shang Period vessels were more independent of the northern prototypes.

But this similarity between the two areas had ceased by the beginning of the early Western Zhou. Bronzes of this date are not, it seems, to be found in the southeast, nor were the northern tomb types employed here. Instead the moment of interest comes at the beginning of the middle Western Zhou.

Attention focuses on a local tomb type, the *dun*. Unlike metropolitan shaft tombs, *dun* were constructed above rather than below ground. They often contained a mixture of metropolitan and locally made bronzes. A tomb at Jiangsu Dantu Dagang Muzidun can be taken as an example (figure 21.7).[11] The vessels in this tomb included the following: two *ding*, a *li*, two *gui*, a *zun*, a *you*, a *hu* with a bird-shaped lid, and a bird-shaped vessel. Apart from the last two vessels, all the containers were based on metropolitan prototypes. One of the *gui*, a vessel on a square base decorated with plumed birds, seems indeed to be a metropolitan piece; its shape, decoration and inscription are entirely metropolitan. It is highly unlikely that at this date the peoples of the southeast could write in perfect characters suitable for bronze inscription. This inscription probably indicates that the bronze was cast at the capital or, if made in the southeast, that it was made by a caster who had moved there from a metropolitan center.

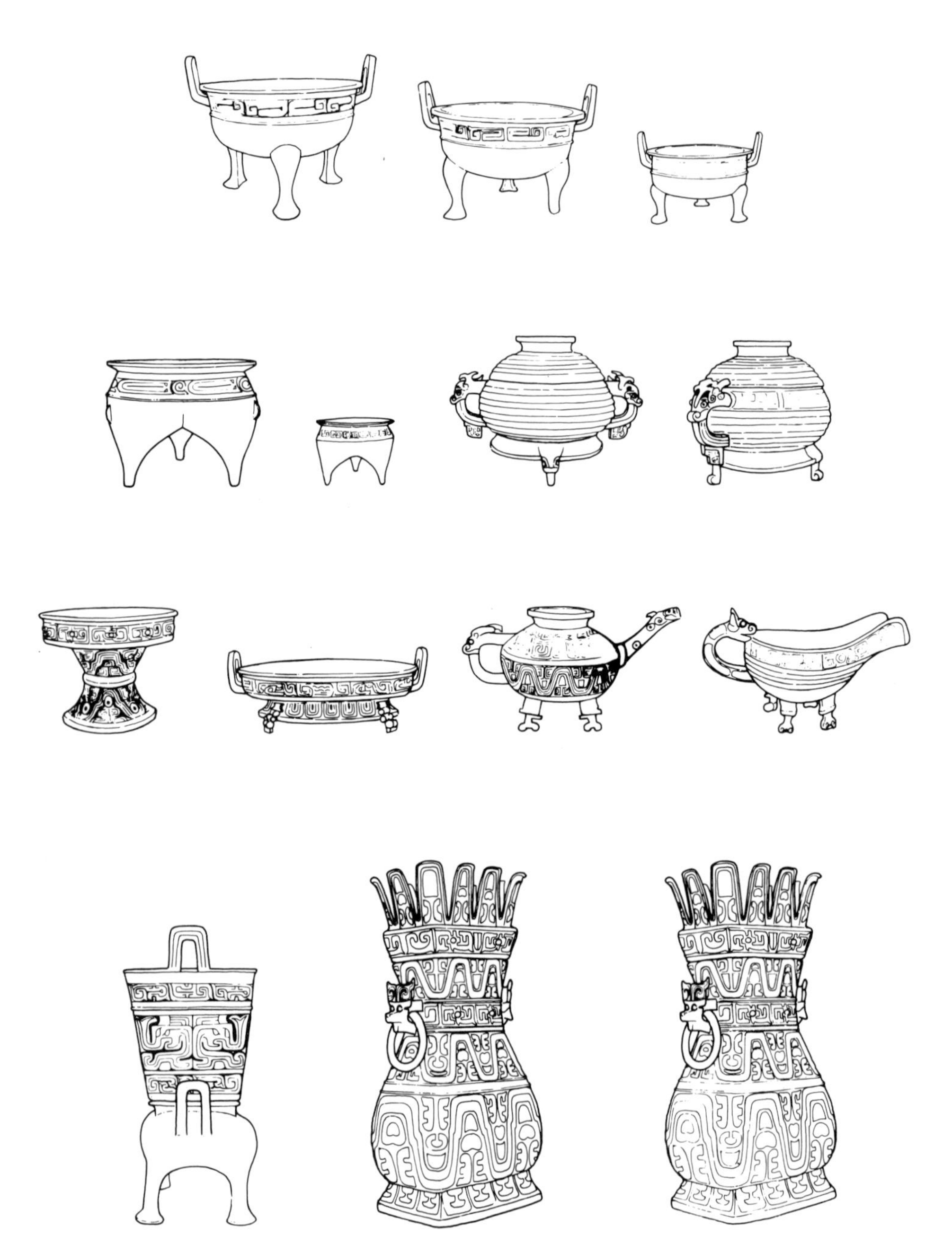

Figure 21.6 Drawings of vessels found in a hoard at Hubei Jingshan Songhequ: three *ding* (three of nine), a large *li* (one of two); a small *li* (one of seven), two *gui* (two of seven), a *yan*, a *dou* (one of two), two *fang hu*, a *pan*, an *yi*. Two of the *ding*, the two *dou*, and the *hu* are inscribed with a dedication by the second son of the Zeng Hou. Early Eastern Zhou. After Hubei Provincial Museum (1972, pls. 9, 10, figs. 4–10).

The Muzidun *gui* can be compared with an almost identical bronze from tomb M17 at Huayuancun in Shaanxi Chang'an Xian (figure 21.8). This second *gui* is inscribed with an account of campaigns against the Chu and the Jing.[12] It is likely that these are the campaigns of the King Zhao Wang, and it is probable that the *gui* dates to Zhao Wang's reign or to the immediately succeeding reign of Mu Wang. The Muzidun bird-decorated *gui* (figure 21.7) is therefore likely to date to the end of the early or beginning of the middle Western Zhou.

Once early middle Western Zhou bronzes became known in the southeast, they stimulated quite a large industry. Excavation of sites in Jiangsu and Anhui provinces, especially at Danyang in Jiangsu and Tunxi in Anhui, illustrate characteristics of these bronzes:[13]

1. Early middle Western Zhou vessel shapes predominated, including *ding*, *gui*, cylindrical *zun*, and S-profile *zun*.
2. These vessels were varied according to the tastes of the region; they were sometimes made in large sizes; they were modified in profile over time and decorated with local motifs combining metropolitan and local taste.
3. Some vessel types remained in use in the southeast long after they were abandoned in metropolitan areas. Thus cylindrical *zun* survived in Eastern Zhou tombs.[14]

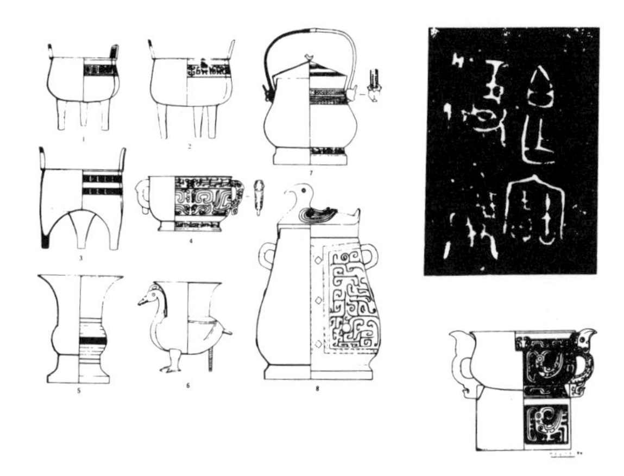

Figure 21.7 Drawings of vessels excavated from tomb at Jiangsu Dantu Dagang Muzidun. Middle Western Zhou. After Zhenjiang Museum et al. (1984, fig. 4).

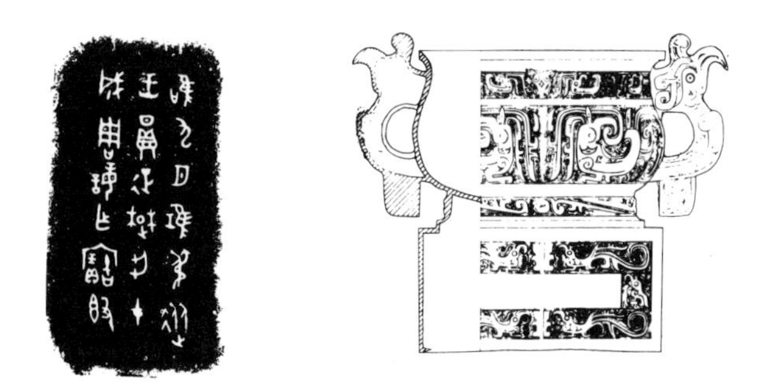

Figure 21.8 Bird-decorated *gui* and its inscription, excavated from Shaanxi Chang'an Huayuancun tomb M17. Early to middle Western Zhou. After Committee for Cultural Relics in Shaanxi Province (1986, figs. 23 and 24).

A *you* in the British Museum appears to be a product of this southeastern tradition. Like other vessels from the southeast it follows the shape of a metropolitan *you* dating to the end of the early or beginning of the middle Western Zhou. In figure 21.9 the provincial *you* has been photographed with a metropolitan *you* of the second half of the early Western Zhou, also in the British Museum. By comparison with this conventional vessel, the provincial *you* is enormous (with its handle it is 62 cm in height). A *zun* from Danyang in Jiangsu Province is similarly inflated.[15]

Its great size notwithstanding, the *you* was cast in piece molds. The mold seams on the two sides, aligned with the handles, can be seen clearly in the discontinuities in the design. Like metropolitan *you*, the southern vessel shows traces of subsidiary mold seams down the center panels of the body. In addition it shares with vessels from the capitals the technique casting the handle onto two loops at the vessel's sides. Thus the technology established at the center had been transferred to the periphery.

On the evidence of the British Museum's *you*, the composition of southern castings differs slightly from the range of bronze alloys known from Shaanxi and Henan (table 21.1). The levels of copper are lower than those for the metropolitan bronze, and the levels of lead and silver are rather higher. Somewhat similar results have been obtained by the team analyzing the bronzes in the Sackler Collections. From its similarity to vessels from Anhui and Jiangsu, a *pan* in that collection can be identified as coming from the south (figure 21.10). Its major components have been analyzed as follows: 64.3% Cu; 13.6% Sn; 11.8% Pb; 0.056% As; 0.313% Ag. The resemblance to the alloy of the British Museum's *you* may be fortuitous; however, the comparison is at least suggestive, especially in light of the relatively high percentage of silver also found in excavated bronzes from Tunxi in Anhui Province.[16]

For its size the *you* is rather thin. In the north bronzes were thicker and more crisply cast. Perhaps the ceramic materials used for the southeastern molds did not allow the sharp definition achieved in Henan and Shaanxi. Above all, the decoration of the *you* is highly eccentric. Large coiled snakes on its belly are probably a local interpretation of the coiled dragon motif already illustrated on the Jingyang bronzes (figure 21.1).[17] A background of dense crisscrosses is based on local ceramic motifs.

Similar provincial bronzes employing local variants of middle Western Zhou vessel shapes and decorated

Figure 21.9 Bronze ritual vessel *you*, late Western Zhou (left). Probably from southern China, shown for comparison with a metropolitan *you* (right) of the early Western Zhou. Height to handle 62 cm. British Museum.

Figure 21.10 Bronze ritual vessel *pan*. Probably from southern China. Late Western or early Eastern Zhou. Height 7.6 cm, diameter 26.6 cm. Sackler Collections.

with local interpretations of northern motifs seem to have persisted in the southeast until the end of the Western Zhou. Furthermore, there seems to have been little or no contact with the center in the late Western Zhou. For this reason vessels, such as the *you* and the *zun*, that were abandoned in the west and eliminated by chance from central-southern China also survived into much later centuries in the southeast. When these vessel types are found in Eastern Zhou contexts, either in central-southern China or in the extreme south and southwest in the Guangdong and Guangxi provinces, they are likely to have been borrowed from southeastern examples.[18]

The industries in central-southern and southeastern China were thus distinct and gave birth to two quite separate early Eastern Zhou traditions. A recognition of the revolution in ritual vessel shapes and of the different effects of this revolution in the provinces is essential background to a study of the spread of casting across southern China.

Notes

1. The contents of Fu Hao's tomb are described in *The Tomb of Fu Hao at Yinxu* (1980).

2. The excavation of the tomb is reported in Ge Jin (1972).

3. The excavation of the tomb at Qijiacun is reported in "Western Zhou tomb no. 19" (1979). Its contents can be dated by comparison with the contents of a tomb excavated at Shaanxi Chang'an Puducun (Cultural Relics Committee, 1957). This latter tomb contained the Chang Fu *he*, whose inscription names Mu Wang, and the vessel and the tomb are therefore dated to one of the immediately succeeding reigns, probably the reign of Gong Wang. Other inscribed bronzes, including the Qiu Wei vessels (Cultural Relics Bureau of Qishan Xian et al., 1976), support the dating of vessels with these proportions to the second half of the middle Western Zhou.

4. The discovery of the Zhuangbai hoard is reported in "Brief report of the excavation" (1978). Li Xueqin suggests that *jue* were retained exceptionally by this family because of their long-standing connections with Shang areas (Li 1980, p. 14).

Table 21.1 Atomic absorption analyses of two Chinese bronze *you* (British Museum)

Registration number	Cu	Sn	Pb	As	Ag	Au	Ni	Zn	Fe	Sb	Bi	Mn	Cd	Co
Provincial *you*														
1983, 2-2, 1	64.9	11.9	13.0	0.19	0.162	<0.003	0.19	0.046	0.07	0.13	0.055	<0.0008	<0.0004	0.009
Metropolitan *you*														
1947, 7-12, 329	85.1	13.5	0.27	0.15	0.063	<0.004	0.039	0.009	0.32	0.06	0.119	<0.0010	<0.0006	0.006

5. The hoard, which contains a number of vessels dedicated by one San Bo Che Fu, is reported by Shi Yan (1972).

6. For alternative terminology for the *dou* and the *fu*, see Gao Ming (1982).

7. The Chang Fu tomb mentioned in note 3 contained *li* of the new type and a set of *zhong*, but otherwise the vessels were of the old type.

8. Contrast the bronzes from Hubei Huangpi Panlongcheng (Hubei Provincial Museum, 1976) with, say, the four-ram *zun* from Hunan Ningxiang (Fong 1980, no. 20).

9. The tombs at Hubei Huangpi Lutaishan are described by the Cultural Relics Office in Huangpi Xian (1983). Additional finds are reported in "Two bronzes" (1965), Cultural Relics Sales Office (1985, p. 103), "Shang and Zhou bronzes" (1982), and Suizhou City Museum (1982, 1984).

10. For the hoard from Hubei Jingshan Songhequ, see Hubei Provincial Museum (1972).

11. The tomb at Muzidun is described in Zhenjiang Museum et al. (1984). For discussion of tomb chronology and bronze development in the area, see Liu and Wu (1982) and Liu (1985).

12. For an account of the *gui* from Huayuancun and a discussion of its inscription, see Committee for Cultural Relics in Shaanxi Province (1986).

13. For vessels from Tunxi, see Cultural Relics Office of Anhui Province (1959) and "A large group of Western Zhou relics" (1965). For references to Danyang, see note 15.

14. Cylindrical *zun* appear in the tomb of the Marquis of Cai, for example (Cultural Relics Committee of Anhui Province and Anhui Provincial Museum, 1956, pl. 9).

15. The *zun* is illustrated in Zhenjiang City Museum et al. (1980, fig. 1.7).

16. I am grateful to E. Sayre and Lore Holmes for this information and for their permission to quote the results. For bronzes from Tunxi, see Cultural Relics Office of Anhui Province (1959). For similar *pan*, see Feng (1981).

17. A small bronze decorated with tight spirals in the Shanghai Museum (Chen 1984, pl. 1) may be an intermediary between the coiled dragon and this southern snake design.

18. S-profile *zun* have been found in both Hunan and Guangxi provinces (Fong 1980, no. 66, fig. 88). These vessels were probably not copied either from the metropolitan areas or from Hubei-Hunan tradition of the late Western or early Eastern Zhou, but derive from the southeastern casting area, the only region in which the vessel shape survived after the ritual revolution at the capital. The vessel from Guangxi is decorated with snakelike creatures. The snakes on the British Museum *you* may indicate that, notwithstanding the similarities of its composition with that of the southeastern vessels, it belongs to the later extension of the southeastern tradition into the extreme south and southwest.

References

"Brief report of the excavation of Western Zhou bronzes from pit no. 1 at Zhuangbai, Fufeng, Shaanxi Province." 1978. *Wenwu* 3:1–18. (In Chinese.)

Chen Peifen. 1984. The Ji Zhong *hu*." *Wenwu* 6:21–23. (In Chinese.)

Chinese Archaeological Research Institute. 1974. "Western Zhou slave burials excavated in the vicinity of Beijing." *Kaogu* 5:309–321.

Committee for Cultural Relics in Shaanxi Province. 1986. "A brief report on the excavation of tombs in the vicinity of the Western Zhou capital of Hao." *Wenwu* 1:1–36. (In Chinese.)

Cultural Relics Bureau of Qishan Xian et al. 1976. "Brief report of the recovery of Western Zhou bronzes from a pit at Dongjiacun, Qishan Xian, Shaanxi Province." *Wenwu* 5:26–44. (In Chinese.)

Cultural Relics Committee of Anhui Province and Anhui Provincial Museum. 1956. *Relics Excavated from the Tomb of the Marquis of Cai*. Beijing: Kexue Chubanshe. (In Chinese.)

Cultural Relics Committee of Shaanxi Province. 1957. "Excavation of Western Zhou tombs at Puducun at Chang'an." *Kaogu Xuebao* 1:75–85. (In Chinese.)

Cultural Relics Office in Huangpi Xian. 1982. "Buildings and tombs of the Western and Eastern Zhou periods found at Lutaishan in Huangpi Xian, Hubei Province." *Jianghan Kaogu* 2:37–60. (In Chinese.)

Cultural Relics Office of Anhui Province. 1959. "Report on the excavation of Western Zhou tombs at Tunxi in Anhui Province." *Kaogu Xuebao* 4:59–90. (In Chinese.)

Cultural Relics Sales Office in Wuhan. 1985. "The *Wei zun* of the Western Zhou." *Jianghan Kaogu* 1:103. (In Chinese.)

Feng Puren. 1981. "Bronzes from Beizhougang at Wuxi." *Kaogu* 4:302–303, 369. (In Chinese.)

Fong Wen, ed. 1980. *The Great Bronze Age of China: An Exhibition from the People's Republic of China*. New York: Metropolitan Museum of Art.

Gao Ming. 1982. "A discussion of the terms *gu* and *fu*." *Wenwu* 6:70–73, 85. (In Chinese.)

Ge Jin. 1972. "Report of the excavation of an early Zhou tomb at Jingyang Gaojiabao." *Wenwu* 7:5–8. (In Chinese.)

Hubei Provincial Museum. 1972. "Zeng state bronzes discovered at Jingshan in Hubei Province." *Wenwu* 2:47–53. (In Chinese.)

Hubei Provincial Museum. 1976. "Bronzes of the Erligang period from Panlongcheng." *Wenwu* 2:26–41. (In Chinese.)

"A large group of Western Zhou relics excavated at Tunxi in Anhui Province." 1965. *Wenwu* 6:52. (In Chinese.)

Li Xueqin. 1980. *The Wonder of Chinese Bronzes*. Beijing: Foreign Languages Press.

Liu Xing. 1985. "Periodization of bronzes from southeastern China." *Kaogu Yu Wenwu* 5:90–101. (In Chinese.)

Liu Xing and Wu Dalin. 1982. "A discussion of the chronology of the *dun* tomb type in the Zhengjiang area." *Wenwu Ziliao Congkan* 6:79–85. (In Chinese.)

Research Institute of Shannxi Province, Cultural Relics Committee of Shaanxi Province, and Shaanxi Provincial Museum. 1979, 1980, 1984. *Shang and Zhou Bronzes Excavated in Shaanxi.* 4 vols. Beijing: Wenwu Chubanshe. (In Chinese.)

"Shang and Zhou bronzes collected at Xiangfan in Hubei Province." 1982. *Wenwu* 9:84–85. (In Chinese.)

Shi Xingbang. 1954. "Report on the excavation of Western Zhou tombs at Punducun at Chang'an." *Kaogu Xuebao* 8: 109–126.

Shi Yan. 1972. "A group of Western Zhou bronzes discovered at Fufeng Zhuangbai." *Wenwu* 6:30–35. (In Chinese.)

Suizhou City Museum. 1982. "Bronzes excavated at Anju in Sui Xian, Hubei Province." *Wenwu* 9:84–85. (In Chinese.)

Suizhou City Museum. 1984. "Shang and Zhou bronzes discovered at Sui Xian in Hubei Province." *Kaogu* 6:510–514. (In Chinese.)

The Tomb of Fu Hao at Yinxu. 1980. Beijing: Wenwu Chubanshe. (In Chinese.)

"Two bronzes found at Xishui in Hubei Province." 1965. *Kaogu* 7:369–370. (In Chinese.)

"Western Zhou tomb no. 19 at Qijiacun in Fufeng, Shaanxi." 1979. *Wenwu* 11:1–11. (In Chinese.)

Zhenjiang City Museum et al. 1980. "Western Zhou bronzes excavated at Danyang in Jiangsu Province." *Wenwu* 8:3–9. (In Chinese.)

Zhenjiang Museum et al. 1984. "Brief report on the excavation of a tomb containing Western Zhou bronzes at Dantu Dagang Muzidun in Jiangsu Province." *Wenwu* 5:1–10. (In Chinese.)

22
The Old Copper Mines of Eastern India

Dilip K. Chakrabarti

Mining and metallurgy have a long history in India. The earliest range of data on indisputably smelted copper is in the fifth to fourth millennium B.C. at the site of Mehrgarh in Baluchistan (figure 22.1). The use of tin as an alloy was known but limited outside the Indus civilization (Chakrabarti 1979). Arsenic alloying was known in the Indus civilization, but it was more prevalent among the "copper hoards" of the Ganges valley, some of which date from the second millennium B.C. (Agrawal et al., n.d.). The fourth millennium B.C. site of Nal in Baluchistan shows the first occurrence of lead and silver. Gold occurs first in period 1 of Jalilpur in Punjab, which belongs to the same general time bracket. Iron occurs in a number of contexts in the second half of the second millennium B.C. but becomes more common from c. 1000 B.C. on (Chakrabarti 1981–1983). The use of zinc processed by a distillation method goes back to the closing centuries of the first millennium B.C. (Willies et al. 1984).

Metal ores are abundant outside the Indo-Ganges system. Not all these occurrences are suitable for modern commercial exploitation, but they were adequate for preindustrial smelters. No significant source of native copper is known in India. Copper ores show concentration in parts of Rajasthan, Madhya Pradesh, Andhra, Gujarat, and Bihar, but there are other areas such as Baluchistan, the Himalayan belt in Uttar Pradesh, etc. that were significant in protohistoric and historic periods. Tin was mined in one locality in eastern India as late as the middle of the nineteenth century and is commercially exploited now in Bastar in Madhya Pradesh. Large lead fields are known in Rajasthan and Andhra. In abundance lead is next to iron in India and occurs in many areas of the Peninsular and Himalayan blocks. Argentiferous galena, although common, does not seem to constitute a rich source of silver in the country. Gold derived from river sands was fairly abundant until the nineteenth century. Gold reefs occur only in Andhra-Karnataka and parts of Bihar–Orissa–West Bengal. Significant sources of zinc ore are in Rajasthan and Orissa. Iron suitable for preindustrial smelting occurs almost everywhere outside the Indo-Ganges alluvium.

Old mines have been located for copper, lead-zinc-silver, and gold. The old copper mines are distributed in Baluchistan, Gujarat, the Uttar Pradesh Himalayas, Rajasthan, Bihar, West Bengal, Sikkim, Madhya Pradesh, Andhra, and Karnataka. The lead-zinc-silver mines are in Rajasthan, Himachal Pradesh, Orissa-Bihar, and Andhra. The old gold mines are in Andhra-Karnataka and Bihar. Research on the history of Indian mining has only begun. Some old shafts in Rajasthan, Gujarat, and Andhra-Karnataka have been intercepted by modern mining shafts, leading to the recovery of some radiocarbon samples (Chakrabarti 1986). The earliest available date goes back to the close of the

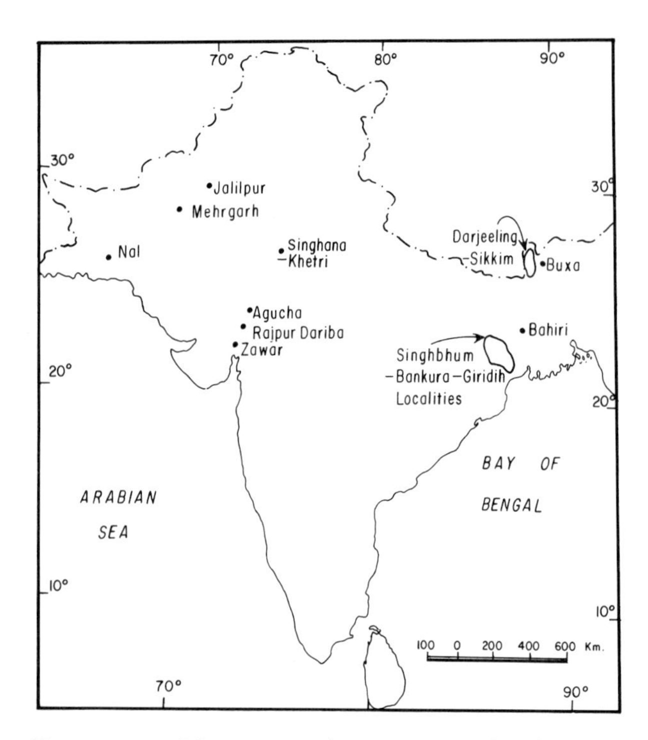

Figure 22.1 Major sites and areas mentioned in the text.

second millennium B.C. in the context of the Rajpur Dariba mines for copper in Rajasthan. The lead-zinc-silver field of Agucha in Rajasthan (Tiwari and Kavdia 1984) has been given a date of 990 ± 100 B.C. from an old mine shaft at a depth of 20 m (PRL-930). It is easily observed that these dates are not at all indicative of an early history of metallurgy in India as reflected in the archaeological record.

The ancient Indian expertise in mining is suggested by the following points. First, wherever ore deposits occur, there are signs of old working. In fact, the geologists of the Geological Survey of India were considerably helped in their task of mineral exploration by the distribution of these workings. Second, the ancient miners showed considerable ability in tracing and following the mineral veins. Third, analyses of slag debris show that the technique of metal extraction was highly efficient. A typical analysis of old copper slag near Musabani in Singhbhum contains only 0.26% copper, which is comparable to the extraction by the modern smelters.

However, in the Indian context a discovery like that of the ancient Tonglushan mines in China (with their wealth of objects used in mining operations of the protohistoric and historic periods) is still missing. Such important discoveries can take place only in the localities of modern open cast mining. In India there is large-scale open cast mining only in one copper-bearing area—the Balaghat district in Madhya Pradesh. This paucity of open cast mining in the modern Indian context has prevented the detailed discoveries of mining procedures and implements.

Some idea of the techniques of ancient mining in India comes from a few nineteenth-century reports on preindustrial mining. These reports are in the contexts of the lead mines of Ajmer in Rajasthan, the copper mines of Singhana, Khetri, and Rajasthan, and the copper mines at Buxa in the northern part of the state of West Bengal (Chakrabarti, forthcoming). There was also a rich preindustrial tradition of metallurgy in India until at least the beginning of the twentieth century. It is possible to infer that there is continuity in different aspects of metallurgy in India from the protohistoric period to the beginning of the twentieth century.

What has been stated so far lends the necessary academic perspective to the study of the old copper mines of eastern India. Old copper mines have been reported from the following areas in eastern India: the Singhbhum district of south Bihar, where one can trace hundreds of old workings roughly between Chakradharpur and Charakmara near the Singhbhum-Mayurbhanj border; the localities of Chhedapathar and Tamkhun in Bankura, which mark the extensions of the Singhbhum copper belt; the locality of Baragunda in the modern Giridih district; Buxa in the Dooars area of the northern part of the state of West Bengal; and a few miscellaneous localities in Darjeeling-Sikkim. The Buxa mines were worked in the nineteenth century by following preindustrial methods, and the same may also be true of the Darjeeling-Sikkim mines. I believe that these mines came into operation only in the medieval period when there was a considerable demand for copper for religious images. From the archaeological viewpoint the copper mining localities of Singhbhum, Bankura, and Giridih are important for a number of reasons. First, the entire area is the home of the so-called east Indian copper hoards, which are generally treated as protohistoric material. Second, there is a wide distribution of black-and-red ware (second millennium B.C.) Chalcolithic sites in the adjacent regions of Bihar and Bengal. Third, east India has a significant art school of copper and bronze images dating from the early medieval period, and it is logical to postulate that these copper mines were the major source of this apparently prolific supply of copper to the image makers of east India. Fourth, no records are available on when these mines fell into disuse, and this fact by itself is suggestive of a general antiquity of these mines.

The Darjeeling-Sikkim and Buxa Mines

The major interest of the Darjeeling-Sikkim and Buxa mines in the present context is a somewhat elaborate account of the preindustrial mining operations at Buxa dating from the early 1870s. In view of the extreme paucity of such records in the Indian context, this account is worth citing in detail.

Mines ... magnified rabbit holes; props to support the roof are only occasionally made use of, and the passages meander with the courses of the ore, and do not exceed a yard square diminishing downwards in the narrowest parts to half this dimension. The tools used are an iron hammer, a gad or chisel held in a split bamboo, and a pick. The light is afforded by thin slips of bamboo, the smoke from which in the confined passages is not so irritating to the eyes as that from other kinds of wood. The ore is carried out in narrow baskets, and picked, crushed and finally pounded with a stone hammer or pounder fixed in a forked stick. It is then subjected to several washings, ground in a handmill, rewashed, and is at last ready for the preliminary smelting.

The furnace is either built of refractory clay, or is simply a hole dug in the ground with a clay rim at the sides and front, and a higher one or a stone at the back; the inside is plastered with refractory clay if it is not sufficiently so naturally. The furnace is 18 inches deep, a foot square at the top, and tapering to the bottom, where charcoal is rammed down so as to form a floor through which the molten metal cannot sink. There is no orifice at the base of the furnace. Two clay tuyeres dip into it vertically and are connected with simple skin bellows by horizontal tuyeres, which are about a yard long. The furnace when filled with lighted charcoal is soon raised to its full heat by the alternate working of the bellows, and the ore is then sprinkled on from time to time until a sufficient quantity of regulus, covered by lighter slag, has collected at the bottom. On the removal of the charcoal the surface of the slag is cooled with a wisp of wet straw, and several cooled and solidified cakes of slag are removed in succession, leaving the heavy regulus behind, which when set is taken out, pounded, and kneaded with cowdung into balls; these are dried in the sun and then roasted with free access of air in a shallow furnace formed of a ring of slag cakes placed on edge. The last process of refining consists in treating the powder produced from these roasted balls in the same furnace, and in precisely the same manner as was the original ore, the result being that a fluid mass of copper is found at the bottom of the furnace which on cooling is removed.... It has to be still further refined before it is available for ordinary purposes. (Ball 1881, pp. 276–277).

The Singhbhum Copper Belt

Jones (1833) first reported the possibility of finding copper in Singhbhum (at Rajdoha). This locality was subsequently mined fruitfully for copper. The existence of copper in the district was ascertained beyond doubt by Haughton (1854) by smelting a small quantity of ore. This led the local landlords to copper mining and smelting, and soon a reasonable quantity of ore was being mined during the dry season every year. In Haughton's estimate the mining operations ceased to exist for about a hundred years until his work revived it in the middle of the nineteenth century.

Haughton refers to the old workings in the Kharsawan, Seraikela, and Dhalbhum divisions of the district. Pits not more than 10 ft deep marked the course of the vein near Kharsawan town for about half a mile. The twelve old shafts noted in the Tama Dungri hill near Narainpur in Seraikela were small and irregular, but the largest of them was about 60 ft deep. The traces of old furnaces could be seen for several miles near the Dhobani hill in Dhalbhum. The old pits at Badia in the same area were 10–40 ft in depth. Old pits were also observed for about 200 yd at Kamerara, presumably in the same area.

There have been several discussions on the old copper mines of Singhbhum since Haughton wrote in the middle of the nineteenth century (Ball 1869, 1870; Stoehr 1870; Dunn 1937; Murray 1940; Chakrabarti and Chattopadhyay 1985). The most comprehensive set of references may be found in Dunn's memoir on the mineral deposits of eastern Singhbhum and surrounding areas (Dunn 1937). I completed the preliminary fieldwork on this problem in 1985 (Chakrabarti and Chattopadhyay 1985).

The distribution of these mines can best be studied in a number of clearly identifiable sections from the northwest to the southeast. The localities in the westernmost section are Duarpuram (24°46′ N, 85°34′ E), the area between the villages of Jaypur (22°45′ N, 85°36′ E) and Itihassa, and the area to the northwest of Roladih (22°45′ N, 85°40′ E). Only small depressions have been reported from this section. The Kharsawan section has two major foci in Dunn's account: Galudih (22°47′ N, 85°44′ E) and Ragadih (22°47′ N, 85°45′ E). At present these workings can be studied in the area of the Lapso Kyanite Complex of the Hindustan Copper Limited. Apart from the more usual type of large shafts in rock, one notices in this area shafts dug in what are now agricultural fields to reach the mineral-bearing rocks below. Further east, the Seraikela region seems to be fairly full of apparently minor workings, especially at Sini (22°47′ N, 85°57′ E), Dugni (22°46′ N, 85°58′ E), Tama Dungri (22°45′ N, 86°02′ E), etc. These areas are close to the Sanjay River, and along its banks one can easily locate an archaeological assemblage with a profusion of Neolithic celts, pottery, burnt rice, microliths, and iron slag. The next major section seems to begin south of the modern steel city of Jamshedpur, and this continues up to Rajdoha where the Garra Nadi, a tributary of the Suvarnarekha, comes out of the hills. The old workings are reasonably dense along the line of ridges in this area, especially around Ramchandrapahar or Chandra Buru (22°43′ N, 86°13′ E). At Rajdoha large shafts on ridge tops are a prominent feature. The Rakha mines are being worked by the Hindustan Copper Limited, but the traces of old mining activities, including old shafts, are still abundant. The geological map of the Ruam-Sidheswar blocks shows seven old workings in an axis of 2.4 km (courtesy: U. K. Sharma, geologist, the Rakha Mines). Both old workings and slag heaps are plentiful on the top of the neighbouring Sidheswar hill.

Between Chapri (22°37′ N, 86°24′ E) and Laukesra

(22°33′ N, 86°27′ E) the old workings, which extend for about 2000 ft along the top of a ridge at Surda, are singularly impressive, both in their density of distribution and apparent depths. The old workings in the section of the Musabani mines are said to be the largest in the copper belt and continue throughout the length of the lodes. For instance, regarding the old workings at Dhobani (22°31′ N, 86°27′ E), a village about a mile west of Musabani, Dunn writes: "The old workings here, especially at the north end, are among the largest in the belt, individual pits being up to 100 feet in width with a depth of 30 to 40 feet. They are very persistent.... It is quite obvious that this was one of the largest mining centres of the ancient miners, and that they were worked lodes of an unusually persistent nature" (Dunn 1937, pp. 118–119).

There are old workings south of the Musabani belt, but the southermost distribution in Singhbhum is in the Baharagora area. Charakmara (22°17′ N, 86°41′ E) in this region demands special attention because of the scatter of apparently medieval pottery around the mouth of the old shafts.

A number of points are striking in the context of the distribution of these old workings. First, the entire length of this belt is dotted with old workings. The modern prospectors only had to study the distribution of old workings to map the broad alignment of the copper lodes. Second, the intensity of the old mineral exploration in this belt is also indicated by the abundance of slag heaps, some material from which was in fact used to lay concrete for King George's Dock in the Calcutta Port between 1926 and 1928. Third, the ancient miners were attentive even to minor deposits. And finally, they showed a remarkable skill not merely in prospecting but also in mining and smelting.

The skill of these ancients is indicated in the manner of their minings. Down to the depth at which they ceased working, usually water level, they have left no workable copper except in the pillars for holding up the walls; they have picked the country as clean as the desert vulture picks a carcass. Looking over some of these old workings it is often remarked that "they must have worked over it with tooth picks." Even their spoil heaps provide no abundant specimens of copper. Not only does this indicate considerable skill, but also it suggests an infinite patience. (Dunn 1937, p. 54)

The Old Workings in Bankura and at Baragunda in Giridih

There are only two localities with old copper mines in the Bankura district, which forms part of the eastern boundary of Singhbhum. These localities, Chhedapathar and Tamkhun, are both within the jurisdiction of the Ranibandh Police Station. At Chhedapathar, the site of a privately owned tungsten mining operation, old workings for copper can be traced for about a kilometer, and the trace of at least one old shaft is still visible at Tamkhun.

Baragunda is a locality off the Giridih-Dumri road. The old workings in this area have been seriously disturbed by the early twentieth-century mining operations in the area. The first notice of the old mines at Baragunda dates from 1850, when forty-eight ancient mines were noticed along a 3/4-mi-long ridge. On the basis of the earlier reports Ball offers a description of these mines:

We are not in possession of any information as to who the ancients were who made the numerous excavations at Baragunda, of which ample evidence is still to be seen. These workings extend all along the outcrop, and from their overlapping in places it would appear that the deposit was not limited to one line of strike.... Along the main line the width of the excavations average from 25 to 30 yards. The miners appear to have thrown the debris behind them as they progressed, the depth to which they could go being limited; thus there are a succession of basin-like pits separated from one another by mounds of debris, and bounded by the faces of rock which form the foot and roof of the deposit. (Ball 1881, p. 255)

The Problem of Dating

The first point to be understood about dating is that modern mining shafts do not intersect the old ones in this region, and thus the possibility of collecting samples for dating (in the form of radiocarbon samples or identifiable historical antiquities) is rather remote without a direct clearance of the shafts themselves. All that one can hope to achieve is to point out the broad chronological spectrum in which these mines were possibly operative. The basic focus will have to be around the Singhbhum mines because elsewhere in the region there is not even any circumstantial evidence.

If Haughton's testimony is to be believed, the chief of the local princely state of Seraikela assured him that there had been no mining operations in the area "during the time his family had been settled there, that is for about a century" (Haughton 1854). This takes one back to the middle of the eighteenth century. The preindustrial copper mining of Singhbhum apparently stopped before that. This is in contrast to the historical situation in the Rajasthan copper belt, where preindustrial mining continued until the middle of the nineteenth century.

Going backward, what may be called the next chronological point is related to Ball's assertion (Ball 1869) that the Seraks or the Jaina laity were responsible for the mining activities in Singhbhum. This inference was based on the occurrence of various Jaina antiquities in the region. The report of a fairly extensive spread of Jaina antiquities leads to an interesting parallel in the existence of medieval Jaina temples and inscriptions in some mining areas of Rajasthan (compare Zawar,

Rajpur Dariba in Udaipur district). In Singhbhum the Jaina antiquities occur notably in the Chandil and Chakradharpur-Chaibasa regions. Such antiquities are also widespread in the neighboring West Bengal districts of Purulia, Bankura, and Midnapur. The chronological range of these antiquities is about tenth to thirteenth century A.D.

This approximate date is also suggested by the large-scale occurrence of smelting hearths and slag heaps at the village of Parihati in Midnapur district (Chakrabarti and Chattopadhyay 1983). The apparently medieval pottery associated with the smelting hearths could not be precisely dated, but a Jaina image in the village belongs to c. thirteenth century A.D., and this may well be the date of the smelting hearths as well. Remains of the same type are also found at Khanamauda at the border of Singhbhum and Midnapur. In this context I would like to put on record my opinion that some of the so-called copper hoards of eastern India, which are found in the Chhotanagpur plateau section of West Bengal, Bihar, and Orissa, may also be medieval in date. This opinion is based on fieldwork at two copper hoard sites of West Bengal—Parihati and Perua, both in Midnapur district. Parihati yielded heavy flat celts, double axes, rings, etc., whereas Perua yielded celts, some of which were shouldered. There are remains of smelting hearths virtually all over the modern village of Parihati, but my work has shown only medieval pottery and large terra-cotta crucibles, apart from slag debris and plans of furnaces (Chakrabarti and Chattopadhyay 1983). The spot where the Perua hoard was found has a clear scatter of potsherds, but these sherds cannot be put earlier than the medieval period. The evidence is not, of course, specific but suggestive.

The immediately earlier chronological point is furnished by the discovery of the Puri-Kushan coins of the sixth to seventh century A.D. in the neighborhood of the old copper workings at Rakha. The discovery was reported in 1919 [for the details, see Chakrabarti and Chattopadhyay (1985)]. One of the Rakha coins was inscribed and paleographically dated before the middle of the seventh century A.D. It is worthwhile to mention here that there is an unconfirmed report of the discovery of Puri-Kushan coins in the old shafts of Chhedapathar in Bankura.

Murray was an old-style prospector in the Rakha area, and in 1940 (Murray 1940) he summed up his impressions of the old mines of the Singhbhum copper-gold belt. His observations on the Kunderkocha gold mines at the border of Singhbhum and Mayurbhanj districts are not relevant here, but two of his findings are important enough to be remembered in the context of the old copper mines of the region. First, the beads of carnelian, agate, onyx, crystal, and glass in some of the copper mining localities, as reported by Murray, may indicate an early historic (early centuries A.D.?) presence in this area. Second, Murray found neoliths in the mining debris around the old copper workings of Chandra Buru or Ramachandra Pahar and those between Talsa and Turumdih. With Murray's observation in mind one may turn to the archaeological assemblage of such sites as Barudih and Dugni on the banks of the Sanjay River. Apart from showing a profusion of Neolithic celts both these sites yield microliths, iron slag and implements, rice, beads of semiprecious stone, and pottery, the most distinctive of which is a coarse black-and-red ware. On the whole the assemblage, which at Barudih has been dated to c. 1000 B.C., is the Chalcolithic–Iron Age assemblage of eastern India. Now, old copper workings have been reported at such places as Barudih, and one can logically suggest that these sites mark the occupation of copper miners in the area and thus take back the antiquity of copper mining in Singhbhum to the second millennium B.C. I believe that Murray's find of Neolithic celts in the mining debris of Chandra Buru and between Talsa and Turumdih makes sense.

That the metal resources of the Chhotanagpur plateau were opened up in the second millennium B.C. is suggested by the find of tin as an alloy in a small finger ring from the "Chalcolithic" level at Bahiri, a black-and-red ware site in Birbhum, West Bengal, dated c. 1000 B.C. [for the site, see Chakrabarti and Hasan (1982); for tin (about 10% in the analyzed finger ring), see the information from P. K. Chattopadhyay of Durgapur Alloy Steels, Durgapur]. The only plausible source of tin in the region is Nurungo, near Giridih (Chakrabarti 1979). The fact that the tin of the Giridih area was reaching the Birbhum plain in West Bengal, more than 200 km to the east, as early as 1000 B.C. is revealing in the metallurgical history of the region.

Summary

With a brief and general summary of the present stage of knowledge in the history of mining and metallurgy in India, I have highlighted what is currently known and understood about the old copper mines in the Darjeeling-Sikkim area and at Buxa in the northern section of West Bengal and a little beyond, the Singhbhum and Giridih areas of south Bihar, and the Bankura area of the southern section of West Bengal. The preindustrial mining tradition of Buxa is illuminated by a late nineteenth-century witness report, but otherwise the historical interest of this and other mines of the region is not inferred to be earlier than the medieval period. The two old working localities of Chhedapathar and Tamkhun in Bankura and the single old working locality of Baragunda in Giridih have been briefly mentioned. There are more data on the old workings of the Singhbhum copper belt, but even here only the localities can be identified along with a discussion identifying the chronological range [from the second millennium B.C. to the late medieval period (?)] in which these mines were operative. Collaborative work among the archaeologists, geologists, and miners is necessary before anything can be said in more detail about these mines. This comment is also valid for the history of mining in India in general.

References

Agrawal, D. P., R. V. Krishnamurthy, and S. Kusumgar. n.d. "Arsenical copper in the Indian Bronze Age." Unpublished.

Ball, V. 1869. "On the ancient copper miners of Singhbhum." *Proceedings of the Asiatic Society of Bengal,* 170–175.

Ball, V. 1870. "On the copper of Dhalbhum and Singhbhum." *Records, Geological Survey of India* 3 : 94–103.

Ball, V. 1881. *A Manual of the Geology of India. Part III. Economic Geology.* Calcutta: Geological Survey of India.

Chakrabarti, D. K. 1979. "The problem of tin in early India: A preliminary survey. *Man and Environment* 3 : 61–74.

Chakrabarti, D. K. 1981–1983. "Study of the Iron Age in India." *Puratattva* 13/14 : 81–85.

Chakrabarti, D. K. 1986. "Mining in ancient and medieval India," in *The Cultural Heritage of India* (Calcutta: Ramakrishna Mission Institute of Culture), vol. 6.

Chakrabarti, D. K. Forthcoming. "The preindustrial mines of India." *Puratattva* 16.

Chakrabarti, D. K., and R. K. Chattopadhyay. 1983. "Parihati: A medieval smelting site in Midnapur district, West Bengal." *Puratattva* 21 : 161–163.

Chakrabarti, D. K., and R. K. Chattopadhyay. 1985. "A preliminary report on the archaeology of Singhbhum with notes on the old mines of the district." *Man and Environment* 9 : 128–150.

Chakrabarti, D. K., and S. J. Hasan. 1982. "The sequence at Bahiri (Chandra Hazrar Danga), District Birbhum, West Bengal." *Man and Environment* 6 : 111–149.

Dunn, J. A. 1937. "The mineral deposits of eastern Singhbhum and surrounding areas." *Memoirs, Geological Survey of India* 59, pt. 1.

Haughton, J. C. 1854. "Memorandum on the geological structure." *Journal of the Asiatic Society of Bengal* 23 : 103–122.

Jones, W. 1833. "Description of the northwest coal district. *Asiatic Researches* 18 : 183–170.

Murray, E. F. O. 1940. "The ancient workers of western Dhalbhum." *Journal of the Royal Asiatic Society of Bengal, Letters* 6 : 79–104.

Stoehr, E. 1870. "The copper mines of Singhbhum." *Records, Geological Survey of India* 3 : 86–93.

Tiwari, R. K., and N. K. Kavdia. 1984. "Ancient mining activity around Agucha village, Bhilwara district, Rajasthan." *Man and Environment* 8 : 81–87.

Willies, L., P. T. Craddock, L. J. Gurjar, and K. T. M. Hegde. 1984. "Ancient lead and zinc mining in Rajasthan, India." *World Archaeology* 16 : 222–233.

23
Science and Magic in African Technology: Traditional Iron Smelting in Malawi

Nikolaas J. van der Merwe and Donald H. Avery

Iron smelting was still widespread in Africa at the beginning of the twentieth century and was recorded in a number of areas by missionaries, administrators, travelers, and the like. A wide variety of technical procedures and associated magic was observed in the process. The quality of these early accounts varies, depending on the technical expertise of the observers and their interest in (or bias against) the ritual aspects. Reviews of the early literature, both technical and ritual, have been provided by Cline (1937) and others, but no attempt in this direction will be made here. Suffice it to say that magic associated with smelting appears to be universal in sub-Saharan Africa, as are certain themes. These include the following:

1. For iron smelting to be successful, the smelters must be technically expert and a variety of supernatural forces must be propitiated. This propitiation takes the form of ceremonies and ritual and involves the application of special materials or medicines. The combination of ritual and medicines is referred to here as magic.
2. The iron smelters, especially the leaders, command the knowledge to carry out both the technical and magical components of smelting. This makes them more than ordinary craftsmen and gives them special status, usually high; alternatively they may be outcasts.
3. Sexual symbolism and taboos accompany smelting, often expressed through the exclusion of women from smelting in progress, compulsory celibacy for the smelters, or female anthropomorphic details added to smelting furnaces.

Explanations for magic associated with smelting or other metallurgical procedures are as varied as their observers. These range from simple dismissals of superstition to elaborate functional explanations. An example of the latter is Bronowski's analysis of the Shinto ritual that accompanies the making of samurai swords. The chants and movements are explained as devices for recipe recall and timing: "When you have no written language, when you have nothing which can be called a chemical formula, then you must have a precise ceremonial which fixes the sequence of operations so that they are exact and memorable" (Bronowski 1973, p. 171).

The case in question involves forging, not smelting, but the same principle could apply. As yet nobody has produced such a functional explanation for African smelting magic, nor has anyone suggested that the medicines have a chemical function. Dismissing smelting magic as unimportant is more common. Thus Kjekshus (1977, p. 91) believes that the "ritualist school of research into ... ironworking" has tended to see "the manufacturing of iron as a by-product of the magico-religious needs of the East African peoples."

Kjekshus reviews iron production in Tanzania at the turn of the century, emphasizing that people in the lakes region (for example, the Fipa) were able to supply their own needs and also export iron, usually in the form of hoes. He rejects as misplaced emphasis the observation of Wyckaert (1914, p. 374) that Fipa smelters "seem to attach greater importance to the supernatural influences than their own work," Robert's (1949, p. 20) characterization of the master smelter as a magician, and Wise's (1958, p. 232) contention that, for smelting to be successful, "the living and the dead, the spirits of the air, wood and rock must work together." Instead Kjekshus sides with those who describe smelting ritual as of secondary importance, at most "a type of trade-union secrecy aimed at maintaining monopoly over an important economic enterprise" (Kjekshus 1977, p. 91). He explains that the demise of the local iron smelting industry, after its traumatic contact with colonial technology, resulted in "the decline of chemical/technical knowledge and the rise of the magical element as a substitute explanation of the metallurgical process." If this is interpreted to mean that those not familiar with smelting technology are likely to emphasize the ritual aspects in their accounts of it, we would tend to agree. This is hardly an alternative explanation, however.

The problem with this debate in any case is that the opposing explanations have been produced from two antagonistic European ideologies. Neither side need be entirely right or wrong; they may simply be missing the point of an African ideology. It is appropriate, therefore, to accept Kjekshus's challenge to contemporary anthropologists to provide a more satisfactory explanation of the magical element in iron smelting. This we have attempted to do by working with iron smelters in Malawi, not far from the area about which Kjekshus writes. In our reporting we should admit to a bias toward technical explanations. We are an anthropologist of the archaeological persuasion (van der Merwe) and a metallurgical engineer (Avery); both of us are metallurgical historians. It is doubly significant, therefore, that our perceptions of the importance of the magical element in smelting increased substantially as we were progressively introduced to its mysteries. The investment of time and effort in smelting magic is large and the expertise required is extensive. It is a technology in its own right, with its own body of knowledge and network of personal and commercial relations. We did not find that magic has a scientific contribution to make to iron smelting technology but nevertheless concluded that smelting could not be conducted without it, because it is too important to the smelters.

Iron Smelting in Malawi: A Reconstruction

In 1982 and again in 1983 we reconstructed traditional iron smelting at two sites in Malawi. One of these reconstructions was at Chulu village, near Kasungu in central Malawi, the other near Phoka Court, in the vicinity of Livingstonia in northern Malawi. The two sites are some 255 km apart.

Chulu is a sizable village with a courthouse and is the seat of a chief, or traditional authority. The village lies on the edge of Kasungu National Park, a wildlife refuge established by the British colonial authorities (figure 23.1). Many of the people of Chulu village, or at least their parents, lived in Kasungu Park before it was proclaimed a park or had access to the area for purposes of agriculture or hunting. Chulu also lies on the periphery of the urban culture of Kasungu, the nearest town (50 km), and Lilongwe, the capital city of Malawi (200 km away). A daily bus connects Chulu with these population centers. The people of Chulu are Cewa speakers, members of the largest linguistic group in Malawi, which is also dominant politically. When asked, they are likely to identify themselves as Cewa but are unlikely to belabor the point. Loyalties to their village government, the Malawi Congress Party, and to the nation cross-cut this linguistic identity. They are also matrilineal and matrilocal; thus many of the men of Chulu come from outside the village. Chulu is consequently subject to a mix of rural influences and traditions, as well as being poised between rural and urban (Western) lifestyles. To avoid any ambiguity, the term "Chulu" is used here to refer to the village, its people, and its way of doing things. It would be a mistake to project Chulu onto a larger canvas as representing, say, the traditions of a tribe or all Cewa speakers.

The mixture of rural tradition and Western urban values in Chulu can be observed in their most basic activity, agriculture. Subsistence farming is the major occupation here, but cash cropping of tobacco and maize also occurs. Flue-cured tobacco is produced on fully mechanized estates in the vicinity, and some of this mechanization has been adopted by successful small farmers. Traditional hoe agriculture is still much in evidence, although hoes are imported from British factories, not locally made. The hoe has a thick shovel-shaped blade, is mounted at a right angle to the handle (which stands about chest height), and is used in a chopping sidestroke like a pick. A variety of iron tools are produced by local blacksmiths from scrap metal (primarily axes and adzes for woodworking, but also some spears, arrow points, and knives), but no iron smelting had been done in the area since about 1930. Because of the expressed interest of President Banda, who grew up near Kasungu, local interest in iron smelting was revived around 1980. One of our informants built a nonworking example of a smelting furnace at this time, and another wrote an account of smelting procedures based on interviews with Vujomo, legendary chief smelter of the district. Vujomo died in 1980, and with him much of the local tradition of smelting. In reconstructing iron smelting at Chulu, it

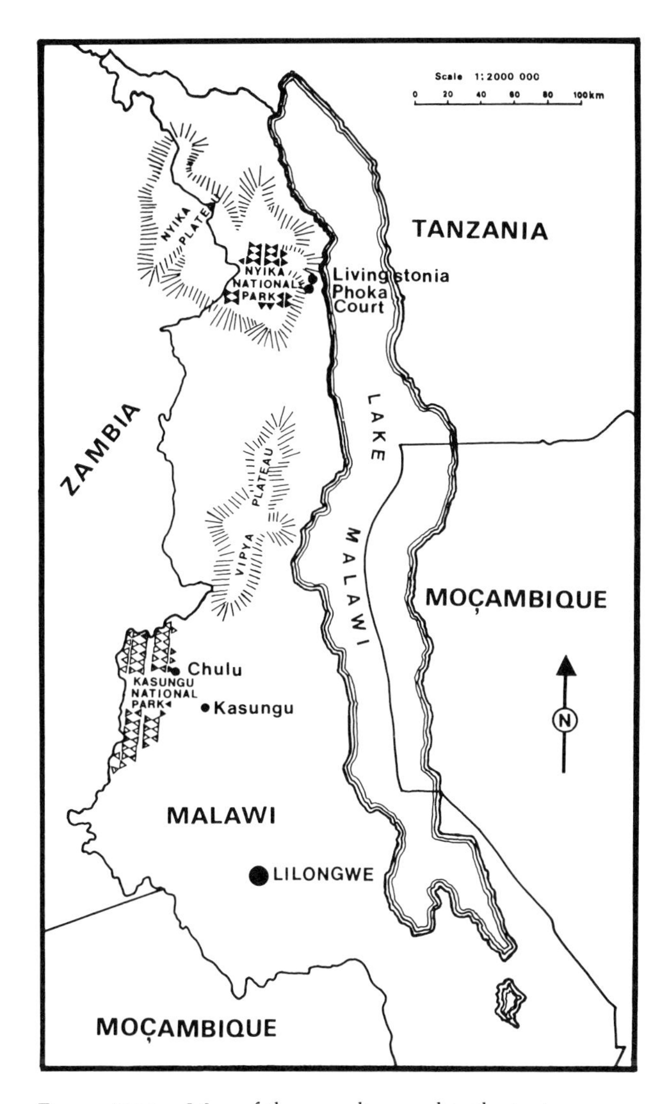

Figure 23.1 Map of the area discussed in the text.

was difficult to assess how much of this tradition we were able to recover, especially because the men came from various villages and there were some differences of opinion about procedures.

The Phoka of northern Malawi are a subgroup of Tumbuka speakers. They identify themselves by speaking the Phoka dialect, although they can all speak "normal" Tumbuka, and by reminding the observer at times of their Phoka identity. The region they live in embraces the steep escarpment slopes and foothills on the eastern and southern sides of the Nyika mountain massif, which is considered inaccessible by most non-Phoka. This eccentricity is commented on by the Phoka, usually when looking at their glorious view of mountains, hills, and Lake Malawi. Who would want to trade all this for the overpopulated flatlands? Many of the Phoka used to live further up the Nyika slopes, some on the Nyika plateau at altitudes as high as 2,500 m.

The center of the plateau, largely unpopulated, was proclaimed as a wildlife refuge in colonial times. Under the current Malawi government, Nyika National Park was enlarged to take up the entire plateau and to spill down the escarpment sides. This proclamation involved the destruction of some 10,000 houses and the removal of their owners, many of them Phoka. In many ways, this was a reenactment in reverse of some recent Phoka history. In the latter half of the nineteenth century they survived the Nguni invasion of Malawi by moving higher into the mountains.

The Phoka are not numerous, but they are stubbornly independent. They tend to live on the ridges in independent family homesteads, with their terraced fields on the steep slopes in between. A young man or woman who feels like being independent may choose to move to his or her own ridge and build a house there. Thus the Phoka tend not to have villages of any size, just a few concentrations of houses at the homes of headmen or subchiefs (Sub-Traditional Authorities, or STAs in the new government structure). Because the Phoka are so insistent on their own ethnic identity, we use the term "Phoka" here to refer to the people we worked with and their traditional ways of doing things.

Note that both the Phoka and the people of Chulu live on the edge of a national park in which they used to own land. Both parks contain a kind of museum of their past activities in the form of village remains, old fields, and especially well-preserved iron smelting furnaces. Here the resemblance between the two groups ends, for the Phoka are much further removed from urban culture. A Scots Presbyterian mission has been active near the Phoka at Livingstonia since 1896, and the level of literacy is fairly high. The Phoka, however, seem to have been preadapted to Scots Calvinism; they added the new religion and value set to their own, instead of replacing one with the other. They are widely regarded as the most traditional of Malawi's peoples.

The basic livelihood of the Phoka is still subsistence hoe agriculture, with minor cash cropping of coffee beans to provide clothes and a few extras, such as sugar and soap. A few of their chief iron smelters were still alive in 1981 when this project started—old men who could contribute advice but not much active assistance. Iron smelting ceased here in the 1930s, but there is still considerable interest in the process. In our reconstruction the smelting was led by men in their late sixties who had been apprenticed to their smelter fathers in their youth. The organizer of this smelting team was the son of another well-known chief smelter, the late Khaura Msiska; the son had also written an account of smelting and kept a collection of smelting memorabilia in an empty hut in his compound. The survival of Phoka tradition was particularly evident in the ritual associated with smelting. All the participants were familiar with the procedures and were keen to instruct us in them. Because they had grown up together, there was also near-unanimity on what was appropriate. An extensive collection of medicines (animal, vegetable, and mineral) was employed; it was firmly based on the local pharmacopoeia and formed part of their daily lives. The smelters were also regular users of the full range of traditional hand tools made by local blacksmiths out of scrap. These included brush hooks, knives, axes, adzes, spears with digging chisels hafted to the tail, and arrow points. A few hoes made from locally smelted iron were proudly displayed as historical items, as were some hoes traded in from Tanzania many years ago. Hoes for everyday use have to be purchased at the farmers' cooperative, however; in 1982 these cost K2.70 (about US $2.50), the price of a live chicken or two days' wages for a skilled laborer.

We interviewed a large number of people in both the Chulu and Phoka areas who had some knowledge of smelting, or claimed to. At Chulu, the interviews took place in the courthouse and were a public event. The selection committee was composed of one of us (van der Merwe) on the bench, flanked by Traditional Authority Chulu (the chief of the area) and his secretary on one side and by two park rangers and an archaeologist from the government department of antiquities on the other. The leaders of the smelting team were selected after a committee assessment of the candidates, based on their knowledge of smelting and tradition. With the Phoka the Malawian assessors included only the local subchiefs and a research assistant from the National Parks department who was a Phoka himself. The selection was much easier here, as knowledge of smelting was more current.

Chulu and Phoka Iron Smelting

The iron smelting practiced by both the Phoka and the people of Chulu is characterized by a two-stage process employing two furnaces. A large, primary furnace (*ng'anjo* in both Cewa and Tumbuka) operates by natural draft and produces a sintered sponge of slag with small pieces of iron in it. The slag is a glassy material, consisting mostly of melted rock and other unwanted material from the ore, in which is suspended the bits of iron that need to be recovered; the whole lot is sintered together (incompletely melted) into an irregular spongelike mass, which also contains ash and charcoal. This product is resmelted in a small secondary furnace (*chiramba* in Cewa, *kathengu* in Tumbuka), which is bellows driven, and produces an iron bloom. The bloom is a solid chunk of iron, so called because it "blooms" when red hot, and contains some trapped slag. Much of this slag is squeezed out when the blacksmith reheats and hammers the bloom during forging, but some slag is always present in bloomery iron (wrought iron) as a useful impurity that imparts some hardness.

The furnaces of the two areas differ in the details of shape, size, and operating characteristics (especially the draft and potential temperature), but the basic process is the same. The tall, natural draught furnaces (*ng'anjo*) used in the first stage are particularly unusual. Most iron bloomeries (that is, furnaces that smelt iron in its solid form) are pumped with bellows. Smelting by natural draft (also called "induced draft") is an unusual process, confined almost entirely to Africa. Its distribution extends from west Africa through the Congo basin to Zambia and Malawi. Little is known about the origins and development of this process, but the general impression is that it was invented in west Africa and spread from there to central Africa. In Malawi, examples of natural draft *ng'anjo* can be seen in both the Kasungu and Nyika national parks, having been built there during the last century or so by the ancestors of the modern day people of Chulu and of the Phoka, respectively.

Although iron smelting is nearly 2,000 years old in Malawi (Mgomezulu 1981), it is unlikely that induced draft smelting is that old. The few iron smelting furnaces of early Iron Age times (first millennium A.D.) that have been excavated in west, central, and southern Africa seem to be of the smaller, forced draft variety and were apparently used in a single-stage smelting process. The Cewa speakers of Malawi are associated with an archaeological tradition of the later Iron Age (second millennium A.D.), which may provide an indication of when natural draft smelting appeared on the local scene. There is no direct evidence for this suggestion, however. The Tumbuka speakers in northern Malawi are thought to have arrived there after about 1500 A.D. (Derricourt 1979), but knowledge of their smelting technology at the time is equally lacking.

The antiquity of natural draft smelting in Africa remains unexplored archaeological territory. The fact is that natural draft furnaces are not readily recognizable from archaeological remains and may have been overlooked in many places, both in Africa and in other parts of the world. They have been observed in historical times in upper Burma. Archaeologically there is a strong suggestion that they were used at Noricum in Austria during Hallstatt times, which may mean that they may yet be recognized elsewhere in the remains of early Iron Age Europe. These disconnected bits of evidence provide tantalizing threads that still need to be unraveled.

In 1864, when David Livingstone traveled through the area that is now Kasungu Park, he noted that "at every third of fourth village, we see a kiln-looking structure ... for smelting iron. No flux is used ... and yet capital metal is produced" (Livingstone and Livingstone 1865, p. 536). Many such furnaces can still be seen in the park, some of them attributed to known smelters, now dead, who lived near Chulu village in their later years. The same is true for the Nyika mountain massif: There are a few well-preserved furnaces in the national park on the plateau and several more in the foothills now occupied by Phoka.

The furnaces that can be seen in Kasungu Park (figure 23.2) vary somewhat in design, but the *ng'anjo* of this area is generally a cylindrical chimney about 2.3 m tall and 1.5 m in outside diameter, constricted to about 0.7 m at the top. The internal diameter is about 1 m, constricted to about 0.5 m at the top and also at a point about one-third of the way up from the floor. The furnace thus has two internal chambers, the upper one about twice the size of the lower, with the internal constriction helping to support the weight of the charge in the stack. The smelting zone is in the lower chamber, which has six or more holes for the insertion of one tuyere (ceramic air pipe) apiece, spaced evenly around the circumference near the base, plus a doorway that is plastered shut during smelting. Some of the furnaces are also waisted on the outside at the same height as the internal constriction (figure 23.3).

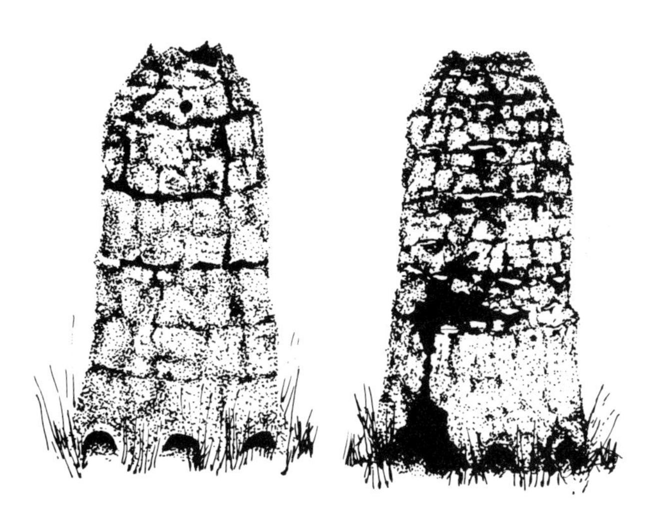

Figure 23.2 Two views of the *ng'anjo* in Kasungu Park, built by an unknown smelter. The front view (right) shows the rake hole and coursed construction, with broken tuyeres in the seams. The rear view (left) shows the inspection hole, or *jojo*, near the top.

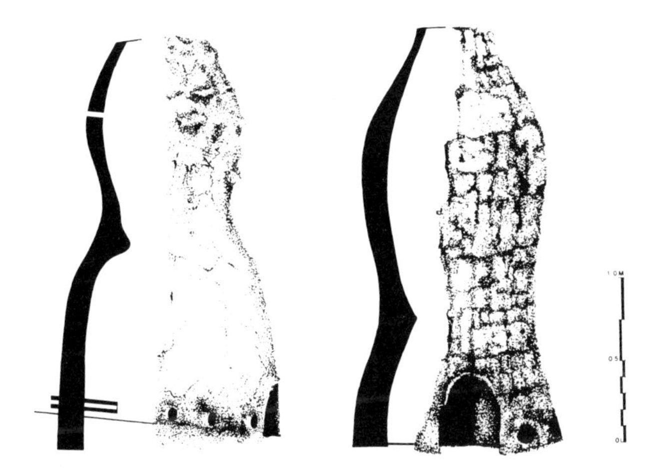

Figure 23.3 The *ng'anjo* (induced draught furnace) in Kasungu Park (right), built by Vujomo around 1910. On the left is an attempted copy built by Chulu smelters in 1982. During smelting each furnace was equipped with a ladder from which the smelters could pour the fuel and ore.

The furnaces still standing on or around the Nyika are a good deal smaller than those of Kasungu (figure 23.4). Here the *ng'anjo* is shaped like a cone with the point cut off, tapering from about 1.8 m in outside diameter at the base to 0.8 m at the top; the height is 1.5 m or less. On the inside the diameter tapers from about 1.3 m at floor level to 0.7 m at the top, the wall being thickest at the base. A dozen or more tuyere holes penetrate the wall near the base (three tuyeres per hole), as does a doorway, which is plastered shut during smelting after the insertion of six tuyeres. The floor is simply the soil, which is dug so that it slopes toward the door; the same is true for the Kasungu furnaces.

The secondary smelting furnaces used in both research areas are much less substantial than a *ng'anjo*. An archaeological survey of Kasungu Park by David J. Killick (personal communication) has failed to turn up a single *chiramba*, and none were known to survive at Chulu village. Near Phoka Court, the seat of STA Kachulu, one *kathengu* built in 1910 by Khaura Msiska still stands (figure 23.5). The *chiramba* used in 1982–1983 by the Chulu smelters was a simple cylinder, about 0.75 m in internal diameter and height, with four tuyere ports; four pumpers with four goatskin bag bellows operated it. The *kathengu* used by the Phoka is an oval-shaped beehive, about 0.5 m long and 0.4 m wide and high, with a chimney hole of 0.2 m diameter in the top. It has a rake hole in one side, which is plastered shut during smelting, and tuyere holes at the ends. Two pumpers, each with one goatskin bag, operate the furnace.

It is not the intention here to give a full technical description of the smelting procedures we observed, as this will be done elsewhere (Avery et al., this volume). It is more important to gain an impression of the

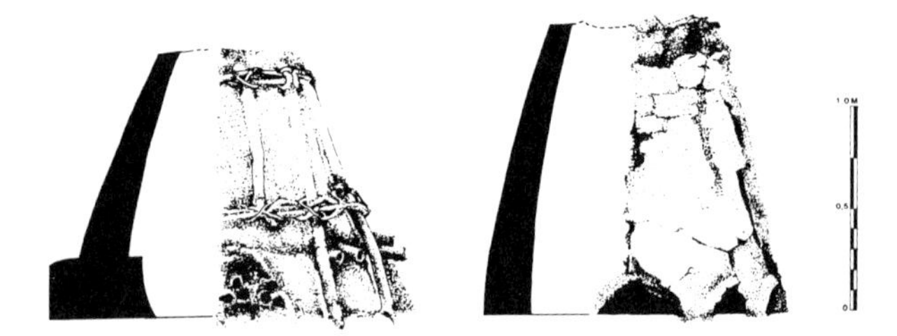

Figure 23.4 (Left) A *ng'anjo* built by a Phoka smelting team in 1982 and 1983. (Right) A *ng'anjo* built by Khaura Msiska near Phoka Court around 1910. During smelting this furnace also had an exterior wooden skeleton for the smelters to stand on. The difference between the two reflects differing family traditions of building the footing.

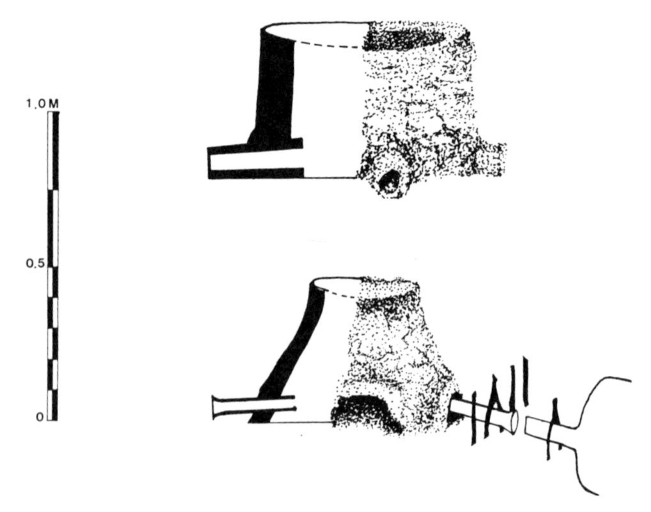

Figure 23.5 Secondary refining furnaces in which the iron-rich slag made in the *ng'anjo* is resmelted to make iron. The *chiramba* built by the Chulu smelters (top) is pumped with four goatskin bellows; the *kathengu* used by Phoka smelters requires only two.

amount of work involved and the wide economic network required for smelting. At Chulu and with the Phoka the full cycle of charcoal preparation, mining of ore, building the furnaces, gathering medicines, making tuyeres, brewing beer, and, finally, smelting took about three months (July to October). This was done in the dry season, between harvest and preparation of the new fields for planting. At times, only a half dozen men were involved; at others, as many as fifty people, including women and children. The latter case was vividly demonstrated by the Phoka in 1982, when they built their furnaces against a steep mountain slope at an altitude of 2,000 m (6,000 ft), about 1,000 m (3,000 ft) above the nearest road. This inaccessible site was chosen because it is near a ready supply of the right tree species for charcoal making and, primarily, because it is near the house of the chief smelter. Iron ore had to be mined on the Nyika plateau and carried to the site over several mountain ridges, about a day's walk for the very fit. Building the furnaces required 6,500 kg (14,300 lb) of wet clay to be carried up about 600 m (2,000 ft) of steep mountain slope. A metric ton of charcoal was consumed in the smelting, requiring the felling of 88 m^3 of trees. The maximum yield that could be expected from this prodigious effort was about four hoes and some smaller tools; this was the unanimous opinion of many informants over a wide area. In the past participants were rewarded with shares of the iron, but later they received primarily food, which is the real end product. In reconstructing the process, we paid government-approved wage rates; this provides a rough measure of the labor involved. On this basis a hoe could be calculated to be worth about $250 (US), that is, a hundred times the price of the one imported from the United Kingdom or the value of about eighteen goats.

Smelting in the Chulu *ng'anjo* consisted of starting a fire on the floor, plastering the door shut, and then feeding charcoal through the chimney until the furnace was half full. Pellets of lateritic iron ore were then poured in, alternating with more charcoal, until the furnace was full. As the fuel was consumed, more charcoal was added to keep the furnace full. After twenty-four hours, the fire had traveled through the stack, and a pale blue plume of carbon monoxide flame erupted from the chimney. The temperature in the zone in front of the tuyeres rose to about 1250°C, a workable smelting temperature, and finally reached a maximum of about 1350°C. Fuel was added for 90 hours, some 1,450 kg of charcoal being consumed in the process of smelting 75 kg of iron ore. When the furnace door was broached 114 hours after the fire was started, 91 kg of sintered sponge was extracted. This material was broken up and the heavier pieces selected for resmelting in the *chiramba*. The latter did not work very well, reaching at best 1250°C for short periods. This was insufficient for draining the slag away from the small pieces of iron in it, so as to allow the latter to come together in a bloom. Sufficient reduction of the ore had taken place in the *ng'anjo*, however, for us to see how the full smelting cycle would have worked if the smelters had been more technically expert.

Smelting in the Phoka *ng'anjo* was more difficult than in the Chulu furnace, because the Phoka *ng'anjo* is shorter and has less draft. Manipulating it to a workable temperature of about 1250°C requires careful adjustment of a number of variables, particularly the particle size of the fuel and charge (fine particles choke the furnace). The Phoka load the entire charge into the furnace at the tuyere level, cover it with a heavy cap of wood made from half a cross section of a big tree trunk, start a fire with burning coals around the charge, and then fill the *ng'anjo* with charcoal. This procedure makes it doubly difficult to heat up the charge and seems technically inexplicable. It was clear, however, that this was the proper, traditional way of going about it. On the other hand, starting the melt in the furnace was made easier by mixing the ore with an equal amount of discarded slag from previous smelts, collected at old smelting sites. On the second try in 1983, all these variables meshed, and a 75-kg lump of sintered sponge was produced from 55 kg of charge. To achieve this required about 100 hours of smelting and 1,000 kg of fuel. The sponge was broken up, and the most promising pieces resmelted in the *kathengu*. The latter worked quite effectively: Two pumpers expending heroic amounts of energy could keep the temperature above 1300°C. Provided that the amount of charge was kept small, the product of the *ng'anjo* could be consolidated into usable lumps of iron. With more expertise and practice the Phoka smelters could produce a larger bloom in this manner.

In both procedures described here the *ng'anjo* is designed to process a relatively large amount of low-grade iron ore into a form that can be resmelted in a smaller furnace. This provides a technical solution to the problem of dealing with low-grade ore but is demanding of labor and fuel. After allowing for knowledge lost over the fifty years when smelting was not carried out, we still find these smelting processes fraught with difficulties and open to failure on many counts. It was not surprising, then, to hear many accounts of furnaces that failed in the past and of what one has to do when this happens. Explanations of failure can be technical, and it was clear that the experienced smelters had a considerable knowledge of how to correct such problems. The explanation was just as likely, however, to involve medicines, ritual, or witchcraft. The amount of attention directed at these items in the course of smelting easily equals that expended on technical problems.

Chulu Ritual and Medicines

In the absence of a recognized chief smelter at Chulu, the required rituals and placement of medicines were conducted by three members of the smelting crew who had knowledge of such things. One of these was an elderly man with considerable *gravitas*; the second, a middle-aged man who had written the reminiscences of the late chief smelter, Vujomo; and the youngest a son (grandson?) of Vujomo. There were some differences of opinion among them about proper procedure and a definite reluctance to discuss medicines in front of others. This information had to be elicited in private interviews.

The medicines involved in smelting are called *msinkho* or *mitsiriko* in Cewa. A full list of the items employed are provided in appendix A, with botanical identifications wherever possible. Twenty-seven items are listed, of which twenty are specific to the smelting process only. The other seven have additional uses. Two are used to counteract bad dreams, considered to be due to evil spirits. A third is used to help cure the "dancing disease," a disorder with symptoms of fever, headache, shouting, and generally carrying on in a disturbed manner; its ultimate cure is a night of energetic drumming and dancing. Yet another is used to treat children subject to epilepsy. This group of medicine can be classed as cures for spirit possession, that is, actions by malevolent ancestor spirits. For protection against the living, poles cut from the *chimphakasa* tree are incorporated at intervals in the fence of a cattle enclosure to protect the livestock against evil spells from jealous neighbors. The multipurpose *musambamfumu* tree provides medicines against ancestor spirits, and its boiled bark is used to anoint the body of a chief before important meetings to enhance his authority; more mundane, it also cures Newcastle disease in chickens! A final item worth noting is a piece of the nest of the *katawa* bird (hamerkop, *Scopus umbretta*), a solitary wader usually seen standing motionless in shallow water, waiting to catch a frog. Known as the lightning bird to many Africans, it is believed to bring rain and, while staring into water, to see the future.

None of our three informants claimed to be experts on the wider uses of medicines, their knowledge having been acquired in connection with smelting. There may be some hidden information in the list of *msinkho* as a result, but enough information is available to observe that the medicines are aimed at increasing the capabilities of the smelters and counteracting the malevolent influences of others, both living and dead. The beneficence of the medicines are furthermore combined with sacrifices, the keeping of sexual taboos, and prayers to God and the ancestors.

We observed a Chulu smelt in 1982. The first ceremony took place when the foundation trench for the *ng'anjo* was dug. The site was chosen on the western side of a large termite hill that could provide good refractory clay for building the furnace. The prevailing wind in this area is from the east, but during the dry season it is in fact quite variable, so that the choice appeared arbitrary. A wooden peg was placed in the ground, a circle drawn around it with the help of string, and the foundation trench dug. The *mitsiriko* ceremony ("protecting the furnace," but also a word for the medicines) then followed. Two pegs were placed to outline the position where the door of the furnace would be, and two more directly opposite them on the circumference. Of the five pegs the central one is not considered to have additional importance, but the four on the circumference are and must be cut from the *chimphakasa* or *msindira* Chulu trees. The first of these, one notes, is of the species used in the fence of cattle enclosures to protect the cattle against witchcraft. After the placing of the pegs, medicines were placed in the foundation trench between the two areas outlined by the pegs. These included pieces of seventeen items listed in appendix A plus some maize flour. The latter had been pounded and ground in the village by an old woman with no husband, that is, a woman both infertile and celibate. As the leader of the ceremony (the oldest smelter) placed these items, he chanted quietly: "We are building this *ng'anjo* here and we do not want evil to befall it, or the jealousy of a person. We want peacemakers only. We are building this *ng'anjo* because we are poor. We pray for success to God, through the ancestral spirits." The leader then ran away for a short distance, to signify the departure of evil spirits, and returned to brush the foundation trench with a green branch, representing the creator and purity. These actions consecrated the furnace, protecting it from witchcraft.

A day went by, to ensure that the foundations of the furnace were not disturbed in any way, and then building commenced. Four days later, with the furnace half finished, it was found in a collapsed state with human excrement in four places around the periphery. Unscrupulous enemies? A cover-up of a botched construction job? We never found out, but the dedication ceremony was repeated with better medicine, and building was restarted with renewed energy and by a different method. About a week later the furnace was finished and the smelting team held a party, consuming maize porridge, chicken, and beer. The furnace was left to dry out for a month or more.

For a new furnace a religious ceremony is held on the evening before smelting begins; it is open to anyone in the village. For this furnace the ceremony was held at the shrines (*kavuwa*) to three deceased chiefs of Chulu village, through whom prayers are directed to God (*Mphambe*) to bless the smelting and to counter evil influences. Each shrine is a small timber hut with a pot inside, into which libations of freshly brewed beer are poured. Prayers were conducted by a senior man from the village, one of the protectors of the shrine, with responses of "*pepa*" ("forgive us") from the participants. They then sang Hymn No. 1 from the hymnal of the local Christian church: *Mzimu wanga wabwera*, "My spirit has come." This song, which everyone knew by heart, is taken to mean that an ancestor spirit is present to intercede with God, a fact that may come as a surprise to the missionaries.

After the ceremony at the *kavuwa*, most of the congregation went home, but some elderly members accompanied the smelters to a *musolo* tree near the furnace. One of the elders, a village headman, led the group in prayer again, repeating the pleas made at the *kavuwa*. The same prayers were repeated a final time at the *ng'anjo* itself, led by one of the smelter leaders. The head of a white rooster was then cut off and its blood spilled over the inside and outside of the *ng'anjo*. The insides of the rooster, plus its head, lower legs, and feet were placed under the *musolo* tree as an offering, along with a small pot of fresh beer from the large supply in the smelting camp. The villagers who were not smelters all left at this point, and the smelters' camp was sealed by lowering a boom over the path to the village. A font consisting of a bowl of tree bark in a forked upright was erected next to this gate. Pieces of bark, wood, and leaves were placed in this font and water was added to them. In all, twenty-two items of *mitsiriko* were used (appendix A), but only six of these were compulsory—without them the smelting could not proceed. Upon lowering the boom over the path, each member of the smelting team (including us—we were accepted as full members) washed his hands in the font; this serves as both an act of purification and one of distancing the smelters from the people outside. The few visitors who were subsequently allowed inside the smelting camp (Chief Chulu, for example) had to wash their hands in the same manner. Finally, the sacrificial rooster was roasted and shared among us, with beer; a pot of beer was sent to Chief Chulu and another was sent to the village. Smelting could then commence, and the smelting team undertook to observe the required taboos for the duration: isolation from outsiders, especially women; celibacy; and no washing of the body.

Before lighting the furnace, a small hole (*chitewerende*) was dug in the middle of its floor, and three items concerned with the fire setting placed in it (appendix A, items 23–25). Some kindling was added. A flame was kindled by twirling a hard stick on a piece of soft wood, a method now practically extinct in the village. Drifted grass (appendix A, item 23) was ignited from the flame and the fire started. Small pieces of the full list of *msinkho*, twenty-five items in all, were added to the fire, and the doorway (*chilema*) was plastered shut. The furnace was fed through the chimney from a scaffolding until half full with charcoal. The ore charge was added and all the leftover bits of *msinkho* thrown in after it.

No further ceremonies took place during smelting, which was a week-long tiring and dirty business. A great deal of beer was consumed, and many songs were sung to keep up morale. The songs, like the religious ceremony before the smelting, were an interesting mixture. The most popular was in Tumbuka and concerned the many Christian denominations striving to lead their adherents to the Kingdom of God, ending with the rousing chorus "Rush! Everybody rush to register your names, so that you may enjoy the fruits of the Kingdom of God." Many of the other songs came from the *Nyao* secret men's society, in which most of the smelters held membership; they were raucous, bawdy, and far from feminist.

After the furnace was broached, villagers were allowed to come into the camp, and the taboos were lifted. No special ritual was attached to the operation of the secondary furnace, the *chiramba*. The smelters were not successful at producing an iron bloom, but had they been, they would have carried it to the chief's house and a celebration would have followed. Villagers are allowed to see the bloom, for a fee, at the party and may place orders for hoes.

Phoka Ritual and Medicines

The Phoka ritual associated with smelting is more elaborate than the Chulu ritual and more firmly grounded in tradition. Two defined persona lead the smelting: the *murungu*, or chief smelter, who pronounces on technical matters; and the *msofi*, or medicine man, who leads in ritual and deals with medicines. *Murungu* is the Tumbuka word for prophet, not to be confused with *musungu*, a white man. Further north, among the Fipa of Lake Tanganyika, *irungu* (plural *marungu*) is the word for a primary iron smelting furnace (Wyckaert 1914); the word *rungu* has a wide distribution in central Africa in connection with iron smelting. During the Phoka iron smelting of 1982, the *murungu* and *msofi* were two different men; in 1983, one man assumed both offices. The latter case is the general rule: a chief smelter is usually a man who knows about medicines, for which one must have a calling.

The metaphor for Phoka smelting is fertility and, in a more restricted sense, human reproduction. This is reflected in the words applied to the primary furnace, *ng'anjo*, in whole or in part. While it is being built, it is simply *chimpani*, a pot. When the basic structure is complete but before holes have been cut in it for the doorway and tuyeres, it becomes *mwali*, a young woman ready for marriage. From this point on, until the furnace has dried and is readied for smelting, it is protected every night by encircling the top with a knobbly sapling from the *kowi* tree (a cypress) thus "tying it together." This sapling, which is eventually incorporated into a protective wooden skeleton around the furnace, is called *nkhwamba* (Tumbuka) or *khwamba* (Phoka), the belt a woman uses to hold up her underwear. When smelting is underway, the smelters refer to the furnace as "our wife."

The tuyere holes near the furnace base are called *mapaa* or feet, and the doorway is *chilema*, which means lame or crippled. The implication of the latter is obscure; no informant could explain it. It is noteworthy that the words *ng'anjo* and *chilema* occur in both Tumbuka and Cewa; further west, in Zambia, the Cewa call a primary furnace *mucelo*, and it is of a quite different design (Phillipson 1968).

In the course of smelting the Phoka make use of an extensive list of medicines, called *sepo*. Items of animal, vegetable, and mineral nature are involved. Some originated from Lake Malawi, others from the Nyika plateau; the two places are separated by 40 km (25 mi) and 2,500 m (7,500 ft) in altitude, with some formidable mountain ridges in between. Collecting the *sepo* was therefore a lengthy process. As a result of the proclamation of an enlarged Nyika National Park and because of the enforcement of Malawi's game laws, some items were difficult or impossible to acquire locally. Although the entire list of *sepo* is not required for smelting, a few items are considered essential, the most important being a piece of hippo hide: It had to be flown in from South Africa. The *sepo* are listed in appendix B. They include thirty-six plants, fifteen animals, fish, or insects, four minerals, and three artifacts. Most of the plants have medicinal uses other than that of assisting with smelting and so do some of the animals and minerals. They constitute most of the Phoka pharmacopoeia for dealing with witchcraft, diseases caused by witchcraft and spirit possession, epidemics that threaten a community, or venereal diseases that reduce fertility. Other items are clear symbols of fertility and plenty: millet, peanuts, and other edibles, schooling fish, a termite queen, a fish net, a beer strainer, and the like. The final group symbolizes the desirable qualities of strength, toughness, hardness, wile, and speed—qualities one wishes to impart to the smelting and its product. Our informants tended to classify the *sepo* as animal, vegetable, or mineral, but they clearly also understood the symbolism involved.

The power of the *sepo* is combined with ritual, prayer, and the observation of taboos. A terrace had to be constructed for the smelting, and its position was chosen for purely practical reasons. Whether the choice was appropriate, however, was determined by divining. Several divining methods can be used, depending on the practitioner. The basic procedure is to manipulate an object and to ask it questions while suggesting answers: It stops at the right answer. The objects that can be used for divining include a decorated tortoise shell descending on a vertical string stretched between the forefinger and the big toe, a bell rattle under an upturned dish rubbed in circles on a flat basket, and a maize pounder stick rubbed back and forth on the ground. After a positive answer was received, the terrace was leveled and a small hole dug at the center of the proposed furnace; the roots of the shrubs *m'pinga viwanda* ("chases off evils") and *kawinga wa zimu* ("chases away ancestor spirits") were placed in it. These are both medicines used against such afflictions as bad dreams and the dancing disease, which are associated with spirit possession. The site of the *ng'anjo* was thus protected against interference by malevolent spirits. A stick was placed upright on the spot and the circular foundation trench dug around it. The central medicine hole is called *chihikiriro*, and it becomes the receptacle for all the medicines associated with the furnace.

When the foundation trench was complete, a brown rooster was sacrificed by cutting off its head and dropping the jerking body in the middle of the circle. Where it eventually flops into the foundation trench determines where the doorway of the furnace will be. Because the doorway has to face the downslope edge of the terrace, the choice is actually fairly limited: A wrong result could involve further ritual. The killing of the rooster was accompanied by the following prayer: "At this place, allow that this pot produce a wealth of good iron. Many people did not want this pot to be built here, but all places belong to you." Various items of *sepo* were then placed at intervals in the foundation trench, with strands of *bopa makala* grass and *mwambula* (maidenhair creeper) completely circling it. The body of the rooster was wind dried for later use; its head eventually wound up in the *chihikiriro*.

When the furnace walls were standing, the ceremony of *kaghoghora ng'anjo* ("we are pleased with the furnace") was held, involving the smelters and women and children from the village. *Nsima*—the stiff, white maize porridge that is the local staple—was prepared, and the sacrificial rooster was roasted. The *murungu* kneaded some of the maize porridge into a sausage around a cooking stick and stuck bits of chicken meat in it. This stick was held out to the participants by the *murungu* as he walked around the *ng'anjo*, so that each could take a small amount to eat. The *murungu* then started a song and led his chorus in a circling dance around the *ng'anjo*, while everyone chewed their small portion of the communal food offering. The song has two verses, which alternately use Tumbuka and Phoka dialect for the words "be pleased," and translates as follows.

1st Song

Leader:	*Kaghogho* [Phoka]	Be pleased
Chorus:	*Kaghoghora ng'anjo*	Pleased with *ng'anjo*
Leader:	*Ine kaghogho*	I am pleased
Chorus:	*Kaghoghora ng'anjo*	Pleased with *ng'anjo*

2nd Song

Leader:	*Sekerera, sekerera* [Tumbuka]	Be pleased
Chorus:	Eeh eeh eeh, eeh eeh eeh, *Henyani chanthemha* [Phoka]	(give) with love from our hearts
Leader:	*Sekerera mwali withu* [Tumbuka]	Be pleased with our young woman
Chorus:	Eeh eeh eeh [repeat]	[Repeat]

The elders were pleased to be able to teach the children the words to these songs and to have them participate in the ceremony, lest they lose touch with the traditions of their people. At its conclusion the participants separated into men, women, and children and ate the rest of the maize porridge and chicken from separate serving dishes.

Shortly thereafter a ceremony took place in preparation for the opening of the furnace door (*chilema*), the tuyere ports (*mapaa*), and the observation hole (*jojo*), which is situated about 0.2 m from the top. The *msofi* and some of the senior smelters knelt at the furnace and sprinkled millet from a basket and water from a pot in front of them. The *msofi* prayed: "Accept this millet; if it sprouts, our furnace will be a success." The millet was then sprouted and beer made from it, a process that takes several days. Some of this beer was carried in a calabash by the *murungu* to the Nyika plateau, for his personal use during mining of the ore. A libation of beer was poured on the furnace wall at the place where the doorway was to be cut, immediately before doing so. The smelters then drank the rest. In contrast to the elaborate ceremonies associated with the *ng'anjo*, no ritual attended the building of the small refining furnace, *kathengu*.

The culminating ceremony took place when everything was in readiness and smelting about to begin. By this time the performance of various younger assistants had been evaluated over months of preparation and a final smelting team of eight men was chosen. In general, spectators had not been encouraged at the smelting site, but there had been a constant stream of people from nearby houses to carry building materials and supplies. In 1982 the smelting camp was about 2 km from the nearest house, but in 1983 several houses were within easy shouting distance, screened from the smelting camp by dense shrubbery. The *msofi* now closed the camp by barring the footpaths to it and by burying three items of *sepo* (see appendix B) in the paths and at intervals around the perimeter. From this point no one could leave or enter the camp until the smelting was complete. Surprisingly two women were included in the team to cook and fetch water. The cook was an elderly woman, hence not fertile; the other was a nursing mother with an infant, hence not sexually available by local custom. A fertile woman, especially a pregnant one, is considered detrimental to the smelting process.

Everyone in the camp, including the women, participated in the ceremony that followed. The audience was subdued, and various prayers were offered for the success of the smelting. The women had pounded together the leaves of all the plants on the list of *sepo* and mixed them with water into a green mulch. A portion of this mixture was placed in a bowl and the rest poured in a wooden trough. The *msofi* now called each participant, us included, in turn to squat in the doorway of the *ng'anjo* and to drink from the bowl. The young mother held her infant to have some poured in his mouth. Again, each participant was called forward to the trough, where they anointed their faces, forearms, lower legs, and feet with the *sepo*. By these actions, the team dedicated themselves to the *ng'anjo* and undertook to observe the vows of celibacy, isolation, peaceful conduct with their fellows, and abstinence from washing their bodies for the duration.

Loading of the *ng'anjo* then commenced, timed to end at sundown. The *msofi* prepared the *sepo* by laying out all the items on flat baskets, adding pieces of root and bark as they were chopped from the available materials. Most of the *sepo*, especially the rare items, was stored in two small gourds for future use. A yellow fist-size *chihimo* tuber was hollowed out and most of the *sepo* placed in it. The remaining items were arranged around it in the *chihikiriro* hole in the furnace floor, which the *msofi* reached by crawling through the doorway. The hole was then capped with charcoal and a heavy piece of slag placed over it. The furnace was ready for loading.

The *murungu* loaded the furnace, first placing tuyeres through the doorway and plastering it shut, then climbing in through the top and supervising the placement of the other tuyeres. As charcoal, ore, and even live coals were handed to him, he poured it around his feet and climbed on top of the furnace contents. With the furnace nearly full, he climbed out of it. All leftover bits of *sepo* were picked up and tossed on top of the charge.

No further ceremony attended the smelting. When the furnace was broached some 100 hours later, the smelters were released from all taboos and went home to wash and see their families. In 1982 this was a glum occasion as the smelting had failed; a divining ceremony was held the next day to determine the cause. The answer was entirely technical, having to do with the type of ore used. This problem was corrected in 1983, when jubilation greeted the fact that "our wife" had produced a large chunk of sponge that could be resmelted in the *kathengu*.

Operating the *kathengu* requires two bellows, which are made from goatskins. The goats are killed by smothering, holding their mouths and noses closed, the skin is pulled off like a garment starting from the head to keep it whole. The skin of a hind leg is fitted with a tuyere and becomes an air outlet. Before firing up the *kathengu*, three items of *sepo* are placed in a small hole in the floor, also called *chihikiriro*, but no further ritual is involved and women can watch. Teams of smelters vigorously pumped the bellows in relay, each group trying to outdo the other and shouting encouragement to their pumper. As flames shot rhythmically into the night sky, women ululated in high keening voices, and even we found ourselves atavistically leaping about and shouting at the top of our lungs.

Discussion

It is clear from our Malawi field data that the magical component of iron smelting is complicated and important to the whole enterprise. The following seven observations follow from the data.

1. The smelters at Chulu were far less expert in the area of medicines than the Phoka. Western technology has adversely affected not only smelting technology but also the associated magic. In the rural-urban continuum, the Phoka are much further from the city than Chulu, hence the difference.

2. The knowledge required to assemble the full range of Phoka medicines for smelting is extensive, particularly regarding plants. The effort required to obtain everything is equally large. This component is a technology in its own right.

3. Ritual and medicines associated with smelting are not secret in the sense that they form a separate, compartmentalized segment of knowledge. Many of the medicines have other uses, the prayers are simply appropriate words for the occasion, and the villagers participate in some of the rituals. The "secrecy" results from the fact that medicine per se is a specialized area of knowledge, which requires "a calling" and apprenticeship, and is not lightly discussed in public. The exclusivity of smelting further contributes to this apparent secrecy.

4. The sexual taboos associated with smelting are elegantly exemplified by the Phoka case. The metaphor for smelting is fertility, conception, birth. The furnace is a woman going through the stages of puberty, marriage, reproduction. Women may participate in the smelting only if the are post-menopausal or sexually unavailable because of breast feeding. The men remain celibate during smelting and do not wash off the layers of grime associated with the furnace, their wife. The prevalent explanation about women being a contaminating force does not fit this situation well; the symbolism is rather that of fidelity in marriage to prevent jealousy and to ensure fertility.

5. The technology of smelting is difficult and insufficiently understood to guarantee success, especially with raw materials that vary considerably in composition. Much of this lack may be due to a loss of knowledge on the part of the smelters during the past fifty years, but it is also our independent assessment of the smelting process and of the information that furnaces often failed when smelting was in its heyday. At the same time, we know that both the Tumbuka and Cewa produced substantial amounts of iron: Their marginal technology worked well enough to invite constant repetition.

6. For smelting to work, both the metallurgical and magical components must work. Failure in smelting is likely to increase ritual activity, but the diagnosis of the problem may be technical or magical.

7. If it is not obvious from the data, we unequivocally state that none of the rituals or medicines have any direct or indirect metallurgical or chemical function. This does not mean that they do not have practical value: They do, but it is of a different kind.

We return now to Kjekshus's contention that the ritual component of smelting is of secondary importance, at most a type of trade union secret, and that it has become an alternative explanation for the smelting process because of the fading of technical knowledge. This viewpoint is simply not supported by the facts. The amount of knowledge and effort involved in smelting magic is a substantial portion of the total effort.

An alternative explanation is provided by Wyckaert (1914), a missionary of the order of White Fathers. Writing of the Fipa, he reflects:

No doubt the reader must have noted with surprise the purism the smelters affect through the whole process . . . . All this is rather surprising, especially for one who is acquainted with these smelters and who knows the ordinary laxness of their morals. . . . It is proof simply that even the most depraved natures recognise perfectly the beauty of purity, suspect its happy influence, and even understand its necessity. It proves that these sad natures know very well that sin displeases the God, attracts punishment, causes many failures even in temporal affairs. (Translated by D. H. Avery)

Wyckaert has a degree of insight into the problem, as he is providing explanations from one system of religious beliefs for behavior shaped in part by the tenets of another. The concept of sin does not fit too well in the African context, however. Thus we find neither the materialist nor the missionary position entirely appropriate.

Malinowski's functional analysis of "productive magic" (Malinowski 1948, pp. 1–71; Firth 1958, p. 129) fits the Malawi case in several instances. He notes that magic may serve to throw a cloak of sanctity over a technical process, buttressed by a system of rewards and punishments, thus increasing the seriousness of the performers. This is an idea similar to Wyckaert's "purism" and is certainly evident in the Malawi smelting rituals. Magic may also provide an organizing force, for a sequence of rituals may order a technical process. This element was not observed in the smelting magic: The rituals merely punctuate the successful completion of various preparatory phases. Particularly apt is Malinowski's observation that magic produces confidence in those who have to cope with the unknown. The smelting technology used by our informants was not very effective, at least not to the point of being able to predict the results. It is not necessary to conclude that they had lost this knowledge during the past fifty years; even when the collective technical expertise of their fathers was at its peak, the magical component played its indispensible role. It is not necessary, of course, that one have a scientific understanding of the solid-state reduction of iron in a fluid slag medium in order to be able to smelt iron. It is a fact that metallurgy did not become scientific in Western industrial economies to the point of producing guaranteed compositions until the later half of the nineteenth century. The vagaries of producing steel before such understanding was achieved are illustrated by Nordic sagas about rare magical swords that brought their owners victory in combat and conferred on them the position of leader or king. In the same vein, Excalibur was probably an exceptional blade of just the right grade of steel, accidentally produced.

A technology that is based on sufficient knowledge to make it work but not enough to predict the outcome falls in the same category as agriculture, hunting, fishing, or love. One does one's level best, but it may still not be enough. There are too many forces at large in the world that may cause failure: God, evil spirits and ghosts, or your jealous neighbors. Iron smelting, which produces implements for agriculture and defense, is as vital to survival as food procurement or reproduction. Under these circumstances it is rational that the practitioners should see iron smelting in terms of sexual reproduction and that they should enlist the aid of a sizable pharmacopoeia to ensure fertility. Magic is as much a component of smelting as it is of life. One could not consider undertaking one without the other because, as Damon Runyon put it, in the affairs of man the odds are always six-to-five against.

Appendix A: *Mitsiriko* or *Msinkho* (Medicine) Used at Chulu for Iron Smelting

This information was obtained in private from two informants who differed on some of the details; comments were obtained from a third informant. The twenty-seven items are categorized, first, according to their use. Twenty-two of them are used in connection with *ng'anjo*, the primary furnace, in the foundation trench during construction or in a hole in the middle of the floor (*chitewerende*). These are subdivided under the headings "Plants" and "Other." Three items used exclusively for setting the fire in *ng'anjo* form a separate category, as do two other items, used only in the font at the gate that closes off the smelting area. Items in each category are arranged alphabetically by their Cewa names; scientific identifications and English equivalents are given as available. The symbols preceding each numbered item indicate its use during smelting in accordance with the key. Other uses of *mitsiriko* items are noted; the absence of such notes mean that usage is restricted to smelting.

Key

× *Chitewerende* of *ng'anjo* (hole in floor of primary furnace)
% in font at gate
< for firesetting in *ng'anjo*
* in foundation trench
! peg in foundation trench

Plants

×%! 1. *Chimphakasa* (*Lonchocarpus capassa*). Tree. Root used in *ng'anjo*. One of compulsory six items for gate font, where leaves are used. Saplings are used here and there in a kraal fence to prevent spells on the livestock. Saplings provide wooden pegs for *ng'anjo* foundation.

× %* 2. *Chimwemwe*. Tree, unidentified. Root used.

× %* 3. *Chipombo* (*Brachystegia* sp.). Tree. Root and leaves used. One of the six items that are compulsory for the gate font.

× %* 4. *Chitongololo* (*Acacia macrothyrsa*). Tree. Root and leaves used. See *Musambamfumu* (no. 10) for other uses.

× %* 5. *Kathyothyo*. Tree, unidentified. Root used. Leaves made into steam for pneumonia.

× %* 6. *Mkonje* (*Manilkara mochisia*). Tree. Root used.

× % 7. *Mpungulira* (*Antidesma venosum*). Tree. Root used.

× %* 8. *M'thumpu*. Tree, unidentified. Bark and root used.

× %* 9. *Mtima umodzi*. Small bush, unidentified. Root used.

× %* 10. *Musambamfumu* (*Afzelia quanzensis*). Tree. Root used. Several others uses: (a) Bark boiled for chief to wash his hands and body in order to enhance his authority at a meeting. (b) Bark mixed with leaves of *Chitongololo*, then soaked or boiled; the liquid is taken to counter bad dreams. (c) Bark of *musambamfumu*, leaves of *chitongololo*, and roots of *mzonono* are boiled into liquid for children with epilepsy.

× %* 11. *Musolo* (*Pseudolachnostylis maprouneifolia*). Tree. Roots and leaves used.

× % 12. *Mzonono* (*Acacia sieberiana*). Tree. Root and leaves used. See *musambamfumu* for other uses.

× %* 13. *N'ngaka*. Tree, unidentified. Root used.

× %* 14. *Nthukumusi*. Mushroomlike plant, unidentified. Tuber used.

× %* 15. *Nthundu* (*Ximenia americana*). Tree with edible fruit. Root used. This is one of six items that are compulsory for the gate font.

× %* 16. *Palijekanthu*. Tree, unidentified. Root used.

Mitsiriko, Other

× %* 17. *Chisa cha katawa* ("Piece of nest of *katawa* bird"). The *katawa* (*Scopus umbretta*, hamerkop) is associated with rain by the Tswana, and called "the lightning bird." Because it stares motionless into water, it is considered capable of seeing into the future. These beliefs are widely held in southern Africa, but it was not established whether the Cewa share them. (The bird could not be identified in the field.)

× %* 18. *Makochezi a madzi* ("any floating object").

× %* 19. *M'pinga njira* ("any root which crosses a path").

× %* 20. *Muzu wayambuka dambo* ["root (of any tree) that crosses a stream"]. One of six items that are compulsory for the gate font. Liquid boiled from such a root is taken to cure dancing disease.

× 21. *Shoni* (*Hystrix africaeaustralis*). Porcupine. Piece of skin and quills used.

× 22. *Tambala*. Domestic chicken, must be white. The blood and some feathers are sprinkled on and in *ng'anjo*.

Firesetting

These three items were wrapped together in a bundle, lit from a flame that was kindled with a fire drill, and then used to light the fire in *ng'anjo*.

< 23. *Makochezi*. Grass that floated down a stream and caught on something like a tree branch.

< 24. *Msindira chulu* (*Diospyros squarrosa*). Tree. Root used. Also provides wooden pegs for *ng'anjo* foundation.

< 25. *Chiputu* (*Panicum deustum*). Clump of grass that grows on a tree (where there is soil).

Gate Font

Twenty-two items of *mitsiriko* were placed in a font at the gate that barred access from the village to the smelting area. The font was simply a piece of curved tree bark resting on a forked upright. Water was added to the pieces of bark, roots, and leaves in the font. All the smelters washed their hands in this water, as did the few visitors (like the chief) who were allowed entry. Of the twenty-two items mentioned, only six are compulsory. Four of these were also involved at *ng'anjo* (nos. 1, 3, 15, and 20), but two are exclusive to the gate font.

% 26. *Mtengo wa vihanda* ("tree for spirits"). Leaves, stems and roots used. Herbalist rubs leaves in hands, chews them, and spits around you on ground to prevent evil spirits troubling you with bad dreams.

% 27. The sixth charm that is compulsory for the gate "has no known name." It was not clear whether this item was actually present and its name unknown, whether it was absent, or whether its name was being withheld. The first alternative appeared most likely.

Appendix B: *Sepo* (Medicine) Used by Phoka for Iron Smelting

This information was obtained during the 1983 smelting cycle and is much more extensive than that obtained in 1982. In a few instances additional information from 1982 has been added and so noted. Although the core of the smelting teams remained the same, there was some change in membership between 1982 and 1983, hence the small differences.

The items of *sepo* are arranged alphabetically in Tumbuka or Phoka vernacular, categorized as plants, animals, or minerals and artifacts. Where available and appropriate, the names of each item are provided in Tumbuka or Phoka, Cewa, scientific Latin, and English. The portion or part of each item used for *sepo* comes next, followed by notes on its uses other than for iron smelting. The symbols that precede each numbered item indicate the places and occasions where it is used as *sepo*, according to the key provided. For trees and shrubs the general rule is that roots, fruits, and bark are buried in the furnaces or along the perimeter of the smelting area, whereas leaves are pounded and mixed with water. The resulting green slurry, or infusion, is ceremonially drunk by the smelters and used to anoint parts of their bodies. Exceptions to the rules are noted.

Key

×	*Chihikiro* of *ng'anjo* (hole in floor of primary furnace)
*	Foundation of *ng'anjo*
°	Infusion for smelters
#	Paths and perimeter of smelting area
+	*Chihikiriro* of *kathengu* (hole in floor of secondary furnace)

Plants

*+ 1. *Bopa makala* (Tumbuka; *Oplismenus* sp.). Grass from Nyika. Circles foundation trench. No other use.

°*× 2. *Chifwaja* (Phoka; *Tabernaemontana angolensis*). Tree with edible fruit from Nyika forests. Bark, leaves used. Cure dancing disease with bark and roots.

* 3. *Chihimo* (Phoka; *Plectranthus esculentus*). Livingstone potato. Bright yellow edible tuber with fingerlike roots. All the *sepo* are placed inside the tuber for placement in *chihikiriro*. No other use.

× 4. *Chipindi* (Tumbuka; *chipanda* in Cewa; *Lagenaria siceraria*). Bottle gourd, large type. Seeds used. Used as a container.

°* 5. *Chipululu* (Phoka; *Disa axicola*). Nyika disa, with red flowers. Stem, flowers, and leaves used. Leaves and roots are mixed with other items to cure malnutrition in a nursing child of a mother who is prematurely pregnant again.

°*× 6. *Chisese* (Tumbuka and Cewa; *Faurea speciosa*). Tree, widespread. Bark and leaves used. Boil roots, mix liquid with lake flies, and drink for bad cough due to witchcraft.

°* 7. *Chiyere* (Tumbuka and Cewa; *Faurea saligna*). Tree, widespread, used for planks. Bark and leaves used. To cure malaria from mosquitoes (as opposed to malaria from witchcraft), the roots are boiled and the mixture poured in a hollow in the ground, from which the patient drinks three times a day.

*× 8. *Dopa* (Phoka; *Casis abbreviata*). Nyika tree that supplies planks. Roots and leaves used. Cures pneumonia and dancing disease.

°*× 9. *Kaŵingaŵazimu* (Tumbuka; *Clerodendrum glabrum*; "chases evil spirits"). Small shrub; stem used for outside sweeping. Roots and leaves used. A new grave is swept with its branches to prevent witches from stealing body parts. To prevent bad dreams about the deceased among children, it is soaked and the liquid is used to wash the hands and to drink.

°*× 10. *Kayunga* (Tumbuka; *kalama* in Cewa; *Combretum apiculatum*). Tree. Roots and leaves used. Cure for pneumonia caused by witchcraft. Root tied to fishnet to improve catch.

*× 11. *Khwewa* (Tumbuka and Cewa; *Dracaena reflexa*). Tree. Roots used. To cure dysentery, boil roots and use the liquid.

°*× 12. *Kowi* (Phoka; *Widdringtonia nodiflora*). Nyika cypress. Roots and leaves used. Its knobbly sapling provides the "belt" of the furnace. No other uses.

°*× 13. *Lufifya* (Phoka; *Securinega virosa*). Nyika tree. Roots and leaves used. Bark, leaves, and roots boiled together into tea to cure heart attack.

°*× 14. *Lupindura* (Tumbuka; *Rhus longipes*). Tree. Roots and leaves used. Cure for pneumonia caused by witchcraft. Root tied to fishnet to improve catch.

°*× 15. *Mandinya* (Phoka; *Flacourtia indica*). Nyika tree. Roots and leaves used. Prevents and cures witchcraft. Taken as preventive when there is an epidemic in the area.

°*× 16. *Marombo/Malombo* (Phoka; *Colocasia antiquorum*). Nyika tree. Roots and leaves used. To prevent witchcraft, the roots and leaves are boiled into medicine.

*× 17. *M'kanda-zovu* (Tumbuka; *mkanga-njobvu* in Cewa; *Mundulea miricea*; "resistant to breaking by elephants"). Tree, widespread, which bends instead of breaking. Roots and leaves used. Before an elephant hunter sets out, he takes liquid boiled from the roots and cleans his body with an infusion made from the leaves. His wife is required to stay inactive and in her hut while he is away.

°*× 18. *M'ndopi* (Tumbuka; *mhihi* in Phoka; *Ficalhoa laurifolia*). Tree, Nyika. Bark and leaves used. Boil roots for bad cough due to witchcraft or for dancing disease.

°*× 19. *Mng'ina/Mung'ina* (Tumbuka; *mung'ona* in Cewa; *Breonadia microcephala*). Tree, widespread, used for strong planks. Roots and leaves used. To cure dancing disease or prevent witchcraft, the roots, bark, and leaves are boiled together into medicine. In 1982 an informant also noted its use to cure rheumatism, miscarriage, and pneumonia.

°*# × 20. *Mnyongoloka* (Tumbuka; *mphinga* in Phoka; *Steganotaenia araliacea*; *Mphinga* means "to bar access"). Small shrub. Roots, stem, and leaves used. Buried in paths leading to village as protection against witchcraft; also used to cure strife in a village.

*# × 21. *M'pinga viŵanda* (Tumbuka; *chitongololo* in Cewa; *Acacia macrothyrsa*; "chases off evils"). Tree. Roots used. Taken to prevent dreams about dead relatives, against witchcraft, and for dancing disease.

°* × 22. *Msolo* (Tumbuka and Cewa; *Pseudolachnostylis maprouneifolia*). Tree, widespread, used for *nsima* stirring sticks, hoe handles etc. Roots and leaves used. Made into charcoal and rubbed in razor blade cuts on side of head or on chest to cure dancing disease and headache.

* × 23. *Mtatana/Mutatani* (Tumbuka; *mtatani* in Cewa; *Garcinia huillense*). Tree, widespread, with edible fruit. Roots used. No other use.

°*# × 24. *Mupotolo* (Cewa; *mnyongoloka* in Tumbuka; *Hymenocardia acida*). Small shrub. Roots, stem, and leaves used. No other use.

°* × 25. *Murungu/Mulunga* (Tumbuka; *Droogmansia pteropus*). Nyika tree. Roots and leaves used. Prevents and cures witchcraft. Taken as preventive when there is an epidemic disease in the area.

°* × 26. *Mvivu, Bokoto* (Tumbuka; *Ocotea usambarensis*). Nyika tree, used for planks. Roots and leaves used. Prevents and cures witchcraft. Taken as preventive when there is an epidemic in the area.

*+ 27. *Mwambula* (Phoka; *Thunbergia kirkiana*). Stringlike creeper, like maiden-hair. Completely circles foundation trench of *ng'anjo* and is also placed in middle of *kathengu*. For coughing, soak leaves and drink the liquid.

°* × 28. *Nkhwaju* (Tumbuka and Cewa; *Cyphomandra betacea*). Tree with edible fruit (tree tomato). Fruit and leaves used. No other use.

* × 29. *Nthesi* (Tumbuka; *chitoŵe* in Cewa). Shrub with edible seeds, which are ground into flour. Seeds used. No other uses. This plant is probably cultivated sesame, *Sesamum indicum*, or the wild species *Sesamum angolense* (Williamson 1975, p. 213).

°* × 30. *Nthundu* (Tumbuka and Cewa; *Ximenia caffra*). Tree with edible fruit. Roots and leaves used. Prevents or cures witchcraft. Cures bad malaria (witchcraft caused) in children when all else fails. Cures epilepsy. In 1982 an informant noted that the roots could be used to cure "child fever," accompanied by enlarged spleen. This may be glandular fever but may also refer to "bad malaria."

° × 31. *Nthutumusi* (Tumbuka; *nthukumusi* in Cewa). A mushroomlike fungus that grows on acacia trees. Prevents or cures witchcraft and pneumonia caused by witchcraft. This may be *Auricularia auricula*, or Jew's ear, an edible tree fungus (Williamson 1975, p. 335).

*+ 32. *Sangazinji* (Tumbuka; *sakazinji* in Cewa). Another creeper that hangs in a bush.

× 33. *Skaŵa* (Tumbuka; *n'tedza* in Cewa; *Arachis hypogaea*). Ground nuts. Thrown into tuyeres from inside before setting fire; one or more in each pipe.

× 34. *Zgama* (Tumbuka; *nzama* in Cewa; *Voandzeia subterranea*). Bambarra groundnut, cultivated plant with edible seeds in underground pods. Seeds used.

Animals

× 35. *Boli* (Tumbuka; *Equus burchelli*). Zebra. Heel piece of hoof. No other use.

× 36. *Chigwere* (Tumbuka; *mvuu* in Cewa; *Hippopotamus amphibius*). Hippo. Piece of skin used. For venereal disease, soak skin with herbs in water and drink liquid.

* × 37. *Chimbuli* (Tumbuka; *chihuli* in Cewa; *Mellivora capensis*). Honeybadger, a tough fighter. Skin used. No other use.

× 38. *Chisa cha masanganavo* (Tumbuka; "nest of wasps"). Paperlike nest of brown wasp. Piece used. No other use.

× 39. *Godi* (Tumbuka and Cewa; *Orycteropus afer*). Antbear, aardvark. A tough animal. Paw used. No other use.

× 40. *Lupemphezi* (Tumbuka; *Erinaceus frontalis*). Hedgehog. Paws used. For venereal disease proceed as with hippo skin (no. 36).

× 41. *Mphumi* (Phoka; *kabenubenu* in Tumbuka and Cewa). Beetle, black, spotted, with hard wing covers. The wing covers are used to help melt hard stones. No other use.

* × 42. *Mphyayi* (Phoka; *Dendrohyrax arboreus*). Tree hyrax. Hard to kill. Paw used. No other use.

× 43. *Njereyere* (Phoka; *sambisambi* in Cewa). Water beetle, zigzag swimmer. Used whole. No other use.

× 44. *Nkhungu* (Tumbuka and Cewa). Lake flies, one or more used. They can be boiled with roots and the liquid taken for a bad cough due to witchcraft. Lake flies occur in dense clouds over Lake Malawi, where they are netted from boats and pressed into dried cakes for subsequent cooking and eating.

× 45. *Nyinawambulika* (Phoka). Termite queen. Whole used. Obvious fertility symbol. A piece can be boiled with appropriate herbs and the liquid used to protect people against an epidemic, as if they are secure in a termite mound.

× 46. *Sato* (Tumbuka; *nsato* in Cewa). Python. Vertebrae used. They can also be used to attract customers to your shop.

× 47. *Tambala* (Tumbuka and Cewa). Brown rooster. Head used. The sacrificial brown rooster figures at several stages of the smelting ritual.

× 48. *Usipa* (Tumbuka and Cewa). Edible sardinelike, small silver fish. One or more used. No other use.

× 49. *Utaka* (Tumbuka and Cewa). Dried fish, about 5 × 2 cm. Edible; no other use.

Minerals and Artifacts

× 50. *Chipunu* (Tumbuka). Piece of head of canoe. No other use.

*× 51. *Chisusu* (Tumbuka and Cewa). Piece of beer container woven from grass. No other use.

× 52. *Lubani* (Tumbuka and Cewa). Myrrh, white and brown. A piece will attract customers to your shop.

× 53. *Lung'alang'ala* (Phoka; *Mbonekera* in Cewa). Mica. No other use.

× 54. *Mkhawo* (Tumbuka and Cewa). Fishnet. No other use.

× 55. *Nkhama* (Tumbuka and Cewa; *kundu* in Phoka). Red ocher. When making any medicine in a pot, a cross is drawn on the inside and outside of its bottom with ocher.

× 56. *Uzira* (Tumbuka; *chiziro* in Cewa). Graphite. Same use as red ocher.

Acknowledgments

This article was originally published in the journal *Africa* (July 1987, no. 2); it is reproduced here by permission of the editors.

Gadi Mgomezulu (Malawi Department of Antiquities) and R. H. V. Bell (Malawi National Parks) provided invaluable support in getting this project going. Many other Malawians helped us: the smelters, traditional authorities, government employees, private citizens—too many to name. Our sponsors were the National Science Foundation (USA), the Human Sciences Research Council (South Africa), the Explorer's Club, and Roberts Construction Company (Malawi). Plant identifications were done by Matthew Matemba and his staff at Nyika National Park. Sharma Saitowitz drew the figures. We thank them all.

References

Bronowski, J. 1973. *Ascent of Man*. London: BBC.

Cline, W. 1937. *Mining and Metallurgy in Negro Africa*. Menasha, Wis.: George Banta.

Derricourt, R. 1979. *People of the Lakes*. Lusaka: Manchester University Press for University of Zambia.

Firth, R. 1958. *Human Types*. New York: New American Library.

Kjekshus, H. 1977. *Ecology Control and Economic Development in East African History*. London: Heinemann.

Livingstone, D. and C. Livingstone. 1865. *Narrative of an Expedition to the Zambesi and Its Tributaries; and of the Discovery of Lakes Shirwa and Nyassa*. London: John Murray.

Malinowski, B. 1948. *Magic, Science and Religion and Other Essays*. Boston: Beacon Press.

Mgomezulu, G. G. Y. 1981. "Recent archaeological research and radiocarbon dates from eastern Africa." *Journal of African History* 22:435–456.

Phillipson, D. W. 1968. "Cewa, Leya, and Lala iron-smelting furnaces." *South African Archaeological Bulletin* 23:102–113.

Robert, R. P. J. M. 1949. *Croyance et Coutumes Magico—Religieuses des Fipa Païens*. Kipalapala, Tobora: Tanganyika Missionary Press.

Williamson, J. 1975. *Useful Plants of Malawi*. Limbe, Malawi: Montford Press.

Wise, R. 1958. "Iron smelting in Ufipa." *Tanganyika Notes and Records* 50:106–111.

Wyckaert, R. P. 1914. "Forgerons païens et forgerons chrétien au Tanganyika." *Anthropos* 9:371–380.

24
The Metallurgy of the Iron Bloomery in Africa

Donald H. Avery, Nikolaas J. van der Merwe, and Sharma Saitowitz

The sub-Saharan African Iron Age differs significantly from the Iron Age in other regions in three significant aspects. First, there is little or no prior pyrotechnology predating the development of bloomery iron. Pottery was and in general continues to be open fired, rather than fired in controlled-atmosphere kilns; although decorated with oxides and pigments, the pottery seldom has a true vitreous glaze. Also, with few exceptions there was no active Copper or Bronze Age preceding the Iron Age. The mining and smelting of tin, copper, and gold would appear to be largely in response to export demand for the product, which did not enter into the local technology except as a measure of currency or wealth and for various decorative, ritual, or magic uses. There also seems to be little extension of the high-temperature technology used in iron smelting and forging to other materials after the iron bloomery was mastered. We do not find a development of kiln-fired pottery or glazes, and in most areas, except west Africa and the Arab-influenced east coast, we do not find extensive development of casting or nonferrous alloys.

Second, in Africa there exists an amazing diversity of iron smelting furnaces (Cline 1937). The African furnaces differ among themselves not only in form but also in the technique used. Furthermore, smelters located in close geographical and temporal proximity often have furnaces and traditions that are different. The furnace types range from open pit furnaces to highly stylized female symbols complete with legs and breasts. There are forced draft furnaces of remarkable technical sophistication capable of attaining temperatures of 1700°C and tall natural draft furnaces up to 7 m high that should be capable of a high level of production. At the other extreme there are complicated small furnaces, such as those of the Phoka, which require an extraordinary expenditure of resources for a low output of iron. These may well serve more spiritual and ritual functions than the economic production of iron. All in all the African Iron Age is unique in that it sprang up without precedents some 2,500 years ago, probably in the Nok region of Nigeria, then spread extremely rapidly through the Congo basin to central and east Africa, where there is an extraordinarily diverse spectrum of furnace types, technology, and efficiency, all virtually unchanged or technologically degenerate. For all the early sophistication of the African bloomery, it did not develop or spread to new materials. It was limited to small-scale forged implements; neither carburization nor heat treatment was recognized or exploited.

The third difference in studies of the African Iron Age is that smelting still continues in some remote sites and was only recently abandoned, within the lifetime of many practicing smelters, in many other areas. Thus we have people who are, or were in their youth or as young men, practicing smelters with a concrete knowledge of how they practiced and how they perceived the smelting of iron. The smelting furnaces were in almost all cases abandoned when cheap scrap iron from Europe became readily available. It is important to note that the demise of the iron smelting industry is generally attributed by the former smelters not to economics but to colonial authorities whose iron was not as good as the traditional iron and to missionaries who feared the spiritual power and respect that the smelters commanded. These smelters remain interested in their skills and trade and are eager to describe and to practice it, given an economic incentive. Most of the African furnaces are prodigal in their use of materials, and even during World War II local attempts to reestablish the iron industry in Malawi as an aid in producing iron, which was temporarily in short supply, proved unfeasible. Local people remain intensely interested and have periodically reconstructed furnaces, starting with the Shona demonstration at the 1930 Imperial Agricultural Show in Johannesburg (Stanley 1931). Most of these projects, however, have been inadequately supported, used the wrong materials, and have not been pursued to completion.

In all of the cases that we will describe and have encountered, the production of iron is strongly associated with spirits and complicated rituals, taboos, and medicine. Among the men who had been associated with smelting as traditional smelters and had learned smelting from their fathers without input of Western schooling or analysis, we found that smelting was almost always associated with birth; the analogies were to planting, hunting, and husbandry.

The Iron Age smelters in Africa were farmers, herders, or hunters who had to master extremely complex technologies but whose results varied greatly despite their efforts. Every year, elaborate planting, hunting, war, and love rituals were carried out and sometimes rewarded with successful crops, increase in herds, successful hunts, and faithful wives; in other years the same rituals and technical efforts produced little or no results. This contrasts with the rational, mechanistic Newtonian view of the Western scientist or engineer who looks on a world of machines in which results are directly related to materials and processing procedures. The furnaces that we describe here were invented or developed in a world totally different from that.

The African Iron Age smelters hold a world view in which good results come out of a harmony with nature; there are many evil spirits, hostile and jealous individuals, and vengeful and neglected ancestors who cast spells and practice witchcraft to oppose successful smelting. If the smelters pay attention to medicine, please and draw support from their ancestors, and follow taboos in order to be clean and purged of witchcraft and spells and if they remain isolated from the outside world after ritual functions of cleansing have been performed, then their efforts will be rewarded with a bountiful production of iron that reflects not only the technology but also the purity and spiritual strength of the smelters. Indeed many of the songs regarding smelting involve the smelters returning from the taboo smelting site bearing their iron while people sing their praise for the purity that overcame evil spirits. For an interesting account, see (Wyckaert 1914).

In my ten years of living with smelters at a variety of sites in central Africa and of building and operating furnaces, I have never encountered a serious characterization or classification of materials in the Western sense. Iron ore, although certainly necessary, is merely one of the ingredients that goes into the furnace and often is not held to be as important to success as many medicines, herbs, and other spiritual taboos that go into the production of iron. In many cases the iron ore is associated not with a specific appearance but with a place. This is usually a place with which various ritual or spiritual properties are associated, but often it is a place that is unique only because of its difficult access. "We must get the ore from that hill where our fathers went to get ore." Divining is often used to select a site for clay or iron ore, and the subsequent success or failure of the smelt is attributed to the diviner and his power over various evil spirits who want the smelt to be unsuccessful. The secondary level of taboo involves purification of the ore from various evil spirits so that it may enter the smelting furnace. These rituals often involve technical operations, such as crushing, washing, and roasting, that have advantageous effects. The attempt, however, is not to modify or upgrade the ore or to concentrate it but to purify the ore from the evil spirits or contamination that it may possess. This aspect of smelting is discussed much more thoroughly by van der Merwe and Avery (this volume).

An interesting account of pre–Great War smelting in German Tanganyika (Wyckaert 1914) makes precisely this point. When discussing the chief's reaction to a failed smelt Wyckaert observes:

What does the chief do then? Does he check his ore to see if it is good quality? Does he check his flux to see if it is suitable to the quality of the ore? Will he try to find out if some natural cause has upset the operation? Occasionally yes; he will ask himself these questions. Ordinarily, no! He will simply say with resignation, "Some spirit does not want this. Let us appease him with a sacrifice." Or else he will say with resignation: "My medicines are not worth anything any more, let us find others." And he will really work very hard at coming up with rarer bones and more extraordinary feathers and the skins of even nastier snakes. Most often he will cry angrily, "Again our women are behaving badly in the village. They are spoiling our work." And he will send spies to spy on them. Heaven help the poor accused or suspected one. She will have to prove her innocence by the poison test. (p. 375)

Throughout our work with traditional iron smelting, we became continually aware of the vast difference between Wyckaert's account and the monolithic Bantu culture as discussed in *Livingstone's Lake* (Rainsford 1966):

The highest ambition is to lead the same lives as their ancestors had led—for these ancestors were the Gods who filled their pantheon. The design of the Amaravi social system was to integrate the individual not unto himself but into a society that felt undivided from the past even by time. In Maravi the dead were just as valid as the living—and sometimes much more powerful. (p. 28)

Thermodynamics of the Bloomery Furnace

Before going into detailed technical descriptions of various bloomery furnaces it is useful to review the thermodynamics of such furnaces.

The iron oxides hematite, magnetite, and limonite, laterite, hydrous deposits from leaching, and bog ore are available almost everywhere. With hindsight it is clear that the late development of iron and the elaborate technology associated with its production arose from the fact that iron was the only metal of antiquity to be smelted below its melting point.

To understand this conclusion, one must analyze how a furnace works. When fuel is burned, heat is released. The adiabatic flame temperature is calculated on the assumption that all the heat released in burning the fuel is available to raise the temperature of the products of combustion. Inert reactants such as nitrogen, water, and ash also absorb heat and must be included. Because the heat for complete oxidation of carbon to CO_2 (96 kcal/mol), is considerably greater than for carbon to CO (26.4 kcal/mol) and because the amount of inert nitrogen is twice as much for CO_2 as for CO, the adiabatic flame temperature is strongly dependent on the CO/CO_2 ratio, as shown in figure 24.1. With wood rather than charcoal the heat of reaction is less, and water vapor, both in the wood and as a product of combustion, lowers the maximum temperature. Because iron requires a strongly reducing atmosphere ($CO/CO_2 > 3$), furnace temperatures in excess of 1250°–1300°C are undesirable; any iron formed would be reoxidized. Heat must also pass out of the flue at the furnace temperature; it is only the sensible heat between the adiabatic flame temperature and the furnace temperature that is available to supply the heat to the furnace and that is balanced by the heat lost through the furnace walls, absorbed in heating the ore charge, or in endothermic reactions such as the reduction of iron ore.

Thus, as the furnace temperature approaches the adiabatic flame temperature, less heat is available and fuel consumption increases exponentially (figure 24.2). Furnace temperature is controlled by the rate of heat

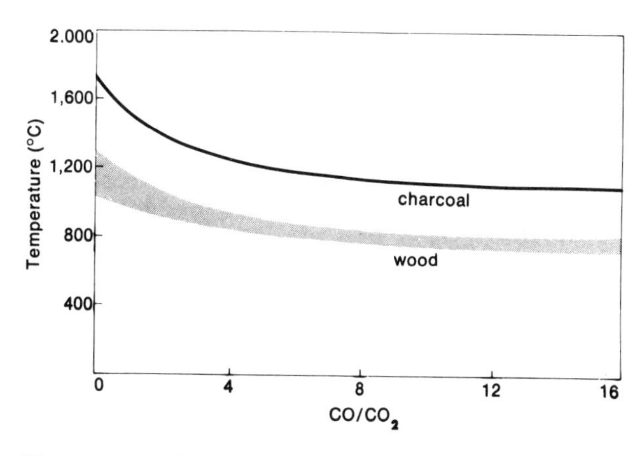

Figure 24.1 Temperature versus CO/CO_2.

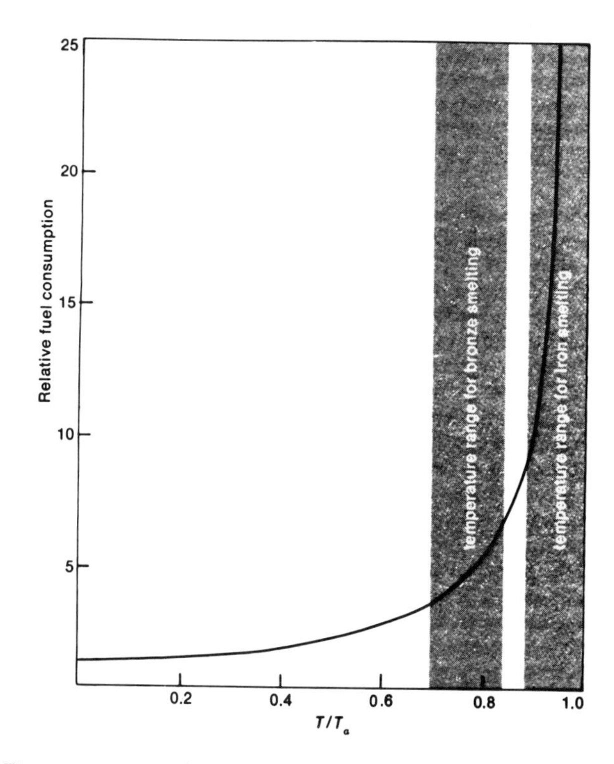

Figure 24.2 Fuel consumption versus T/T_0.

generation and loss. In the simplest case of an open bonfire there are large heat losses from radiation, and the maximum temperature is limited to about 850°C, independent of the amount of fuel burned. Placing walls around the open fire reflects most of the radiant heat losses back into the furnace, and temperatures increase to the 1000°–1100°C range. (The importance of the African wife's three cooking stones reflects this.) With better insulating materials for the wall, heat losses can be decreased and temperatures raised.

It must be appreciated that the temperature and oxidation state vary throughout the furnace. Immediately in front of the tuyeres the atmosphere is oxidizing. As air enters the furnace, it reacts with charcoal and forms CO_2, reaching the highest temperature in an oxidizing zone of pure CO_2. Carbon dioxide then reacts with additional carbon to form two CO and a reducing zone of increasing CO/CO_2 and decreasing temperature. As gases rise up the stack, the temperature drops and the atmosphere remains mainly CO.

The rate of fuel burning is dependent on the draft and the aerodynamic resistance to flow of the furnace shaft. The draft is usually measured in inches of water. Hot gases in the furnace have a lower density than the outside air so that there is a pressure difference between the top and the bottom of the furnace. In the short furnaces of the ancient world, about 1–2 m high, the pressure difference is only 0.03–0.06 in of water; this does not produce a sufficient flow of air for the furnace to attain temperatures much greater than 1000°C.

Higher temperatures can be produced in a forced draft furnace where air pressure at the base of the furnace is increased by bellows forcing air into the furnace through ceramic blowpipes called "tuyeres." This permits a high enough rate of combustion to produce temperatures in the 1200°C range. If the temperature were to get much higher, the furnace would become oxidizing and would not produce iron. In the Mediterranean-European tradition, from the early Iron Age to the Roman Period, furnaces become larger and higher, going from bowl to shaft to domed designs. They all require forced drafts to reach smelting temperatures.

During the operation of such a forced draft furnace, ore particles and charcoal are charged at the top. As the ore particles descend and are heated, they lose any hydrated water, becoming porous and internally fissured. Hydrated ores such as limonite or bog iron are the easiest to smelt, as they shrink and crack and thus have a large surface area after heating. During smelting, CO, at about 800°C in the furnace stack, reduces the iron oxide to small flakes of finely divided iron on the exposed free surfaces. As the charge passes down the stack, the temperature rises and the flakes begin to coarsen and agglomerate. At about 1150°C wüstite reacts with silica to form a glassy wüstite-fayalite slag, which seals the iron particles from further contact with the furnace gases and permits them to pass into the high-temperature oxidizing zone above the tuyeres. With high-grade ores some silica must be added to form this protective slag.

Below the tuyeres the atmosphere is again reducing. The most fluid slag (80% FeO, 20% SiO_2) drains down into the bowl or is tapped by periodically opening a channel from the furnace hearth. The solid iron particles and some fluid slag are filtered on the solid bed of charcoal and stay above the bowl and sinter together in a pasty mass called the "bloom." The bloom picks up some carbon from local contact with charcoal, but the continual washing with oxide slag draining from the shaft keeps the carbon content relatively low.

After four to twelve hours of firing, the furnace is allowed to cool down and the bloom is removed. It must then be reheated in a forge to a high temperature and hammered to expel the fluid slag and to weld the porous iron into a dense solid. In order to remove as much of the refactory slag as possible, initial forging must be above 1300°C. This requires that the bloom be close to the tuyeres in an oxidizing zone. During this process any carbon that might be in the bloom is oxidized, and the forged, consolidated bloom is consequently a low-carbon iron with stringers of included slag.

Technical Description of Four Smelting Furnaces in Central Africa

In figure 24.3, four quite different furnaces are presented; their operation is described later. Figures 24.3a and 24.3b show two rather inefficient natural draft furnaces in central and northern Malawi. Figure 24.3c shows a sophisticated forced draft furnace operated in Buhaya in northwestern Tanzania close to the Ugandan border. Figure 24.3d depicts again a remarkably sophisticated forced draft furnace operated by the Barungu in western Tanzania slightly south of the Ermin Pasha Gulf at the southwest corner of Lake Victoria. In all of these furnaces the smelts were managed by old men who no longer smelted but who had been directly involved as youths or young men up to the cessation of extensive smelting during the 1930s.

In all cases we acted as observers; we took no part in directing the furnace or indicating approval or disapproval of what the smelters were doing. During an actual smelt we did necessarily interfere as we made temperature and gas measurements, and we occasionally raised a question of the nonperformance of specific individuals when they ceased following the procedures set out by the senior smelters and being followed by the other men. In this case, however, we attempted to do this in the form of a question as to why one individual was not operating his bellows, rather than ordering

a

b

c

d

Figure 24.3 Four African furnaces. (a) Cewa furnace, reconstructed at Chulu, Malawi, 1982. (b) Phoka furnace, reconstructed at Nyika, Malawi, 1982. (c) Buhaya furnace, reconstructed at Nyungue village, West Lake Province, Tanzania, 1976; (d) Barungu furnace, reconstructed at Bwanga, West Lake Province, Tanzania, 1979.

or directing the process. Many times this procedure was frustrating, as earlier mistakes were repeated and work continued on projects that everyone agreed were not likely to succeed. It was important, however, because we were trying to investigate the technical procedures and problem solving of traditional smelters, not our own. Without a fairly rigorous adherence to all aspects of the traditional taboos and life, we found a tendency to slack off and to lose direction, particularly among the younger men. The making and wearing of traditional bark cloth and the isolation of the smelting site from visitors, in particular women, produced a much more intense involvement and commitment to the furnace.

Natural Draft Furnaces, Malawi

In Malawi we studied two forms of natural draft furnaces. A short truncated cone furnace was operated by the Phoka in the north, on the side of the rift valley leading up to the Nyika escarpment. In central Malawi we dealt with a tall 8-ft shaft furnace that was operated by the Cewa. Remains of both of these furnaces were abundant everywhere and were associated with most villages. Both of our sites in Malawi were operated on the edge of national parks that had not been extensively exploited agriculturally; thus furnace remnants were easily identifiable, and many furnaces remained standing whole and undamaged. These were extremely widespread, and in a short walk of a few kilometers half a dozen furnaces or slag heaps would invariably be encountered.

When David Livingstone traveled through the area west of Kasungu in central Malawi in 1864, he witnessed a flourishing local iron industry (Livingstone and Livingstone 1865, p. 536):

Here at every third or fourth village we see a kiln looking structure about six feet high by 2 or 3 feet in diameter. It is a clay fire hardened furnace for smelting iron. No flux is used whether the specular iron, the yellow haematite or magnetic iron ore is fused and yet capital metal is produced. The native manufactured iron is so good that the natives declare English iron to be 'rotten' in comparison and specimens of African hoes were pronounced at Birmingham to be nearly the equal of the best Swedish iron. As we passed along men sometimes ran from the fields they were working in and offered for sale new hoes, axes, and spears of their own workmanship. It is certainly the iron age here.

The furnaces Livingstone saw in Malawi were built of clay that baked during smelting to produce a permanent structure usable for many years. Even today, over fifty years after the end of active smelting and furnace building, many smelting furnaces still remain. Some good examples can be seen in Kasungu National Park. They are a bit larger than Livingstone's recollection but otherwise can be recognized from his description.

One of the best surviving examples was built (supposedly in 1910) by Vujomo, who lived in Chulu village at the edge of the park until his death in 1981. He was the last remaining master or chief smelter from the region. Much smelting knowledge died with him, although some of his reminiscences have been recorded (Mwaza 1976) The Kasungu furnaces are cylindrical structures, often built in courses of about 0.2 m (8 in) high with broken half-round pieces of tuyeres in the seams. Details vary, but they are about 2.3 m ($7\frac{1}{2}$ ft) high and 1.4 m ($4\frac{1}{2}$ ft) in outside diameter. From a shoulder about 1.9 m (6.2 ft) above the base, the furnace tapers inward to a chimney hole of 0.5 m (1.5 ft) in diameter. The internal diameter varies between 0.9 m (3 ft) at the base to about 0.7 m (2.25 ft) below the shoulder, but it is constricted to about 0.5 m (1.5 ft) at a point about 0.8 m (2.5 ft) above the floor. Thus the furnace has two internal chambers much like the modern blast furnace, with the constriction helping to support the weight of the charge in the stack and to maintain burning space around the tuyeres. Some furnaces are waisted on the outside at the height of the internal constriction, giving it an hourglass appearance; Vujomo's furnace is an example of this style.

Among the Cewa-speaking people of Chulu village are some men, now in their sixties, who participated in smelting in their youth. They call their tall furnaces *ng'anjo* and explain that they functioned without the use of bellows. The product of the furnace could not be forged directly, but had to be resmelted in a small secondary furnace called a *chiramba* to which four goatskin bellows were applied. No examples of *chiramba* have survived. Further west in Zambia, Cewa people used a domed furnace called *mucelo,* which also functioned by natural draft, and resmelted its product in a forced draft *kathengu* (Phillipson 1968).

In northern Malawi a similar two-stage smelting process was used by the Phoka, a Tumbuka-speaking group that lives on the south and east sides of the Nyika mountain massif. Many Phoka lived on top of the Nyika plateau before it became a national park. Examples of their iron smelting furnaces can still be found there and in uninhabited forest patches along the eastern escarpment. Smelting also ended here about 1930, but a few chief smelters were still alive when we worked in the area in 1981–1984. They, and other informants one generation younger, recount a smelting process that uses a natural draft *ng'anjo* and a secondary forced draft *kathengu.*

The *ng'anjo* in this case looks like a truncated cone about 1.5 m (4.8 ft) high tapering from about 1.8 m (5.75 ft) on the outside diameter at the base to about 0.8 m (2.5 ft) at the top. On the inside the diameter is about 1.3 m (4 ft) at floor level and 0.7 m (2.25 ft) at the chimney. A rake hole and several tuyere holes (from eight to thirteen in known examples) occur around the base. One example of a *kathengu* survives a few kilometers from the courthouse of the subtraditional authority of Kachulu. It stands about 10 m from the *ng'anjo* with which it was paired and is an oval-shaped beehive about 0.5 m (1.5 ft) long and 0.4 m (1.25 ft) wide and high with a chimney hole of 0.2 m (8.0 in) at the top. It has a rake hole at the side that is plastered shut during smelting and tuyere ports at the ends, each for a pumper with a single goatskin bellows. The *ng'anjo* and *kathengu* referred to here were built in 1910 by Kahaura Msiska, a well-known smelter of the region. His son, Chikotawalala Chapasi Msiska, has a collection of iron blooms, forged items (such as hoes), and assorted smelting paraphernalia in a private smelting museum. He has also produced a written account of Phoka smelting (Msiska, n.d.).

Induced draft smelting is an unusual technique confined almost entirely to Africa. Here it has been reported along a band that stretches from west Africa, through the tropical forests of the Congo basin, to central Africa (Cline 1937). This distribution is not continuous: forced draft furnaces can also be found throughout the area in question, and outlying examples of induced draft furnaces can be found among people like the Galla of the African Horn. Most induced draft furnaces produce forgeable iron in a single step; they are usually tall and thus able to heat a considerable amount of ore to high temperatures. Examples from west Africa include the Dedougou furnace [6.7 m (20 ft)] and many others in the 3–5-m range (Cline 1937). An impressive example of the product of induced draft smelting was that of the Yoruba, who routinely produced 35-kg blooms of 1.7% carbon content (Bellamy and Harboard 1904).

Two-stage smelting, in which the primary furnace works by induced draft, can be found in the area of Lakes Malawi, Tanganyika, and Rukwa. Wright (personal communication) calls this the *ilungu* zone, after the work most commonly associated with iron smelting in this area. So far as productivity is concerned, the two-stage smelting in central Africa should not be compared with the large induced draft furnaces of west Africa but rather with the ubiquitous short shaft furnaces that operate by forced draft. This furnace type, which is described later, is widely distributed in Africa spatially and temporally.

The productivity of the Malawi natural draft furnaces does not approach that of the Buhaya forced draft example, even when the most optimistic expectations of informants are considered. The combined abnormalites of furnaces that seem too short to have sufficient draft for iron smelting and low productivity reported by informants prompted us to arrange a reconstruction of traditional smelting in Malawi in 1982 and 1983.

At Chulu the smelting in 1982 was organized and led by Potiphar Mvula and Addison Mwaza. In 1983 Wilfred Kalemba was added as a leader; he is a son or grandson of the late chief smelter Vujomo. The smelting crew was composed of nine men but was augmented by five women during the building of the furnace.

The Phoka smelting was organized by Chikolawalala (C. C.) Msiska, son of the late chief smelter Kahaura Msiska. He insisted that others more knowledgeable than himself could be found to lead the smelting and appointed such men. In 1982 the furnace was built near Chakaka village in the area of STA Kachulu. The chief smelter, or *murungu*, was Peter Nyirenda; his brother, Waterha Nyirenda, acted as *msofi*, or medicine man. This furnace, although it produced substantial microscopic iron, failed to produce the massive consolidated bloom associated with forgeable iron; because of this failure, the site was moved. The 1983 smelt occurred near the home of C. C. Msiska. On this occasion an aged chief smelter, Thomas Chivanda, was brought out of retirement to act as both *murungu* and *msofi*, but he was assisted at all times by C. C. Msiska and Watherha Nyirenda.

The two natural draft furnaces studied differed in almost all metallurgical aspects. Both furnaces were built by a group of men who had been involved in smelting in their youth but who had not been master smelters. The furnaces were constructed based on models of surviving furnaces: the Chulu furnace modeled after Vujomo's furnace from Kasungu National Park and the Phoka furnace copied from the remains of a furnace built by C. C. Msiska's father. Although the furnaces are not exact duplicates, they were built closely to scale and lie in the middle of the range of surviving furnaces.

The Cewa Furnace The Cewa *ng'anjo* was approximately 8 ft tall, and the shaft was constricted at mid-height. The lower chamber was 1.1 m (42 in) in diameter at the base, decreasing to 0.9 m (36 in) at 4 ft, mid-height. The constriction then pulled in to 0.5 m (19 in), with a 10-cm- (3-in-) thick lip around the radius. Figure 24.3a shows a scale drawing of the Chulu furnace. The constriction divides the furnace into two chambers. We believe that the purpose of the constriction is to support the charge in the upper chamber while providing a smelting zone below the constriction in which slag melting and draining occur.

The Chulu furnace contained nine tuyeres at the base and was built on sloping ground on the west side of termite hill, as were all traditional furnace remains in the area. The purpose of the west side of the termite hill does not seem to be connected to prevailing winds, which during the dry season are light and variable.

Construction commenced with soaking and wedging clay dug directly from the termite mound and letting it stand overnight. A trench was then dug 6 in wide and approximately 8 in deep. Medicine was placed in the bottom of the trench where magic stakes had been driven in. The chief smelter then ran away from the furnace, signifying the fleeing of evil spirits. The spirits of everyone who stayed, the smelting crew, thus stay at the furnace even when they are not physically present and protect it from evil spirits. After the furnace had been consecrated, the trench was filled with large balls of wet clay that were forcefully flung into the trench. Above ground the wall was built up by throwing handfuls of clay. The walls were then scraped to 15–20 cm (6–8 in) thick. When the walls were up to 23 cm (9 in), nine 8-cm-wide slits were cut in the walls with a mean spacing of 28 cm (11 in). Short lengths of sapling 5 cm (2 in) in diameter were plastered in at a mean height above the floor of 15 cm (6 in). A 28-cm (11-in) door was also cut at the lowest (western) point of the furnace floor.

The furnace construction required approximately one month. The furnace was built of clay in layers of about 6 in per day. This slow rate of construction was required to permit sufficient drying and strengthening of the large furnace so that the walls could support the construction as it progressed. On several occasions when construction exceeded this pace, the walls slumped and sections fell in and had to be replaced. When the smelters were questioned about the furnace collapses and excessive drying cracks that developed, their reply was invariably, "We entertain not any pessimism." There was some argument regarding the constriction that, when first built, almost completely choked the furnace off. This was subsequently enlarged to 48 cm (19 in), approximately half the furnace diameter. Further construction then took place, with the upper chamber of the same diameter as the furnace constriction, like a smokestack and totally unlike any remaining traditional furnace. The smelters agreed that this was wrong but resisted changing it for several days. Finally, after extensive argument the "smokestack" was torn down and the upper chamber enlarged to the original diameter, so as to be similar to Vujomo's standing furnaces in the park. At the top (2.5 m; 8 ft) the furnace was again constricted to a 43-cm opening (17 in), approximately the diameter of the constriction. A monitoring hole, the *jojo* hole, was made at a height of 2.2 m (7.2 ft), 0.3 m (9 in) below the rim for visual monitoring of the smelt. Six thermocouple holes were also made at even intervals of 1 ft to monitor temperature.

The furnace walls were 6 in thick and dried slowly. When the door was cut halfway through the wall three weeks after starting, the clay 2 in in from the surface still had the same moisture content as the wet construction clay. Without openings in the furnace bottom there was little circulation in the furnace, and it was drying from the outside in. This created large shrinkage cracks on the outside [approximately 0.6 cm (0.25 in)] and compressed the clay on the inside. As the clay on the inside dried, it also shrunk, pulling the exterior cracks together and leaving mosaic blocks neatly fitted together but cracked along a 10-cm (4-in) craze pattern.

With the furnace at Chulu some attention was paid to medicine, but not a good deal compared with the Phoka furnace. The men at Chulu also had had more contact with modern life and did not take the traditional aspects of the furnace rituals, taboos, and medicine terribly seriously. Later on, when they began to be worried about success, they became much more concerned and attempted to follow the rituals, including the taboo restrictions and the wearing of traditional bark cloth.

The other two tasks consisted of the mining of ore and the production of charcoal. The smelters at Chulu were adamant that traditional ore be mined from alluvial laterite sills or deposits at the edge of *dambos*. *Dambos* are iron-rich sandy deposits formed in shallow grassy valleys during the rainy season and subsequently exposed when later erosion cuts new gullies through the silted up old valleys.

The ore that the Cewa smelters chose was hard and required intensive physical labor with picks and hammers to break it out and into small cubes a few centimeters on a side. Good or "first-class" ore shows specular fractures when broken and contains muscovite or mica without too much sand or large pebble inclusions. Most pieces seem to have substantial silica or sand inclusions comprising a significant portion of the fracture. The "good ore" is hard and gives off a clear ring when struck with the pick. The broken pieces show extensive embedded sand or quartz grains cemented with hydrous iron oxides and appear to be a marginal ore. Chemical analysis by X-ray fluorescence of the ore is given in table 24.1. It is clear that both ores have a high silica content. In the Chulu ore the silica content is more than twice the amount required to form the minimal melting point for fayalite-wüstite slag. Thus the formation of a fluid tap slag would

Table 24.1 Chemical analysis of Malawi ores

Component	Chulu ore	Phoka ore
SiO_2	40.70	23.10
TiO_2	0.36	5.07
Al_2O_3	5.84	14.92
Fe_2O_3	41.58	47.48
MnO	0.06	0.26
MgO	0.17	0.42
CaO	0.10	0.01
Na_2O	0.10	0.00
K_2O	1.13	2.68
P_2O_5	0.14	0.13
H_2O	0.84	0.57
ignition loss	8.83	4.97

consume all of the iron, leaving substantial amounts of undissolved silica. The explanation must be that the large quartz grains do not enter the slag and should not be included in the analysis.

From the start of the smelting project to opening the furnace, nine men and four women worked to various degrees for fifteen weeks in 1982. The results of this labor were approximately 800 kg of ore and 2,000 kg of charcoal. They also built the furnace and a smelting house and made two dozen tuyeres, bark cloth for twelve people, and numerous enormous pots of beer. Had the smelt been successful, they would have expected to have produced 30–40 kg of bloom and perhaps half that weight of implements after forging. Admittedly the building of the furnace would not have to be repeated in the following years, and the efficiency of mining and charcoal making would be expected to be improved; still the process, if successful, represents approximately 0.9 kg of iron per worker per month at a cost of 1,500 kg of charcoal or approximately 15,000 kg of hardwood.

For charcoal making the smelters were all agreed on the use of the *mwanga* tree (*Pericopsis angolenesis*). This is a large dense hardwood akin to Rhodesian teak. The trees cut ranged from 40 to 60 ft tall and up to 16 to 20 in in diameter. The wood is very dense: A 6-ft log 16 in in diameter weights approximately 500 lb, or almost 60 lb/ft^3, just barely able to float. Carrying the logs to the charcoal burning pile was laborious; the smelters laid down saplings to form a road and rolled the log with pry bars. Moving a 6-ft log 200 yd took 6 men an hour and a half. Smaller logs were carried by six men using poles and going about 30 yd between stops, accompanied by rhythmic singing.

To make the charcoal, the smelters normally used two trees per pile. The wood was piled up, with the logs generally parallel, and covered with earth, leaving a hole on the windward side. The hole was then filled with brush and meilie (corn stalks) and lit. In half an hour, when the fire was burning well, the hole was bricked up with clods of earth. The dirt was not thrown on the fire but was built up as a wall, with clods approximately a foot in front of the fire. The whole pile was then covered with more earth, and two men were left to watch it. Watching continued in shifts of two for 24 hours, and the men covered any open flames that broke through cracks in the earth as the pile settled. After 24 hours the pile was again covered with earth to seal any vents and left to distill, smoking gently through the soil for at least a week. Nine days after burning the first pile was opened. It was still hot inside but no longer burning, and the earth was burnt to a deep red. The men pulled the pile apart carefully and separated the charcoal by hand. Any hot pieces were covered with soil to quench them. The charcoal was dense and in fact broke brittly across the grain with a conchoidal fracture. One pile started six days earlier was still smoking, and we gathered from the smelters that opening a pile too soon causes it to burst into flame; little charcoal is recovered from this all too frequent disaster. As usual with disasters, the primal cause of the fires is attributed to witchcraft and only secondarily if at all to the premature opening.

The tuyeres or blowpipes were made from a different clay than the furnace. The smelters said that this was the clay used in traditional times. It was a mixture of the termite earth being used for the furnace and a sticky ball clay, seemingly lacking in quartz or granular refractory material, which seemed more suitable for low-fired pottery than refractory fire-clay tuyeres. The clay was broken up, mixed with water, shaped into wedges, and then left for several days. The initial tuyeres were only about 12 in long, formed on a sapling approximately 2 in in diameter and partially stripped of its bark. At the junction of the stripped section and the bark a cone was formed; the tuyeres were left to dry on the sapling. In several days the tuyeres all cracked radially, as they could not contract and slide freely on the saplings. Repeated attempts to suggest removing the tuyeres from the saplings while they were still soft and let them dry slowly were resisted, and a sufficient quantity of unbroken tuyeres was obtained only just before the smelting.

Before smelting purification rites were carried out by Chief Chulu and much of the village at several shrines to the ancestors. The smelting site was then sealed off, bark cloth donned, and the taboos of washing, women, sex, and fighting were put into effect. Small medicine sticks and various charms and a gate served to protect the site from unclean visitors.

Smelting was started by Tonga, the smallest smelter, who crawled in the rake hole and placed medicine and charms in the furnace. The chief smelter then cut off the head of a cock and held it over the top of the furnace. The blood flowed into the furnace freely, signifying acceptance by the spirits and a propitious beginning of the smelt.

The furnace was also surrounded by a scaffold with a ladder for loading from the top. Nine tuyeres were then placed through the tuyere holes and extended between 6 and 10 in into the furnace. The tuyeres lay on the ground, and the tuyere doors were then sealed up with mud. Fifteen kilograms of ore was then passed into the furnace and placed at the center of the furnace just past the end of the circle of tuyeres. Some charcoal was also packed around this, and finally an additional 368 kg of charcoal and an additional 15 kg of ore were added from the scaffold at the top. Coals from a fire that had been started in the traditional manner by the friction rubbing of a stick were then inserted into the tuyeres; these special iron-making coals lit the charcoal quite readily, and the furnace began to burn.

The ore was a low-grade laterite with a typical analysis given in table 24.1. There was a high concentration of old furnace sites in the area with extensive deposits of free running tap slag surrounding them. During the 1982 and 1983 campaigns we formed fluid slag inside the furnace but did not get a sufficient amount or high enough temperatures for it to run out through the tuyeres or the tap hole. The furnace was started at 3 P.M., and within a few hours temperatures at the end of the tuyeres were about 1040°C. There was reasonable draft through the tuyeres, at this point estimated to be 1–2 knots. The charcoal consumption rate was initally rather low and rose over the course of the next day from about 5 kg/hr to a maximum rate of about 10 kg/hr. At 9 P.M., six hours after firing began, a CO flame ignited at the top of the furnace. At this point an additional 45 kg of ore was added. The temperature in the lower chamber was measured using thermocouples in mullite protection tubes in the furnace and an optical pyrometer sighted through the tuyeres. After 24 hours the temperature in the lower chamber was up to 1000°–1040°C, and at this point there was also evidence of liquid slag dripping and also flowing into the tuyeres themselves.

The tuyeres also began to soften and bend under the load, flattening out and reducing the air velocity and the combustion temperature. The presence of free-flowing slag while the bulk furnace temperature was at 1000°C is indicative of local temperature variations, with the hottest zones being just above the tuyeres, which were not accessible for temperature measurement. The temperatures at the end of the tuyeres rose to 1200°–1240°C and remained relatively constant until the end, when they rose to 1360°C as the charcoal burned down in the furnace, providing less constriction and a higher air flow.

Most of the ore wound up below the tuyeres in the cooler zone of the furnace before the furnace reached its highest temperatures. We believed that the ore should have been put in the furnace later in the run, when temperatures had climbed in both the upper and lower sections and also that the furnace should not have been filled so high. It is clear that the furnace was kept overfilled with charcoal and that this reduced the draft. Flow rates through the tuyeres were calculated based on charcoal consumption to be 0.43 knot, which is basically undetectable by drafts or dry dust dropped in front of them. With a lower level of charge in the furnace there would be less restriction of flow and a higher rate of burning and more heat. We believe that this practice was followed in traditional times and raised the temperature sufficiently to produce free-flowing wüstite-fayalite tap slag from the bulk of the ore charge rather than from a small portion. There is also an observation hole built halfway up the upper chamber, called a *jojo* hole, that is used to observe the charcoal burning. Most of the time the charcoal level was filled well above the observation port, which thus could not be used for its purpose and indicates overcharging of the furnace.

It was clear that the furnace was not operating as well as it could, but it was close. The temperatures were a bit lower than expected on the basis of the fluidity of traditional tap slags. The center of the furnace seemed to be operational, however; the ore at the bottom was below the hot zone and did not contribute significantly to the bloom. Firing continued for five days with a total charcoal consumption of 1,455 kg.

On opening the furnace there was a substantial mass of slag and bloom, totaling 91 kg, 16 kg more than the 75 kg of ore that had been added to the furnace. Because the recovery of slag and bloom was less than complete and some slag remained fused to the inside of the furnace, a substantial amount of ash must have incorporated itself into the slag. The product was separated into four classes: The first consisted of about 12 kg of strongly magnetic material containing a molten fluid slag that appeared partially drained; an additional 9 kg was weakly magnetic; the largest amount of slag was 34 kg of nonmagnetic fused ore and fayalite that had not reached the glassy free-flowing molten slag state; there was also 36 kg of a green glassy slag.

Metallographic analysis showed a well-defined nucleation and growth mechanism for the iron and similar features in both the traditional blooms and in the material from the 1982 and 1983 smelts. Furthermore, the metallurgical analysis of the Phoka furnace, which is described later, and material recovered from other traditional sites all show evidence of the same stages.

The iron had been largely reduced at this point, but the bulk of the bloom was fragile and the smelters declared that it was not forgeable until it had been further refined in the small furnace. The purpose of the small furnace appears to be to increase the temperature causing the further formation of fluid tap slag and its draining away to produce a material that has a continuous network of iron, whereas most of the material recovered from the *ng'anjo* had a continuous network of slag that would break if hammered at lower temperatures. At elevated temperatures material recovered from the *ng'anjo* should be forgeable, although many of the small flakes could well be lost in expelling the slag.

The Phoka Furnace The Phoka furnace consisted of a much shorter cone about 1.5 m high, less than half the height of the Chulu furnace, with a base diameter of 1.8 m (5.7 ft). The furnace was built on a terrace halfway up the Nyika massif at a site chosen by divination. The Nyika is part of the rift escarpment rising to 8,500 ft above Lake Malawi at 1,600 ft in less than 10 mi. The land is ridged and everywhere steep, normally on the order of 30°. Goods and materials were transported by the Phoka on their heads; their paths took the shortest route up the steepest contours. Gardens, dwellings, and other sites of activity generally have to be built on terraces excavated into the mountainside. The furnace site was remote from both the clay source 2,000 ft below and the ore source 2,000 ft above and three mountains and 10 mi away. The furnace was constructed from 6,500 kg (14,300 lb) of wet clay fetched from the valley at the base of the escarpment. This was carried by porters at 20 kg per porter per day. Individual head loads ranged from 6 to 20 kg, the size of the load governing the number of trips to make the 20 kg daily norm. Once the clay had been gathered, the furnace was built in three days and left to dry for a month. The furnace was generally in the clouds and covered with a thatched roof, which was removed when the sun shone. Before smelting the furnace was dried by burning wood that was incompletely charred during the charcoal production.

The Phoka use a mixture of new ore, which is again a high-silica low-iron ore (see table 24.1) fluxed by adding fluid tap slag or glassy slag and partially fused material from previous smelts. The mines are located in rolling grassland on top of the plateau and consist of a series of deep shafts 9–15 ft deep, with interconnecting stopes (Wener and van der Merwe 1986). It is estimated that surface weathering and filling in of the pits has reduced the depth by 5 or 6 ft and that the mining was originally conducted at a Precambrian basement granite–gneiss contact some 15–20 ft below the surface. The weathered surface gneisses consist of quartz, feldspar, kaolin and iron oxides, including limonite, hematite, and some magnetite. Whole rock analysis shows 40–68% SiO_2, 20–30% Al_2O_3, and 11–22% Fe_2O_3, with other elements restricted to traces amounts. This ore seems uncommonly ill-suited for smelting, particularly considering the great distance over which it was transported—15 mi of mountainous terrain. Whole rock analysis can be misleading, however: The ore was highly sorted, discarding quartz and kaolin; furthermore, much of the massive quartz is found unchanged in the slags and thus should not be counted; also the alumina reacts directly with silica to produce refractory mullite, which will not enter into the reaction because of its high melting temperature. In fact we used mullite protection tubes for our thermocouples with no problems of slag erosion during the smelt. Discounting the refractories and the inert materials, what is left is an iron silicate, probably not far off the minimum melting eutectic for wüstite-fayalite ($2FeO{-}SiO_2$), about 80% iron oxide. In actual smelting the Phoka did not smelt their ore directly but fluxed it with previously smelted old slags, which act as solvents to extract selectively the lowest melting components. The Phoka furnaces, and indeed all the African furnaces, depend on a fluid slag solvent, thus requiring about 20% silica. Modern high-grade ores have proven singularly refractory and nonproductive when used in experimental furnace reconstructions.

Charcoal burning by the Phoka was much less efficient than that done by the Cewa. A slow-growing, hard shrub, *Phillipia benguelensis* (an *Erica* or heather type) was used. These trees are small (less than 6 in) and widely scattered. Eighty-eight cubic meters of wood was collected at various sites and stacked and dried for a month, often below the cloud level. Pits approximately 1 m deep and 10 m^2 in area were dug and stacked with wood to a height of 2 m, 1 m above the ground. The pile was then lit and allowed to burn as an open bonfire until the level dropped 50% to ground level. The fire was then smothered with large quantities of large wet leaves and immediately covered with sod. Afterward tenders walked over the surface watching for cracks and smoke and covering them with more earth. After twenty-four hours the fire had cooled and the pit was uncovered. The charcoal and incompletely charred logs were separated. In all 1,500 kg of charcoal was obtained, of which 1,000 kg was used in the smelt.

The Phoka procedure involves much more attention to ritual than the Cewa and is discussed by van der Merwe and Avery (this volume). The ore charge consists of a purple magnetite plus a softer clay-looking pink stone that is quite magnetic; a third part is a glassy slag from old smelts. The Phoka procedure of using the glassy slag and pieces of old bloom from previous smelts seems sensible from a kinetic point of view, as the premelted slag would much more readily wet and selectively dissolve the ore, forming a fluid solvent slag.

The Phoka are also quite determined in their placement of the ore at the bottom of the furnace below the tuyeres. The main aerodynamic resistance to convection consists of the charcoal body in the furnace, which effectively chokes the flow and restricts the air intake to only a fraction of what would be expected to flow through the tuyeres. Charcoal was added continually over five days with a consumption rate that varied from 6 kg to 12 kg per hour. Flow through the tuyeres was estimated at approximately 1 knot. The calculated air flow based on an average charcoal consumption rate of 9 kg per hour was 1.1 knots. The temperature levels of the Phoka kiln are much the same as the Chulu kiln with flame temperatures at the tip of the tuyere in the range of 1220°C. At the end, when we allowed the charcoal level to drop, creating less resistance, the temperature rose to 1340°–1360°C. Because the ore was not mixed with charcoal and was relatively low at the bottom of the furnace, the ore temperature rose slowly compared to the flame temperature. By the end of the first day the temperature in the ore body measured by thermocouple had risen to 774°C. After the second day it reached 897°C; the third day added 25°, to 922°C. During the afternoon of the fourth day the charcoal level was allowed to drop and the ore temperature rose 125°C within four hours. During the early morning hours of the fifth day, as the furnace burned low, the ore reached over 1200°C and formed both fluid and glassy slags. It is clear that the ore placed below the tuyeres was held at a temperature below which a fluid slag would occur and in which growth of iron would be expected to form a sponge or fine particles without agglomeration and/or separation. Thus, most of the time the ore was in the furnace, we do not believe that there was molten slag that would cause the fine particles of ore to agglomerate. Only during the final stage, when charcoal additions had stopped and the furnace was being allowed to burn out, did the temperature rise above 1200°C. Although the furnace burned for five days, we believe that the bulk of the bloom formation and consolidation occurred in the final twelve hours.

On the morning of the fifth day the furnace had burned down, and a large mass of fused material, weighing 35.5 kg was removed. Most of the charge was sintered together in a single mass. There was dripping fluid slag at the bottom that had penetrated into the charcoal bed. There were also numerous other pieces of fused material. The smelters separated this into two main lots: massive iron bloom suitable for direct forging, and fused material suitable for refining by resmelting in the *kathengu*, which was further subdivided into three lots (table 24.2). The fused material was subsequently graded as 1, 2, and 3, the criteria being weight, the degree of fusion, and appearance. Grades 1 and 2 were resmelted, and grade 3 material was discarded. Finally, the remainder, a greenish glassy slag, was retained for fluxing the next smelt.

Refining took place in a small furnace, the *kathengu*, which had a squarish base of 0.5 m (20 in) on a side, a height of 0.4 m (15 in), and a round hole top 0.2 m (8 in) in diameter at the top. There were two blowing ports on opposite sides, each with a single tuyere mudded in, and a rake hole on the third side. The tuyeres were supported on forked sticks that passed through a clay-covered wattle heat shield. The bellows were skins from goats that had been strangled and skinned from the mouth, leaving the entire skin intact, except for holes at the feet and the head. The skin was softened with oil from the castor bean. The tuyere was tied to one rear leg of the skin; the other legs were tied shut, and the skin was trimmed at the neck. Blowing commenced by pulling the neck open and up, then closing it and pushing down to force air through the tuyeres. The furnace was loaded with charcoal and grades 1 and 2 slags from the *ng'anjo*. Five minutes after starting, the temperature below the tuyeres where the bloom forms was up to 750°C. In 15 min the temperature had reached 900°C, and after 20 min, 1100°C. In approximately an hour the bed temperature was over 1300°C, and the furnace was allowed to cool down. The top temperature at the end of the tuyere was 1400°C. The resulting bloom was small but well-consolidated iron with few voids; the bloom can be filed to a flat metallic surface.

Table 24.2 Furnace products

Product	Magnetic property	Amount
Bloom	strongly magnetic	0.5 kg
Grade 1 material	magnetic	17 kg
Grade 2 material	slightly magnetic	21 kg
Grade 3 material	nonmagnetic	12.5 kg
Slag	nonmagnetic greenish glass	6 kg

The Buhaya Forced Draft Furnace

The Haya live in West Lake Province, northwestern Tanzania, near the lake shore on relatively low sloping land between the lake and the mountains of Ruwanda. The land receives extensive rainfall throughout the year because of its proximity to the lake; it originally supported a terminal rain forest. The soil is lateritic and is largely infertile now as a result of extensive deforestation, which appears to be associated with iron smelting. Again, as with Malawi, the existence of iron slag is extensive; in the clearing of fields for banana planting, 50% of the rocks piled up along the edge may well consist of broken iron slag as well as large cartwheels or furnace bottoms of iron slag weighing 50–100 lb. The Haya furnaces were investigated in 1976 and 1979 by D. H. Avery (a metallurgist) and Peter R. Schmidt (an archaeologist, anthropologist, and oral historian with extensive field experience in the Haya region). The work has been described previously (Schmidt and Avery 1978; Avery and Schmidt 1979; Avery 1981). We review it here in less detail but not with less emphasis than the Malawi furnaces. The Haya furnace is a bowl shaft furnace, which is much more typical of African furnaces than the natural draft furnaces studied in Malawi.

Mining, as in so much of Africa, consisted of shallow pit excavations to weathered laterite with water-deposited secondary deposits lying on top of the basement rock. In this case the ore was a banded ironstone with alternating bands of leached quartzite and water-deposited iron oxides; in many cases there were also large veins of clay. The fractured samples are high in quartz and would appear to be totally unsmeltable if all the quartz enters into the slag.

Charcoal making was not sophisticated; great effort was expended in searching the swamps looking for *muchwesi* trees, associated with the *bachwesi* spirit mediums, traditionally the exclusive practitioners of iron smelting. When an area was finally selected at a remote and inaccessible area of the swamp, wood cutting began. The area was clear cut, felling both tall and short, broad-leafted and narrow-leafed *muchwesi* trees. The wood was split and piled up on islands formed from dredged debris from the swamp. It was then burned in a bonfire, and when the flames died down, the embers were quenched with water. The charcoal was then sifted with two baskets. One of these was a traditional iron smelting basket with holes approximately 2 cm^2, which retained about 60% of the charcoal and allowed the fines to be discarded. At the same time, other charcoal was being sifted with a new tightly woven basket, which retained essentially all the charcoal. Subsequently, from previous smelting experience, it was decided that charcoal from the traditional basket made better iron. The size of the charcoal and the presence of fines was never commented on.

The furnace consisted of a deep bowl (1 m in diameter and 60 cm deep) lined with a refractory mixture of termite earth, which dries over an extended period of time. The bowl was cleaned between firings and used for many smelts; the shaft was built and pulled down on the day of firing. The shaft was a tapered cone with six to nine ports for blowpipes in the base, approximately 4 ft high and tapering from a 3-ft diameter at the base to 1.5 ft at the top. The lower section of the furnace was built from large oval plates, called *empega* blocks, which were cut from iron slag from old furnace bottoms. These slag blocks were used to rebuild the furnace shaft after each firing. The large *empega* blocks were placed on the bowl rim, long axis vertical and with spaces in between that formed the openings for the tuyeres. The *empega* blocks have to be trimmed to the circumference of the bowl, with space for the proper number of tuyeres and bellows.

On the day of smelting the furnace bowl was filled with hollow reeds and grass cut from a nearby swamp. This was burned under conditions of limited oxygen access, charring the reeds and forming a charcoal consisting of a large area of fine carbonaceous tubes. There is also sufficient salt in these reeds to form an alkali flux, which is subsequently wet by iron ore that is drawn into the hollow reeds and along their surface by capillary action, leaving long cylinders and holes in the slag which faithfully reproduce the surface of the reeds. Thus the furnace is completely filled with charcoal capillary tubes that are well wet by iron slag. The reed burning continues until the pile of grass char forms a conical peak (approximately 30° slope) arising from the bowl. Next the *empega* blocks are propped up with small slag on their ends with gaps in between to insert the blowpipes or tuyeres. A typical furnace may have between six and nine bellows and tuyeres surrounding it, the number being generally governed by the availability of the bellows.

The bellows are the property of the individual smelters and their ownership entitles people to participate in the smelt and thus in some of the production of iron. The bellows are valveless and cut from a single large log, approximately 3 ft long, with two bowls and associated holes proceeding to a nozzle. Skins are tied over the bowls and tied to long poles to form a drum that can be driven at a relatively high frequency, up to 400 strokes per minute. This produces an alternate pulsing of an air jet from the end of the tuyeres and the drawing of air back in through the nozzle into the drum.

The tuyeres are approximately 2 ft long with 1-in inner diameter and $1\frac{3}{4}$-in outer diameter. They are formed on mandrels from a refractory clay that is tempered with approximately 40–50% crushed tuyeres, which had previously been successfully used in smelting. Thus the tuyeres are relatively refractory, and the selection of temper from tuyeres that did not melt in prior smelts helps maintain a highly refractory blowpipe. The blowpipes are inserted into the furnace to almost their full length with only the bell-shaped mouth projecting out beyond the furnace. Thus they initially lie on the conical pile of grass char pointing up at a 30° angle, with their ends forming a circle approximately 15 cm in diameter.

The shaft was then continued by chinking the *empega* blocks above the tuyeres with smaller pieces of slag and plastering them together with termite earth. The upper part of the shaft was formed from small pieces of slag trimmed to fit and mortared with termite mud. The construction of the whole furnace took approximately 4 hours. At this point the grass char had usually started to burn; also, some charcoal was generally added before the shaft was completed. The furnace was left, and the smelters generally engaged in a special meal containing delicacies such as palm nuts, dried fish, or possibly some chicken cooked with the normal diet of boiled plantains. Salt, normally in extremely short supply, had also been hoarded and was freely passed about.

After the meal the smelters returned to the furnace, which was normally glowing red at the tip of the tuyeres, and charcoal and ore were poured in through the top. Bellowing started at this point and the temperature rose quite rapidly. The chief smelter carefully positioned the bellows approximately one to two bellows diameters back from the bell-shaped opening of the tuyeres. Thus when the bellow was pumped, it alternately forced a high-velocity jet of air into the convergent opening of the bell-shaped bellows. This acted as an aspirator and drew additional air from outside the bellow into the tube. The calculated total air consumption based on the charcoal burning rate seems to be more than twice the estimated volume blown from the bellows. Thus nearly half of the air entering the tuyeres must have been secondary aspirated air, rather than the primary jet from the bellows. When the stick was lifted, air then flowed back into the nozzle of the bellow, but most of this air was drawn from outside the tuyere. Thus the aspiration of the bell-shaped bellows, and the gap between the aspirator and the bellows provided a valve forcing air into the tuyere. On the reverse cycle the flow does the reverse, however, and at later stages in smelting molten drops of slag or softened tuyere hanging from the end of the tuyere could be seen to move in and out with each stroke of the bellow. Thus the air flow in the tuyere was not straight through but rather cycled back and forth. This had a significant effect on the rate of heat transfer from the tuyere to the air entering the furnace.

The air blast split in the oxidizing region at the center of the furnace, with part of the forced air blast being forced out into a central column that rose through the furnace; another part flowed back along the outside of the tuyere, passing out of the furnace through chinks near the tuyere entrance and rising along the inner wall of the furnace to the top. The air flow along the tuyere produces extensive heating of the outer tuyere surface, which in all cases was wet with molten slag, meaning that the outer surface temperature of the tuyere was in excess of 1250°C. Sections of the tuyere after smelting showed fluid slag penetration through the wall of the tuyere to approximately the midpoint. Iron slag will wet clay and penetrate so long as the temperature is above its flow point. Thus the internal boundary of the slag penetration represents the freezing point of the slag, approximately 1200°C depending on composition. The interior of the tuyere almost never, except at the end, reached a high enough temperature to be measured with an optical pyrometer or to glow in the dark. Thus the interior temperature remained below 700°–800°C. Therefore the temperature gradient across the tuyere wall is nonlinear, being shallow through the region above 1200°C, where the clay is wet with slag. The inner surface is at least 400°C cooler, demonstrating substantial cooling and heat transfer from the inner wall.

Thermocouple scans of the tuyeres showed significant heating of the air blast. It is difficult to measure air temperatures with thermocouples because of the effect of radiant heating from the hot zone at the open end of the tuyere. Various heat shields were used and, based on all the evidence, we estimated a temperature rise of the air passing through the tuyere in the region of 500°–600°C. This measured heat rise is considerably larger (Rehder 1986) than one would expect in normal once-through flow in heat exchangers. Heat flow calculations taking into effect the longer path and dwell time caused by reverse flow indicate that temperature rises were in reasonable agreement with calculated heat transfer (Avery and Schmidt 1986). It should be noted that the nominal inefficiency of the valveless bellows and uncoupled tuyeres was responsible for the high level of air preheating, producing the periodic (200 cycle min) reverse flow that lengthens the dwell time, increases mixing, and peels off and reduces the thickness of the boundary layer that would develop in steady-state flow.

The preheating of the air blast permitted high temperatures to be reached in the central zone. The combustion front temperature at the end of the tuyere could be measured through the tuyeres with an optical pyrometer and a platinum-rhodium thermocouple and reached approximately 1700°C. This temperature normally could not be reached without preheating or having a completely oxidizing environment. The gas measurements of furnace samples extracted further up the stack show almost pure CO with negligible CO_2 and indicate a fully reducing atmosphere in most of the furnace.

The ore was, in general, considered to be contaminated or impure, owing to its taboo but unavoidable contact with women. Thus it was necessary to purify the ore before it was added to the furnace. This was done before smelting by filling the bowl halfway with wet wood, which was induced to burn in a smoldery manner. The ore was then piled on the smoldering green wood and covered with more green wood. After this operation the temperature in the bowl rose rapidly as the high CO atmosphere in the bowl began to reduce iron oxides in the ore. Thus in the roasting operation the ore supplied oxygen to the wood, causing the temperature to rise to approximately 800°C.

The mechanism at the Haya furnace was as follows. The ore was pretreated by roasting, which produces a fine internal distribution of carbon and extensive fissuring and permeability to the furnace atmosphere. During the firing ore was periodically added from the top of the shaft, initially forming finely dispersed iron sponge platelets as the porous reoxidized ore was again reduced by carbon monoxide. As the temperature rose, however, the ore melted; the glassy slag then contained finely dispersed carbon remaining from the roasting operation. The melted slag then dropped down past the tuyeres, with the glassy slag sealing the internal iron sponge from reoxidation as it passed by the oxidizing zone of the tuyeres. Below the tuyeres the iron slag fell onto the bed of grass char, which it wet, and was drawn into the tubular capillaries providing an extensive slag carbon interface. This puddle of slag could be observed through the tuyeres, where it undergoes a carbon boil, slowly bubbling CO as the oxides and the molten slag are reduced both by reaction with the grass char and by the internal carbon deposited in the ore.

As the oxygen content was reduced, large quite perfect crystals of iron formed and grew in the liquid slag. The growth interface of this material had small dendrites sticking out and was basically continuous, consisting of a ferrite surface growing in an iron oxide, iron silicate, or wüstite-fayalite slag. As the oxygen concentration was reduced, the slag was reduced, its melting temperature rose, and it ultimately solidified around the iron as a silica aluminous slag with relatively low iron content and a very high softening temperature measured in the laboratory as being in excess of

1400°C. This refractory slag was then used to build further furnaces.

Some of the grass char is incorporated into the iron behind the growing carbon-free alpha iron interface, and, as the interface (which is in equilibrium with the oxygen-rich slag) progresses past these carbon particles, they begin to carburize the bloom. On removal from the furnace the bloom has a medium carbon content of 0.4–0.6%. During subsequent forging without preheating, the same bellows produced a highly oxidizing blast in a small bowl forge that had a high temperature but was highly oxidizing, and almost all the carbon that was in the bloom was burned out, producing final implements that consisted of a low-carbon ferrite with entrained slag.

The Barungu Forced Draft Furnace

The Barungu live in the south of West Lake Province inland of the Gulf of Speke at the southwest corner of Lake Victoria. In precolonial times they were noted as suppliers of iron all the way to Burundi and Lake Tanganyika. They stopped smelting perhaps twenty-five years ago. The furnace they built did not quite work when I observed it with Schmidt in 1979, because of inadequate refractory tuyeres, which melted before the furnace reached the high temperatures of which it was obviously capable.

Again the source, history, and medicine surrounding the materials were of utmost importance. Charcoal was made from *mounga* logs, which were gathered and stacked in a conical pile about 5 ft high. A complicated *dowa* (medicine), requiring some forty different plants and substances, was collected by the senior smelter, Mzee Musaka, and this was placed on the wood to make it "iron charcoal." Ordinary charcoal will not work, only "iron charcoal" made with the iron medicine.

The Barungu furnace is the most ingenious and thermodynamically sophisticated furnaces that I have seen either in Africa or in modern industry. In many ways the technology is more akin to multistage rocket design than to early furnaces.

The Barungu furnace was again a shaft bowl furnace in which the bowl was approximately 2 ft in diameter and $1\frac{1}{2}$ ft deep with a medicine hole in the center. The Barungu place much more emphasis on medicine than the Buhaya, and every aspect of furnace construction was intimately tied to medicines and special rituals that purified the material and prepared it for iron smelting. The Barungu furnace did not use the grass char reaction bed of the Buhaya furnace. An extremely large and complicated collection of medicine was placed in the medicine hole in the bowl floor and covered with a medicine log. There were also complicated charms and purification of the bowl, the smelting hut, and the smelters. The walls of the shaft were built of slabs of termite earth, leaving straight slots the full height of the shaft between them. The furnace bowl was then filled with a thick layer of green leaves and coarse charcoal to the level of the tuyeres. The tuyeres were inserted through the slots of the refractory termite earth and rested on the charcoal where they almost met in the center. Other tuyeres were then broken to form semicircular ceramic vanes that were then mudded into the slots and angled up at approximately 45°. Thus above each tuyere was a ceramic ladder of vanes, much like a venetian blind angled up, preventing direct radiation out through the slots.

The furnace was then loaded with pure charcoal placed around the circumference, and the central core above the tuyere structure was built up with alternating layers of charcoal and ore and old slag, both iron containing and glassy. The furnace was then lit by pushing embers down the tuyeres, and bellowing began. The bellows were again bowl bellows, although this time the bowls were made of pottery and the nozzles were clay tuyeres reinforced externally with bamboo. Two bellows were used at each tuyere. The bellows were also valveless, producing aspirated reversing flow like the Buhaya bellows. As the furnace began to heat, the outside charcoal caught on fire and produced an outer annulus that was a natural draft furnace, with the air supplied through the slots and directed upward by the ceramic louvers. This outer annulus rapidly reached a temperature of 1000°C from the natural draft burning. The air in the blowpipes passed through this outer annulus, which preheated it, although the tuyeres were shorter and less refractory than the Buhaya furnaces and melted and collapsed, terminating the central core blowing before the maximum temperature was reached. It was quite evident that with more refractory tuyeres, which would not collapse, we would have had considerable preheating in the blast that would rise up through the central column, which was surrounded by the outer 1000°C annular natural draft furnace. Thus we would expect high temperatures equal to or exceeding that of the Buhaya furnace.

It is an elegant concept to put a furnace in a furnace. If you have a relatively inefficient furnace that is limited to 1000°C, a simple way to get higher temperatures is to put a second furnace inside the first so that its environment is 1000°C and then to blow that and get a similar temperature rise, 2000°C at most. As smelting went on, the smelters continued to add ore, old slag, and charcoal to the center of the furnace and charcoal alone to the outer annulus.

Microscopic examination of the material showed extensive development of iron flakes, plates, and nodules; however, the early failure of the tuyeres in these runs prevented the iron flakes from sintering together to form the massive solid blooms that the Barungu were accustomed to.

Metallurgical Mechanisms and Kinetics

We have now performed extensive metallography on a variety of samples from all four furnaces, and a rather clear but complex picture of the formation of iron bloom has appeared. Examination of traditional slags and bloom shows the same structures and stages that we obtained from our experimental smelts. The main difference with traditional blooms and slags and the experimental furnaces that we built with people who were knowledgeable but had not practiced for some time was that the temperatures tended to be lower, the slags were less fluid, and the iron, although extensive, had not sintered together to form massive pieces of iron that could be easily forged and recognized as good bloom.

It should be noted that the material removed from the furnace consisted of material that had been subjected to a wide range of temperatures. There were some pieces of ore that had remained near cold spots, which had not heated sufficiently to melt or appreciably change the initial state of the ore. Other materials in the hot zone had reached relatively high temperatures and had been there for an extended period; thus the inside of the furnace was not homogeneous, and pieces representing the whole thermal history of smelting could be found. The presence of one particular feature should not be interpreted as indicating that the furnace in fact operated at the temperature associated with that feature but merely that it is part of the product of the furnace.

The metallographic analysis showed a well-defined nucleation and growth mechanism for the iron and similar features in both the traditional blooms and in the material from the four furnaces. The first appearance of iron occurred in the cracked but unmelted pieces of ore. Small spheres, approximately 0.01 mm in diameter, formed in cracks in ore samples (figure 24.4a) At a later stage of smelting we found areas of ore containing iron precipitate clusters. These consisted of a rather uniform array of closely spaced spheres, 0.05–0.1 mm in diameter (figure 24.4b). In all cases the spheres appeared to be approximately the same size. They did not grow at the expense of the smaller spheres. In a few areas the spherical precipitates were replaced by fine iron dendrites.

As the temperature increased, flakes, approximately 0.1 mm long, appeared. At this stage the slag was glassy with fayalite needles, and we presumed that the slag was partially liquid and acting as a strong flux. The iron flakes grew rapidly, forming a "Chinese script" pattern (figure 24.4c), and the spheres and dendrites dissolved. There was no slow ripening of the Ostwald type. A *Widdmanstätten* array of plates swept into the clusters of small spheres with a sharp boundary where the spheres were dissolved (figure 24.4d). In front of the boundary there were only spheres, and beyond the boundary there were only plates. Etching revealed the plates to be of pure ferrite with a low carbon content (figure 24.5a).

At higher temperatures the "Chinese script" structure began to organize itself into islands (figure 24.5b). These islands gradually assumed a concentric arrangement of platelets parallel to the surface, and the inner plates dissolved (figure 24.5c), leaving dense iron nodules of a few millimeters in diameter with hollow or internally fissured interiors (figure 24.5d). Depending on time and the temperature of the furnace all of these features were seen.

In the Barungu furnace, operated at a higher temperature than the Malawi natural draft furnaces, the plates grew lengthwise much more rapidly and formed thin continuous sheets extending over an inch in many cases (figure 24.6b). The Barungu furnace again shows all the stages observed in the Malawi furnaces of fine precipitate, platelet, and nodule development. At higher temperatures the platelets grew to form much larger continous sheets, as opposed to the "Chinese script" typical of the Phoka furnace. These sheets became thicker and formed a network structure that was high in iron. As the network structure became continuous, the slag could be cold-formed with a hammer without breaking; this was deemed "first class," acceptable for forging. The material that breaks when cold-formed still has considerable iron in it.

The Barungu practice was to add the "first class" iron-containing slag to the new ore, where it provided iron nuclei on which additional iron could grow, greatly hastening the kinetics by providing a fluid slag that fluxed the new ore and formed a fluid iron-rich slag that then dripped down over the plates developed in the previous smelt. Iron precipitated out of the slag on the plates while the fluid low-melting high-silica tap slag drained through to the bottom of the furnace bowl.

At the highest temperatures the sheets and nodules dissolved and reprecipitated out on massive pieces, yielding a material with a continuous iron network (figure 24.6b) that is malleable at ambient temperature. The bloom from the slags that were examined showed that the coarsening and the platelet growth and nodule formation occurred in the presence of the free-flowing tap slag. In the glassy slag region there were extensive clusters of fine spherical precipitates, but the glassy slag did not operate as a solvent and did not agglomerate into platelets, clusters, and nodules.

We believe that the slag dissolved the finely divided iron particles and transported iron by convection in the slag to the platelets, which then grew as the slag precipitated iron onto them. We believe that the fluid slag was responsible for the increasing density and deposition of iron at the surface of the clusters.

The Buhaya furnace, which operated at the highest temperature and in which the slag underwent direct

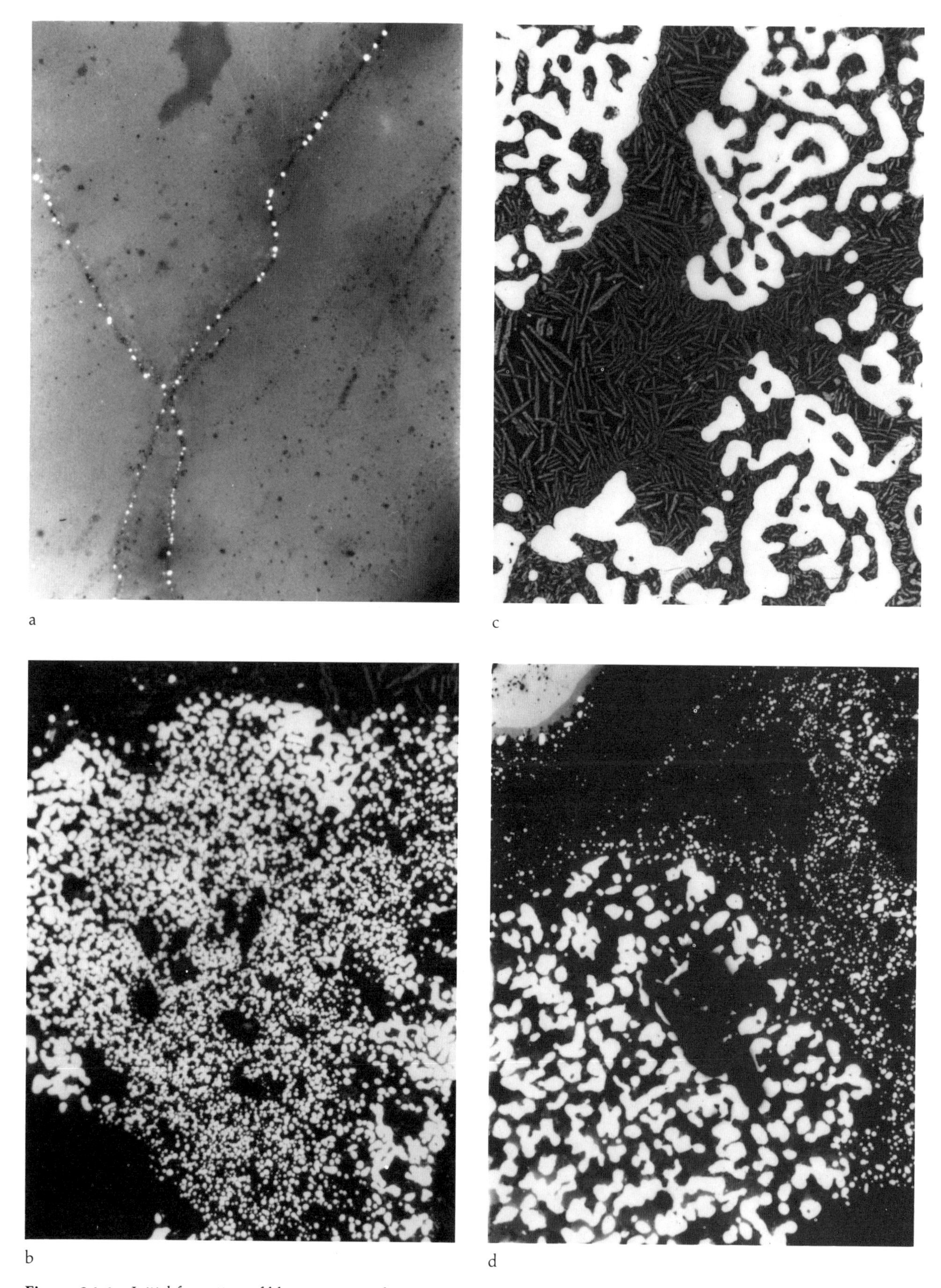

Figure 24.4 Initial formation of bloomery iron (from 1982 smelt). (a) Small spheres in cracks, Chulu *ng'anjo*, unetched, ×500. (b) Clusters of spheroids, Chulu *ng'anjo*, unetched ×1100. (c) "Chinese script" platelet structure, traditional Cewa slag, unetched, ×225. (d) Boundary of "Chinese script" growing into spheroids, Phoka *ng'anjo*, unetched, ×225.

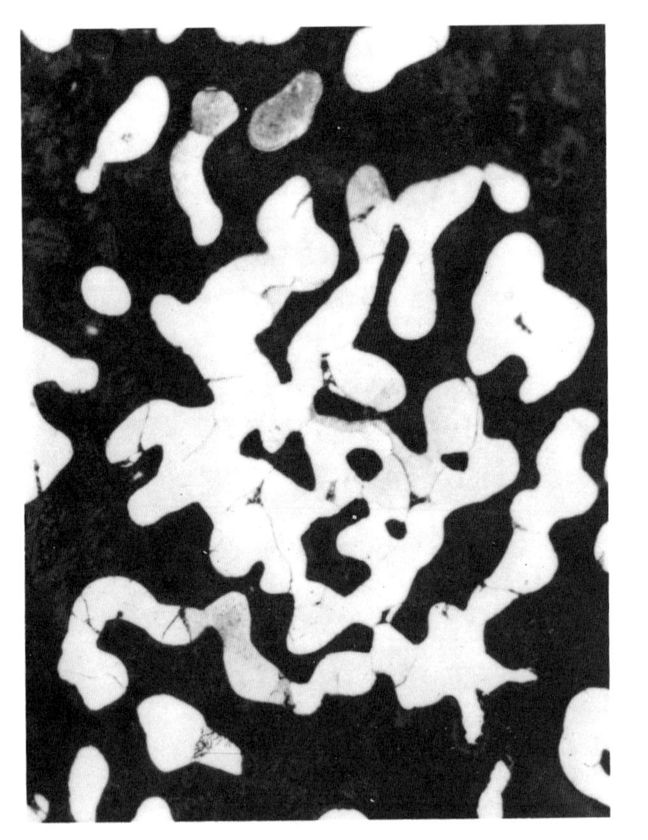

a

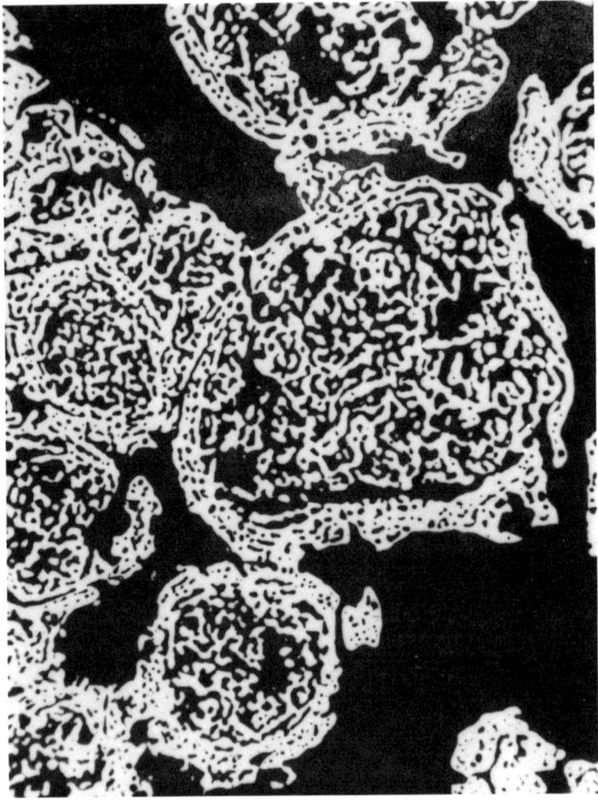

c

b

d

Figure 24.5 Growth stages of bloomery iron (from 1983 smelt). (a) Etched ferrite plates, Chulu *ng'anjo*, 3% nital, × 225. (b) Platelet islands, traditional Angoni (Cewa region) slags, unetched, × 225. (c) Concentric formation of platelets into protonodules, Chulu *ng'anjo*, × 112. (d) Hollow nodule, traditional Angoni, × 56.

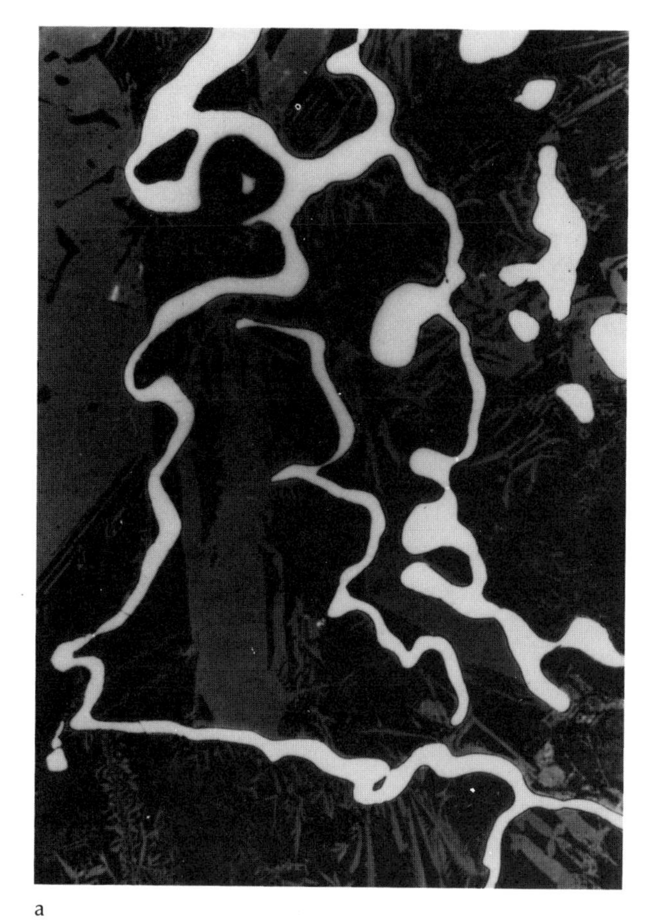

a

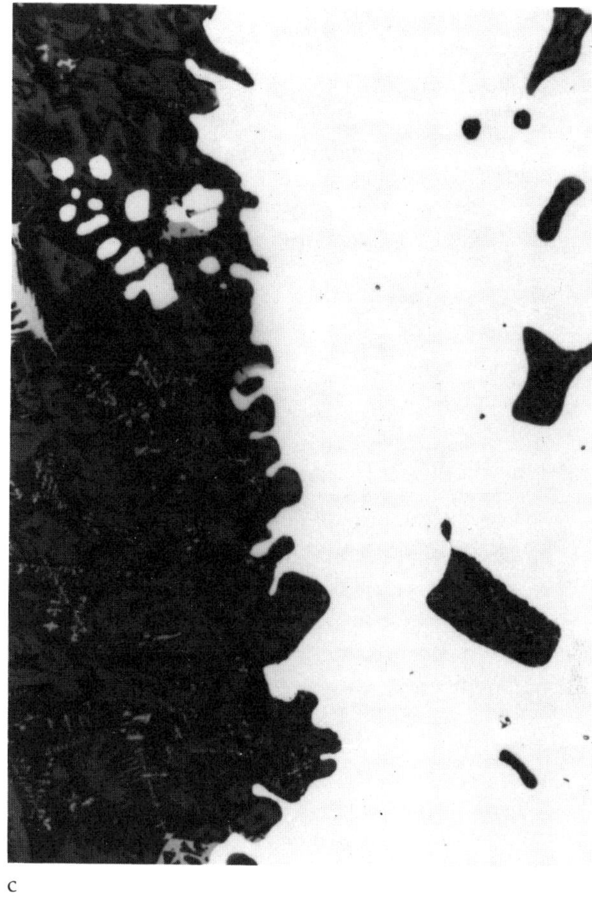

c

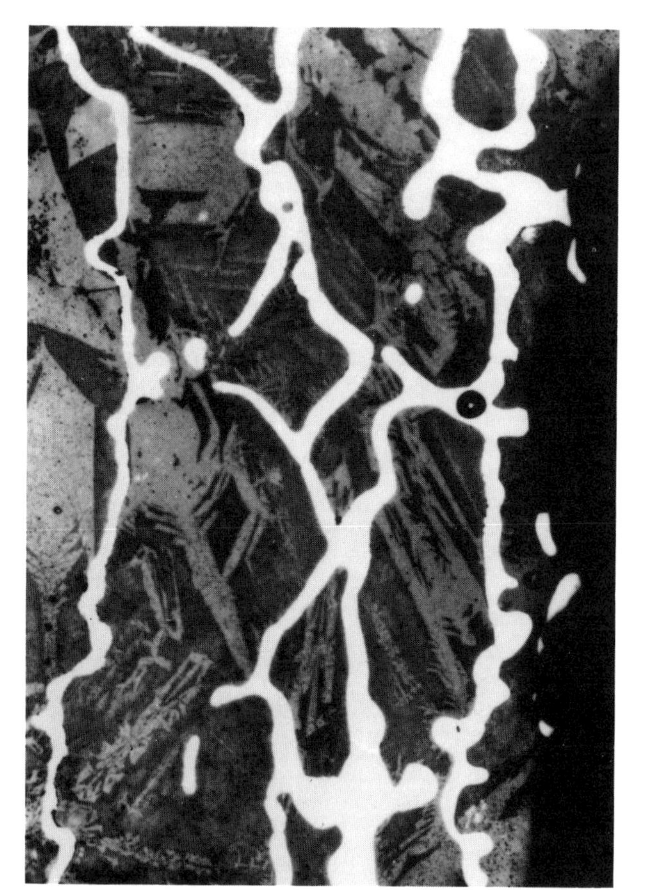

b

Figure 24.6 Forced draft structures. (a, b) Iron sheets, Barungu furnace (1979), unetched, × 225. (c) Massive iron network, Barungu furnace (1979), unetched, × 56.

reduction with a carbon boil in the grass char reaction bed, showed a different growth mechanism. We believe that all the other morphologies—spheres, script, plates, and nodules—existed before the formation of large crystals in the reaction bed but were dissolved and reprecipitated onto the bulk crystals.

In the case of the forced draft Buhaya and Barungu furnaces, very high temperatures, above 1600°C, were measured. In the Buhaya furnace, where there was an active reaction bed, we saw a different growth mechanism: Internally precipitated, finely divided carbon (soot) caused direct reduction of a highly fluid slag. The molten slag that entered the reaction bed undoubtedly contained the flakes and precipitates that were encountered in the lower-temperature Malawi slags. In this case, however, additional reduction of oxygen in the slag caused an active carbon boil, which produced supersaturation of metallic iron in the slag. In this case large crystals developed in the fluid slag and grew by dendritic and nodular projections into the slag from a solid ferritic growth front (figure 24.7a). This bloom, when fractured, showed large eqiaxed grains, several millimeters in diameter, with straight internal twins (figure 24.7b). Slag from cooler regions of the Buhaya furnace also showed the precipitate to platelet to nodule sequence observed in the other bloomeries.

The efficiency of the Buhaya furnace with the carbon boil reduction of the slag and extraction of most of the iron left a highly refractory slag with little residual iron. Thus the slag remaining from the Buhaya process is not readded to the furnace as with the Barungu and Phoka but is rather used as a highly refractory and well-insulating material for constructing furnaces. In fact, although temperatures of 1700°C were being measured inside the furnace, ants were freely climbing over the outside walls, some 6 in thick.

We believe that the fluid slag dissolved each successive stage and transported it by convection and gravity flow down the stack. The lowest melting and most fluid slag was formed near the eutectic with about 23% SiO_2. Thus successful operation of the traditional bloomery required rather low-grade ores with high silica contents and a large production of slag. One of us (van der Merwe) has worked extensively in the Phalabora region of the low veld. Here a high-grade magnetite ore (0.95% Fe_2O_3) was used; it required the deliberate addition of crushed quartz pebbles from a different source.

A second corollary arising from the slag solution-transport-reprecipitation mechanism is the efficacy and indeed necessity of fluctuation temperatures in the consolidation of the bloom. Thus the seeming inefficiency of these bloomeries—where the temperature rises and falls with the waxing and waning attention of the smelters, the enthusiasm of the bellows operators, the drinking of beer and the singing of smelting songs, the force and direction of the wind, the height and

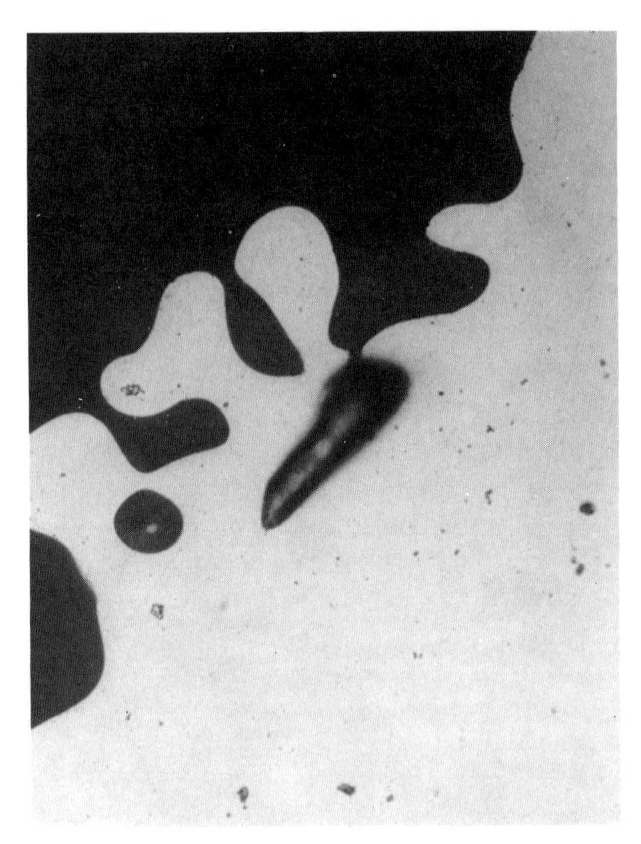

a

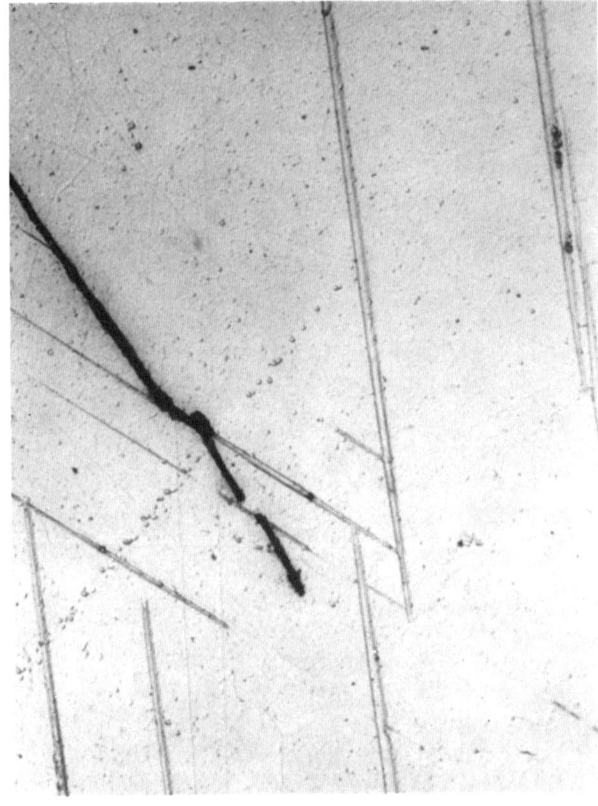

b

Figure 24.7 Massive iron bloom growth. (a) Ferritic growth front, Phoka *katanga* (1983), unetched, × 225. (b) Equiaxed grains with twins, traditional Buhaya bloom, × 56.

density of the charcoal in the furnace stack, and the diligence in keeping the tuyeres open and clean of slag—all serve to produce a cyclic variation in temperature, with the small precipitates being dissolved as the temperature rises and solubility increases and subsequently precipitating out on larger and more planar plates as the temperature falls. It is also noteworthy that it is the inefficiency of the Buhaya are Barungu valveless bellows that reduces the boundary layer, promotes mixing, increases dwell time, and results in the increased heat transfer that produces the high temperatures that these furnaces attain with eight elderly smelters.

In contrast, the history of academic reconstructions of traditional furnaces (of which we have contributed our share) have been notably unsuccessful, as we controlled our temperatures with thermocouples, our air blast with blowers, and smelted our high-quality iron ore obtained from steel companies' benefication plants. We seldom, however, produced more than a minuscule quantity of iron.

Summary

We have described four iron smelting furnaces and their operation by men who had practiced this tradition, which they learned from their fathers in the recent past. The furnaces differed greatly in their technology, and yet all produced iron with varying degrees of economy. In many cases the success and efficiency of the operation is dependent on factors that seem to modern technological eyes as primitive or grossly inefficient—low-grade iron ore and valveless bellows. Some, notably the Phoka and Chulu natural draft furnaces, put extraordinary demands on fuel and could not have been economically feasible for large-scale manufacture of implements. We also know that in the Phoka areas particularly, many remaining traditional hoes were not in fact locally manufactured but were imported from more efficient producers in Tanzania.

In all these cases we find that there is a large spiritual component in the making of iron and that the people who are successful in smelting iron are considered to have strong powers over evil spirits, to have pure souls, and to be in harmony with nature. Success does not come to the bad spirited, to the breaker of taboos, to the unclean, or to the unpure. Songs of smelting all celebrate the success of the smelters. It is a great badge of achievement to have made iron. The iron is made not in the workshop or in a businesslike environment but in a secluded semireligious setting from which outsiders and the unpure are barred and in which rigid taboos are enforced.

As for the origins of the African Iron Age, we do not believe that one can say with any certainty that it diffused or was strongly influenced by outside technology. Certainly the existence of iron came into the African consciousness in the first millennium A.D. and started various groups producing it.

That the production of iron had important ritual and magical significance is apparent, both in songs and legends and as seen in the importance of iron for various ritual uses, such as king stools. The actual economics seem to be secondary to the ritual importance of producing iron, and thus relatively uneconomic processes could exist side by side with those that were more efficient. Although the preheating of the Buhaya and the furnace in a furnace of the Barungu in Tanzania were highly sophisticated, they do not appear to have spread or to have been copied extensively anymore than were the inefficient and wasteful furnaces of the Phoka. Furthermore, we have never encountered a traditional technical description of smelting; rather, smelting has always been associated with ritual and purity, taboos, witchcraft, jealous ancestors, and medicine. The processes developed for smelting are limited to iron and have not been adapted to the production of pottery or other pyrometallurgical processes.

Acknowledgments

This work represents forty years of involvement with the Iron Age for us and thanks are due to all the people of Africa who recounted their history, pushed our stalled vehicles, and worked with great enthusiasm at mining, charcoal making, furnace building, and smelting; to the people who supplied us with shelter and food and who drank beer with us and sang and danced; and to the White Fathers, game rangers, and settlers who preserved and recounted changing customs and aided with supplies and repairs.

We also thank the chief smelters Mzee Herman Kariba in Nyungwe village, and Mzee Musaka in Bwanga village, Tanzania; and Potiphar Mvula in Chulu, and C. C. Msiska, Peter Nyerenda, Thomas Chilbanda, Walthera Nyerenda in Phoka, Malawi.

Our thanks also go to the Department of Antiquities of Tanzania and the Departments of Antiquities and Parks in Malawi for assistance and the privilege of sharing their cultural history. In particular, Gadi Mgomezulu and Fidelius Morrocco of the Department of Antiquities, Malawi, and Richard Bell, Mathew Matemba, Hetherwick Msiska, and David Elias of the Department of Parks deserve thanks for their active participation in the research.

We are grateful to the National Science Foundation (USA) who supported many of these expeditions, the Human Resources Council (Republic of South Africa), The Explorer's Club, Roberts Construction, and the University of Capetown and Brown University for financial support and laboratory facilities. Gratitude is also due, in the case of the universities, for permitting the outer limits of leave policies to be explored.

References

Avery, D. H. 1981. "The iron bloomery," in *A History of Pyrometallurgy*, T. E. Wertime, ed. (Washington, D.C.: Smithsonian Press), 205–215.

Avery, D. H., and P. R. Schmidt. 1979. "A metallurgical study of the iron bloomery, particularly as practiced in Buhaya." *Journal of Metals* 31:14–20.

Avery, D. H., and P. R. Schmidt. 1986. "The use of preheated air in ancient and recent furnaces: A reply to Rehder." *Journal of Field Archaeology* 13:355–357.

Bellamy, C. V., and F. W. Harboard. 1904. "A west African smelting house." *Journal of the Iron and Steel Institute* 56:99–127.

Cline, W. 1937. *Mining and Metallurgy in Negro Africa.* Menasha, Wis.: George Banta.

Livingstone, D., and C. Livingstone. 1865. *Narrative of an Expedition to the Zambesi and Its Tributaries; and the Discovery of Lakes Shirwa and Nyassa.* London: John Murray.

Msiska, C. C. n.d. *The Phoka Kiln.* Available from 1982 Malawi field notes. Available from D. H. Avery, Brown University.

Mwaza, A. A. C. 1976. "Things of olden times." Available from D. H. Avery.

Phillipson, D. W. 1968. "Cewa, Leya, and Laba iron smelting furnaces." *South African Archaeological Bulletin* 23:102–113.

Rainsford, O. 1966. *Livingston's Lake: The Dramma of Nyassa.* London: John Murray.

Rehder, C. F. 1986. "Use of preheated air in primitive furnaces: Comment on views of Avery and Schmidt." *Journal of Field Archaeology* 13:351–354.

Schmidt, P. R., and D. H. Avery. 1978. "Complex iron smelting and prehistoric culture in Tanzania." *Science* 201:1085–1089.

Stanley, G. H. 1931. "Some products of native iron smelting." *South African Journal of Science* 28:131–134.

Wener, D. B., and N. J. van der Merwe. 1986. "Mining for the lowest grade ore: Traditional iron production in Malawi." *Geoarchaeology* 1.

Wyckaert, R. P. 1914. "Forgerons paiens et forgerons chretien au Tanganyika" *Anthropos* 9:371–380. Annotated English translation available from D. H. Avery.

25 Characteristics of Casting Revealed by the Study of Ancient Chinese Bronzes

Pieter Meyers

Among the surviving materials from Bronze Age China, bronze ritual vessels provide an extensive and fascinating record of the artistic and technological accomplishments during the Shang and Zhou dynasties. Careful examination of these vessels has contributed significantly to the understanding of the society that produced them. The most useful information is obtained, of course, from archaeological excavations, where a study of the bronzes and associated materials has allowed vessels to be defined as to purpose, date, context, provenance, etc. Close observation of these and other unprovenanced vessels has permitted investigators to deduce various developments in metal technology and to reconstruct many of the procedures employed in the ancient bronze foundries.

Ideally the study of ceremonial vessels should be a combined effort by archaeologist, art historian, and physical scientist, because it is only by combining information from various different specialities that these objects can properly be analyzed and placed in their historical and cultural context.

The Piece-Mold Casting Process

The development of methods of manufacture of bronze objects in ancient China is unique as compared to that of other civilizations. The earliest known vessels, such as those excavated at Erlitou, are already advanced products of a complicated piece-mold casting procedure. They appear to have mostly been made of a ternary alloy of copper, lead, and tin. Although there have been occasional finds of copper and copper-tin alloy materials predating the Erlitou period, they do not as yet provide sufficient evidence to make reliable conclusions on likely metallurgical developments, such as copper and copper-tin (bronze) phases and the extent, if any, of sheet metal production. Considerably more archaeological work is needed to be able to reconstruct the metallurgical developments in the third and early second millennia B.C.

Fortunately a large number of bronzes and various associated materials (for example, mold fragments) from the Shang and Zhou dynasties have been available to study bronze technology in China during the second half of the second millennium and the first millennium B.C. Thanks primarily to the work of Karlbeck, Shi Zhangru, Barnard, Gettens, and others, it has become quite clear that the piece-mold casting method was the only technique used for the manufacture of bronze vessels (Karlbeck 1935; Shi 1955; Barnard 1961; Barnard and Tamotsu 1975; Gettens 1969). There is no evidence of shaping by hammering. Based on examinations of recently excavated bronzes, the lost-wax casting method was not used until the late Spring and Autumn Period.

The piece-mold or section-mold casting technique used for Chinese bronzes usually starts with a model made of clay. The model is identical in shape with the intended final product or any part thereof. Decoration may be added to the model. After the clay has dried or has been fired, impressions are made from the mold using a suitable clay. For obvious practical reasons these impressions are made in sections. After withdrawal from the model, these mold sections are dried and possibly fired. Decoration may be added to the mold sections, which carry the "inverted" or "negative" image of the model. The mold sections are then incorporated into the casting assembly. A ceramic core is also included. This core may be specially prepared or, more likely, consists of the model, reduced in size by removing 1–2 mm from the entire surface. The hollow space between the core and the assembled mold is defined through strategically placed bronze spacers.

The molten bronze is poured into the casting assembly and flows into the areas between the core and the assembled mold. After cooling the casting assembly is broken, the bronze is cleaned, and the surface of the bronze is abraded and polished to remove the casting skin and most of the surface irregularities.

Although there is general agreement in the literature on the use of the piece-mold casting technique, uncertainties exist on specific details of this method. For example, it is not certain that a model was consistently used. In theory, most objects could have been made without a model. Furthermore, the published literature does not clearly describe how the decoration was executed: Was it applied to the model or to the mold sections or partly to the model and completed in the mold? The methods used to repeat design elements seen especially in Eastern Zhou bronzes, need more extensive consideration. There are also questions on the positioning of the casting assembly. Were certain objects cast right side up or upside down or even sideways? Joining techniques, such as the use of solder, have not been adequately studied. Differences in casting techniques between various centers of bronze production cannot yet be recognized, and a detailed chronological development in casting technology has not been established.

Detailed technical studies of individual vessels, especially of objects from controlled excavations, may provide answers to these and other questions. Such information can then be used to identify developments in bronze casting technology, such as chronological changes; technological differences may allow the identification of bronze casting centers.

Little is known about the production and distribution of the copper, tin, and lead used in the manufacture of bronzes. Except for the copper mines at Tonglushan, the location of the ore sources are for the most part not known (Huangshi Museum 1980). Smelting and purification techniques have not yet been studied, and provenance studies based on the elemental compositions of bronzes have not been successful.

A comprehensive technical study of more than 400 bronzes from the Arthur M. Sackler Collections is underway at the Metropolitan Museum of Art and Brookhaven National Laboratory.[1] So far, 120 bronzes from the Shang and early Western Zhou dynasties have been examined.[2] Each of the vessels has been carefully studied visually and microscopically. Corrosion products and other mineral components were identified by X-ray powder diffraction. Metallographic studies were carried out on available metal cross sections. Important information on casting techniques and on ancient and modern repairs were deduced from X-ray radiographs. Elemental analyses using atomic absorption spectrometry and neutron activation analysis were performed on metal and ceramic core samples. Lead isotope ratio determinations are being processed by mass spectrometry at the National Bureau of Standards.

The immediate aim of this study is to provide accurate descriptions for each of the vessels, including details of the casting technique, condition, ancient and modern repairs, and elemental composition. It was also anticipated that this project would contribute information on certain aspects of the technology related to the ancient Chinese bronze caster, such as the identification and location of metal ore sources in China, smelting, and purification and alloying procedures and developments in the piece-mold casting techniques.

Some of the findings obtained so far are reported in what follows.

The Use of Decorated Models

It has been possible by careful observation of characteristic details in the decoration to conclude that with a few exceptions all bronzes examined would have required a model in their manufacture and that for most bronzes all decoration had been applied to the model.[3] The only exception is the earliest group of bronzes with thread relief design, defined as Style I by Loehr, and a few later vessels (Loehr 1953).

The evidence for this conclusion is based on the characteristic differences in the design as it occurs on the bronze: whether this design was executed in the mold sections ("negative" design) or on the model ("positive" design). If the decoration is added to the mold sections, incised patterns in the mold will result in raised patterns on the bronze. However, if the decoration is added to the model, an incised pattern on the model will appear as an incised pattern on the bronze; in mold impressions made from the model this pattern will, of course, be in relief (raised). If it were possible to compare the designs on the model, the mold, and the object, one might be able to determine at what stage the design was applied, based on such characteristics as line width, line depth, undercutting, and sharp angular protrusions that result when two incised lines meet. Of course, mold and model are not available to be compared with the bronze vessel, but this comparison can be closely approximated. It can be assumed that any design on the model would be duplicated almost exactly on the bronze (except for alterations or additions in the mold and changes from polishing and burnishing of the bronze surface). For all practical purposes, characteristic features of the design on the model and on the bronze vessels are identical. An accurate impression in a plastic material made from the design of a bronze vessel will then closely approximate the design in the original mold. Examination of such impressions allows the study of characteristic features of the design as they would have appeared in the mold. As will be demonstrated, careful examination of the design on a bronze and of its negative image in impressions taken from it will allow the determination for most objects of whether the decorating was executed on the model, on the mold, or on both.

This idea can be demonstrated using the *jia* shown in figure 25.1 as an example. Figure 25.2 presents a detail of this vessel, and figure 25.3 is an impression of this area made in Polyform, a flexible synthetic product often used to make seal impressions. The design on this vessel was clearly achieved by incising lines, but were they drawn on the model or on the mold? The most convincing evidence for the conclusion that this design was executed on the model (positive, as in figure 25.2) is the presence of angular protrusions formed when two incised lines meet. They often occur in the corners of the swirls in the *leiwen* designs (some are indicated in figure 25.2 by arrows). These irregularly shaped segments are logically formed when the design is executed by drawing lines on the model, but if the corresponding areas are examined in the "negative" design, shown in figure 25.3, it can be seen that extensive modeling and carving would be required in addition to drawing lines to achieve the same decoration in the mold. Other characteristic features, such as even line width, confirm the conclusion that the entire decoration for this vessel was executed on the model.

Another example is the *gu*, shown in figure 25.4. A detail of the decoration on the lower part of this vessel is shown in figure 25.5, and an impression from it is shown in figure 25.6. Although a number of typical angular protrusions can be observed in figure 25.5, it is more the uniformity of the line width that indicates the use of a decorative model. The argument becomes especially convincing if it is realized that the entire design can be completed on the model solely by drawing lines (figure 25.5), although considerably more complicated techniques would be needed to execute the decoration in the mold (figure 25.6). It appears well justified to conclude that the manufacture of the *gu* included a fully decorated model.

Figure 25.1 *Jia*, bronze, twelfth century B.C., Arthur M. Sackler Collections, New York, V 325.

Figure 25.2 Detail of design on *jia*, V 325.

Figure 25.3 Impressions made of area shown in figure 25.2. Comparison of the decoration seen in figure 25.2, which is comparable to the decoration that would have been on the model, with that shown here, representative of the decoration on the mold, leads to the conclusion that all decoration was applied to the model. This conclusion is predominantly based on the presence of angular protrusions, such as those indicated by arrows (figure 25.2), that could have been formed only when decorating had taken place on the model.

Figure 25.4 *Gu*, bronze, thirteenth century B.C., Arthur M. Sackler Collections, New York, V 346.

Figure 25.5 Detail of design on *gu*, V 346.

An interesting complication arises because there is another *gu* in the Arthur M. Sackler Collections that is virtually identical in shape, weight, and decoration with the *gu* shown in figure 25.4. There are, however, clearly observable differences in the details of the design. The manufacture of the second *gu* must have also included a decorated model, but because of the differences in design details, it could not have been the same one as used for the first *gu*. It is not likely that two models, identical in shape and size, would have been made independently. It appears reasonable to assume that two undecorated models were produced from a primary mold, consisting of various mold sections. These mold sections, in turn, were probably made using a primary model.[5]

There are various examples of two or more bronze vessels of Shang or Western Zhou dates with identical shape, size, and decoration.[6] Their method of manufacture may well have included the following steps:

1. Manufacture of a primary undecorated model
2. Production of a primary mold from the primary model

Figure 25.6 Impression made of area shown in figure 25.5. Comparison of the "positive" (figure 25.5) and the "negative" (here) images, representing the designs on the model and mold sections, respectively, indicates that the entire design had been executed on the model.

Figure 25.7 *Lei*, bronze, c. fifteenth century B.C., Arthur M. Sackler Collections, New York, V 59. The design appearing in relief on this vessel, was incised in the mold.

3. Production of multiple undecorated models from primary mold, followed by adding decoration to each model
4. Production of molds from each of the decorated models
5. Casting of bronze vessels from multiple casting assemblies

This procedure appears to be the first indication of a mass production process.[7] Although the piece-mold casting process lends itself very well for mass production and even though certain aspects of mass production gradually became part of the bronze casting industry, the character of ceremonial vessels retained its individuality and never suffered from the multiplicity of mass production.

Careful examination of Shang and Western Zhou bronzes has indicated that as a rule fully decorated models were employed in their production. There are some exceptions. Vessels with Style I thread relief decoration, such as the *lei* shown in figure 25.7, are the oldest vessels with decoration. There should be little argument that their design was incised in the mold sections. There is in fact no evidence that a model was used in the manufacture of Style I vessels. However, because of the complexity of some vessel shapes, the use of a model would have considerably simplified the manufacturing process.

Starting with Style II vessels, such as the *jia* shown in figure 25.8, the use of a decorated model became standard practice.[8] Any decorating on the mold was largely restricted to small adjustments, repairs, and relatively minor additions to the decoration, such as circles and possibly extremities such as the *taotie* eyes. There may be some Eastern Zhou bronzes with a design produced in the mold, but otherwise the only other exceptions known to me are a group of bronzes exemplified by the *pou* shown in figure 25.9. This vessel has three different registers of decoration. For the reasons described before, it can be shown that the designs on the shoulder and on the foot were applied on the model. The design covering the lower two-thirds of the bronze vessel, as shown in figure 25.10, has different characteristics. A number of vertical lines can be seen in the decorated band; their positions are indicated by arrows in figure 25.10. The first line on the left is a mold line; the others are lines in relief interrupted only by the relief elements of the design itself. There are twenty-one vertical lines, three of which are mold lines. An explanation for the presence of these lines can be deduced from the impression of part of this area, shown in figure 25.11. Lines were drawn in the mold sections as guidelines to assist the artisan in the execution of the geometric and swirl patterns. All characteristics of the design are consis-

Figure 25.8 *Jia,* bronze, fifteenth to fourteenth century B.C., Arthur M. Sackler Collections, New York, V 108. The manufacture of this vessel included the use of a decorated model.

Figure 25.9 *Pou,* bronze, thirteenth century B.C., Arthur M. Sackler Collections, New York, V 380.

Figure 25.10 Detail of decoration band on lower part of *pou,* V 380. Of particular significance are a number of vertical lines, indicated by arrows.

Figure 25.11 Impressions made of area shown in figure 25.10. A comparison of the "positive" and "negative" images of this design indicates that the design was executed on the mold. The vertical lines are guide lines drawn into the mold before the design was executed.

tent with the conclusion that this decoration band was executed on the mold section.[9]

Among the objects examined, two vessels show additional and accidental evidence of decorated models. A *jia* of pre-Anyang date, illustrated in figure 25.12 exhibits on its interior wall a faint decoration pattern (figure 25.13). The presence of this design can only be explained satisfactorily as resulting from a decorated core. This core consisted of the upper part of a decorated model from which a thin surface layer not deep enough to remove the decoration was scraped off. Not only does this object confirm the existence of a decorated model, but it also proves the use of a scraped-down model as a casting core. It is generally accepted that this practice took place, even though there had been, until now, no direct evidence. It is remarkable that a model for a different vessel was used, because the bronze itself is not decorated and the decorative pattern on the interior is not seen on the exterior surface.

The *jue* illustrated in figure 25.14 appears to have significant areas of its decoration missing (figure 25.15).[10] These flat and smooth areas of missing decoration are on the same level as the top surface of the decorated areas. This observation is significant because, if the decoration was simply worn away or if the molten metal during casting had not flowed in the decorative pattern of the mold, the areas of missing decoration would have been at a level below that of the decorated areas. Therefore these effects did not contribute to the observed phenomenon. As it is likely that the cast metal surface reflects accurately what was on the mold sections, the missing decoration must be the result of inaccurate registration or incomplete transfer of the design when the mold impressions were made from the decorated model. Why the bronze caster would have allowed the use of an imperfect mold is not immediately obvious.

The consistent use of fully decorated models implies that the Shang artisans at an early stage had mastered the skill of making mold impressions from models, without losing even the finest details. This feat is even more remarkable if it is realized that usually only three or four and occasionally six vertical mold divisions were used. This could have been accomplished only if the mold material was sufficiently flexible to allow it to be withdrawn, even from models with high relief decoration such as seen in Loehr's Style V vessels. It appears that nearly all decoration on bronzes was conceived and executed in the "positive" and that decorative patterns already during the Shang Period could easily be transferred from model to mold to bronze without much noticeable loss of detail. This understanding may be of considerable value in the understanding of the development of bronze casting technology and of decorative patterns.

Figure 25.12 *Jia*, bronze, fourteenth century B.C., Arthur M. Sackler Collections, New York, V 319.

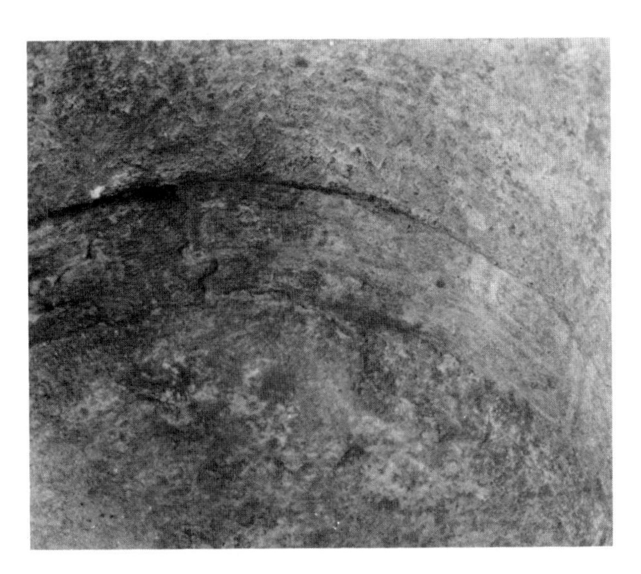

Figure 25.13 Detail of the interior wall of *jia*, V 319. The faintly visible decoration indicates that the core used in the casting process of this vessel consisted of the upper part of a decorated model, its surface scraped down, but not far enough to remove all decoration.

Figure 25.14 *Jue*, bronze, twelfth to eleventh century B.C., Arthur M. Sackler Collections, New York, V 249.

Figure 25.15 Detail of decorative band on *jue*, V 249. The areas with missing decoration are believed to result from incomplete transfer of the design when mold impressions were taken from the decorated model.

An Iron Spacer and Copper Cores

Nearly every bronze vessel examined in this study contained one or more bronze spacers. The function of these spacers is to define and maintain before and during casting a constant space between the core and the mold sections and between the core and the investment. Spacers ensure an even wall thickness of the vessel. Invariably spacers are made of bronze. They sometimes carry part of a decoration, indicating that at least some are made from "scrap" bronze.

It was rather surprising to discover that one of the spacers of the *pou* shown in figure 25.16 proved to be made of iron. It is located in the undecorated area between the upper and middle registers (figure 25.17). Its dimensions are 1.5 cm × 1.1 cm. Because there is no evidence of a repair and because the iron is extensively corroded, the iron spacer could be an original part of the vessel. Energy dispersive X-ray fluorescence spectrometry failed to detect nickel. Therefore the iron appears to be manufactured iron, not meteoritic. This may be a significant observation, as this vessel predates the earliest recorded occurrence of terrestrial iron in China by approximately 500 years.

In the casting process of bronzes, difficulties were often encountered with the legs, expecially of *ding* and *fang ding* type vessels. Ancient repairs and recast legs can often be observed on bronze vessels. Common casting problems were augmented because the vessels were usually cast upside down. This resulted in such problems as casting flaws resulting from trapped air in the legs and premature solidification of the molten alloy.

Various methods were employed by the Shang and Zhou artisans in the manufacture of vessels with legs. They include (1) casting of legs integrally with the main part of the vessel with solid legs, with legs containing a ceramic core extending from the main core, or with legs with a separate ceramic core; (2) precast legs incorporated in the casting assembly; and (3) casting of legs onto a precast vessel.

The present study has revealed that another previously unknown method can be added to this list: Solid copper cores for the legs of *ding* vessels were incorporated in the casting assembly; vessels and legs were cast with one pour of the molten alloy.

Figure 25.18 shows a *ding*, with an X-ray radiograph illustrated in figure 25.19. The legs appear to be solid, but close examination of the X-ray radiograph reveals the existence of a metal core that has not fused with the surrounding metal (see arrows in figure 25.20). Similarly cast legs were observed in two other roughly contemporary *ding* vessels.[11] Elemental analyses of drilled samples from the underside of one of the legs shows that the core consists of unalloyed (pure) copper. This observation may be of some significance because not only is this the first reported occurrence of

Figure 25.16 *Pou,* bronze, twelfth to eleventh century B.C., Arthur M. Sackler Collections, New York, V 120.

Figure 25.18 *Ding,* bronze, twelfth century B.C., Arthur M. Sackler Collections, New York, V 89.

Figure 25.17 Detail of *pou,* V 120, showing iron spacer.

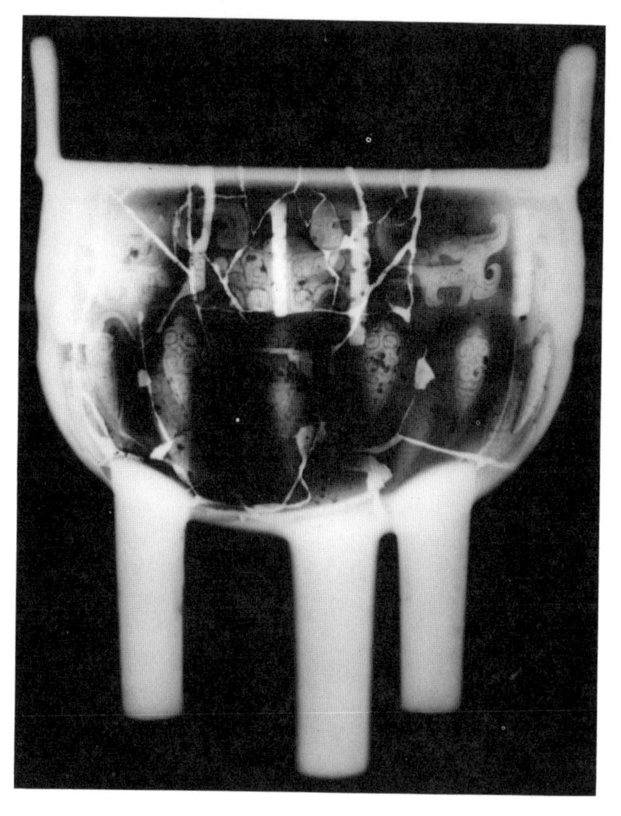

Figure 25.19 X-ray radiograph of *ding* V 89, showing extensive solder repairs and seemingly solid legs.

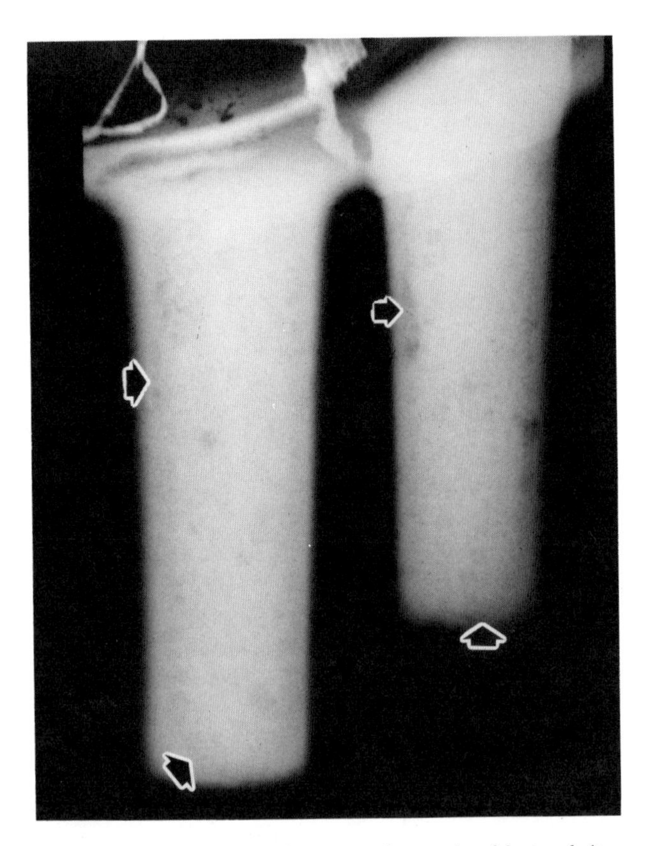

Figure 25.20 Detail of X-ray radiograph of legs of *ding* V 89. The legs are cast around a solid metal core, identified as unalloyed copper.

unalloyed copper in Shang ceremonial vessels but it also may indicate an advanced knowledge of melting points of copper and bronze alloys. The use of a copper core with a melting point considerably higher than the one of the alloy cast around it would have less chance to cause casting failures than the use of a metal alloy core. The metal alloy core would have a melting point sufficiently low that contact with the molten bronze cast around it could cause partial melting of the core, resulting in probable casting failures. More evidence of casting methods of vessel legs is needed to substantiate the speculation on the metallurgical knowledge of the Shang bronze caster.

Elemental Analysis and Lead Isotope Ratios Analysis

Elemental analyses have been carried out so far on about 160 metal samples from 120 vessels. The purposes of this analytical program are to characterize individual bronzes, to gain information on the alloy compositions used, and to investigate the possibility of provenance determinations.

The details of the analytical methods used, the analytical data, and the interpretations of the data are presented elsewhere, together with a comprehensive description of art historical, stylistic and technical properties of each of the vessels examined (Bagley 1987). However, to stimulate interest in analytical programs of Chinese bronzes, I present here a summary of the work currently completed together with the elemental compositions of vessels illustrated in this chapter.

Metal alloy samples were obtained by drilling, mostly using a $\frac{1}{8}$-inch tungsten carbide drill bit rotated in a small hand drill. Corrosion and surface drillings were discarded. Usually, about 50–60 mg of metal was collected. This amount represents a compromise between the need for a representative sample and the wish to minimize disfiguration of the object.

After careful evaluation of available analytical methods, we decided to employ two techniques: atomic absorption spectrometry to determine the concentration of the major components (copper, tin, lead, and iron) and instrumental neutron activation analysis for minor and trace elements (iron, cobalt, nickel, zinc, arsenic, silver, antimony, mercury, and gold).

Elemental analyses were carried out at Brookhaven National Laboratory. Elemental compositions of the objects illustrated are given in table 25.1.

In order to be able to interpret the analytical data, all analyzed vessels were tentatively divided into four groups. This grouping is based on stylistic properties. Because no objects were from known controlled excavations, exact provenances could not be established and any grouping had to be considered preliminary. Four groups were considered: (1) objects predating the Anyang Period, (2) objects probably made in Anyang, (3) Shang bronzes probably not made in Anyang ("provincial"), and (4) vessels from the Western Zhou Period.

Copper, tin, and lead were found to be major components in all the objects analyzed, indicating the consistent use of a ternary alloy. For the Anyang group, the tin and lead contents average 13.3% and 4.0%, respectively. The variations in tin and lead contents appear to follow standard distribution laws, and no correlation is evident between the two elements. This observation is consistent with the theory that lead and tin were independent deliberate additions. Provincial vessels show on average lower tin and higher lead contents with a negative correlation between lead and tin. Apparently lead was often substituted for tin in bronze manufacturing centers outside Anyang.

In attempting to observe differences between the minor and trace element concentrations of objects in each of the four groups, I found that simple graphical representations, such as correlation diagrams, were not fully satisfactory. Better results were obtained using computer-aided multivariate statistical methods. With the objects believed to have been produced in Anyang as a reference group, it could be demonstrated that significant differences occur between the elemental compositions of the Anyang group and each of the other three groups. It was also observed that a number

Table 25.1 Elemental composition of illustrated bronzes

Vessel type	Inventory number	Sample location	Cu (%)	Sn (%)	Pb (%)	Fe (%)	Zn (ppm)	Au (ppm)	Ag (ppm)	Hg (ppm)	As (ppm)	Sb (ppm)	Co (ppm)	Ni (ppm)	Total (%)
Jia	V 325	inside of leg	82.7	11	3.8	≤0.14	19	4.7	650	6.3	280	40	6.6	69	98
Gu	V 346	inside rim of foot	78.4	15	5.9	0.026	56	21	600	20	610	43	32	160	99
Gu	V 345	bottom edge foot	77.6	17	4.1	≤0.062	29	5.1	380	5.8	620	34	7.3	45	99
Lei	V 59	bottom edge foot	85.1	8.6	5.3	<0.05	<28	9.8	2300	10	1700	490	8.8	420	99
Jia	V 108	underside bottom	72.8	6.3	19	<0.04	<17	11	990	4.4	2200	220	38	470	99
Pou	V 380	underside bottom	72.6	16	11	0.023	<14	0.28	130	<1	400	84	0.88	<13	99
Jia	V 319	underside leg	95.0	2.2	2.7	<0.04	<37	9.4	760	<2	2400	550	46	530	100
Jue	V 249	underside leg	77.2	14	8.0	0.33	<68	36	1400	7.6	4600	1400	62	270	100
Pou	V 120	bottom edge foot	70.0	8.0	22	0.21	150	60	970	7.6	6200	250	180	<38	101
Ding	V 89	underside leg	83.8	9.5	3.3	0.089	54	11	1200	6.6	860	120	28	92	97
		underside leg, core	98.8	0.31	0.95	0.044	120	2.8	330	13	540	38	8.7	62	100

Sample size: 50–60 mg, obtained by drilling.
Method of analysis: For major components (copper, tin, lead, and iron), atomic absorption spectrometry on aqueous solutions containing 10 mg bronze. For minor and trace elements, instrumental neutron activation analysis on 20 mg bronze.

of Anyang vessels, especially several believed to have been produced during the early Anyang Period, differed considerably in elemental composition from the average or typical Anyang composition. These vessels were perhaps not manufactured in Anyang. It is also possible that they were made in Anyang but that the metals used in their manufacture were obtained from different ore sources.

Because of the uncertainties in date and provenance of the objects analyzed, it is not possible to describe accurately the potential and limitation of provenance studies based on elemental compositions of the metal alloys. However, these initial results are promising. They appear to justify a systematic analytical study on provenanced vessels.[12] Of course, such a project can be undertaken only in China. I hope that our Chinese colleagues will seriously consider the undertaking of such a study.

Trace elemental compositions of ceramic materials, as determined by neutron activation analysis, have been informative in provenance studies. Encouraged by these successes, I carried out a pilot study on samples of core material available from approximately sixty bronzes. The elemental compositions of these samples were determined using standard neutron activation procedures. The aim was to investigate whether bronzes with common stylistic properties (common provenance) could be grouped together and separated from unrelated objects. The results indicate a considerable uniformity in elemental compositions of all samples with no detectable differences in core material compositions of unrelated objects. Apparently this method is not suitable for provenance determinations. This observation is easily understood if it is realized that the core material in Chinese bronzes is predominantly yellow earth, an eolian deposit that covers most of northern China (Gettens 1969, pp. 107–114). Its composition may show little variation. Except for vessels for which yellow earth was either not available (for example, southern vessels) or consistently not used, trace element analysis of the core material may not provide further information on provenance.

Lead isotope analysis is a technique of considerable potential for provenance studies (Brill et al. 1979). This method is based on the fact that a lead ore deposit can be characterized by ratios of the lead isotopes ^{204}Pb, ^{206}Pb, ^{207}Pb, and ^{208}Pb. For a single deposit these ratios are constant. From one ore deposit to another the ratios are often different, depending on the geological history of each deposit. Because lead isotope ratios do not change during smelting and casting, it is expected that the lead contained in bronze vessels will have the same lead isotope ratios as the lead ores from which the lead was produced.

Lead isotope ratios can be accurately determined by mass spectrometry of small metal samples. Vessels containing lead from the same source will have identical lead isotope ratios, whereas vessels produced in different manufacturing centers will often exhibit different lead isotope ratios.

A program for lead isotope ratios analysis of samples from Chinese bronzes is presently being carried out by the Freer Gallery of Art and the National Bureau of Standards. Metal samples of all vessels examined in the Arthur M. Sackler Collections have been submitted for lead isotope ratios analysis, but results are not yet available.

An accurate evaluation of this technique may soon be possible. However, to employ this technique to its full potential, we need detailed information on the geology of lead deposits in China and lead isotope data on documented ore source materials and excavated bronzes.

Notes

I am most grateful to Arthur M. Sackler for his permission and encouragement to examine the Chinese bronzes from his collections. I would like to thank Lois Katz for her advice and suggestions. Without her enthusiastic support this project could not have been undertaken. I am greatly indebted to Lore L. Holmes for her assistance in the examination and elemental analysis of the bronze vessels. I am also grateful to Robert W. Bagley. The many long discussions with him have contributed greatly to my understanding of Chinese bronzes. All photography, except X-ray radiography, was performed by Otto E. Nelson. Thanks to his skill I am able to present the visual evidence needed in this contribution. Parts of this research were performed at Brookhaven National Laboratory under contract to the United States Department of Energy, supported by its Division of Basic Energy Sciences.

1. This project is undertaken in conjunction with a more general study of the Chinese bronzes in this collection. A multi-volume catalogue of these objects is currently in preparation. Bagley (1987) includes full descriptions of the following objects discussed in this chapter. The objects are identified by inventory number and catalogue number, respectively: *jia*, V 325, cat. no. 6; *gu*, V 346, cat. no. 35; *gu*, V 345, cat. no. 34; *lei*, V 59, cat. no. 1; *jia*, V 108, cat. no. 2; *pou*, V 380, cat. no. 53; *jia*, V 319, cat. no. 3; *jue*, V 249, cat. no. 20; *pou*, V 120, cat. no. 57; *ding*, V 89, cat. no. 82.

2. Preliminary results of this study, including some of those described in this contribution have been reported in Meyers and Holmes (1982).

3. Statements in the published literature on whether decorations were applied to the model or to the mold are often ambiguous and unsubstantiated. See, for example, Gettens (1969, pp. 23–31, 57–59), Bagley (1980, pp. 70–73, 98–102, 182), Smith (1972), and Wan Chiao-pao and Li Chi (1964, pp. 121, 125, 126).

4. *Gu*, bronze, thirteenth century B.C., Arthur M. Sackler Collections, New York, V 345, not illustrated.

5. There is no evidence for the use of a primary model other than that it would be consistent with the general practice during the Shang and Western Zhou periods to start the manufacturing process of a bronze with a positive form. Incidentally it would be quite possible in the manufacturing process of multiple objects for the undecorated primary model to receive a decoration and be used as one of the secondary models.

6. For example, the tomb of Fu Hao at Anyang contained several groups of nearly identical *gu* vessels; see Institute of Archaeology (1980).

7. This procedure is not dissimilar to the mass production system of iron implements, approximately one thousand years later, as described by Li Jinghua.

8. It may be of interest to note that, in the manufacture of a decorated model for the *jia*, shown in figure 25.8, a band of clay was added to the model into which the design was then executed.

9. There are other vessels that appear to have parts of their decoration executed in the mold, for example, two vessels in the Freer Gallery of Art (a *pou*, accession no. 09.334, and a *hu*, accession no. 49.5) [see Pope et al. (1967, pp. 20–24, 40–44)] and a *fang ding* in the Fogg Art Museum, accession no. 1944.57.19 [see Chen Mengjia (1977, p. A69)].

10. The phenomenon of missing decoration is not unique. During my travels in China, following the Symposium on Ancient Metallurgy in China (1981), I noticed two more examples, both *jue* vessels, one in the Museum of Luoyang, the other in the Historical Museum, Beijing, with large plain areas in the decoration band.

11. *Ding*, bronze, Arthur M. Sackler Collections, New York, V 125, and *ding*, bronze, Arthur M. Sackler Collections, New York, V 337. The catalogue numbers (see note 4) are 83 and 87, respectively. Objects not illustrated.

12. Previous attempts to use metal alloy compositions for provenance studies have not been successful. See, for example, Gettens (1969, p. 46) and Barnard (1981).

References

Bagley, R. W. 1980. *The Great Bronze Age of China: An Exhibition from the People's Republic of China*, Wen Fong, ed. New York: Metropolitan Museum of Art.

Bagley, R. W. 1987. *Shang Ritual Bronzes in the Arthur M. Sackler Collections*. Volume 1 of *Ancient Chinese Bronzes in the Arthur M. Sackler Collections*. Cambridge, Mass.: Harvard University Press.

Barnard, N. 1961. *Bronze Casting and Bronze Alloys in Ancient China*. Monumenta Serica Monograph 14. Canberra, Australia: Australian National University and Monumenta Serica.

Barnard, N. 1981. "Aspects of the application of science in the study of Chinese archaeology." Unpublished.

Barnard, N., and Sato Tamotsu. 1975. *Metallurgical Remains of Ancient China.* Tokyo: Nichiosha.

Brill, R. H., K. Yamasaki, I. L. Barnes, K. J. R. Rosman, and M. Diaz. 1979. "Lead isotopes in some Japanese and Chinese glasses." *Ars Orientalis* 11:88–109.

Chen Mengjia. 1977. *A Corpus of Chinese Bronzes in American Collections.* Tokyo: Kyuko Shoin. (In Japanese.)

Gettens, R. J. 1969. *The Freer Chinese Bronzes. Vol. 2. Technical Studies.* Washington, D.C.: Smithsonian Institution.

Huangshi Museum, Hubei. 1980. *Tonglushan (Mt. Verdigris Daye): A Pearl among Ancient Mines.* Beijing: Chinese Society of Metals, Publications Committee, and Archaeometallurgy Group of Beijing University of Iron and Steel Technology. (In Chinese.)

Institute of Archaeology. 1980. *The Shang Tomb of Fu Hao.* Beijing: Wenwu Chubanshe. (In Chinese.)

Karlbeck, O. 1935. "Anyang moulds." *Bulletin of the Museum of Far Eastern Antiquities* 7:39–60.

Loehr, M. 1953. "The bronze styles of the Anyang period (1300–1028 B.C.)." *Archives of the Chinese Art Society of America* 7:42–53.

Meyers, P., and L. L. Holmes. 1982. "Technical studies of ancient Chinese bronzes: Some observations." *The Great Bronze Age of China: A Symposium*, G. Kuwayama, ed. (Los Angeles, Calif.: LA County Museum of Art), 124–136.

Pope, J. A., R. J. Gettens, J. Cahill, and N. Barnard. 1967. *The Freer Chinese Bronzes. Vol. 1. Catalogue.* Washington, D.C.: Smithsonian Institution.

Shi Zhangru. 1955. "Open discussion of Shang Dynasty bronze and jade." *Bulletin of the Academia Sinica* 26:95–129. (In Chinese.)

Smith, C. S. 1972. "Metallurgical footnotes to the history of art." *Proceedings of the American Philosophical Society* 116(2): 107–113.

Wan Chiao-pao and Li Chi. 1964. "Studies of the bronze Ku-beaker." *Archaeologica Sinica*, n.s., 1:119–130.

26
The Technical Examination, Lead Isotope Determination, and Elemental Analysis of Some Shang and Zhou Dynasty Bronze Vessels

I. L. Barnes, W. T. Chase,
L. L. Holmes, E. C. Joel,
P. Meyers, and
E. V. Sayre

The Arthur M. Sackler Collections contain approximately 400 early Chinese ritual bronze vessels. Most of these vessels (106 attributed to the Shang dynasty, 127 to the Western Zhou, and 94 to the Eastern Zhou) have been given a thorough technical examination. The examination has included X-ray radiography, viewing under ultraviolet light, overall visual examination, and microscopic examination of details. In some instances impressions were made of surface details in order to study them in reverse relief. The metal of most of these vessels also has been analyzed for lead isotope ratios by means of mass spectrometry, for their major component composition by atomic absorption spectrometry, and for some minor and trace elemental concentrations by instrumental neutron activation analysis. The aim of the project was to obtain detailed characteristics for each object, such as composition and structural details from which the method of fabrication, including casting and joining techniques, could be inferred, and to evaluate these characteristics in conjunction with art historical evidence to establish or confirm details of Chinese bronze casting technology and to help resolve questions of attribution. Figures 26.1 through 26.3 show examples in the Arthur M. Sackler Collections of vessels from these three periods.

The examination and elemental analysis of these objects was initiated by Pieter Meyers, then with the Metropolitan Museum of Art, with the assistance of Lore L. Holmes, Lambertus van Zelst, and Louis Sangermano. When Meyers left the Metropolitan Museum for the Los Angeles County Museum of Art, the examination and analysis of most of the Shang Dynasty vessels had been completed (Meyers, this volume). Edward V. Sayre then assumed the responsibility of supervising the project. The examination and elemental analysis of the remaining objects was subsequently completed by Holmes and Sayre, with Holmes actually performing the analyses. All of the elemental analyses were carried out in the Chemistry Department of Brookhaven National Laboratory with the support of the US Department of Energy, the Metropolitan Museum of Art, and the Arthur M. Sackler Foundation. Before the inception of the Sackler bronze project, W. T. Chase of the Freer Gallery of Art and I. Lynus Barnes of the National Bureau of Standards were collaborating on a study of lead isotope ratios of some of the Freer bronzes and related material. They agreed to include the determination of lead isotope ratios on the Sackler bronzes as part of their project. The measurements were carried out at the National Bureau of Standards with support from the Smithsonian Scholarly Research Opportunities Fund and the Smithsonian Conservation Analytical Laboratory, the measurements themselves being performed by Emile C. Joel of the Conservation Analytical Laboratory.

Although the project was conceived as an indepen-

Figure 26.1 A *fang ding* attributed to the Shang Dynasty, twelfth to eleventh century B.C. (Sackler, V 18). The vessel conforms to lead isotope ratio group 7 and composition group C2 and is judged to have Anyang style.

Figure 26.3 A *hu* attributed to the Eastern Zhou Dynasty, late sixth to fifth century B.C. (Sackler, V 326, now in the Arthur M. Sackler Gallery, Smithsonian Institution, no. S 87.0011). The vessel conforms to lead isotope ratio group 3 and compositional group C3 and is judged to have metropolitan style.

Figure 26.2 A *guang* attributed to the Western Zhou Dynasty, eleventh to tenth century B.C. (Sackler, V 307, now in the Art Museum, Princeton University, no. 65.3). The vessel conforms to lead isotope ratio group 2 and compositional group C3.

dent study, concerned with the general characterization of early Chinese bronzes, it has been carried out with close collaboration in the preparation of three catalogs describing the Shang, Western Zhou, and Eastern Zhou vessels in the Sackler Collection (Bagley 1987; Rawson 1987; So 1987). The accumulated technical information has been submitted to these authors to be used in their respective catalogs at their discretion. Accordingly the catalogs will include comments on the structure and condition of the vessels and tables listing lead isotope ratios and elemental compositions. Thus the determinations of both isotope ratios and compositions for individual vessels, together with descriptions of the vessels, will be available when the catalogs are published. It is anticipated that a fourth volume devoted to the technical study of the vessels will be prepared.

The technical study is greatly strengthened by having the art historical expertise of Robert W. Bagley, Jessica Rawson, and Fenny F. So available. The value of the analyses and the technical observations will be significantly enhanced if the technical data can be interpreted within the stylistic and iconographic framework of these bronze vessels. Only when completely incorporated with the available art historical and archaeological information can the inferences from the technical study be fully assessed. Unfortunately the analytical work on these vessels was just recently completed and there has been far too little time for a proper joint consideration of the data with the art historians. The multivariate statistical analysis of the data presented here was derived from the numerical values themselves, essentially without regard to historical information. This is a reasonable first step in the evaluation of such data, but a definitive evaluation must include consideration of all the information available. Fortunately an encouraging degree of positive correlation between this initial group of specimens and the art historical information already available to us has emerged.

The Determination of Lead Isotope Ratios

The relative proportions of the stable isotopes of the metallic element lead tend to be different in different natural sources of lead and hence can serve to characterize individual lead sources. The reason for the differences is that three of the naturally occurring stable isotopes, ^{206}Pb, ^{207}Pb, and ^{208}Pb, are continuously being formed in nature through separate chain sequences of radioactive decay that start, respectively, with the radioisotopes ^{238}U, ^{235}U, and ^{232}Th. The fourth stable isotope of lead, ^{204}Pb, is no longer being formed on earth; the amounts encountered today were all formed in the primordial generation of matter. The differences in the ratios of lead isotopes from different sources are small, but they are reasonably consistent and can be measured with high precision using modern mass spectrometers.

Only 0.5 μg of lead is required for an isotope ratio determination. Because one does not expect fractionation of the lead isotopes to occur on corrosion, it would be possible to use lead from a corrosion layer on a bronze for the determination. However, one would still be concerned as to whether lead from outside of the vessel might have been deposited in the corrosion layer, thus affecting the isotope ratios. We were fortunate in being allowed to take drilled samples of pure metal (uncorroded) for analysis. The drilling for samples was done with a hard tungsten carbide drill, the samples being taken from inconspicuous and undecorated locations, usually from the underside of the vessel. The small clean drill holes could later be filled in such a manner as to render them nearly undetectable. The same drilling produced samples for isotope ratio determinations and for elemental analyses.

Each sample was prepared for analysis by first dissolving it in 10 ml of one part of concentrated nitric acid to one part of water. This solution was dried and the residue redissolved in 20 ml of 0.025 *N* perchloric acid solution. The lead was then separated out and purified by electrolytic deposition. The deposited lead was rinsed, dried, and weighed and then redissolved in 0.1 *N* nitric acid containing 5% hydrogen peroxide. This solution was dried and the residue redissolved in 0.1 *N* nitric acid to form as nearly as possible a solution containing 50 μg/ml of lead. All reagents, including water, were specially purified to remove the possibility of contamination from outside sources of lead.

A rhenium filament, onto which this solution was to be deposited, was first degassed by passing a 3-amp electric current through it for 15 min. Five microliters of a concentrated suspension of silica gel was deposited on the filament and dried under a heat lamp while a 1-amp current was passed through the filament for 5 min. Five microliters of the lead solution was then added to the silica gel and similarly dried. Then 5 μl of 0.75 *N* phosphoric acid were added to the sample, which was again dried under a heat lamp with a 1.5-amp current being passed through the filament for 10 min. Finally the filament was transferred to a bell jar filled with nitrogen and heated to a temperature of 950°C for 5 sec. It was then ready for loading into the mass spectrometer. In the mass spectrometer the filament was slowly heated to 1200°C, at which temperature the baseline measurements of backgrounds adjacent to the isotope peaks were made, followed by measurement of the $^{208}Pb/^{206}Pb$, $^{207}Pb/^{206}Pb$, and $^{204}Pb/^{206}Pb$ isotopic ratios. The runs were standardized by comparison to readings obtained with the National Bureau of Standards standard SRM 981.

Evaluation of the Lead Isotope Ratios

Some 80% of the vessels in the Sackler Collections were found to contain between 2% and 25% lead. This is a sufficiently high level to conclude that lead was probably an intentional addition to the metal alloy. Thus the lead isotope measurements would relate to the sources of lead used in the formulation of the vessels. For the remaining 20% a variety of materials must be considered as possible lead sources.

The overall lead isotope ratio data obtained on the Sackler vessels are remarkable in terms of their diversity and spread in magnitude. Figure 26.4 shows a conventional correlation plot relating the $^{208}Pb/^{206}Pb$ ratios to the $^{207}Pb/^{206}Pb$ ratios. One can see a large cluster of low ratio specimens ($^{208}Pb/^{206}Pb$ ratios in the range 1.9–2.0 and $^{207}Pb/^{206}Pb$ ratios in the range 0.73–0.77) and another large cluster of high ratio specimens ($^{208}Pb/^{206}Pb$ ratios in the range 2.1–2.2 and $^{207}Pb/^{206}Pb$ ratios in the range 0.85–0.90) with relatively few specimens lying beyond or between these ranges. One can argue that the specimens lying between the two main groups might be objects in which a mixture of a high ratio lead and a low ratio lead was used, resulting in a lead with intermediate characteristics. This would be a possibility, but considerations to be discussed later will demonstrate that this was probably not the case for many intermediate specimens.

It is interesting to compare figure 26.4 to a similar plot published by Barnes et al. (1987) in a paper dealing with early Chinese lead-containing glasses, shown in figure 26.5. They found the leads in these glasses to separate into high and low isotope ratio groups that to a good degree overlap in their ranges with those we have found in the leads from bronze vessels. In figure 26.5 the intermediate ovals indicate the location of isotope ratios that have characterized leads from European and Mediterranean sites. The ellipse marked "Egypt" corresponds to leads in early Egyptian glasses and glazes; the ellipse marked "L" indicates leads from the ancient mines at Laurion in Greece; group E contains leads from England, Italy, Turkey, and perhaps elsewhere, and group S contains leads from Spain, Wales, and Sardinia. Barnes et al. comment that other leads, for example, some from Mesopotamia, also fall into this central region. Their general conclusion is that the Chinese leads tend to lie, both high and low, beyond the ranges that have been encountered in most more Western leads. Our observations on lead from early Chinese bronze vessels tend to confirm this observation, as have data published by Ch'en et al. (1980) in China and by Mabuchi et al. (1982) in Japan.

When these data are divided according to the three periods to which the vessels are attributed, one observes some pronounced chronological changes. Figure 26.6 shows the distribution of points for the Shang Dynasty vessels, which spread over nearly the entire range of isotope ratios encountered. Clearly a great variety of leads was used in the Shang Period. The scatter plot for the Western Zhou Dynasty (figure 26.7) shows that the use of the low ratio leads was dying out. Relatively few vessels have low or intermediate lead isotope ratios. By the time of the Eastern Zhou Dynasty (figure 26.8) the occurrence of lead with low or intermediate isotope ratios within the Sackler vessels becomes almost negligible; only one vessel with a low ratio and two with intermediate ratios were encountered. Also the clustering in the high ratio range is quite tight. One might infer that during the Eastern Zhou lead from only a few related sources tended to be used. Clearly there is a good degree of correlation between the lead isotope results and chronology.

There is, at times, more to be learned from considering all three ratios determined in the laboratory than just the two ratios presented in the preceding figures. Accordingly a three-dimensional multivariate approach was used for the further analysis of these data. It was decided first to work with the ratios relative to ^{204}Pb because these three ratios are roughly of the same magnitude. By using them, one avoids certain computational problems that arise when one is trying to combine large- and small-valued variates. Next the entire body of data was considered as a core group population, with only the deletion of a few specimens

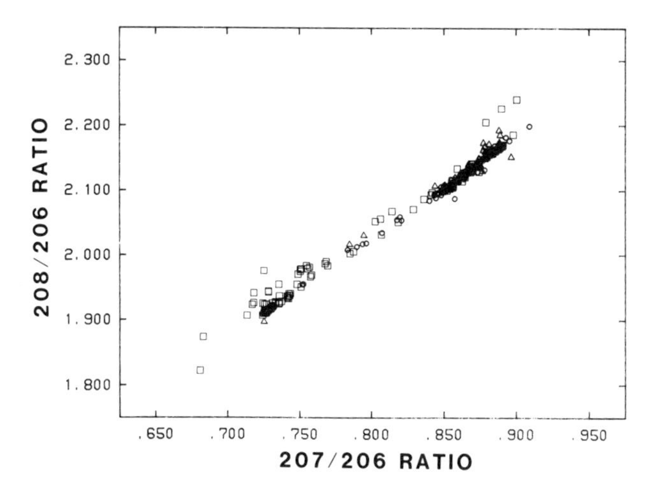

Figure 26.4 Lead isotope ratio distribution for the bronze vessels analyzed. □ indicates Shang vessels, o indicates Western Zhou vessels, and △ indicates Eastern Zhou vessels.

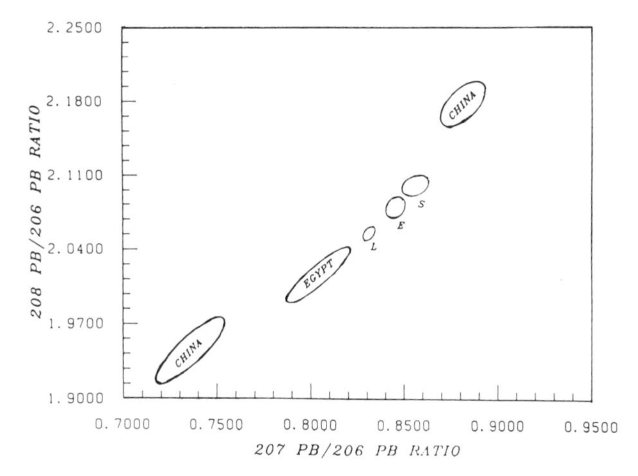

Figure 26.5 Comparison of lead isotope distribution in ancient Chinese glasses with that in European and Near Eastern leads. Reproduced from Barnes et al. (1987, fig. 1).

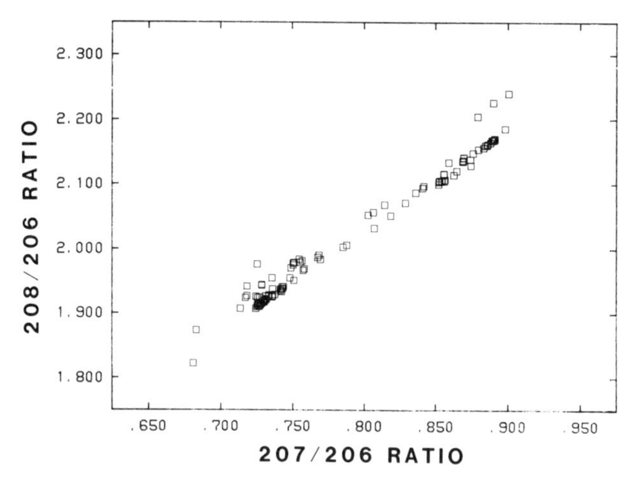

Figure 26.6 Lead isotope ratio distribution for Shang Dynasty vessels in the Sackler Collections.

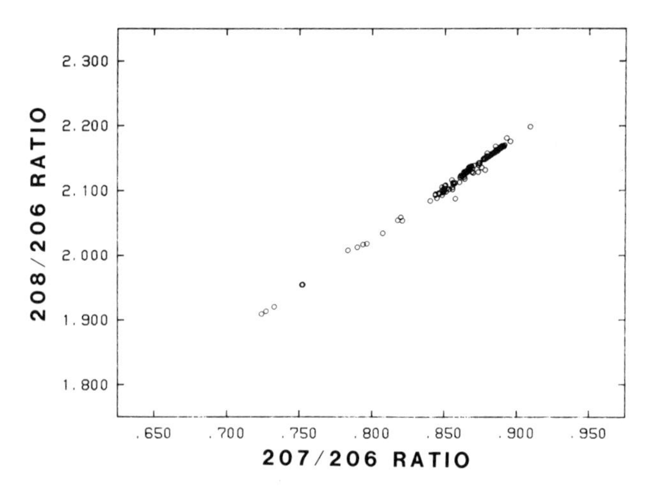

Figure 26.7 Lead isotope ratio distribution for Western Zhou Dynasty vessels in the Sackler Collections.

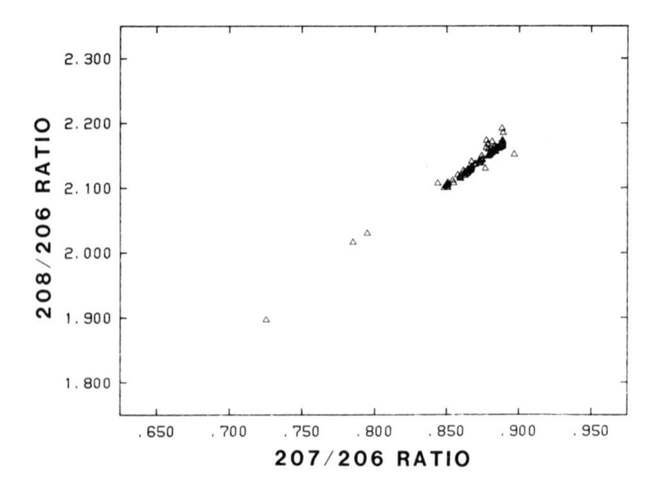

Figure 26.8 Lead isotope ratio distribution for Eastern Zhou Dynasty vessels in the Sackler Collections.

whose values were so deviant from the rest that they would unduly affect the results, and the variance-covariance matrix for this group was calculated. The characteristic vectors for this matrix were determined, and the coordinates for the individual specimens were transformed from the determined isotope ratios into standardized coordinates along the characteristic vectors. This is essentially a principal components procedure, except that we have retained and used all the characteristic vectors, not just those explaining most of the variance. The separation of the specimens into related groups was then undertaken.

Several procedures were used in arriving at these groups, including the calculation of ordered similarity matrices, the generation of clustering dendrograms and of correlation plots, and the calculation of multivariate probabilities of specimens belonging to groups. In all we have arrived at eleven groups for these specimens that would seem to merit consideration. This separation must still be considered to be tentative as the art historical information has not been fully involved in its generation. However, as we shall see, there is an encouraging degree of positive correlation among the groups as they stand and according to the art historical information that we have. We expect, however, that some alteration and clarification of the analysis will occur when we have the opportunity to discuss it in detail with our art historical advisers and to refine our statistical treatment.

It is not possible to show the separation of all eleven of these groups in a single diagram, but figure 26.9, of vector 3 (the vector explaining most variance) versus vector 2 (the vector explaining the intermediate amount of variance), shows good separation between eight of these groups. In this figure the wide separation of groups 6, 7, and 8 is immediately apparent. In figure 26.10, which is just the expanded upper left corner of figure 26.9, the separation of the first five groups is quite convincingly apparent. In these and in most subsequent similar figures the ellipses are two-dimensional probability limits defining containment of 90% of the specimens belonging to a group, assuming the group to contain a large number of specimens. Hotelling's T, which adjusts the probability limits for a small number of specimens within a group, could have been applied in these plots, but doing so for the small groups spreads the ellipses so much that they become confusing.

Viewed in other directions, the groups shift much in relative positions, sometimes appearing to overlap but remaining compact in all views, as they are indeed compact groups in the three-dimensional space. Figures 26.11 and 26.12 are a complete and an expanded plot of vector 3 versus vector 1, vector 1 being the direction of least variance. Along this vector groups 7 and 8, which were widely separated in figure 26.9, now appear to overlap, as do groups 1, 2, 3, and 4. In a plot of vector 2 versus vector 1 (figure 26.13), groups 6 and 7 now appear to overlap groups 1 and 2. A separate similar plot of just the first five groups (figure 26.14) shows that they are well separated in this direction.

The remaining three groups (9, 10, and 11) all lie close to groups 2, 3, and 4 in the three-dimensional space. However, figure 26.15 shows that they are definitely separated from each other and from group 3, and figure 26.16 shows that they are definitely separate from groups 1, 2, and 4. It is interesting to note that group 9 contains no late Eastern Zhou specimens, and that groups 10 and 11 contain no early Shang Dynasty specimens.

Returning our attention to the vector 3 versus vector 2 plot in figure 26.9, it is worth noting that the widely separated groups 1, 6, 7, and 8 are composed almost totally of early samples. All of the specimens comprising group 8 are attributed to the Shang Dynasty, as are more than 90% of those in group 7. Groups

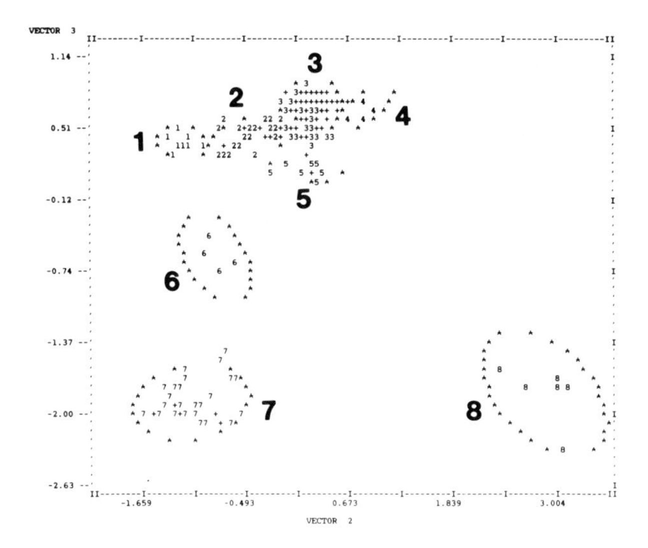

Figure 26.9 Standardized characteristic vector correlation plot (vector 3 versus vector 2) for isotope ratio groups 1–8.

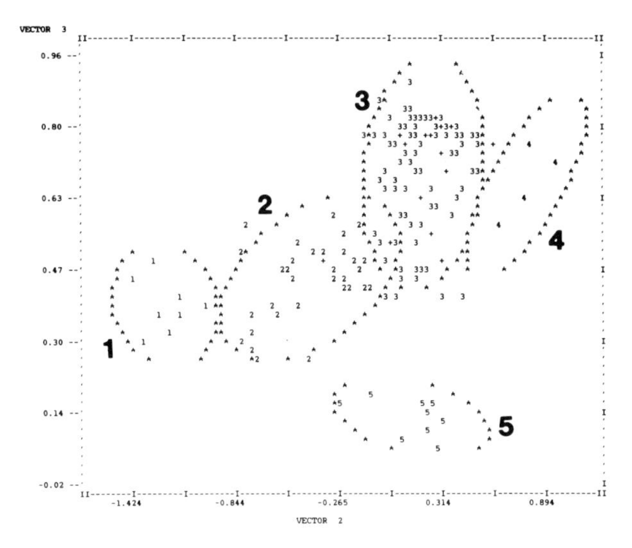

Figure 26.10 Standardized characteristic vector correlation plot (vector 3 versus vector 2) for isotope ratio groups 1–5.

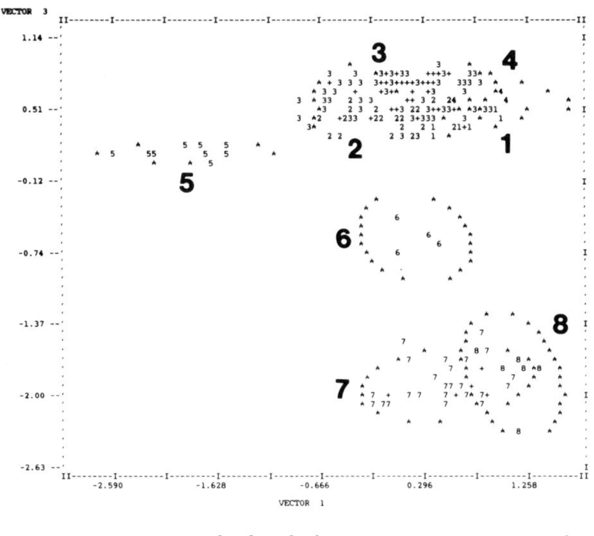

Figure 26.11 Standardized characteristic vector correlation plot (vector 3 versus vector 1) for isotope ratio groups 1–8.

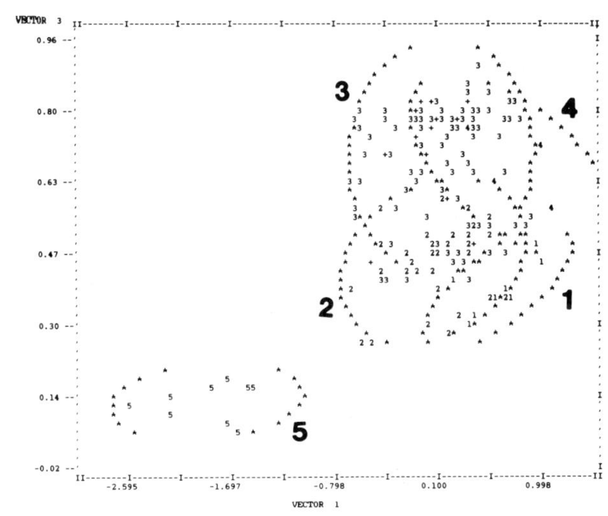

Figure 26.12 Standardized characteristic vector correlation plot (vector 3 versus vector 1) for isotope ratio groups 1–5.

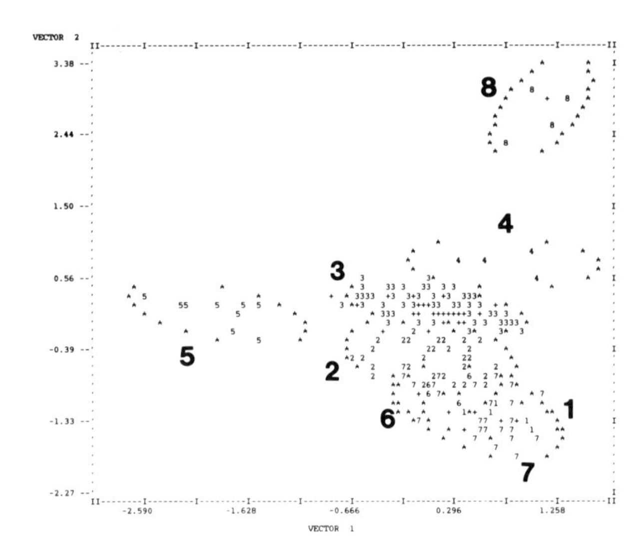

Figure 26.13 Standardized characteristic vector correlation plot (vector 2 versus vector 1) for isotope ratio groups 1–8.

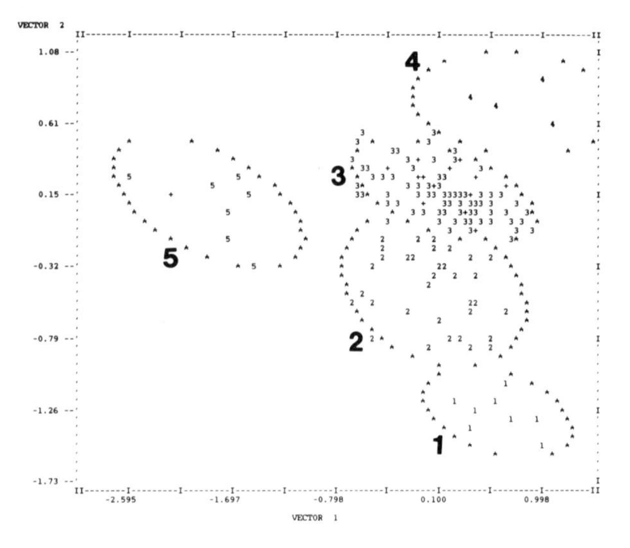

Figure 26.14 Standardized characteristic vector correlation plot (vector 2 versus vector 1) for isotope ratio groups 1–5.

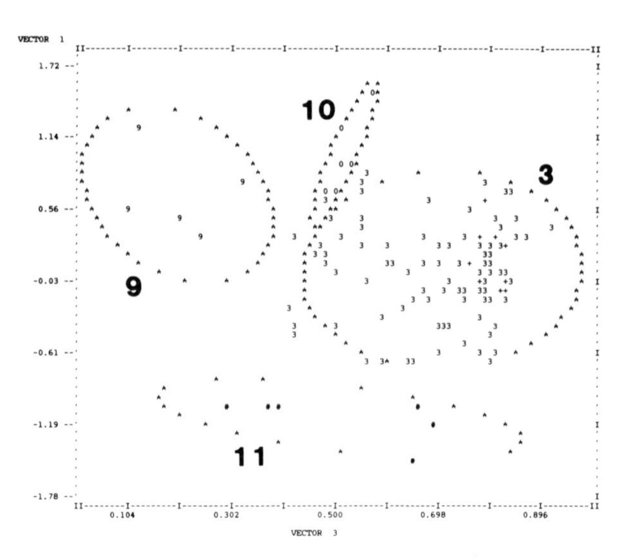

Figure 26.15 Standardized characteristic vector correlation plot (vector 1 versus vector 3) for isotope ratio groups 3, 9, 10, and 11.

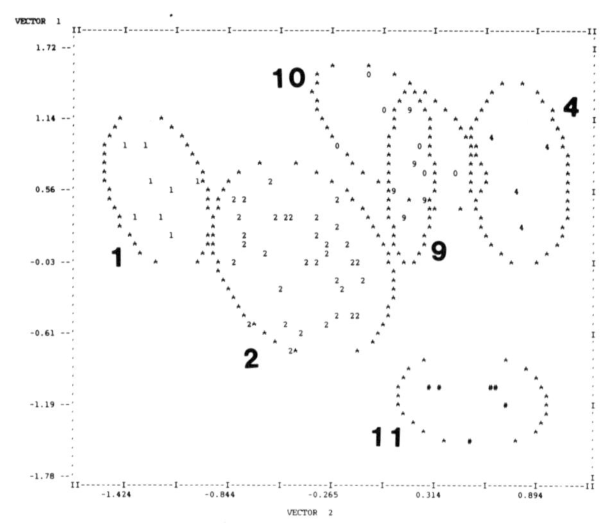

Figure 26.16 Standardized characteristic vector correlation plot (vector 1 versus vector 3) for isotope ratio groups 1, 2, 4, 9, 10, and 11.

1 and 6 are mixtures of Shang and Western Zhou specimens. No Eastern Zhou specimens occur in any of these four groups.

Moreover, all of the members of group 8 are considered by Bagley to be of either pre-Anyang or early Anyang style. Of the pre-Anyang vessels he comments that they are earlier than Anyang bronzes but their geographic distribution is not necessarily any different. However, he would like us to make it quite clear that his attribution of Anyang style to a group of vessels should not be interpreted to mean that he believes that these vessels necessarily were made at Anyang. He notes that, although similar vessels have been found at Anyang, similar vessels have been found at many other places as well. The fact that vessels of this style have been found at Anyang suggests only the possibility that Anyang-style Sackler Collection vessels may have been made there (Bagley, personal communication).

Thirty-one of the thirty-four vessels in group 7 have been attributed to the Shang Dynasty, the remaining three to Western Zhou. Rawson has noted that these three Western Zhou vessels are all of early date. Bagley considered roughly two-thirds of these Shang vessels to be of Anyang style. Of the remaining one-third, five were of styles that have been found at such widespread locations that inferences of their possible origin cannot be drawn; Bagley considered four to be predynastic Zhou, and one he considered to be of a type unlikely to have been made at Anyang. Despite the few exceptions there appears to be a good degree of correlation between the Anyang style and group 7 isotope ratios.

The fact that the leads in the vessels of group 7 are related to Anyang is confirmed by our analysis of a lead ingot in the collection of the Academia Sinica, Nangang. The ingot was found in the Anyang excavations in the 1930s and was misidentified as tin. It has a surface of pure lead carbonate, and the lead isotope ratios of this surface material fall solidly into group 7.

Groups 1, 6, and 9 also contain Shang Dynasty vessels of Anyang style, although the number of vessels in these groups is so small that no general conclusions concerning them are warranted. We have two additional samples from the Academia Sinica taken from helmets from tomb HPK 1004 at Anyang. The lead isotope ratios for both of these helmets are consistent with group 6.

It is quite clear that leads from a number of sources were used in creating Anyang-style vessels and that the use of the particular leads of the groups just discussed did not seem to persist into the Eastern Zhou Period.

The largest of the lead isotope ratio groups, with more than 100 specimens, and by far the most complex one, is group 3. This group is the only one containing a sizable number of vessels from all three dynasties. Of the Shang vessels in this group approximately one-third are thought to be of Anyang style, one-third to be of styles found at such diverse sites that their origin cannot be evaluated, and one-third to be of predynastic Zhou style. Bagley has said that there is a good chance that the predynastic Zhou-style vessels might have been made in the Wei River valley of Shaanxi Province. Rawson believes that a fair number of the Western Zhou vessels in this group may also have come from Shaanxi Province. Thus there is a fair indication that some of the group 3 vessels may have come from a relatively western location.

However, So has told us that the majority of the Eastern Zhou vessels in this group are of metropolitan style, which she would tend to relate to the more eastern Henan or Shanxi provinces. Many of them relate directly in design or construction details to vessels produced at the Eastern Zhou foundry near Houma. Also, Mabuchi et al. (1982) have found some *bu* coins, attributed to Hebei or Henan province during the Warring States Period of the late Eastern Zhou Dynasty, to have lead isotope ratios compatible with group 3. Thus there is an indication that some of the group 3 vessels may have come from a more eastern source.

It is possible that further consideration might permit a subdivision of group 3. It is also possible that lead from a specific source, possibly a centrally located one, was transported to a number of casting sites during these three dynasties. It is even conceivable that there was a manufacturing region that continued to supply vessels to the various capitals throughout the three dynasties. Whatever the case may be, group 3 deserves further serious consideration.

Groups 2, 4, and 5 are almost exclusively composed of relatively late specimens. Only one Shang Dynasty vessel occurs among them; it is in group 2, where it constitutes but one out of thirty-four specimens. Little more can be said about the members of these groups at this time, except that So has expressed the opinion that the Eastern Zhou vessels in group 2 are mostly not of metropolitan style. She doubts, however, that they all came from a common source, as there are some that she would relate to more southern or more northern sites.

It should be considered whether any of these groups could have been formed by combining leads used in other of the groups. If this has happened, the combination group's isotope ratios should lie in the column of three-dimensional space directly joining the two groups whose leads were mixed to form it. Thus in all possible projections the combination group should lie somewhere directly between the parent groups. A review of the previous figures shows that only group 2 might conceivably have been formed by mixing the leads of groups 1 and 3 and that group 6 might have been formed by mixing groups 2 and 7. Because groups 1 and 3 represents leads largely used at different periods, as do groups 2 and 7, it seems unlikely that even they were used in combination. There is little evidence within the groups that mixing of leads was a common practice. However, a number of specimens were not included in any group because their isotope ratios were unique and either were not close to those of any other specimens or, at most, were close to those of only one or two other specimens. A fair number of these specimens lie at intermediate positions between the groups, and their unique isotope ratios might have resulted from the mixing of leads.

It is also possible that the entire lead of a later vessel might have been derived from the remelting of an earlier vessel. Our data indicate that this did not occur often, if at all, in the fabrication of these vessels over long periods of time, as many of the leads that occur in early vessels, for example, group 1, 6, 7, and 8 leads, simply have not occurred in later vessels.

In summary, there are many reasons why objects made at the same site and time might not have consistent lead isotope ratios. Hence inconsistent ratios should not be accepted as proof that objects were fabricated at different sites. However, the variety and range of ratios encountered is such that accidental close agreement between specimens is unlikely. The results obtained to date would encourage one to believe that a close agreement in lead isotope ratios between specimens does indeed indicate that the probability is high that they were made from lead from related sources.

The Determination of Elemental Compositions

The concentrations of the major components in the bronzes—copper, tin, and lead—were determined by atomic absorption spectroscopy. In this method 10-mg samples of bronze were dissolved in 10 ml of a 6 *N* hydrochloric acid solution containing 3% hydrogen peroxide. After dissolution the excess peroxide was removed by gentle heating, and the solution was diluted to 50 ml. The solution was aspirated into an acetylene-air gas stream (nitrous oxide and air for tin) that fed a high-temperature burner. The light characteristic of the element was sent through the flame of this burner and the fraction of the absorbed light measured. The runs were calibrated with standard solutions of the elements determined.

The concentrations of the minor and trace components in the bronzes—iron, cobalt, nickel, zinc, arsenic, antimony, silver, gold, and mercury—were determined by instrumental neutron activation analysis. It was also possible to determine tin by neutron activation. The tin concentrations determined by the two methods were in good agreement, and in some instances the mean of the values obtained by the two methods or the neutron activation results were used instead of the values determined by atomic absorption.

In the neutron activation method 20-mg samples of bronze were sealed in small tubes of highly pure silica. These samples were activated in the Brookhaven National Laboratory High Flux Beam Reactor by exposing them to a flux of thermal neutrons of 2.7×10^{14} neutrons per square centimeter per second intensity for a period of 20 min. The activated samples were allowed to decay for a week and then were counted with a 35-cm^3, 15% efficient lithium–drifted germanium detector for 30–40 min. After a second decay period of four days, the samples were again counted for 60–90 min. Each run was separately calibrated by including within them samples of the following standards, U.S. National Bureauof Standards silicon bronze 158a, tin alloy 431, flyash 1633, and phosphorized copper alloys C 1251, C 1252, and C 1253, plus an in-house multielement standard, Ohio red clay.

The Evaluation of Elemental Compositions

The relative concentrations of the major components —copper, tin, and lead—are primarily determined by the proportions of these metals that were mixed at the foundry in formulating bronze alloys. They only occasionally provide information relative to provenance. There is much to be learned from them regarding ancient metallurgical procedures but, because we are at this time concerned with questions of provenance, we have not as yet made use of them in our data analysis.

Because of the levels of concentration of nickel, zinc, and mercury encountered, it was not possible to determine these metals in all specimens. Therefore, although the occurrence of these elements in some instances has been significant, they have not been used in our present statistical analysis. Also, although iron has been well determined in all specimens, its concentration has varied greatly and inconsistently with other elements. We suspect that iron is being introduced in a somewhat random manner in the metallurgical processes of preparing and handling the bronze. Accordingly, at present iron has been excluded from our statistical analysis.

The statistical analysis was undertaken therefore with silver, gold, arsenic, antimony, and cobalt, all of which, although present only in trace amounts were well determined in almost all the specimens. The approach was essentially the same as that used to analyze the isotope ratio data. With the exception of a few extreme outliers, all of the specimens were combined into a single group for which characteristic vectors of the group's variance-covariance matrix were determined. Standardized coordinates along these characteristic vectors were calculated for the individual specimens and used for further analysis.

Groups have been resolved for the elemental concentration data that are similar to those found for the lead isotope data, but they are not as detailed or as widely spread out. Four groups (C1, C2, C3, and C4) that seem to be significantly different from one another are shown in figure 26.17, a vector 3 versus vector 5 scatter plot. Figure 26.18, a vector 5 versus vector 4 plot, shows the separation between groups C1 and C2 a little more clearly. The groups show a correlation with time similar to that of the isotope ratio groups. Groups C1, C2, and C4 are almost totally composed of Shang and Western Zhou vessels. Group C4 contains no Eastern Zhou vessels, group C1 contains only one late vessel, and only two of the fifty-seven vessels comprising group C2 were Eastern Zhou. In contrast to this 93% of the vessels in group C3 were either Western or Eastern Zhou.

Moreover, there was a good degree of correlation between membership in isotope ratio groups and in compositional groups. For example, there were fifteen Shang Dynasty vessels that were simultaneously members of compositional group C2 and of isotope group 7. This coincidence meant that nearly half of the Shang vessels in either one of the groups was also part of the other group. Only 7% of compositional group C3 vessels were attributed to the Shang Dynasty; however, eight out of the ten Shang vessels that did occur in compositional group C3 were also in isotope group 3. In fact, of the seventy-nine vessels in compositional group C3 that also were contained in isotope groups, forty-nine, nearly two-thirds, were also contained in isotope group 3. There is also, however, a significant degree of overlap between compositional group C3 and isotope group 2. More than two-thirds of the members of isotope group 2 are also to be found in compositional group C3.

On the basis of the assumption that the trace impurities used for this analysis predominantly relate to the copper used in the bronze and that the lead isotope ratios predominately relate to the lead, the overlap of both lead isotope ratio groups 2 and 3 with compositional group C3 would suggest that two different leads had been used with the same copper.

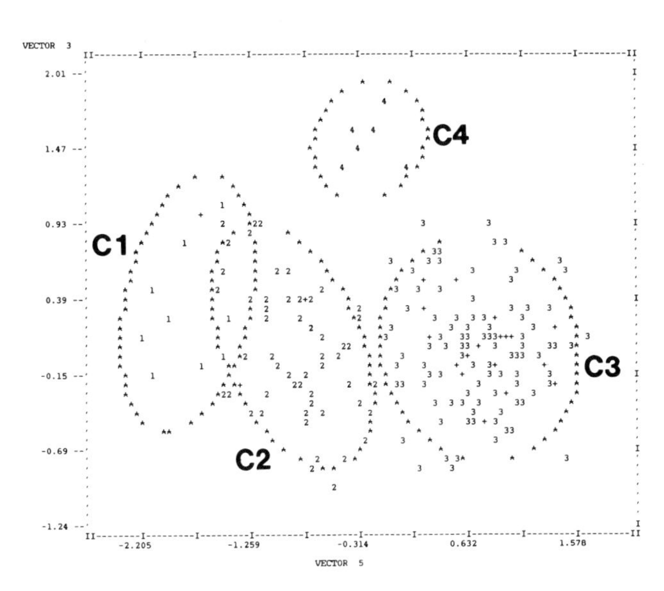

Figure 26.17 Standardized characteristic vector correlation plot (vector 3 versus vector 5) for compositional groups C1–C4.

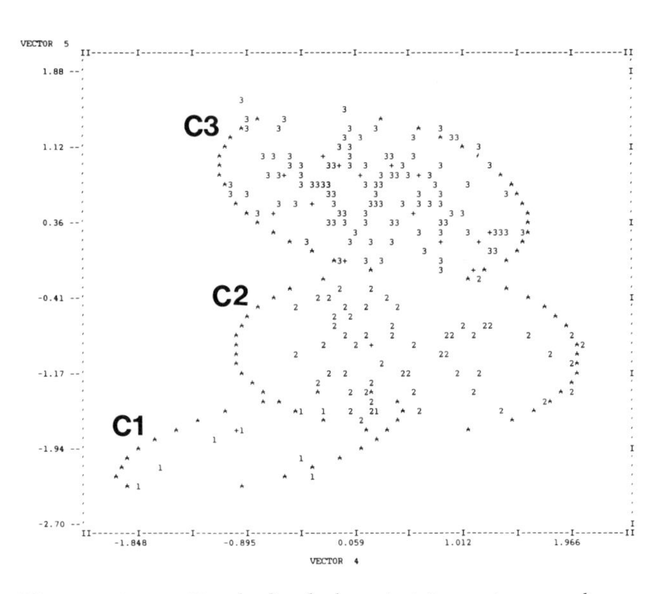

Figure 26.18 Standardized characteristic vector correlation plot (vector 5 versus vector 4) for compositional groups C1–C3.

Conclusion

The data presented will yield many more interesting conclusions as they are considered in detail and as further collaboration with art historians and archaeologists enables more interpretation. The geological implications have not yet been fully explored, and a close consideration of the implications in terms of commerce and transportation of metal should be enlightening. In the near future we hope to explore these areas and to extend our measurements to include vessels of the Han and later dynasties. We also would like to encourage and, to the extent possible, support the application of these techniques on archaeologically attested material.

The results obtained to date, however, do provide good evidence that lead isotope and compositional measurements can significantly supplement structural and stylistic evaluation of the inferred origins and the relationships between Chinese ritual bronze vessels. Such complicating factors as mixing of metals from different sources and remelting of older vessels to form new ones may indeed be responsible for the unique, nonmatching compositions or isotope ratios of some of the vessels. However, the majority of vessels group into well-defined regions of compositional or isotope ratio space, and there was such a high degree of correlation among these groups and between them that one can infer that such complications had not arisen for these vessels.

Acknowledgments

We are most grateful to Arthur M. Sackler for generously making available the Chinese ritual bronze vessels in his collections for examination and analysis and for his encouragement and support of this research. We greatly appreciate the constructive evaluation of our data and the correlation of them with archaeological and art historical information provided by Robert Bagley, Jessica Rawson, and Jenny So. Without their well-considered attributions for the vessels we have analyzed, our analytical data would be of little use. We also want to express our appreciation to Lois Katz for the help and encouragement she has furnished.

References

Bagley, R. W. 1987. *Shang Ritual Bronzes in the Arthur M. Sackler Collections.* Volume 1 of *Ancient Chinese Bronzes in the Arthur M. Sackler Collections.* Cambridge, Mass.: Harvard University Press.

Barnes, I. L., R. H. Brill, and E. C. Joel. 1987. "Lead isotope studies of early Chinese glasses." *Proceedings of the International Symposium on Glass* (Columbus, Ohio: American Ceramic Society).

Chen Yuwei, Mao Cunxiao, and Zhu Bingquan. 1980. "Lead isotopic composition and genesis of Phanerozoic lead deposits in China." *Geochimica* 3:215–229.

Mabuchi, H., Y. Hirao, S. Sato, N. Midorikawa, and K. Igaki. 1982. "Lead isotope ratios in ancient Asian coins." *Archaeology and Natural Science* 15:23–29.

Rawson, J. 1987. *Western Zhou Ritual Bronzes in the Arthur M. Sackler Collections.* Volume 2 of *Ancient Chinese Bronzes in the Arthur M. Sackler Collections.* Cambridge, Mass.: Harvard University Press.

So, J. F. 1987. *Eastern Zhou Ritual Bronzes in the Arthur M. Sackler Collections.* Volume 3 of *Ancient Chinese Bronzes in the Arthur M. Sackler Collections.* Cambridge, Mass.: Harvard University Press.

27
The Fabrication of Gold Foil in Japan

Muneaki Shimura

Fabrication of gold foil is one of the most ancient and finest techniques of metallurgy in the world. The fabrication process of gold foil at Kanazawa, Japan, which preserves some characteristics of the ancient technique, has been surveyed and analyzed from the viewpoint of modern technology. The results suggest that the traditional technique for gold foil fabrication is a wonderful, sophisticated technique of metallurgy and that ancient gold foil artisans had a high degree of skill.

Historical Background

Gold is one of the earliest metals found by humans. It was an ideal material for ornaments, because of its easy workability and because the artisans could lavish their skill on the making of beautiful objects. Much of the early gold work is made from thin sheet or leaf. For example, the figurines of bulls found at the Royal Cemetery at Ur were covered with gold varying in thickness from $\frac{1}{2}$ mm to 2 mm. Some goldsmiths at work in ancient Egypt are shown in the tomb of Ti at Sakkarah (2400 B.C.); the gold was thinned into foil about 10 μm thick and into leaf about 1 μm thick (Farag 1981, p. 15). It is interesting to note that such a thickness was reached in Europe only in the eighteenth century (Aitchison 1960, p. 178). This foil was made by hammering out a small ingot into a sheet. This sheet was cut into squares, which were placed on top of each other with thin animal skin in between. The resulting packet was hammered down again to make the sheets even thinner; the skin layers prevented the sheets from sticking to one another. This process was repeated many times until each packet of gold consisted of thin leaves a few hundredths of a millimeter thick (Tylecote 1979, p. 37). Ancient goldsmiths had a high degree of skill with this "ancient science of technology."

The fabrication technique for making gold foil in Japan was introduced from China, perhaps with the introduction of Buddhism in A.D. 600. Figure 27.1 shows a goldsmith (a foil beater) at work in the sixteenth century [*71 ban Shokunin Utaawase Emaki*; (Picture Scroll of Seventy-One Kinds of Craftsmen Prepared for a Poetry Contest), sixteenth century]. In those days the foil was called *haku*, which means "thin one." It was used for decorations of temples and of sculptures of the Tenpyo and Asuka cultures in Japan.

The Chinese technique for making gold foil was depicted in the *Tiangong Kaiwu* in 1637. The comparison between Japanese and Chinese techniques for making gold foil as deduced from the historical literature shows that the preparation of the paper used between the gold leaf layers during beating is important in both processes, but the preparation process of the paper is different for the two cultures. This suggests the separate development of techniques for making gold foil in China and Japan.

Figure 27.1 Craftsman working on gold foil in sixteenth-century Japan. From *71 ban Shokunin Utaawase Emaki* (A Picture Scroll of 71 Kinds of Craftsmen Prepared for a Poetry Contest).

Nowadays more than 90% of gold foil in Japan is made at Kanazawa. Although foil-beating machines with air-pressure systems were introduced from Germany in 1920, the fundamental process has not changed. The gold foil made industrially by the traditional technique is about 0.2–0.3 μm thick and about 18 cm^2. Although the fabrication of gold foil is apparently easy, because gold is malleable and simple to hammer, the fabrication technique is a typical, sophisticated technique of manual industry, and hence it is not easy to interpret the process in terms of the modern theory of deformation processing of metals. By the way, the thickness of copper foils that are produced industrially is about 2–3 μm, about ten times thicker than the gold foil made by traditional techniques.

Process for Making Gold Foil at Kanazawa

Table 27.1 shows the composition of gold alloys for foil at Kanazawa (Shimoide 1972, p. 48). Silver and copper are added for color. They also increase the deformability (extensionability) of the gold sheet. The composition of Japanese gold is similar to the composition of those found in ancient Egypt (Yasuda and Ota 1981, p. 61). Figure 27.2 shows the whole process for gold foil fabrication in Japan.

The process can be divided into three subprocesses: (1) the *nobekin* process, (2) the *uwazumi* process, and (3) the *hakuuchi* process. *Nobekin* means the "extension of gold." The gold is made into 5-mm squares that are 0.03–0.05 mm thick. The square is called *koppe*. The *koppe* are placed on top of each other with sheets of a special paper called *zumiuchigami* in between. About 200 sets are stacked, and about 30 papers called *furuya* are placed on the top and on the bottom of the stack. Then the stack is wrapped in bag skin. This packet is hammered down to make the gold sheet, which is 3–4 μm thick. This sheet is called *uwazumi*.

Table 27.1 Kinds of gold foil and their composition (%)

Specimen	Au	Ag	Cu
Junkin 5 moshuku	97.087	1.942	0.971
Junkin 1 shoku	97.666	1.358	0.976
Junkin 2 shoku	96.721	2.408	0.870
Junkin 3 shoku	95.795	3.343	0.862
Junkin 4 shoku	94.438	4.901	0.661
Chu shoku	90.909	9.091	0
Sanbu shoku	75.586	24.414	0

The third process is *hakuuchi*, which means "foil beating." The kind of paper is changed, depending on the thickness of the gold sheet in the second process. The quality of the paper plays a definitive role in the third process. The *uwazumi* becomes a foil that is 110 mm^2 and 0.2–0.3 μm thick through nine stages. The most important stages are *komauchi* and *hakuuchi* in the third process. *Koma* is the sheet that is 55–77 mm^2 and 3–4 μm thick. The stack of about 1,500 sheets of special papers called *komagami*, between which the *koma* are drawn, is placed between two sheaves of about fifty sheets of special papers called *megami*. Then the two sheaves of about fifty sheets of special paper called *shirobuta* are placed on the top and on the bottom of them, and further pad skins are placed on the top and on the bottom. Finally, the total stack is wrapped in bag skin. The packet (figure 27.3) is hammered down to make the gold sheet even thinner.

Papers are important tools in the *hakuuchi* process, and the quality of the beating papers depends on the quality of the raw paper and on the preparation technique. The raw paper is composed mainly of *ganpi* (*Wikstromia sikokiana*), *mitsumata* (*Edgeworthia papyrifern*), and *kozo* (*Broussonetia kazinoki*) and some special clay. The raw paper is called *shitazigami*. It is not ready to use as beating paper. Great care and experience are necessary for the special treatment of beating papers. The craftsmen have devised subtle techniques independently, and so the special treatment is a secret, even to this day.

The special treatment is roughly as follows. The first step is *aku-zukuri*, which means "ash soup making." *Aku* ("ash soup") plays a leading role in the process of the special treatment of the papers. *Aku* is made from rice straw ash and hot water, which is mixed with persimmon tannin and eggs. The next step is *aku-zuke*, which means "leaching in ash soup." In this process the raw papers are leached in the ash soup. Then the water is squeezed out. When the papers are dry to some degree, they are hammered. This process is repeated many times in a process called *shoaku* ("the first leaching"). *Hutaaku* ("the second leaching") is the third process and it is similar to the previous step. After the *hutaaku* process the raw papers are known as

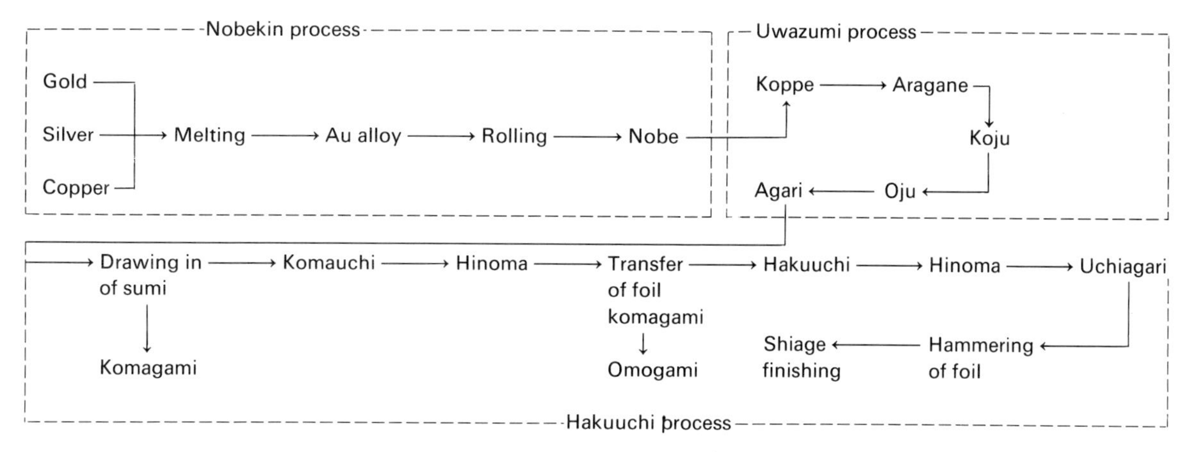

Figure 27.2 Chart showing the process of gold foil fabrication.

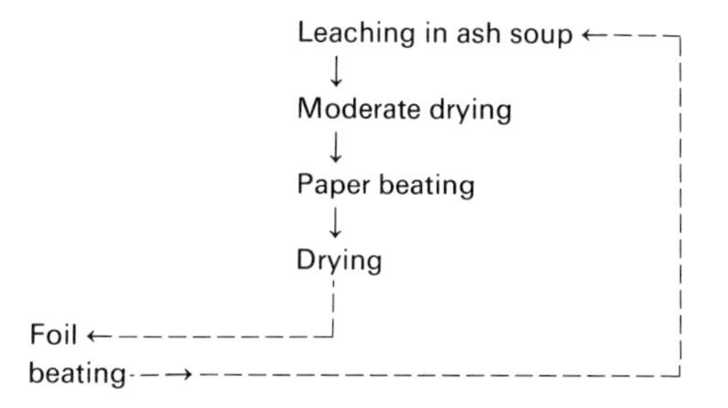

Figure 27.3 Packet for hammering gold foil.

komagami. Komagami is used only in the early stage of hammering the gold sheet. After hammering the gold sheet three or four times using *komagami*, the gold sheet cannot be extended further. The papers are leached again in the ash soup to recover their extension properties. Figure 27.4 shows the process of the special treatment of the papers. The fourth step in the special treatment is *itsuaku* ("fifth leaching"). After the leaching process has been repeated five times, the paper becomes known as *omogami. Omogami* is the beating paper with the highest efficiency for the extension of gold sheets. The last step is *furuya*. After about the fifteenth leaching the paper does not recover and can no longer be used. This paper is called *furuya*.

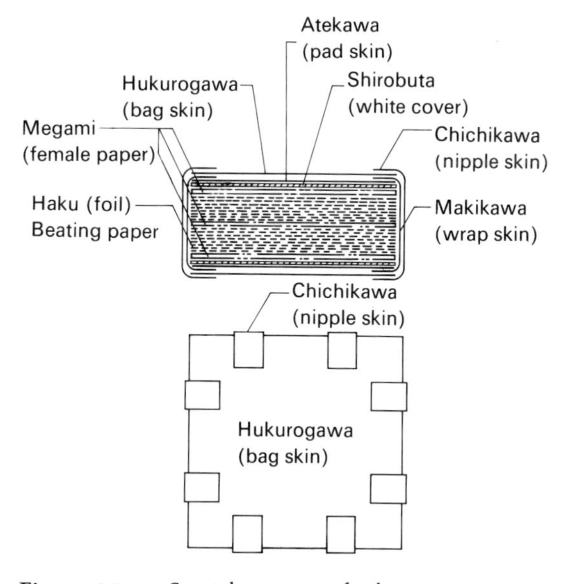

Figure 27.4 Special treatment for beating papers.

Technical Terms with the Traditional Technique

The names of the gold sheet and the paper are changed according to the state in the fabrication process. For example, the name of the gold sheet changes from *koppe* ("little soldier") to *uwazumi* ("surface affected person") and then to *haku*, which means "foil" or "thin one." The extensionability of the gold sheet is influenced strongly by the state of the beating paper. The raw paper is necessary for the most effective extension of the gold sheet. The process for the preparation is called the "process for improvement of *nari*" ("figure of paper") or the "process for bringing up papers" by artisans in Kanazawa.

The name of the paper is changed depending on the process of special treatment, from *shitazigami* ("basal paper") to *komagami* ("a little space paper"). The artisans say, "Komagami is too young to work and he can work only at the half man's rate." After *itsuaku* the paper is called *omogami* ("master paper"). The artisans say, "*Omogami* has the highest might for extension." Time-worn *omogami* is called *furuya* ("old man"). That is, makers of gold foil in ancient Japan likened the beating process to the lifetime of a human and regarded the beating paper as a might for extension of the gold sheet. Such a personification of nature is an idea that has survived in the world of traditional goldsmiths.

Modern Analysis of the Process for Making Gold Foil

Figure 27.3 is an illustration of a typical packet in the *hakuuchi* stage of the fabrication process. The gold sheets that are 55–77 mm^2 and 3–4 μm thick are placed on top of each other with special papers in between. The packet of about 1,700–1,800 leaves of gold is wrapped with thin animal skin and is hammered down to make the sheets even thinner. Such a complicated system may be regarded simply as the compression of metal between two flat plates. When there is friction between the paper and the gold sheet, the incremental increase in pressure is estimated by the friction-hill theory. As the sheet becomes thinner, the pressure required for plastic deformation increases rapidly, and then further compression becomes impossible.

The theory of strip rolling makes it possible to understand the interaction of several factors: strip thickness and hardness, reduction of area or draft, roll diameter, roll gap, and speed and coefficient of friction. When a thin hard strip is rolled, it is not possible to reduce the thickness below a certain limit. Attempts to do so result in greater deformation of the roll, without any plastic deformation of the strip. The limiting thickness is important for thin strip and foil production. It has been found that the limiting thickness is proportional to several parameters:

$$H_{\lim} \propto c\mu RS. \qquad (27.1)$$

Here c is the elastic deformation parameter of the roll, given as $8(1-\gamma)/\pi E$, from Hitchcock (1933, p. 5), where γ is the Poisson ratio and E is Young's modulus of the roll. μ is a coefficient of friction; R is the radius of the roll, and S is the plane-strain yield stress of the metal. Consequently, to obtain the thinnest possible strip, the metal should be annealed.

The coefficient of friction should be as low as possible; this can be accomplished by polishing the rolls and by using a good lubricant. Small diameter rolls should be used (Rowe 1977, p. 277). When the yield stress is low, as in aluminum, larger roll diameters can be used. In this case foils thinner than the theoretical limiting thickness can be made practically by a four-high mill.

Orowan (1943) measured the shape of the work roll during rolling and observed that the contact arc was divided into two "dead parts." The concept of the effective part roll was supported by calculations of the elastic deformation of the work roll during rolling by Troost (1966). That is, the work roll changes its shape in work and forms an effective part roll that has an effective radius that is smaller than the original radius; compression then becomes possible. What is the effective part roll in the process for gold foil fabrication? What effect does the special treatment of the beating papers have on the processing of gold foil? I examined these problems using modern technological methods, especially scanning electron microscopy (SEM).

Figure 27.5 shows a scanning electron micrograph of the surface of *shitazigami*. Salients on the surface are parts of thick fibers, and the rough parts between the fibers are regions of filler, which is added to flatten the surface of the paper in the process of papermaking. The salient parts, the sizes of which are about 2 μm, are distributed lengthwise and crosswise.

Figure 27.6 shows a stereo pair of scanning electron micrographs of the surface of *omogami*. It was difficult to obtain clean photographs because of degassing from the surface of the paper from heating by the electron beam. The salients and the region of filler are flattened, and the salient sink more deeply in the region of filler than in the case of *shitazigami*. It can be considered that the sailent parts act as effective part rolls in the hammering of the packet of gold. The goldsmiths can judge the quality of the beating papers by observing with the unaided eye the surface state of the paper. I find this remarkable.

Figure 27.7 shows a stereo pair of scanning electron micrographs of the surface of the gold foil. Depressed parts on the surface are pressure markings made by paper fibers. The salient parts of *omogami* are transcribed on the surface of the gold foil. Such markings are observed everywhere on the surface of gold foil. Hammering of the packet is not random but is even from end to end. The depressed parts move in response to the pressure of the hammering. It is probable that the formation of the depressed parts by the salient

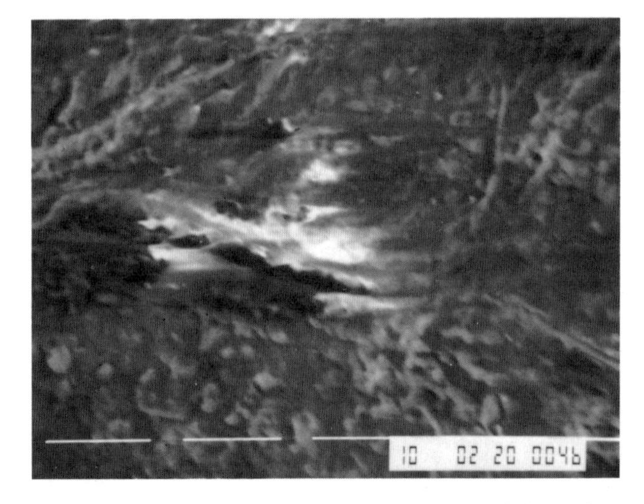

Figure 27.5 Scanning electron micrograph of the surface of *shitazigami*. Scale mark, 10 μm. Acceleration voltage, 2 kV.

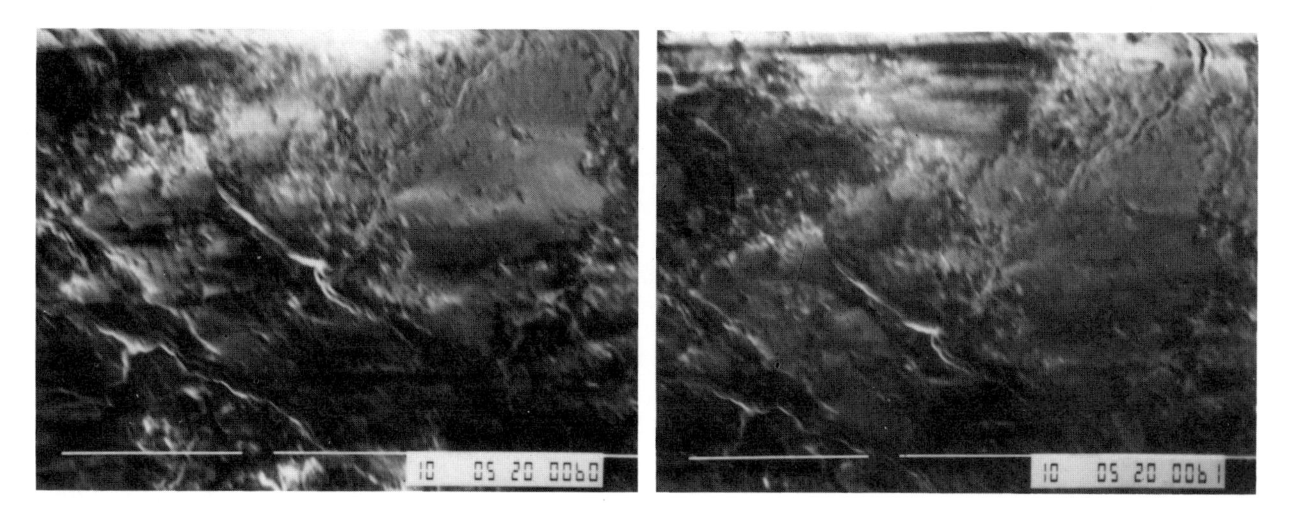

Figure 27.6 Stereo pair of scanning electron micrographs of the surface of *omogami*. Scale mark, 10 μm. Acceleration voltage, 5 kV.

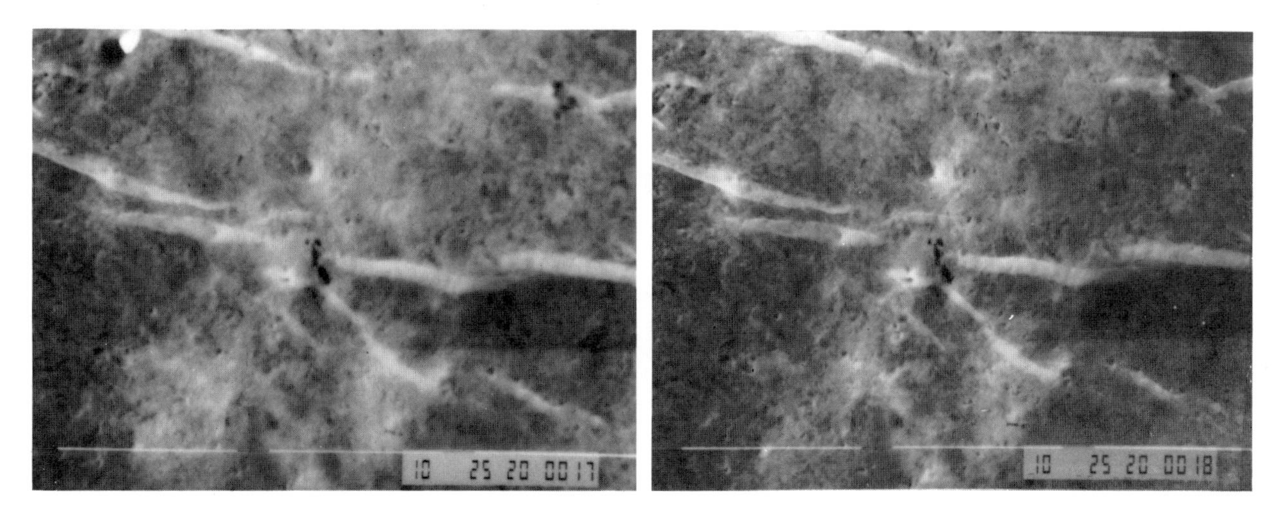

Figure 27.7 Stereo pair of scanning electron micrographs of the surface of gold foil. Scale mark, 10 μm. Acceleration voltage, 25 kV.

parts acts as an effective part roll of about 2 μm in diameter.

Figure 27.8 shows the surface of *furuya*. The surface is flatter than the surface of *omogami*, and the salient parts are flattened and depressed into the region of the filler. Special attention should be paid to the distribution of fatigue cracks along the fibers on the surface. Paper with many fatigue cracks can no longer be used for foil making.

When the packet of gold sheet is hammered, the temperature of the packet rises over 100°C. However, it seems probable that the temperature of the gold sheet becomes higher than 100°C and that recovery and recrystallization of the gold alloy occur in the process. After the *hakuuchi* stage the packet is subjected to *hinoma*. *Hinoma* is a kind of heat treatment performed to anneal the gold sheet, remove water that is absorbed during the work, and prevent the gold sheet from adhering to the paper. The treatment lowers the yield strength of the gold sheet [that is, S in equation (1)]. After *hinoma*, of course, extensionability of the gold sheet increases.

Because it is difficult to measure directly the elastic property of the fibers, I performed compression tests on stacked papers at the stress level of the ultimate tensile strength of gold alloy. Figure 27.9 shows compressive stress versus compressive strain curves of *shitazigami* and *omogami*. It seems that *omogami* has a much higher resilience than the *shitazigami*. Because cellulose fibers, which are the base of paper, are regarded as a compressive viscoelastic/plastic solid, the elastic modulus cannot be obtained easily, but comparison of the stress/strain ratio was performed near the ultimate tensile strength of gold alloy ($\sim$200 MNm^{-2}). It seems that *omogami* has a much higher ratio than *shitazigami*. The stress-strain curve of *shitazigami* shows a decrease of the stress/strain ratio at high compressive stresses. The strain-rate dependence of the stress-strain curve is small in both *omogami* and *shitazigami*, but the stress/strain ratio of *shitazigami* in the high-stress level increases with the strain rate. It seems from these results that the special treatment of the papers raises the resilience of the fibers. That is, it decreases c in equation (27.1) and decreases the limiting thickness of the metal.

Next I measured μ in equation (27.1), that is, the average coefficient of surface of the papers, by the ring test (Male and Cockcroft 1964, p. 38). A flat ring of lead with outside diameter D_0, inside diameter D_i, and height h was continuously compressed axially. In this case the change in D_i is sensitive to the average value of the coefficient of friction. The changes in the inside diameter with compression were plotted on a calibration chart (figure 27.10). The average coefficient of friction of *shitazigami* is about 0.05 in the range of small deformation, but the coefficient of friction increases rapidly after 30% compressive reduction and reaches 0.20 at 70% reduction. This is because the metal adheres strongly to the paper near 20% reduction, and the paper is torn at the region of adhesion.

Omogami after the special treatment shows an average coefficient of friction of about 0.03–0.04. Adhesion still occurs over 30% reduction and increases rapidly the coefficient of friction. *Omogami* after the *hakuuchi* stage shows a lower coefficient of friction than *shitazigami*, but the adhesion occurs near 30% reduction, increasing rapidly the average coefficient of friction; then cracks develop in the region of adhesion.

The results of the ring test are summarized as follows. A small value of the coefficient of friction correlates with a small reduction in compression. Adhesion is enhanced with an increase of reduction until the paper fractures. The special treatment of the paper tends to decrease the coefficient of friction of the paper only

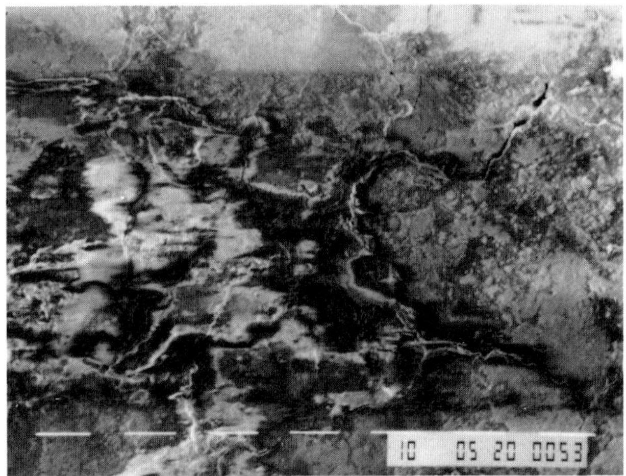

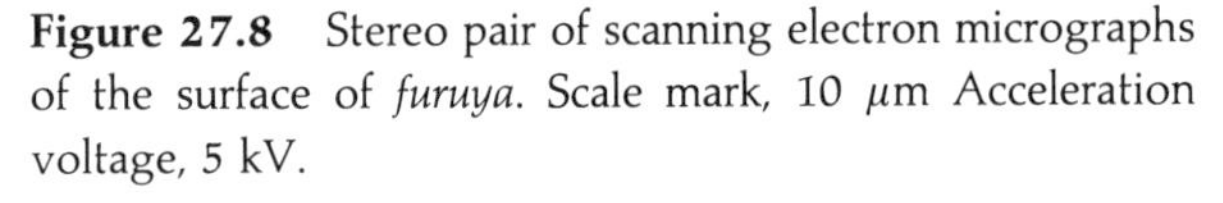
Figure 27.8 Stereo pair of scanning electron micrographs of the surface of *furuya*. Scale mark, 10 μm Acceleration voltage, 5 kV.

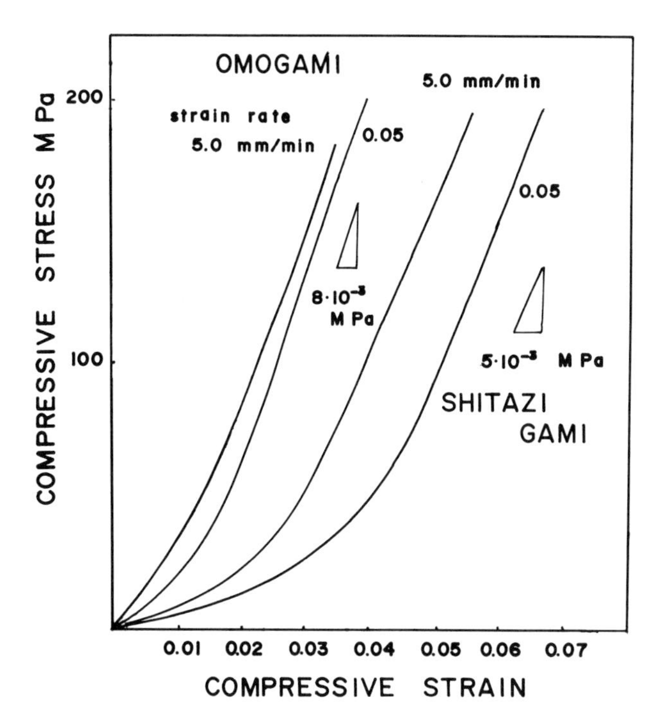

Figure 27.9 Compressive stress versus compressive strain curves for *shitazigami* and *omogami*.

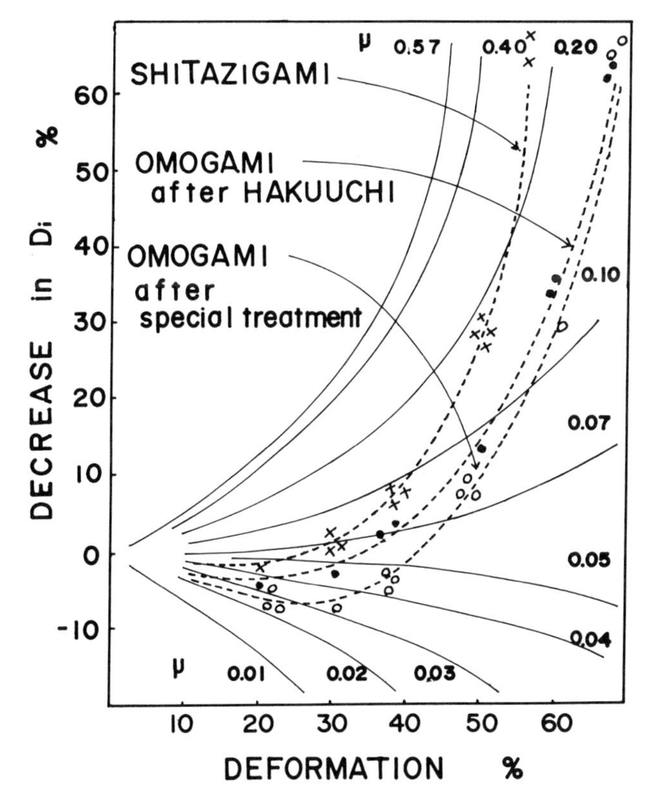

Figure 27.10 Results of ring test plotted on a calibration chart.

for small reductions in continuous compression. The coefficient of friction of the surface of *omogami* increases in the *hakuuchi* stage, making necessary additional special treatments. The results of the ring test apply to continuous compression of metal. A film on the surface of *omogami*, which has a lubricating effect, thins with displacement of the surface layer of the metal, until fracturing and hence adhesion occur. In practice, however, the foil is beaten intermittently and the compressive reduction per beat is small so that the effective coefficient of friction is small and adhesion of *omogami* does not result.

Discussion and Summary

The traditional fabrication technique of making gold foil in Japan is complicated and sophisticated. A modern analysis of the *hakuuchi* stage, which is the main process, was attempted. The existence of an effective roll and its size and resilience, the coefficient of friction of papers, and the yield stress of gold alloys, as shown in equation (27.1), were investigated. According to the SEM results for the surface of *omogami*, cellulose fibers stand out from the flat region, which is composed of filler. These salient parts act as effective rolls, as shown in figure 27.7, which shows pressure markings of fibers of the beating paper on the surface of the gold foil after the *hakuuchi* stage. The diameter of this effective roll is 2 μm. It seems probable that the formation of the depressed parts and their migration on the surface of the gold by hammering are an extension mechanism in the traditional process of gold foil beating.

As a result of cyclic loading of high compressive stress on the fibers in hammering, the salient parts are flattened and fatigue cracks develop along the fibers in regions where there is filler supporting the fibers. When the foundation that supports the fibers is broken, the fibers can no longer act as effective rolls and, however the papers are treated, they can no longer be used as beating papers.

The special treatment of the beating papers is important in the fabrication of gold foil. According to the results of the compression test of stacked papers, the compressive resilience of the paper is increased by the special treatment and the stress/strain ratio becomes high at the flow stress level of gold alloy. That is, the resilience of the fibers that act as effective rolls is enhanced. According to the SEM results for the surfaces of *shitazigami* and *omogami*, the state of the surface was changed by the special treatment. The main change of topography is flattening of the regions composed of filler and flattening of the surface of the fibers. These results agree with the results of measurements of surface roughness of the papers (Ueda and Tanimura 1979, p. 15).

One of the effects of the special treatment is the lowering of the coefficient of friction. The coefficient of friction of *omogami* is low in the case of low reduction. However, the difference between *omogami* and *shitazigami* is not large. The effects of ash, persimmon tannin, and egg on the lubrication are not clear, but it is easily supposed that they act to form a film of basic metallic soap. In any case the effect of lowering the coefficient of friction of the surface of the paper and the effect of preventing the adhesion of the paper to the gold are important for making foils. The goldsmiths in China used a special paper call *wujingzhi*, described in the *Tiangong Kaiwu*. *Wujingzhi* was a film of bamboo coated with lampblack, which is composed of fat and graphite, both good lubricants for deformation processing of metals.

The effects of changes in c, μ, and R in equation (27.1) were discussed earlier. The effect of S, the yield stress of gold alloy, is explained by the lowering of the flow stress of the alloy by the recovery and the recrystallization resulting from the temperature rise in the *hakuuchi* stage and by the annealing during *hinoma*. The extensionability of the gold alloy is enhanced remarkably after *hinoma*.

The main effect of the special treatment of the papers is on the compressive strength of the fibers. The effect can be understood from the mechanism of beating in papermaking. The process of beating changes the quality of the paper. Beating in papermaking is a complex process performed by passing the pulp between the teeth of the beater in water.

The mechanism of beating in papermaking is roughly as follows. The membranes of the fibers are broken, facilitating the swelling of the fibers. Then the fibers become ductile because of the fracturing of the bonds between the fibrils. As the water is removed from the solution of the fibrillated fibers, the space between the fibers and the fibrils becomes small, and hydrogen bonds form between the cellulose molecules. The strength of the paper increases in proportion to the number of bonds (Okajima and Migita 1976, p. 77). Such a process of beating in papermaking is similar to the special treatment of beating papers, especially in the similarity between the effects of beating in papermaking and hammering of the wet paper in the special treatment.

The results of the survey of the traditional technique, the analysis from the viewpoint of modern technology, and the discussions suggest that the traditional technique for making gold foil in Japan is a wonderfully sophisticated technique and that ancient goldsmiths had a high degree of skill. The ancient artisans in Japan likened the sophisticated process to the lifetime of a human, and there is a close relation between their "science of technology" and their high degree of skill.

References

Aitchison, L. 1960. *The History of Metals*, vol. 1.

Farag, M. M. 1981. "Metallurgy in ancient Egypt: Some aspects of techniques and materials." *Bulletin of the Metal Museum of the Japan Institute of Metals* 6.

Hitchcock, J. H., and W. Trinks. 1933. "Strength of roll necks." *Transactions of the ASME* RP55-5 : 67–74.

Rowe, G. W. 1977. *Principles of Industrial Metalworking Processes*. New York: Crane-Russak.

Male, A. T., and M. G. Cockcroft. 1964. "A method for the determination of the coefficient of friction of metals under conditions of bulk plastic deformation." *Journal of the Institute of Metals* 93(2) : 38–45.

Okajima, S., and N. Migita. 1976. *Paper and Natural Fiber*. Tokyo: Dainippon-Tosho. (In Japanese.)

Orowan, E. 1943. "The calculation of roll pressure in hot and cold flat rolling." *Proceedings of the Institute of Mechanical Engineering* 150 : 140–167.

Shimoide, T. 1972. *Gold Foil in Kaga Kanazawa*. Kanazawa: Kitaguni-shuppansha. (In Japanese.)

Troost, A. 1966. In *Grundlagen der bildsamen Formgebung*. Düsseldorf: Verlag Stahleisen. (Japanese edition, 1972).

Tylecote, R. F. 1979. *A History of Metallurgy*. London: Metals Society.

Ueda, M., and K. Tanimura. 1979. "The study of extension mechanism of gold foil." *Bulletin of the Japan Sea Research Institute, Kanazawa University* 11 : 1–24. (In Japanese.)

Yasuda, K., and M. Ota. 1981. "Behind brilliance and calmness: History of precious metals and alloys." *Metals and Technology* (October), 51 : 58–61.

28
The Japanese Alloy *Shakudo:* Its History and Its Patination

Michael R. Notis

Surface treatment of metals to produce a variety of colors and textures has long been a technique used in the Orient. Chinese artisans, possibly as early as the Shang Dynasty, used techniques developed through alchemy to cast and color the beautiful cast bronzes that remain as enduring artifacts of early Chinese civilization. In medieval Japan the ability to color metals was in all probability better than it is in today's modern coatings industries, which are still largely empirically based.

At least as early as the fourteenth century A.D., Japanese metalsmiths had developed methods to produce a deep purple-black patina on copper-gold alloys. These coatings have survived and have been of keen interest to historians of science and as art objects prized for their unequaled beauty.

Here I review the literature concerning the history of the development of the Japanese copper-gold alloy called *shakudo* and the possible relations between the patination process for *shakudo* and the earlier development of "purple sheen" gold in China. The research that has been performed and documented in modern scientific literature concerning corrosion studies on similar copper alloys is described and compared to the traditional techniques for patination of *shakudo*. Finally, patination of a copper-gold alloy has been performed under laboratory conditions, and the resultant coatings have been examined by X-ray diffraction, scanning electron microscopy (SEM), and analytical electron microscopy (AEM).

The Japanese Alloy *Shakudo*

W. Gowland, an Englishman who worked at the Imperial Japanese Mint, has documented many of his studies concerning the alloy *shakudo*. Gowland states:

The alloy has been long known to the Japanese, but there are no records of its first use, and the date of its origin cannot be even approximately determined. Perhaps the least doubtful of the earliest specimens known to us are the mounts of the sword of Ashikaga Takauji, who held the position of shōgun from 1335 to 1337, which is preserved in the temple of Itsukushima. There may be earlier examples, but it was certainly not known in the ninth century. The oldest specimen of Buddhist art-metal work in the decoration of which shakudo appears, so far as I have been able to trace, is a reliquary containing fragments of the bones of St. Nichiren in the famous temple of Minobu (dated the eighth year of Tensho, i.e., 1580). In many temples there are statues of divinities and saints which are said to be composed of this alloy, but those I have had the opportunity of examining were all of ordinary copper-tin-lead bronze. In the seventeenth century it was extensively employed, but the finest examples of it as a decorative alloy are found in the guards and other furniture of the swords of the last century and the first half of the present. The addition of gold to bronze in order to obtain a black patina has been long known to the Chinese. It is hence possible that the Japanese may have learned from them this peculiar property of gold; but the pure alloy of copper and gold, of the true shakudo, is essentially Japanese, and is unapproached in the beauty and richness of its patina by any alloy of the Chinese, either of old or recent times. Its rich deep tones of black, and the splendid polish which it is capable of receiving, render it alike a perfect ground for inlaid designs of gold, silver, and copper, and for being similarly inlaid in them. This alloy, too, possesses physical properties which are of extreme importance to the worker in metals, and enable him to manipulate and fashion it as he desires. It can be cast into any form; can be hammered into sheets and drawn into wire. No large castings, however, have been made of it. (Gowland 1896, p. 412)

A later paper by Gowland (1914–1915) reports a *shakudo* sword guard to have been found together with a sword blade made by Mitsusada dated to the middle of the thirteenth century, and Greey (n.d) reports a colossal solid *shakudo* (?) casting in the temple of Yaku-shi at Nara cast about A.D. 700 and attributed to Gio-gi.

Based on evidence in pictorial scrolls and documents from the time, Sasano (1979) believes that the alloy *shakudo* was in use as early as the Heian Period (794–1183), in disagreement with the earlier belief that the alloy was first produced during the middle Muromachi Period (1393–1573) in the time of the great master Yujo. In his book Sasano (1979) shows a knife (*kozuka*) made of *shakudo* and dated to the late Kamakura Period (1184–1333), as well as a drawing of a *shakudo* skewer (*kogai*) taken from a manuscript, the *Token-zuko*. The *kogai* is said to have belonged to Machimantaro Yoshiie (Heian Period, 794–1183).

The treasury of the Shoso-in (reliquary) of the Todaiji Temple in Nara contains a collection of treasures consisting of various items, such as clothing, furniture, musical instruments, stationary, weapons, medicines, and Buddhist texts or paintings, all of the eighth to ninth century. One of the items in the collection is a cast *shakudo* container and cover, about 11.5 cm high and 7.3 cm wide, indicated to be part of the original collection (*Exhibition Catalog*, 1979). This would appear to be consistent with the opinion of Sasano (1979) that *shakudo* was already in use during the Heian Period.

The Composition and Patination of *Shakudo*

The earliest Western scientific description of *shakudo* appears to be that given by Pumpelly (1866), who describes it as a copper alloy with gold varying between 1 and 10 parts per hundred. The patina is said to be formed by boiling in a solution of copper sulfate, alum, and verdigris (copper acetate). Pumpelly explains the development of the bluish black patina as caused by the superficial removal of copper with the subsequent exposure of a thin film of gold; the bluish color is produced as a result of the action of light on this gold film.

Roberts-Austin (1888, p. 1137; 1893, p. 1022) describes the alloy *shakudo* as approximately 95% copper, with about 4% gold and gives two analyses (table 28.1). The alloy is described as acquiring a beautiful purple-black patina when boiled in certain pickling solutions, given in table 28.2. Hiorns (1892, p. 153; 1901, p. 271; 1912, p. 289) and Huish (1889, p. 184) report the work of Roberts-Austin and describe the use of the pickling solutions given by Roberts-Austin in order to produce a patina on copper alloys in the laboratory.

Gowland (1896, 1914–1915) tabulated chemical analysis for the alloy (table 28.3). He reports the development of a contemporary alloy developed to produce medals from captured Chinese bronze cannon. This alloy, made by adding small quantities of a 60 wt. % iron–arsenic speiss, termed *bajirome*, to the bronze, develops a black patina similar to *shakudo* when boiled in 2.5 g of copper sulfate, 8.3 g of verdigris, and 2 l water.

Gowland (1896, 1914–1915) also describes the method by which the black patina is produced:

The object is first boiled in a lye prepared by lixiviating wood ashes (concentrated potassium hydroxide solution);

Table 28.1 Analyses of *shakudo* given by Roberts-Austin

Component	Analysis 1[a]	Analysis 2[b]
Copper	94.50	95.77
Silver	1.55	0.08
Gold	3.73	4.16
Lead	0.11	
Iron and arsenic	traces	
Total	99.89	100.01

a. Analyses by Gowland, Imperial Japanese Mint at Osaka.
b. Analysis by Kalischer in *Dingl. Polytech. Journal* 215:93.

Table 28.2 Pickling solutions to produce purple-black patina of *shakudo*

Component	Solution 1	Solution 2	Solution 3
Verdigris[a]	438 grains	87 grains	220 grains
Copper sulfate	292 grains	437 grains	540 grains
Niter	–	87 grains	–
Common salt	–	146 grains	–
Sulfur	–	233 grains	–
Water	1 gallon	–	1 gallon
Vinegar	–	1 gallon	5 fluid drachmas

a. Iron sulfate sometimes replaces verdigris.

Table 28.3 Analyses of the alloy *shakudo* (%)

Analysis and analyst	Gold	Silver	Copper	Lead	Iron	Arsenic	Total
1. Kalischer	4.16	0.08	95.77	–	–	–	100.01
2. Gowland	3.73	1.55	94.50	0.11	trace	trace	99.89
3. Gowland	2.67	2.06	94.90	0.11	–	–	99.74
4. Gowland	2.45	1.24	96.00	0.06	–	–	99.75
5. Gowland	1.52	2.01	96.10	0.08	–	–	99.71
6. Gowland	1.00	1.37	97.40	0.07	–	–	99.84
7. Atkinson	0.49	0.29	99.04	–	–	–	99.82

after which it is carefully polished, if necessary, with charcoal powder. It is then immersed in plum-vinegar containing common salt in solution, and, after being washed with a weak lye, is placed in a tub of water to remove all traces of alkali. After this treatment it is digested in a boiling solution of copper sulphate, verdigris, and water, to which sometimes potassium nitrate is added, and the desired patina is produced.

The work of Gowland is later quoted in a number of texts and journals (Brinkley 1903–1904; Robinson 1961; Joly 1909; Wiest 1978).

Lyman (1890–1891) quotes a Japanese reference that describes *shakudo* to be 100 parts (*momme*) of antimonial (?) copper and 4 parts gold. It was patinated in a solution made by mixing 1 qt vinegar with 4 *momme* of *rokusho* (identified *incorrectly* by Lyman as blue vitriol, that is, copper sulfate) and 2 qt water.

Wilson (1902) describes the patination of *shakudo* in the laboratory as being achieved by boiling in one of two different solution compositions (table 28.4).

A modern technique in use in Japan for the aqueous patination of *shakudo* is as follows (Hiroshi Miyajima, personal communication): To 2 l H_2O at 60°C, add 4 g $CuSO_4$ and 4 g Cu acetate. Boil piece for about 10 min. Before patination the piece should be degreased and soaked in a slurry from grated *diakon* (radish) for about 2–3 min. This process is still quite similar to the traditional solution compositions.

I have recently had the good fortune to visit in Tokyo with Satsuo Ando, a master craftsman/goldsmith who was twice (1953, 1973) selected as the artisan to design new sword fittings for the Ise shrine, the holiest Shinto shrine in Japan. Ando, who has retired from his position as executive director of Mikimoto Pearl Co., is an artisan of international repute who still works with and teaches traditional metalworking techniques.

Ando's techniques, both for casting of the alloy *shakudo* and for its patination, show both traditional sensitivity and modern insight. For example, nearly pure copper is extremely difficult to cast without the formation of blow holes from gas evolution on solidification of the molten copper. *Shakudo* must be extensively mechanically worked to prepare the surface for patination, and hence imperfections in the cast material

Table 28.4 Two solutions for patination of *shakudo* given by Wilson

Component	Solution 1	Solution 2
Verdigris	1 dram, $\frac{1}{7}$ oz.	–
Salt	–	$\frac{1}{3}$ scruple, $\frac{1}{4}$ dram
Copper sulfate	1 scruple less 1 dram	1 dram, $\frac{1}{7}$ oz.
Water	1 gallon	$\frac{1}{3}$ cup

pose a particularly difficult problem. Ando uses a casting method in which the molten *shakudo* is poured into an ingot mold that has been covered with cheesecloth and submerged in hot water. *Shakudo* cast in this manner appears to be significantly more workable than the same material cast in air. A similar water casting technique is described by C. S. Smith (1981, p. 151) to have been in use in Japan before the end of the eighteenth century (Masuda and Kudo 1801).

Ando's approach to *irotsuke* ("coloring" or patination) is equally insightful. Ando indicates that the chemical known as *rokusho* (Niiyama 1969) and available commercially in Japan is *not* the copper acetate $Cu(C_2H_3O_2)_2 \cdot H_2O$ identified in the West as verdigris (also, "neutral verdigris") but rather $Cu(C_2H_3O_2)_2 \cdot 2Cu(OH)_2$. This would correspond fairly closely to basic copper acetate, or "blue verdigris" $[Cu(C_2H_3O_2)_2 \cdot Cu(OH)_2 \cdot 5H_2O = Cu(C_2H_3O_2)_2 \cdot CuO \cdot 6H_2O]$ and would thus produce a patination solution with a higher divalent copper concentration. The formulation used by Ando consists of 4 g *rokusho* and 3.2 g copper sulfate ground together and dissolved in 2 l of warm water. The metal object to be colored is first etched in sulfuric or acetic acid, finely polished after surface mechanical working, cleaned using a sodium bicarbonate paste to remove surface oils, and then heated in the coloring solution for about 30 min.

A number of contemporary metalsmiths in the United States produce jewelry by traditional Japanese metalworking techniqes (Pijanowski and Pijanowski 1977, 1978, 1979). Pijanowski and Pijanowski (1977) describe a procedure for the preparation of *rokusho* given by Niiyama (1969). This consists of dissolving 30 g of copper acetate $[Cu(C_2H_3O_2)_2 \cdot H_2O]$, 10 g of calcium carbonate, and 10 g of sodium hydroxide in 750 ml of water. The solution must stand for about a week to allow the precipitate to settle; the remaining liquid ($\sim$500 ml) is then poured off and diluted further with 4.5 l of water. This *rokusho* solution is mixed with 20 g of copper sulfate for use in patinating *shakudo*. In a later paper Pijanowski and Pijanowski (1979) report better results with a solution similar to that published earlier by Wilson (1902) and consisting of 24 g copper sulfate, 6 g sodium chloride, and 90 ml water. A number of other metalsmiths (Vada C. Beetler, Columbus, Ohio; George N. Sawyer; John Paul Miller, Cleveland, Ohio; Mitsuko Kombe Soellner, California) have adapted variations of more traditional Japanese techniques to their own designs.

C. S. Smith (1973, 1977) has proposed that the patina on *shakudo* is produced by a colloidal dispersion of gold in copper oxide. Smith indicates that this patina is the richest black ever produced on a metal and provides the patinated metal surfaces with a high resistance to atmospheric tarnish. Savage and Smith (1979) indicate, as noted in some earlier work by Roberts-Austin and Gowland, that substitute alloys, such as *kuromido* (a dark etching alloy of copper with about 1% arsenic), have been developed that approximate the black color of *shakudo*. These authors also note that the original surface of cast *shakudo* remained relatively unaffected by patination treatments but that a good black patina was developed on abrasively polished surfaces by the solutions described by Roberts-Austin or Gowland or even by standard chromate metallographer's etchants. This would appear to indicate that mechanical deformation plays a significant role in the surface activity of these alloys, as has been reported by Pickering and Swann (1963).

Influence of China

A thorough description of the Chinese alchemical manuscripts concerning the production of "purple sheen gold" is given by Needham (1974). He concludes that the development of "purple sheen gold" started as early as the fifth to third century B.C. and that the manuscripts present a clear picture of the continuing use of a particular alloy over many centuries. It appears to have been first used on a large scale by the Emperor Han Wu Ti (ruled 86–40 B.C.), who was a great patron of Taoists and alchemists. The alloy is presumed to be of much copper and little gold, and Han Wu Ti would have appreciated the economy of its use in bullion. A comment by Kuo Pho in the third century A.D. indicates great admiration for the purple-surfaced alloy. All through the medieval period, purple was one of the most characteristic colors of religious Taoism. It signified lofty empyreal abstraction, like the mystery of dawn and sunset clouds. Needham indicates that, unlike the Mediterranean region, for example, the Murex and senatorial or imperial purple, this color in China had never been one of kingship but was rather tied to Taoist belief and to alchemy. Needham gives a number of examples of thirteenth- to sixteenth-century manuscripts that describe the use of gold added to copper to produce the purple-surfaced alloy.

Needham describes the work of Gowland and Roberts-Austin concerning the Japanese processes for patinating *shakudo* and then compares this with a Chinese recipe of about A.D. 640, described by Sun Ssu-Mo in his manuscript *Tan Ching Yao Chueh*, where use is made of a "plum bath" made from unripe *Prunus mume*. This would have contained a strong mixture of malic, oxalic, and other organic acids, if not acetic acid, in brine and is quite similar to that described by Gowland (1896, 1914–1915).

Recent discussion with Chinese colleagues in archaeometallurgy (Conference on Early Metallurgy, Beijing, October 1981) reveals that not a single piece of the seemingly mythical "purple sheen gold" exists in any of the museum collections in China. There are reports of traditional Chinese artisans currently working in Shanxi Province who produce a black patina on an alloy called *wutong* (brass or copper); this material is being used for the manufacture of small containers or jewelry boxes, the surfaces of which are patinated black and then engraved with a variety of geometric or scenic patterns that highlight the color of the matrix material against the black patinated background. I was also shown photographs of a lacquer box from a recently unearthed late Western Han Dynasty tomb in Hunan Province; it was covered with "black gold" foil that appears to be engraved, uncovering areas of underlying gold. This piece has not been studied in detail, and the workers at the Hunan Provincial Museum in Changsha concede that the piece could have been overlaid with silver that has naturally blackened with age.

Other Alloys Similar to *Shakudo*

Alloys similar to *shakudo* have been found in use in a number of different archaeological contexts. Depletion gilding (gold coloration, or mise-en-couleur, caused by treating the surface of a dilute gold alloy to cause gold enrichment at the surface) has been used by the Indians of Precolumbian Central America (Lechtman 1971, 1973, 1979).

In the European and Mediterranean areas, quite rare occurrences of a similar patina have been noted. Craddock (1982a, b) indicates that the purple-black surface of a small inlaid Roman plaque was found to consist of copper oxide with some gold and silver. The black surface appeared to be developed in situ and not applied. Craddock claims that this material could have been processed by an alchemical transmutation process called "iosis" by Zosimus of Alexandria and said to be carried out with rhubarb (a source of organic acid) and salts. Furthermore, Craddock proposed that this may be an example of the alloy described in detail by Pliny as the prized "Corinthian bronze," of which no examples have been found in modern times.

Finally, there is a reference to the "purple gold" of Tutankhamun as early as 1934 (Wood 1934, pl. 11). Fishman and Fleming (1980) have noted that the most striking feature of the analysis made on a bronze figure of Tutankhamun, having a deep black lustrous patina, was the presence of 4.7% Au.

Oxidation and Aqueous Corrosion of Copper and Copper Alloys

Two different oxides of copper are found to occur in nature: cuprite (Cu_2O, or cuprous oxide), which is deep red and the most common form, and the much more uncommon tenorite (CuO, cupric oxide) which is black (Dana and Ford 1958).

The most recently available equilibrium phase diagram for the $Cu-Cu_2O-CuO$ system as a function of temperature and oxygen partial pressure has been drawn by Schmidt-Whitley et al. (1974) after O'Keeffe and Moore (1962). When metallic copper is oxidized by thermal oxidation, CuO tends to form at lower temperatures and higher oxygen pressures (more oxidizing conditions), whereas Cu_2O forms at higher temperatures and lower oxygen pressures (more reducing conditions). Typically copper oxidizes according to a logarithmic law up to around 200°C. Above that temperature there appears to be a cubic law that is operative over a small range, and then a parabolic law is found. Above 550°C the value of the activation energy, obtained from the variation of the rate constant, corresponds to the activation energy for the diffusion of cuprous ions in Cu_2O. Below this temperature the activation energy is considerably lower and is thought to be associated with a reaction involving CuO (Valensi 1948). The Cu_2O/CuO ratio found experimentally is pressure sensitive, and the amount of CuO decreases with increasing temperature and decreasing oxygen pressure (Kubachewski and Hopkins 1962, p. 106). Typical of the behavior of many binary copper alloys, when Cu-Al alloys with Al $< 10\%$ are oxidized at 800°C, cuprous oxide (Cu_2O) forms rapidly but then forms a continuous protective oxide layer (Dennison and Preece 1953). At lower temperatures Cu_2O forms first, but with time this is consumed by the formation of a CuO outer oxide layer (de Brouckère and Hubrecht 1952a, b). More recent studies using electron microprobe analysis, X-ray diffraction, and optical metallography show that the appearance of a thin inner layer of Cu_2O and a thicker CuO outer layer are typical of Cu-Zn, Cu-Al, and Cu-Ni alpha phase copper alloys (Gabel et al. 1976).

The aqueous oxidation of copper has been studied in a series of papers by Miller and co-workers (Miller and Lawless 1959; Guthrow and Miller 1966; Topham and Miller 1966; Porterfield and Miller 1966). The Cu_2O oxidation product formed in $CuSO_4$ aqueous solutions tends to form initially as a thin film but then grows in the form of well-defined oriented polyhedra of the oxide (Miller and Lawless 1959). The observed behavior is interpreted in terms of the electrochemical theory of corrosion and the dislocation theory of crystal growth. The role of the aqueous salt solution is described to provide both an easy path for the flow of electrons and ions and a readily available source of copper ions. Lower oxidation rates obtained in aqueous K_2SO_4 solutions indicate the importance of the copper ions in solution. Deaerated and aerated $CuSO_4$ solutions yield approximately the same oxidation rates, indicating the lack of importance of dissolved oxygen; it is therefore assumed that oxidation occurs primarily through the reactions

$$2Cu^+ + 2OH^- \rightarrow Cu_2O + H_2O,$$

$$2Cu^{++} + 2e^- + H_2O \rightarrow CuO + 2H^+.$$

The pH of the solution is observed to decrease from the 3.8–4.3 range to about 3.2 over a period of about 24 hours, with an attendant decrease in $[Cu^{1+}]$ and decreased oxide growth rate. Under the same initial conditions of Cu^{2+} ion concentration, bulk pH, and temperature, oxide growth in $Cu(OAc)_2$ solutions produces relatively smooth and continuum films (Guthrow and Miller 1966; Porterfield and Miller 1966; Scott and Miller 1966), yielding improved corrosion resistance. These $Cu(OAc)_2$ solutions are produced by the dissolution of $Cu(OAc)_2 \cdot H_2O$ to a pH of 5.6 (0.05 *N*) and then by adding glacial acetic acid to reduce the pH to 4.3, the same pH as the $CuSO_4$ solutions. In contrast to the previous results using $CuSO_4$, the pH in the $Cu(OAc)_2$ solution remains essentially constant. The constancy is expected because the acetic acid added to adjust the initial pH to 4.3 would allow the solution to act as a buffer system. Referencing the work of Yamada et al. (1957), which indicates that in concentrated Cu^{2+} formate solutions there are infinite chains of $Cu(CHO_2)_2 \cdot 2H_2O$ joined together by hydrogen bonding through the two water molecules and that in copper acetate solutions there are binuclear complexes of the form $Cu_2(CH_3COO)_4 \cdot 2H_2O$, Topham and Miller (1966) proposed that aqueous corrosion in formate and acetate solutions should be similar. Topham and Miller also indicate that the Cu^{2+} ions in solutions of $CuSO_4$ and $Cu(ClO_4)_2$ would exist predominantly as $Cu(H_2O)_4^{2+}$ and that the perchlorate and sulfate ions would exist in a noncomplexed form, and hence sulfate and perchlorate solutions should produce similar corrosion behavior. Their experimental corrosion studies confirm the predictions both in terms of similarity of corrosion products and of pH behavior.

Lichter and Wagner (1960) found that, during sulfidation of copper-gold alloys at 500°C, the copper-rich alloys form an outer homogeneous sulfide layer and an inner two-phase layer consisting of sulfide and a spongelike network of nearly pure gold in accordance with preferential sulfidation of copper with interdiffusion of the components in the alloy.

Aqueous oxidation of copper alloys has been intensively studied by Pickering and co-workers (Pickering and Swann 1963; Pickering and Wagner 1967; Pickering and Byrne 1971; Pickering 1968a, 1968b, 1969; Swann 1969). Pickering and Swann (1963) first studied the nature of chemical attack of Cu−5 atomic % Au, Cu−25 atomic % Au, Cu−40 atomic % Au, copper-aluminum, and copper-zinc alloys using transmission electron microscopy. In copper-gold alloys they observed increased chemical attack with increased deformation, noting that, with the addition of gold to copper, the cross slip of dislocations becomes more difficult, causing the cellular arrangement of dislocation tangles characteristic of deformed pure copper to be replaced by a planar dislocation array with increasing gold content. Deformation introduced into bulk material in the form of dislocations or the presence of an internal stress has also been shown to enhance surface reactivity in iron (Gulbransen and Copan 1960) and iron-chromium alloys (Frankenthal 1968).

Copper−5 atomic % Au exposed to ferric chloride or dilute aqua regia solutions produced a pitted corrosion, with pits approximately 5−10 nm in diameter and with the pitted regions being of enhanced gold content (Pickering and Swann 1963). Potassium cyanide−hydrogen peroxide solutions appeared to cause the formation of a fine gold precipitate on the attached surface. It was concluded that under certain conditions there appear to be three stages involved with the initiation and propagation of pits or tabular attack: (1) formation of a protective film over the major portion of the surface, (2) acceleration of attack at remaining uncovered areas or at points of local breakdown of the protective film, and (3) development of a local electrochemical cell capable of preventing film repair and sustaining anodic dissolution.

In later papers Pickering and Wagner (1967) and Pickering and Byrne (1971) describe the anodic dissolution of copper-gold alloys as occurring such that a copper-depleted layer forms over the surface, limiting the rate of copper dissolution to that of copper diffusion through the layer. Initially ionization of the less noble metal occurs concurrently with movement of both metals. However, only Cu^{2+} ions enter into solution, leaving an enriched gold layer behind. This is similar to an earlier model used by Gerischer and Rickert (1955) to describe dissolution of a Cu−10 atomic % Au alloy in aqueous H_2SO_4. Pickering first proposed a model that emphasized volume diffusion as an important feature of the dissolution mechanism (Pickering 1968a, b); later work, however, did not support the initial model but rather demonstrated that surface diffusion and redeposition may be operative (Pickering 1969; Swann 1969).

More recently, Forty (1979, 1981) and Forty and Rowlands (1981) have made essentially similar observations to those of Pickering and Swann (1963), but in addition have used transmission electron microscopy (TEM) and a quantitative model to follow the development of this island-channel structure. For the high gold compositions studied, the island growth was found to occur rapidly such that pits arising from incomplete merger of the gold islands are formed after only short exposures to corrosive environments. Furthermore, Forty (1981) reports the probable formation of Au_2O_3 (Jones et al. 1980) within the near-surface disordered zone during corrosion of silver-rich gold alloys by strong acids.

Experimental Studies

Because Pickering (1968b) and Swann (1969) were unable to obtain a clear indication of the composition of the gold-rich second phase formed on copper-gold alloy surfaces and because it was desired to examine the patination of *shakudo* both in early fabricated pieces and on layers produced under laboratory conditions, the present study was initiated.

Specimens of Cu−3 wt. % Au and Cu−5 wt. % Au *shakudo* alloys were purchased from Komorimiya Kingen Kogyo Co., Tokyo, in the form of strips ~1.6 mm thick in the as-rolled (full hard) condition. A number of *shakudo* artifacts were also studied; these are described in appropriate places in the text.

Metallographic specimens of the purchased alloys were mounted, polished, and then etched chemically with freshly made 1:1 mixture of 10% Na_2HSO_4 and 10% KCN. The solution was immediately applied to the sample; etching times were typically a few minutes (Ntukogu 1979). The best structure was obtained by slightly overetching the sample and subsequently lightly polishing the stained surface with 0.3 μm alumina powder.

All traditional metallographic etchants for copper alloys tended to produce a heavy black or purple-black stain on the as-rolled (deformed) material. One specimen that had been annealed for a short time at ~500°C was etched without staining and with little difficulty in the same etching solutions. In both cases the microstructures appeared typical of those observed in most single-phase copper alloys, that is, mechanical slip traces and elongated grains in the as-rolled specimen and equiaxial grains showing annealing twins in the annealed specimen.

Small samples of the as-rolled 3 wt. % and 5 wt. % Au alloys were polished on a flat silk cloth using a water slurry containing 0.3 μm alumina powder.

The specimens were supported using cotton thread and placed in glass beakers containing the appropriate patination solutions as described previously. Several other patination experiments, each with only a single chemical component in aqueous solution, were also tried. Table 28.5 lists each of the solutions used for the patination experiments.

X-ray diffraction patterns were obtained from the surface of each of the patinated alloys by measurements made on a Philips APD-3600 Automated Powder Diffractometer. Diffraction patterns from the majority of the specimens consisted of strong Cu_2O peaks and weak Au peaks (figure 28.1), as expected, as the volume fraction of gold particles in Cu_2O would be small. These Au peaks are much weaker than those observed in the work of Pickering (1968b, 1969) and Pickering and Wagner (1967) because in these studies the nature of the corrosion-producing solution was such as to preferentially attack the surface by pitting corrosion, thereby depleting it of copper and enriching it in gold. Also, the present X-ray results indicate little or no shift from the peak position expected for pure gold. In the earlier work of Pickering, shifts in the Au peak were interpreted as indicating the metallic phase to be gold with copper in solid solution as a result of a volume diffusion contribution during the growth process. Finally, some of the patinated *shakudo* layers (especially those etched with dichromate) indicated the presence of CuO as well as the presence of a few weak and unidentified peaks.

Figure 28.2 shows a piece of decorative metalwork containing inlaid sections of *shakudo* alloy; the piece shown is the pommel (*koshira*) found at the rear of the hilt of a Japanese sword and is typical of such pieces fabricated in the eighteenth or nineteenth century. The items shown on this *koshira* consist of a war helmet and war wand or fan (used to relay battle signals) in the folded position. The wand is shown at 10 × magnification in the scanning electron microscope picture of figure 28.3 (ETEC SEM); it is fabricated from a small inlaid piece of *shakudo* that has been patinated and then possibly surface chipped to produce a beautiful woodlike surface texture (figure 28.4). Higher magnification SEM photographs within the chipped regions (figure 28.5) indicate the presence of a second phase consisting of lighter dispersed particles; typically higher atomic number elements produce enhanced electron intensity, resulting in enhanced or "atomic number" contrast. SEM pictures taken using backscattered rather than secondary electrons give even better "atomic number" contrast, and a higher magnification picture taken in the backscatter mode is shown in figure 28.6.

The scanned image of a CRT also allows images to be constructed from the characteristic X-rays emitted locally as a result of electron bombardment. Figure 28.7 shows an image of the same area shown in figure 28.6 but constructed from the Au-L line radiation; it

Table 28.5 Patination solutions for *shakudo*

Roberts-Austin, solution 1 (table 28.2)
Roberts-Austin, solution 2 (table 28.2)
Roberts-Austin, solution 3 (table 28.2)
Roberts-Austin, solution 3, vinegar deleted
Gowland
Wilson, solution 1 (table 28.4)
Wilson, solution 2 (table 28.4)
0.25 *M* NaOH
$CuSO_4$ (50g/l)
Standard dichromate etch
Dichromate etch + 20% NaCl
Dichromate etch + 50% NaCl
Dichromate etch + 20% H_2SO_4

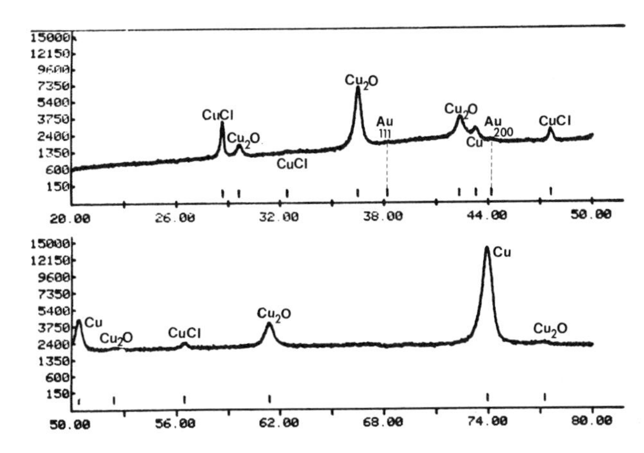

Figure 28.1 X-ray diffractometer trace of patina developed on Cu–3 wt. % Au alloy.

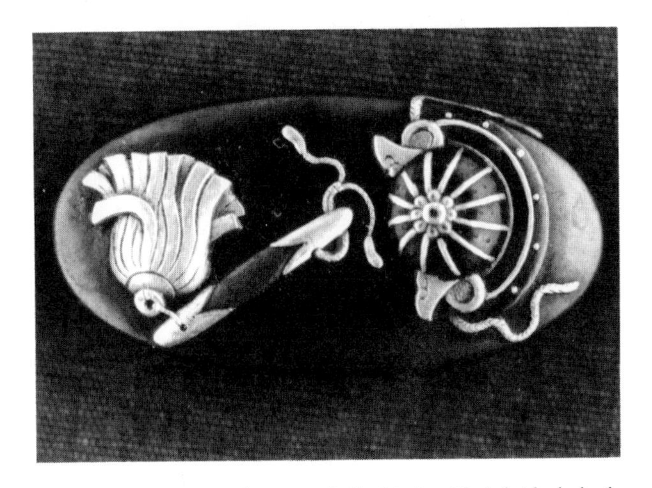

Figure 28.2 Sword pommel (*koshira*) with inlaid *shakudo* war fan and war helmet. Late eighteenth or early nineteenth century.

Figure 28.3 SEM photomicrograph of war fan shown in figure 28.2 (× 10).

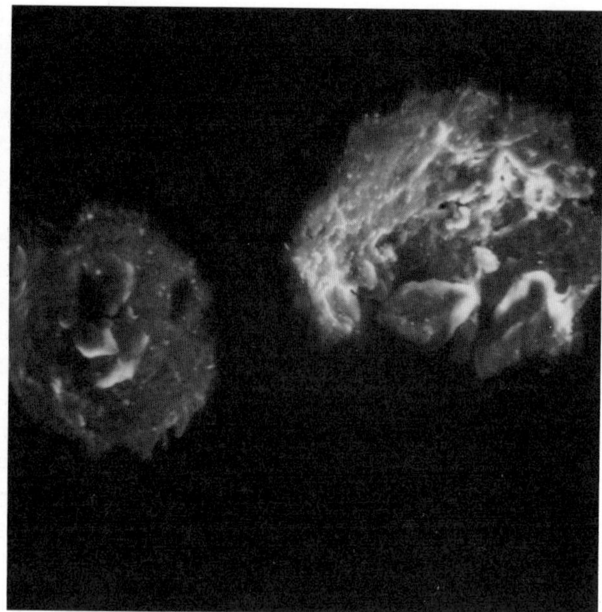

Figure 28.5 SEM photomicrograph of chipped area of fan in figure 28.2 (× 200). Note small white particles within chipped area.

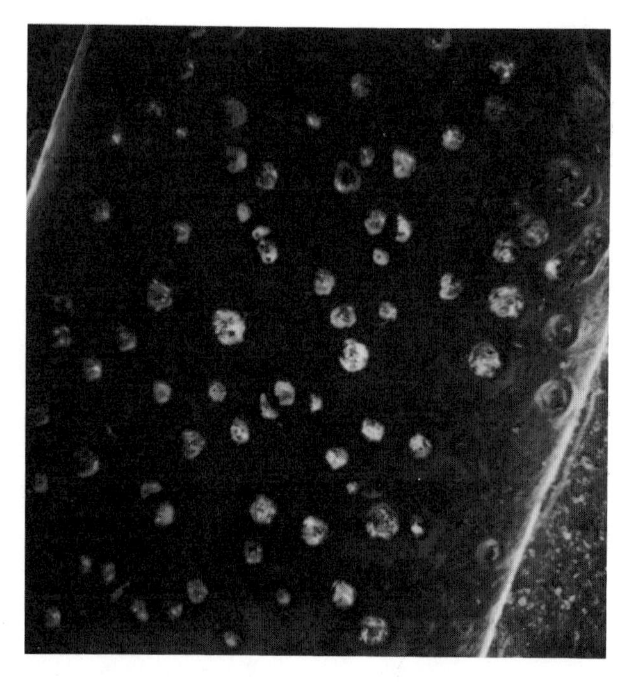

Figure 28.4 SEM photomicrograph of detail in war fan of figure 28.2 (× 50).

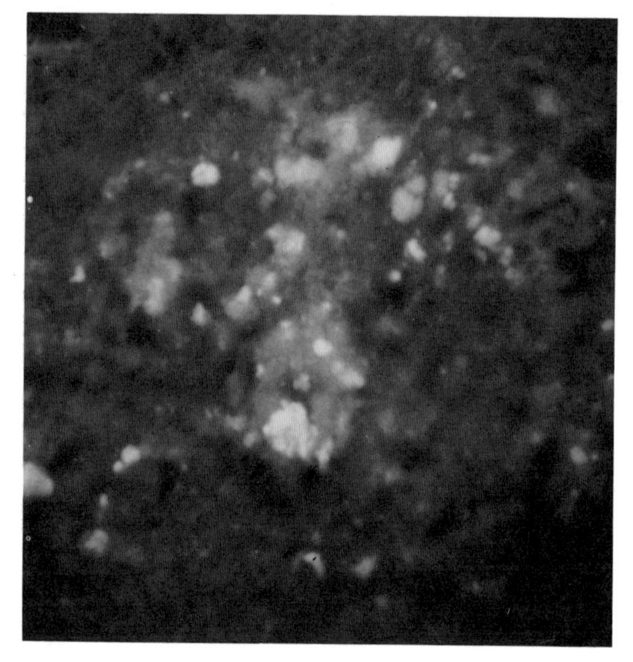

Figure 28.6 Backscattered electron SEM image of chipped area of fan in figure 28.2 (× 3000). Total counts, Au = 5436, Cu = 578 (60 sec).

indicates the particles to contain a significantly higher gold content than the surrounding matrix. Rather than scanning the electron beam back and forth over an entire area, the electron beam may be positioned at a particular point and the energy dispersive X-ray detector can be used to obtain an X-ray spectrum giving semiquantitative chemical information from a localized region. Such a spectrum obtained from one of the lighter particles is shown in figure 28.8 and indicates the particles to consist almost entirely of gold. However, because of the limited spatial resolution of the electron beam in SEM ($\geqslant 1\ \mu m$) and the possible effects of the measurement of excess copper X-radiation from scattering from the surrounding matrix and given the small size of the particles, an exact analysis could not be obtained; analytical techniques with better spatial resolution are required for direct verification of the particle chemistry.

Thin foils for electron microscope examination were obtained from the patinated *shakudo* surfaces by applying a few drops of acetone and an acetate tape to the specimen surface, peeling back the dried tape, and floating the small extracted flakes onto microscope grids immersed in acetone and alcohol. The foils were examined in a Philips 400 TEM/STEM equipped with an EDAX X-ray detector. A thin foil of the patinated layer, typical of those examined, is shown in figure 28.9. It consists of a dispersion of dark spherical particles (the largest of which is ~8000 Å) in an electron-transparent matrix. Electron diffraction patterns obtained from the matrix regions indicated the matrix to be Cu_2O.

Examination using electron diffraction was also helpful in determining the nature of the second phase particles. Figure 28.10 shows the diffraction ring pattern obtained for a thin foil gold standard; figure 28.11 shows the diffraction pattern obtained near the edge of a thin foil from a patinated *shakudo* layer (figure 28.12). The diffraction pattern obtained from the *shakudo* foil consists mainly of spots lying on Au diffraction rings; a dark field image (figure 28.13) obtained from these diffraction spots indicates that the spots originate from the second phase particles, that is, the particles are metallic gold. STEM/X-ray spectra indicate the particles to be gold enriched, but X-ray information from the matrix is also included in the spectra (figure 28.14).

In the cases shown here, the X-ray diffraction, SEM/EDS, TEM, electron diffraction, and STEM/X-ray analyses are all consistent with the identification of the *shakudo* patina as nearly pure metallic gold particles in a Cu_2O matrix. However, examination of the patinated layers on a few other artifacts visually identified as *shakudo* indicated the second phase particles to be Ag_2S, Ag_2O, or $Ag_2S_2O_7$ and sometimes to consist of particles in a CuO rather than a Cu_2O matrix. Be-

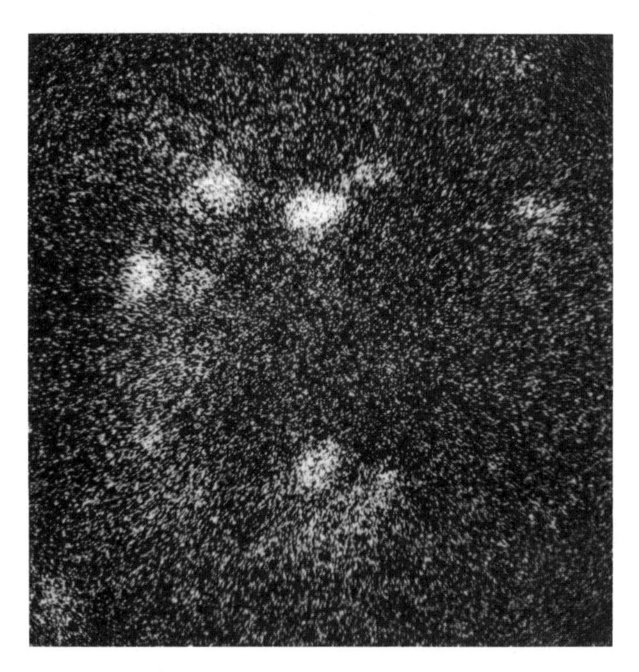

Figure 28.7 Au-L X-ray image of chipped area of fan in figure 28.2 (× 3000).

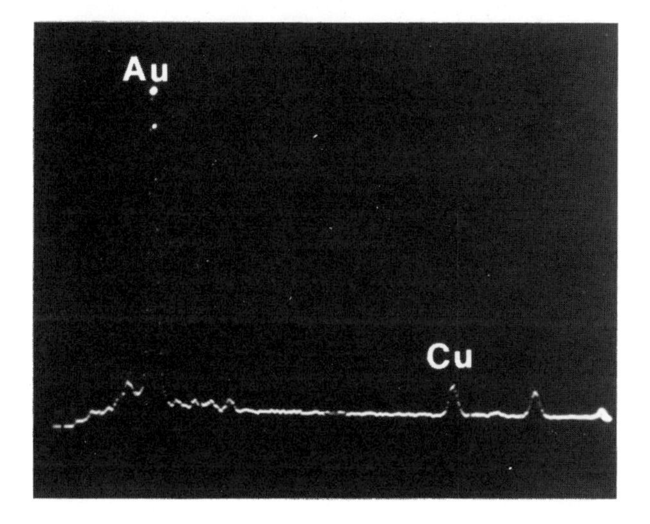

Figure 28.8 Energy dispersive X-ray spectrum from a large second phase particle in figure 28.7.

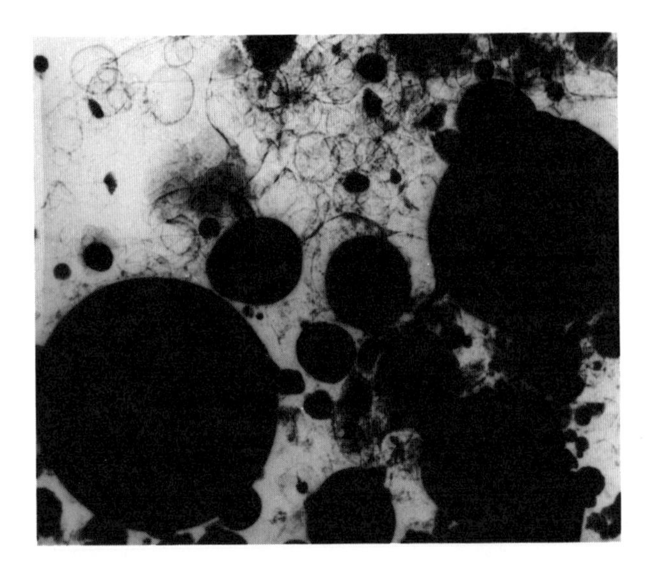

Figure 28.9 TEM photomicrograph of patinated layer extracted from *shakudo* alloy (× 50,000).

Figure 28.10 Electron diffraction ring pattern for Au standard.

Figure 28.11 Electron diffraction pattern of patinated *shakudo* layer.

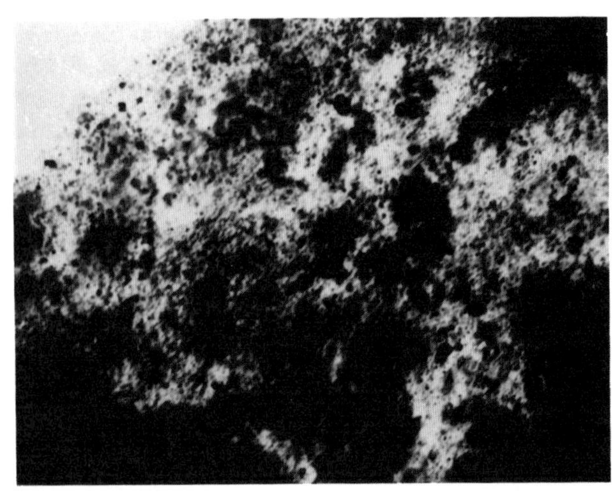

Figure 28.12 Bright field TEM photomicrograph of patinated *shakudo* layer.

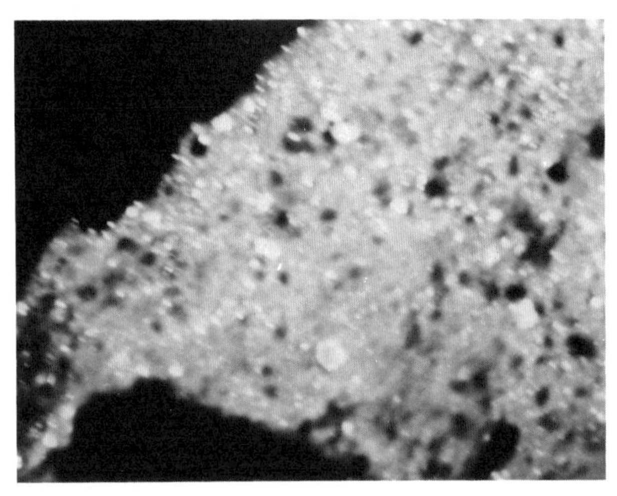

Figure 28.13 Dark field TEM photomicrograph of patinated *shakudo* layer of figure 28.12 using diffraction spot identified as originating from Au reflection.

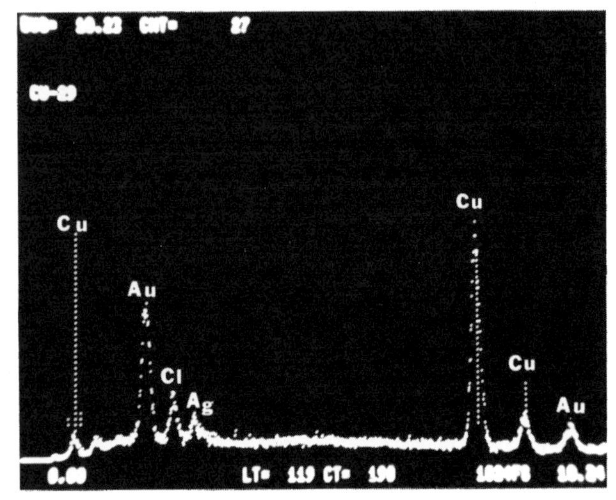

Figure 28.14 STEM/EDS X-ray spectrum obtained at second phase particle in patinated *shakudo* layer.

Figure 28.15 SEM photomicrograph of morphology of patina formed on Cu–3 wt. % *shakudo* alloy, placed in $CuSO_4$ solution.

cause many of the pieces obtained for analysis tended to be of more recent manufacture and/or from less expensive artifacts, this could indicate either the use of a variety of patination techniques or the use of less expensive alloys with lower gold content for many of the pieces.

It appears that, although all the patinating solutions produce a deep black color, the surface appearance ranges from highly lustrous, found to be associated with a continuous patina layer, to a dull matte texture related to the growth of discrete polyhedra at the surface (figure 28.15) similar to that observed previously by Miller and Lawless (1959) under similar corrosion environments but on pure copper.

Acknowledgments

I wish to express gratitude to the Mellon Foundation for its support through a faculty support grant to Lehigh University. Thanks are due to Jeff Binns and Gillian Parry, undergraduate summer assistants from Manchester University, England, for much of the SEM work and to Barry Bender for the electron microscopy.

References

Brinkley, F. 1903–1904. *Japan, Its History, Art and Literature.* London.

Craddock, P. T. 1982a. "Corinthian bronze: Rome's purple sheen gold." *MASCA Journal* 2(2):40–41.

Craddock, P. T. 1982b. "Gold in antique copper alloys." *Gold Bulletin* 15(2):69–72.

Dana, E. S., and W. E. Ford. 1958. *A Textbook of Mineralogy.* New York: Wiley.

de Brouckere, L., and L. Hubrecht. 1952a. "Dry oxidation of copper and its alloys." *Bulletin de la Société Chimie Belgique* 60:311.

de Brouckere, L., and L. Hubrecht. 1952b. "Dry oxidation of copper and its alloys." *Bulletin de la Société Chimie Belgique* 61:101–123, 205.

Dennison, J. P., and A. Preece. 1953. "High-temperature oxidation characteristics of a group of oxidation-resistant copper-base alloys." *Journal of the Institute of Metals* 81: 229–234.

Exhibition Catalog of the Shoso-in Treasures—1979. Nara: Nara National Museum, item 7.

Fishman, B., and S. J. Fleming. 1980. "A bronze figure of Tutankhamun: Technical studies." *Archaeometry* 22(1): 81–86.

Forty, A. J. 1979. "Corrosion micromorphology of noble metal alloys and depletion gilding." *Nature* 282:597–598.

Forty, A. J. 1981. "Micromorphological studies of the corrosion of gold alloys." *Gold Bulletin* 14(1):25–35.

Forty, A. J., and G. Rowlands. 1981. "A possible model for corrosion pitting and tunnelling in noble metal alloys." *Philosophical Magazine,* sec. A, 43(1):171–188.

Frankenthal, R. P. 1968. "Effect of surface preparation and deformation on anodic dissolution of Fe-Cr alloys." *Corrosion Science* 8:491–498.

Gabel, H., J. A. Beavers, J. B. Woodhouse, and E. N. Pugh. 1976. "The structure and composition of thick tarnish films on alpha-phase copper alloys." *Corrosion* 32(6):253–257.

Gerischer, H., and H. Rickert. 1955. "Electrochemical behavior of copper-gold alloys and the mechanism of stress corrosion." *Zeitschrift für Metallkunde* 46:681–689. (In German.)

Gowland, W. 1896. "Japanese metallurgy I. Gold and silver and their alloys." *Journal of the Society of the Chemical Industry* 15:404–414.

Gowland, W. 1914–1915. "Metals and metalworking in old Japan." *Transactions of the Japan Society* (London) 13:20–94. Reprinted in *Japanese Crafts: Materials and Their Applications,* B. Hickman, ed. (London: Fine Books Oriental, 1977).

Greey, E. n.d. "Description of a superb collection of Japanese tsuba." New York.

Gulbransen, E. A., and T. P. Copan. 1960. "Microtopology of the surface reaction of oxygen and water vapor with metals." 229.

Guthrow, C. E., and G. T. Miller. 1966. "Oxidation of copper single crystals in aqueous solutions of inorganic salts I." *Journal of the Electrochemical Society* 113(5):415–424.

Hiorns, A. H. 1892. *Metal Colouring and Bronzing*. London: Macmillan.

Hiorns, A. H. 1901. *Mixed Metals or Metallic Alloys*. London: Macmillan. Second edition, 1912.

Huish, M. B. 1889. *Japan and Its Art*. London: Fine Arts Society.

Joly, H. L. 1909. "Introduction to the study of sword mounts." *Bulletin de la Société Franco-Japonaise de Paris* 14:51.

Jones, P., H. Rumpel, G. M. Sheldrick, and E. Schwartzmann. 1980. "Gold (III) oxide and oxychloride." *Gold Bulletin* 13(2):56.

Kubachewski, O., and B. E. Hopkins. 1962. *Oxidation of Metals and Alloys*, second edition. London: Butterworths.

Lechtman, H. N. 1971. "Ancient methods of gilding silver: Examples from the Old and New worlds," in *Science and Archaeology*, R. H. Brill, ed. (Cambridge, Mass.: MIT Press), 2–30.

Lechtman, H. N. 1973. "The gilding of metals in Pre-Columbian Peru," in *Applications of Science in Examination of Works of Art*, W. J. Young, ed. (Boston, Mass.: Museum of Fine Arts), 38–52.

Lechtman, H. N. 1979. "A Pre-Columbian technique for electrochemical replacement plating of gold and silver on copper objects." *Journal of Metals* 31(12):154–160.

Lichter, B. D., and C. Wagner. 1960. "The attack of copper-gold, silver-gold, nickel-copper, and silver-copper alloys by sulfur at elevated temperatures." *Journal of the Electrochemical Society* 107:168–180.

Lyman, B. S. 1890–1891. "Japanese swords." *Numismatic and Antiquarian Society Proceedings*, 23–60.

Masuda, T., and Kudo Zuroko. c. 1801. *Picture Book on the Smelting of Copper*. Osaka: Zenryu Shinakawa. (In Japanese.) English translation published by Bundy Library (Norwalk, Conn.) in 1983.

Miller, G. T., and K. R. Lawless. 1959. "An electron microscope study of the formation of oxide on Cu single crystals immersed in an aqueous solution of copper sulphate." *Journal of the Electrochemical Society* 106(10):854–860.

Needham, J. 1974. *Science and Civilization in China*. Cambridge: Cambridge University Press, vol. 5, pt. 2, 257–265.

Niiyama, 1969. *Wrought Gold Chasing Techniques*, vol. 1. Nihon Kinko Sakka Kyokai. Reprinted in 1982. (In Japanese.)

Ntukogu, A. O. 1979. *Solid Solution Miscibility in Ternary Copper-Silver-Gold System*. Ph.D. Dissertation, Polytechnic Institute of New York.

O'Keeffe, M., and W. J. Moore. 1962. "Thermodynamics of the formation and migration of defects in cuprous oxide." *Journal of Chemical Physics* 36(11):3009–3013.

Pickering, H. W. 1968a. "The surface roughening of a Cu-Au alloy during electrolytic dissolution." *Journal of the Electrochemical Society* 115:690–694.

Pickering, H. W. 1968b. "Volume diffusion during anodic dissolution of a binary alloy." *Journal of the Electrochemical Society* 115:143–147.

Pickering, H. W. 1969. "Stress corrosion via localized anodic dissolution in Cu-Au alloys." *Corrosion* 25(7):289–290.

Pickering, H. W., and P. J. Byrne. 1971. "Preferential anodic dissolution of alloys." *Journal of the Electrochemical Society* 118(2):209–215.

Pickering, H. W., and P. R. Swann. 1963. "Electron metallography of chemical attack upon some alloys susceptible to stress corrosion cracking." *Corrosion* 19:373–389.

Pickering, H. W., and C. Wagner. 1967. "Electrolytic dissolution of binary alloys containing a noble metal." *Journal of the Electrochemical Society* 114(7):698–706.

Pijanowski, H. S., and E. M. Pijanowski. 1977. "Lamination of nonferrous metals by diffusion: Adaptions of the traditional Japanese techniques of mokume-gane." *Goldsmith's Journal* 3(4):20.

Pijanowski, H. S., and E. M. Pijanowski. 1978. "Workshop: Mokume-gane." *Crafts Horizons* (February), 32–35.

Pijanowski, H. S., and E. M. Pijanowski. 1979. "Update: Mokume-gane." *Goldsmith's Journal* 5(1):16–20.

Porterfield, W. W., and G. T. Miller. 1966. "Oxidation of copper single crystals in aqueous solutions of inorganic salts III. Kinetics of oxidation in $Cu(OAc)_2$ solution." *Journal of the Electrochemical Society* 113(6):528–531.

Pumpelly, R. 1866. "Notes on Japanese alloys." *American Journal of Science* 42:43–45.

Roberts-Austin, W. C. 1888. "Cantor lectures on alloys. Lecture III." *Journal of the Society of Arts* 36:1137.

Roberts-Austin, W. C. 1893. "Cantor lectures on alloys. Lecture IV." *Journal of the Society of Arts* 41:1022.

Robinson, B. W. 1961. *The Arts of the Japanese Sword*. London. Second edition, 1970.

Sasano, M. 1979. *The Origins of Sword Fittings*. Tokyo. (In Japanese.)

Savage, E. I., and C. S. Smith. 1979. "Techniques of the Japanese tsuba maker." *Ars Orientalis* 11:291–328.

Schmidt-Whitley, R. D., M. Martinez-Clemente, and A. Revcolevschi. 1974. "Growth and microstructural control of single-crystal cuprous oxide." *Journal of Crystal Growth* 23(2):113–120.

Scott, D. B., and G. T. Miller. 1966. "Oxidation of copper single crystals in aqueous solutions of inorganic salts IV." *Journal of the Electrochemical Society* 113(9):883–886.

Smith, C. S. 1973. "Historical notes on the colouring of metals," in *Recent Advances in Science and Technology of Materials*, A. Bishay, ed. (New York: Plenum), vol. 3, 157–167.

Smith, C. S. 1977. "Some constructive corrodings," in *Corrosion and Metal Artifacts*. NBS Publication 479 (Gaithersberg, Md.: National Bureau of Standards), 143.

Smith, C. S. 1981. *A Search for Structure*. Cambridge, Mass.: MIT Press.

Swann, P. R. 1969. "Mechanism for corrosion tunnelling with special reference to Cu_3Au." *Corrosion* 25(4):147–150.

Topham, R. W., and G. T. Miller. 1966. "Oxidation of copper single crystals in aqueous solutions of inorganic salts II." *Journal of the Electrochemical Society* 113(5):421.

Valensi, G. 1948. "New applications of the theory of oxidation of metals having two oxides." *Review of Metallurgy* 45(7):205–210.

Wakansandsaidzue, vol. 59, folio 15 [as given in Lyman (1890–1891)].

Wiest, F. K. 1978. *Newsletter of the Japanese Sword Society of the United States* 10(1):9.

Wilson, H. 1902. *Silverwork and Jewelry*. London: Pitman. Reprinted in 1978.

Wood, R. W. 1934. "The purple gold of Tutankhamun." *Journal of Egyptian Archaeology* (London) 20:62–65.

Yamada, S., H. Nakamura, and R. Tsuchida. 1957. "Spectrochemical study of microscopic crystals XVI. Structure of cupric formate, acetate, and propionate." *Bulletin of the Chemical Society of Japan* 30:953–957.

29
The Metallurgy of Ancient West Mexico

Dorothy Hosler

The Context

A vigorous and innovative copper-based metallurgy developed in Mexico between approximately A.D. 800 and the sixteenth-century Spanish invasion in the area encompassed by the modern states of Michoacan, Colima, Nayarit, and Jalisco (figure 29.1). The technology was closely related to the indigenous metallurgies of Central and South America, which had emerged hundreds of years earlier. Metallurgy appeared relatively late in the history of ancient Mexico; plant domestication had begun by 3500 B.C., and sedentary agricultural communities were common at about 1800 B.C. By 1100 B.C. complex society was emerging among the Olmec, in the tropical lowlands of the southeast.

One of the most striking features of the metallurgy of ancient Mexico is that the technology appeared long after the advent of complex sociopolitical organization. By A.D. 800 Mesoamerica's most powerful Classic Period states (A.D. 300–900) had reached their apogee and were declining. Teotihuacan, for example, a large urban center in the valley of Mexico near Mexico City, had reached its maximum population between A.D. 550 and 600. At that time the city covered 8 mi^2 and supported a population estimated at 125,000 people (Millon 1981). The florescence and decline of Mayan civilization, centered in the lowland regions of Guatemala and Mexico to the southeast, had likewise occurred before the appearance of metal.

The western region of Mexico, in which metallurgy developed, is rich in mineral resources and environmentally extremely diverse. The area is defined by the rugged Sierra Madre Occidental, a north-south range

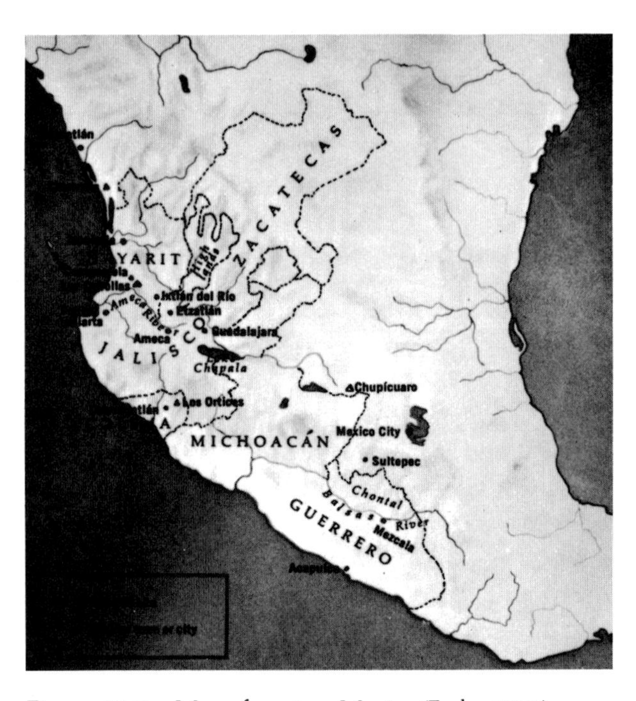

Figure 29.1 Map of western Mexico (Easby 1970).

of mountains, and the transverse volcanic axis with peaks reaching to 17,000 ft. Varied ecological zones occur within a fairly circumscribed geographical area as the central Mexican plateau, averaging some 7,000 ft in altitude, drops off to the Pacific Ocean. These zones range from the cool highland oak-conifer forests at the higher elevations to the tropical deciduous and semideciduous vegetation of the hot coastal plain. The region is heavily mineralized, containing significant deposits of copper, silver, lead, zinc, and other metals. Malachite and chalcopyrite are the most usual sources of copper; silver occurs as metallic silver, argentite, and silver sulfosalts. Arsenopyrite is the most common source of arsenic.

Unlike central Mexico, state-level organization did not emerge in western Mexico until approximately A.D. 1200. The Tarascan empire centered at Tzintzuntzan in central Michoacan constituted the largest political unit in Mexico apart from the Aztec at the time of the Spanish invasion. It consisted of a centralized kingdom surrounded by tributary and buffer areas that encompassed at least the modern state of Michoacan, parts of Jalisco, and surrounding areas. Smaller-scale settlements also flourished throughout the western Mexican region in the Prehispanic era. These communities, like the Tarascans, were agricultural, raising maize, beans, squash, and other crops adapted to the local soils, climatic conditions, and altitude. Some settlements were tiny hamlets populated by only a handful of people. Others, for example, Amapa on the coastal plain of Nayarit, became large. Amapa extends over 1 km^2 (Meighan 1976), has planned residential structures, a plaza and mound groups, and a ball court. Centers such as these were inhabited by a sizable population of farmers, specialized religious and political functionaries, artisans, and craftspeople.

The Study

The metallurgy that developed in ancient West Mexico was investigated through laboratory studies of Prehispanic metal artifacts from that region. The artifacts are part of a museum collection in Guadalajara, Mexico, containing approximately 2,300 metal objects that reflect both the range of artifact types fabricated in the Prehispanic era—from tools to status objects—and their relative abundance.[1] The artifacts, housed in the collection of the Regional Museum of Guadalajara (RMG), are identical with metal artifacts found in archaeological contexts dating from approximately A.D. 800 to the Spanish conquest. A sample of 400, representing distinct functional types, was selected for intensive laboratory analyses. The studies included qualitative and quantitative analyses of chemical composition, metallographic examinations to identify fabrication technique, microhardness tests, and the examination of alloy-property relations. The research also investigated the ore geology of the western Mexican region to identify the particular ore types most likely to have been the parent material for these objects. Ethnohistoric and historic sources served to amplify and corroborate the data about the cultural function of the artifacts established through the laboratory studies.

The Cultural Focus of West Mexican Metallurgy

The study results indicate (Hosler 1986b, "Style in Materials") that the primary use of metal in ancient West Mexico was to fashion objects that symbolically marked elite and sacred or religious domains of culture. These objects were principally made from copper and alloys of copper and silver, copper and arsenic, and copper and tin. The use of copper-arsenic and copper-tin bronzes was widespread. Despite the predominantly religious or sacred emphasis of the technology, West Mexican smiths also made tools and household implements. The metalworkers utilized a variety of techniques, including lost-wax casting, open mold casting, hot and cold working, and so forth.

The most frequent use of metal was for bells, which comprise approximately 60% of the objects in the corpus studied. West Mexican bells exhibit a wide range of forms. Twenty-nine discrete subtypes were identified in the study, a few of which are illustrated in figure 29.2. All West Mexican bells are relatively small. None exceeds approximately 10 cm in height. They contain a loose pebble clapper (occasionally of metal) and have a narrow slit opening at their base and a suspension ring at the top. As figure 29.2 indicates, the resonators of some bells exhibit wirework surfaces, which in some cases are also distinguished by raised zigzag designs.

The metallographic studies show that all the bells, including those exhibiting elaborate external wirework, were lost-wax cast in a single piece. Figure 29.3 is a photomicrograph of the suspension ring and the upper portion of the resonator chamber of a cast wirework copper-tin alloy bell (2.81% Sn). As the photomicrograph indicates, the structure of the suspension ring and the upper body of the bell is continuous. The grain boundaries are indistinct, but the size of the dendrites indicates that the grains are very large. The heavily cored dendritic structure is unusual for a copper-tin alloy with such a low tin concentration. These structural features and the grain size indicate that the metal cooled slowly in the mold.

West Mexican smiths made bells from several different metals and alloys. In addition to copper-tin bronze, they utilized unalloyed copper, copper-arsenic bronze, and occasionally silver and gold. Certain bell subtypes identified on the basis of form and dimension are consistently made from alloys—of either copper-tin or copper-arsenic bronze. The concentration of the alloying element varies. The mean concentration for the

Figure 29.2 West Mexican bells.

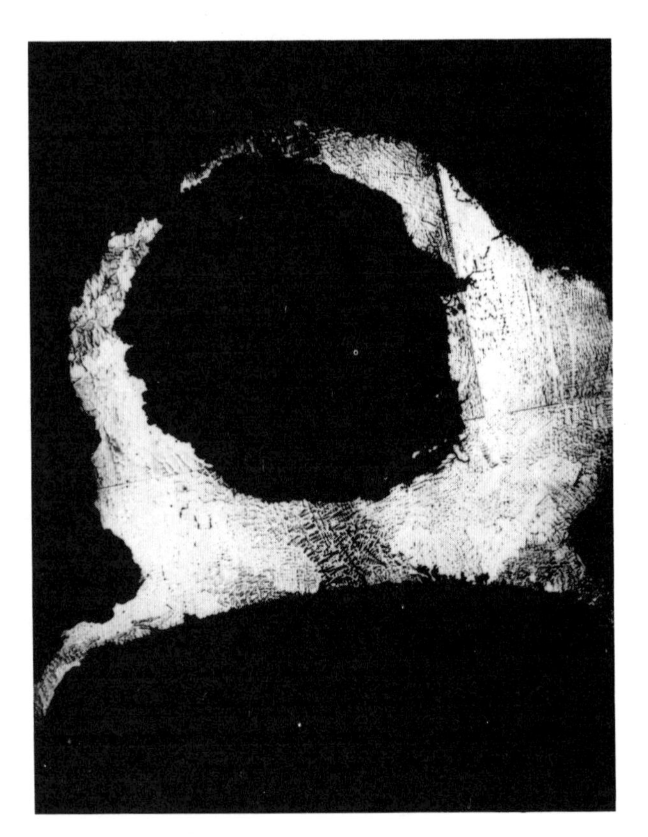

Figure 29.3 Section of bell suspension ring and upper resonator (2.81% Sn). × 100. Potassium dichromate etch.

copper-tin bronze bells is about 8% tin by weight; for bells made from the copper-arsenic alloy, arsenic is present at a level of approximately 3 wt. %, although there are some bells that contain arsenic in concentrations as high as 23%.

Bells made from the bronze alloys tend to be relatively thin walled, often with intricate wirework decorations. Figure 29.4a shows a copper-arsenic bronze wirework bell (1.94% As) with a raised zigzag design. The photomicrograph in figure 29.4b is a section through the wirework and zigzag design showing the individual raised wirework "threads" and the superimposed zigzag design. The bell wall measures only 0.05 cm in thickness. It is clear from the continuous dendritic structure across both the wall and the zigzag design that these zones were cast in a single piece. When design is considered with respect to properties conferred by the alloys, it is clear that alloys were used intentionally for their increased fluidity and their strength, both of which are necessary in casting these complex thin-walled objects.

Other bell types, by contrast, are consistently made from unalloyed copper. Such bells are usually undecorated and their walls are thick. The smiths capitalized on the increased fluidity afforded by the alloys to produce the intricate castings, reserving unalloyed copper for designs whose mechanical requirements allowed the use of that material. Figure 29.5a illustrates a bell made from copper; figure 29.5b is a photomicrograph of a section of the bell's suspension ring showing a typical cast structure. The metal contains substantial oxygen.

Bells became extremely important in ancient West Mexico as symbols of status, power, and divinity.[2] Figure 29.6 illustrates Tezcatzoncatl, a powerful god of the Aztecs, as depicted by local informants at the time

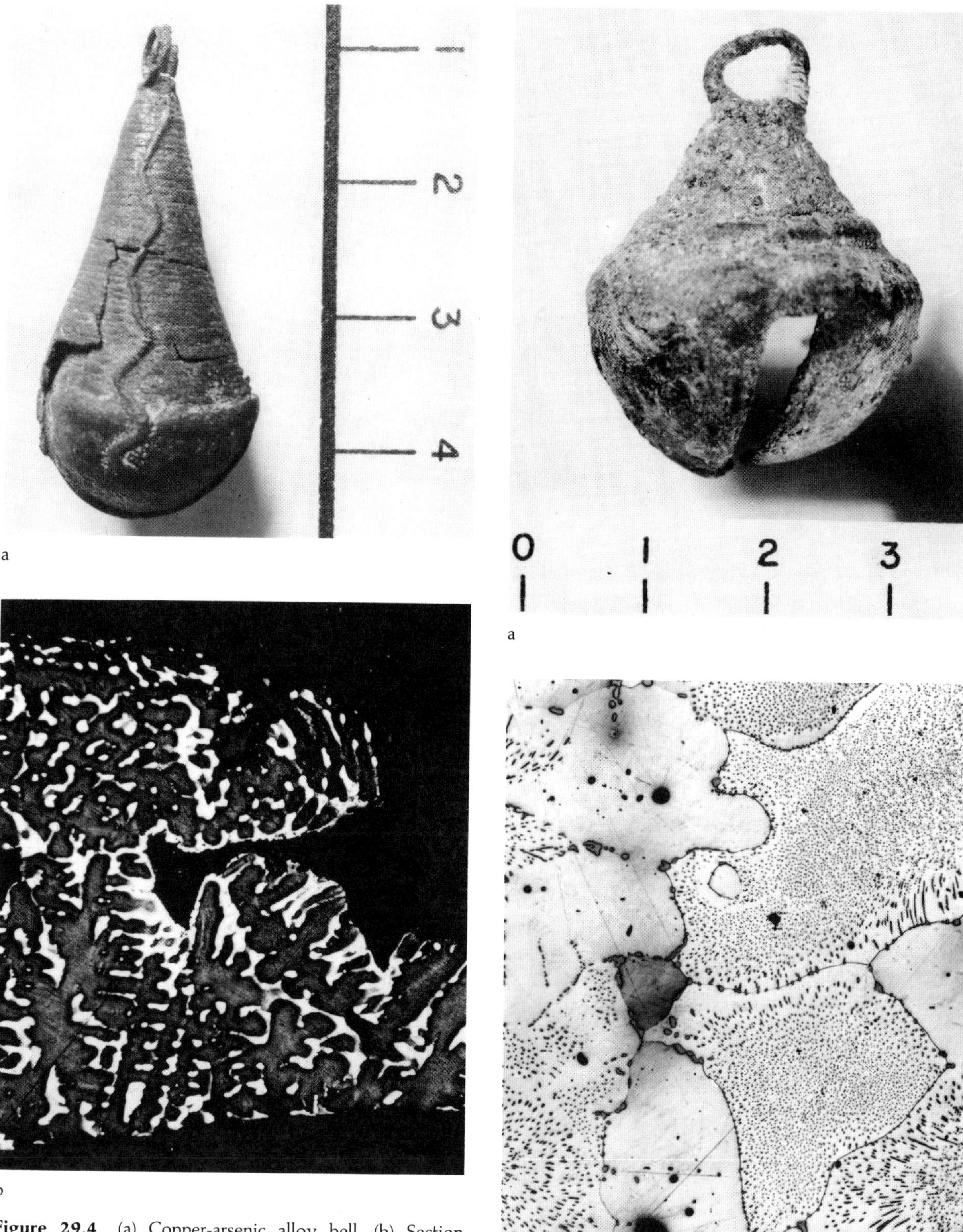

Figure 29.4 (a) Copper-arsenic alloy bell. (b) Section through wirework bell wall and raised zigzag design. Copper-arsenic alloy (1.94% As). ×50. Potassium dichromate and hydrochloric acid etch.

Figure 29.5 (a) Copper bell. (b) Section of bell suspension ring. Copper. ×100. Potassium dichromate etch.

of the Spanish conquest (Sahagún 1951–1978, vol. 1, fig. 21). He wears bells at his ankles. Deities frequently are portrayed wearing bells in conquest period and preconquest documents. Rulers and elites also wore bells around their ankles or wrists and sewn onto their clothing (Hosler 1986b). Musicians and dancers wore or shook bells in ceremonies. Figure 29.7 (Martí 1961, fig. 116) shows a dancer with a rattle stick in each hand. The resonators of the rattle sticks were made from copper or gold; they contained bells or metal rattles. A sacred version of this instrument was designed so that bells could be strung on it (Stevenson 1968); it could be shaken only in religious ceremonies. Ethnohistoric evidence indicates that the sounds of instruments with bells and of bells themselves were extremely important culturally. The bell sounds, it was believed, brought power, strength, and sanctity to the individuals who wore the bells and to ritual and ceremonial activities (Hosler 1986b, "Style in Materials").

West Mexican bells were not tuned to particular pitches. Pitch was determined by the internal volume of the resonator cavity of the bell. Although the material composition varied among the bells, composition was a much less significant factor to pitch than bell size and shape. The diverse shapes and sizes of West Mexican bells determined that, when sounded together, the bells produced many different pitches. The simultaneous production of different pitches, the realization of "textured sound," is a typical characteristic of the indigenous musical systems of the Americas. The fabrication of a great diversity of bell sizes and shapes to produce that textured quality is consistent with the norms of those musical systems.

Bells were thus important in ancient West Mexico for the sounds they produced, which were symbols of the strength and power that are inherent in the elite and in the divine. Eliteness and divinity were also communicated visually through color. The bells that are alloys of copper-tin bronze, for example, contain tin in concentrations that vary from approximately 2% to 23%. The color of the metal becomes increasingly yellow as the tin concentration rises. Figure 29.8 is a longitudinal section of the wall of a tin-bronze bell containing approximately 18% tin by weight. The color of this bell was golden. Figure 29.9 shows a longitudinal section through a raised wirework design from the wall of a copper-arsenic alloy bell. The bell contains arsenic at a concentration of 22%, and its color was silvery. The two bronze alloys were generally utilized in concentrations far higher than necessary to provide the fluidity and strength sufficient to achieve the thin-walled delicate castings. The use of these alloys in high concentrations was for color—the brilliant yellow, almost golden color was accomplished using the high-tin bronzes, and the silvery color was accomplished using the high-arsenic alloys. In Mexico, as in the Andes, the colors of gold and silver were

Figure 29.6 Tezcatzoncatl wearing bells (Sahagún 1951, vol. 1, fig. 21).

Figure 29.7 Dancer with rattle sticks (Martí 1961, fig. 116).

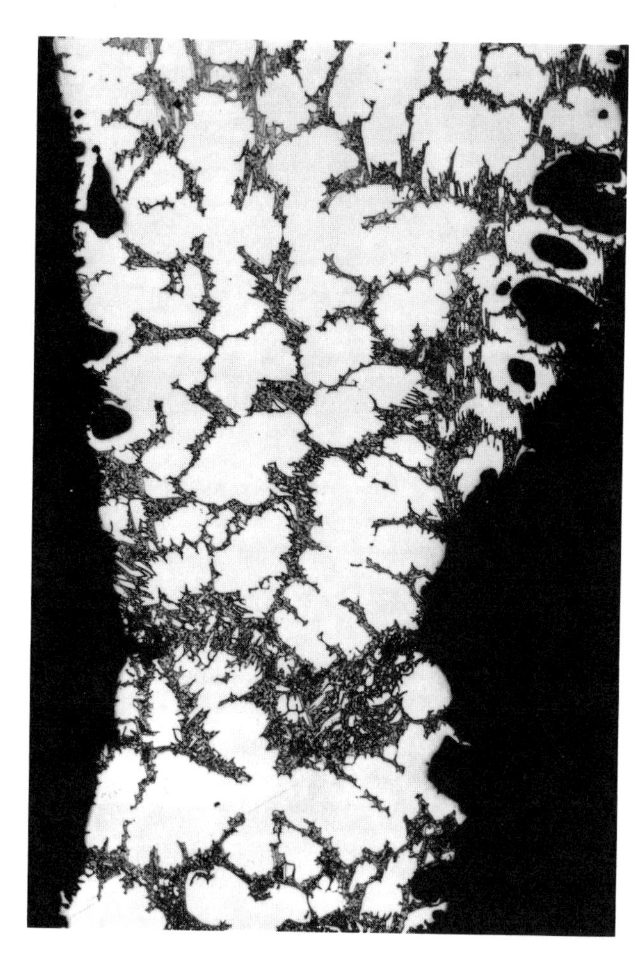

Figure 29.8 Longitudinal section of bell wall. Copper-tin alloy (18.1% Sn). × 100. Ferric chloride and ferric nitrate etch.

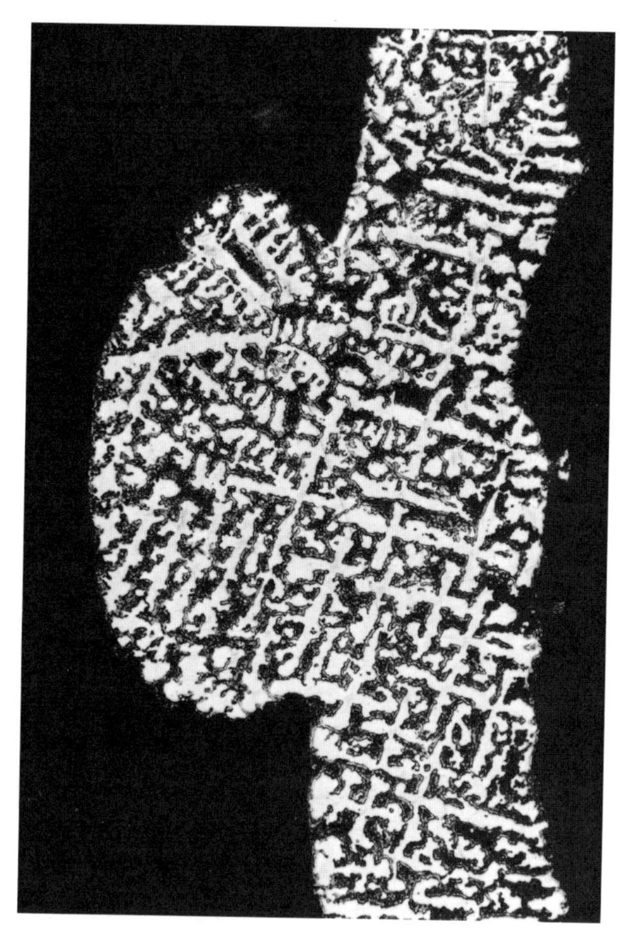

Figure 29.9 Longitudinal section of bell wall. Copper-arsenic alloy (22.0% As). × 50. Potassium dichromate and hydrochloric acid etch.

associated with the solar and lunar deities (Hosler 1986b, "Style in Materials").

The second most frequent type of artifact made from metal in ancient West Mexico was open loops or rings, as shown in figure 29.10. Archaeological evidence suggests that these loops formed part of a hair ornament (Hosler 1986b). Loops make up 20% of the metal artifacts in the RMG collection. They range in diameter from approximately 1 cm to 4 cm and are made from unalloyed copper, copper-tin bronze, and occasionally copper-arsenic alloys. Two different formal types appear. The cross section of one (shown in figure 29.10) is round. Loops made with round cross sections are made from copper or copper-tin bronze. Those made from copper were cold-worked to shape, annealed, and left in the annealed condition. Figure 29.11, a transverse section through such a copper loop, shows strung out cuprous oxide inclusions reflecting extensive initial cold working to shape the metal and equiaxed grains with annealing twins, indicating one or more sequences of annealing.

The cross section of the other type of loop is rectangular, forming a thin band of metal. Loops made with rectangular cross sections tend to be made of copper-tin bronze, with the alloying element present in concentrations up to 20%. These high-tin bronze loops were hot-worked to shape, as the microstructural evidence given in figure 29.12 indicates. The photomicrograph shows a longitudinal section of a loop containing 20% tin by weight. The etched section contains very small equiaxed grains. Strain markings are visible in some grains, and elongated pools of eutectoid strung out in the direction of the flow of the metal is present throughout the section. Microstructural evidence for hot working is suggested by the presence of the strain markings, indicating that in some regions shaping continued below the recrystallization temperature of the metal. The elongated and oriented eutectoid pools indicate plastic deformation as the beta phase above 586°C.

The use of the copper-tin alloy in the case of the rectangular cross section loops appears to have been mechanically necessary to execute the design. These loops are hammered to a thickness of at most 0.05 cm. However, the use of the alloy in such high concentrations—these loops rarely contain less than 8% tin—is for the golden color of the metal. The copper-tin alloy provided the requisite strength to accomplish the

Figure 29.10 Open loops.

Figure 29.11 Transverse section of round cross section loop. × 200. Potassium dichromate etch.

Figure 29.12 Longitudinal section of rectangular cross section loop (20.0% Sn) showing elongated pools of eutectoid. × 200. Ferric chloride and ferric nitrate etch.

design and at the same time dramatically altered the color of the object.

Thus the two types of metal artifacts that appear most frequently in ancient West Mexico and that comprise almost 80% of the objects fabricated are symbols of elite status. Elite status was conveyed through sounds and through the golden color accomplished with the copper-tin bronze alloy and, less frequently, through the silvery color imparted by the copper-arsenic alloy. In both artifact types the alloy, although mechanically necessary to the design, was used in concentrations far higher than necessary to confer mechanical advantage.

The other artifact types made from metal in ancient West Mexico occur in smaller number. They include tweezers, sheet metal ornaments, axes, sewing needles, awls, and miscellaneous ornaments.

The sheet metal objects present in the RMG collection are usually made from alloys of copper and silver. The concentration of the alloying element, either silver or copper, ranges from 14% to 50% by weight. Some are made from nearly pure silver. The artifacts include neck pieces, disks, headbands, sheet metal pendants, and wire coils. Figure 29.13 illustrates a disk, and figure 29.14 a wire coil.

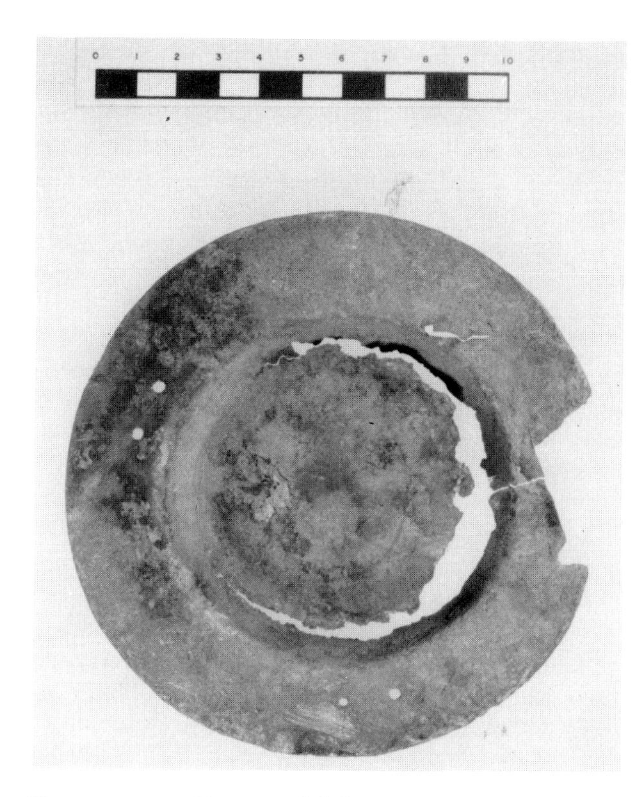

Figure 29.13 Sheet metal disk.

Figure 29.14 Wire coil.

Metallographic examination of the sheet metal objects revealed that they were worked to shape. Lechtman (1980) has shown that copper-silver alloys are especially appropriate for such objects because they become hard but not brittle when hammered into thin sheets. The alloy retains both flexibility and toughness when worked.

The sheet metal objects, regardless of the specific alloy concentration, developed enriched silver surfaces as they were shaped and would have looked like silver. Silver was symbolically important in Mesoamerican societies. The word for silver in Nahuatl, the widely spoken language of the Aztecs, was associated lexically with the moon, and the moon was a powerful Mesoamerican deity. Historical sources report that silver objects were worn by elites and by the king and that the king guarded chests filled with silver disks in honor of the moon (Craine and Reindorp 1970; Hosler 1986b). The individuals who wore these objects identified themselves with that deity. Copper-silver alloys thus were used for both the silvery surfaces they could confer and their exceptional properties, which facilitated the fabrication of thin sheet metal objects.

Tweezers, like the sheet ornaments, bells, and loops, were important status symbols among the rulers and nobility. Ethnohistoric sources show that they were worn by the priests and leaders on ceremonial occasions as symbols of authority (Craine and Reindorp 1970; Hosler 1986b) and that they were exchanged by elites as gifts. Figure 29.15 shows a Tarascan priest wearing a pair of tweezers (Craine and Reindorp 1970, plate 19), and figure 29.16 shows a pair of archaeological tweezers of the same design. Tweezers were also functional depilatory implements.[3] West Mexican tweezers were made from unalloyed copper and from alloys of copper and tin, copper and arsenic, and occasionally copper and silver. There is a large number of visually distinct types in the West Mexican corpus, several of which are shown in figure 29.17.

Despite their variety, all tweezers can be sorted into two design types (Hosler 1986b). In profile the blade of one displays a domelike or shell-like curvature. The metal is extremely thin, measuring only 0.05 cm or less. The blades of the other design type lack this characteristic curvature; they are flat and substantially thicker. All but one of the shell design tweezers were fabricated from copper-tin, copper-arsenic, or copper-silver alloys. Those that contain tin in concentrations higher than about 7% were hot-worked to shape. All tweezers of the flat design are made from unalloyed copper and were shaped through cold working.

A computer simulation study of the stress and deformation behavior of these two tweezer designs (given the differing metal compositions of individual tweezers) showed that the increased strength conferred by the alloys was necessary for the domed shell design and to the functionality of the tweezer (Hosler, "Style

Figure 29.15 Tarascan priest wearing spiral tweezers (Craine and Reindorp 1970, pl. 19).

Figure 29.16 Prehispanic spiral tweezers.

in Materials"). The flat design was fully functional when made from unalloyed copper because the extra thickness of the metal provided the strength that the unalloyed metal lacked. The alloys, particularly the copper-tin alloy, were used not only for the increased strength they imparted but also for their color. Tin, for example, occurs in concentrations that average about 9% by weight, making the color of these tweezers golden.

West Mexican smiths also fashioned household implements and tools from metal. These objects comprise only a small proportion of the total. They include axes, awls, and sewing needles. The axes are made from unalloyed copper and from copper-tin and copper-arsenic bronze alloys. The metallographic studies show that some of the unalloyed copper axes were left in the annealed state and lack the requisite strength and hardness to have been used as tools. The photomicrograph in figure 29.18 shows the cutting edge of the blade of a copper ax. The etchant revealed a fully recrystallized, annealed structure in which none of the grains is deformed. The microstructure shows no evidence that the ax was used. Hardness values are predictably low, ranging from 60–80 VHN. Ethnohistoric sources indicate that copper axes were powerful symbols of authority in ancient Mexico; deities shown in Prehispanic codices were often depicted wielding metal axes (Hosler 1986b). Figure 29.19 from the Codex Kingsborough (1964–1967, vol. 4, pl. 1) for example, shows a figure identified as "The Lord of the Earth" brandishing a metal ax over a captive. In view of the symbolic importance of copper axes, it is likely that some copper axes were never used as tools; nor were they intended to serve as such but functioned rather in ceremonial and religious contexts.

By contrast, the axes made from copper-tin or copper-arsenic bronze were designed for cutting. They are substantially thinner than axes made from unalloyed copper and were cold-worked at the cutting edge of the blade to increase hardness. Figure 29.20 is a photomicrograph of a longitudinal section from an ax blade. The ax contains 8.06% tin by weight. Metallographic examination reveals severe cold working at the blade edge. Strain markings indicating final cold working after the last annealing occur within the grains throughout the entire section but become more dense at the edge of the blade. Hardness values range from 156 VHN to 210 VHN and increase toward the blade edge. This ax would have served as a highly efficient cutting tool.

Ethnohistorical evidence indicates that metal axes were used for woodcutting and woodworking in ancient West Mexico and elsewhere in Mesoamerica (Craine and Reindorp 1970; Hosler 1986b). The Tarascan king maintained a guild of woodcutters whose primary task was to gather wood for ritual temple fires. Their insignia was a hafted metal ax (Craine and Rein-

Figure 29.17 West Mexican tweezer types.

Figure 29.18 Longitudinal section from blade of copper ax. × 200. Potassium dichromate etch.

Figure 29.19 Lord of the Earth with metal ax (Codex Kingsborough 1964–1967, vol. 4, pl. 1).

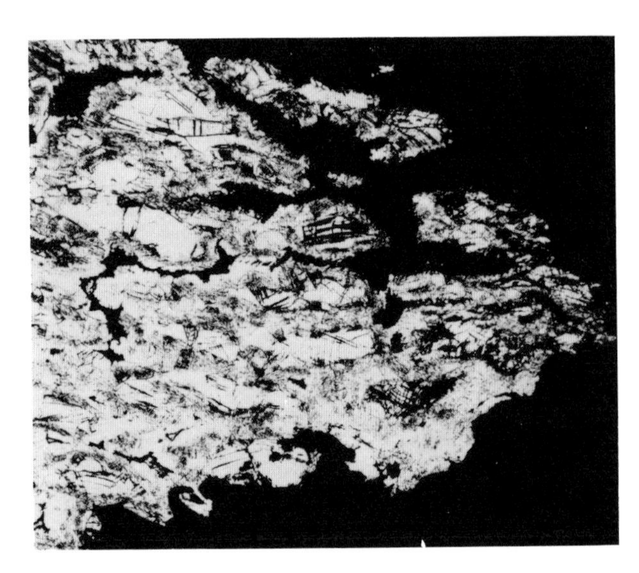

Figure 29.20 Longitudinal section from blade of copper-tin alloy ax (8.06% Sn). × 138. Potassium dichromate and ferric chloride etch.

dorp 1970, pl. 3). Individuals employing hafted metal axes for cutting and working wood are illustrated in Durán (1967, fig. 61), Sahagún (1951–1978, vol. 11, fig. 371), and other sources.[4]

Awls and eyed sewing needles figure among the only strictly utilitarian implements in the West Mexican corpus. The sewing needles measure from 7 cm to 10 cm in length, although some are much longer. They are made from unalloyed copper and from copper-arsenic alloys containing the alloying element in concentrations of about 2%. The copper-arsenic alloy needles are finer, that is, they have smaller diameters, than those made from unalloyed copper. The bronze alloy was again used for the increased strength it lends to the material.

The West Mexican smiths thus employed unalloyed copper and copper-tin, copper-arsenic, and copper-silver alloys to produce a constellation of cultural artifacts whose primary function was to enhance the status of the elites. The parent materials from which these metals and alloys were produced can be reconstructed with reasonable certainty by evaluating information from artifact chemical compositions in light of infomation concerning the ore geology of western Mexico.

Ore Deposits, Metals, and Alloys

Western Mexico is a zone in which copper deposits are plentiful (Gonzalez Reyna 1956; Salas 1966, 1975; Hosler 1986b), as figure 29.21 (Hosler 1986b) indicates. The most common copper ores are malachite, chalcopyrite, and chalcocite. Native copper is present but is far less frequent. On the basis of the presence in the RMG artifacts of certain suites of trace elements (determined by chemical analysis), it can be argued that all these copper sources were probably used. Some objects were fashioned from a fairly pure metal, derived from either an oxide ore or native copper; others were smelted from more complex ores, such as chalcopyrite, as suggested by the presence of trace elements that commonly characterize these ores. For example, a typical suite of elements in an object assumed to be derived from a chalcopyrite ore included copper as the major element and traces of tin, bismuth, and nickel.[5] This suite of elements is particularly characteristic of chalcopyrite ores in which tin in trace amounts sometimes occurs in solid solution. Chalcopyrite is commonly found in metallogenic association with ores of nickel and bismuth, which would account for the presence of these two elements in the chemical analyses (Hosler 1986b).

Tin, in contrast to copper, occurs infrequently in western Mexico proper. The only source of tin in Mexico is cassiterite, which occurs in relatively small pockets scattered throughout the north-central region of the country (Foshag and Fries 1942; Hosler 1986b; Pan 1974). The evidence for the intentional alloying of copper with tin to produce bronze is unequivocal. On the one hand, Mexico contains no known deposits of stannite, the tin-bearing copper ore that, when smelted, directly yields a copper-tin alloy. On the other hand, Mexican cassiterite ores contain the trace element indium (Caley and Easby 1967; Hosler 1986b; Pan 1974). Indium was detected only in RMG artifacts that are made of copper-tin bronze and that contain tin in concentrations greater than approximately 1% by weight (Hosler 1986b).

Arsenic-bearing copper ores, such as enargite and tennantite, are infrequent in western Mexico. Arsenopyrite, by contrast, is abundant and is often associated with chalcopyrite (Hosler 1986b). Based on the frequency of these ores in the western Mexican region, the case for the production of copper-arsenic alloys from the smelting of a mixed ore of chalcopyrite and arsenopyrite is a good one. It is strengthened by the fact that nickel and tin are present in trace amounts in many of the analyses. These two elements generally do not occur in enargite or tennanite ores and normally should not be present in objects made from them. As previously mentioned, they do tend to be present in chalcopyrite ores.

Silver is abundant throughout the western Mexican region, as figure 29.22 shows (Gonzalez Reyna 1956; Hosler 1986b; Salas 1966, 1975). Argentite is by far the most common silver mineral, but silver ores such as proustite and native silver are not uncommon. These silver ores co-occur in many deposits with chalcopyrite (Hosler 1986b). Silver-bearing copper ores that contain silver in concentrations as high as those found in the artifacts analyzed do not exist in Mexico. The likelihood is great, therefore, that the copper-silver alloys were made by winning silver from a silver ore and copper from a copper ore and then melting the two together to form the alloy.

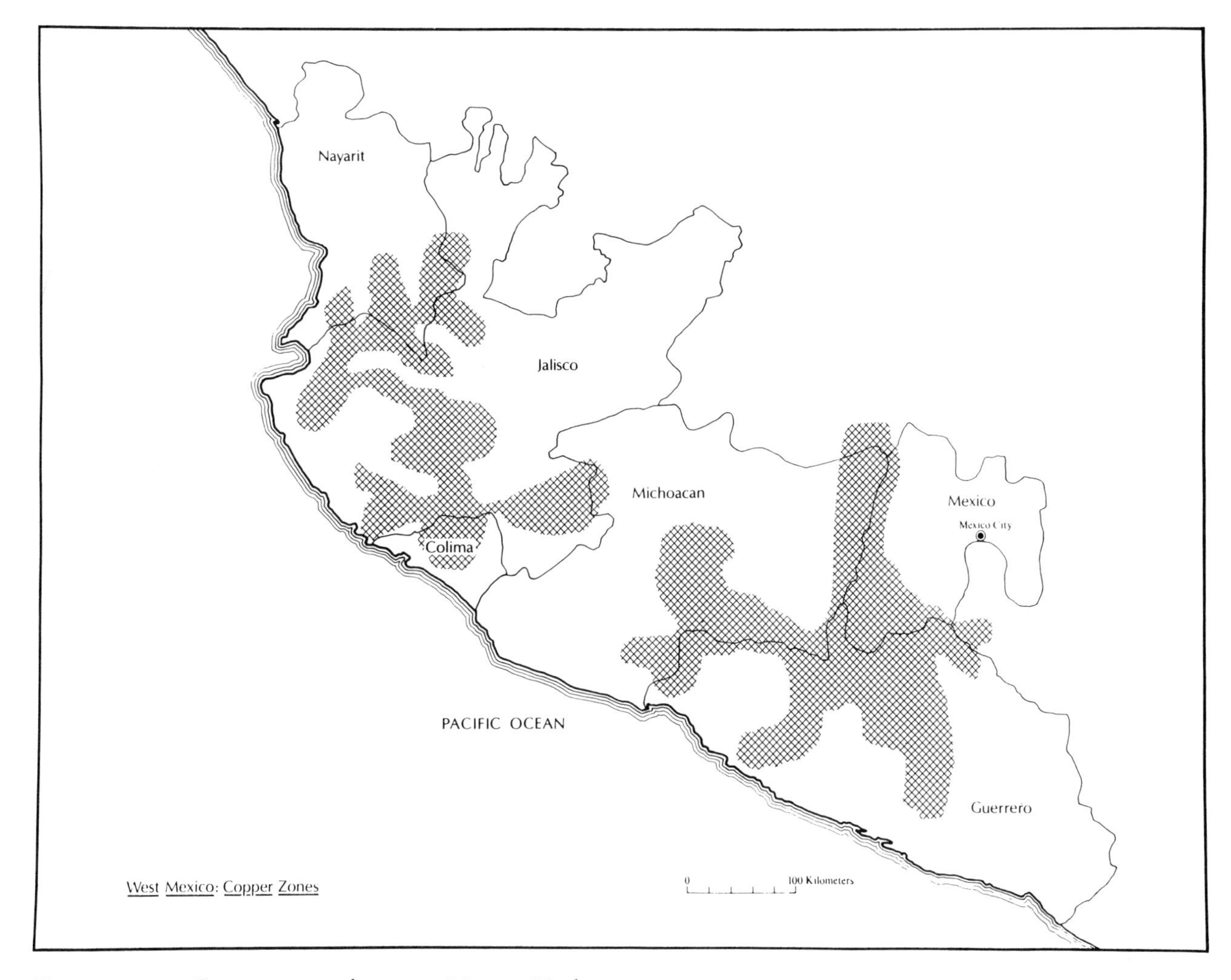

Figure 29.21 Copper zones of western Mexico (Hosler 1986b).

Metal As an Elite Material: Explanations

The metal that these ores yielded was used to produce a constellation of artifacts that functioned primarily in elite domains of culture. It is clear, however, that this particular developmental course for West Mexican metallurgy was not inevitable; the resources and technical expertise were available to develop a metallurgy with a strongly utilitarian emphasis. The elite cultural focus of the technology can be explained in several ways.

This particular emphasis of West Mexican metallurgy was related to the time when metallurgy developed in Mexico and to the social context in which it arose. Metallurgy emerged in West Mexico long after ancient Mexican societies had met their basic societal needs using nonmetal materials. Those technologies, which involved the fabrication of tools, weapons, and agricultural implements from bone, stone, and wood, were highly sophisticated, the result of hundreds and in some cases thousands of years of progressive refinement. Once locked into such technological traditions and the social apparatus that supported them, it is unlikely that objects made from metal would quickly replace others made from such materials. If we were dealing with the introduction of iron, whose mechanical advantages surpass those of copper and its alloys for household tools and implements, the result might have been different.

The social context in which West Mexican metallurgy emerged is another factor influencing its development. Metallurgy appeared after complex societies had been established in the region for many centuries. Those societies were already hierarchical, ruled by elites whose social legitimacy required continuous symbolic affirmation. Although a metal production site has yet to be excavated in western Mexico, I suspect that the development of the technology was controlled by local rulers who oversaw the exploitation and processing of metals and ores and the fabrication of metal objects. Metal was a material that in West Mexico served elite interests—it reinforced, expressed, and asserted their power by means of a constellation of metal objects that symbolically identified those individuals with

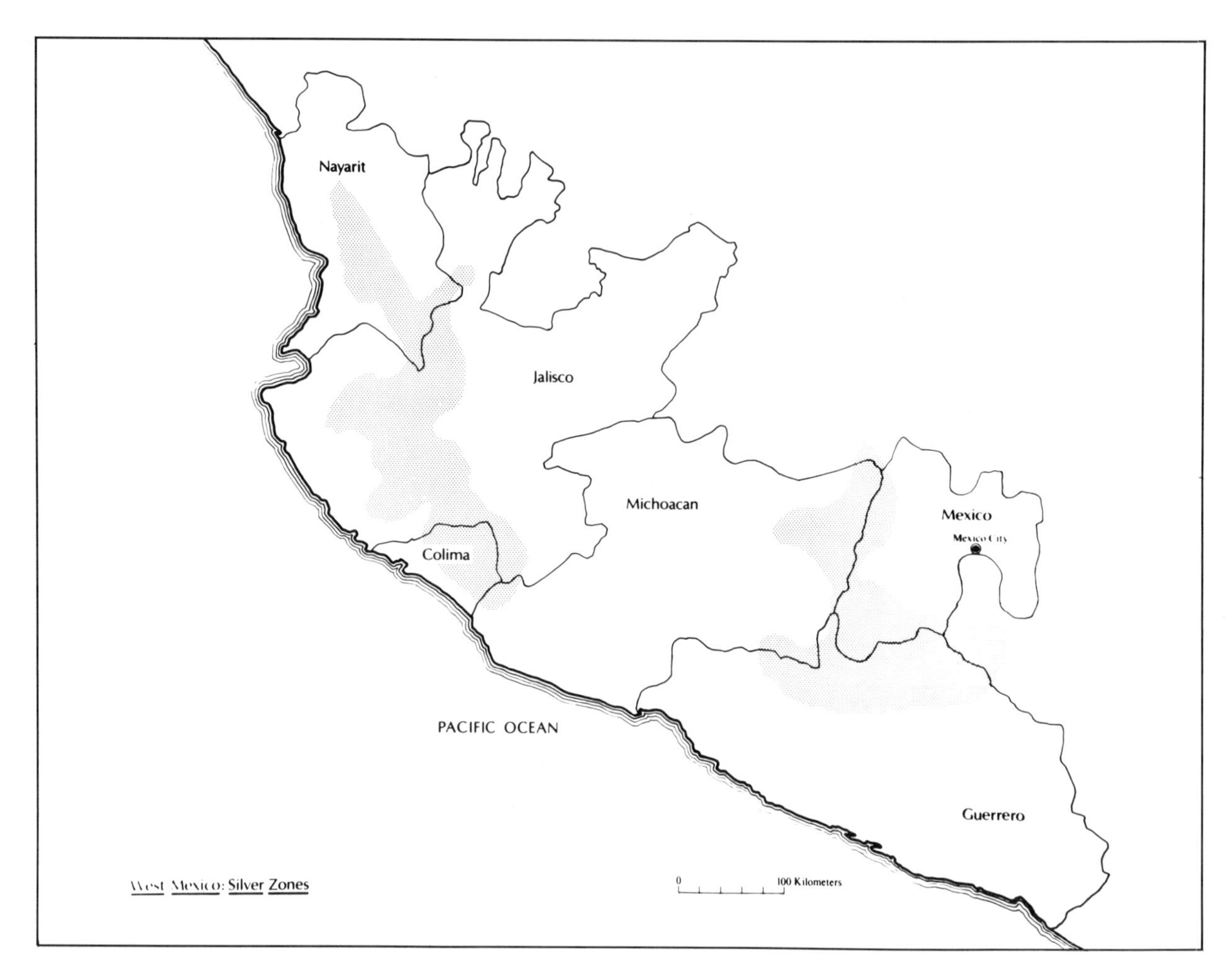

Figure 29.22 Silver zones of western Mexico (Hosler 1986b).

deities and with the sacred and religious realm of experience.

Origins

The metallurgical technology that emerged in West Mexico was strongly influenced by the metallurgies of lower Central America and Colombia on the one hand and the Andean area on the other, as mentioned previously. Scholars have long recognized the probable South and or Central American origins of the technology (Meighan 1969; Mountjoy 1969; Pendergast 1962). However, until now, the data to investigate the issue thoroughly had been unavailable. Such an investigation requires identifying the technical characteristics of West Mexican metallurgy, placing them temporally (Hosler 1986b, "Ancient West Mexican Metallurgy"), and then comparing them to the characteristics of other New World metallurgical technologies during specific time periods. It is now possible to specify particular components of West Mexican metallurgy that were derived from Central and South America and the regions from which they originated. I treat the question briefly here because I have discussed it in detail elsewhere [Hosler 1986b, in press (b)].

The metallurgies of Colombia, Ecuador, and Peru, to which the West Mexican technology is related, took shape significantly earlier than that of West Mexico. Initial developments occurred in Peru by about 1500 B.C., in Ecuador during the Regional Developmental Period (500 B.C.–A.D. 500), and in Colombia by about 100 B.C. Colombian (and lower Central American) metallurgy is characterized by the extensive use of lost-wax casting. The vast majority of metal objects are status objects made from gold-copper alloys. By contrast, in southern Ecuador and Peru metal was primarily shaped by working it into thin metal sheet. Lost-wax casting was rare. The preference for shaping objects by cold working rather than by casting was so pronounced that three-dimensional objects were made by joining pieces of metal sheet rather than by casting a hollow or solid form. These objects were made from gold, silver, and copper-silver and copper-gold alloys. Peruvian smiths also utilized copper-tin and copper-arsenic bronze extensively, primarily for agricultural tools and work implements. Copper-arsenic bronze was common in northern Peru (and probably southern Ecuador) where arsenical copper ores are abundant. In southern Peru and Bolivia, where tin deposits occur, copper-tin bronze was widely employed (Lechtman 1980).

West Mexican metallurgy developed over a period of some 800 years, during which the technical repertoire progressively expanded. Some—by no means all—of those technical elements were introduced from the metallurgies of the Andean and Central American region. During an initial phase, between approximately A.D. 800 and A.D. 1200 (Hosler 1986b, "Ancient West Mexican Metallurgy"), West Mexican metallurgy was most directly related to the metallurgy of coastal Ecuador and Colombia. For example, among the earliest metal objects in West Mexico are tweezers, open loops, sewing needles, and bells made from unalloyed copper. The tweezers are the flat design, the loops are those with a round cross section, and the needles are the type made with a perforated eye (Hosler 1986b, "Ancient West Mexican Metallurgy"). Tweezers, needles, and loops of the same design, fabrication method, and materials are found on the coast of Ecuador (Hosler 1986b, "Ancient West Mexican Metallurgy") several hundred years before they appear in West Mexico. Turning to the bells, the earliest West Mexican bells are identical in fabrication method, overall design (a suspension ring and resonator enclosing a loose clapper), and specific attributes with several bell types from Colombia, Panama, and Costa Rica that have earlier dates. The probability is high that these bells, tweezers, needles, and loops were introduced to West Mexico from their respective geographical areas of manufacture. Nonetheless there is no evidence for wholesale import of such objects; the artifacts that appear in West Mexico served as prototypes that were later copied by West Mexican smiths.

During the subsequent phase of West Mexican metallurgy, from A.D. 1200–1300 up to the Spanish Conquest, the copper alloys—copper and silver, copper and tin, and copper and arsenic—and new artifact types appeared in West Mexico (Hosler 1986b, "Style in Materials"). The data indicate that some of these elements were derived from the south—in this case from Peruvian metallurgy—although contacts continued with other regions [Hosler 1986b, in press (b)].[6] For example, the copper-silver alloy objects that are found in West Mexico were identical with or similar to artifact types made from that alloy in the Andes, specifically the coastal regions of northern Peru and southern Ecuador. In both West Mexico and these regions copper-silver alloys were commonly used for thin sheet metal objects [Hosler, in press (a); 1986b]. Copper-tin bronze, which also appears in West Mexico during this time, seems to derive from central and southern Peru, where the alloy was widely used [Hosler 1986b, in press (a), in press (b)].

The most plausible mechanism for the introduction of these and other elements of the technology is by a maritime route. On one hand, many of these technological elements are distributed discontinuously; that is, they appear in western Mexico and the Andean or Central American region but are absent in the intervening area. This distribution of traits, particularly as found in western Mexico and the northern Andes, has been noted in other realms as well; a style of tomb known as a shaft tomb occurs only in western Mexico and the Northern Andes, and specific ceramic vessel types and design motifs are found only in those two regions. The idea of maritime introduction for such traits has been discussed by many researchers [see, for example, Kelly (1947), Marcos (1978), Meighan (1969), Mountjoy (1969), and Paulsen (1977)]. At the same time the idea is corroborated by extensive ethnohistoric documentation for a system of maritime trade along the coasts of Peru and Ecuador (Murra 1982) in the Prehispanic era, that may also have extended as far north as Mesoamerica (Marcos 1978; Paulsen 1977). A reconstruction of Ecuadorian sailing technology shows that the large sailing rafts used by these people were capable of nonstop voyages in the open sea (Edwards 1969) and thus could have made the trip north (Hosler 1986b).

Certain technological complexes that were fundamentally important in West Mexican metallurgy—artifact types, alloy systems, and so forth—were thus derived from the metallurgies of lower Central America, Colombia, Ecuador, and Peru. However, West Mexican smiths rapidly incorporated and then inventively reconfigured many technological elements that had been introduced from those regions. For example, whereas several bell types were introduced into West Mexico from lower Central America and Colombia, the West Mexican predilection for bells, manifest in the astonishing proliferation of bell designs made from a variety of metals and alloys, occurs nowhere else in the New World. The salient characteristics of West Mexican metallurgy—the cultural interest in sound expressed in the vast number of bells and the use of bronze alloys, sometimes in extraordinarily high concentrations for status objects (Hosler 1986b, "Style in Materials")—were unique to West Mexico. The metallurgy of ancient West Mexico thus does not resemble that of the Andes, lower Central America, or Colombia. Rather, the technology constitutes an original regional experiment in metallurgy stimulated to be sure by metallurgies of other areas but ultimately expressing the particular cultural realities and requirements of West Mexican societies of the time.

Notes

The research was supported by a grant-in-aid from the Wenner-Gren Foundation for Anthropological Research, a Doctoral Disseration Improvement Grant (BNS8025822) from the National Science Foundation, and a predoctoral fellowship from Industrial Minera México S.A., Mexico. The Consejo de Arqueología, Instituto Nacional de Antropología e Historia, Mexico, issued a permit for the study, and Otto Schöndube and Frederico Solórzano (INAH-SEP Museo Regional de Guadalajara) facilitated the work in every way possible. I would like to thank each of these individuals and institutions.

1. The collection was originally assembled by Frederico Solórzano (director of the Regional Museum of Guadalajara at the time the study was carried out) and is now the property of the Mexican government in accordance with Mexico's antiquities laws. The research was performed under a permit issued in 1981 by the Consejo de Arqueología, Instituto Nacional de Antropología e Historia, México D. F., following Mexican government regulations concerning archaeological research. The laboratory analyses were carried out at the Center for Materials Research in Archaeology and Ethnology at the Massachusetts Institute of Technology.

2. The ethnohistoric data cited here are from central Mexico; however, the musical cultures throughout preconquest Mexico were fundamentally similar (Stanford 1966).

3. See the extended discussion in Hosler (1986b, "Style in Materials").

4. Metal axes can be distinguished from Mesoamerican stone axes by shape.

5. Silver and iron were present in trace concentrations in many artifacts and were thus nondiagnostic.

6. Copper-arsenic alloy objects containing arsenic in extremely low concentrations are present before A.D. 1200, but the use of the alloy became more common later.

References

Caley, E., and D. Easby. 1967. "Indium as an impurity in ancient West Mexican tin and bronze artifacts and in local tin ore." *Science* 155:686–687.

Craine, E. R., and R. C. Reindorp. 1970. *The Chronicles of Michoacan*. Norman, Okla.: The University of Oklahoma Press.

Durán, Fray Diego. 1967. *Historia de las Indias de Nueva España e Islas de la Tierra Firma*, vol. 2, A. M. Garibay, ed. Mexico City: Editorial Porrua, S.A.

Easby, E., and J. Scott. 1970. *Before Cortes: Sculpture of Middle America*. New York: Metropolitan Museum of Art.

Edwards, C. 1969. "Possibilities of Pre-Columbian maritime contacts among New World civilizations," in *Pre-Columbian Contact within Nuclear America*, J. C. Kelly and C. Riley, eds. (Carbondale, Ill.: Southern Illinois University Museum), 3–10.

Foshag, W. F., and C. Fries, Jr. 1942. "Tin deposits of the Republic of Mexico." *US Geological Survey Bulletin* 935C: 99–176.

Gonzalez Reyna, J. 1956. *Riqueza Minera y Yacimientos Minerales de México*, third edition. XX Sesión del Congreso Geológico Internacional. Mexico City: Banco de Mexico, S.A.

Hosler, D. In press (a). "External influences in the metallurgy of ancient West Mexico: A technological report," in *18th Reunion Mesa Redonda* (Mexico City: Sociedad Mexicana de Antropología).

Hosler, D. In press (b). "Ancient West Mexican metallurgy: South and Central American origins and West Mexican transformations." *American Anthropologist*.

Hosler, D. 1986a. "The cultural organization of technology: Copper alloys in ancient West Mexico," in *Precolumbian American Metallurgy*, Special Publication of the New World Metallurgy Symposium, Forty-Fifth International Congress of Americanists, Bogotá (Bogotá, Colombia: Banco de La República), 67–86.

Hosler, D. 1986b. *The Origins, Technology and Social Construction of Ancient West Mexican Metallurgy*. Ph.D. Dissertation, Department of Anthropology, University of California Santa Barbara.

Hosler, D. n.d. "Ancient West Mexican metallurgy: A technological chronology." Unpublished.

Hosler, D. n.d. "Style in materials: The metallurgy of ancient West Mexico." Unpublished.

Kelly, I. 1947. *Excavations at Apatzingan, Michoacan*. Viking Fund Publications in Anthropology 7. New York: Johnson Reprint.

Kingsborough, Viscount Edward King. 1964–1967. *Antiguedades de México: Basadas en la Recopilación de Lord Kingsborough*, 4 vols. Interpreted by José Corona Nuñez. Mexico City: Secretaria de Hacienda y Crédito Público.

Lechtman, H. 1980. "The Central Andes: Metallurgy without iron," in *The Coming of the Age of Iron*, T. A. Wertime and J. O. Muhly, eds. (New Haven, Conn.: Yale University Press), 267–334.

Marcos, J. G. 1978. "Cruising to Acapulco and back with the thorny oyster set: A model for a lineal exchange system." *Journal of the Steward Anthropological Society* 9(1–2):99–131.

Martí, S. 1961. *Canto, Danza, y Música Precortesianos*. Mexico City: Fondo de Cultura Económica.

Meighan, C. 1969. "Cultural similarities between western Mexico and Andean regions." *Mesoamerican Studies* 4:11–25.

Meighan, C. 1976. *The Archaeology of Amapa, Nayarit*. Los Angeles, Calif.: Institute of Archaeology, University of California.

Millon, R. 1981. "Teotihuacán: City, state, and civilization," in *Archaeology*, J. Sabloff, ed. Supplement to the *Handbook of Middle American Indians*, vol. 1, V. Bricker, general editor (Austin, Tex.: University of Texas Press), 198–243.

Mountjoy, J. B. 1969. "On the origin of West Mexican metallurgy." *Mesoamerican Studies* 4:26–42.

Murra, J. V. 1982. "El Tráfico de Mullu en la Costa del Pacífico," in *Primer Simposio de Correlaciones Antropológicas Andino-Mesoamericano* (Guayaquil, Ecuador: Escuela Superior Politécnica del Litoral), 265–273.

Pan, Yuh-Shyi. 1974. *The Genesis of the Mexican-Type Tin Deposits in Acidic Volcanics*. Ph.D. Dissertation, Columbia University.

Paulsen, A. C. 1977. "Patterns in maritime trade between south coastal Ecuador and western Mesoamerica, 1500 B.C.–A.D. 600," in *Dumbarton Oaks Conference on the Sea in the Pre-Columbian World* (Washington, D.C.: Dumbarton Oaks Research Library and Collections), 141–160.

Pendergast, D. M. 1962. "Metal artifacts in Prehispanic Mesoamerica." *American Antiquity* 27(4):520–545.

Sahagún, Fray Bernardino de. 1951–1978. *Florentine Codex: General History of the Things of New Spain*, 14 vols., A. J. O. Anderson and C. E. Dibble, trans. Monographs of the School of American Research and the Museum of New Mexico. Santa Fe, N.M.: Museum of New Mexico.

Salas, Ing, Guillermo P. 1966. *Cartas Metalogénicas de La República Mexicana*, M-11 (Yacimientos de Plata de La República Mexicana), M-17 (Yacimientos de Estaño de La República Mexicana), and M-19 (Yacimientos de Cobre de La República Mexicana). Mexico City: Instituto de Geología, Universidad Nacional Autónoma de México.

Salas, Ing. Guillermo P. 1975. *Carta y Provincias Metalogenéticas de la República Mexicana*, no. 21E. Mexico City: Consejo de Recursos Minerales.

Stanford, E. T. 1966. "A linguistic analysis of music and dance terms for three sixteenth-century dictionaries of Mexican Indian languages," in *Yearbook*, G. Chase, ed. (New Orleans, La.: The Inter-American Institute of Musical Research, Tulane University), vol. 2, 101–159.

Stevenson, R. 1968. *Music in Aztec and Inca Territory*. Berkeley, Calif.: University of California Press.

30
Traditions and Styles in Central Andean Metalworking

Heather Lechtman

Archaeologists of the southern Americas tend to discuss the prehistory of the continental land mass from Mexico to the southern limits of Chile and Argentina in terms of three large culture-historical regions: Mesoamerica, including modern Mexico, Guatemala, and portions of the republics to their south as far as northern Costa Rica; Intermediate America, which includes the Central American states of Nicaragua, Costa Rica, and Panama and Colombia and northwest Venezuela; and Andean America, or the western portion of South America dominated by the Andes mountain range from northern Ecuador to central Chile and including most of Bolivia and northwest Argentina (figure 30.1). Andean America, in turn, has traditionally been divided into three zones: northern, central, and southern Andes, reflecting broad differences in ecology and culture-historical development. For the purposes of this discussion I consider the central Andean metalworking area as subsuming the modern southern border of Colombia, Ecuador, and Peru. It is this area within which metallurgy was invented in the Americas and from which metallurgical technologies spread both north and south. Within the area styles and traditions of metalworking exhibited strong continuity over millennia, but the products of these technologies—the metal objects themselves—are as varied as the villages, the religious bodies, or the states that made and used them.

The most striking feature of the central Andean metalworking tradition is its commitment to shaping metal by working rather than by casting. Most objects, regardless of size or function, are worked to shape—cold-worked and annealed or forged hot—and many are made of sheet metal, preformed pieces of which are mechanically or metallurgically joined to produce the desired form. From the earliest appearance of metal objects in the Andes (Grossman 1972; Lechtman 1980, 1984b) right through to the Inka use of metal in affairs of state (Lechtman, in press), this approach to the handling of the material never altered. Metal was meant to be treated as a solid, not as a liquid, and Andean smiths pushed the working of metal to its limits, avoiding other possibilities for solving what we today recognize as engineering problems of design.

In no way is this overriding interest in the plastic deformation of metal more clearly demonstrated than by comparison with the equally strong but conceptually different tradition of lost-wax casting so impressively developed by the peoples of the Intermediate Area, especially Colombia—neighbors to the north. Figure 30.2 illustrates a miniature gold trophy head—the head of a victim taken in battle—from the Peruvian south coast (Lechtman 1984a). It measures only 1.95 cm in height. Figure 30.3 shows a small gilt *tumbaga* object excavated in the northwest corner of Venezuela but most likely made by Colombian craftspeople

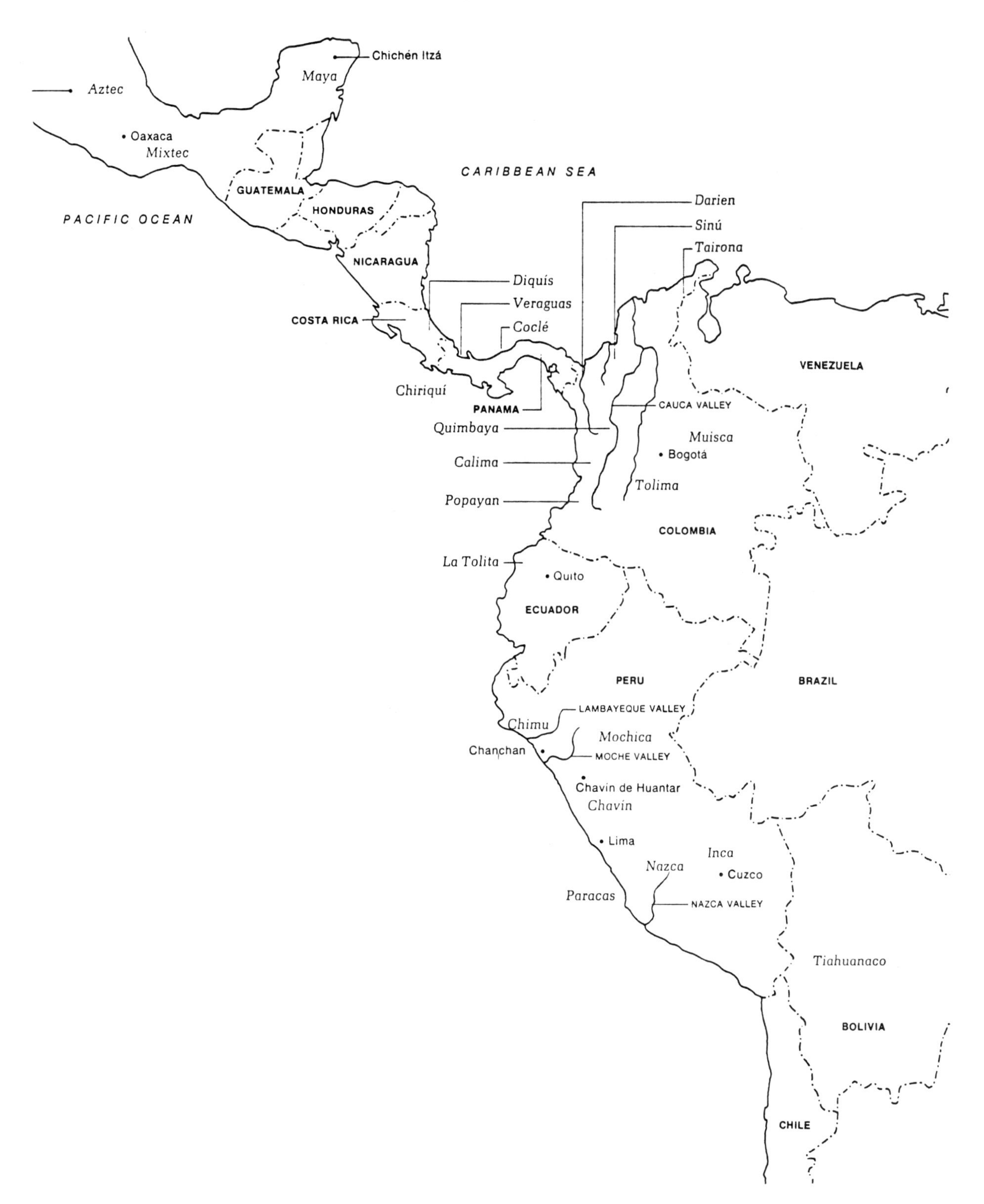

Figure 30.1 Map of the central Andean culture area and portions of the Intermediate Area.

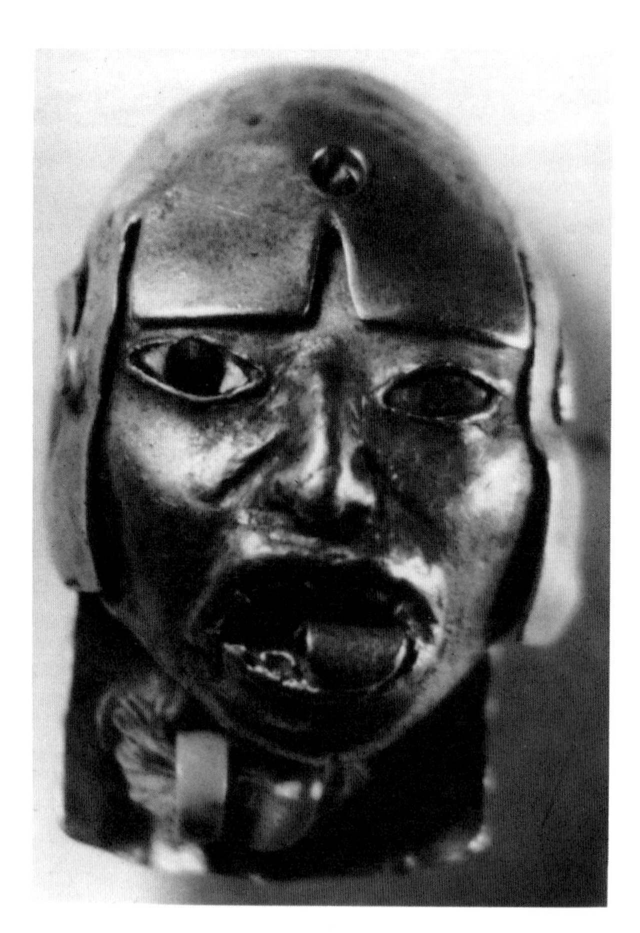

Figure 30.2 One of a pair of hollow gold trophy heads of Nasca style. Ica valley, Peruvian south coast. Private collection. Height: 1.95 cm.

Figure 30.3 Gilt *tumbaga* object from La Era Nueva, Mérida, Venezuela (Wagner 1970).

(Wagner 1970; Lechtman 1973b). It too is a miniature —a sequin or pendant—1.3 cm high and 0.012–0.02 cm thick. The trophy head arrived at the laboratory with credentials as a solid casting; the *tumbaga* object was thought to be an extremely thin piece of hammered sheet. Faithful to the traditions that conceived and made them, however, the Andean product was worked and joined; the Venezuelan object is a casting.

The trophy head has been assembled from nineteen individual pieces of sheet gold, hammered to shape and meticulously soldered or welded together. Every tooth, the tongue, and the nose are separate units. An X-radiograph of the head, given in figure 30.4b, clearly indicates the join between the face at the front and the back of the head, with its separate attachment representing hair. On the other hand, in the photomicrograph of figure 30.5, which represents a cross section through the sequin, we see the unmistakable evidence of a cast structure, with large, highly cored dendrites of the *tumbaga* alloy (45.3% Cu, 46.7% Au, 7.4% Ag). The object was hammered only around its periphery, to perfect its shape. The thinning of the edges through final and localized cold working accounts for the elongation and orientation of the dendrites at the extreme edge of the section.

Each of these objects is equally audacious in its method of manufacture. But culturally neither was audacious in the sense that each manifests an approach to the handling of metal as a material that was deeply rooted and supported by centuries of highly skilled performance. They represent attitudes about what metal is, which properties are to be selected and developed, how the management of form is to be achieved. I have said elsewhere that the metalworking traditions of Andean and Intermediate Area artisans were almost diametrically opposed, that the Andean metalworkers were exquisite forgers of metal and the Intermediate Area artisans superb founders (Lechtman 1979a, p. 31). Of course, each tradition utilized a wide variety of fabrication procedures. As a mature and sophisticated technology, neither was exclusive in its familiarity with and exploitation of the normal range of metal properties—mechanical, visual, and acoustical. But it would be fair to say that appropriate metallurgical performance lay poles apart for these two culture areas. The styles of metalworking that stimulated the most sophisticated and inventive aspects of the technologies developed along quite disparate routes.

Casting

Although casting was not the preferred means by which metal objects were produced by central Andean smiths, it is nevertheless of interest to document the repertoire of casting techniques that has so far been recognized within the corpus of artifacts.

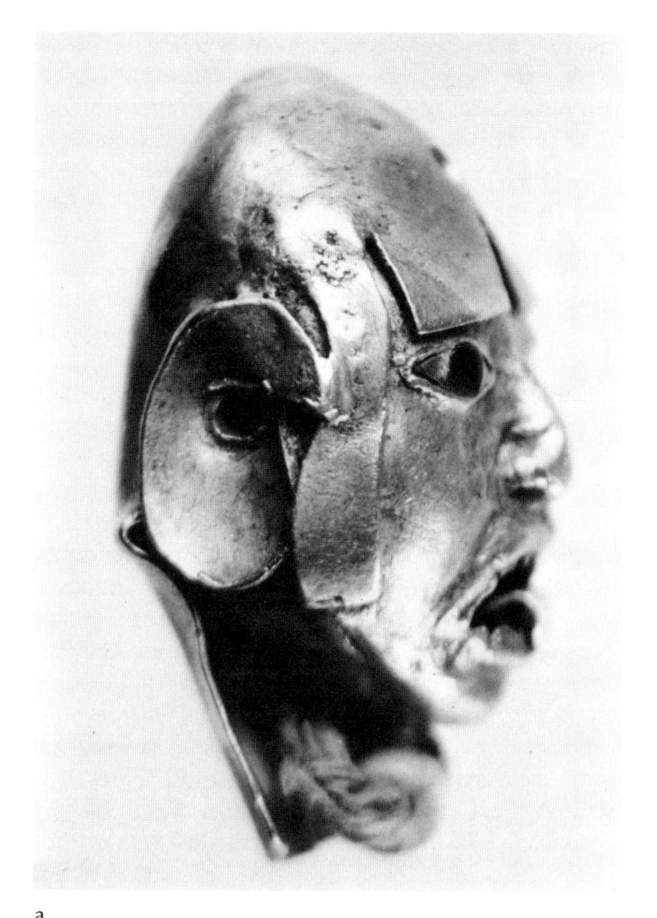

a

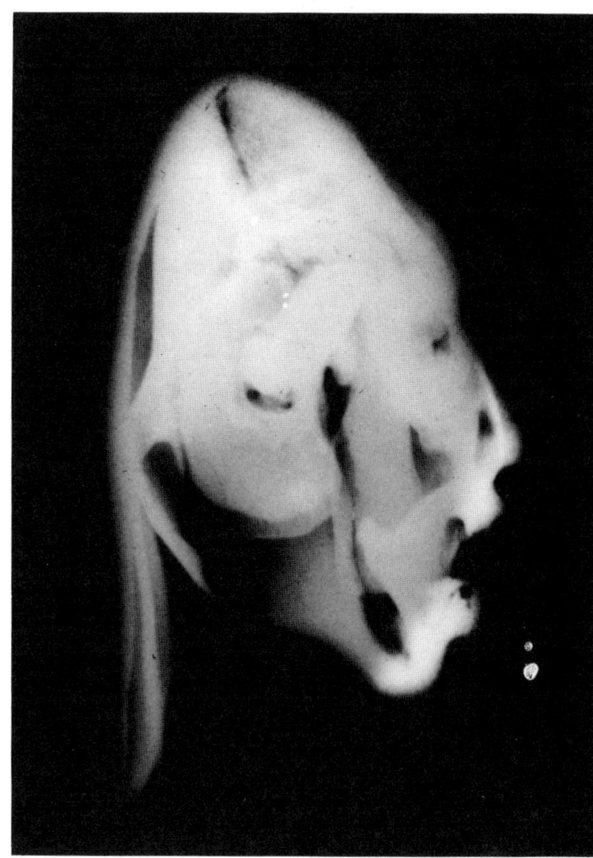

b

Figure 30.4 (a) Side view of the trophy head illustrated in figure 30.2. The seam between the face and the back of the head is just visible under the chin. The large flat ear is joined to a hollow tube that projects into the interior of the head. Solder or weld joins are visible between the side lock of hair and the ear and between the side lock and the cap or headdress. (b) X-radiograph of the trophy head oriented as shown in (a). Note the seam between the face and the back of the head. The tongue passes through a plate at the rear of the mouth that holds it in place.

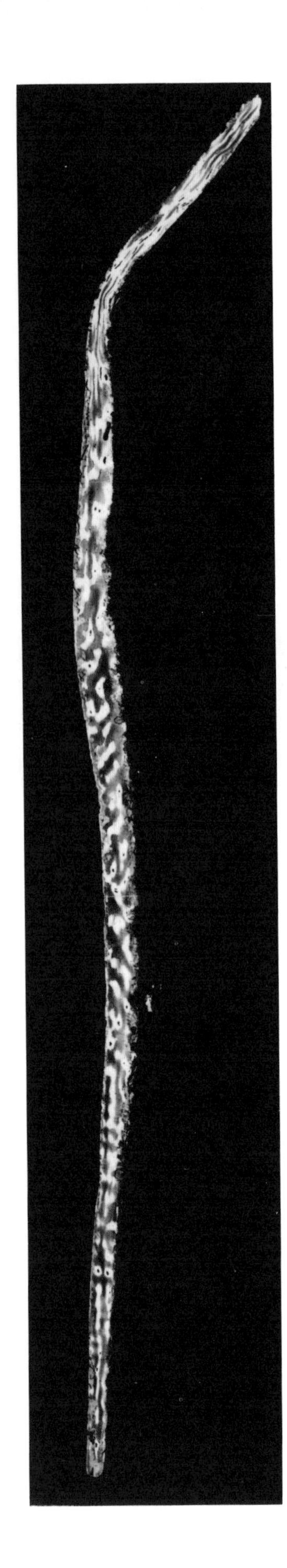

Figure 30.5 Photomicrograph of a complete section through the *tumbaga* object shown in figure 30.3. The cast structure exhibits highly cored dendrites that have become elongated at the extreme edges of the object where it was hammered to reduce its thickness. × 500. Potassium dichromate etch.

Lost-wax casting is rare, and the known examples tend to be from the north coast of Peru. They appear in the region between the Moche and Lambayeque river valleys. Although unexcavated and without association, they can probably be assigned to the Late Intermediate Period (see table 30.1). Speculating on the virtual absence of lost-wax casting in the Central Andes, Bird (1979) suggested that it may have been a result of the scarcity of wax derived from the stingless honey bee (*Meliponidae* family). Whereas the native habitat of these bees includes the tropical and subtropical regions of Mexico, Central America, and South America as far south as northern Argentina, they are absent from the cold highland areas and from the desert valleys of Peru (Bird 1979, p. 50). On the other hand, such wax could have been obtained by exchange from the Amazonian lowlands. Bird also notes that, even in those cases in which castings appear to have been made by a process similar to the lost-wax method —objects such as those illustrated in figures 30.6 and 30.7—the model may have been made of some other material that was later burned out of the mold. He bases this observation on the quality of the castings themselves, which are not as delicately wrought as those achieved with beeswax (Bird 1979, p. 51).

The heavy staff finial illustrated in figure 30.6 is typical of these north Peruvian castings [see also Antze (1965)]. Many have pendant bells. Smaller, more delicate castings do appear to have been made utilizing wax or some similar highly plastic material. Figure 30.7 illustrates an example collected by Henri Reichlen in the 1940s from a site he called Batanes (undoubtedly Batán Grande) in the Lambayeque valley (Reichlen, personal communication).[1] Although the object appears to have been assembled from a number of loops and thick "wires" of metal joined metallurgically, it is a single-piece casting of arsenic bronze (1.9% As by weight). The photomicrograph (figure 30.8) reveals the highly cored dendrites typical of several sections removed from the object at locations that appear to be characterized by joins. Each shows the same cast microstructure.

The north coast of modern Peru, between the Moche and Piura river valleys, seems to have been something of an experimental laboratory where there was constant attention to new techniques of handling metal, to alloy development, and to property management. This was particularly the case during the Early Intermediate Period, when the Moche peoples held sway over most of the north Peruvian coast, but the region continued

Table 30.1 Chronology for Central Andean prehistory

Years	Relative Chronology	Historical Events
1534	Colonial Period	Spanish Empire
1500	Late Horizon	Inka Empire
1476		
1000	Late Intermediate Period	Kingdom of Chimor; ← Sicán Culture →; ← Chimú Culture →
	Middle Horizon	Huari Empire
500		
A.D. 0 B.C.	Early Intermediate Period	← Nasca Influence →; ← Moche Culture →
500		
1000	Early Horizon	Paracas-Nasca Tradition; ← Chavín Influence →
1500		
2000	Initial Period	
	Preceramic Period	

Figure 30.6 Staff ornament with pendant bells, cast in copper or copper alloy. Collected by Adolph Bandelier in Chimbote, Peru. Collection: American Museum of Natural History, New York (Cat. no. B/4758). Height: 6.6 cm. Photograph courtesy of American Museum of Natural History.

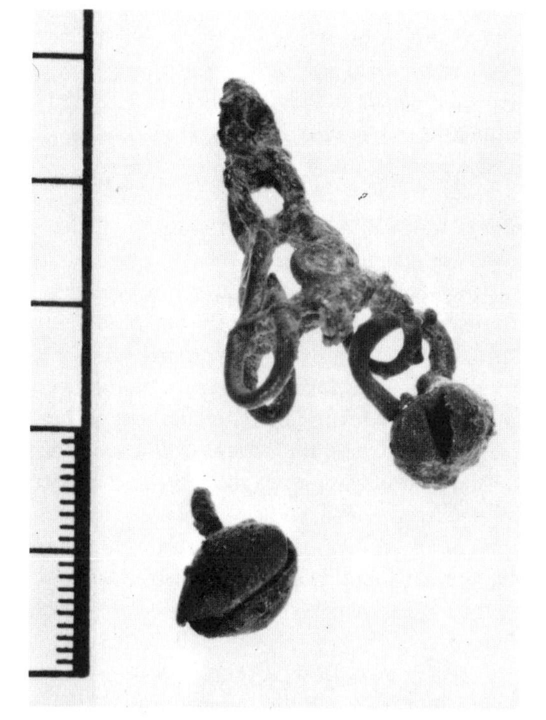

Figure 30.7 Pendant with bells. One-piece casting from Batán Grande, Peru (Cu-As bronze, containing 1.9% As). Collected by Henri Reichlen. Collection: Laboratory for Research on Archaeological Materials, MIT.

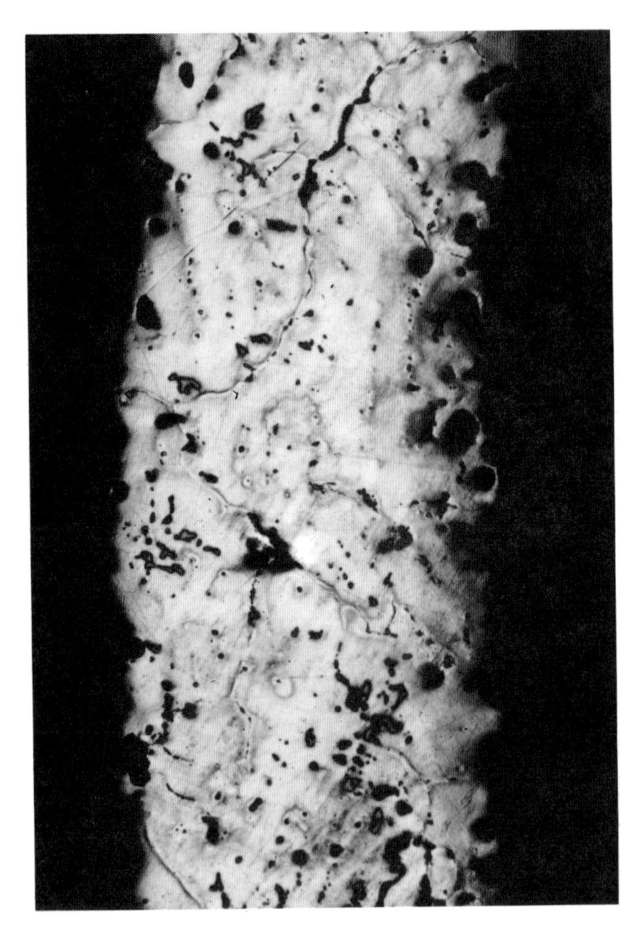

Figure 30.8 Photomicrograph of a section through a loop on the object illustrated in figure 30.7. The structure exhibits large, highly cored dendrites with tiny pools of eutecticlike material in the arsenic-rich areas at grain boundaries. ×50. Potassium dichromate etch.

Figure 30.9 Group of arsenic bronze implements made by a slush casting technique. The four at the left resemble blades for agricultural digging sticks. Lambayeque valley, Peru. Collection: Museo Arqueológico Brüning, Lambayeque, Peru.

as the center of innovative metallurgy until the Chimor kingdom was conquered by the Inka in A.D. 1470.

Slush casting, one of the techniques developed by coastal smiths, enjoyed limited and special use by the lords of Chimor during Chimú domination of the coast, but the technique does not seem to have been adopted outside the Moche-Lambayeque area (Lechtman 1981). At Chan Chan, the capital of the kingdom of Chimor, and also in the Lambayeque valley, a large number of socketed metal tools—probably agricultural digging stick points or hoes (figures 30.9 and 30.10)—have been found buried in hoards. Although some were extensively worked, the majority was slush cast from a low-arsenic copper-arsenic alloy (0.5–4.1% As by weight). Figure 30.11 presents a proposed reconstruction of the stages in the manufacture of these implements by a slush casting process. The castings tend to be extremely porous, and many of these artifacts probably were never used or meant to be functional as working tools. I have pointed out elsewhere (Lechtman 1981, pp. 96–97) that their characteristics—the weight of metal they contain, their porosity, their burial in large numbers—suggest that their primary purpose was as concentrators of metal that constituted real wealth. Whether it was the agricultural aspect of the tool, its arsenic-bronze metal, or both that imparted special significance to these objects, we cannot say.

Most castings in the Central Andes were made in piece molds, often of a simple two-piece variety. An interesting and more intricately designed example is given by an Inka copper-tin bronze weapon (10.02% Sn by weight) in the form of a star-shaped mace with an axlike blade (figures 30.12 and 30.13) (Bandelier 1910; Mead 1915). Examination of its surface revealed traces of the parting line where the two original abutting mold edges were slightly misaligned or where metal had seeped out along the seam between the mold parts (figure 30.14). To explore the design of the original piece mold in which this object was cast, a silicone rubber two-piece mold was made directly from the mace and a plaster replica of the object was cast from the mold.[2] The position of the seam between the rubber mold parts was determined by following the orig-

Figure 30.10 The undersides of the objects illustrated in figure 30.9 display the long sockets for hafting. A protuberance at the junction of socket and blade is characteristic of many of these implements. The object at the right has no socket; its use is unknown.

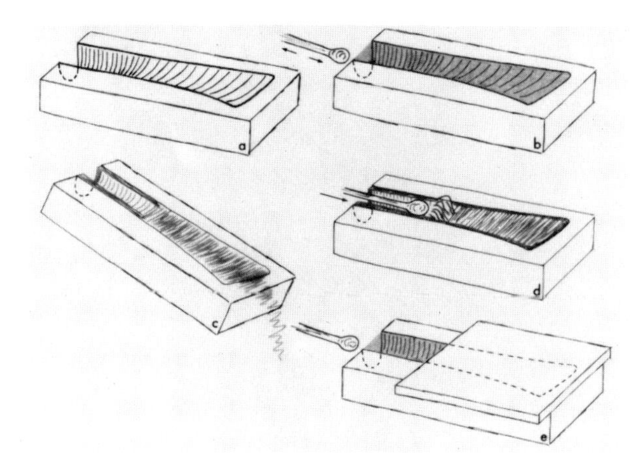

Figure 30.11 Sketch of the stages in a proposed reconstruction of the slush casting method used to produce the solid socketed tool castings from the Peruvian north coast. (a) Open mold. (b) Molten metal fills mold, and socket end is kept open by rod movement. (c) When desired socket thickness is achieved, mold is tipped and excess metal runs off. (d) While metal is still mushy and semisolid, rod is inserted through the socket and pushes up the protuberance. (e) Diagram of partly open, partly closed mold such as may have been used to cast some of the objects.

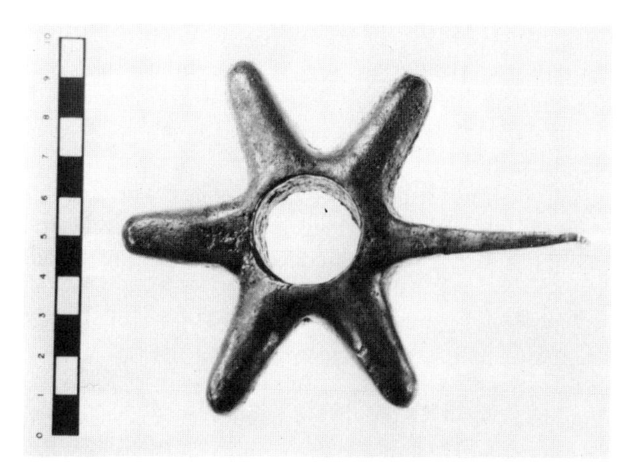

Figure 30.12 Top view of Inka style mace-ax of copper-tin bronze. Collected by Adolph Bandelier on the Island of Titicaca, Bolivia. Collection: American Museum of Natural History, New York (Cat. no. B/1846).

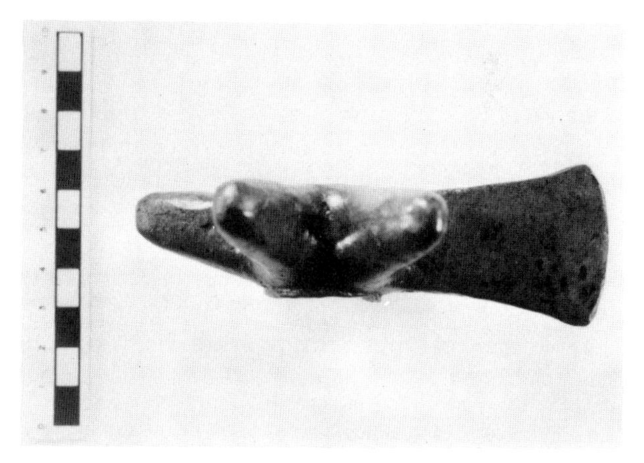

Figure 30.13 Side view of Inka mace-ax.

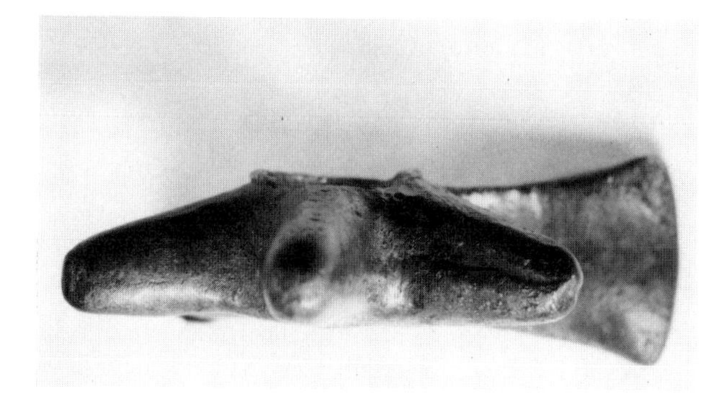

Figure 30.14 Detail of the Inka mace-ax showing traces of the mold line on mace points at left (faint) and right (pronounced). The line registers the location of abutting edges of a two-piece mold.

inal parting line on the mace; where no traces of this line were evident, the most obvious line was taken to complete the short distances between two known points. The cast plaster replica of the object clearly demonstrates the viability of the technique. Figure 30.15 illustrates a portion of this replica. The flash or fin was deliberately left untrimmed to highlight the location and changing direction of the original mold seam.

The most complex piece-mold casting we have examined to date is the small solid silver figurine shown in figure 30.16 (Lapiner 1976, fig. 689). Originally described by Bird (1979) as having been cast in an eleven-piece mold, more recent study of the object has revealed that the mold consisted of nineteen pieces.[3] Because the smiths who cast and finished the figure did not remove the flashes of metal that had seeped out along the mold seams, these raised ridges delineate the boundaries between adjacent mold pieces (figure 30.17). Figure 30.18 plots the position of each mold piece on a simplified line drawing of the figurine and registers the location of the mold lines. We have no evidence of how the mold parts were held together or about the existence of possible vents or risers. The figure was probably cast upside down, as a rather large shrinkage cavity underneath one foot suggests that the pouring sprue was located nearby. Nor is there provenance available for this unique object. Other researchers have reported a small number of similar figurines that hold cobs of maize or are decorated with maize cobs and plants. Some are cast in silver-copper alloy, others in silver (Jones 1964, figs. 58, 59; Emmerich 1965, fig. 57; *Peru durch die Jahrtausende*, 1984, p. 227, fig. 1.58).[4] They are described as Inka in style.

Molds used for casting metal artifacts are rare in the Central Andes. Those that have survived are usually open molds made of fired clay.

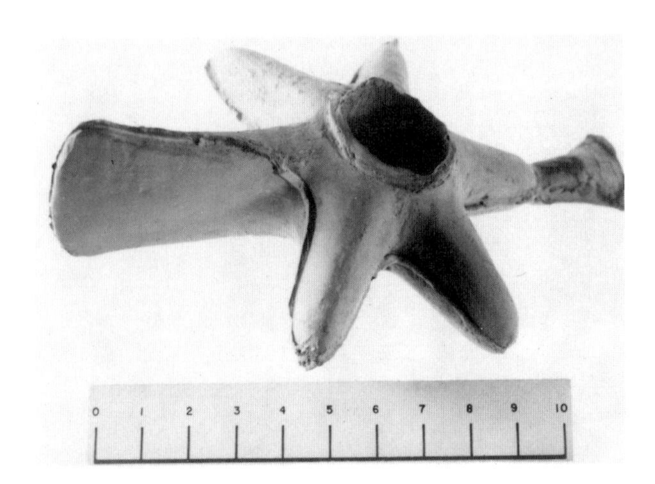

Figure 30.15 Cast plaster replica of the Inka-style ax blade shown in figures 30.12 and 30.13. Note the change in direction of the mold line (raised fin or flash of plaster) from transverse on the points to longitudinal on the blade. The pouring sprue can be observed at the far right.

Figure 30.16 Silver figurine representing a man holding cobs of maize. Cast in a multiple piece mold. Peru. Collection: American Museum of Natural History, New York (Cat. no. T-86/1).

Working

A tradition of shaping metal that relies heavily on the use of sheet to render three-dimensional form places certain requirements on the metals and alloys it uses. They must be malleable but tough, sufficiently plastic for ease in shaping and strong and tough enough to resist fracture and to retain the shape achieved. A further consequence of a sheet metal tradition for the manufacture of objects is the development of joining procedures, both mechanical and metallurgical, for uniting an array of two-dimensional parts into a three-dimensional whole.

Metals and Alloys

Copper From today's perspective copper appears to have been the most important metal in the Central Andes both in terms of the value accorded it and as the base material for manufacture of all the significant

Figure 30.17 Detail of silver figurine illustrated in figure 30.16 showing raised ridges or flashes of metal that seeped out between adjacent mold pieces during casting. The flashes visible here on the maize correspond to the boundaries of mold parts 2, 5, and 6 on the drawing in figure 30.18; the ridges on the underside of the arms indicate where parts 12 and 13 abutted mold part 14.

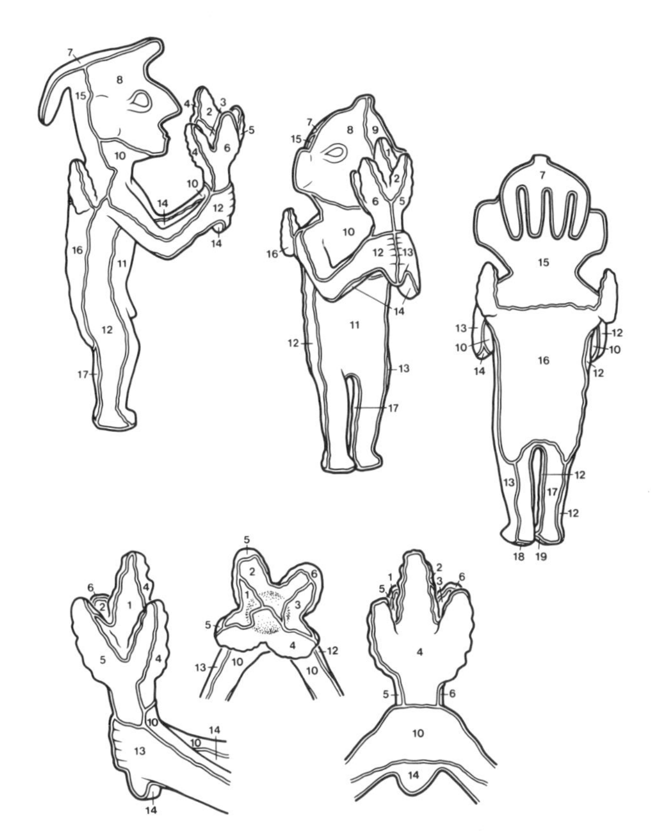

Figure 30.18 Drawing of the silver figurine illustrated in figure 30.16. Each of the mold parts identified through examination of the flash lines on the figure has been given a number. Nineteen mold pieces were assembled to cast the figure. The position of the mold joins is indicated by a thin double line in the drawings. Drawing by S. Whitney Powell.

alloys developed there (Lechtman 1980). Moche peoples in particular made heavy use of copper, most frequently for the production of metal sheet. The Moche rattles (figure 30.19) and the seated figure from the site of Loma Negra (figure 30.20)—perhaps the most northerly outpost of Moche cultural influence—are typical examples of the use of copper sheet of quite different thicknesses. Moche interest in copper as a material of choice for objects hammered and assembled to shape never abated, even in face of the new alloys they themselves invented or developed into materials of common use and broad dispersion.

Copper-Silver Alloy The alloy systems developed during the Early Intermediate Period are extremely important, for they continued to be used in later periods and some of them dominated the Andean metallurgical scene up to the time of the Spanish invasion. The first of these was the alloy of copper and silver. A small bead from Malpaso, in the Lurín valley, is perhaps the earliest object we have of copper-silver alloy, dating to approximately 700 B.C. (Lechtman 1979a, 1980). By Moche times metalworkers used this alloy over a wide range of silver concentrations, from just a few percent to over 40% by weight. The alloy became particularly popular both on the north coast, within Chimú territory, and on the south coast, where it was used extensively by Chincha metalworkers, during the Late Intermediate Period (Root 1949).

Copper-silver alloys have two properties that were important for Andean craftspeople: their toughness when hammered and their development of enriched silver surfaces when hammered and annealed. These alloys were used almost exclusively for the manufacture of objects made of sheet metal, although their use as solders was also of primary importance.

The cup illustrated in figure 30.21 is typically Chimú, and, like many others of its type, it has been fashioned from two thin sheets of copper-silver alloy (figure 30.22). Its surfaces are silver in color. An ingot cast of this alloy will solidify through the formation of primary beta or copper-rich dendrites in a groundmass of eutectic microconstituent. As the ingot is hammered to reduce its thickness and increase its area, the primary dendrites and the duplex eutectic become increasingly deformed and elongated in the direction of metal flow. Eventually, after sufficient deformation the silver-rich and copper-rich phases separate from each other, and the microstructure of the sheet pro-

Figure 30.19 Two copper rattles attributed to the Moche culture, north coastal Peru. The one at the right is intact; the example at the left is damaged and shows the three units of its construction: rattle box, ferrule, and blade. Collection: American Museum of Natural History, New York (Cat. nos. 41.2/5429 a–c and 41.2/5430).

Figure 30.20 Seated man of gilt sheet copper, Loma Negra, Peru. The front, back, and arms of the figure are separate pieces fastened mechanically by tab-and-slot joins. Collection: Metropolitan Museum of Art, New York (Cat. no. 69.82).

duced will resemble that in figure 30.23: thin lamellae of alternating silver-rich and copper-rich phases intermeshed throughout the thickness of the metal (Lechtman 1971, 1979a). It is precisely this laminated structure that imparts strength and toughness to the metal sheet, inhibiting fracture during working and helping to maintain the form achieved.

The surface enrichment properties of copper-silver alloys have been described in some detail (Lechtman 1971, 1973a). Suffice it to say that, particularly for alloys containing at least about 10 wt. % silver, the repeated sequences of hammering, annealing, and removal of the surface copper oxide scale formed on annealing—sequences necessary to the fabrication of sheet metal from the alloy ingot—produce enriched silver surfaces on the resulting sheet as the surface copper is removed. Thus metal made from such alloys, of mottled copper color when cast, is bright silver in color after having been hammered into sheet. The formation of silvered surfaces on objects hammered from these alloys is an inescapable consequence of annealing and of the attendant loss of surface copper through its oxidation. There is essentially no way of preventing it.

Copper-silver alloys were used by the Mochica to produce objects of sheet metal because the sheet was

Figure 30.21 Cup made of silver-copper alloy sheet, Chimú culture, Peru. Collection: Peabody Museum of Archaeology and Ethnology, Harvard University (Cat. no. 42-12-30/3174).

Figure 30.22 Photomicrograph of a cross section of metal removed from the base of the cup where the circular base and vertical walls meet. The join between the two has been achieved by crimping—bending the edge of the base plate up and over onto the bottom edge of the vessel wall. × 40. As polished.

hard and tough and because objects made of such sheet looked like silver. In later periods these alloys continued to be used for the same reasons, and the well-known *vasos retratos*, or effigy beakers, of Chimú and Chincha origin, said to be of silver, are sometimes made of copper-silver alloy (Root 1949; Easby 1955).

One is led to speculate on the genesis of copper-silver alloys in the Central Andes. Copper ores that contain silver are found abundantly in the central highlands. They are sulfide ores. The smelting of copper ores containing argentiferous tetrahedrite (freibergite), for example, a mineral type extremely common to the Peruvian Andes, could easily have resulted in the direct production of alloys of copper and silver. Analyses of samples of such ores being mined in Peru today indicate silver concentrations of about 6–18 wt. %. The ratio of copper to silver in such an ore is about 4:1, and one would expect the smelted alloy to contain these two metals in approximately the same proportion, giving an alloy of roughly 80% copper and 20% silver by weight. As a sulfide, however, it could not be smelted by reduction techniques without at least some prior roasting. By the Early Intermediate Period, it is much more likely that the Mochica manufactured their copper-silver alloys by melting together metallic copper and metallic silver, either the native metals or, far more commonly, metals smelted from copper and silver ores, which abound in the northern and central sierra.

Copper-Gold and Copper-Gold-Silver Alloys By far the most important alloy system developed during the Early Intermediate Period and often used by the Mochica in the manufacture of sheet metal objects was that of copper-gold. This alloy, like the copper-silver alloy, was also widely used as a solder and weld metal. Copper and gold, when melted together, form a complete solid solution series throughout the entire range of possible alloy compositions, and objects varying widely in composition have been encountered. Some silver is also often found in these Andean alloys either because the gold used contained some silver, as placer gold from the Andes often does, or because gold was added to a copper-silver alloy. Patterson (1971) analyzed an ingot of Mochica origin excavated by Max Uhle in the Moche valley. It contains 60% copper, 31% gold, and 10% silver. By contrast, Kroeber (1944) reports the compositions of several pieces of Mochica "sheet gold," also excavated by Uhle, as containing, in one case, 68% gold, 13% copper, 19% silver, and in another case, 67% gold, 11% copper, and 22% silver by weight. Copper-gold alloys such as these are often referred to in the literature of the New World as *tumbaga*, a term more often associated with the metal as it came to be used by the peoples of Colombia in the remarkable lost-wax castings they made from it. The Mochica ingot analyzed by Patterson is a copper-rich *tumbaga*. The alloys described by Kroeber may be considered gold-rich *tumbagas* with a high concentration of silver. In fact, such metal is equivalent to 16 karat gold.

Copper-gold alloys, like copper-silver alloys, become hard on hammering but retain their flexibility. They were therefore perfectly suited to the sheet metal tradition already characteristic of north Peruvian metal-

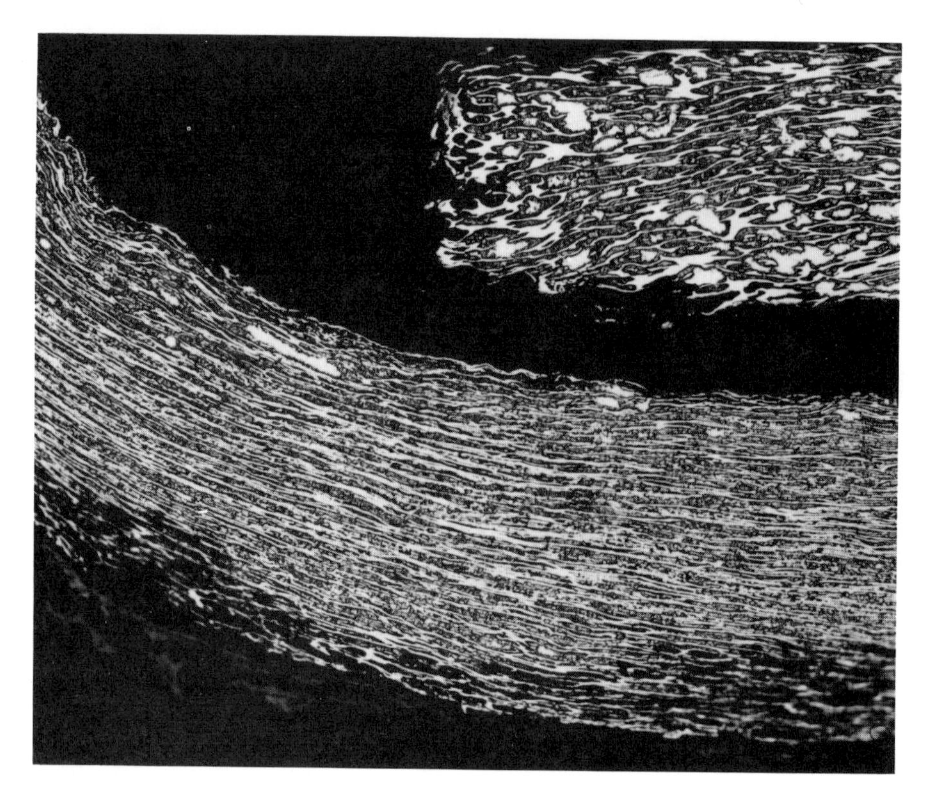

Figure 30.23 More highly magnified and etched detail of the cross section of the Chimú cup shown in figure 30.22. The etchant has revealed the typical lamellar microstructure of a heavily worked silver-copper alloy. White areas are silver-rich alpha phase; dark areas are copper-rich beta phase. The extreme elongation of these phases has resulted from extensive plastic deformation of the alloy in forming the metal sheet. × 200. Potassium dichromate and hydrochloric acid etch.

working. But these alloys were used and subsequently highly developed primarily for another property—the gold color that they confer on articles made from them once the surfaces of such objects are suitably treated. The color-enhancement properties of *tumbaga* alloys are discussed later.

Copper-Arsenic Alloys Toward the end of the Middle Horizon, at about A.D. 900, a new alloy was developed by metalworkers on the north Peruvian coast—copper-arsenic bronze (Lechtman 1979a, 1980, 1981; Epstein and Shimada 1984; Shimada 1985). We are still uncertain about the specific types of ore that were smelted to produce the alloy, but the north Peruvian highlands abound in copper sulfarsenide ores, such as enargite and tennantite, and on the coast arsenopyrite is not uncommonly found in association with chalcopyrite and other primary sulfides of copper (Lechtman 1976, 1979a). From analyses of north coast artifacts that date from the end of the Middle Horizon up to the conquest of the kingdom of Chimor by the Inka in 1470, it is clear that such ores were being smelted directly to win the metal in alloy form. The concentration of arsenic in these objects varies from about 0.5 wt. % to 5 wt. %, the majority falling within the 1–3% As range (Lechtman 1981). It is also quite evident that initially arsenical copper was tantamount to "dirty copper"; on smelting the arsenic-bearing ores, arsenic remained with the copper to constitute a major impurity in the smelted metal (Lechtman 1981). There was no way of removing it. Later, with experience in using the alloy and appreciation of its excellent mechanical properties, it was deliberately produced and became one of the primary alloys of the Chimú state. Only conquest of that state by the Inka caused abandonment of production of copper-arsenic bronze in the northern zone of the Central Andes.

Arsenic bronze was used regularly in casting ornaments and heavy work tools (see figures 30.7 and 30.9) and in hammering both small and large tools to shape. Heavy axes, such as the Ecuadorian example from the site of La Compañía, illustrated in figure 30.24 (1.5% As), were cast roughly to shape and then extensively cold-worked to render the final form and to harden them for use. Figure 30.25, a photomicrograph of a longitudinal section through the tip of this ax blade, reveals the residual center line porosity of the casting that at the tip of the tool has been highly compressed through hammering. At the work-hardened tip, the grains are severely distorted and full of deformation lines.

On the other hand, arsenic bronze seems to have seldom served in sheet metal production. Evidently

Figure 30.24 Cast and worked ax from the site of La Compañía, Los Rios Province, Ecuador. Milagro culture. Height: 12.5 cm. Collection: National Museum of Natural History, Smithsonian Institution, Washington, D.C. (Inventory no. 1260-2). Photograph courtesy of Betty Meggers.

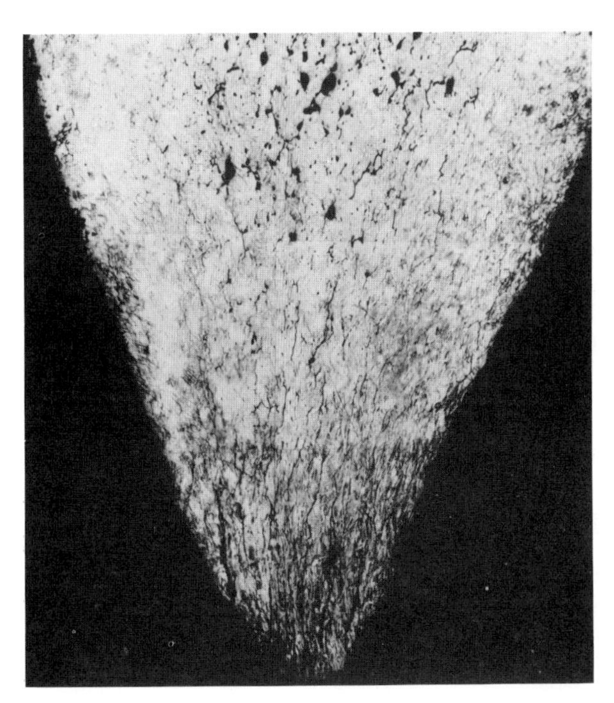

Figure 30.25 Photomicrograph of a longitudinal cross section removed from the tip of the blade of the ax shown in figure 30.24. The pores at the center of the casting are fairly round toward the body of the ax but have been severely compressed at the blade tip from extensive working of the metal there. The microstructure at the work-hardened tip consists of highly distorted grains replete with deformation lines. × 13. Ammonium hydroxide and hydrogen peroxide etch.

copper, copper-silver, and copper-gold-silver alloys performed adequately for this purpose. Instead, household implements such as needles and spindle whorls (see figures 30.26 and 30.27) were commonly fashioned by hot or cold working the alloy (Lechtman 1981; Arnold 1986), and even the so-called architectural cramps (figure 30.28) found at the site of Tiwanaku, Bolivia (Mead 1915; Posnansky 1945, vol. 1, figs. 14, 14a) and probably dating to Tiwanaku Phase IV (c. A.D. 375–700) are made of copper-arsenic bronze.[5] The cramp shown in figure 30.28 is one of five such artifacts from Tiwanaku, and their use there for architectural construction purposes is a rare use of metal in the Andes. The one illustrated here was cast roughly to its present shape, then hammered extensively at both ends to form the flanged extremities of the cramp. Figure 30.29 shows the heavily worked microstructure of the highly segregated alloy (2.34 wt. % As), its grains distorted and full of deformation lines. Whether the cramp was set into wood or masonry, it was hammered along one surface until it fitted into the slot provided.[6]

It should come as no great surprise that items such as these cramps, chronologically placed at the end of

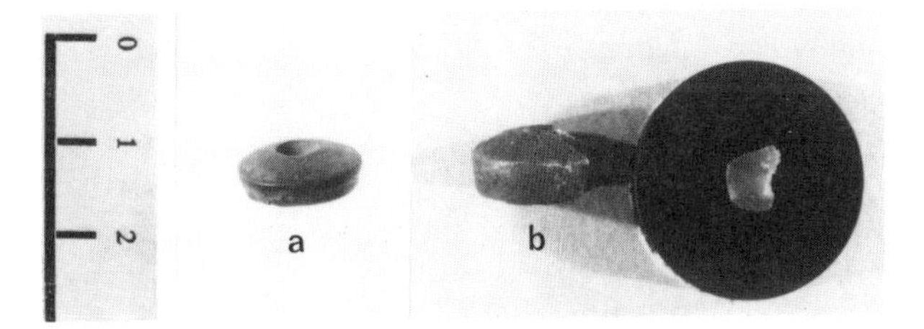

Figure 30.26 Spindle whorls of copper-arsenic bronze. (a) Chan Chan, Moche valley. (b) Batán Grande, Lambayeque valley, Peru. Both whorls were made in the identical manner. The example from Chan Chan is reproduced intact; the Batán Grande specimen has been sampled and is shown with the mounted and polished cross section. This section gave rise to the photomicrograph illustrated in figure 30.27.

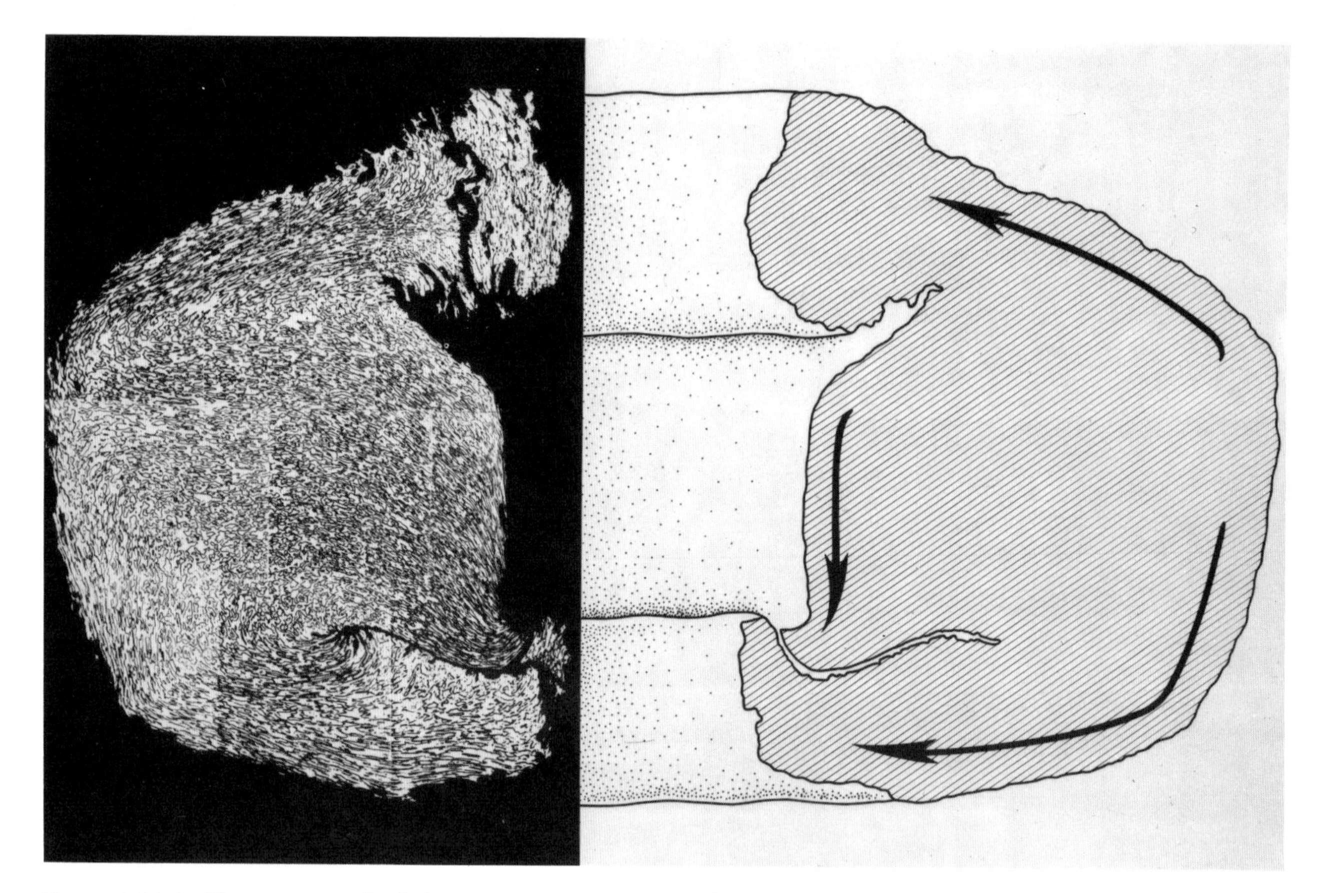

Figure 30.27 Photomicrograph of the cross section removed from the spindle whorl shown in figure 30.26b. The line drawing at the right completes the whorl in the plane of the section, with arrows to indicate the direction of metal flow so clearly indicated by elongation and distortion of the highly cored dendrites (1.90% As). A hole was partly cut, partly punched through the center of the original lump of metal, the blunt tool dragging the metal down along with it. Afterward, the whorl was shaped by hammering metal from the edges of the lump inward, toward the central hole. × 13. Potassium dichromate etch. Drawing by S. Whitney Powell.

the Early Intermediate Period or at the onset of the Middle Horizon and located at the southern end of the central Andean zone, should be made of arsenic bronze. Although northern Bolivia presents some of the richest tin fields in the world, which were heavily exploited by the Inka, objects of copper-tin bronze are not abundant in the region until the Inka presence there. Furthermore, local outcrops of copper sulfarsenide ores are known from northwest Argentina, where arsenic bronze artifacts have been found that belong to the Aguada phase, or the beginning of the ninth century (Fester 1962; González 1979). Thus, although we rightly attribute the development and intensive use of arsenic bronze to the metalworkers of the Peruvian north coast, it is clear that there was widespread experimentation in the smelting of arsenic-bearing ores of copper in many central Andean locales, wherever such ores were a serious source of metal.

Copper-Tin Alloys Just as the copper-arsenic alloy typified bronze production in the northern reaches of the central Andean zone, so did copper-tin alloy become the hallmark of bronze manufacture in the southern portion of that zone (Lechtman 1979a, 1980).

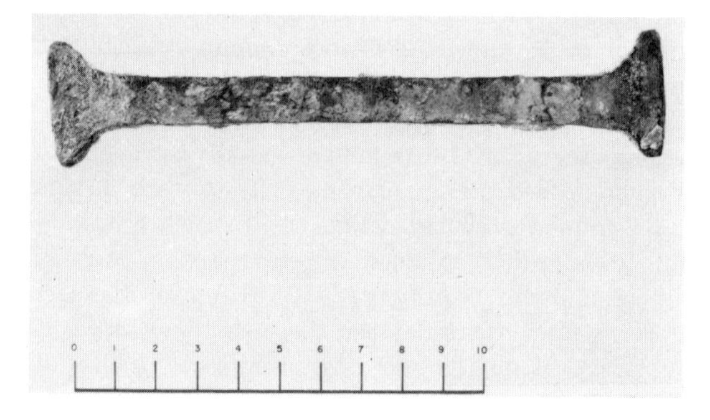

Figure 30.28 Architectural cramp of copper-arsenic bronze from the site of Tiwanaku, Bolivia. Collection: American Museum of Natural History, New York (Cat. no. B/2792-c).

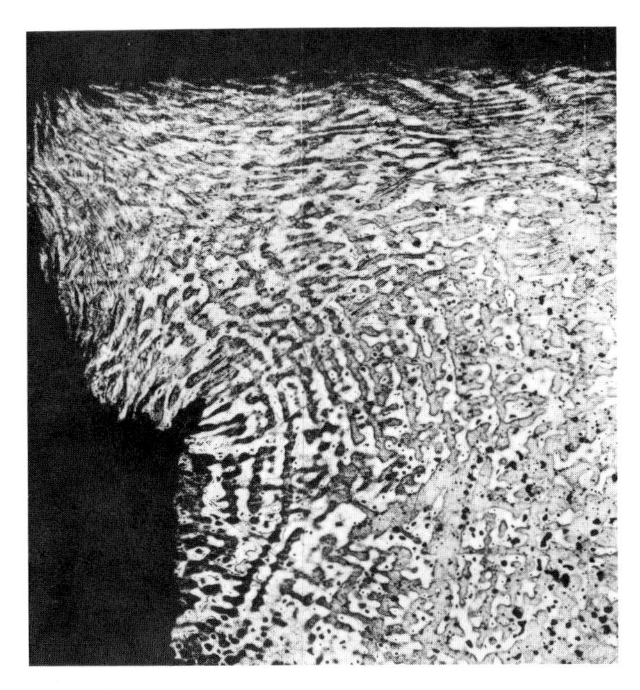

Figure 30.29 Photomicrograph of a longitudinal cross section removed from one of the widened ends of the cramp illustrated in figure 30.28. The microstructure is characterized by marked coring of the dendrites in the as-cast blank of copper-arsenic alloy (2.34% As). These dendrites have been severely elongated at the upper edge of the section (slightly upset end of the cramp) as a result of appreciable cold working. The grains here contain a high density of deformation lines.

The alloying element, tin, was readily available as cassiterite (the oxide of that metal), which abounds in southern Peru, northern Bolivia, and northern Argentina both as placer deposits and in veins. It is most likely that cassiterite was smelted to produce metallic tin, which was then melted with copper metal to form bronze. Support for this conjecture is given by several lumps and sheets of quite pure metallic tin found at the Inka site Machu Picchu and interpreted by Gordon as stock material used in the preparation of copper-tin bronze (Mathewson 1915; Gordon and Rutledge, 1987). Furthermore, I am not aware of a mixed source of copper-tin ores in this region that could have been mined and smelted directly to give tin bronze in the composition range characteristic of the artifacts from the area—normally 5–15 wt. % tin. The mineral stannite (Cu_2FeSnS_4) is common in Bolivia and frequently accompanies minerals of copper and silver, but its inclusion in a copper ore could be expected to account for the presence of tin in artifacts in concentrations of only about 1.5% (Lechtman 1980, fn. 126).

Copper-tin bronze was used for every conceivable kind of object—ornaments, implements, weapons (see figure 30.12)—whether they were cast or worked. In his superb study of the Late Horizon bronzes from Machu Picchu, Mathewson (1915) establishes the mastery of the alloy by Inka metalworkers and their appreciation and control of the properties and working characteristics of the metal [see also Gordon and Rutledge (1984)]. High-tin bronzes (generally between about 10% and 13% tin) were used in castings, taking advantage of the two main qualities of such alloys: strength and superior castability. On the other hand, low-tin bronzes (about 5% tin) were used for objects that were largely worked to shape, including those made of sheet metal, such as axes, chisels, depilatory tweezers, *tumis,* and *tupus,* because these alloys are ductile and easily worked cold without becoming brittle. When hardness was required above and beyond that naturally conferred on the alloy by the addition of these relatively modest amounts of tin, the smiths work-hardened the edges of *tumis,* chisels, and other cutting tools to give them added durability and sharpness.

In a more recent study, Gordon (1985) suggests that at Machu Picchu the common T-shaped ax, broadly disseminated throughout *Tawantinsuyu* by the Inka, may have been used as a stone cutting tool. The average tin content of these Machu Picchu artifacts is 5%, and many have severely deformed blade edges. An example of such an Inka ax is illustrated in figure 30.30, and figure 30.31 presents a photomicrograph of a longitudinal section removed from the tip of the blade. The microstructure of this bronze (4.31% Sn) exhibits fully recrystallized grains with annealing twins. Left in the annealed condition, there is no evidence that the ax was ever used. Many never were, nor were they intended to be. They served often as symbols of state, hafted and borne as emblems of the imperial presence. It is especially interesting to compare the microstructure of this tin-bronze ax blade with that of the Ecuadorian arsenic-bronze blade in figure 30.25. The Ecuadorian blade was a heavy working tool; the tin-bronze ax probably was a lighter weight bearer of symbolic message.

By the end of the Middle Horizon, copper-arsenic bronze had established itself in the northern portion of our central Andean metalworking area, and copper-tin bronze was coming into its own in the south (Lechtman 1979a, 1980). They existed side by side and served regional needs well. Only the Inka expansion to the north and their conquest of the kingdom of Chimor in 1470 put an end to arsenic bronze production in the Central Andes. Thereafter tin bronze became the preferred alloy of the state (Lechtman 1979a, 1980; Owen 1987).

The chief difference between the two bronzes lay in their method of manufacture. Arsenic bronze involved the roasting and smelting of sulfarsenide ores of copper or the direct smelting of a mixture of oxide and arsenic-bearing sulfide ores.[7] Tin bronze, on the other hand, was produced by melting together metallic copper and metallic tin, each metal won from ores typical of the southern region. Otherwise, both bronzes were

Figure 30.30 Inka ax of tin bronze, cast and hammered. Provenance may be Cuzco, Peru. Height: 8.0 cm. Collection: American Museum of Natural History, New York (Cat. no. B/9188).

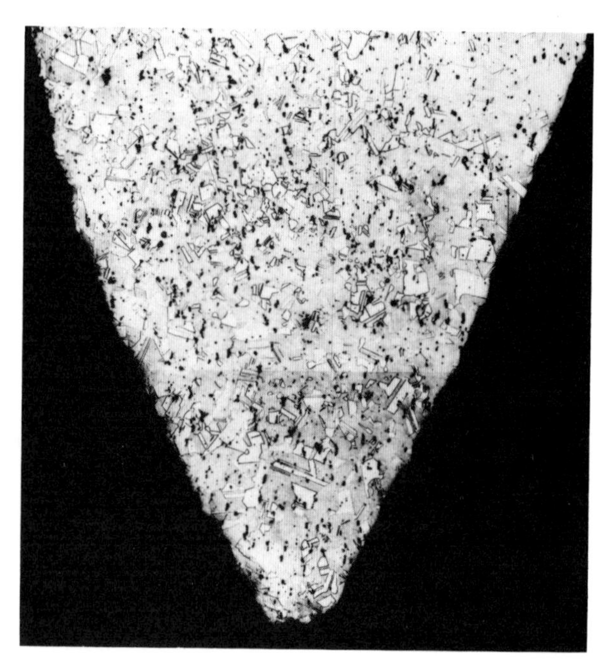

Figure 30.31 Photomicrograph of a longitudinal section removed from the tip of the blade of the ax illustrated in figure 30.30. The microstructure reveals fully equiaxed grains with annealing twins. The ax was probably worked and annealed several times in the process of shaping the blade, but it was left in the annealed condition. There is no evidence that the blade was used, and the tip has not been work hardened. × 25. Potassium dichromate etch.

used for similar purposes, although copper-tin bronze was more frequently worked into sheet for the manufacture of items such as tweezers than was copper-arsenic bronze.

Thus we see that the range of metals and alloys used and developed by central Andean metalsmiths—copper, copper-silver, copper-gold, copper-gold-silver, copper-arsenic, and copper-tin—in addition to gold and silver alone, was carefully tailored to accommodate the thrust of the metalworking tradition. Emphasis was placed on alloys that balanced plasticity with toughness, alloys that would serve as good solders or weld materials, and alloys whose surfaces could undergo transformations in color.

Joining

It is convenient to distinguish two types of metal join: mechanical joins and metallurgical joins. Metallurgical joins require heat to achieve union between the metal parts (as in soldering, brazing, welding). Both types of join were common to metalworking practice among smiths whose major concern was often the realization of three-dimensional form through the assembly of many individual pieces of metal [see Tushingham et al. (1979)]. The joins had to be strong, permanent, and sometimes invisible. I have long held the opinion that one of the stimuli to the development of alloys in the Central Andes was precisely the need to produce solders and brazing metals that would meet the demands of a metalworking tradition deeply rooted in the art of assembly (Lechtman 1980).

Mechanical Joins The seated figure reproduced in figures 30.20 and 30.32, from the site of Loma Negra on the far north coast of Peru (Lechtman 1979b; Lechtman et al. 1982), can be considered as representative of the northernmost reaches of Moche cultural influence in the area at about A.D. 300. It illustrates the most common Andean mechanical joining procedures used in sheet metal work. The object is assembled from various pieces of preformed and gilt copper sheet. The front and back halves of the figure, fashioned separately, fit tightly together along an overlapped seam. The arms enter snugly into sockets and are secured to the lap of the seated man by short tabs that project from the hands into narrow slots at the knees. Tabs at the base of the figure hold the man rigidly to his seat, and the seat, in turn, has a number of tabs located underneath that originally secured the ensemble to some other artifact, now lost. There is no solder anywhere on this object.

The front and back of the seated figure form a close fit; undoubtedly the outer sheet of copper (the front half) was hammered down along the join to force the two overlapped edges together. Nowhere is the metal crimped, however. A good example of crimping—the bending of the edge of one sheet of metal tightly over the edge of its neighbor—is given by the Chimú cup

Figure 30.32 Seated man, Loma Negra, Peru. Side view (see also figure 30.20). The join made by tightly overlapping the edges of the front and back halves of the figure is clear in this view. Note the tab at the bottom of the figure projecting down into a slot in the base and the arm fitted closely into its socket.

illustrated in figure 30.21. The base of the cup is a circular sheet fashioned separately from the flaring wall of the vessel. It has been attached to the wall by simple mechanical crimping together of their common edges, as is clearly demonstrated in figure 30.22.

The tab-and-slot technique for joining metal sheet was extremely popular and is found on objects almost irrespective of geography or chronology, although it exhibits greatest versatility along the north coast from the Early Intermediate Period through the end of the Chimú ascendancy there.

An unusual and extremely fine pendant of white shell from the south-central Peruvian coast (figures 30.33 and 30.34) illustrates the tab-and-slot technique as used to make a seam, a fairly common practice in sheet metal work. The head and face of the pendant consist of a single piece of sheet gold wrapped around the narrow end of the shell and closed at the rear by a running tab-and-slot seam. Figure 30.35 shows the seam in greater detail. The head is capped with a slab of purple shell and is fixed to the shell body with a dark resinous material that acts as a cement.

Another gilt sheet copper object from Loma Negra, portrayed in figure 30.36, displays the penchant of Andean peoples, particularly from northern Peru and Ecuador, for the shimmering effect of dangling and

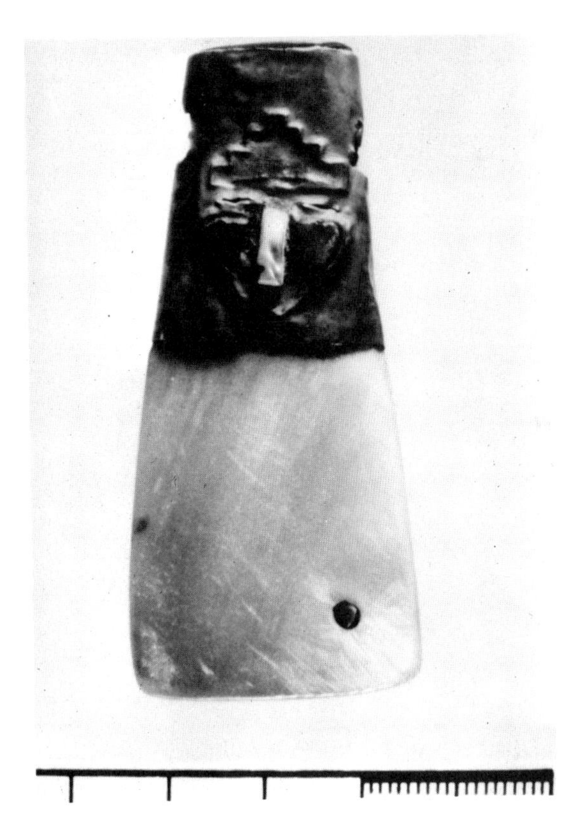

Figure 30.33 Shell pendant with gold head, Nasca style, from the Ica valley, Peru. Inlaid eyes of green stone, nose of Spondylus shell. Front view. Private collection.

Figure 30.34 Shell pendant, back view. Note the tab-and-slot join that secures the two edges of the gold sheet.

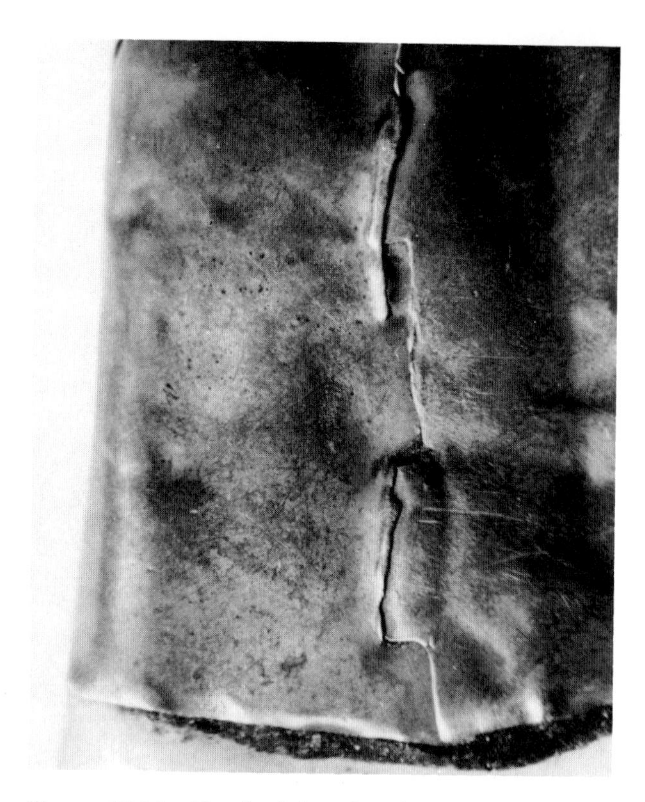

Figure 30.35 Detail of the tab-and-slot seam shown in figure 30.34.

Figure 30.36 Head of gilt sheet copper, Loma Negra, Peru. The headdress attaches to the face with a tab-and-slot join. Flat metal wire straps secure the dangling sequins to the metal sheet.

highly reflecting metal sequins or precious stones appended to metal objects. They were part of the metallic aesthetic. Such danglers were suspended and fastened by metal wires, most often thin flat straps of metal of rectangular section. Not only was such wire used to fasten pendant features, but it also often virtually sewed together overlapping sheets of metal [see Tushingham et al. (1979) and Carcedo and Shimada (1985)].

The photomicrograph (figure 30.37) shows a section through a piece of flat strap wire from a Loma Negra object similar to the mask shown in figure 30.36. Wire was never drawn in the Andes. Whether round or flat, it was hammered to shape from thin stock rod or from sheet, as the example illustrated here demonstrates.

Thicker wire was also made into artifacts in their own right. Ecuadorian smiths were especially adept at making earrings and nose ornaments from heavy square, round, or oblate wire (Zevallos 1956). Here I illustrate just one example of a small Ecuadorian copper-silver alloy earring (c. 15% Ag) made of wire that has been fashioned into a spiral (figure 30.38). A transverse section cut through the spiral (figure 30.39) shows the changing contour of the wire as it was hammered to shape from the original thin cast rod. A more highly magnified detail (figure 30.40) illustrates clearly that it was plastic deformation of the metal that resulted in the desired contour.

As Tushingham et al. (1979) have amply demonstrated, crimping, lacing, tabbing, slotting, and other such techniques were all appropriate to a sheet metal tradition, and all were widely used in the area under discussion. Some of these joining procedures are akin to the way in which cloth is joined, and those who have studied Andean sheet metal work have often considered the closeness in certain aspects of the handling of two-dimensional sheets of metal and of webs of woven cloth.

Metallurgical Joins Metallurgical joins require heat to effect a bond between the metal parts being joined. Soldering, brazing, and welding are the three most common techniques that fall within this category. The first two require a filler metal that is introduced between the parts. Solders and brazing metal must melt at a temperature lower than those parts so that they can run freely at the joint and, on freezing, bond the parts together. There is no concurrent local melting of the parts themselves. According to the definition set by the American Welding Society, soldered joins are those made at temperatures below 800°F (427°C), whereas brazed joins take place at higher temperatures. When a join is made by welding, melting occurs along the edges of two contiguous pieces of metal; the edges fuse and solidify together. If a filler metal is inserted at the joint to facilitate the weld, the filler is of the same composition as the parts being joined (see, for example, the discussion later of the sweat-welded joins on the Peruvian jaguars).

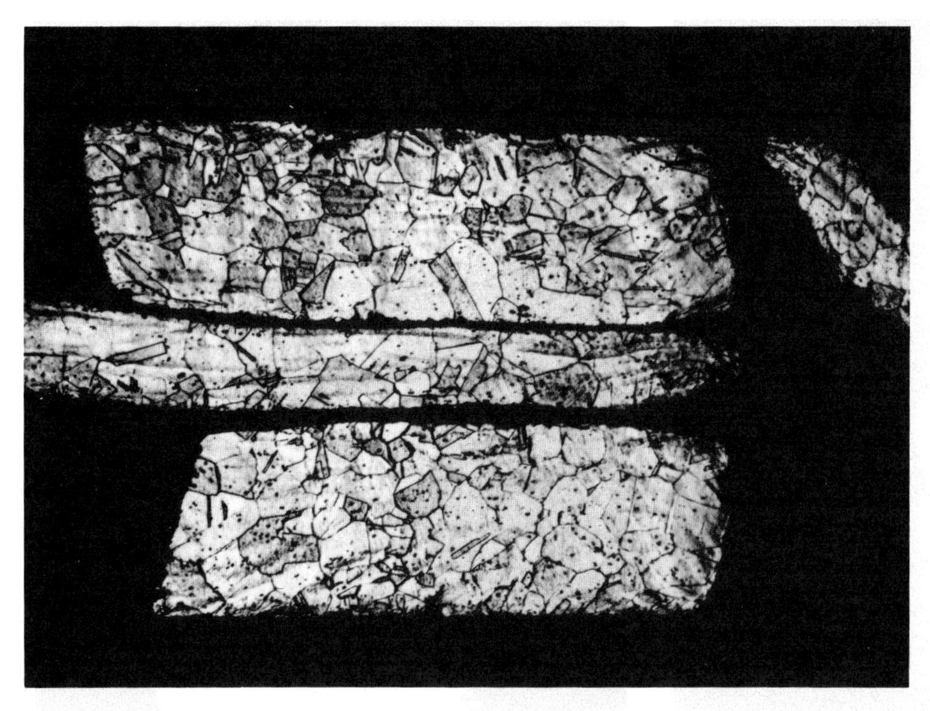

Figure 30.37 Photomicrograph of a transverse section through a flat wire strap similar to those that secure the danglers on the Loma Negra object illustrated in figure 30.36. The thick wire surrounds the thin copper sheet on its front and back surfaces. Both sheet and wire have been hammered and annealed and left in the annealed condition. × 50. Potassium dichromate and ferric chloride etch.

Figure 30.38 Spiral earring of thick wire, from the site of Peñon del Río, Guayas Province, Ecuador. Height: 2.0 cm. The copper alloy contains both silver and arsenic. The earring is on loan to MIT from the Escuela Superior Politécnica del Litoral (ESPOL), Guayaquil, Ecuador.

Soldering and welding were mastered by goldsmiths during the period of spread and influence of the Chavín cult, and by the end of the Early Horizon both techniques were expertly used for joining pieces of sheet gold (Lothrop 1941, 1951; Lechtman 1984a). It is also probable that a joining process akin to granulation was current by about 600 B.C. (Lechtman 1980, 1984a). In all of this early material the solders or brazing metals appear to have been gold-copper or gold-silver alloys.

By the Early Intermediate Period, accompanying the serious use of copper and its alloys, other solders were developed. Figure 30.19 illustrates two of a set of eight copper objects of Moche origin, probably dating to the latter part of Moche Phase IV (c. A.D. 550–650) (Donnan 1978, fig. 254, and personal communication). They have been identified by Junius Bird as rattles (Friedman et al. 1972) because the upper boxlike portion is hollow and contains several pellets, probably of stone.[8] This upper chamber is composed of three pieces of thick sheet copper: (1) the vertical walls, fashioned from a single sheet that has been bent to form four corners and closed by an overlapping joint, riveted and soldered together; (2) an upper and (3) a lower plate soldered to the top and bottom of the box (figure 30.41). In most of the soldered joints the solder occurs rather evenly, but at times it appears as individual flat pieces of metal that have not melted sufficiently to lose their shape. Figure 30.42 shows a polished section of a sample cut diagonally through the lower right corner of the box, as seen in figure 30.41. The solder is porous and spongy in appearance, but it has flowed around the overlapping vertical edges of the box and has effectively joined the lower plate to the walls.

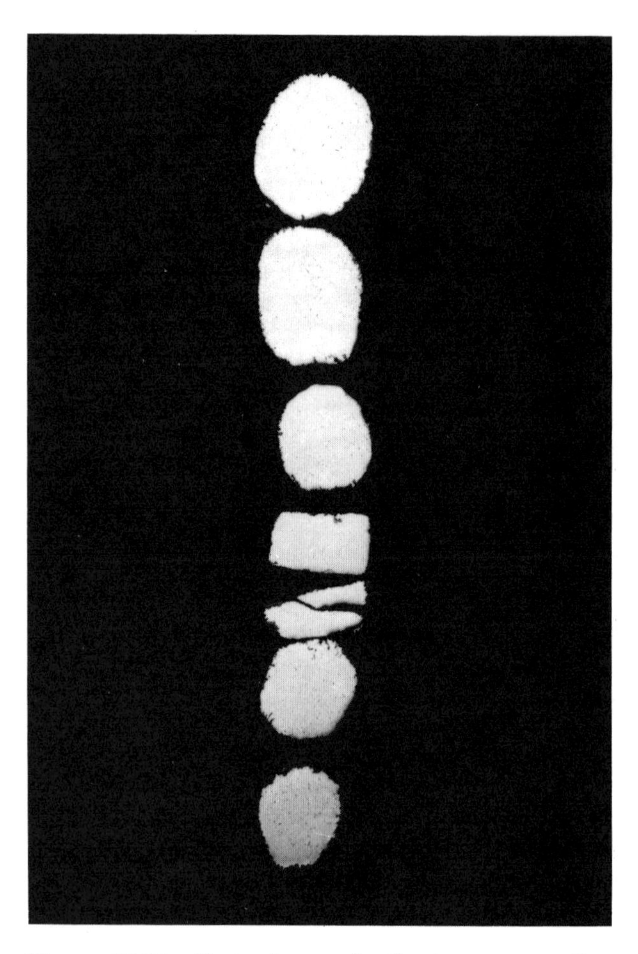

Figure 30.39 Photomicrograph of a transverse section through the spiral portion of the earring illustrated in figure 30.38. Note the dramatic change in contour of the wire, from oval to rectangular. × 11. As polished.

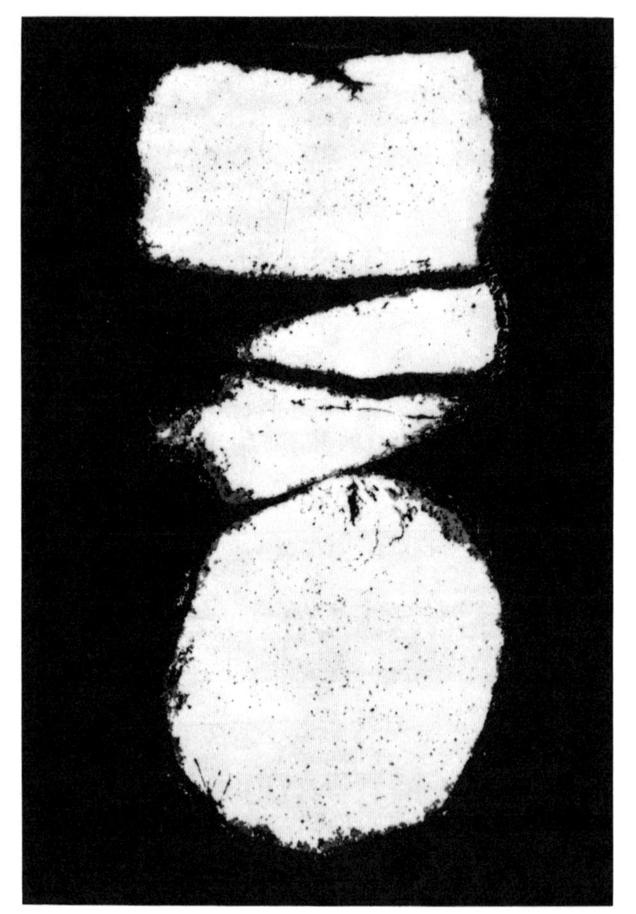

Figure 30.40 Detail of the photomicrograph of figure 30.39 showing the three central turns of wire in the spiral. The hammering of the wire to change its shape is especially clear in the rectangular portion. There surface fissures have formed where metal overlapped and abutted as a result of plastic flow. × 40. As polished.

Figure 30.41 Detail of the upper box portion of the Moche rattle illustrated in figure 30.19 (left). Note the copper rivets along the right vertical edge that secure the overlapping ends of the metal sheet there. Solder, now partially corroded, joins the top and bottom plates to the walls.

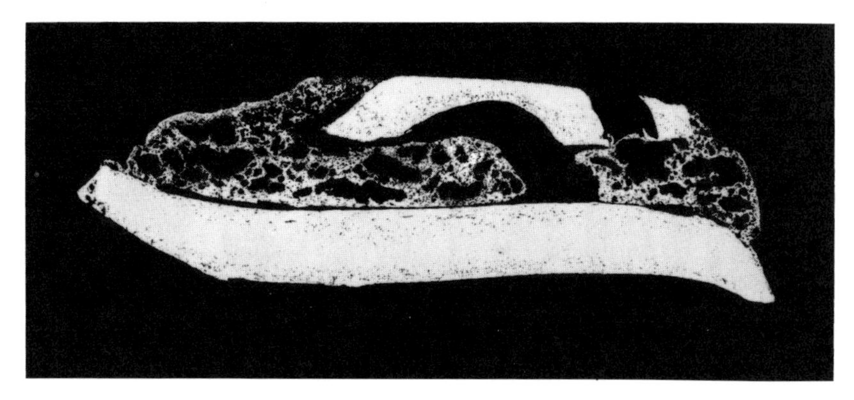

Figure 30.42 Oblique section through the soldered joint at the lower right-hand corner of the rattle box, as oriented in figure 30.41. The broad horizontal band of metal (white) at the bottom of the micrograph corresponds to the bottom plate of the box; the upper bands of metal of similar appearance are portions of the two overlapping ends of the wall. The spongy, porous material that connects wall and plate is a copper-arsenic alloy solder. × 5. As polished.

Electron microanalysis of the solder revealed the presence of a copper-arsenic alloy containing 5.4% arsenic by weight. The melting point of the solder (about 850°C) is 180° lower than that of the copper sheet (about 1030°C), which the analysis proved to be copper with an average of 0.8% As. This is the only Andean example I known of that uses a copper-arsenic alloy as a solder. It is especially intriguing for two reasons: first, because it was the Mochica who began experimenting with copper-arsenic alloys on the north Andean coast (Lechtman 1980), and this is an imaginative and rather precocious use of an alloy with which smiths had relatively recent experience; and, second, because most central Andean artifacts of arsenic bronze contain between about 1% and 3% As, rarely 5% or higher (Lechtman 1981). Moche smiths were quick to recognize the range of uses of this new alloy, and the rattles are a surprising instance of how they tested its properties.

One of the most elegant and sophisticated examples of soldered and welded joins on a single object is given by the group of seven matched hollow gold jaguars from the site of Pampa Grande on the Peruvian north coast (Lechtman et al. 1975). These small felines, one of which is reproduced in figure 30.43, were made at the end of the Moche presence (Moche V) and the beginning of the Middle Horizon (Lechtman et al. 1975, fn. 74), at some time between c. A.D. 560 and 900. They are hollow, as the X-radiograph of figure 30.44 demonstrates, and each is a composite of twelve shaped pieces of hammered sheet gold whose composition is 75.9% Au, 14.9% Ag, and 8.9% Cu (M.P. = 930°C). The body has an upper and a lower portion, joined along their common edges. Each extremity—tail, ears, front legs—is made of halves, joined along their midlines. The jaguars were assembled by joining the extremities to the body after the top and bottom body parts themselves had been united

The halves of each tail, ear, and leg were joined by a sweat-welding technique that can be appreciated with reference to figure 30.45, a diagram of the steps in the assembly of the tail. Originally the edges of each tail half were fitted with a kind of strip clip that ran like a thin ribbon along the entire length of the edge (figures 30.45a, d). The clip was made of extremely thin sheet metal (0.002 cm)—of the same composition as the metal of the tail—whose ends were rolled into a tubular shape so that the ribbon would clamp onto the tail edge when pressed into place. The two halves of the tail were brought together, registered, and held mechanically under some pressure before and during heating. Figure 30.45b illustrates how a cross section through the seam would appear at this stage of the procedure. The clips provide a substantially increased surface area over which bonding can be achieved, and good contact is made at all the surfaces that eventually must fuse to effect the join. Figure 30.45c illustrates in cross section three such possible types of join achieved by this design. In each case the left and right halves of the tail are connected through the ribbon clips, but the zone of fusion is different for each.

A look at several photomicrographs of sections through the tail seam affords further insights into the nature of the bond achieved. Figures 30.46 and 30.47 demonstrate that the melting of the ribbon clips was never complete, nor was it meant to be. In all cases they retain their original shapes. The seam was never brought to a high enough temperature nor sustained at that temperature long enough to melt the added metal so that it ran like a true weld filler. Instead, the seam was heated until only its surfaces—the surfaces of the tail metal and the surfaces of the ribbon clips—began to melt or to sweat. The clips, being of much thinner metal than the tail, began to sweat first, thus initiating the bonding action. As soon as enough surface metal became molten and could fuse, the heat source was removed and the join allowed to cool and solidify.

The photomicrograph of figure 30.46 indicates that complete bonding did not occur between the right and

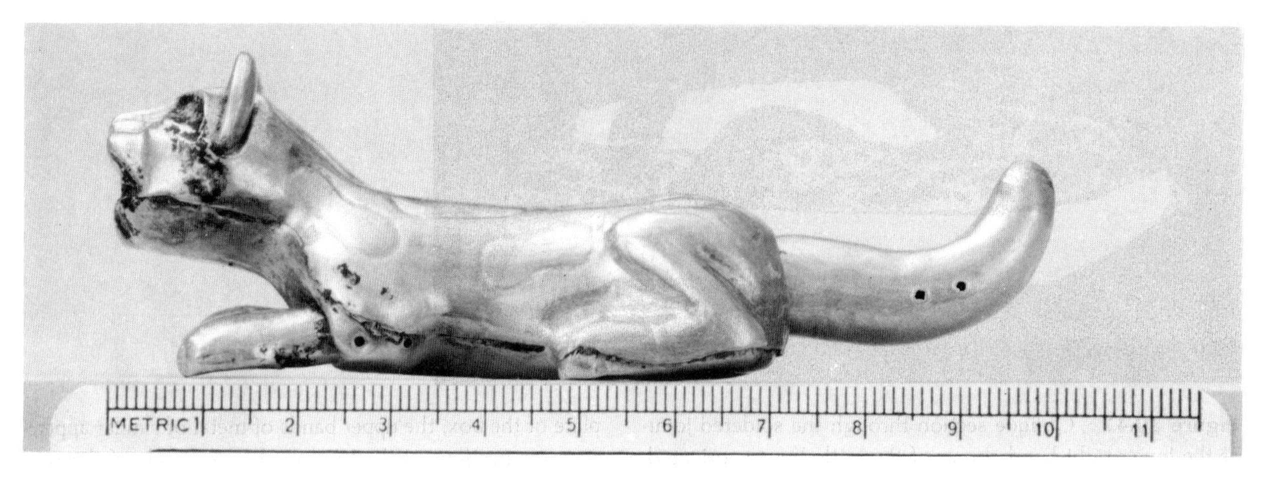

Figure 30.43 Jaguar of sheet gold, Pampa Grande, Lambayque valley, Peru. Collection: Chicago Art Institute (Cat. no. 70.420). Photograph courtesy of the Chicago Art Institute.

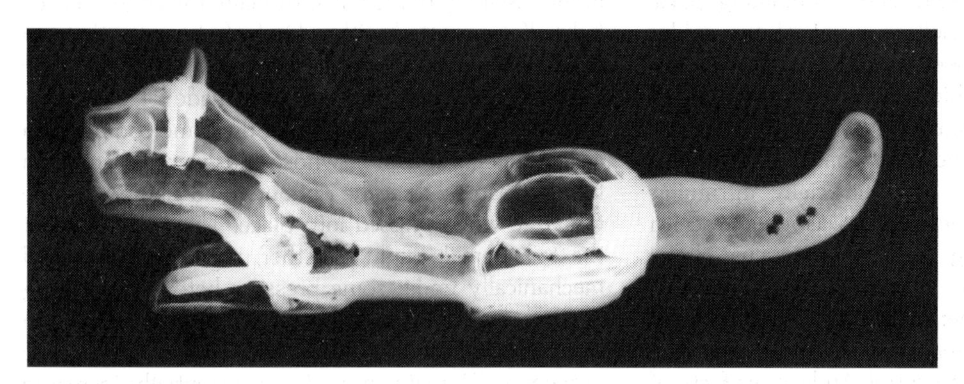

Figure 30.44 X-radiograph of the jaguar illustrated in figure 30.43. Note pallions of silver-copper alloy solder all along the body seam.

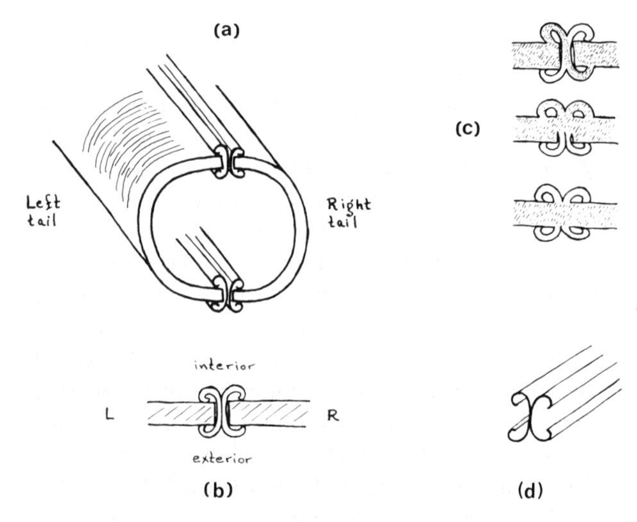

Figure 30.45 Diagram illustrating the configuration of ribbon clips at the seam between the halves of the jaguar tail. (a) Clips pressed along length of seam, following seam contour outside and inside. (b) Cross section of seam shown in (a). (c) Variations in the bond configuration across the join. (d) Clips joined along their middle before mounting on seam edges.

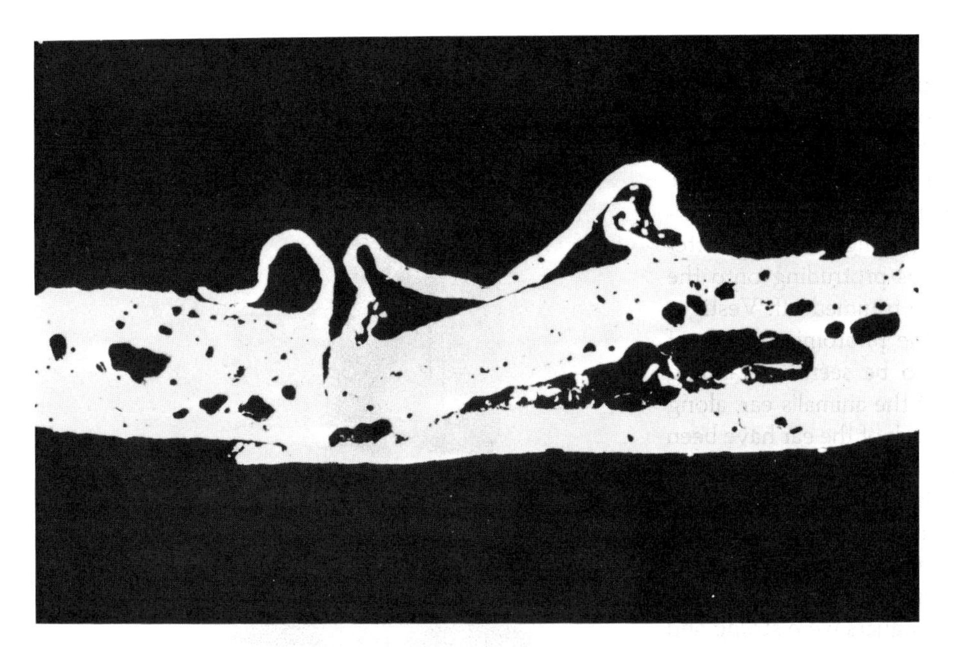

Figure 30.46 Jaguar, photomicrograph of a transverse section through the tail join. Note the tubular clips intact on the interior surface but ground away on the exterior. × 100. As polished.

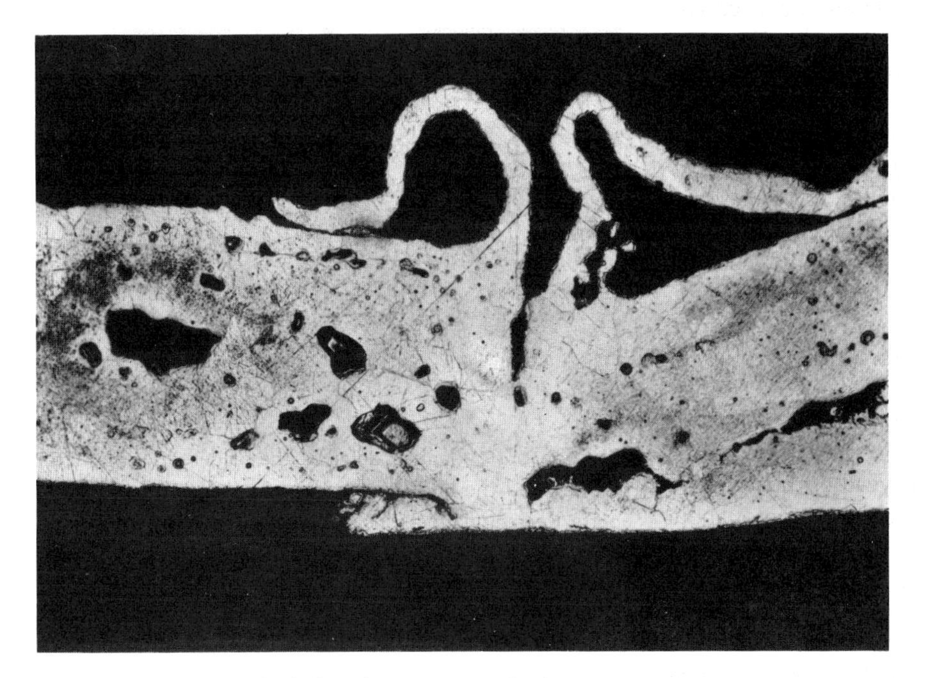

Figure 30.47 Detail of the photomicrograph shown in figure 30.46. Note the small equiaxed grains containing annealing twins that are continuous across the join. × 200. Potassium cyanide and ammonium persulfate etch.

left tail halves at the location of this section; yet all along the seam one continuous bond, even if only a partial bond, is made at each position, creating a sturdy join. Where fusion did occur, the worked and annealed structure of the tail metal and the clips is continuous, the small grains with annealing twins running uninterrupted across the bonded zone (see figure 30.47). After the halves were joined, the clips protruding onto the exterior surface of the tail were trimmed off. Vestiges of these clips are evident in the photomicrograph of figure 30.46, and they can also be seen as standing slightly proud of the surface of the animal's ear, along the seam where the front and back of the ear have been sweat welded together (figure 30.48).

The beauty of this system lies not only in its effectiveness in joining extremely thin sheets of metal but also in the virtual invisibility of the seam, because the color of the filler clips is the same as that of the halves of each extremity. Because the jaguars were strung and probably worn around the neck as a kind of chest ornament, with their backs uppermost, facing the viewer, the soldered seam so prominent in the side view given in figure 30.43 would not have been visible. And the sweat-welded joins, neatly trimmed and polished and fully golden in color, would have escaped notice.

The upper and lower body parts were joined with a silver-copper solder in the form of short rectangular pallions of metal inserted at intervals along the overlapping seam (see figure 30.43). These are easy to recognize in the X-radiograph (figure 30.44) by their marked porosity. Later, each extremity was united with the body using this same soldering technique. Figure 30.49 presents a detail of a solder pallion positioned at the joint where one front leg is inserted into the body socket prepared to receive it. Nearby is another pallion, fitted at the seam between the upper and lower body parts. A sample of a pallion of solder removed from the joint between the tail and the body (figure 30.50) was analyzed by quantitative metallography, yielding an alloy composition of about 52% Ag and 48% Cu (M.P. = 860°C). A detail of the bubbly porous structure of primary copper-rich dendrites surrounded by eutectic is given in the photomicrograph of figure 30.51.

Whereas the sweat welding of the extremities was probably achieved by first heating the entire extremity almost to the sweating point and then concentrating the heat at the seam, the soldering entailed heating only the local area covered by each solder pallion. This must have necessitated a fine blowpipe, whose jet would have allowed each pallion to be heated while the rest of the animal remained relatively cold. Given the rather small difference in melting temperature between the sweat-weld metal and the solder (approximately 70°C), the spot-solder technique was imperative to concentrate the heat at discrete areas from which it was rapidly dissipated.

Figure 30.48 Jaguar, detail of the seam at the midline join of the right ear. Portions of the sweat-welded ribbon clips stand slightly proud of the surface.

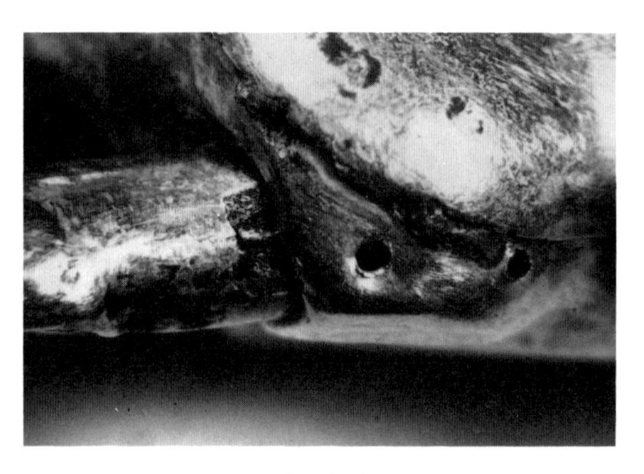

Figure 30.49 Jaguar, detail of the soldered join between the leg and the body and between upper and lower body parts. Rectangular pallions of silver-copper alloy solder were fitted into the seams and heated there until they partially melted.

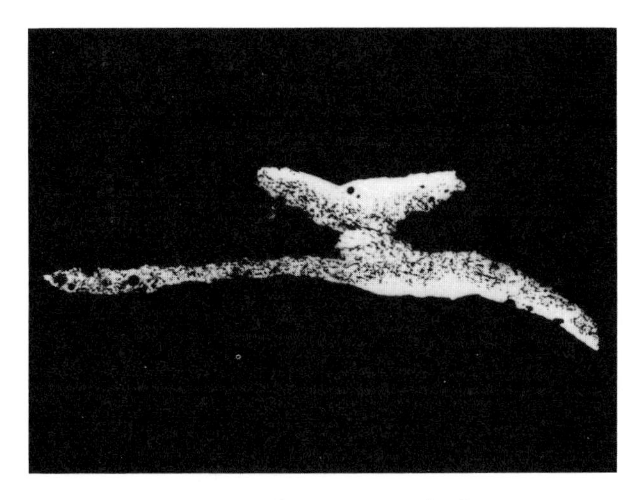

Figure 30.50 Jaguar, photomicrograph of a cross section through a silver solder pallion removed from the tail. The upper surface originally bonded to the body metal, the lower surface to the tail metal. × 39. Ferric chloride etch.

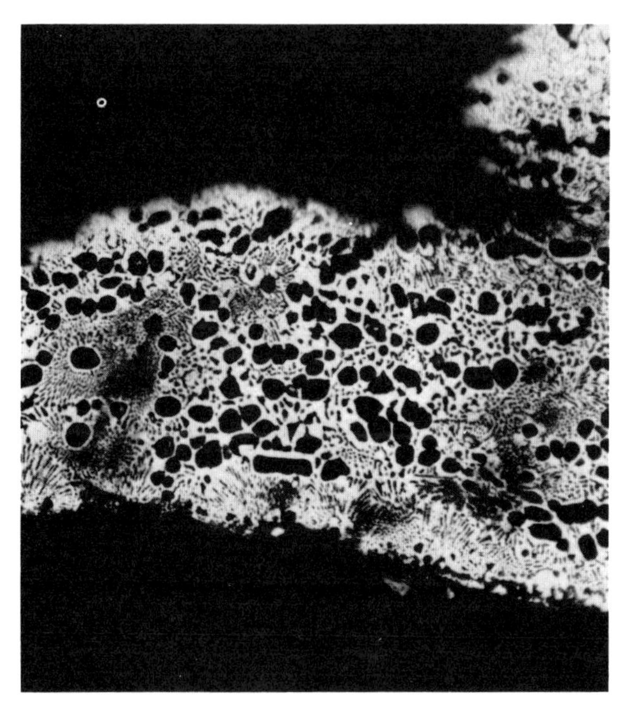

Figure 30.51 Etched detail of a portion of the silver-copper solder pallion shown in figure 30.50. The primary copper-rich phase (dark etching) is surrounded by eutectic (light etching). × 500. Ferric chloride etch.

Tushingham et al. (1979) discuss other kinds of welds on Andean artifacts made without benefit of any filler metal. Throughout the entire course of the central Andean metalworking tradition, metallurgical joins and the alloys required for their accomplishment were a focal point of the smith's skill and inventiveness [for example, see Griffin (1986)]. Nowhere is that artistry more accomplished than in the Nasca-style trophy head shown in figure 30.2 or in another miniature figure that accompanied it, illustrated here in figures 30.52–30.54. Because sampling of the minute areas where joins are visible is extremely difficult and has not been attempted for either of these objects, it remains unclear whether those joins were made by soldering or welding. In either case, they are masterfully done.

Color

Color was the single property of metal whose achievement and manipulation stimulated the most innovative and sophisticated developments of Andean metallurgical technologies. Except for iron, which was not manufactured in the prehistoric era, Andean metalworkers produced and used most of the same metals and alloys we associate with the Old World: copper, copper-arsenic bronze and copper-tin bronze, silver, lead, and gold (Lechtman 1980). In addition, copper-silver and copper-gold alloys were uniquely developed for their ability to confer the colors of silver and gold on objects made from them. The mechanical properties of metals were not neglected by Andean smiths, as we have seen. But if we look for the locus of attention of the metallurgy, for that arena in which both the expectations of metal performance and the realization of such performance were highest, we find it not in the realm of utility but in the realm of the symbolic.

In the Andes, metals carried and displayed the content or message of status, wealth, and political power and reinforced the affective power of religious objects (Lechtman 1984a). Andean metallurgy thus received its greatest stimulus in the arena dominated by status and political display. A glance at the inventory of surviving Precolumbian metal objects that functioned as symbols of rank and of political and religious power—earrings, nose rings, death masks, religious cult objects, and so forth—indicates that it was their color as much as, if not more than, their form that acted as the carrier of such symbolic information. From the earliest involvement of Andean peoples with metal up to the time of the Spanish conquest of the Inka empire, the two colors that were paramount in the metallurgical spectrum were gold and silver. Thus a concern for the color of metals rather than for their mechanical properties became a central and an early concern in Andean metallurgical developments.

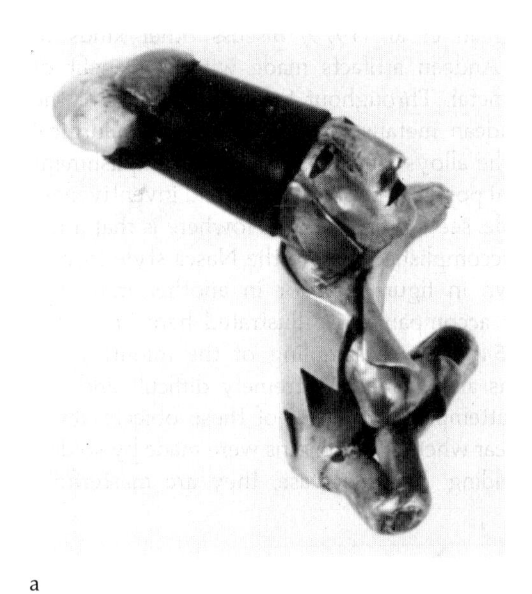

a

b

c

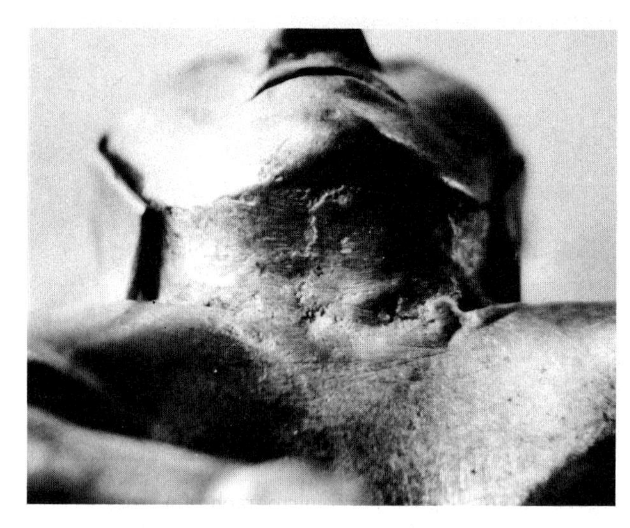

Figure 30.53 Detail of the miniature shown in figure 30.52. The neck consists of a short cylinder of sheet gold whose edges are joined under the chin. The cylinder, in turn, connects the face and upper torso. In this view three seams are visible: (1) a vertical seam at the location of the two edges of the sheet cylinder, (2) a horizontal seam (only partially filled with metal) between cylinder and chin, and (3) another horizontal seam between cylinder and torso.

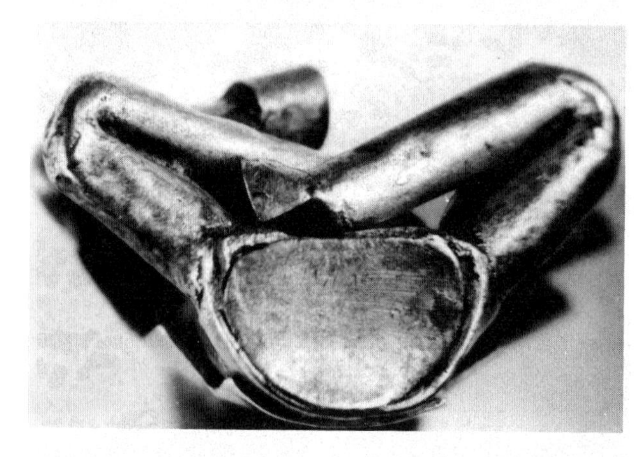

Figure 30.54 Base of the seated man of figure 30.52. Each leg consists of four parts: foot, calf, knee cap, and thigh. The legs were soldered or welded to the semicylindrical trunk of the figure, and its seat was closed with a base plate tacked in place.

Figure 30.52 Miniature seated man of sheet gold with fezlike hat of silver crowned with Spondylus shell. Nasca style. Ica valley, Peru. Private collection. Height: 2.75 cm.

Once color becomes the focus of property development, we are dealing with the metallurgy of surfaces, because the color of a metal object resides at its surface. The object may have one color at the surface and a totally different color underneath, corresponding to differences in metal or alloy composition at the surface and in the underlying material. Because Andean metallurgy was color oriented, it was also surface oriented.

The most creative and interesting aspects of Andean metallurgy arose as a response of the technology to an overriding system of visual communication, as smiths attempted to produce metallic gold and metallic silver surfaces on metal objects that were made of other metals. These efforts resulted in the purposeful manufacture of binary and ternary alloys of copper, silver, and gold and a remarkable set of metallurgical and electrochemical procedures for gilding and silvering objects made of copper.

The two most sophisticated of Andean gilding procedures—electrochemical replacement plating and depletion gilding—can now be credited to the metalsmiths of Moche society, which flourished along the north coast of present-day Peru from c. 100 B.C. to A.D. 800 (Lechtman et al. 1982). In many respects Moche metallurgy epitomizes the central Andean tradition of metal manufacture. The virtuosity of its technology and the quality of its products were unequaled by those of succeeding cultures, including the Inka.

A large group of objects of Moche style has been found recently at the site of Loma Negra, on the far north coast of Peru, near the Ecuadorian border (Lechtman et al. 1982). Two are illustrated in figures 30.20, 30.32, and 30.36. Most are made of hammered sheet copper, individual shaped pieces of which were often joined to produce three-dimensional forms. The outside and occasionally the inside surfaces of the copper sheet were covered with extremely thin coatings of gold or silver so that the objects appeared to be made of those metals.

Laboratory study of the Loma Negra artifacts together with experiments designed to reproduce the gross characteristics and the microstructures of these remarkably thin (0.5–2 μm) and uniform gold and silver coatings revealed that the copper had been plated (gilt or silvered) by some system of electrochemical deposition (Lechtman 1979b). In other words, Moche smiths were able to deposit gold or silver onto copper without the use of modern chemicals, such as cyanide or aqua regia for dissolving the gold, and without the use of an external source of current. They apparently dissolved gold and silver in mixtures of corrosive minerals, such as common salt, potassium nitrate, and potassium aluminum sulphate, which were readily available in the coastal desert environment and which earlier studies have shown were probably used by them (Lechtman 1973a). Once in solution, the gold or silver plated directly onto the surfaces of copper objects dipped into the solution, providing those surfaces with infinitesimally thin and regular coatings of the precious metals.

Many metallographic examinations of small samples removed from the objects defined the characteristics of these coatings: (1) On any given object they are of a relatively uniform thickness and cover all surfaces, including the often paper thin edges of the object; (2) there is a solid-state diffusion zone between the gold and the copper, indicating that at some stage of the coating process heat was applied; (3) there is no evidence of mercury gilding, the use of gold leaf or foil, or the flushing-on of molten gold.

To reproduce the gross characteristics and the microstructures of these gold and silver coatings using systems of electrochemical deposition, we dissolved gold or alloys of gold and silver in an aqueous solution of equal parts of common salt (NaCl), potassium nitrate (KNO_3), and potassium aluminum sulfate [$KAl(SO_4)_2 \cdot 12H_2O$]. This solution contains inter alia the same ions as those present in aqua regia needed to dissolve the gold in the form of trivalent ions. Chloroauric acid, $H(AuCl_4) \cdot 3H_2O$, would crystallize from the solution. Once this highly acidic solution was neutralized to a pH of 9, copper sheet dipped into it was uniformly coated on all its surfaces (including the edges) with a film of gold approximately 1 μm thick after 5 min of gentle boiling.

Depositing a thin film of gold onto a copper substrate by simple electrochemical plating was only occasionally sufficient to bond the gold permanently to the copper. A more durable bond could be effected, however, by heating the gilt copper at a temperature high enough—between about 500° and 800°C—to produce solid-state diffusion of the gold and copper across their common interface. Ultimately we were successful in achieving plates that were quite close in their visual characteristics and in their microstructures to the platings on the Loma Negra objects. The Loma Negra microstructures (figure 30.55) are typified by large annealed grains within the copper sheet and, at the surface of the sheet, by heat-induced solid-state diffusion of the plated coating along the grain boundaries. Our copper sheet, plated by immersion with an 85 Au:15 Ag coating, then annealed in a bunsen flame for 15 sec, closely reproduced the Loma Negra microstructures and formed a well-bonded golden surface (figure 30.56).

Plating by electrochemical replacement occurs when a metal, such as copper, high in the electromotive series (at the "negative" or "base" end of the series) is placed in an electrolyte containing ions of a metal, such as gold, lower in the series (at the "positive" or "noble" end). Chemically a simple replacement reaction occurs, such as

$$2AuCl_3 + 3Cu \rightarrow 2Au + 3CuCl_2.$$

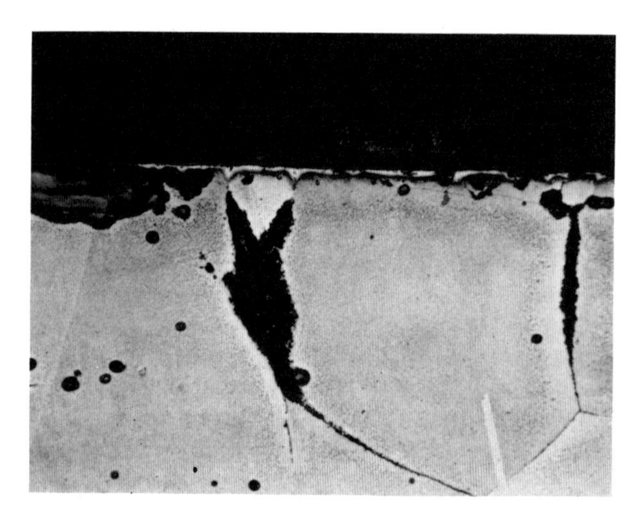

Figure 30.55 Photomicrograph of an etched cross section removed from a tab underneath the base of the Loma Negra seated man illustrated in figure 30.20. Note the large grain size of the annealed copper, the spherical oxide inclusions, and the grain boundary corrosion. At the surface of the copper sheet the electrochemically deposited gold layer is just visible. Annealing of the gilt sheet caused solid-state diffusion between the copper and the gold. That diffusion proceeded preferentially along grain boundaries and is recognized in the micrograph as small dips or inverted peaks in the gold layer where the gold has entered along a grain boundary. × 380. Potassium dichromate etch.

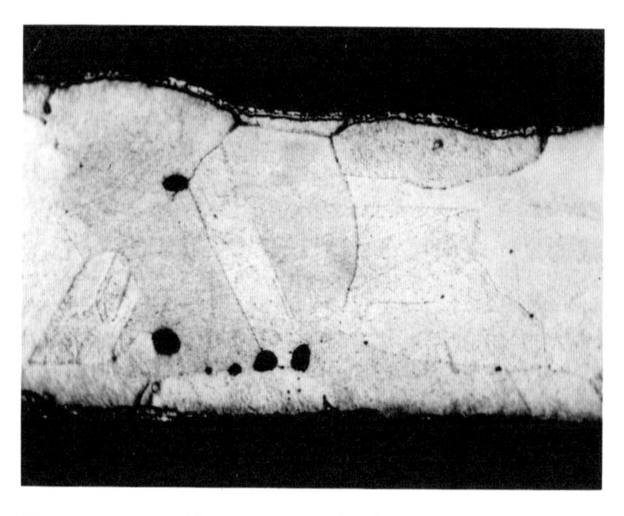

Figure 30.56 Photomicrograph of an etched cross section through a piece of hammered sheet copper gilt in the laboratory by immersion into an aqueous salt solution/suspension of 85% gold, 15% silver. After electrochemical replacement plating the sheet was flame annealed for 15 sec. The structures produced in the laboratory are almost identical with those in the sample removed from the Loma Negra seated figure (figure 30.55). × 760. Potassium dichromate etch.

But such an equation does not reveal the mechanism that is identical with that in the simple cells used by the first electroplaters. It is necessary to have anode and cathode areas, both of which must be in contact with an electrolyte, and a complete circuit for electrons to flow from the anode through the metal to the cathode area, balanced by the return flow of ions through the electrolyte.

In replacement plating different parts of the same metal surface provide both the anode and the cathode. In the case of the Loma Negra objects, small pits or irregularities on the surface of the copper sheet, such as those shown in the photomicrograph of figure 30.57, initially acted as anodes and continued their anodic activity until they were completely blocked or protected from the electrolyte by the deposit—the gold—that plated onto adjacent cathodic surfaces (Lechtman 1979b).

Moche metalsmiths were not limited solely to gilding and silvering objects made of copper. Their desire to achieve culturally valued color effects was played out in the alloy systems they developed, some of which have come to be considered the hallmark of Andean prehistoric metallurgy.

One of the most interesting and exciting aspects of any indigeneous metallurgy is the way in which alloys develop—the reasons for which particular alloy systems emerge from the range of possible choices and the relations between those alloy systems and other aspects of culture. In the Andes there is at least one trend that we can follow from its earliest circumscribed appearance to its ultimate far-flung dispersion from present-day Peru to Mexico: the development of a series of alloys whose primary use was the creation of a silver or gold color on the surface of metal objects made of the alloys.

The earliest of these alloys that has been identified archaeologically is that of copper and silver (Lechtman 1979a), discussed previously. But by far the most important alloy system developed by the Moche was *tumbaga*, the alloy of copper and gold or of copper, gold, and silver. An ingot of copper-gold alloy may be coppery red or pink or golden, depending on the amount of gold it contains. When such an ingot is hammered to produce thin metal sheet, copper is lost from the surfaces of the alloy through oxidation on annealing, just as in the case of copper-silver alloys. Objects made from reddish copper-rich *tumbagas* soon develop golden surfaces as increasing amounts of copper are lost.

Many of the large golden mummy masks from Peru familiar to museum goers are of Sicán or Chimú origin (Carcedo and Shimada 1985). The largest known example of these, presently in the collection of the Metropolitan Museum of Art (New York) and reproduced here in figure 30.58, measures 29 in wide by 16 in high. Its golden surface is dazzling, yet the mask is

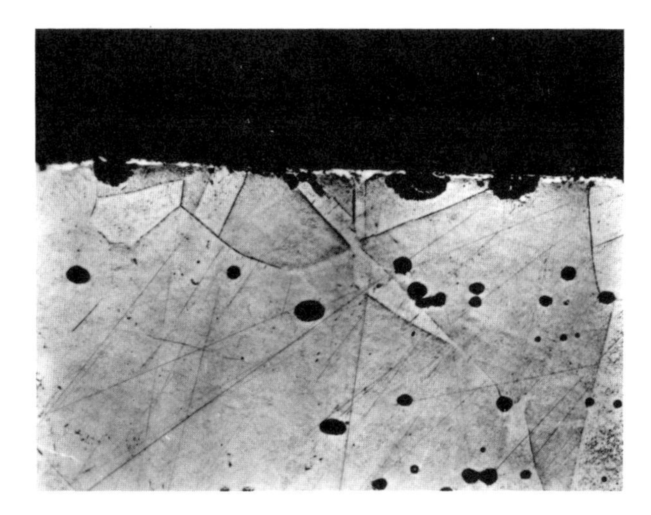

Figure 30.57 Photomicrograph of an etched cross section through the surface of a fragment removed from a gilt copper object from Loma Negra. The surface anodic pores are partially covered by a thin gold-silver plate electrochemically deposited in the laboratory. ×380. Potassium dichromate and ferric chloride etch.

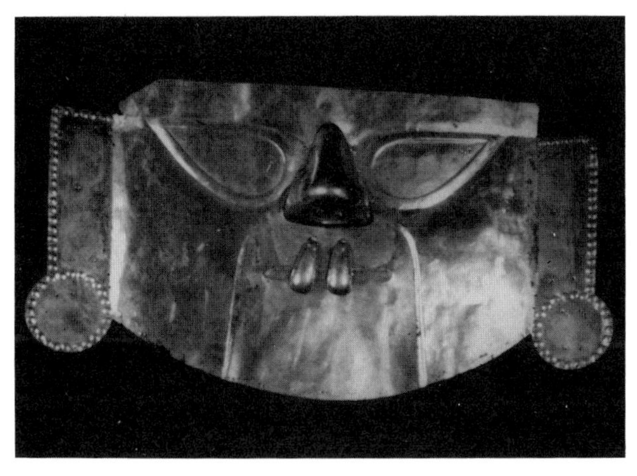

Figure 30.58 Largest golden mummy mask of Chimú origin extant, measuring 74 cm wide by 40 cm high. Originally some of its surfaces were painted with cinnabar, and the holes in the eyes, the ears and ear spools, and along the bottom edge of the mask indicate that other decorations, such as plaques of copper, precious stones, or fibers, were attached when the mummy bundle was interred. Collection: Metropolitan Museum of Art, New York (Cat. no. 57. 161), Michael C. Rockefeller Memorial Collection, bequest of Nelson A. Rockefeller (1979).

made of a ternary alloy containing only 39% gold, the remainder being silver (49%) and copper (12%) (Lechtman 1973a).

In manufacturing the sheet metal for a mummy mask, periodic annealing keeps the alloy malleable and formable. On each anneal, a brown to black copper oxide scale coats the surface and must be removed before hammering can continue. Andean smiths could have used certain acid plant juices or stale urine, which degrades to ammonia, to pickle off the copper oxide scale. After many sequences of hammering, annealing, and pickling, the resulting thin metal sheet loses enough surface copper to enrich the surfaces in silver and gold. A mummy mask at this stage of manufacture would appear as if made of silver, because the binary silver-gold alloy remaining at the surface is silvery white (see stage 1 in figure 30.59, and figure 30.60).

But the mummy masks are brightly golden. Sicán and Chimú smiths, approaching the task of gilding from the same point of view as that of silvering, were faced with the problem of removing the silver in the surface silver-gold binary alloy in order to leave only the gold in place. A modern chemist might use various cyanide solutions or distilled nitric acid to part silver from gold, but we can assume that such distilled acids were not available to Andean smiths. Laboratory studies suggest (Lechtman 1973a) that what they did have at hand and apparently used were combinations of naturally occurring acid minerals, such as ferric sulfate and sodium chloride (common table salt), which in aqueous solution effectively remove silver from a silver-gold alloy, leaving the gold behind (see stage 2 in figure 30.59). This discovery had already been made by the Moche, although it was most widely adopted by Sicán smiths working in the Lambayeque valley and by the Chimú, who several centuries later dominated the entire north coast of Peru.

Depletion gilding—the term used to denote this set of procedures—relies on the removal from the surface of an alloy of its baser metal constituents in order to leave the noblest metal in place. It was used effectively by central Andean metalworkers to gild sheet metal objects that contained as little as 12% gold by weight.

The *tumbaga* alloys, with their inherent gold enrichment properties, swept through the Americas from Peru to Mexico and were in common use in that entire region when the Spaniards invaded Mexico, Central America, and South America in the sixteenth century. They constitute the most significant contribution of the New World to the repertoire of alloy systems developed among ancient societies.

Electrochemical replacement and depletion silvering or gilding represent two opposite approaches to the production of a colored metal surface on a metal substrate. Both involve sophisticated chemistry, and Precolumbian surface metallurgy is surely as much chemistry as it is metallurgy. But replacement plating

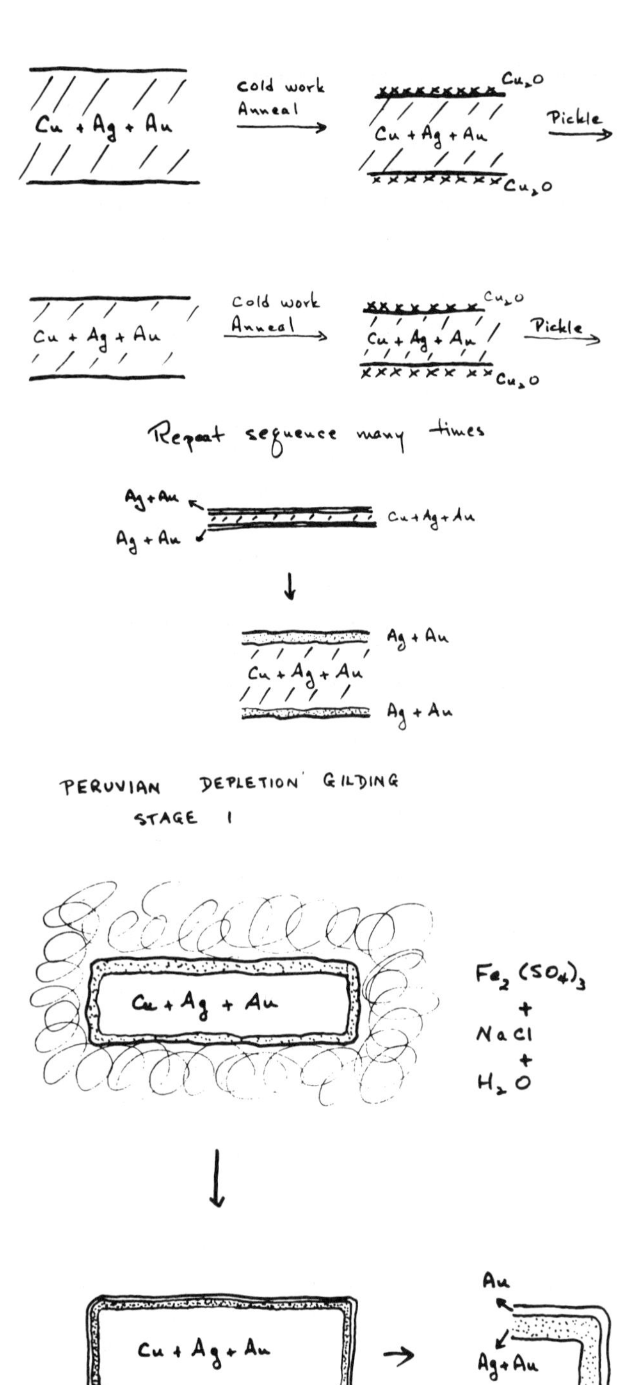

Figure 30.59 Diagram of the mechanisms by which Andean metalworkers gilt by depletion sheet metal objects hammered from ingots of copper-silver-gold alloys.

Stage 1 The ingot, containing copper, silver, and gold, is cold-worked to reduce its thickness and annealed. At each anneal a scale of cuprous oxide (Cu_2O) forms on the surface. This is pickled off, and the hammering continues. After many sequences of hammering, annealing, and pickling, the thin sheet has lost sufficient copper at its surfaces so that those surfaces become binary silver-gold alloys and appear silver in reflected light.

Stage 2 To remove silver from the surface of the sheet, the object is covered completely with an aqueous paste of ferric sulfate and salt. This corrosive material dissolves the silver from the surface silver-gold alloy, leaving a thin skin of gold in place. The object is left with a three-layered structure: an inner core of ternary copper-silver-gold alloy, a surface zone of binary silver-gold alloy, and a visible surface skin of gold.

Figure 30.60 Photomicrograph of an etched cross section of a tiny fragment removed from a broken edge of the mummy mask illustrated in figure 30.58. The structure is typical of severely worked metal, with tiny grains and alternating phases of the ternary alloy strung out in the direction of working. The white zones at both surfaces represent regions of surface enrichment in silver and gold resulting from depletion of the surfaces in copper. × 250. Potassium cyanide and ammonium persulfate etch.

achieves color change by covering one metal with another, hiding what is underneath. Depletion systems, on the other hand, transform surfaces by selectively developing at the surface certain properties of the material that lies beneath. One system covers and hides; the other system develops and enhances.

It is significant that of the various surface covering or coating systems identified thus far in a general inventory of Andean objects—electrochemical replacement plating, the use of gold foil, the flushing on of molten gold—none had more than a local and brief impact on the development of Precolumbian metallurgical technologies. Depletion surface metallurgy, on the other hand, dominated the metallurgical scene in the New World for between one and two millennia.

The metallurgy of surfaces was important in the New World because Precolumbian metallurgy was color oriented. The increasing attention of craftspeople to surface phenomena and their sophistication in developing systems to transform metal surfaces reflects the importance of the surface as the visible manifestation of an inner state or an inner property. The surface became the location where aspects of the inner state were realized and enhanced. Surfaces are, after all, boundaries between an inner condition and an external reality or environment. The surface is where the two meet. It is the place of communication, the seat of greatest information content. When information is conveyed visually, the surface assumes special importance.

Although enrichment systems, whether of silver or gold, have been used by metalworkers in other areas of the world, the role they played in the Andes is unique. Almost from the earliest appearance of metallurgy there, depletion and enrichment processes assumed a special place that persisted throughout the entire course of Andean metallurgical development; they were responsible for stimulating some of its most interesting achievements. Although ideological considerations may have had little to do with the initial working out of these procedures, it seems certain that the way in which Andean peoples perceived such processes (or at least the objects that resulted from their use) had a great deal to do with the way in which the technology emerged and matured (Lechtman 1984a). Belief systems and attitudes toward materials supported the technology and gave rise to further developments along similar lines.

Traditions and Styles

In reviewing the central themes of Andean prehistoric metalworking, we see that two long-standing and deeply rooted traditions dominated the enterprise. First is an overwhelming commitment to the working of metal, to its formability as a solid material. Andean objects convey a fascination with the plasticity of metal, that property above all others that rendered form. The central Andean metalworking tradition was indeed a tradition of working, not of casting, metal. Second we note, at least in items destined for elite use, complete involvement with the visual aspect of metals, with the property of color. Thus the working tradition was joined with a tradition focused on surfaces, emphasizing the metal surface as the carrier of information or the seat of visual communication.

These two traditions were played out in certain styles of technological performance, certain technological ways of meeting the issues of color and of form and of handling those issues (Lechtman 1977). The style of assembly was a response to the problem of creating three-dimensional form from essentially two-dimensional building blocks—the thin metal sheet that served as the basic raw material of the smith. Objects such as the Nasca-style miniatures from the Peruvian south-central coast embody this unwavering approach to the realization of form in metal. It is technological style successfully and boldly pushed to its limit.

The issue of color was met with a wide variety of techniques, all of which essentially "treated" the metal surface to achieve the desired effect. Unlike the Intermediate Area or West Mexico (Hosler 1986), where color was usually rendered by varying alloy composition, central Andean smiths focused on the ultimate surface itself as the carrier of color. Depletion and enrichment processes transformed metal surfaces by enhancing them in culturally valued aspects of what lay beneath. It was this approach to the achievement of surface color that eventually spread to the Intermediate Area and to parts of southern Mexico and that dominated the American metalworking scene for millennia.

The analogy between Andean metalworking and Andean cloth production is not an idle one (Lechtman, in press, 1984a). In both cases it is the flexibility or pliability of the material being shaped and manipulated that counts. And form takes shape under one's fingers. Furthermore, the weaver or the smith builds form in planes of materials, and symbolic information carried at the surface is generated there by managing the very structure of the material. I have suggested (Lechtman 1984a, p. 33) that key technologies associated with cloth and metal production in the Andes shared certain stylistic modes, perhaps because those modes are expressions of cultural ideals, that is, they incorporate ideological concerns of the society at large. Both systems of technology seem to insist that cultural message be embodied in and expressed by structure.

Notes

Many of the objects discussed here were studied by former graduate students at the Center for Materials Research in Archaeology and Ethnology at MIT and by participants in the Center's Summer Institute program. I wish to thank them for their excellent work: Judith Levinson, Janet Levy, Gerald Macomber, Stephen Mellor, Billie Milam, Donna Strahan. Special thanks are due Craig Morris of the American Museum of Natural History (AMNH), New York, for lending many of the objects from the museum's collections for laboratory study and for encouraging such study of Andean material culture and technology. Thanks also to Peter Kvietok of the Department of Anthropology at the AMNH for his kind assistance in providing background information about certain objects, and to Dana Brown for her patient work in reconstructing the casting procedures used in the manufacture of several interesting pieces. A final acknowledgement to S. Whitney Powell for her excellent drawings and to Edward Barry and Christopher Craig for help with photographic documentation.

1. The form and construction of this and other similar objects from the Lambayeque valley is atypical for Peru. On examination of this piece, Dorothy Hosler remarked that the form of the pendant bells is almost identical with certain varieties of West Mexican bells, varieties that are seldom found outside Mesoamerica. This is an important observation in view of Hosler's demonstration of close ties between the metallurgies of northern Peru and Ecuador and that of West Mexico [Dorothy Hosler, personal communication; see also Hosler (1986, this volume).

2. Dana Brown made the flexible mold and the plaster casting of this object in 1986 while working as conservator to the new Hall of South American Peoples at the American Museum of Natural History, New York. She later constructed a plaster two-piece mold similar to the one in which the Inka mace was originally cast. The object, the plaster mold, and the plaster replica are presently on exhibition at the American Museum of Natural History.

3. Joining me in the examination of this figurine were Dana Brown of the American Museum of Natural History, S. Whitney Powell, science illustrator, and Christopher Craig of the Center for Materials Research in Archaeology and Ethnology (CMRAE) at MIT. Brown made the final determination of the number of pieces that constituted the mold; Powell rendered the drawing of the pieces shown in figure 30.18, and Craig took numerous detail photographs of the object to help with the drawing.

4. The American Museum of Natural History has on loan a similar figurine (Loan T364–443) made of silver-copper alloy. It wears a headdress consisting of three maize cobs that extend down the back of the head. Two additional maize cobs are on the chest, and three maize plants decorate the back of the figure. This standing male was probably holding a maize cob or plant, but both arms have been broken off at the wrists. Another example, in silver, published by Jones (1964, fig. 58), is said to be from the north or central coast of Peru.

5. In a 1979 article (Lechtman 1979a) I reported the cramps from Tiwanaku as made of copper, on the basis of Wissler's analysis of one of the five (B/2792-C), which showed a complete absence of tin (Mead 1915). Recently we analyzed the same cramp by atomic absorption spectrometry and determined the presence of arsenic in a concentration of 2.34 wt. %.

6. Bennett (1960) discusses the stone masonry techniques used to construct the massive architectural units at Tiwanaku: "Stones are fitted by means of notches, joints, and copper cramps placed in cut-out T and I-shaped grooves (Fig. 38)" (Bennett 1960, p. 188, fig. 38 on p. 189). His figure 38 illustrates a stone pavement with I-shaped "sockets" or slots meant to receive metal cramps or, as he calls them, "tie bolts." Although no scale is given in the photograph, it is clear that the sockets are much larger than the metal cramps we are discussing here. The cramp shown in figure 30.28 could not have functioned as a tie mechanism between stone blocks of the size Bennett describes and illustrates. It is possible, however, that they may have been used to register much smaller stones or possibly even wooden beams, if such were used at the site. Alan Kolata, who has been excavating at Tiwanaku, has also suggested that cramps such as these may have had a decorative function. "There are a number of examples of carved stone lintels with mold-like impressions that were clearly intended to accept some type of inlay—most likely metal. This seems to be a more likely function for this rather gracile artifact" (Alan L. Kolata, personal communication, February 1987).

7. Smelting experiments I have conducted with a variety of copper ores from the Central Andes indicate that it is not difficult to smelt a charge consisting of copper oxide and copper sulfarsenide ores to produce metallic copper-arsenic alloy.

8. In 1982 we made an X-radiograph of rattle 41.2/5429a–c. It showed three spherical pellets within the hollow box, each measuring between 0.6 cm and 0.9 cm in diameter. Elizabeth Fitzhugh West analyzed the copper from this same rattle by emission spectrography at the Freer Gallery laboratories in 1963. She reported the following results in a letter to Junius Bird dated March 25, 1963:

Analysis of copper of rattle 41.2/5429a–c

Sample location	Ag	Au	Zn	Fe	Ni	As	Sb	Al	Mg	Si	Sn	Pb
Handle	0.002	n.d.	n.d.	<0.001	0.03	n.d.	n.d.	n.d.	<0.001	0.002	n.d.	0.03
Ferrel	0.002	n.d.	n.d.	0.001	0.03	0.3	0.3	n.d.	<0.001	0.002	n.d.	0.03
Base panel	<0.001	n.d.	n.d.	0.001	0.03	0.3	0.3	n.d.	<0.001	0.1	0.02	0.05

Christopher B. Donnan reports another similar rattle in the collections of the Larco Hoyle Museum in Lima, Peru (Donnan, personal communication, February 1987).

References

Antze, G. 1965. *Trabajos en metal en el norte del Perú.* Lima, Peru: Universidad Nacional Mayor de San Marcos.

Arnold, S. V. 1986. "Aspects of casting practice in Prehispanic Peru," in *Perspectives on Andean Prehistory and Protohistory,*" D. Sandweiss and P. Kvietok, eds. (Ithaca, N.Y.: Latin American Studies Program, Cornell University), 131–143.

Bandelier, A. F. 1910. *The Islands of Titicaca and Koati.* New York: Hispanic Society of America.

Bennett, W. C. 1960. "The Central Andes," in *Andean Culture History,* W. C. Bennett and J. B. Bird, eds. (New York: American Museum of Natural History), 95–244.

Bird, J. 1979. "Legacy of the stingless bee." *Natural History* 88(9):49–51.

Carcedo Muro, P., and I. Shimada. 1985. "Behind the golden mask: The Sicán gold artifacts from Batán Grande, Peru," in *The Art of Precolumbian Gold: The Jan Mitchell Collection,* E. P. Benson, ed. (New York: Metropolitan Museum of Art), 60–75.

Donnan, C. 1978. *Moche Art of Peru.* Los Angeles, Calif.: Museum of Cultural History, University of California.

Easby, D. T., Jr. 1955. "Los vasos retratos de metal del Perú," *Revista del Museo Nacional* (Lima), 24:137–153.

Emmerich, A. 1965. *Sweat of the Sun and Tears of the Moon.* Seattle, Wash.: University of Washington Press.

Epstein, S. M., and I. Shimada. 1984. "Metalurgia de Sicán." *Beiträge zur Allgemeinen und Vergleichenden Archäologie* 5: 379–430.

Fester, G. A. 1962. "Copper and copper alloys in ancient Argentina." *Chymia* 8:21–31.

Friedman, A. M., E. Olsen, and J. B. Bird. 1972. "Moche copper analyses: Early New World metal technology." *American Antiquity* 37(2):254–258.

González, A. R. 1979. "Precolumbian metallurgy of northwest Argentina," in *Pre-Columbian Metallurgy of South America,* E. P. Benson, ed. (Washington, D.C.: Dumbarton Oaks), 133–207.

Gordon, R. B. 1985. "Laboratory evidence of the use of metal tools at Machu Picchu (Peru) and environs." *Journal of Archaeological Science* 12:311–327.

Gordon, R. B., and J. W. Rutledge. 1984. "Bismuth bronze from Machu Picchu, Peru." *Science* 223:585–586.

Gordon, R. B., and J. W. Rutledge. 1987. "The work of metallurgical artificers at Machu Picchu, Peru." *American Antiquity* 52:578–594.

Griffin, J. A. 1986. "An investigation in joining metals," in *Metalurgia de América Precolombina,* C. Plazas, ed. (Bogotá: Museo del Oro), 367–377.

Grossman, J. W. 1972. "An ancient gold worker's tool kit." *Archaeology* 25:270–275.

Hosler, D. 1986. *The Origins, Technology, and Social Construction of Ancient West Mexican Metallurgy,* Ph.D. dissertation, University of California, Santa Barbara.

Jones, J. 1964. *Art of Empire: The Inca of Peru.* New York: Museum of Primitive Art.

Kroeber, A. L. 1944. *Peruvian Archeology in 1942.* New York: Wenner-Gren Foundation.

Lapiner, A. 1976. *Pre-Columbian Art of South America.* New York: Abrams.

Lechtman, H. 1971. "Ancient methods of gilding silver: Examples from the Old and the New worlds," in *Science and Archaeology,* R. H. Brill, ed. (Cambridge, Mass.: MIT Press), 2–30.

Lechtman, H. 1973a. "The gilding of metals in Pre-Columbian Peru," in *Application of Science in Examination of Works of Art,* W. J. Young, ed. (Boston, Mass.: Museum of Fine Arts), 38–52.

Lechtman, H. 1973b. "A tumbaga object from the high Andes of Venezuela." *American Antiquity* 38:473–482.

Lechtman, H. 1976. "A metallurgical site survey in the Peruvian Andes." *Journal of Field Archaeology* 3:1–42.

Lechtman, H. 1977. "Style in technology: Some early thoughts," in *Material Culture: Styles, Organization, and Dynamics of Technology,* H. Lechtman and R. S. Merrill, eds. (St. Paul, Minn.: West Publishing Co.), 3–20.

Lechtman, H. 1979a. "Issues in Andean metallurgy," in *Pre-Columbian Metallurgy of South America,* E. P. Benson, ed. (Washington, D.C.: Dumbarton Oaks), 1–40.

Lechtman, H. 1979b. "A Precolumbian technique for electrochemical replacement plating of gold and silver on objects of copper." *Journal of Metals* 31:154–160.

Lechtman, H. 1980. "The Central Andes: Metallurgy without iron," in *The Coming of the Age of Iron,* T. Wertime and J. Muhly, eds. (New Haven, Conn.: Yale University Press), 267–334.

Lechtman, H. 1981. "Copper-arsenic bronzes from the north coast of Peru." in *Annals of the New York Academy of Sciences: The Research Potential of Anthropological Museum Collections* 376:77–121.

Lechtman, H. 1984a. "Andean value systems and the development of prehistoric metallurgy." *Technology and Culture* 25(1):1–36.

Lechtman, H. 1984b. "Technical examination of a gold-alloy object from Chavín de Huántar," in *The Prehistoric Occupation of Chavín de Huántar, Peru,* R. L. Burger, ed. (Berkeley, Calif.: University of California Press), 271–276.

Lechtman, H. In press. "Technologies of power: The Andean case," in *Configurations of Power in Complex Society,* P. J. Netherly and J. S. Henderson, eds. (Ithaca, N.Y.: Cornell University Press).

Lechtman, H., A. Erlij, and E. J. Barry, Jr. 1982. "New perspectives on Moche metallurgy: Techniques of gilding copper at Loma Negra, N. Peru." *American Antiquity* 47:3–30.

Lechtman, H., L. Parsons, and W. Young. 1975. "Seven matched hollow gold jaguars from Peru's Early Horizon." *Studies in Precolumbian Art and Archaeology* 16:1–45.

Lothrop, S. K., 1941. "Gold ornaments of Chavín style from Chongoyape, Peru." *American Antiquity* 6:250–262.

Lothrop, S. K. 1951. "Gold artifacts of Chavín style." *American Antiquity* 16:226–240.

Mathewson, C. H. 1915. "A metallographic description of some ancient Peruvian bronzes from Machu Picchu." *American Journal of Science* 40:525–616.

Mead, C. W. 1915. "Prehistoric bronze in South America." *Anthropological Papers of the American Museum of Natural History* 12:15–51.

Owen, B. 1987. *The Role of Common Metal Objects in the Inka State.* Master's thesis, University of California, Los Angeles.

Patterson, C. C. 1971. "Native copper, silver, and gold accessible to early metallurgists." *American Antiquity* 36: 286–321.

Peru durch die Jahrtausende: Kunst und Kultur im Lande der Inka. Exhibition catalog. Recklinghausen: Verlag Aurel Bongers.

Posnansky, A. 1945. *Tihuanacu.* New York: J. J. Augustin.

Root, W. C. 1949. "The metallurgy of the southern coast of Peru." *American Antiquity* 15(1):10–37.

Shimada, I. 1985. "Perception, procurement, and management of resources: Archaeological perspective," in *Andean Ecology and Civilization,* S. Masuda, I. Shimada, and C. Morris, eds. (Tokyo: University of Tokyo Press), 357–399.

Tushingham, A. D., U. M. Franklin, and C. Toogood. 1979. *Studies in Ancient Peruvian Metalworking.* History, Technology, and Art Monograph 3. Toronto: Royal Ontario Museum.

Wagner, E. 1970. "Arqueología de la Región de Mucuchiés en los Andes Venezolanos." *Acta Científica Venezolana* 21: 180–185.

Zevallos M., C. 1956. "Tecnología metalúrgica arqueológica." *Cuadernos de Historia y Arqueología* (Guayaquil, Ecuador), 6:209–215.

Contributors

Donald H. Avery
Department of Engineering
Brown University
Providence, Rhode Island

Hans Gert Bachmann
Degussa AG
Frankfurt/Main, West Germany

I. Lynus Barnes
US National Bureau of Standards
Washington, D.C.

Emma C. Bunker
Asian Art Museum
San Francisco, California

Dilip K. Chakrabarti
Archaeology Unit
Department of History
Delhi University
Delhi, India

W. Thomas Chase
Freer Gallery of Art
Smithsonian Institution
Washington, D.C.

Christopher J. Davey
Merchant Bank, First National Ltd.
Melbourne, Australia

Janos Gömöri
Liszt Ferenc Müzeum
Sopron, Hungary

Andreas Hauptmann
Deutsches Bergbau Museum
Bochum, West Germany

Charles Higham
Department of Anthropology
University of Otago
Dunedin, New Zealand

Lore L. Holmes
The Arthur M. Sackler Foundation
New York, New York

Dorothy Hosler
Center for Materials Research in Archaeology and Ethnology
Massachusetts Institute of Technology
Cambridge, Massachusetts

Hu Youyan
City Museum of Huangshi
Hubei, China

Gunborg O. Janzon
Department of Archaeology
Stockholm University
Stockholm, Sweden

E. C. Joel
Conservation Analytical Laboratory
Smithsonian Institution
Washington, D.C.

Borislav Jovanović
Institut Archéologique
Belgrade, Yugoslavia

Heather Lechtman
Center for Materials Research in Archaeology and Ethnology
Massachusetts Institute of Technology
Cambridge, Massachusetts

Fulvia LoSchiavo
Soprintendente Archeologo per le province de Sassari e Nuoro
Sassari, Italy

Lu Benshan
City Museum of Huangshi
Hubei, China

Robert Maddin
Peabody Museum
Harvard University
Cambridge, Massachusetts

Gert Magnusson
Riksantivarieambetet
Stockholm, Sweden

Pieter Meyers
Conservation Center
Los Angeles County Museum of Art
Los Angeles, California

P. R. S. Moorey
Ashmolean Museum
Oxford, England

James D. Muhly
Department of History
University of Pennsylvania
Philadelphia, Pennsylvania

Robert E. Murowchick
Department of Anthropology
Harvard University
Cambridge, Massachusetts

Surapol Natapintu
Archaeology Division
Thai Fine Arts Department
Bangkok, Thailand

Michael R. Notis
Department of Metallurgy and Materials Engineering
Lehigh University
Bethlehem, Pennsylvania

Vincent C. Pigott
Museum Applied Science Center for Archaeology
The University Museum
University of Pennsylvania
Philadelphia, Pennsylvania

George Rapp, Jr.
Archaeometry Laboratory
University of Minnesota
Duluth, Minnesota

Jessica Rawson
Department of Oriental Antiquities
British Museum
London, England

William Rostoker
Department of Civil Engineering, Mechanics, and Metallurgy
University of Illinois at Chicago
Chicago, Illinois

Sharma Saitowitz
Archaeology and Materials Science
University of Cape Town
Cape Town, South Africa

Edward V. Sayre
Conservation Analytical Laboratory
Smithsonian Institution
Washington, D.C.

C. A. Shell
Archaeology Department
University of Cambridge
Cambridge, England

Muneaki Shimura
Research Institute for Iron, Steel, and Other Metals
Tohoku University
Sendai, Japan

Tamara Stech
Department of Materials Science and Engineering
University of Pennsylvania
Philadelphia, Pennsylvania

Sarah J. Taylor
Archaeology Department
University of Cambridge
Cambridge, England

Nikolaas J. van der Merwe
Department of Archaeology
University of Cape Town
Cape Town, South Africa

Gerd Weisgerber
Deutsches Bergbau Museum
Bochum, West Germany

Joyce C. White
The University Museum
University of Pennsylvania
Philadelphia, Pennsylvania

Zhou Baoquan
City Museum of Huangshi
Hubei, China

Index